HYPERSPECTRAL REMOTE SENSING
OF VEGETATION
SECOND EDITION
VOLUME I

Fundamentals, Sensor Systems, Spectral Libraries, and Data Mining for Vegetation

Hyperspectral Remote Sensing of Vegetation
Second Edition

Volume I: Fundamentals, Sensor Systems, Spectral Libraries, and Data Mining for Vegetation

Volume II: Hyperspectral Indices and Image Classifications for Agriculture and Vegetation

Volume III: Biophysical and Biochemical Characterization and Plant Species Studies

Volume IV: Advanced Applications in Remote Sensing of Agricultural Crops and Natural Vegetation

HYPERSPECTRAL REMOTE SENSING OF VEGETATION
SECOND EDITION
VOLUME I

Fundamentals, Sensor Systems, Spectral Libraries, and Data Mining for Vegetation

Edited by
Prasad S. Thenkabail
John G. Lyon
Alfredo Huete

CRC Press
Taylor & Francis Group
Boca Raton London New York

CRC Press is an imprint of the
Taylor & Francis Group, an **informa** business

CRC Press
Taylor & Francis Group
6000 Broken Sound Parkway NW, Suite 300
Boca Raton, FL 33487-2742

First issued in paperback 2022

ISBN 13: 978-1-03-247588-2 (pbk)
ISBN 13: 978-1-138-05854-5 (hbk)

DOI: 10.1201/9781315164151

Publisher's Note
The publisher has gone to great lengths to ensure the quality of this reprint but points out that some imperfections in the original copies may be apparent.

Visit the Taylor & Francis Web site at
http://www.taylorandfrancis.com

and the CRC Press Web site at
http://www.crcpress.com

Dr. Prasad S. Thenkabail, *Editor-in-Chief of these four volumes would like to dedicate the four volumes to three of his professors at the Ohio State University during his PhD days:*

1. **Late Prof. Andrew D. Ward**, *former professor of The Department of Food, Agricultural, and Biological Engineering (FABE) at The Ohio State University,*
2. **Prof. John G. Lyon**, *former professor of the Department of Civil, Environmental and Geodetic Engineering at the Ohio State University, and*
3. **Late Prof. Carolyn Merry**, *former Professor Emerita and former Chair of the Department of Civil, Environmental and Geodetic Engineering at the Ohio State University.*

Contents

Foreword to the First Edition

The publication of this book, *Hyperspectral Remote Sensing of Vegetation*, marks a milestone in the application of imaging spectrometry to study 70% of the Earth's landmass which is vegetated. This book shows not only the breadth of international involvement in the use of hyperspectral data but also in the breadth of innovative application of mathematical techniques to extract information from the image data.

Imaging spectrometry evolved from the combination of insights from the vast heterogeneity of reflectance signatures from the Earth's surface seen in the ERTS-1 (Landsat-1) 4-band images and the field spectra that were acquired to help fully understand the causes of the signatures. It wasn't until 1979 when the first hybrid area-array detectors, mercury-cadmium-telluride on silicon CCD's, became available that it was possible to build an imaging spectrometer capable of operating at wavelengths beyond 1.0 μm. The AIS (airborne imaging spectrometer), developed at NASA/JPL, had only 32 cross-track pixels but that was enough for the geologists clamoring for this development to see *between* the bushes to determine the mineralogy of the substrate. In those early years, vegetation cover was just a nuisance!

In the early 1980s, spectroscopic analysis was driven by the interest to identify mineralogical composition by exploiting absorptions found in the SWIR region from overtone and combination bands of fundamental vibrations found in the mid-IR region beyond 3 μm and the electronic transitions in transition elements appearing, primarily, short of 1.0 μm. The interests of the geologists had been incorporated in the Landsat TM sensor in the form of the add-on, band 7 in the 2.2 μm region based on field spectroscopic measurements. However, one band, even in combination with the other six, did not meet the needs for mineral identification. A summary of mineralogical analyses is presented by Vaughan et al. in this volume. A summary of the historical development of hyperspectral imaging can be found in Goetz (2009).

At the time of the first major publication of the AIS results (Goetz et al., 1985), very little work on vegetation analysis using imaging spectroscopy had been undertaken. The primary interest was in identifying the relationship of the chlorophyll absorption red-edge to stress and substrate composition that had been seen in airborne profiling and in field spectral reflectance measurements. Most of the published literature concerned analyzing NDVI, which only required two spectral bands.

In the time leading up to the 1985 publication, we had only an inkling of the potential information content in the hundreds of contiguous spectral bands that would be available to us with the advent of AVIRIS (airborne visible and infrared imaging spectrometer). One of the authors, Jerry Solomon, presciently added the term "hyperspectral" to the text of the paper to describe the "...multidimensional character of the spectral data set," or, in other words, the mathematically, over-determined nature of hyperspectral data sets. The term hyperspectral as opposed to multispectral data moved into the remote sensing vernacular and was additionally popularized by the military and intelligence community.

In the early 1990s, as higher quality AVIRIS data became available, and the first analyses of vegetation using statistical techniques borrowed from chemometrics, also known as NIRS analysis used in the food and grain industry, were undertaken by John Aber and Mary Martin of the University of New Hampshire. Here, nitrogen contents of tree canopies were predicted from reflectance spectra by regression techniques using reference measurements from laboratory wet chemical analyses of needle and leaf samples acquired by shooting down branches. At the same time, the remote sensing community began to recognize the value of "too many" spectral bands and the concomitant wealth of spatial information that was amenable to information extraction by statistical techniques. One of them was Eyal Ben-Dor who pioneered soil analyses using hyperspectral imaging and who is one of the contributors to this volume.

As the quality of AVIRIS data grew, manifested in increasing SNR, an ever-increasing amount of information could be extracted from the data. This quality was reflected in the increasing number of nearly noiseless principal components that could be obtained from the data or, in other words, its dimensionality. The explosive advances in desktop computing made possible the application of image processing and statistical analyses that revolutionized the uses of hyperspectral imaging. Joe Boardman and others at the University of Colorado developed what has become the ENVI software package to make possible the routine analysis of hyperspectral image data using "unmixing techniques" to derive the relative abundance of surface materials on a pixel-by-pixel basis.

Many of the analysis techniques discussed in this volume, such as band selection and various indices, are rooted in principal components analysis. The eigenvector loadings or factors indicate which spectral bands are the most heavily weighted allowing others to be discarded to reduce the noise contribution. As sensors become better, more information will be extractable and fewer bands will be discarded. This is the beauty of hyperspectral imaging, allowing the choice of the number of eigenvectors to be used for a particular problem. Computing power has reached such a high level that it is no longer necessary to choose a subset of bands just to minimize the computational time.

As regression techniques such as PLS (partial least squares) become increasingly adopted to relate a particular vegetation parameter to reflectance spectra, it must be remembered that the quality of the calibration model is a function of both the spectra and the reference measurement. With spectral measurements of organic and inorganic compounds under laboratory conditions, we have found that a poor model with a low coefficient of determination (r^2) is most often associated with inaccurate reference measurements, leading to the previously intuitive conclusion that "spectra don't lie."

Up to this point, AVIRIS has provided the bulk of high-quality hyperspectral image data but on an infrequent basis. Although Hyperion has provided some time series data, there is no hyperspectral imager yet in orbit that is capable of providing routine, high-quality images of the whole Earth on a consistent basis. The hope is that in the next decade, HyspIRI will be providing VNIR and SWIR hyperspectral images every 3 weeks and multispectral thermal data every week. This resource will revolutionize the field of vegetation remote sensing since so much of the useful information is bound up in the seasonal growth cycle. The combination of the spectral, spatial, and temporal dimensions will be ripe for the application of statistical techniques and the results will be extraordinary.

Dr. Alexander F. H. Goetz PhD
Former Chairman and Chief Scientist
ASD Inc.
2555 55th St. #100
Boulder, CO 80301, USA
303-444-6522 ext. 108
Fax 303-444-6825
www.asdi.com

REFERENCES

Goetz, A. F. H., 2009, Three decades of hyperspectral imaging of the Earth: A personal view, *Remote Sensing of Environment*, 113, S5–S16.

Goetz, A.F.H., G. Vane, J. Solomon and B.N. Rock, 1985, Imaging spectrometry for Earth remote sensing, *Science*, 228, 1147–1153.

BIOGRAPHICAL SKETCH

Dr. Goetz is one of the pioneers in hyperspectral remote sensing and certainly needs no introduction. Dr. Goetz started his career working on spectroscopic reflectance and emittance studies of the Moon and Mars. He was a principal investigator of Apollo-8 and Apollo-12 multispectral photography studies. Later, he turned his attention to remote sensing of Planet Earth working in collaboration with Dr. Gene Shoemaker to map geology of Coconino County (Arizona) using Landsat-1 data and went on to be an investigator in further Landsat, Skylab, Shuttle, and EO-1 missions. At NASA/JPL he pioneered field spectral measurements and initiated the development of hyperspectral imaging. He spent 21 years on the faculty of the University of Colorado, Boulder, and retired in 2006 as an Emeritus Professor of Geological Sciences and an Emeritus Director of Center for the Study of Earth from Space. Since then, he has been Chairman and Chief Scientist of ASD Inc. a company that has provided more than 850 research laboratories in over 60 countries with field spectrometers. Dr. Goetz is now retired. His foreword was written for the first edition and I have retained it in consultation with him to get a good perspective on the development of hyperspectral remote sensing.

Foreword to the Second Edition

The publication of the four-volume set, *Hyperspectral Remote Sensing of Vegetation*, second edition, is a landmark effort in providing an important, valuable, and timely contribution that summarizes the state of spectroscopy-based understanding of the Earth's terrestrial and near shore environments. Imaging spectroscopy has had 35 years of development in data processing and analysis methods. Today's researchers are eager to use data produced by hyperspectral imagers and address important scientific issues from agricultural management to global environmental stewardship. The field started with development of the Jet Propulsion Lab's Airborne Imaging Spectrometer in 1983 that measured across the reflected solar infrared spectrum with 128 spectral bands. This technology was quickly followed in 1987 by the more capable Advanced Visible Infrared Imaging Spectrometer (AVIRIS), which has flown continuously since this time (albeit with multiple upgrades). It has 224 spectral bands covering the 400–2500 nm range with 10 nm wavelength bands and represents the "gold standard" of this technology. In the years since then, progress toward a hyperspectral satellite has been disappointingly slow. Nonetheless, important and significant progress in understanding how to analyze and understand spectral data has been achieved, with researchers focused on developing the concepts, analytical methods, and spectroscopic understanding, as described throughout these four volumes. Much of the work up to the present has been based on theoretical analysis or from experimental studies at the leaf level from spectrometer measurements and at the canopy level from airborne hyperspectral imagers.

Although a few hyperspectral satellites have operated over various periods in the 2000s, none have provided systematic continuous coverage required for global mapping and time series analysis. An EnMap document compiled the past and near-term future hyperspectral satellites and those on International Space Station missions (EnMap and GRSS Technical Committee 2017). Of the hyperspectral imagers that have been flown, the European Space Agency's CHRIS (Compact High Resolution Imaging Spectrometer) instrument on the PROBA-1 (Project for On-Board Autonomy) satellite and the Hyperion sensor on the NASA technology demonstrator, Earth Observing-1 platform (terminated in 2017). Each has operated for 17 years and have received the most attention from the science community. Both collect a limited number of images per day, and have low data quality relative to today's capability, but both have open data availability. Other hyperspectral satellites with more limited access and duration include missions from China, Russia, India, and the United States.

We are at a threshold in the availability of hyperspectral imagery. There are many hyperspectral missions planned for launch in the next 5 years from China, Italy, Germany, India, Japan, Israel, and the United States, some with open data access. The analysis of the data volumes from this proliferation of hyperspectral imagers requires a comprehensive reference resource for professionals and students to turn to in order to understand and correctly and efficiently use these data. This four-volume set is unique in compiling in-depth understanding of calibration, visualization, and analysis of data from hyperspectral sensors. The interest in this technology is now widespread, thus, applications of hyperspectral imaging cross many disciplines, which are truly international, as is evident by the list of authors of the chapters in these volumes, and the number of countries planning to operate a hyperspectral satellite. At least some of the hyperspectral satellites announced and expected to be launched in this decade (such as the HyspIRI-like satellite approved for development by NASA with a launch in the 2023 period) will provide high-fidelity narrow-wavelength bands, covering the full reflected solar spectrum, at moderate (30 m pixels) to high spatial resolution. These instruments will have greater radiometric range, better SNR, pointing accuracy, and reflectance calibration than past instruments, and will collect data from many countries and parts of the world that have not previously been available. Together, these satellites will produce an unprecedented flow of information about the physiological functioning (net primary production, evapotranspiration, and even direct measurements related to respiration), biochemical characteristics (from spectral indices and from radiative transfer first principle methods), and direct measurements of the distributions of plant and soil biodiversity of the terrestrial and coastal environments of the Earth.

This four-volume set presents an unprecedented range and scope of information on hyperspectral data analysis and applications written by leading authors in the field. Topics range from sensor characteristics from ground-based platforms to satellites, methods of data analysis to characterize plant functional properties related to exchange of gases CO_2, H_2O, O_2, and biochemistry for pigments, N cycle, and other molecules. How these data are used in applications range from precision agriculture to global change research. Because the hundreds of bands in the full spectrum includes information to drive detection of these properties, the data is useful at scales from field applications to global studies.

Volume I has three sections and starts with an introduction to hyperspectral sensor systems. Section II focuses on sensor characteristics from ground-based platforms to satellites, and how these data are used in global change research, particularly in relation to agricultural crop monitoring and health of natural vegetation. Section III provides five chapters that deal with the concept of spectral libraries to identify crops and spectral traits, and for phenotyping for plant breeding. It addresses the development of spectral libraries, especially for agricultural crops and one for soils.

Volume II expands on the first volume, focusing on use of hyperspectral indices and image classification. The volume begins with an explanation of how narrow-band hyperspectral indices are determined, often from individual spectral absorption bands but also from correlation matrices and from derivative spectra. These are followed by chapters on statistical approaches to image classification and a chapter on methods for dealing with "big data." The last half of this volume provides five chapters focused on use of vegetation indices for quantifying and characterizing photosynthetic pigments, leaf nitrogen concentrations or contents, and foliar water content measurements. These chapters are particularly focused on applications for agriculture, although a chapter addresses more heterogeneous forest conditions and how these patterns relate to monitoring health and production.

The first half of Volume III focuses on biophysical and biochemical characterization of vegetation properties that are derived from hyperspectral data. Topics include ecophysiological functioning and biomass estimates of crops and grasses, indicators of photosynthetic efficiency, and stress detection. The chapter addresses biophysical characteristics across different spatial scales while another chapter examines spectral and spatial methods for retrieving biochemical and biophysical properties of crops. The chapters in the second half of this volume are focused on identification and discrimination of species from hyperspectral data and use of these methods for rapid phenotyping of plant breeding trials. Lastly, two chapters evaluate tree species identification, and another provides examples of mapping invasive species.

Volume IV focuses on six areas of advanced applications in agricultural crops. The first considers detection of plant stressors including nitrogen deficiency and excess heavy metals and crop disease detection in precision farming. The second addresses global patterns of crop water productivity and quantifying litter and invasive species in arid ecosystems. Phenological patterns are examined while others focus on multitemporal data for mapping patterns of phenology. The third area is focused on applications of land cover mapping in different forest, wetland, and urban applications. The fourth topic addresses hyperspectral measurements of wildfires, and the fifth evaluates use of continuity vegetation index data in global change applications. And lastly, the sixth area examines use of hyperspectral data to understand the geologic surfaces of other planets.

Susan L. Ustin
Professor and Vice Chair, Dept. Land, Air and Water Resources
Associate Director, John Muir Institute
University of California
Davis California, USA

REFERENCE

EnMap Ground Segment Team and GSIS GRSS Technical committee, December, 2017. *Spaceborne Imaging Spectroscopy Mission Compilation*. DLR Space Administration and the German Federal Ministry of Economic Affairs and Technology. http://www.enmap.org/sites/default/files/pdf/Hyperspectral_EO_Missions_2017_12_21_FINAL4.pdf

BIOGRAPHICAL SKETCH

Dr. Susan L. Ustin is currently a Distinguished Professor of Environmental and Resource Sciences in the Department of Land, Air, and Water Resources, University of California Davis, Associate Director of the John Muir Institute, and is Head of the Center for Spatial Technologies and Remote Sensing (CSTARS) at the same university. She was trained as a plant physiological ecologist but began working with hyperspectral imagery as a post-doc in 1983 with JPL's AIS program. She became one of the early adopters of hyperspectral remote sensing which has now extended over her entire academic career. She was a pioneer in the development of vegetation analysis using imaging spectrometery, and is an expert on ecological applications of this data. She has served on numerous NASA, NSF, DOE, and the National Research Council committees related to spectroscopy and remote sensing. Among recognitions for her work, she is a Fellow of the American Geophysical Union and received an honorary doctorate from the University of Zurich. She has published more than 200 scientific papers related to ecological remote sensing and has worked with most of the Earth-observing U.S. airborne and spaceborne systems.

Preface

This seminal book on *Hyperspectral Remote Sensing of Vegetation* (Second Edition, 4 Volume Set), published by Taylor and Francis Inc.\CRC Press is an outcome of over 2 years of effort by the editors and authors. In 2011, the first edition of *Hyperspectral Remote Sensing of Vegetation* was published. The book became a standard reference on hyperspectral remote sensing of vegetation amongst the remote sensing community across the world. This need and resulting popularity demanded a second edition with more recent as well as more comprehensive coverage of the subject. Many advances have taken place since the first edition. Further, the first edition was limited in scope in the sense it covered some very important topics and missed equally important topics (e.g., hyperspectral library of agricultural crops, hyperspectral pre-processing steps and algorithms, and many others). As a result, a second edition that brings us up-to-date advances in hyperspectral remote sensing of vegetation was required. Equally important was the need to make the book more comprehensive, covering an array of subjects not covered in the first edition. So, my coeditors and myself did a careful research on what should go into the second edition. Quickly, the scope of the second edition expanded resulting in an increasing number of chapters. All of this led to developing the seminal book: *Hyperspectral Remote Sensing of Vegetation*, Second Edition, 4 Volume Set. The four volumes are:

Volume I: Fundamentals, Sensor Systems, Spectral Libraries, and Data Mining for Vegetation
Volume II: Hyperspectral Indices and Image Classifications for Agriculture and Vegetation
Volume III: Biophysical and Biochemical Characterization and Plant Species Studies
Volume IV: Advanced Applications in Remote Sensing of Agricultural Crops and Natural Vegetation

The goal of the book was to bring in one place collective knowledge of the last 50 years of advances in hyperspectral remote sensing of vegetation with a target audience of wide spectrum of scientific community, students, and professional application practitioners. The book documents knowledge advances made in applying hyperspectral remote sensing technology in the study of terrestrial vegetation that include agricultural crops, forests, rangelands, and wetlands. This is a very practical offering about a complex subject that is rapidly advancing its knowledge-base. In a very practical way, the book demonstrates the experience, utility, methods, and models used in studying terrestrial vegetation using hyperspectral data. The four volumes, with a total of 48 chapters, are divided into distinct themes.

- **Volume I**: There are 14 chapters focusing on hyperspectral instruments, spectral libraries, and methods and approaches of data handling. The chapters extensively address various preprocessing steps and data mining issues such as the Hughes phenomenon and overcoming the "curse of high dimensionality" of hyperspectral data. Developing spectral libraries of crops, vegetation, and soils with data gathered from hyperspectral data from various platforms (ground-based, airborne, spaceborne), study of spectral traits of crops, and proximal sensing at field for phenotyping are extensively discussed. Strengths and limitations of hyperspectral data of agricultural crops and vegetation acquired from different platforms are discussed. It is evident from these chapters that the hyperspectral data provides opportunities for great advances in study of agricultural crops and vegetation. However, it is also clear from these chapters that hyperspectral data should not be treated as panacea to every limitation of multispectral broadband data such as from Landsat or Sentinel series of satellites. The hundreds or thousands of hyperspectral narrowbands (HNBs) as well as carefully selected hyperspectral vegetation indices (HVIs) will help us make significant advances in characterizing, modeling, mapping, and monitoring vegetation biophysical, biochemical, and structural quantities. However, it is also important to properly understand hyperspectral data and eliminate redundant bands that exist for every application and to optimize computing as well as human resources to enable seamless and efficient handling enormous volumes of hyperspectral data. Special emphasis is also put on preprocessing and processing of Earth Observing-1 (EO-1) Hyperion, the first publicly available hyperspectral data from space. These methods, approaches, and algorithms, and protocols set the stage for upcoming satellite hyperspectral sensors such as NASA's HyspIRI and Germany's EnMAP.

- **Volume II**: There are 10 chapters focusing on hyperspectral vegetation indices (HVIs) and image classification methods and techniques. The HVIs are of several types such as: (i) two-band derived, (ii) multi-band-derived, and (iii) derivative indices derived. The strength of the HVIs lies in the fact that specific indices can be derived for specific biophysical, biochemical, and plant structural quantities. For example, you have carotenoid HVI, anthocyanin HVI, moisture or water HVI, lignin HVI, cellulose HVI, biomass or LAI or other biophysical HVIs, red-edge based HVIs, and so on. Further, since these are narrowband indices, they are better targeted and centered at specific sensitive wavelength portions of the spectrum. The strengths and limitations of HVIs in a wide array of applications such as leaf nitrogen content (LNC), vegetation water content, nitrogen content in vegetation, leaf and plant pigments, anthocyanin's, carotenoids, and chlorophyll are thoroughly studied. Image classification using hyperspectral data provides great strengths in deriving more classes (e.g., crop species within a crop as opposed to just crop types) and increasing classification accuracies. In earlier years and decades, hyperspectral data classification and analysis was a challenge due to computing and data handling issues. However, with the availability of machine learning algorithms on cloud computing (e.g., Google Earth Engine) platforms, these challenges have been overcome in the last 2–3 years. Pixel-based supervised machine learning algorithms like the random forest, and support vector machines as well as object-based algorithms like the recursive hierarchical segmentation, and numerous others methods (e.g., unsupervised approaches) are extensively discussed. The ability to process petabyte volume data of the planet takes us to a new level of sophistication and makes use of data such as from hyperspectral sensors feasible over large areas. The cloud computing architecture involved with handling massively large petabyte-scale data volumes are presented and discussed.
- **Volume III**: There are 11 chapters focusing on biophysical and biochemical characterization and plant species studies. A number of chapters in this volume are focused on separating and discriminating agricultural crops and vegetation of various types or species using hyperspectral data. Plant species discrimination and classification to separate them are the focus of study using vegetation such as forests, invasive species in different ecosystems, and agricultural crops. Performance of hyperspectral narrowbands (HNBs) and hyperspectral vegetation indices (HVIs) when compared with multispectral broadbands (MBBs) and multispectral broadband vegetation indices (BVIs) are presented and discussed. The vegetation and agricultural crops are studied at various scales, and their vegetation functional properties diagnosed. The value of digital surface models in study of plant traits as complementary\supplementary to hyperspectral data has been highlighted. Hyperspectral bio-indicators to study photosynthetic efficiency and vegetation stress are presented and discussed. Studies are conducted using hyperspectral data across wavelengths (e.g., visible, near-infrared, shortwave-infrared, mid-infrared, and thermal-infrared).
- **Volume IV**: There are 15 chapters focusing on specific advanced applications of hyperspectral data in study of agricultural crops and natural vegetation. Specific agricultural crop applications include crop management practices, crop stress, crop disease, nitrogen application, and presence of heavy metals in soils and related stress factors. These studies discuss biophysical and biochemical quantities modeled and mapped for precision farming,

hyperspectral narrowbands (HNBs), and hyperspectral vegetation indices (HVIs) involved in assessing nitrogen in plants, and the study of the impact of heavy metals on crop health and stress. Vegetation functional studies using hyperspectral data presented and discussed include crop water use (actual evapotranspiration), net primary productivity (NPP), gross primary productivity (GPP), phenological applications, and light use efficiency (LUE). Specific applications discussed under vegetation functional studies using hyperspectral data include agricultural crop classifications, machine learning, forest management studies, pasture studies, and wetland studies. Applications in fire assessment, modeling, and mapping using hyperspectral data in the optical and thermal portions of the spectrum are presented and discussed. Hyperspectral data in global change studies as well as in outer planet studies have also been discussed. Much of the outer planet remote sensing is conducted using imaging spectrometer and hence the data preprocessing and processing methods of Earth and that of outer planets have much in common and needs further examination.

The chapters are written by leading experts in the global arena with each chapter: (a) focusing on specific applications, (b) reviewing existing "state-of-art" knowledge, (c) highlighting the advances made, and (d) providing guidance for appropriate use of hyperspectral data in study of vegetation and its numerous applications such as crop yield modeling, crop biophysical and biochemical property characterization, and crop moisture assessment.

The four-volume book is specifically targeted on hyperspectral remote sensing as applied to terrestrial vegetation applications. This is a big market area that includes agricultural croplands, study of crop moisture, forests, and numerous applications such as droughts, crop stress, crop productivity, and water productivity. To the knowledge of the editors, there is no comparable book, source, and/or organization that can bring this body of knowledge together in one place, making this a "must buy" for professionals. This is clearly a unique contribution whose time is now. The book highlights include:

1. Best global expertise on hyperspectral remote sensing of vegetation, agricultural crops, crop water use, plant species detection, crop productivity and water productivity mapping, and modeling;
2. Clear articulation of methods to conduct the work. Very practical;
3. Comprehensive review of the existing technology and clear guidance on how best to use hyperspectral data for various applications;
4. Case studies from a variety of continents with their own subtle requirements; and
5. Complete solutions from methods to applications inventory and modeling.

Hyperspectral narrowband spectral data, as discussed in various chapters of this book, are fast emerging as practical most advanced solutions in modeling and mapping vegetation. Recent research has demonstrated the advances and great value made by hyperspectral data, as discussed in various chapters in: (a) quantifying agricultural crops as to their biophysical and harvest yield characteristics, (b) modeling forest canopy biochemical properties, (c) establishing plant and soil moisture conditions, (d) detecting crop stress and disease, (e) mapping leaf chlorophyll content as it influences crop production, (f) identifying plants affected by contaminants such as arsenic, and (g) demonstrating sensitivity to plant nitrogen content, and (h) invasive species mapping. The ability to significantly better quantify, model, and map plant chemical, physical, and water properties is well established and has great utility.

Even though these accomplishments and capabilities have been reported in various places, the need for a collective "knowledge bank" that links these various advances in one place is missing. Further, most scientific papers address specific aspects of research, failing to provide a comprehensive assessment of advances that have been made nor how the professional can bring those advances to their work. For example, deep scientific journals report practical applications of hyperspectral narrowbands yet one has to canvass the literature broadly to obtain the pertinent facts. Since several papers report this, there is a need to synthesize these findings so that the reader gets the correct picture of the best wavebands for their practical applications. Also, studies do differ in exact methods most suited for detecting parameters such as crop moisture variability, chlorophyll content, and stress levels. The professional needs this sort of synthesis and detail to adopt best practices for their own work.

In years and decades past, use of hyperspectral data had its challenges especially in handling large data volumes. That limitation is now overcome through cloud-computing, machine learning, deep learning, artificial intelligence, and advances in knowledge in processing and applying hyperspectral data.

This book can be used by anyone interested in hyperspectral remote sensing that includes advanced research and applications, such as graduate students, undergraduates, professors, practicing professionals, policy makers, governments, and research organizations.

Dr. Prasad S. Thenkabail, PhD
Editor-in-Chief
Hyperspectral Remote Sensing of Vegetation, Second Edition, Four Volume Set

Acknowledgments

This four-volume *Hyperspectral Remote Sensing of Vegetation* book (second edition) was made possible by sterling contributions from leading professionals from around the world in the area of hyperspectral remote sensing of vegetation and agricultural crops. As you will see from list of authors and coauthors, we have an assembly of "**who is who**" in hyperspectral remote sensing of vegetation who have contributed to this book. They wrote insightful chapters, that are an outcome of years of careful research and dedication, to make the book appealing to a broad section of readers dealing with remote sensing. My gratitude goes to (mentioned in no particular order; names of lead authors of the chapters are shown in bold): **Drs. Fred Ortenberg** (Technion–Israel Institute of Technology, Israel), **Jiaguo Qi** (Michigan State University, USA), **Angela Lausch** (Helmholtz Centre for Environmental Research, Leipzig, Germany), **Andries B. Potgieter** (University of Queensland, Australia), **Muhammad Al-Amin Hoque** (University of Queensland, Australia), **Andreas Hueni** (University of Zurich, Switzerland), **Eyal Ben-Dor** (Tel Aviv University, Israel), **Itiya Aneece** (United States Geological Survey, USA), **Sreekala Bajwa** (University of Arkansas, USA), **Antonio Plaza** (University of Extremadura, Spain), **Jessica J. Mitchell** (Appalachian State University, USA), **Dar Roberts** (University of California at Santa Barbara, USA), **Quan Wang** (Shizuoka University, Japan), **Edoardo Pasolli** (University of Trento, Italy), (Nanjing University of Science and Technology, China), **Anatoly Gitelson** (University of Nebraska- Lincoln, USA), **Tao Cheng** (Nanjing Agricultural University, China), **Roberto Colombo** (University of Milan-Bicocca, Italy), **Daniela Stroppiana** (Institute for Electromagnetic Sensing of the Environment, Italy), **Yongqin Zhang** (Delta State University, USA), **Yoshio Inoue** (National Institute for Agro-Environmental Sciences, Japan), Yafit Cohen (Institute of Agricultural Engineering, Israel), **Helge Aasen** (Institute of Agricultural Sciences, ETH Zurich), **Elizabeth M. Middleton** (NASA, USA), **Yongqin Zhang** (University of Toronto, Canada), **Yan Zhu** (Nanjing Agricultural University, China), **Lênio Soares Galvão** (Instituto Nacional de Pesquisas Espaciais [INPE], Brazil), **Matthew L. Clark** (Sonoma State University, USA), **Matheus Pinheiro Ferreira** (University of Paraná, Curitiba, Brazil), **Ruiliang Pu** (University of South Florida, USA), **Scott C. Chapman** (CSIRO, Australia), **Haibo Yao** (Mississippi State University, USA), **Jianlong Li** (Nanjing University, China), **Terry Slonecker** (USGS, USA), **Tobias Landmann** (International Centre of Insect Physiology and Ecology, Kenya), **Michael Marshall** (University of Twente, Netherlands), **Pamela Nagler** (USGS, USA), **Alfredo Huete** (University of Technology Sydney, Australia), **Prem Chandra Pandey** (Banaras Hindu University, India), **Valerie Thomas** (Virginia Tech., USA), **Izaya Numata** (South Dakota State University, USA), **Elijah W. Ramsey III** (USGS, USA), **Sander Veraverbeke** (Vrije Universiteit Amsterdam and University of California, Irvine), **Tomoaki Miura** (University of Hawaii, USA), **R. G. Vaughan** (U.S. Geological Survey, USA), Victor Alchanatis (Agricultural research Organization, Volcani Center, Israel), Dr. Narumon Wiangwang (Royal Thai Government, Thailand), Pedro J. Leitão (Humboldt University of Berlin, Department of Geography, Berlin, Germany), James Watson (University of Queensland, Australia), Barbara George-Jaeggli (ETH Zuerich, Switzerland), Gregory McLean (University of Queensland, Australia), Mark Eldridge (University of Queensland, Australia), Scott C. Chapman (University of Queensland, Australia), Kenneth Laws (University of Queensland, Australia), Jack Christopher (University of Queensland, Australia), Karine Chenu (University of Queensland, Australia), Andrew Borrell (University of Queensland, Australia), Graeme L. Hammer (University of Queensland, Australia), David R. Jordan (University of Queensland, Australia), Stuart Phinn (University of Queensland, Australia), Lola Suarez (University of Melbourne, Australia), Laurie A. Chisholm (University of Wollongong, Australia), Alex Held (CSIRO, Australia), S. Chabrillant (GFZ German Research Center for Geosciences, Germany), José A. M. Demattê (University of São Paulo, Brazil), Yu Zhang (North Dakota State University, USA), Ali Shirzadifar (North Dakota State University, USA), Nancy F. Glenn (Boise State University, USA), Kyla M. Dahlin (Michigan State

University, USA), Nayani Ilangakoon (Boise State University, USA), Hamid Dashti (Boise State University, USA), Megan C. Maloney (Appalachian State University, USA), Subodh Kulkarni (University of Arkansas, USA), Javier Plaza (University of Extremadura, Spain), Gabriel Martin (University of Extremadura, Spain), Segio Sánchez (University of Extremadura, Spain), Wei Wang (Nanjing Agricultural University, China), Xia Yao (Nanjing Agricultural University, China), Busetto Lorenzo (Università Milano-Bicocca), Meroni Michele (Università Milano-Bicocca), Rossini Micol (Università Milano-Bicocca), Panigada Cinzia (Università Milano-Bicocca), F. Fava (Università degli Studi di Sassari, Italy), M. Boschetti (Institute for Electromagnetic Sensing of the Environment, Italy), P. A. Brivio (Institute for Electromagnetic Sensing of the Environment, Italy), K. Fred Huemmrich (University of Maryland, Baltimore County, USA), Yen-Ben Cheng (Earth Resources Technology, Inc., USA), Hank A. Margolis (Centre d'Études de la Forêt, Canada), Yafit Cohen (Agricultural research Organization, Volcani Center, Israel), Kelly Roth (University of California at Santa Barbara, USA), Ryan Perroy (University of Wisconsin-La Crosse, USA), Ms. Wei Wang (Nanjing Agricultural University, China), Dr. Xia Yao (Nanjing Agricultural University, China), Keely L. Roth (University of California, Santa Barbara, USA), Erin B. Wetherley (University of California at Santa Barbara, USA), Susan K. Meerdink (University of California at Santa Barbara, USA), Ryan L. Perroy (University of Wisconsin-La Crosse, USA), B. B. Marithi Sridhar (Bowling Green University, USA), Aaryan Dyami Olsson (Northern Arizona University, USA), Willem Van Leeuwen (University of Arizona, USA), Edward Glenn (University of Arizona, USA), José Carlos Neves Epiphanio (Instituto Nacional de Pesquisas Espaciais [INPE], Brazil), Fábio Marcelo Breunig (Instituto Nacional de Pesquisas Espaciais [INPE], Brazil), Antônio Roberto Formaggio (Instituto Nacional de Pesquisas Espaciais [INPE], Brazil), Amina Rangoonwala (IAP World Services, Lafayette, LA), Cheryl Li (Nanjing University, China), Deghua Zhao (Nanjing University, China), Chengcheng Gang (Nanjing University, China), Lie Tang (Mississippi State University, USA), Lei Tian (Mississippi State University, USA), Robert Brown (Mississippi State University, USA), Deepak Bhatnagar (Mississippi State University, USA), Thomas Cleveland (Mississippi State University, USA), Hiroki Yoshioka (Aichi Prefectural University, Japan), T. N. Titus (U.S. Geological Survey, USA), J. R. Johnson (U.S. Geological Survey, USA), J. J. Hagerty (U.S. Geological Survey, USA), L. Gaddis (U.S. Geological Survey, USA), L. A. Soderblom (U.S. Geological Survey, USA), and P. Geissler (U.S. Geological Survey, USA), Jua Jin (Shizuoka University, Japan), Rei Sonobe (Shizuoka University, Japan), Jin Ming Chen (Shizuoka University, Japan), Saurabh Prasad (University of Houston, USA), Melba M. Crawford (Purdue University, USA), James C. Tilton (NASA Goddard Space Flight Center, USA), Jin Sun (Nanjing University of Science and Technology, China), Yi Zhang (Nanjing University of Science and Technology, China), Alexei Solovchenko (Moscow State University, Moscow), Yan Zhu, (Nanjing Agricultural University, China), Dong Li (Nanjing Agricultural University, China), Kai Zhou (Nanjing Agricultural University, China), Roshanak Darvishzadeh (University of Twente, Enschede, The Netherlands), Andrew Skidmore (University of Twente, Enschede, The Netherlands), Victor Alchanatis (Institute of Agricultural Engineering, The Netherlands), Georg Bareth (University of Cologne, Germany), Qingyuan Zhang (Universities Space Research Association, USA), Petya K. E. Campbell (University of Maryland Baltimore County, USA), and David R. Landis (Global Science & Technology, Inc., USA), José Carlos Neves Epiphanio (Instituto Nacional de Pesquisas Espaciais [INPE], Brazil), Fábio Marcelo Breunig (Universidade Federal de Santa Maria [UFSM], Brazil), and Antônio Roberto Formaggio (Instituto Nacional de Pesquisas Espaciais [INPE], Brazil), Cibele Hummel do Amaral (Federal University of Viçosa, in Brazil), Gaia Vaglio Laurin (Tuscia University, Italy), Raymond Kokaly (U.S. Geological Survey, USA), Carlos Roberto de Souza Filho (University of Ouro Preto, Brazil), Yosio Edemir Shimabukuro (Federal Rural University of Rio de Janeiro, Brazil), Bangyou Zheng (CSIRO, Australia), Wei Guo (The University of Tokyo, Japan), Frederic Baret (INRA, France), Shouyang Liu (INRA, France), Simon Madec (INRA, France), Benoit Solan (ARVALIS, France), Barbara George-Jaeggli (University of Queensland, Australia), Graeme L. Hammer (University of Queensland, Australia), David R. Jordan (University of Queensland, Australia), Yanbo Huang (USDA, USA), Lie Tang (Iowa State

University, USA), Lei Tian (University of Illinois. USA), Deepak Bhatnagar (USDA, USA), Thomas E. Cleveland (USDA, USA), Dehua ZHAO (Nanjing University, USA), Hannes Feilhauer (University of Erlangen-Nuremberg, Germany), Miaogen Shen (Institute of Tibetan Plateau Research, Chinese Academy of Sciences, Beijing, China), Jin Chen (College of Remote Sensing Science and Engineering, Faculty of Geographical Science, Beijing Normal University, Beijing, China), Suresh Raina (International Centre of Insect Physiology and Ecology, Kenya and Pollination services, India), Danny Foley (Northern Arizona University, USA), Cai Xueliang (UNESCO-IHE, Netherlands), Trent Biggs (San Diego State University, USA), Werapong Koedsin (Prince of Songkla University, Thailand), Jin Wu (University of Hong Kong, China), Kiril Manevski (Aarhus University, Denmark), Prashant K. Srivastava (Banaras Hindu University, India), George P. Petropoulos (Technical University of Crete, Greece), Philip Dennison (University of Utah, USA), Ioannis Gitas (University of Thessaloniki, Greece), Glynn Hulley (NASA Jet Propulsion Laboratory, California Institute of Technology, USA), Olga Kalashnikova, (NASA Jet Propulsion Laboratory, California Institute of Technology, USA), Thomas Katagis (University of Thessaloniki, Greece), Le Kuai (University of California, USA), Ran Meng (Brookhaven National Laboratory, USA), Natasha Stavros (California Institute of Technology, USA).

Hiroki Yoshioka (Aichi Prefectural University, Japan), My two coeditors, **Professor John G. Lyon** and **Professor Alfredo Huete**, have made outstanding contribution to this four-volume *Hyperspectral Remote Sensing of Vegetation* book (second edition). Their knowledge of hyperspectral remote sensing is enormous. Vastness and depth of their understanding of remote sensing in general and hyperspectral remote sensing in particular made my job that much easier. I have learnt a lot from them and continue to do so. Both of them edited some or all of the 48 chapters of the book and also helped structure chapters for a flawless reading. They also significantly contributed to the synthesis chapter of each volume. I am indebted to their insights, guidance, support, motivation, and encouragement throughout the book project.

My coeditors and myself are grateful to **Dr. Alexander F. H. Goetz** and **Prof. Susan L. Ustin** for writing the foreword for the book. Please refer to their biographical sketch under the respective foreword written by these two leaders of Hyperspectral Remote Sensing.

Both the forewords are a must read to anyone studying this four-volume *Hyperspectral Remote Sensing of Vegetation* book (second edition). They are written by two giants who have made immense contribution to the subject and I highly recommend that the readers read them.

I am blessed to have had the support and encouragement (professional and personal) of my U.S. Geological Survey and other colleagues. In particular, I would like to mention Mr. Edwin Pfeifer (late), Dr. Susan Benjamin, Dr. Dennis Dye, and Mr. Larry Gaffney. Special thanks to Dr. Terrence Slonecker, Dr. Michael Marshall, Dr. Isabella Mariotto, and Dr. Itiya Aneece who have worked closely with me on hyperspectral research over the years. Special thanks are also due to Dr. Pardhasaradhi Teluguntla, Mr. Adam Oliphant, and Dr. Muralikrishna Gumma who have contributed to my various research efforts and have helped me during this book project directly or indirectly. I am grateful to Prof. Ronald B. Smith, professor at Yale University who was instrumental in supporting my early hyperspectral research at the Yale Center for Earth Observation (YCEO), Yale University. Opportunities and guidance I received in my early years of remote sensing from Prof. Andrew D. Ward, professor at the Ohio State University, Prof. John G. Lyon, former professor at the Ohio State University, and Mr. Thiruvengadachari, former Scientist at the National Remote Sensing Center (NRSC), Indian Space Research Organization, India, is gratefully acknowledged.

My wife (Sharmila Prasad) and daughter (Spandana Thenkabail) are two great pillars of my life. I am always indebted to their patience, support, and love.

Finally, kindly bear with me for sharing a personal story. When I started editing the first edition in the year 2010, I was diagnosed with colon cancer. I was not even sure what the future was and how long I would be here. I edited much of the first edition soon after the colon cancer surgery and during and after the 6 months of chemotherapy—one way of keeping my mind off the negative thoughts. When you are hit by such news, there is nothing one can do, but to be positive, trust your doctors, be thankful to support and love of the family, and have firm belief in the higher spiritual being (whatever your beliefs are). I am so very grateful to some extraordinary people who helped me through this difficult life event: Dr. Parvasthu Ramanujam (surgeon), Dr. Paramjeet K. Bangar (Oncologist), Dr. Harnath Sigh (my primary doctor), Dr. Ram Krishna (Orthopedic Surgeon and family friend), three great nurses (Ms. Irene, Becky, Maryam) at Banner Boswell Hospital (Sun City, Arizona, USA), courage-love-patience-prayers from my wife, daughter, and several family members, friends, and colleagues, and support from numerous others that I have not named here. During this phase, I learnt a lot about cancer and it gave me an enlightened perspective of life. My prayers were answered by the higher power. I learnt a great deal about life—good and bad. I pray for all those with cancer and other patients that diseases one day will become history or, in the least, always curable without suffering and pain. Now, after 8 years, I am fully free of colon cancer and was able to edit the four-volume *Hyperspectral Remote Sensing of Vegetation* book (second edition) without the pain and suffering that I went through when editing the first edition. What a blessing. These blessings help us give back in our own little ways. To realize that it is indeed profound to see the beautiful sunrise every day, the day go by with every little event (each with a story of their own), see the beauty of the sunset, look up to the infinite universe and imagine on its many wonders, and just to breathe fresh air every day and enjoy the breeze. These are all many wonders of life that we need to enjoy, cherish, and contemplate.

Dr. Prasad S. Thenkabail, PhD
Editor-in-Chief
Hyperspectral Remote Sensing of Vegetation

Editors

Prasad S. Thenkabail, Research Geographer-15, U.S. Geological Survey (USGS), is a world-recognized expert in remote sensing science with multiple major contributions in the field sustained over more than 30 years. He obtained his PhD from the Ohio State University in 1992 and has over 140+ peer-reviewed scientific publications, mostly in major international journals.

Dr. Thenkabail has conducted pioneering research in the area of hyperspectral remote sensing of vegetation and in that of global croplands and their water use in the context of food security. In hyperspectral remote sensing he has done cutting-edge research with wide implications in advancing remote sensing science in application to agriculture and vegetation. This body of work led to more than ten peer-reviewed research publications with high impact. For example, a single paper [1] has received 1000+ citations as at the time of writing (October 4, 2018). Numerous other papers, book chapters, and books (as we will learn below) are also related to this work, with two other papers [2,3] having 350+ to 425+ citations each.

In studies of global croplands in the context of food and water security, he has led the release of the world's first Landsat 30-m derived global cropland extent product. This work demonstrates a "paradigm shift" in how remote sensing science is conducted. The product can be viewed in full resolution at the web location www.croplands.org. The data is already widely used worldwide and is downloadable from the NASA\USGS LP DAAC site [4]. There are numerous major publication in this area (e.g. [5,6]).

Dr. Thenkabail's contributions to series of leading edited books on remote sensing science places him as a world leader in remote sensing science advances. He edited three-volume *Remote Sensing Handbook* published by Taylor and Francis, with 82 chapters and more than 2000 pages, widely considered a "magnus opus" standard reference for students, scholars, practitioners, and major experts in remote sensing science. Links to these volumes along with endorsements from leading global remote sensing scientists can be found at the location give in note [7]. He has recently completed editing *Hyperspectral Remote Sensing of Vegetation* published by Taylor and Francis in four volumes with 50 chapters. This is the second edition is a follow-up on the earlier single-volume *Hyperspectral Remote Sensing of Vegetation* [8]. He has also edited a book on *Remote Sensing of Global Croplands for Food Security* (Taylor and Francis) [9]. These books are widely used and widely referenced in institutions worldwide.

Dr. Thenkabail's service to remote sensing community is second to none. He is currently an editor-in-chief of the *Remote Sensing* open access journal published by MDPI; an associate editor of the journal *Photogrammetric Engineering and Remote Sensing* (PERS) of the American Society of Photogrammetry and Remote Sensing (ASPRS); and an editorial advisory board member of the International Society of Photogrammetry and Remote Sensing (ISPRS) *Journal of Photogrammetry and Remote Sensing*. Earlier, he served on the editorial board of *Remote Sensing of Environment* for many years (2007–2017). As an editor-in-chief of the open access *Remote Sensing* MDPI journal from 2013 to date he has been instrumental in providing leadership for an online publication that did not even have a impact factor when he took over but is now one of the five leading remote sensing international journals, with an impact factor of 3.244.

Dr. Thenkabail has led remote sensing programs in three international organizations: International Water Management Institute (IWMI), 2003–2008; International Center for Integrated Mountain Development (ICIMOD), 1995–1997; and International Institute of Tropical Agriculture (IITA),

1992–1995. He has worked in more than 25+ countries on several continents, including East Asia (China), S-E Asia (Cambodia, Indonesia, Myanmar, Thailand, Vietnam), Middle East (Israel, Syria), North America (United States, Canada), South America (Brazil), Central Asia (Uzbekistan), South Asia (Bangladesh, India, Nepal, and Sri Lanka), West Africa (Republic of Benin, Burkina Faso, Cameroon, Central African Republic, Cote d'Ivoire, Gambia, Ghana, Mali, Nigeria, Senegal, and Togo), and Southern Africa (Mozambique, South Africa). During this period he has made major contributions and written seminal papers on remote sensing of agriculture, water resources, inland valley wetlands, global irrigated and rain-fed croplands, characterization of African rainforests and savannas, and drought monitoring systems.

The quality of Dr. Thenkabail's research is evidenced in the many awards, which include, in 2015, the American Society of Photogrammetry and Remote Sensing (ASPRS) ERDAS award for best scientific paper in remote sensing (Marshall and Thenkabail); in 2008, the ASPRS President's Award for practical papers, second place (Thenkabail and coauthors); and in 1994, the ASPRS Autometric Award for outstanding paper (Thenkabail and coauthors). His team was recognized by the Environmental System Research Institute (ESRI) for "special achievement in GIS" (SAG award) for their Indian Ocean tsunami work. The USGS and NASA selected him to be on the Landsat Science Team for a period of five years (2007–2011).

Dr. Thenkabail is regularly invited as keynote speaker or invited speaker at major international conferences and at other important national and international forums every year. He has been principal investigator and/or has had lead roles of many pathfinding projects, including the ~5 million over five years (2014–2018) for the global food security support analysis data in the 30-m (GFSAD) project (https://geography.wr.usgs.gov/science/croplands/) funded by NASA MEaSUREs (Making Earth System Data Records for Use in Research Environments), and projects such as Sustain and Manage America's Resources for Tomorrow (waterSMART) and characterization of Eco-Regions in Africa (CERA).

REFERENCES

1. Thenkabail, P.S., Smith, R.B., and De-Pauw, E. 2000b. Hyperspectral vegetation indices for determining agricultural crop characteristics. *Remote Sensing of Environment*, 71:158–182.
2. Thenkabail, P.S., Enclona, E.A., Ashton, M.S., Legg, C., and Jean De Dieu, M. 2004. Hyperion, IKONOS, ALI, and ETM+ sensors in the study of African rainforests. *Remote Sensing of Environment*, 90:23–43.
3. Thenkabail, P.S., Enclona, E.A., Ashton, M.S., and Van Der Meer, V. 2004. Accuracy assessments of hyperspectral waveband performance for vegetation analysis applications. *Remote Sensing of Environment*, 91(2–3):354–376.
4. https://lpdaac.usgs.gov/about/news_archive/release_gfsad_30_meter_cropland_extent_products
5. Thenkabail, P.S. 2012. Guest Editor for Global Croplands Special Issue. *Photogrammetric Engineering and Remote Sensing*, 78(8).
6. Thenkabail, P.S., Knox, J.W., Ozdogan, M., Gumma, M.K., Congalton, R.G., Wu, Z., Milesi, C., Finkral, A., Marshall, M., Mariotto, I., You, S. Giri, C. and Nagler, P. 2012. Assessing future risks to agricultural productivity, water resources and food security: how can remote sensing help? *Photogrammetric Engineering and Remote Sensing*, August 2012 Special Issue on Global Croplands: Highlight Article. 78(8):773–782. IP-035587.
7. https://www.crcpress.com/Remote-Sensing-Handbook---Three-Volume-Set/Thenkabail/p/book/9781482218015
8. https://www.crcpress.com/Hyperspectral-Remote-Sensing-of-Vegetation/Thenkabail-Lyon/p/book/9781439845370
9. https://www.crcpress.com/Remote-Sensing-of-Global-Croplands-for-Food-Security/Thenkabail-Lyon-Turral-Biradar/p/book/9781138116559

John G. Lyon, educated at Reed College in Portland, OR and the University of Michigan in Ann Arbor, has conducted scientific and engineering research and carried out administrative functions throughout his career. He was formerly the Senior Physical Scientist (ST) in the US Environmental Protection Agency's Office of Research and Development (ORD) and Office of the Science Advisor in Washington, DC, where he co-led work on the Group on Earth Observations and the USGEO subcommittee of the Committee on Environment and Natural Resources and research on geospatial issues in the agency. For approximately eight years, he was director of ORD's Environmental Sciences Division, which conducted research on remote sensing and geographical information system (GIS) technologies as applied to environmental issues including landscape characterization and ecology, as well as analytical chemistry of hazardous wastes, sediments, and ground water. He previously served as professor of civil engineering and natural resources at Ohio State University (1981–1999). Professor Lyon's own research has led to authorship or editorship of a number of books on wetlands, watershed, and environmental applications of GIS, and accuracy assessment of remote sensor technologies.

Alfredo Huete leads the Ecosystem Dynamics Health and Resilience research program within the Climate Change Cluster (C3) at the University of Technology Sydney, Australia. His main research interest is in using remote sensing to study and analyze vegetation processes, health, and functioning, and he uses satellite data to observe land surface responses and interactions with climate, land use activities, and extreme events. He has more than 200 peer-reviewed journal articles, including publication in such prestigious journals as *Science* and *Nature*. He has over 25 years' experience working on NASA and JAXA mission teams, including the NASA-EOS MODIS Science Team, the EO-1 Hyperion Team, the JAXA GCOM-SGLI Science Team, and the NPOESS-VIIRS advisory group. Some of his past research involved the development of the soil-adjusted vegetation index (SAVI) and the enhanced vegetation index (EVI), which became operational satellite products on MODIS and VIIRS sensors. He has also studied tropical forest phenology and Amazon forest greening in the dry season, and his work was featured in a *National Geographic* television special entitled "The Big Picture." Currently, he is involved with the Australian Terrestrial Ecosystem Research Network (TERN), helping to produce national operational phenology products; as well as the AusPollen network, which couples satellite sensing to better understand and predict pollen phenology from allergenic grasses and trees.

Contributors

Itiya P. Aneece
Western Geographic Science Center
United States Geological Survey
Flagstaff, Arizona

Sreekala G. Bajwa
Department of Agricultural and Biosystems Engineering
North Dakota State University
Fargo, North Dakota

E. Ben-Dor
Department of Geography
Porter School of Environment and Earth Science
Tel-Aviv University
Tel-Aviv, Israel

Andrew Borrell
Queensland Alliance for Agriculture and Food Innovation (QAAFI)
University of Queensland
Brisbane, Australia

S. Chabrillat
Remote Sensing
Helmholtz Centre Potsdam GFZ German Research Centre for Geosciences
Potsdam, Germany

Scott C. Chapman
CSIRO Agriculture Flagship
and
School of Agriculture and Food Sciences
The University of Queensland
Queensland, Australia

Karine Chenu
Queensland Alliance for Agriculture and Food Innovation (QAAFI)
University of Queensland
Brisbane, Australia

Laurie A. Chisholm
School of Earth an Environmental Sciences
University of Wollongong
Wollongong, Australia

Jack Christopher
Queensland Alliance for Agriculture and Food Innovation (QAAFI)
University of Queensland
Brisbane, Australia

Kyla M. Dahlin
Department of Geography, Environment, and Spatial Sciences
Michigan State University
East Lansing, Michigan

Hamid Dashti
Department of Geosciences
Boise State University
Boise, Idaho

José A. M. Demattê
Department of Soil Science
College of Agriculture "Luiz de Queiroz"
University of São Paulo
São Paulo, Brazil

Mark Eldridge
Queensland Alliance for Agriculture and Food Innovation (QAAFI)
University of Queensland
Brisbane, Australia

Barbara George-Jaeggli
Queensland Alliance for Agriculture and Food Innovation (QAAFI)
University of Queensland
Brisbane, Australia

and

Department of Agriculture and Fisheries
Agri-Science Queensland
Queensland, Australia

Nancy F. Glenn
Department of Geosciences
Boise State University
Boise, Idaho

Graeme L. Hammer
Queensland Alliance for Agriculture and Food Innovation (QAAFI)
University of Queensland
Brisbane, Australia

Alex Held
CSIRO-Land and Water
Commonwealth Scientific and Industrial Research Organisation
Canberra, Australia

Muhammad Al-Amin Hoque
Remote Sensing Research Centre
School of Earth and Environmental Sciences
University of Queensland
Brisbane, Australia

and

Department of Geography and Environment
Jagannath University
Dhaka, Bangladesh

Andreas Hueni
Geography Department
Remote Sensing Laboratories
University of Zurich
Zurich, Switzerland

Alfredo Huete
School of Life Sciences
University of Technology Sydney
New South Wales, Australia

Nayani Ilangakoon
Department of Geosciences
Boise State University
Boise, Idaho

Yoshio Inoue
Agro-Ecosystem Informatics Research
National Institute for Agro-Environmental Sciences
Ibaraki, Japan

David R. Jordan
Queensland Alliance for Agriculture and Food Innovation (QAAFI)
University of Queensland
Brisbane, Australia

Angela Lausch
Department Computational Landscape Ecology
Helmholtz Centre for Environmental Research—UFZ
Leipzig, Germany

and

Department of Geography
Humboldt University of Berlin
Berlin, Germany

Kenneth Laws
Department of Agriculture and Fisheries
Agri-Science Queensland
Queensland, Australia

Pedro J. Leitão
Department of Geography
Humboldt University of Berlin
Berlin, Germany

and

Department of Landscape Ecology and Environmental System Analysis
Technische Universität Braunschweig
Braunschweig, Germany

John G. Lyon
American Society for Photogrammetry and Remote Sensing
Chantilly, Virginia

Megan C. Maloney
Department of Geography
Appalachian State University
Boone, North Carolina

Gabriel Martín
Instituto Superior Técnico
Technical University of Lisbon
Lisbon, Portugal

Gregory McLean
Department of Agriculture and Fisheries
Agri-Science Queensland
Queensland, Australia

Jessica J. Mitchell
Montana Natural Heritage Program
Spatial Analysis Lab
University of Montana
Missoula, Montana

Fred Ortenberg
Technion
Israel Institute of Technology
Haifa, Israel

Stuart Phinn
Remote Sensing Research Centre
School of Earth and Environmental Sciences
University of Queensland
Brisbane, Australia

Antonio Plaza
Department of Technology of Computers and Communications
Hyperspectral Computing Laboratory
University of Extremadura
Cáceres, Spain

Javier Plaza
Department of Technology of Computers and Communications
University of Extremadura
Badajoz (UNEX), Spain

Andries B. Potgieter
Queensland Alliance for Agriculture and Food Innovation (QAAFI)
University of Queensland
Brisbane, Australia

Jiaguo Qi
Department of Geography
Center for Global Change and Earth Observations
Michigan State University
East Lansing, Michigan

Sergio Sánchez
Department of Technology of Computers and Communications
Hyperspectral Computing Laboratory
University of Extremadura
Cáceres, Spain

Alimohammad Shirzadifar
Department of Agricultural and Biosystems Engineering
North Dakota State University
Fargo, North Dakota

Terrance Slonecker
Eastern Geographic Science Center
United States Geological Survey
Reston, Virginia

Lola Suarez
Department of Infrastructure Engineering
University of Melbourne
Melbourne, Australia

Prasad S. Thenkabail
Western Geographic Science Center
United States Geological Survey
Flagstaff, Arizona

James Watson
Queensland Alliance for Agriculture and Food Innovation (QAAFI)
University of Queensland
Brisbane, Australia

Narumon Wiangwang
Royal Thai Government
Department of Fisheries
Information Technology Center
Bangkok, Thailand

Yu Zhang
Department of Agricultural and Biosystems Engineering
North Dakota State University
Fargo, North Dakota

Acronyms and Abbreviations

ρ	Correlation
σ	Variance
δ	Bending Vibration
ν	Stretching Vibration
R^2	Coefficient of Determination
A/CHEAT	Auscover/Curtin Hyperion Enhancement and Atmospheric Correction Technique
AC	Atmospheric Corrector
ACORN	Atmospheric Correction Now
ADEOS	Advanced Earth Observing Satellite
AISA	Airborne Image Spectrometer for Applications
ALI	Advanced Land Imager
ALS	Airborne Laser Scanning
ANN	Artificial Neural Network
AOP	Airborne Observation Platform
API	Application Programming Interface
ASCII	American Standard Code for Information Interchange
ASD	Analytical Spectral Devices Inc.
ASI	Agenzia Spaziale Italiana
ASO	Airborne Snow Observatory
ASTER	Advanced Spaceborne Thermal Emission and Reflection Radiometer
ASU	Arizona State University
ATCOR	Atmospheric and Topographic Correction
ATLAS	Advanced Topographic Laser Altimeter System
ATREM	Atmospheric Removal
AVHRR	Advanced Very High-Resolution Radiometer
AVIRIS	Airborne Visible and Infrared Imaging Spectrometer
BRDF	Bidirectional Reflectance Distribution Function
BSC	Biological species concept
C	Carbon
COV	Covariance
CAO	Carnegie Airborne Observatory
CBERS-2	China-Brazil Earth Resources Satellite
CDA	Canonical Discriminant Analysis
CEOS	Committee on Earth Observation Satellite
CHRIS	Compact High Resolution Imaging Spectrometer
CHV	Convex Hull Volume
CLM	Community Land Model
CMF	Color Matching Functions
CNES	Centre National d'Etudes Spatiales
CNN	Convoluted Neural Network
CO_2	Carbon Dioxide
CoETP	ARC Centre of Excellence for Translational Photosynthesis
COVER%	Crop cover percentage
CP	Crude protein (%)
CPU	Central Processing Unit

CSR	Competitive-, stress, and ruderal strategy types of plants
CTIC	Cross-Track Illumination Correction
CWA	Continuous Wavelet Analysis
CWC	Canopy Water Content
CWSI	Crop Water Stress Index
D	Derivative
DAAC	Distributed Active Archive Center
DAIS	Digital Airborne Imaging Spectrometer
DART	Discrete Anisotropic Radiative Transfer
DBH	Diameter at Breast Height
DESIS DLR	Earth Sensing Imaging Spectrometer
DISORT	Discrete Ordinate Radiative Transfer
DN	Digital Numbers
DNA	Deoxyribonucleic acid
DOAS	Differential Optical Absorption Spectroscopy
DoD	Department of Defense
DSM	Digital Surface Model
DTM	Digital Terrain Model
DWEL	Dual-Wavelength Echidna LiDAR
E	Radiance
EBVs	Essential Biodiversity Variables
ECOSTRESS ECOsystem	Spaceborne Thermal Radiometer Experiment on Space Station
ED	Ecosystem Demography
EFFORT	Empirical Flag Field Optimal Reflectance Transformation
EGU	European Geoscience Union
ELD	Effective Leaf Density
EMP	Extended Morphological Profile
EnMAP	Environmental Mapping and Analysis Programme
ENVI-IDL	Environment for Visualizing Images-Interactive Data Language
ENVI	Environment for Visualizing Images
ENVISAT	Environmental Satellite
EO-1	Earth Observing-1
EO	Earth Observations
EOS	Earth Observing System
EOSDIS	Earth Observing System Data and Information System
ERDAS	Earth Resource Data Analysis System
ERDAS	Earth Resource Development Assessment System
EROS	Earth Resources Observation and Science
ESA	European Space Agency
ESDIS	Earth Science Data and Information System
ESMO	Earth Science Mission Operations
ESMs	Earth System Models
ETM+	Enhanced Thematic Mapper Plus
EUFAR	European Facility for Airborne Research
EVI	Echidna Validation Instrument
FAO	Food and agriculture organization of the United Nations
FEDM	Frequent Domain Electro Magnetic
FLAASH	Fast-Line-of-Sight Atmospheric Analysis of Hypercubes
FLDA	Fisher's Linear Discriminant Analysis
FLEX	Fluorescence Explorer
FLiES	Forest Light Environment Simulator

FloX	Fluorescence Box—dual field of view spectroradiometer
FNAI	Florida Natural Areas Inventory
FNIR	Far near-infrared (1100–1300 nm)
FORMOSAT	Satellite operated by Taiwanese National Space Organization NSPO Data marketed by SPOT Image
FOV	Field of View
FTHSI	Fourier Transform Hyperspectral Imager
FTIR	Fourier Transform Infrared Spectroscopy
FWF	Full Waveform
FWHM	Full Width at Half Maximum (pulse width)
G-LiHT	Goddard's LiDAR, Hyperspectral and Thermal Airborne Imager
GA	Genetic Algorithm
GB	Gigabyte
GECKO	Genetic by environment characterization (via) kinetic observation
GEDI	Global Ecosystem Dynamics Investigation
GEE	Google Earth Engine
GEO	BON Group on Earth Observation Biodiversity Observation Network
GEOEYE-1 and 2	Providing data in 0.25–1.65 m resolution
GERIS	Geophysical and Environmental Research Imaging Spectrometer
GIS	Geographic Information System
GLAS	Geoscience Laser Altimeter System
GLI	Global Imager
GOME	Global Ozone Monitoring Experiment
GORT	Geometric Optical-Radiative Transfer
GPR	Ground Penetrating Radar
GPS	Global Positioning System
GxExM	Genotype by environment by management
H	Entropy
HDF	Hierarchical Data Format
HICO	Hyperspectral Imager for the Coastal Ocean
HISUI	Hyperspectral Imager SUIte
HRS	Hyperspectral remote sensing
HSI	Hyperspectral Imagers
HSR	High Spatial Resolution
HRS	Hyperspectral Remote Sensing
HTTP	High-throughput phenotyping
HTV	H-2 Transfer Vehicle
HVI	Hyperspectral vegetation indices
HyMAP	Hyperspectral MAPping sensor
Hyperion	First spaceborne hyperspectral sensor onboard Earth Observing-1(EO-1)
HyPLANT	Airborne imaging spectrometer with a fluorescence module
HyspIRI	Hyperspectral InfraRed Imager
ICA	Independent Component Analysis
ICAMM	Independent Component Analysis-based Mixed Model
ICARE	International Conference on Airborne Research for the Environment
ICESat	Ice, Cloud, and Land Elevation Satellite
IKONOS	Very High Spatial Resolution (sub-meter to 5-m) satellite operated by GeoEye

INSAR	Interferometric Synthetic-Aperture Radar
IR	Infrared
IRS-1C/D-LISS	Indian Remote Sensing Satellite/Linear Imaging Self Scanner
IRS-P6/AWiFS	Indian Remote Sensing Satellite/Advanced Wide Field Sensor
IS	Imaging Spectroscopy
ISS	International Space Station
JAXA	Japan Aerospace Exploration Agency
JHU	Johns Hopkins University
JPL	Jet Propulsion Laboratory
JRBP	Jasper Ridge Biological Preserve
KFD	Kernel Fisher Discriminant
KOMFOSAT	Korean multipurpose satellite. Data marketed by SPOT Image
KPLSR	Kernel-based Partial Least Square Regression
L	Irradiance
LA	Leaf Area
LAI	Leaf area index (m^2/m^2)
LAN	Leaf Angle
Landsat-1, 2, 3 MSS	Multispectral Scanner
Landsat-4, 5 TM	Thematic Mapper
Landsat-7 ETM+	Enhanced Thematic Mapper Plus
LANDSAT-TM	Land Satellite Thematic Mapper Sensor
LANDSAT	Land Satellite Multispectral Sensor
LDA	Linear Discriminant Analysis
LiDAR	Light Detection and Ranging
LMA	Leaf Mass per Area
LNA	Leaf nitrogen accumulation
LOD	Linked open data
LOWTRAN LOW	Resolution model for predicting atmosphere Transitions
LOWTRAN	Low Resolution Atmospheric Transmittance
LSMs	Land Surface Models
LUE	Light use efficiency
LUI	Land-use intensity
LULC	Land Use, Land Cover
LVIS	Land, Vegetation, and Ice Sensor
LWC	Leaf Water Content
LWIR	Longwave Infrared
MEI	Management, equipment, and infrastructure
MESMA	Multiple End-member Spectral Mixture Analysis
MF	Matched Filter
MI	Mutual Information
MIA	Mutual Information Analysis
MLC	Maximum Likelihood Classification
MLC	Maximum Likelihood Classifier
MLP	Multi-Layer Perceptron
MLR	Multiple Linear Regression
MLRA	Machine Learning Regression Algorithms
MNDVI	Modified Normalized Differential Vegetation Index
MNF	Minimum Noise Fraction
MODIS	Moderate Resolution Imaging Spectroradiometer
MODTRAN	Moderate Resolution Atmospheric Transmittance
Mrad	Milliradians

MS	Multispectral
MSC	Morphological species concept
MWIR	Mid-Wave Infrared
N	Nitrogen (%)
N	Number (of pixels, bands, and so on)
NANO	Headwall Nano-Hyperspec camera
NASA JPL NASA	Jet Propulsion Laboratory
NASA	National Aeronautics and Space Administration
NDI	Normalized Difference Index
nDSM	Normalized Digital Surface Model
NDVI	Normalized Difference Vegetation Index
NDWI	Normalized Difference Water Index
NeCTAR	National eResearch collaboration tools and resources project
NEON	National Ecological Observatory Network
NIR	Near-Infrared
NIRS	Near-Infrared Spectroscopy
NN	Neural Network
NOAA	National Oceanic and Atmospheric Administration
NPP	Net Primary Productivity
NRL	Naval Research Laboratory
NSA	Normalized Spectral Area
NSC	Non-Structural Carbohydrates
NSMI	Normalized Soil Moisture Index
OLS	Ordinary Least Squares
OM	Organic matter
OMI	Ozone Monitoring Instrument
OO	Ocean Optics USB2000+ point spectrometer
OSP	Orthogonal Subspace Projection
P	Phosphorous
PAI	Plant Area Index
PC	Principal Component
PCA	Principal Components Analysis
PFTs	Plant Functional Types
PLNTHT	Plant height (mm)
PLS	Partial Least Squares
PLSR	Partial Least Square Regression
POS	Penetrating Optical Sensor
PP	Projection Pursuit
PRF	Pulse Repetition Rate
PRI	Photochemical reflectance index
PRISMA	PRecursore IperSpettrale della Missione Applicativa
PROBA	Project for On-Board Autonomy
PSC	Phylogenetic species concept
QSM	Quantitative Structure Models
QSMT	Quantitative Spectral Matching Technique
QUICKBIRD	Satellite from DigitalGlobe, a private company in the USA
RAPID EYE	Satellite constellation from Rapideye, a German company
REP	Red-edge Position Determination
RESOURCESAT	Satellite launched by India
RF	Random Forest
RGB	Red-Green-Blue

RMSE	Root Mean Square Error
RPAS	Remotely Piloted Airborne System
RPD	Ratio of Prediction to Deviation
RS-ST/STV-C	Remote sensing-spectral trait/spectral trait variation-concept
RS	Remote sensing
RSRC	Remote Sensing Research Centre
RTM	Radiative Transfer Model
S	Scatter Matrix
SALCA	Salford Advanced Laser Canopy Analyser
SAM	Spectral Angle Mapper
SAVI	Soil-Adjusted Vegetating Index
SBS	Sequential Backward Selection
SBUV	Solar Backscatter Ultraviolet
SCIAMACHY	Scanning Imaging Absorption Spectrometer for Atmospheric Cartography
SCP	Sorghum conversion program
SDAR	Spectral Diversity Area Relationships
SFF	Spectral Feature Fitting
SfM	Structure from Motion
SFS	Sequential Forward Selection
SID	Spectral Information Divergence
SIF	Solar induced chlorophyll fluorescence
SIS	Spectral Information System
SMA	Spectral Mixture Analysis
SMARTS	Simple Model of the Radiative Transfer of Sunshine
SMGM	Soil Moisture Gaussian Model
SNR	Signal-to-Noise Ratio
SOC	Soil Organic Carbon
SPEAR (ENVI tool)	Spectral Processing Exploitation and Analysis Resource
SPECIM	SPECtral IMaging
SPECPR	SPECtrum Processing Routines
SPOT	Système Pour l'Observation de la Terre
ST	Spectral traits
STV	Spectral trait variations
SVAT	Soil-Vegetation-Atmosphere-Transfer
SVH	Spectral Variation Hypothesis
SVM	Support Vector Machine
SVR	Support Vector Regression
SWIR	Shortwave Infrared
TAU	Tel Aviv University
TCARI	Transformed Chlorophyll Absorption in Reflectance Index
TES	Tropospheric Emission Spectrometer
TIR	Thermal Infrared
TLS	Terrestrial Laser Scanning
TOMS	Total Ozone Mapping Spectrometer
UAS	Unmanned Aircraft System
UAV	Unmanned Aerial Vehicle
UFL	Ultraviolet Fluorescence LiDAR
UMV	Unmanned Vehicle
UNFCCC	United Nations Framework Convention on Climate Change
UQ	University of Queensland

USAD	United States Department of Agriculture
USGS	United States Geological Survey
UV	Ultraviolet
Vc	Rate that rubisco works in chloroplasts
VCA	Vertex Component Analysis
Vcmax	Maximum rate for a given leaf (temperature dependent)
VDMs	Vegetation Dynamical Models
VI	Vegetation Indices
VIS-NIR	Visible and Near-Infrared
VIS	Visible
VMEMSMA	Variable Multiple EndMember Spectral Mixture Analysis
VNIR	Visible and Near-Infrared
VSL	Vegetation Spectral Library
VSWIR	Visible and shortwave infrared
W	Water
WASMA	Wavelength Adaptive Spectral Mixture Analysis
WBM	Wet biomass (kg\m^2)
WFIS	Wide Field-of-view Imaging Spectrometer
WORLDVIEW	DigitalGlobe's Earth imaging satellite
WR	White reference
WSC	World Soil Congress
WSN	Wireless sensor networks
WT	Wavelet Transform
YI	Yellowness Index

Section I

Introduction to Hyperspectral Remote Sensing of Agricultural Crops and Vegetation

1 Advances in Hyperspectral Remote Sensing of Vegetation and Agricultural Crops

Prasad S. Thenkabail, John G. Lyon, and Alfredo Huete

CONTENTS

1.1 INTRODUCTION AND RATIONALE

In the recent past, hyperspectral remote sensing has gained great utility for a variety of applications. At last, one can do remotely what chemists have done in laboratories with spectra. It is now possible to be diagnostic in sensing species and plant communities and do so in a direct and informed manner. The true promise of remote sensing for plant studies is realized with application of modern tools and analysis.

Hyperspectral data can provide significant improvements in spectral information content when compared with traditional broad-band analysis. We now can address a variety of phenomena and processes only dreamed of. A number of thoughtful investigators have demonstrated robust applications, and a list of these includes: (a) detecting plant stress (Thenkabail et al., 2004b), (b) measuring the chlorophyll content of plants (Blackburn and Ferwerda, 2008), (c) identifying small differences in percent of green vegetation cover (Chen et al., 2008), (d) extracting biochemical variables such as nitrogen and lignin (Blackburn and Ferwerda, 2008, Chan and Paelinckx, 2008, Dalponte et al., 2009; Houborg and Boegh, 2008), (e) discriminating land cover types (Thenkabail et al., 2004a), (f) detecting crop moisture variations (Colombo et al., 2008), (g) sensing subtle variations in leaf pigment concentrations (Yang et al., 2009, Zhao et al., 2007, Blackburn and Ferwerda, 2008), (h) modeling biophysical and yield characteristics of agricultural crops (Thenkabail et al., 2000, 2002, Houborg et al., 2009), (i) improving the detection of changes in sparse vegetation (Lyon et al., 1998), and (j) assessing absolute water content in plant leaves (Jollineau and Howarth, 2008). This is a fairly detailed list but not exhaustive. It gives the reader a measure of the current, proven experimental capabilities and operational applications. It also elevates the mind in thinking of new, ambitious applications that are attainable by driven investigators.

The spectral properties of vegetation are strongly determined by their biophysical and biochemical attributes such as leaf area index (LAI), the amount of live biomass and senesced biomass, moisture content, pigments (e.g., chlorophyll), and spatial arrangement of structures (Asner, 1998, Hill, 2004). We are now capable of measuring those phenomena and processes, and harness them in testing hypotheses and valuable applications on a variety of ecosystems. For example, the assessment of biophysical and biochemical properties of vegetation such as rangelands (Darvishzadeh et al., 2008b, c), agricultural crops (Thenkabail et al., 2000, Thenkabail et al., 2002, 2004a,b, Darvishzadeh et al., 2008a), and weeds (Thenkabail et al., 2004a) are essential for evaluating productivity, providing information needed for local farmers and institutions, and assessing grazing potential for livestock.

Even though remote sensing has been recognized as a reliable method for estimating various biophysical and biochemical vegetation variables, the existing broad-band sensors were inadequate or supplied limited information for the purpose (Thenkabail, 2003, Thenkabail et al., 2004a,b, Vaiphasa et al., 2007, Maire et al., 2008, Yao et al., 2010). Clearly, broad-bands have known limitations in providing adequate information on properties such as crop growth stage, crop type differentiation, generation of agricultural crop statistics, identification of forest type and species identification, characterizing complex-forest versus non-forest interactions, and detailed mapping of land-cover classes and all to serve diverse scientific and other user communities (Thenkabail et al., 2002, 2004a).

These limitations of broad-band analysis are illustrated by vegetation indices (VIs) which saturate beyond a certain level of biomass and leaf area index (LAI) (Thenkabail et al., 2000). For example, VIs typically increase over an LAI range from 0 to about 3–5 before an asymptote is reached. While extremely useful over the years, the upper limit of this sensitivity apparently differs among vegetation types and can only be driven so far to a solution for a given application. Saturation is more pronounced for planophile canopies (Atzberger, 2004). However, compared with erectophile canopies of the same LAI, planophile canopies are less influenced by soil brightness variations (Darvishzadeh et al., 2008a). In contrast, hyperspectral datasets allow identification of features, allow direct measurement of canopy variables such as biochemical content (e.g., chlorophyll, nitrogen, lignin), forest species, chemistry distribution, timber volumes, water conditions (Gong et al. 1995, Carter, 1998, Blackburn and Ferwerda, 2008), as well as biophysical (e.g., leaf area index, biomass) and yield characteristics (Thenkabail et al., 2002, Thenkabail et al., 2004a, Zhao et al., 2007, Galvão et al., 2009).

Hyperspectral sensors gather near-continuous spectra from imaging spectrometers such as the National Aeronautics and Space Administration (NASA) designed Airborne Visible-InfraRed Imaging Spectrometer (AVIRIS) and Compact Airborne Spectrographic Imager (CASI). This new generation of spaceborne sensors offer tremendous improvements in spatial, spectral, radiometric, and temporal resolutions as well as improvements in optics and mechanics when compared with older generations of sensors (Table 1.1). The promise and potential of hyperspectral narrow-band sensors for a wide array of Earth resource applications motivated the design and launch of spaceborne

TABLE 1.1
Broad-Band and Narrow-Band Satellite Sensor Spatial, Spectral, Radiometric, Waveband, and Other Data Characteristics

Sensor	Spatial (meters)	Spectral (#)	Radiometric (bit)	Band Range (μm)	Band Widths (μm)	Irradiance (W m^{-2} sr^{-1} μm^{-1})	Data Points (# per hectares)	Frequency of Revisit (days)
A. Coarse Resolution Broad-Band Sensors								
1. AVHRR	1000	4	11	0.58–0.68	0.10	1390	0.01	daily
				0.725–1.1	0.375	1410		
				3.55–3.93	0.38	1510		
				10.30–10.95	0.65	0		
				10.95–11.65	0.7	0		
B. Coarse Resolution Narrow-Band Sensors								
2. MODIS	250, 500, 1000	36/7	12	0.62–0.67	0.05	1528.2	0.16, 0.04, 0.01	daily
				0.84–0.876	0.036	974.3	0.16, 0.04, 0.01	
				0.459–0.479	0.02	2053		
				0.545–0.565	0.02	1719.8		
				1.23–1.25	0.02	447.4		
				1.63–1.65	0.02	227.4		
				2.11–2.16	0.05	86.7		
C. Multispectral Broad-Band Sensors								
3. Landsat-1, 2, 3 MSS	56 × 79	4	6	0.5–0.6	0.1	1970	2.26	16
				0.6–0.7	0.1	1843		
				0.7–0.8	0.1	1555		
				0.8–1.1	0.3	1047		
4. Landsat-4, 5 TM	30	7	8	0.45–0.52	0.07	1970	11.1	16
				0.52–0.60	0.80	1843		
				0.63–0.69	0.60	1555		
				0.76–0.90	0.14	1047		

(*Continued*)

TABLE 1.1 ***(Continued)***
Broad-Band and Narrow-Band Satellite Sensor Spatial, Spectral, Radiometric, Waveband, and Other Data Characteristics

Sensor	Spatial (meters)	Spectral (#)	Radiometric (bit)	Band Range (μm)	Band Widths (μm)	Irradiance (W m^{-2} sr^{-1} μm^{-1})	Data Points (# per hectares)	Frequency of Revisit (days)
				1.55–1.74	0.19	227.1		
				10.4–12.5	2.10	0		
				2.08–2.35	0.25	80.53		
5. Landsat-7 ETM+	30	8	8	0.45–0.52	0.65	1970	44.4, 11.1	16
				0.52–0.60	0.80	1843		
				0.63–0.69	0.60	1555		
				0.50–0.75	0.150	1047		
				0.75–0.90	0.200	227.1		
				10.0–12.5	2.5	0		
				1.75–1.55	0.2	1368		
				0.52–0.90 (p)	0.38	1352.71		
6. ASTER	15, 30, 90	15	8	0.52–0.63	0.11	1846.9	44.4, 11.1, 1.23	16
				0.63–0.69	0.06	1546.0		
				0.76–0.86	0.1	1117.6		
				0.76–0.86	0.1	1117.6		
				1.60–1.70	0.1	232.5		
				2.145–2.185	0.04	80.32		
				2.185–2.225	0.04	74.96		
				2.235–2.285	0.05	69.20		
				2.295–2.365	0.07	59.82		
				2.360–2.430	0.07	57.32		
			12	8.125–8.475	0.35	0		
				8.475–8.825	0.35	0		
				8.925–9.275	0.35	0		
				10.25–10.95	0.7	0		
				10.95–11.65	0.7	0		

(Continued)

TABLE 1.1 (*Continued*)
Broad-Band and Narrow-Band Satellite Sensor Spatial, Spectral, Radiometric, Waveband, and Other Data Characteristics

Sensor	Spatial (meters)	Spectral (#)	Radiometric (bit)	Band Range (μm)	Band Widths (μm)	Irradiance (W m^{-2} sr^{-1} μm^{-1})	Data Points (# per hectares)	Frequency of Revisit (days)
7. ALI	30	10	12	0.048–0.69 (p)	0.64	1747.8600		
				0.433–0.453	0.20	1849.5	11.1	16
				0.450–0.515	0.65	1985.0714		
				0.425–0.605	0.80	1732.1765		
				0.633–0.690	0.57	1485.2308		
				0.775–0.805	0.30	1134.2857		
				0.845–0.890	0.45	948.36364		
				1.200–1.300	1.00	439.61905		
				1.550–1.750	2.00	223.39024		
				2.080–2.350	2.70	78.072727		
8. SPOT-1	2.5–20	15	16	0.50–0.59	0.09	1858	1600, 25	3–5
-2				0.61–0.68	0.07	1575		
-3				0.79–0.89	0.1	1047		
-4				1.5–1.75	0.25	234		
				0.51–0.73 (p)	0.22	1773		
9. IRS-1C	23.5	15	8	0.52–0.59	0.07	1851.1	18.1	16
				0.62–0.68	0.06	1583.8		
				0.77–0.86	0.09	1102.5		
				1.55–1.70	0.15	240.4		
				0.5–0.75 (P)	0.25	1627.1		
10. IRS-1	23.5	15	8	0.52–0.59	0.07	1852.1	18.1	16
				0.62–0.68	0.06	1577.38		
				0.77–0.86	0.09	1096.7		
				1.55–1.70	0.15	240.4		
				0.5–0.75 (P)	0.25	1603.9		
11. IRS-P6-AWiFS	56	4	10	0.52–0.59	0.07	1857.7	3.19	16
				0.62–0.68	0.06	1556.4		

(*Continued*)

TABLE 1.1 (*Continued*)
Broad-Band and Narrow-Band Satellite Sensor Spatial, Spectral, Radiometric, Waveband, and Other Data Characteristics

Sensor	Spatial (meters)	Spectral (#)	Radiometric (bit)	Band Range (μm)	Band Widths (μm)	Irradiance ($W\ m^{-2}\ sr^{-1}\ \mu m^{-1}$)	Data Points (# per hectares)	Frequency of Revisit (days)
				0.77–0.86	0.09	1082.4		
				1.55–1.70	0.15	239.84		
12. CBERS –2, –3B →	20 m pan, 20 m MS		11	0.51–0.73	0.22	1934.03	25, 25	
				0.45–0.52	0.07	1787.10		
–3, –4 →	5 m pan, 20 m MS			0.52–0.59	0.07	1587.97	400, 25	
				0.63–0.69	0.06	1069.21		
				0.77–0.89	0.12	1664.3		
D. Hyperspectral Narrow-Band Sensors								
13. Hyperion	30	220 (196[a])	16	196 effective Calibrated bands VNIR (band 8–57) 427.55–925.85 nm SWIR (band 79–224) 932.72–2395.53 nm	10 nm wide (approx.) for all 196 bands	See data in Neckel and Labs (1984). Plot it and obtain values for Hyperion bands	11.1	16
14. ASD spectroradiometer	1134 cm² @ 1.2 m Nadir view 18° Field of view	2100 bands 1 nm width between 400–2500 nm	16	2100 effective bands	1 nm wide (approx.) in 400–2500 nm	See data in Neckel and Labs (1984). Plot it and obtain values for Hyperion bands	88,183	5–16
15. HyspIRI VSWIR	60	210	16	210 bands in 380–2500 nm	10 nm wide (approx.) for all 210 bands	See data in Neckel and Labs (1984). Plot it	2.77	19
16. HyspIRI TIR	60	8	16	7 bands in 7500–12,000 nm 3000–5000 nm (3980 nm center)	7 bands in 7500–12,000 nm and 1 band in	See data in Neckel and Labs (1984). Plot it	2.77	5

(*Continued*)

TABLE 1.1 (*Continued*)
Broad-Band and Narrow-Band Satellite Sensor Spatial, Spectral, Radiometric, Waveband, and Other Data Characteristics

Sensor	Spatial (meters)	Spectral (#)	Radiometric (bit)	Band Range (μm)	Band Widths (μm)	Irradiance (W m^{-2} sr^{-1} μm^{-1})	Data Points (# per hectares)	Frequency of Revisit (days)
				E. Hyperspatial Broad-Band Sensors				
17. IKONOS	1–4	4	11	0.445–0.516	0.71	1930.9	10,000, 625	5
				0.506–0.595	0.89	1854.8		
				0.632–0.698	0.66	1156.5		
				0.757–0.853	0.96	1156.9		
18. QUICKBIRD	0.61–2.44	4	11	0.45–0.52	0.07	1381.79	14,872, 625	5
				0.52–0.60	0.08	1924.59		
				0.63–0.69	0.06	1843.08		
				0.76–0.89	0.13	1574.77		
19. RESOURSESAT	5.8	3	10	0.52–0.59	0.07	1853.6	33.64	24
				0.62–0.68	0.06	1581.6		
				0.77–0.86	0.09	1114.3		
20. RAPID EYE -A -E	6.5	5	12	0.44–0.51	0.07	1979.33	236.7	1–2
				0.52–0.59	0.07	1752.33		
				0.63–0.68	0.05	1499.18		
				0.69–0.73	0.04	1343.67		
				0.77–0.89	0.12	1039.88		
21. WORLDVIEW	0.55	1	11	0.45–0.51	0.06	1996.77	40,000	1.7–5.9
22. FORMOSAT-2	2–8	5	11	0.45–0.52	0.07	1974.93	2500, 156.25	daily
				0.52–0.60	0.08	1743.12		
				0.63–0.69	0.06	1485.23		
				0.76–0.90	0.14	1041.28		
				0.45–0.90 (p)	0.45	1450		

(*Continued*)

TABLE 1.1 (*Continued*)
Broad-Band and Narrow-Band Satellite Sensor Spatial, Spectral, Radiometric, Waveband, and Other Data Characteristics

Sensor	Spatial (meters)	Spectral (#)	Radiometric (bit)	Band Range (μm)	Band Widths (μm)	Irradiance (W m^{-2} sr^{-1} μm^{-1})	Data Points (# per hectares)	Frequency of Revisit (days)
23. KOMPSAT-2	1–4	5	10	0.5–0.9	0.4	1379.46	10,000, 625	3–28
				0.45–0.52	0.07	1974.93		
				0.52–0.6	0.08	1743.12		
				0.63–0.59	0.04	1485.23		
				0.76–0.90	0.14	1041.28		

Source: Adapted and edited from Melesse et al., 2007; Thenkabail P.S. et al. 2010. *Journal Remote Sensing* 2(1): 211–261.

[a] Of the 242 bands, 196 are unique and calibrated. These are: (A) Band 8 (427.55 nm) to band 57 (925.85 nm) that are acquired by visible and near-infrared (VNIR) sensor; and (B) Band 79 (932.72 nm) to band 224 (2395.53 nm) that are acquired by shortwave infrared (SWIR) sensor.

sensors such as Hyperion on-board the Earth Observing-1 (EO-1) satellite (Pearlman et al., 2003, Ungar et al., 2003), and upcoming Hyperspectral Imaging Spectrometer and Infrared Imager (HypspIRI). These sensors gather data in 210–220 narrow-bands from 380 to 2500 nanometers at 60 m resolution or better (Table 1.1). The HyspIRI's Thermal Infrared (TIR) sensor has 7 bands in the 7500–12,000 nm range that saturates at 400 K and 1 band in 3000–5000 nm (centered at 3980 nm) that saturates at 1400 K and acquires data in 60 m pixel size.

However, it must be noted that using hyperspectral data are much more complex than using multispectral data. Hyperspectral systems collect large volumes of data in a short time leading to a number of issues that need to be addressed. For example, Hyperion, the first spaceborne hyperspectral sensor launched by NASA's New Millennium Program (NMP) gathers near-continuous data in 220 discrete narrow-bands along the 400–2500 nanometer spectral range at a 30-meter spatial resolution and in 12-bit bytes with 12-bit radiometry. Each image is 7.5 kilometers in swath by 100 kilometers along track. The volume of data collected using Hyperion for an area equivalent to the Landsat TM image area will increase by about 37 times.

Such increases in data volume pose great challenges in data handling. The issues include data storage volume, data storage rate, downlink or transmission bandwidth, real-time analog to digital bandwidth and resolution, computing bottle necks in data analysis, and new algorithms for data utilization (e.g., atmospheric correction is more complicated; Thenkabail et al., 2002, 2004b). These issues make it imperative that newer methods and techniques be developed to handle these high-dimensional datasets.

Future generations of satellites may carry specially optimized sensors to gather data for targeted applications. Or they may carry a narrow-waveband hyperspectral sensor like Hyperion and HyspIRI from which users with different application needs can extract appropriate optimal wavebands. However, having too many narrow-bands does not necessarily mean more information. Indeed, most of these bands, and especially the ones that are close to one another, provide redundant information. This will make users devote unnecessary time in data mining, complex processing to identify and remove redundant bands, and put a heavy burden on computing, processing, and storage resources.

A far better option will be to focus on the design of an optimal sensor for a given application such as for vegetation studies and by dropping redundant bands. Even when the data are acquired in full range of hundreds or thousands of hyperspectral narrow-bands, knowledge of optimal bands for a particular application will help. Investigators can quickly select these bands, and spend time and expertise resources in applying selected optimal bands for the required application. Optimal hyperspectral sensors will help reduce data volumes, eliminate the problems of high-dimensionality of hyperspectral datasets, and make it feasible to apply traditional classification methods on a few selected bands (optimal bands) that capture most of the crop characteristics (Thenkabail et al., 2002, Thenkabail et al., 2004a, Chan and Paelinckx, 2008, Dalponte et al., 2009, Yang et al., 2009). Thereby, knowledge of application specific "optimal bands" for high-dimensional datasets such as Hyperion and HyspIRI is mandatory to reduce costs in data analysis and computer resources.

Table 1.1 compares the spectral and spatial resolution of narrow-band and broad-band data that are currently in use or are soon to be put into use. A number of recent studies have indicated the advantages of using discrete narrow-band data from a specific portion of the spectrum when compared with broad-band data to arrive at optimal quantitative or qualitative information on crop or vegetation characteristics (e.g., Thenkabail et al., 2000, Thenkabail et al., 2002, 2004a, Yang et al., 2009, Blackburn and Ferwerda, 2008, Chan and Paelinckx, 2008, Dalponte et al., 2009). Hence, this approach or direction has caught the attention of investigators and the community is moving in this direction.

Given this background, the overarching goal is to explore and determine the optimal hyperspectral narrow-bands in the study of vegetation and agricultural crops and to enumerate on methods and approaches. This is partially to overcome the "curse" of high-dimensionality or Hughes phenomenon where the ratio of the number of pixels with known class identity (i.e., training pixels) and the number of bands must be maintained at or above minimum value to achieve statistical confidence and functionality. In hyperspectral data with hundreds or even thousands of wavebands, the training pixels needed can grow exponentially (Hughes phenomenon), making it very difficult to address this spectral diversity.

First, it is necessary to identify hyperspectral narrow-bands that are best suited for studying natural vegetation and agricultural croplands. In the process, one detects and eliminates redundant bands or examples that supply little knowledge to the application. This will lead to highlighting optimal hyperspectral wavebands, in the 400–2500 nanometer range, best suited to study vegetation and agricultural crops. There are a number of studies (e.g., Blackburn, 1998, Nolin and Dozier, 2000, Thenkabail et al., 2000, 2002, 2004a,b, Chan and Paelinckx, 2008, Dalponte et al., 2009, Yang et al., 2009, Yao et al., 2010) that indicate that the narrow wavebands located in specific portions of the spectrum have the capability to provide required optimal information sought for a given application.

However, there is a clear need for synthesis of these various studies to see a general consensus of optimal wavebands based on studies conducted in different parts of the world for varying vegetation types and agricultural practices. Optimal Hyperion wavebands determined for vegetation studies established in this study will help reduce data volumes, eliminate the problems of high-dimensionality of hyperspectral datasets, and make it feasible to apply traditional classification methods on a few selected bands (optimal bands) that capture most of the information of vegetation and agricultural croplands.

Secondly, it is important to present and discuss methods and approaches of hyperspectral data analysis. This is to quickly eliminate redundant bands and identify the most useful bands. For example, when a user receives hyperspectral data and their application is vegetation and agricultural studies, they can directly select wavebands recommended and ignore all other wavebands which are redundant.

The process establishes categorization approaches to achieve highest accuracies, develop indices and wavebands that best model biophysical and biochemical quantities, and identify and eliminate redundant bands. The process involves an exhaustive review of the performance of hyperspectral vegetation indices (HVIs) that will help establish the utility of wavebands associated with these indices in the study of vegetation and agricultural crops.

1.2 HYPERSPECTRAL REMOTE SENSING OF VEGETATION AND AGRICULTURAL CROPS

Agricultural crops are significantly better characterized, classified, modeled, and mapped using hyperspectral data. There are many studies supporting this, conducted on a wide array of crops and their biophysical and biochemical variables such as yield (Wang et al., 2008), chlorophyll a and b (Zhu et al., 2007, Delegido et al., 2010), total chlorophyll (Haboudane et al., 2004), nitrogen content (Rao et al., 2007), carotenoid pigments (Blackburn, 1998), plant stress (Zhao et al., 2005), plant moisture (Penuelas et al., 1995), above ground biomass (Shen et al., 2009), and biophysical variables (Darvishzadeh et al., 2008a,b,c).

Hyperspectral remote sensing is well suited for early detection of nitrogen (N) as an important determinant of crop productivity and its quality. Knowing nitrogen levels early in the growing season will help remedy any deficiency. Leaf nitrogen is a key element in monitoring crop growth stage, its fertilization status, and assessing productivity. Leaf N accumulation, as a product of leaf N content and leaf weight, reflect not only information on leaf N status, but also vegetation coverage during crop growth (Yao et al., 2010). For example, nitrogen deficiency can be evaluated using ratios such as R_{743}/R_{1316} (Abdel-Rahman et al., 2010), where "R" is radiance at the given narrowband in nanometers.

Thenkabail et al. (2000, 2002) studied numerous crops on marginal lands (see Figure 1.1a) that included cotton (*Gossypium*), potato (*Solanum Erianthum*), soybeans (*Glycine max),* corn (*Zea mays*), sunflower (*Helianthus*), barley (*Hordeum vulgare L.*), wheat (*Triticum aestivum L. or Triticum durum Desf.*), lentil (*Lens esculenta Moench.* or *Lens orientale* [*Boiss.*] *Schmalh.* or

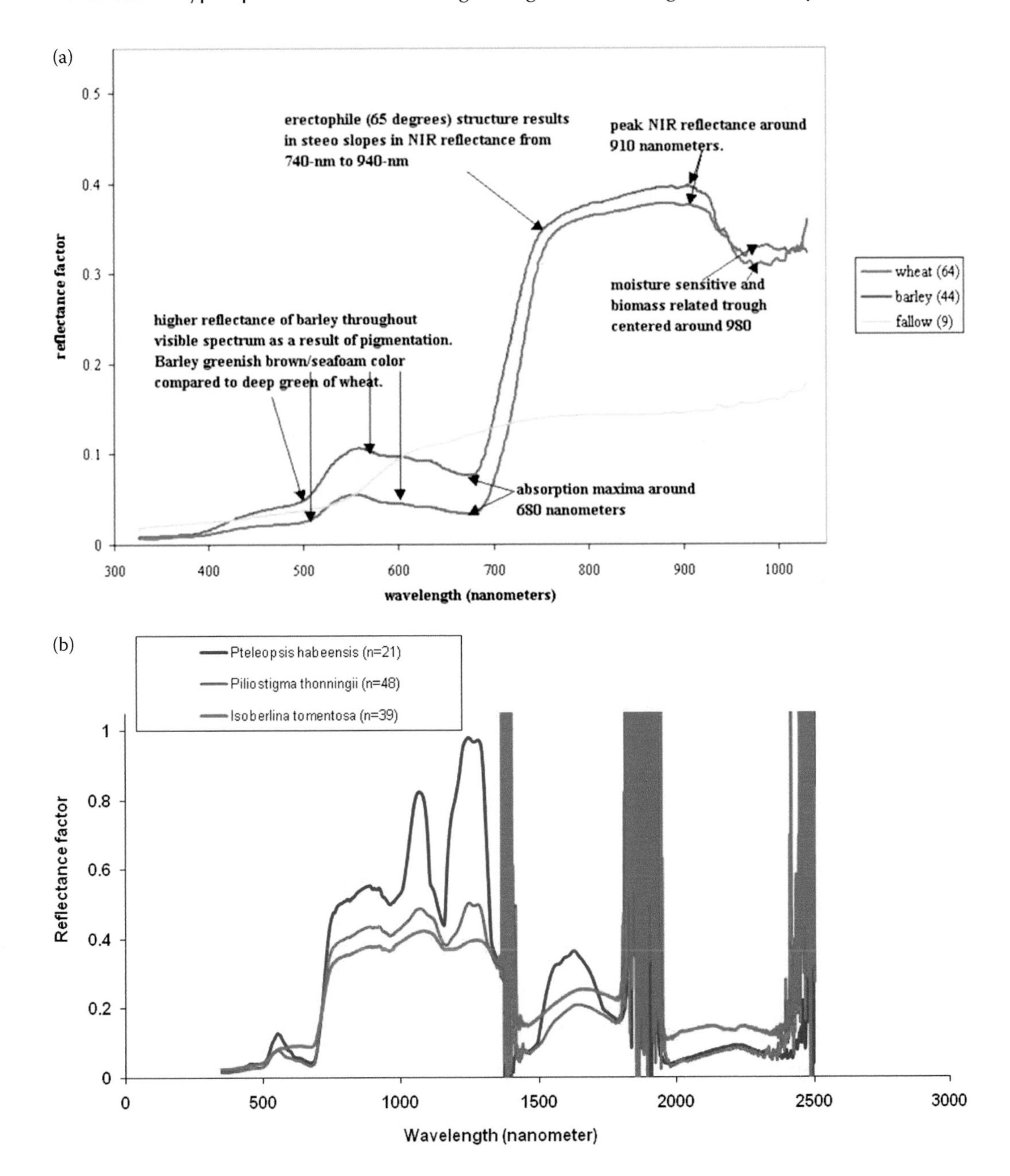

FIGURE 1.1 Hyperspectral characteristics illustrated for few vegetation and agricultural crops. Hyperspectral narrow-band data obtained from ASD spectroradiometer illustrated for: (a) agricultural crops, (b) shrub species.

(Continued)

Lens culinaris Medikus), cumin (*Cuminum cyminum L.*), chickpea (*Cicer arietinum L.*), and vetch (*Vicia narbonensis L.*). The parameters studied and found useful included: LAI or the leaf area index (m^2/m^2), WBM or wet biomass (kg/m^2), the above ground plant height (meters), CP or plant crude protein (%), N or nitrogen (%), and CC or mean canopy cover (%). Thousands of spectral measurements were made. These comprehensive studies led to identifying 12 optimal narrow-bands in 400–1050 nm (Thenkabail et al., 2000, 2002).

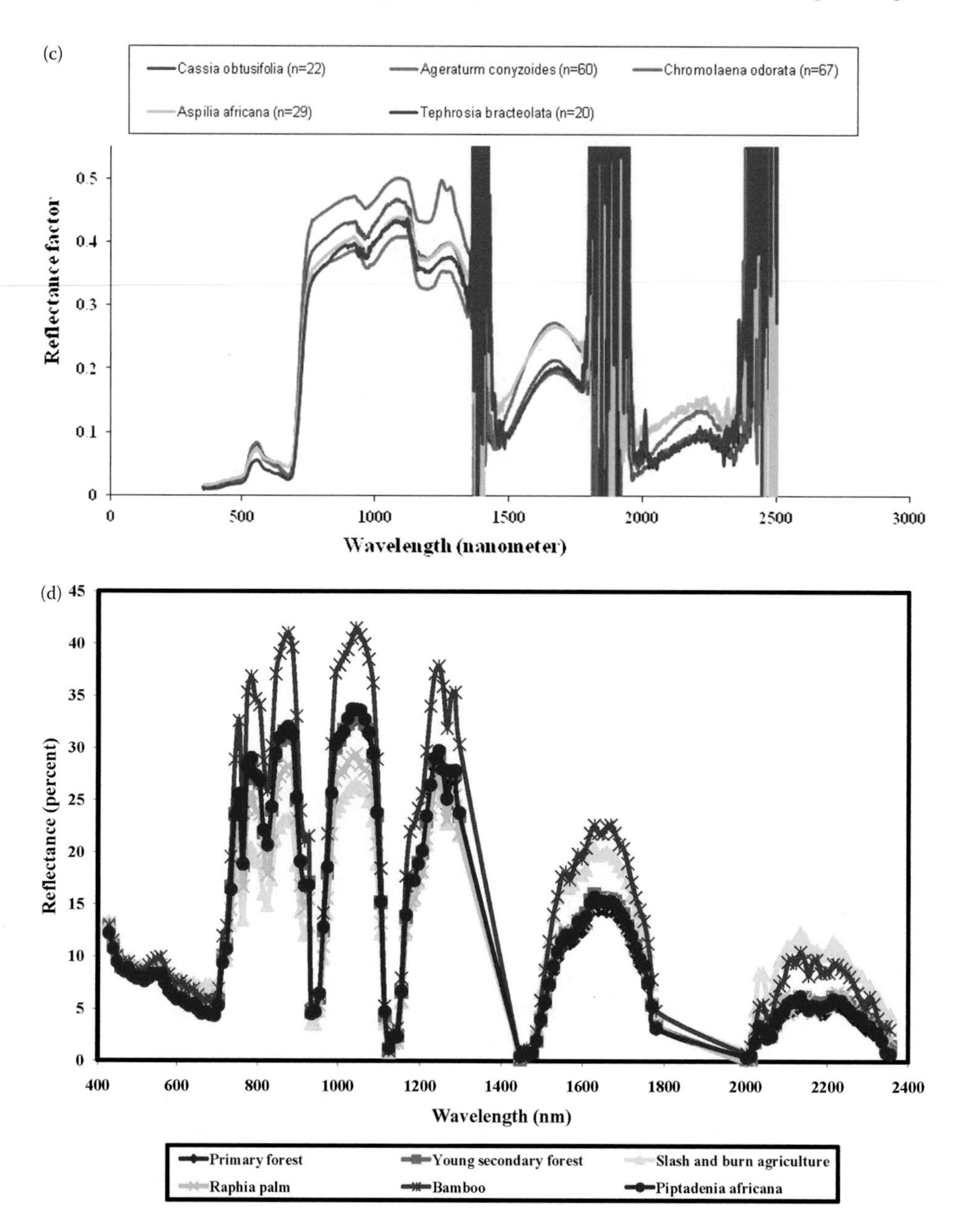

FIGURE 1.1 (Continued) Hyperspectral characteristics illustrated for few vegetation and agricultural crops. Hyperspectral narrow-band data obtained from ASD spectroradiometer illustrated for: (c) weed species. Hyperspectral narrow-band data obtained from Hyperion sensor onboard Earth Observing-1 (EO-1) satellite illustrated for: (d) tropical rainforest vegetation.

Forest and savanna biotic variables (e.g., height, basal area, biomass, and LAI) are modeled and mapped with significantly greater accuracies using hyperspectral data (Schlerf et al., 2005, Kalacska et al., 2007, Asner and Martin, 2009, Asner et al., 2009, Stagakis et al., 2010; e.g., Figure 1.1b–d) when compared to broad-band data. Ability to distinguish species type, assess age, structure (e.g., leaf angle), and phenology (e.g., leaf pigmentation, flowering), and map shade trees and understory crops is becoming increasingly possible as a result of hyperspectral narrow-band data (Oldeland et al., 2010, Papeş et al., 2010).

When there are significant topographic effects such as shadows cast from ridges, LiDAR data can be a great addition to hyperspectral data (Anderson et al., 2008). Thenkabail et al. (2004a,b) conducted a hyperspectral study of vegetation in African savannas and rainforests using Hyperion and an ASD spectroradiometer—both collecting data in the 400–2500 nm range (Table 1.1). They collected spectra from thousands of locations (see the few examples in Figure 1.1) that included an extensive and comprehensive series of vegetation types such as: (a) weeds (*Imperata cylindrical* the most prominent weed of African savannas and *Chromolenea Odorata* the most prominent weed in African rainforests), (b) shrubs (e.g., *Pteleopsis habeensis, Piliostigma thonningii, Isoberlinia tomentosa;* Figure 1.1b), (c) grasses (e.g., *Brachiaria jubata, Brachiaria stigmatisata, Digitaria sp.*), (d) agricultural crops (corn, rice, cowpea, groundnut, soybean, cassava), (e) other weeds (e.g., *Ageraturm conyzoids, Aspilia Africana, Tephrosia bracteolate, Cassia obtusifolia*), (f) agricultural fallows (e.g., of different age years), (g) primary forest (e.g., pristine, degraded), (h) secondary forest (e.g., young, mature, mixed), (i) regrowth forest (e.g., <3 years, 3–5 years, 5–8 years, >8 years), (j) slash and burn agriculture, and (k) forest tree species (e.g., *Piptadenia africana, Discoglypremna coloneura, Antrocaryon klaineanum, Pycnantus angolensis, Rauwolfia macrophylla, Alstonia congensis, Cissus spp., Lacospermas secundii, Haumania delkelmaniana, Alchornea floribunda, Lacospermas secundii, Alchornea cordifolia, Elaeis guineensis, Elaeis guineensis, Pteredium aquilinium, Megaphrenium spp., Pteredium aquilinium, Megaphrenium spp., Pteredium aquilinium, Epatorium adoratum, Aframomum giganteum, Pteredium acquilinium*).

The opportunities to distinguish various species using hyperspectral data are many. For example, the shrub species *Isoberlinia tomentosa* has high reflectivity in the red and NIR bands when compared with *Piliostigma thonningii*, but in the SWIR bands, it is just the opposite (Figure 1.1b). The shrub species *Pteleopsis habeensis* has dramatically high reflectivity relative to the other two shrub species (*Piliostigma thonningii, Isoberlinia tomentosa)* but beyond 2000 nm, *Isoberlinia tomentosa* has higher reflectivity than the other two species (Figure 1.1b). Similarly, the weed species *Aspilia Africana* has higher reflectivity relative to the other weed species *(Ageraturm conyzoids, Tephrosia bracteolate, Cassia obtusifolia)* only beyond 2000 nm. These results clearly imply the value of using data from distinct parts of the electromagnetic spectrum. These studies resulted in recommending 22–23 optimal bands to study vegetation in 400–2500 nm (Thenkabail et al., 2004a,b).

The relationship between vegetation characteristics and spectral indices can vary considerably based on vegetation types, crop types, and species types (Papeş et al., 2010). Biomass and NDVI relationships within and between agricultural crops and various vegetation types are very distinct and are useful discriminators. These differences occur as a result of distinctive features such as (Thenkabail et al., 2002): (a) plant structure (e.g., erectophile and planophile, Figure 1.1a), (b) plant composition (e.g., nitrogen, lignin, chlorophyll a and b), and (c) quantitative characteristics (e.g., biomass per unit area, LAI, and plant height). Thereby, more specific and accurate attempts at biomass estimations using remotely sensed data of various characteristics will need to take into consideration the specific vegetation, land cover, crop type, and species types (Thenkabail et al., 2000).

Specific advances are using radiative transfer models to derive quantities such as LAI and biomass of forests (e.g., Papeş et al., 2010, Schull et al., 2010). These models require surface reflectance and land-cover products. Accuracy of modeled quantities will depend on the number of land covers that can be mapped. Recent studies have also established the possibilities of direct biomass measurements using LiDAR that will provide within-canopy structure, ground topography, and tree height, thus enabling the indirect estimation of biomass and leaf area indices (Anderson et al., 2008).

From the literature and hard work of investigators, it has been established that hyperspectral narrow-bands provide significant additional information when compared with similar information obtained from broad-band sensors in estimating biophysical characteristics (Wang et al., 2008, Asner and Martin, 2009), biochemical properties (Blackburn, 1998, Cho, 2007), forest classification (le Maire et al., 2008), and in other vegetation studies involving forests, grasses, shrubs, and weeds (Thenkabail et al., 2004a,b). This is due to many limitations of the broad-band data.

For example, most of the indices derived using the older generation of sensors tend to saturate (Thenkabail et al., 2000), resulting in significant limitations in distinguishing the species or modeling biophysical and biochemical quantities. Nondestructive measurements of vegetation characteristics such as canopy chemical content (e.g., chlorophyll, nitrogen), forest species, chemistry distribution, timber volumes, and plant moisture (Haboudane et al., 2002, 2004, Zarco-Tejada et al., 2005, Galvão et al., 2009), and biophysical and yield characteristics (Blackburn, 1998, Thenkabail et al., 2000, 2002, 2004a,b) are best measured using hyperspectral data. Hyperspectral narrow-band data have also the potential to separate vegetation into taxonomic levels (Ustin et al., 2004) and forest ecotypes (Chan and Paelinckx, 2008). But, this aspect still needs considerably more study.

These efforts cover a wide array of agricultural crops and vegetation classes from forests, rangelands, savannas, and other ecosystems. They also cover a wide range of agro-ecologies and ecosystems spread across the world, resulting in a more holistic view of hyperspectral data performance in the study of agricultural crops across agro-ecological systems as well as forests and other vegetation across varied ecosystems (Hamilton, 2005, Gillespie et al., 2008).

During this process, we also took into consideration that the high-dimensionality of hyperspectral data can, at times, lead to over-fitting of statistical models (Thenkabail et al., 2000) leading to an over-optimistic view of their power (Lee et al., 2004). This can be unfortunate, but once the issue is understood, approaches can be elucidated and employed to avoid difficulties.

Taking all these factors into consideration, we set forth two key objectives, which were to: (a) establish optimal hyperspectral narrow-bands best suited to study vegetation and agricultural crops by pooling collective findings from a comprehensive set of investigations, and (b) identify and discuss a suite of methods and approaches that help to rapidly identify hyperspectral narrow-bands best suited to study through data mining of hundreds or thousands of wavebands. These objectives were met through the cumulative efforts of the co-editors and co-authors of the Second Edition,

1.3 HYPERSPECTRAL DATA COMPOSITION FOR STUDY OF VEGETATION AND AGRICULTURAL CROPS

Hyperspectral narrow-bands, typically, contain 100–1500 wavebands and collect data in a near-continuous spectra from ultraviolet, visible, near-, mid-, and far-infrared; Figure 1.1. Resulting datasets offer many opportunities to study specific vegetation variables such as leaf area index (LAI) (Koger et al., 2003), biomass (Lee et al., 2004, Darvishzadeh et al., 2008a), vegetation fraction (Asner and Heidebrecht, 2002), canopy biochemistry (Asner and Heidebrecht, 2002), forest structure and characteristics (e.g., diameter at breast height, tree height, crown size, tree density) (Kalacska et al., 2007, White et al., 2010), and plant species (Martin et al., 1998, Oldeland et al., 2010). These data in hundreds or even thousands of bands are composited into a single hyperspectral data cube (HDC; Figure 1.2). For example, a click on any pixel of Hyperion HDC (see Figure 1.2), will provide a continuous spectrum describing the composition of that pixel (see Figure 1.1d).

The principles of hyperspectral data composition and analysis can also be applied to time-series multispectral data. For example, a MODIS NDVI monthly maximum value composite for one year is composed as a single mega-file data cube (MFDC) (see Figure 1.3). Time-series spectra of a few classes extracted from this simple 12-band MFDC is illustrated in Figure 1.3, which is akin to hyperspectral data (Figure 1.1d) extracted from HDC (Figure 1.2).

1.4 METHODS AND APPROACHES OF HYPERSPECTRAL DATA ANALYSIS FOR VEGETATION AND AGRICULTURAL CROPS

The goal of dimension reduction is to reduce the number of features substantially without sacrificing significant information. However, care should be taken to retain all key wavebands, so as not to: (a)

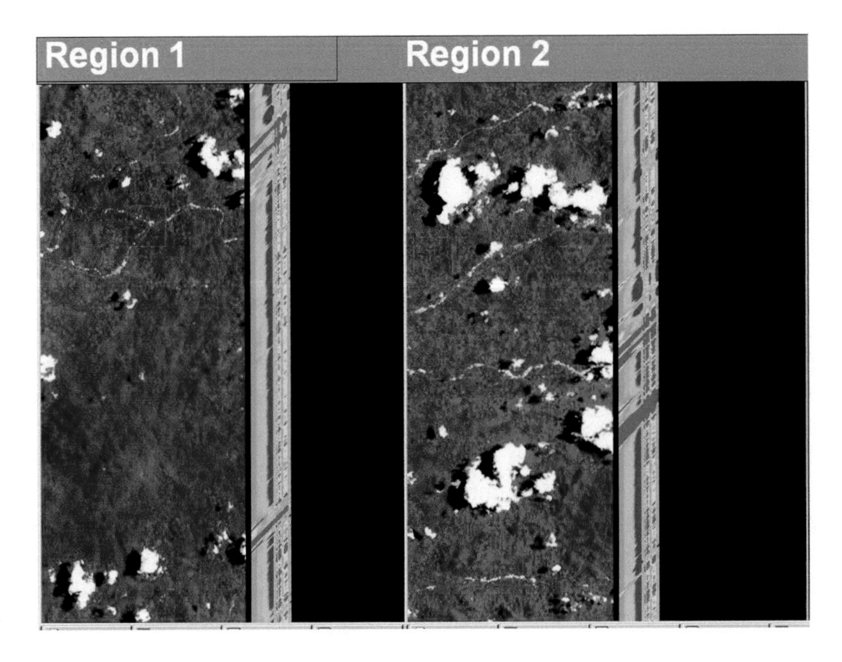

FIGURE 1.2 Hyperspectral data cube (HDC). The 242 band Hyperion image composed as HDC for 2 areas of African rainforests. Spectral signatures derived for a few classes from this Figure are illustrated in Figure 1.1d.

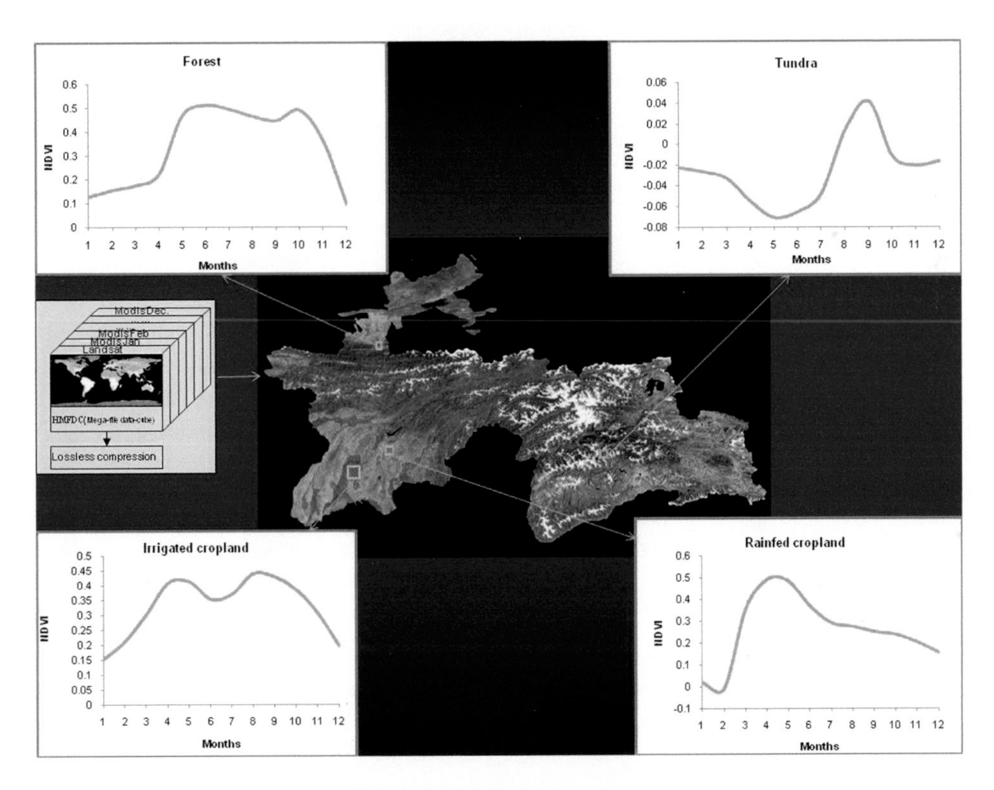

FIGURE 1.3 Mega-file data cube (MFDC). Spectral signatures extracted for a few classes from the 12 band MODIS monthly NDVI maximum value composite time-series MFDC. Akin to hyperspectral data (see Figures 1.1 and 1.2), this multispectral time-series is composed as a hyperspectral data cube. All hyperspectral data analysis techniques can be applied here. Note: illustrated for the country of Tajikistan.

reduce the discrimination power, (b) result in models that lead to lower accuracies, and (c) lead to models that fail to explain maximum variability.

1.4.1 Lambda (λ_1) versus Lambda (λ_2) Plots

Hyperspectral narrow-band data has high inter-band correlation leading to multiple measurements of the same quantity (Melendez-Pastor et al., 2010). Thereby, one of the key steps in hyperspectral data analysis is to identify and remove redundant bands. Using all the bands will only add burden to computing resources and does not add to additional information content.

For example, wavebands that are close to one another (e.g., 680 nm vs. 690 nm), typically, are highly correlated (e.g., R^2-square value of 0.99) and provide similar information. Thereby, using one of the two bands will suffice. Lambda versus Lambda plots determines (Figure 1.4): (a) redundant bands and (b) unique bands. Figure 1.4 shows inter-band correlation of rainforest vegetation gathered using 242 Hyperion bands (in 400–2500 nm with each band having 10 nm band-width) plotted as Lambda 1 (λ_1 = 400–2500 nm) versus Lambda 2 (λ_2 = 400–2500 nm).

In Figure 1.4, the least redundant bands (R^2-square values of <0.004) are shown in a magenta color. For example, band 680 nm vs. band 690 nm (Figure 1.4) was highly correlated and hence redundant (meaning we can use one of the 2 bands). However, 680 nm is significantly different from 890 and 920 nm and 2050 nm that are distinctly different bands. This means it will be valuable to have bands: 890, 920, 2050 nm and either 680 nm or 690 nm. Thereby, it will suffice to select the

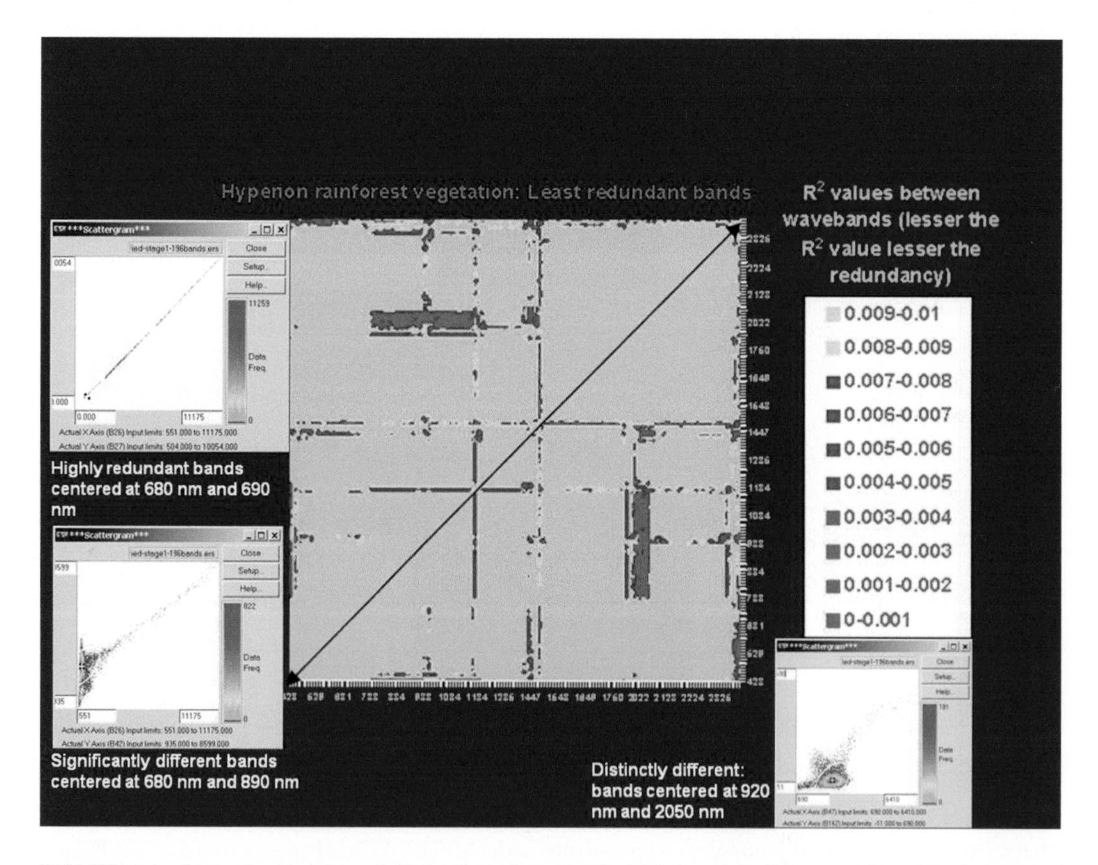

FIGURE 1.4 Redundant bands and distinctly unique bands. This Lambda (λ_1) versus Lambda (λ_2) plot of Hyperion bands show redundant bands (high correlation between bands) and distinctly unique bands (low correlation between bands).

least correlated bands (or most distinct bands; e.g., magenta, green, and yellow in Figure 1.4) for hyperspectral analysis, eliminating all redundant bands.

1.4.2 Principal Component Analysis

Principal component analysis (PCA) plays two key roles in hyperspectral data analysis for vegetation and agricultural crops. It helps in:

1. Selecting best wavebands to model biophysical and biochemical quantities; and
2. Eliminating redundant bands (by highlighting key wavebands).

Through this process, we are left with the best bands, able to eliminate redundant bands, and help reduce data volumes.

PCA is a method in which original data is transformed into a new coordinate system, which acts to condense the information found in the original inter-correlated variables into a few uncorrelated variables, called principal components (Zhao and Maclean, 2000, Tsai et al., 2007). Typically, the first few PCs explain an overwhelming proportion of variability (explained by the eigenvalue) in data. Often, adjacent hyperspectral wavebands contain redundant information (Thenkabail et al., 2002). The original high-dimensional data are thus transformed to a few bands that contain most of the information in the original bands.

The importance of the hyperspectral wavebands in each principal component (PC) are determined based on the magnitude of eigenvectors or factor loadings (higher the eigenvector, higher is the importance of the band) for every vegetation and crop biophysical and biochemical variables. It is important to take these PCs and look at eigenvectors of the wavebands. Again, for each PC, the first few bands possess high eigenvectors. The importance of hyperspectral wavebands is established based on high factor loadings (or eigenvectors) associated with these wavebands on the principal component axes of PCA (Thenkabail et al., 2004a,b, Ferwerda et al., 2005). The importance of the wavebands will vary depending on biophysical and biochemical quantities. Thus, PCA helps in determining: (a) wavebands that have greatest influence in PCA1, PCA2, and so on (based on factor loadings or eigenvectors of each PC); and (b) percent variability explained by each PC (eigenvalues).

1.4.3 Other Hyperspectral Data Mining Algorithms

There are numerous other methods of hyperspectral data analysis to reduce the large number of wavebands to a manageable number while still retaining optimal information. They are discussed in other chapters and will not be discussed here. But it is worthwhile to mention the other key data reduction approaches that still retain optimal information for a particular application. These are:

A. *Uniform feature design* (Filippi and Jensen, 2006): The main goal of the uniform feature design (UFD) is to reduce the dimensionality of the data set while still retaining as much spectral shape information in hyperspectral data. However, UFD averages bands in not an optimal feature extraction method, and loses some of the physical meaning (e.g., spectral shape).
B. *Wavelet transforms* (Sakamoto et al., 2005, Hsu, 2007, Martínez and Gilabert, 2009): The basic idea behind wavelet transform is to analyze data at different scales or resolutions. Wavelet analysis is especially suited for phenology studies such as in: (a) identifying the start of the growing season, duration of crop growth, and the time of harvest, and (b) removing the high-frequency noise caused by the frequent cloud-cover in a highly automated way, especially in the tropics. Wavelet analysis is optimal in terms of detecting transient events in data and adapts well to conditions where responses change significantly in amplitude during experiments. However, some of the wavelet methods are based on the best approximation for data representation and this is a limitation in determining actual conditions.

C. *Artificial neural networks* (Ingram et al., 2005, Trombetti et al., 2008, Liu et al., 2010): An artificial neural network (ANN) was one of the first methods used in data mining and consists of a system of simple, interconnected neurons, or nodes. It is a model representing a nonlinear mapping between an input vector and an output vector. The biggest difficulty in ANN is "training the nodes" based on a substantial and accurate knowledge base. Since hyperspectral data consists of hundreds or thousands of bands, training large number of nodes can be painstaking.

All of them have their own advantages and disadvantages. In a study of vegetation biophysical and biochemical properties, the most useful band reduction approach that still provides optimal information by retaining key bands and removing redundant bands is the Lambda (λ_1) versus Lambda (λ_2) plots (Section 1.4.1). The main advantage is in retaining the identity of original bands which can then be linked clearly to biophysical and biochemical attributes and the physical basis of sensitivity of these variables to wavebands clearly explained.

1.4.4 Optimal Hyperspectral Narrow-Bands: Hyperspectral Vegetation Indices (HVIs) to Study Vegetation and Crop Biophysical and Biochemical Properties

Spectral vegetation Indices (VIs) are computed by using broad-band data as well as narrow-band data. Thereby, even though VIs are computed using the same algebraic manipulations, their calculated values are different, thus affecting their stability in predicting agronomic variables such as total green leaf area index (Zhao et al., 2007). Previous research has demonstrated their usefulness and potential for agricultural applications such as estimating and forecasting crop yields, monitoring crop conditions, classifying and mapping crop types, and assisting precision farming activities (Yang et al., 2009).

However, there are two significant limitations of broad-band (e.g., Landsat, SPOT, IRS) derived VIs. First, the VIs saturate at high vegetation levels (Jiang and Huete, 2010). So, beyond a certain LAI or biomass or canopy cover they are asymptotic (Thenkabail et al., 2000, Jiang and Huete, 2010). Secondly, an overwhelming proportion of the broad-band VIs are constructed with red and NIR spectral measurements or even otherwise offer only a few VIs. The broad-band VIs are significantly correlated with crop agronomic variables such as LAI, above ground biomass, and chlorophyll content. However, a large proportion of variability in modeling biophysical and biochemical quantities are not explained by broad-band VIs.

The hyperspectral vegetation indices (HVIs) overcome both these limitations to a significant degree (see Thenkabail et al., 2000, 2002, 2004a,b, Cho, 2007, Chan and Paelinckx, 2008, Darvishzadeh et al., 2008c). The HVIs have greater dynamic range, providing an ability to better model crop biophysical and biochemical properties and explaining a significantly higher proportion of their variability. Further, a large number of HVIs offer greater opportunity in finding an appropriate index for studying a vegetation variable (e.g., moisture sensitivity using an index involving 970 nm, stress index involving a red-edge band around 720 nm, and nitrogen deficiency using a band that involves 1316 nm).

Given the above facts, four distinct types of HVIs are recommended to study any vegetation or agricultural crop biophysical and biochemical quantities. These indices are discussed as follows.

1.4.4.1 Hyperspectral Two-Band Vegetation Index (HTBVI)

The HTBVIs for narrow-bands i and j will be (Thenkabail et al., 2000, 2002):

$$\text{Narrow-band HTBVI}_{ij} = \frac{(R_j - R_i)}{(R_j + R_i)} \tag{1.1}$$

where, i, j = 1, N, with N = number of narrow-bands. Hyperion 220 bands (N), for example, the total number of possible HTBVIs will be 48,400 (N × N). Since the indices above and below the

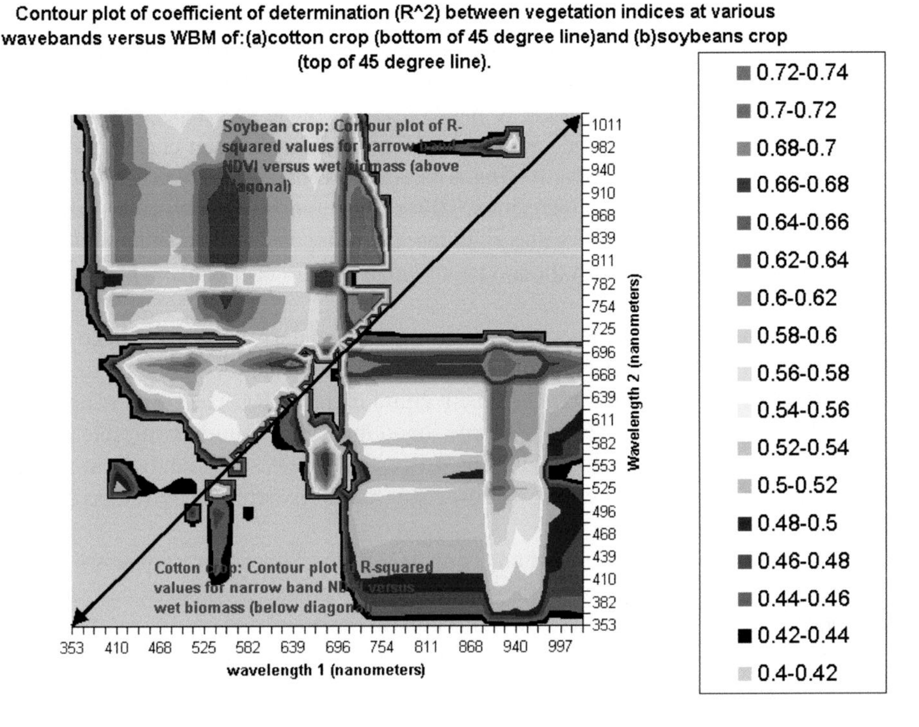

FIGURE 1.5 Hyperspectral two-band vegetation index (HTBVI) versus crop biophysical variable depicting areas of rich information content. The HTBVIs are correlated with crop wet biomass (WBM) and contour plot of R^2-values depicted for soybean crop (above diagonal) and cotton crop (below diagonal). The "bulls-eye" features help us determine waveband center and waveband width with highest R^2-values. These are the best bands to model crop biophysical and biochemical quantities.

diagonal in a N × N matrix "mirror" each other, it will suffice to consider indices that are either below or above the diagonal of a matrix. The number of unique indices will be 24,090 [(48,400–220)/2]. In comparison, the 6 non-thermal Landsat bands will have a meager 36 VIs (N × N), of which only 15 are unique. Such a large number of indices (24,090) provides many opportunities to study vegetation and agricultural crop biophysical and biochemical properties, however the best information is contained in a few selected bands or indices with the rest becoming redundant.

For example, 490 narrow-bands each of 1.43 nm wide spread across 350–1050 nm were available for computing HTBVIs (Figure 1.5). These 490 bands were reduced to 49 bands (each band of 10 nm wide) that resulted in 1176 unique HTBVIs. Each of these 1176 unique indices were correlated with the wet biomass (WMB) of soybean and cotton crops. The resulting R^2-values are plotted in a contour plot (Figure 1.5) for the soybean crop (above diagonal) and cotton (below diagonal). Such a contour plot of Lambda 1 (λ_1 = 350–1050 nm) versus Lambda 2 (λ_2 = 350–1050 nm) depict "bulls-eye" of wavebands centers and waveband widths that provide highest R^2-values (Figure 1.5). This will help us identify hyperspectral narrow-band centers and widths that are best suited to model crop biophysical and biochemical variables.

1.4.4.2 Hyperspectral Multiple-Band Models (HMBMs)

The HMBMs are computed as follows (Thenkabail et al., 2000, 2002):

$$\mathrm{HMBM_i} = \sum_{j=1}^{N} a_{ij} R_j \tag{1.2}$$

where, HMBMs = crop variable i, R = reflectance in bands j (j = 1 to N with N = 220 for Hyperion); a = the coefficient for reflectance in band j for the i variable. The process involves running stepwise linear regression models (e.g., using MAXR algorithm in Statistical Analysis System or SAS) with any one biophysical or biochemical variable as a dependent variable and the numerous hyperspectral narrow-bands as independent variables (Guisan et al., 2002, Fava et al., 2009). The MAXR method begins by finding a narrow-waveband variable (R_j) that produces the highest coefficient of determination (R^2) (SAS, 2010). Then another narrow-band variable, the one that yields the greatest increase in R^2 value, is added. Once the two narrow-band model is obtained, each of the narrow-band variables in the model are compared to each narrow-band variable not in the model.

For each comparison, MAXR determines if removing one narrow-band variable and replacing it with the other narrow-band variable increases R^2. Comparisons begin again, and the process continues until MAXR finds that no replacement could increase R^2. The two narrow-band model thus achieved is considered the best two narrow-band model fit. Another variable is then added to the model, and the comparing-and-switching process is repeated to find the best three narrow-band model, and so forth until the best narrow-band model is determined (Guisan et al., 2002, SAS, 2010).

There is a chance of over-fitting when using HMBMs. This is overcome when additional narrow-bands are considered only when they significantly increase R^2 value from a previous model (Thenkabail et al., 2000). For example, if R^2 value of a two narrow-band model was 0.75, we consider a three narrow-band model only if the R^2 value goes equal to or beyond 0.80 (an R^2 value increase of 0.05). In a study involving many biophysical quantities of several crops, Thenkabail et al. (2002) demonstrated the following distinct advantages of the using HMBVIs. These were:

A. Variability of several crop biophysical variables were explained at 95% or above using HMBMs;
B. HMBMs explained up to 27% greater variability in crop biophysical variables than similar broad-band models;
C. HMBMs explained up to 11% greater variability in crop biophysical variables than HTBVI models;

1.4.4.3 Hyperspectral Derivative Greenness Vegetation Indices (HDGVIs)

When a difference index is formulated using two closely spaced band centers, it is indicative of the slope of the reflectivity with respect to wavelength and is often referred to as a hyperspectral derivative vegetation greenness index (HDGVI, Elvidge and Chen, 1995, Broge and Leblanc, 2001). First derivatives (HDGVI1) are approximated by dividing the difference in reflectance value between spectrally adjacent bands by the corresponding difference in band center wavelength (Lucas and Carter, 2008). Second derivatives (HDGVI2) are determined likewise from *d*1 spectra (Lucas and Carter, 2008). The performance of two indices was shown to be practically identical and thus it will suffice to calculate only one of these. HDGVI1 measure the amplitude of the chlorophyll and are computed by taking near-continuous spectra such as 626–795 nm (Elvidge and Chen, 1995) or along the chlorophyll red-edge (0.700–0.740 μm). The chlorophyll red-edge portion is considered to have maximum sensitivity to changes in green vegetation per unit change in wavelength in the electromagnetic spectrum (Thenkabail et al., 2002). The first-order (HDBVI1) are computed as follows:

$$\mathrm{HDGVI} = \sum_{\lambda_1}^{\lambda_n} \frac{(\rho'(\lambda_i) - (\rho'(\lambda_j)}{\Delta\lambda_I} \tag{1.3}$$

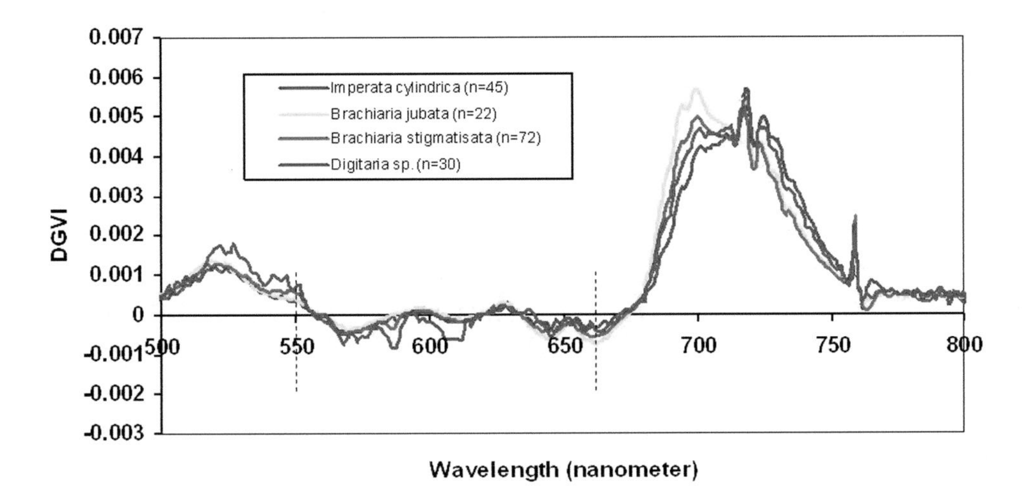

FIGURE 1.6 First-order hyperspectral derivative greenness vegetation index (HDGVI1) computed along 500–800 nm range for certain weed species.

where, i and j are band numbers, λ = center of wavelength, Lambda 1 (e.g., $\lambda_1 = 626$ nm) versus Lambda 2 (e.g., $\lambda_2 = 795$ nm). ρ' = first derivative reflectance. One can integrate HDGVIs using different waveband ranges (e.g., $\lambda_1 = 700$ nm, $\lambda_2 = 740$ nm or $\lambda_1 = 940$ nm, $\lambda_2 = 980$ nm). HDGVI1 computed for 500–800 nm is illustrated for certain weed species in Figure 1.6.

Indications are that derivative indices are of particular importance in monitoring plant stress (Thenkabail et al., 2002), complex vegetation conditions (mixture of green and brown; Broge and Leblanc, 2001), grassland, or weed canopies (Elvidge and Chen, 1995, Curran et al., 1997).

1.4.4.4 Hyperspectral Hybrid Vegetation Indices

1.4.4.4.1 Soil-Adjusted Hyperspectral Two-Band Vegetation Indices (SA HTBVIs)

Soil-adjusted vegetation indices (SAVIs) were first proposed by Huete (1988, 1989) and were developed to account for changes in vegetated canopy spectra due to multiple scattering interactions with background soil optical properties. Soil influences were minimized through alignment of VI isolines with biophysical greenness isolines (usually expressed in terms of LAI) over the entire dynamic range of the greenness measure (Broge and Leblanc, 2001). SAVI is defined as (Huete, 1988):

$$\mathrm{SAVI} = \frac{\mathrm{NIR} - \mathrm{RED}}{\mathrm{NIR} + \mathrm{RED} + \mathrm{L}}(1 + \mathrm{L}) \tag{1.4}$$

where, NIR = near-infrared band reflectance, RED = red band reflectance, L = soil adjustment factor which varies according to vegetation cover or LAI. L = 0.5 is taken as a default global value, however, ideally L approaches zero for full canopy cover (here, SAVI = NDVI), while L values greater than 10 best represent zero canopy cover. Thus, NDVI is a subset of SAVI, for the special case of L = 0, while $L > 10$ mimics linear mixing of soil and vegetation, as in linear mixture models.

It is possible to compute SA HTBVIs for each of the 24,090 HTBVIs computed in Section 1.4.2. To reduce the computing time and optimize resources, SA HTBVIs can be computed only for the best HTBVIs (Figure 1.5). The value of computing SA HTBVIs is to enhance the sensitivity of HTBVIs by normalizing soil background influences; thus, helping to explain greater variability in crop biophysical and biochemical variables such as biomass, LAI, nitrogen, and chlorophyll a.

1.4.4.4.2 Atmospherically Resistant Hyperspectral Two-Band Vegetation Indices (AR HTBVIs)

The atmospherically resistant vegetation index (ARVI) was first proposed by Kaufman and Tanre (1996) and was developed to account for aerosol effects by using the difference in blue and red reflectances to derive the surface red reflectance. This index is especially useful to study leaf pigment. Galvão et al. (2009) defined ARVI for Hyperion data as:

$$((\lambda 864 - (2 * \lambda 671 - \lambda 467))/(\lambda 864 + (2 * \lambda 671 - \lambda 467)) \tag{1.5}$$

Another approach to minimize atmospheric effects on NDVI is to use the middle-infrared wavelength region (1.3–2.5 μm) as a substitute for the red band, since longer wavelengths are much less sensitive to smoke and aerosols (Jiang et al., 2008).

Again, it is possible to compute AR HTBVIs for each of the 24,090 HTBVIs computed in Section 1.4.2. To reduce the computing time and optimize resources, AR HTBVIs can be computed only for the best HTBVIs (Figure 1.5). The value of computing AR HTBVIs is to enhance the sensitivity of HTBVIs by normalizing atmospheric aerosol and moisture effects; thus, helping to explain greater variability in crop biophysical and biochemical variables such as biomass, LAI, nitrogen, and chlorophyll a.

1.4.4.4.3 Hyperspectral Vegetation Indices of SWIR and TIR Bands (HVIST)

Obtaining narrow-band data beyond 1100 nm, from the shortwave infrared (SWIR; 1100–2500 nm) and thermal infrared (TIR; 3000–14,000 nm), are important unique and/or complementing and/or supplementing information in addition to information on vegetation gained in the visible and near infrared (VNIR) discussed in previous paragraphs. The non-photosynthetic vegetation (NPV) such as litter, senesced leaves, and other dry vegetation are best differentiated based on lingo-cellulose bands in the SWIR (Numata et al., 2007, 2008) but are not spectrally separable from soil in the visible and near-infrared wavelength region (Asner and Lobell, 2000). Separating forest types and cropland classes, categorizing forest age classes or croplands at various stages of growth, modeling forest and cropland biotic factors such as canopy height, basal area, biomass, and LAI are often best predicted through a combination of visible and shortwave infrared (SWIR) (Cho, 2007, White et al., 2010). Moisture and plant stress properties are best quantified by including TIR bands along with SWIR, and VNIR (Kalacska et al., 2007, Galvão et al., 2009).

1.4.5 Other Methods of Hyperspectral Data Analysis

There are numerous other methods of hyperspectral data analysis for studying vegetation and agricultural crop biophysical and biochemical properties. These include:

A. *Independent component analysis* (Huang et al., 2018) which is an unsupervised temporal unmixing methodology that helps de-compose spectra. For example, it can recover both the time profile and area distribution of different crop types (Ozdogan, 2010);
B. *Wavelet transform* can detect automatically the local energy variation of a spectrum in different spectral bands at each scale and provide useful information for hyperspectral image classification (Hsu, 2007);
C. *Radiative transfer models* such as PROSPECT (le Maire et al., 2004) and LIBERTY (Coops and Stone, 2005) are used to simulate wide ranges of hyperspectral signatures and their impact on vegetation and crop variables;
D. *Minimum noise fraction (MNF) transformation* separates noise from the data by using only the coherent portions, thus improving spectral processing results; and
E. *Spectral unmixing analysis* (SMA; Shippert, 2001, Pacheco and McNairn, 2010) uses reference spectra (referred to as "endmembers") in order to "unmix" characteristics of

spectra within each pixel. The biotic and abiotic characteristics of the constituents within a pixel will be gathered through ground truth data. The spectra of the individual constituents within the pixel can be obtained using a spectroradiometer and/or from spectral libraries. SMA classifications are known to map land cover more accurately than maximum likelihood classification (Okin et al., 2001, Pacheco and McNairn, 2010) and have the advantage of assessing the within pixel composition. Once all the materials in the image are identified, then it is possible to use linear spectral unmixing to find out how much of each material is in each pixel (Shippert, 2001).

1.4.6 Broad-Band Vegetation Index Models

It was determined by Thenkabail et al. (1999) that the vegetation indices computed for a wide range of existing broad-band sensors such as Landsat MSS, Landsat TM, SPOT HRV, NOAA AVHRR, and IRS-1C, using simulated data for crops from a spectroradiometer, were highly correlated (R^2 value = 0.95 or higher). Hence, computing vegetation indices for any one of these sensors will provide nearly the same information as similar indices computed for other sensors. The most common categories of broad-band indices are:

a. NIR and red-based indices (Tucker, 1977);
b. Soil-adjusted indices (Huete, 1988, 1989);
c. Atmospheric resistant indices (Kaufman and Tanre, 1996); and
d. Mid-infrared-based indices (Thenkabail et al., 1994a, 1995).

1.5 SEPARATING VEGETATION CLASSES AND AGRICULTURAL CROPS USING HYPERSPECTRAL NARROW-BAND DATA

1.5.1 Class Separability Using Unique Hyperspectral Narrow-Bands

There are fewer opportunities in separating vegetation classes using broad-bands from sensors such as Landsat TM (e.g., Figure 1.7a) when compared to numerous opportunities offered by narrow-bands from sensors such as EO-1 Hyperion (e.g., Figure 1.7b). For example, the Landsat TM broad-bands of near-infrared (TM4) versus red (TM3) could not separate a wheat crop from a barley crop (Figure 1.7a). However, when 2 narrow-bands involving the near-infrared centered at 910 nm with a narrow-band-width of 10 nm, versus red centered at 675 nm with narrow-band-width of 10 nm were used, an overwhelming proportion of the wheat crop fields were separated from barley crop fields (Figure 1.7b). Since hyperspectral sensors have 100 s of wavebands, the likelihood of finding wavebands that can separate vegetation/crop types (e.g., Figure 1.7b) or various groups of vegetation/crop biophysical and biochemical quantities increases drastically.

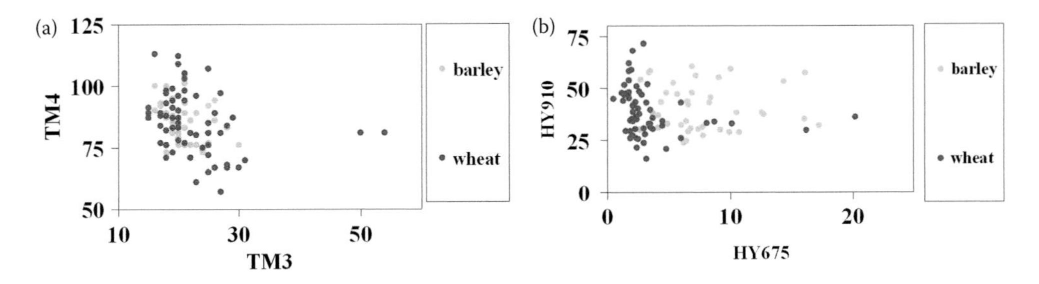

FIGURE 1.7 Separating 2 agricultural crops using broad-bands versus narrow-bands. The 2 broad-bands fail to separate wheat from barley whereas 2 distinct narrow-bands separate wheat from barley. The Landsat TM (a) and Hyperion (b) imaged the same surface at 30 m spatial resolution.

1.5.2 Class Separability Using Statistical Methods

Stepwise discriminant analysis (SDA) is a powerful statistical tool for discriminating vegetation and agricultural crop types based on their quantitative and qualitative characteristics such as biophysical quantities, biochemical compositions, structural properties, and species type. The independent hyperspectral waveband variables are chosen to enter or leave the model using: (a) the significance level of the F test analysis of covariance, where the variables already chosen act as covariates and the variable under consideration is the dependent variable, or (b) the squared partial correlation for predicting the variable under consideration from the CLASS variable, controlling for the influences of the variables were selected for the model (SAS, 2010). Stepwise selection begins with no variable in the model. At each step, if a variable already in the model fails to meet the criterion to stay, the worst such variable is removed. Otherwise, the variable that contributes most to the discriminatory power of the model is entered. When all variables in the model meet the criterion to stay, and more of the other variables meet the criterion not to enter, the stepwise selection process stops (SAS, 2010). The stepwise discriminant analysis (Draper and Smith, 1981) was performed using the PROC STEPDISC algorithm of SAS (Klecka, 1980; SAS, 2010).

Class separability in SDA can be expressed, most powerfully and lucidly, using: (a) Wilks' Lambda, and (b) Pillai's trace. Wilks' Lambda and Pillai's trace are based on the eigenvalues Γ of A^*W^{-1} where A is the among the "sum-of-squares" (SS) and cross-products matrix, and W is the pooled SS and cross-products matrix:

$$\text{Wilks' } \Lambda = \Pi\ 1/(1 + \lambda_i)$$

$$\text{Pillai's trace} = \Sigma\ \lambda_i/(1 + \lambda_i) \quad (1.6)$$

Determinants (variance) of the S matrices are found. Wilks' Lambda is the test statistic preferred for multivariate analysis of variance (MANOVA) and is found through a ratio of the determinants:

$$\Lambda = \frac{|S_{error}|}{|S_{effect} + S_{error}|} \quad (1.7)$$

where, S is a matrix which is also known as: SS and cross-products, cross-products, or sum-of-products matrices.

Wilks' Lambda is the most commonly available and reported, however Pillai's criterion is more robust and therefore more appropriate when there are small or unequal sample sizes. When separating 2 classes:

A. The lower the value of Wilks' Lambda, the greater is the separability between two classes.
B. Higher the value of Pillai's trace, greater is the separability between 2 classes.

Class separability using hyperspectral data is illustrated for a few vegetation types in Figure 1.8. In each case, with the increase in number of wavebands, the separability also increases (Figure 1.8; note: lower the Wilks' Lambda, greater the separability). For example, vegetation types and shrubs vs. grasses vs. weeds are best separated using around 17 hyperspectral narrow-bands (Figure 1.8), whereas, the 2 weeds (*Chromolenea odorata* vs. *Imperata cylindrica*) are separated best using 11 hyperspectral narrow-bands. It is possible to have a small incremental increase in spectral separability beyond 11 bands for weeds and beyond 17 bands for vegetation types, but those increases are very small and statistically insignificant. Figure 1.8 also shows that the fallows (1-year fallow

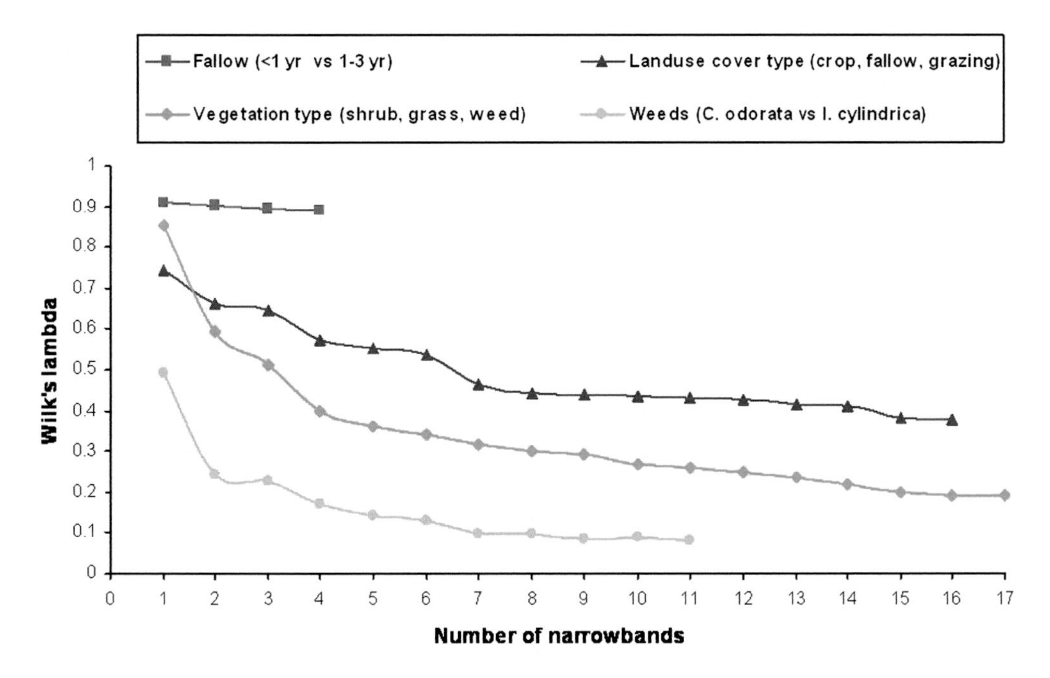

FIGURE 1.8 Class separability using hyperspectral narrow-bands determined based on Wilks' Lambda. Lower the Wilks' Lambda, greater the separability. So, with the addition of wavebands, the separability increases, reaching an optimal point beyond which the addition of wavebands does not make a difference.

vs. 1–3-year fallow) are not well separated by even hyperspectral narrow-bands since the Wilks' Lambda remains high.

In each case, it is possible to determine the exact wavebands that help increase separability (see Thenkabail et al., 2000, 2002, 2004a,b).

1.5.3 Accuracy Assessments of Vegetation and Crop Classification Using Hyperspectral Narrow-Bands

Classification accuracies of vegetation and agricultural crops attained using hyperspectral narrow-band data are substantially higher than broad-bands (Bork et al., 1999, Thenkabail et al., 2004b, Papeş et al., 2010, Schull et al., 2010). For example, greater than 90% classification accuracies have been obtained in classifying 5 agricultural crops using about 20 selected hyperspectral narrow-band data, relative to just about 60% accuracies obtained using 6 non-thermal broad-bands (e.g., Figure 1.9). Similarly, Dalponte et al. (2009) studied a forest area in Italy characterized by 23 different classes reaching accuracies of about 90% with hyperspectral data acquired at a spectral resolution of 4.6 nm in 126 bands. Clark et al. (2005) studied seven deciduous tree species with the HYDICE sensor, using three different classifiers, reaching accuracies to the order of 90%.

Also, sensitivity to above ground biomass and ability to map forest successional stages are achieved using hyperspectral narrow-bands (Kalacska et al., 2007, Asner et al., 2009). Earlier, we established that selected hyperspectral narrow-bands can play a crucial role in separating 2 vegetation categories (e.g., Figure 1.7b) when 2 broad-bands fail to do so (e.g., Figure 1.7a). Also, the hyperspectral narrow-bands explain about 10%–30% greater variability in modeling quantitative biophysical quantities (e.g., LAI, biomass) relative to broad-band data (Thenkabail et al., 2002, Chan and Paelinckx, 2008). The specific narrow-bands that contribute to increased accuracies were also identified.

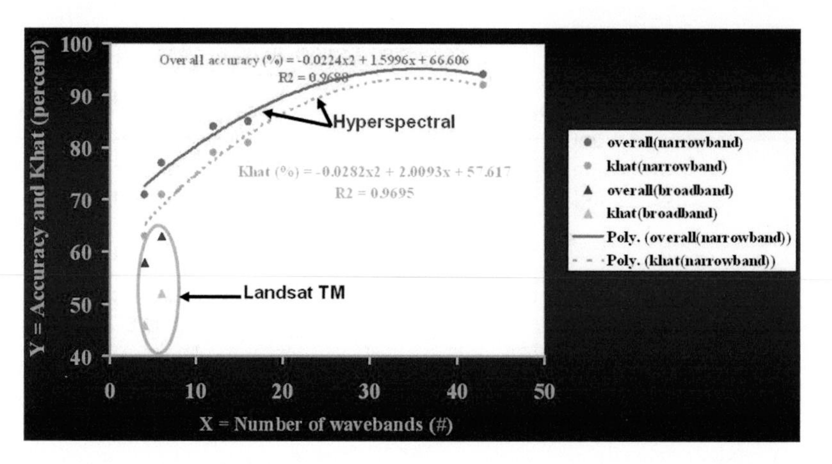

FIGURE 1.9 Classification accuracies using hyperspectral narrow-bands versus Landsat broad-bands. The accuracies increased by 25%–30% when about 25 hyperspectral narrow-bands are used to classify 5 agricultural crops when compared to 6 Landsat broad-bands. Classification accuracies reach about 95% with 30 bands, beyond which accuracies do not increase. (Thenkabail P. S. et al. 2004b. *Remote Sensing of Environment* 91(2–3): 354–376.)

1.6 OPTIMAL HYPERSPECTRAL NARROW-BANDS IN THE STUDY OF VEGETATION AND AGRICULTURAL CROPS

Optimal hyperspectral narrow-bands in the study of vegetation and agricultural croplands (Table 1.2) were determined based on an exhaustive review of literature that included: (a) identifying redundant bands (e.g. Figure 1.4), (b) modeling by linkage of crop biophysical and biochemical variables with hyperspectral indices and wavebands (e.g., Figure 1.5), (c) establishing wavebands that best help separate vegetation and crop types (e.g., Figure 1.7), (d) identifying wavebands, through statistical and other approaches, that best separate vegetation and crop types and characteristics (e.g., Figure 1.8), and (e) establishing classification accuracies of vegetation and crop classes and identifying wavebands that best help enhance these accuracies (e.g., Figure 1.9). The 28 wavebands listed in Table 1.2 are based on their frequency of occurrence in numerous studies discussed here. These prominent wavebands include: (a) two red bands centered at 675 nm (chlorophyll absorption maxima) and 682 nm (most sensitive to biophysical quantities and yield); (b) three NIR bands at 845 nm (mid-point of NIR "shoulder" that is most sensitive to biophysical quantities and yield), 915 nm (peak spectrum in NIR, useful for computing crop moisture sensitivity index with 975 nm); and 975 nm (moisture and biomass sensitive band). The NIR portion of the spectrum is highly sensitive to changes in biophysical quantities and plant structure. For example, the more horizontal, planophile structure plant leaf contributes significantly to greater reflectance in NIR and greater absorption in the red when compared with the more vertical erectophile structure (Thenkabail et al., 2000). The erectophile structure leads to significant slope changes in spectra in the region of 740 nm to 940 nm; (c) five green bands: 515 nm, 520 nm, 525 nm, 550 nm, and 575 nm. These bands are overwhelmingly sensitive to biochemical properties. Band $\lambda = 550$ nm is strongly correlated with total chlorophyll. The green band λ centered at 520 nm provides the most rapid positive change in reflectance per unit change in wavelength anywhere in the visible portion of the spectrum. The green band λ centered at 575 nm provides the most rapid negative change in reflectance per unit change in wavelength anywhere in the visible portion of the spectrum. Overall, the green bands are very sensitive to plant/leaf nitrogen and pigment. There are dramatic shifts in crop-soil spectral behavior around 568 and 520 nm when there is change in pigment content and chloroplast structure for different crop types, growth stages, and growing conditions; and for (d) three red-edge bands at 700 and 720 nm (sensitive to vegetation stress), and 740 nm (sensitive to vegetation nitrogen content). The red-edge bands are

TABLE 1.2
Optimal Hyperspectral Narrow-Bands Recommended in the Study of Vegetation and Agricultural Crops

Waveband Number #	Waveband Center nm	Importance in Vegetation and Agricultural Cropland Studies	Reference
		A. Blue Bands	
1	375	*fPAR, leaf water*: fraction of photosynthetically active radiation (fPAR), leaf water content	Thenkabail et al. (2000, 2002, 2004a,b)
2	466	*Chlorophyll*: Chlorophyll a and b in vegetation	Thenkabail et al. (2000, 2002, 2004a,b)
3	490	*Senescing and loss of chlorophyll\browning, ripening, crop yield*: Sensitive to loss of chlorophyll, browning, ripening, senescing, and soil background effects.	Thenkabail et al. (2004a,b)
		B. Green Bands	
4	515	*Nitrogen*: leaf nitrogen, wetland vegetation studies	Thenkabail et al. (2004a,b), Curran (1989)
5	520	*Pigment, biomass changes*: positive change in reflectance per unit change in wavelength of this visible spectrum is maximum around this green waveband. Sensitive to pigment.	Galvão, L.S. et al. (2011), Thenkabail et al. (2004a), Blackburn (1998, 1999)
6	525	*Vegetation vigor, pigment, nitrogen*: positive change in reflectance per unit change in wavelength in maximum as a result of vegetation vigor, pigment, and nitrogen.	Curran (1989)
7	550	*Chlorophyll*: Total chlorophyll; Chlorophyll/carotenoid ratio, vegetation nutritional and fertility level; vegetation discrimination; vegetation classification.	Yang et al. (2009), Chan and Paelinck (2008), Thenkabail et al. (2002)
8	575	*Vegetation vigor, pigment, and nitrogen*: negative change in reflectance per unit change in wavelength is maximum as a result of sensitivity to vegetation vigor, pigment, and N.	
		C. Red Bands	
9	675	*Chlorophyll absorption maxima*: greatest crop-soil contrast is around this band for most crops in most growing conditions. Strong correlations with chlorophyll a and b	Chan and Paelinck (2008), Thenkabail et al. (2004a,b)
10	682	*Biophysical quantities and yield*: leaf area index, wet and dry biomass, plant height, grain yield, crop type, crop discrimination	Galvão et al. (2011), Thenkabail et al. (2004a,b)
		D. Red-Edge Bands	
11	700	*Stress and chlorophyll*: Nitrogen stress, crop stress, crop growth stage studies	Chan and Paelinck (2008), Thenkabail et al. (2004a,b), Jensen (2000)
12	720	*Stress and chlorophyll*: Nitrogen stress, crop stress, crop growth stage studies	Maire et al. (2008), Thenkabail et al. (2002), Penuelas et al. (1995)
13	740	*Nitrogen accumulation*: Leaf nitrogen accumulation	Chan and Paelinck (2008), Thenkabail et al. (2004a,b)

(*Continued*)

TABLE 1.2 (*Continued*)
Optimal Hyperspectral Narrow-Bands Recommended in the Study of Vegetation and Agricultural Crops

Waveband Number #	Waveband Center nm	Importance in Vegetation and Agricultural Cropland Studies	Reference
		E. Near-Infrared (NIR) Bands	
14	845	*Biophysical quantities and yield*: leaf area index, wet and dry biomass, plant height, grain yield, crop type, crop discrimination, total chlorophyll	Maire et al. (2008)
15	915	*Moisture, biomass, and protein*: peak NIR reflectance. Useful for computing crop moisture sensitivity index.	Thenkabail et al. (2002), Penuelas et al. (1995)
16	975	*Moisture and biomass*: Center of moisture sensitive "trough"; water band index, leaf water, biomass	Yao et al. (2010), Adam and Mutanga (2009), Müller et al. (2008)
		F. Far NIR (FNIR) Bands	
17	1100	*Biophysical quantities*: sensitive to biomass and leaf area index. A point of most rapid rise in spectra with unit change in wavelength in far near-infrared (FNIR)	Abdel-Rahman et al. (2010)
18	1215	*Moisture and biomass*: A point of most rapid fall in spectra with unit change in wavelength in FNIR. Sensitive to plant moisture.	Yao et al. (2010), Becker et al. (2005)
19	1245	*Water sensitivity*: water band index, leaf water, biomass. Reflectance peak in 1050–1300 nm.	Thenkabail et al. (2002), Nichol et al. (2000), Elvidge and Chen (1995)
		G. Shortwave Infrared (SWIR) Bands	
20	1316	*Nitrogen*: leaf nitrogen content of crops	Curran (1989)
21	1445	*Vegetation classification and discrimination*: ecotype classification; plant moisture sensitivity. Moisture absorption trough in early shortwave infrared (ESWIR)	Thenkabail et al. (2002)
22	1518	*Moisture and biomass*: A point of most rapid rise in spectra with unit change in wavelength in SWIR. Sensitive to plant moisture.	Zhao et al. (2007), Thenkabail et al. (2004a,b), Gitelson and Merzlyak (1997)
23	1725	*Lignin, biomass, starch, moisture*: sensitive to lignin, biomass, starch. Discriminating crops and vegetation.	Yao et al. (2010), Maire et al. (2008), Vaiphasa et al. (2007), Thenkabail et al. (2002)
24	2035	*Moisture and biomass*: moisture absorption trough in far shortwave infrared (FSWIR)	Thenkabail et al. (2004b)
25	2173	*Protein, nitrogen*	Chan and Paelinck (2008), Thenkabail et al. (2004a,b)
26	2260	*Moisture and biomass*: moisture absorption trough in far shortwave infrared (FSWIR). A point of most rapid change in slope of spectra based on vegetation vigor, biomass.	Thenkabail et al. (2002), Nichol et al. (2000), Elvidge and Chen (1995)
27	2295	*Stress*: sensitive to soil background and plant stress	Yao et al. (2010), Chan and Paelinck (2008), Thenkabail (2000)
28	2359	*Cellulose, protein, nitrogen*: sensitive to crop stress, lignin, and starch	Galvão et al. (2009), Thenkabail et al. (2004a,b), Penuelas et al. (1995)

especially sensitive to crop stress and changes in total chlorophyll and have potential to form a useful drought index. These are sensitive to senescing rates, chlorophyll changes, browning, ripening, carotenoids, and soil background influences; (e) as are three FNIR bands at 1100 nm (sensitive to biophysical quantities), 1215 nm (sensitive to moisture), and at 1245 nm (sensitive to water), and (f) nine SWIR bands (1316, 1445, 1518, 1725, 2035, 2173, 2260, 2295, and 2359 nm). These wavebands are overwhelmingly sensitive to moisture and biochemical properties; as are (g) three blue bands (375, 466, and 490 nm) that are especially sensitive to fraction of photosynthetically active radiation (fPAR), and senescing. Band $\lambda = 490$ nm is especially sensitive to carotenoids, leaf chlorophyll, and senescing conditions.

A nominal bandwidth ($\Delta\lambda$) of 10 nm can be used for all wavebands. Too narrow a waveband will lead to lower signal to noise ratios. Keeping a band-width ($\Delta\lambda$) of 10 nm will ensure that optimal information on a particular feature is captured rather than average conditions captured in broader bands.

In general, there is overwhelming evidence that the hyperspectral wavebands and indices involving specific narrow wavebands provide significantly better estimates on vegetation and crop biophysical quantities (e.g., LAI, biomass, plant height, canopy cover), biogeochemistry (e.g., lignin, chlorophyll a and b, photosynthesis, respiration, nutrient use (e.g., through LAI, and fPAR), canopy water content, and improved discrimination of species and land cover.

1.7 CONCLUSIONS

An overwhelming proportion of the hyperspectral narrow-bands can be redundant. So, it will be extremely important to identify and remove the redundant bands from further analysis to ensure best use for resources and to overcome the curse of high-dimensionality or Hughes phenomenon. This led us to identify 28 optimal hyperspectral narrow-bands (Table 1.2) best suited in the study of vegetation and agricultural crops. The wavebands were identified based on their ability to: (a) best model biophysical and biochemical properties, (b) distinctly separate vegetation and crops based on their species type, structure, and composition, and (c) accurately classify vegetation and crop classes. Some caution is needed in eliminating redundant bands, since some of the redundant bands in one application may be useful in some other. Taking this into consideration, methods and approaches of data mining to quickly identify key wavebands important for a given application (and as a result eliminate redundant bands) are highlighted and these key wavebands are then used in more detail studies of that particular application.

References cited here highlighted computation and use of various hyperspectral vegetation indices (HVIs) in studying and determining the most valuable wavebands (Table 1.2). The specific importance of each of the wavebands to biophysical and biochemical properties of vegetation and agricultural crops are also called out in Table 1.2. Typically, there is more than one waveband for each crop variable. For example, biophysical properties and yield are best modeled using wavebands centered at 682, 845, and 1100 nm. Similarly, moisture sensitivity in leaf/plants can be studied using several wavebands centered at 915, 975, 1215, 1518, 2035, and 2260 nm. The 28 wavebands (Table 1.2) will also allow us to compute 378 unique hyperspectral two-band vegetation indices (HTBVIs). Overall, there is clear evidence that the combination of the 28 identified bands (Table 1.2) and various HVIs computed from them will suffice to best characterize, classify, model, and map a wide array of vegetation and agricultural crop types and their biophysical and biochemical properties.

ACKNOWLEDGMENT

The authors would like to thank Ms. Zhouting Wu of Northern Arizona University (NAU) for preparing Figure 1.3. Comments from reviewers were much appreciated.

REFERENCES

Abdel-Rahman, E.M., Ahmed, F.B., and Van den Berg, M. 2010. Estimation of sugarcane leaf nitrogen concentration using *in situ* spectroscopy, *International Journal of Applied Earth Observation and Geoinformation* 12(1): S52–S57.

Adam, E., and Mutanga, O. 2009. Spectral discrimination of papyrus vegetation (*Cyperus papyrus L.*) in swamp wetlands using field spectrometry, *ISPRS Journal of Photogrammetry and Remote Sensing* 64(6): 612–620.

Anderson, J.E., Plourde, L.C., Martin, M.E., Braswell, B.H., Smith, M.-L., Dubayah, R.O., Hofton, M.A., and Blair, J.B. 2008. Integrating waveform LiDAR with hyperspectral imagery for inventory of a northern temperate forest, *Remote Sensing of Environment* 112: 1856–1870.

Asner, G.P. 1998. Biophysical and biochemical sources of variability in canopy reflectance, *Remote Sensing of Environment* 64: 234–253.

Asner, G.P., and Heidebrecht, K.B. 2002. Spectral unmixing of vegetation, soil and dry carbon cover in arid regions: Comparing multispectral and hyperspectral observations, *International Journal of Remote Sensing* 23: 3939–3958.

Asner, G.P., and Lobell, D.B. 2000. A biogeophysical approach for automated SWIR unmixing of soils and vegetation, *Remote Sensing of Environment* 74: 99–112.

Asner, G.P., and Martin, R.E. 2009. Airborne spectranomics: Mapping canopy chemical and taxonomic diversity in tropical forests, *Frontiers in Ecology and the Environment* 7: 269–276.

Asner, G.P., Martin, R.E., Ford, A.J., Metcalfe, D.J., and Liddell, M.J. 2009. Leaf chemical and spectral diversity in Australian tropical forests, *Ecological Applications* 19: 236–253.

Atzberger, C. 2004. Object-based retrieval of biophysical canopy variables using artificial neural nets and radiative transfer models, *Remote Sensing of Environment* 93: 53–67.

Becker, B.L., Lusch, D.P., and Qi, J. 2005. Identifying optimal spectral bands from *in situ* measurements of Great Lakes coastal wetlands using second-derivative analysis, *Remote Sensing of Environment* 97(2): 238–248.

Blackburn, G.A. 1998. Quantifying chlorophylls and carotenoids at leaf and canopy scales: An evaluation of some hyperspectral approaches, *Remote Sensing of Environment* 66: 273–285.

Blackburn, G. 1999. Relationships between spectral reflectance and pigment concentrations in stacks of broadleaves, *Remote Sensing of Environment* 70: 224–237.

Blackburn, J.A. 2006. Hyperspectral remote sensing of plant pigments, *Journal of Experimental Botany* 58(4): 855–867.

Blackburn, A.G. and Ferwerda, J.G. 2008. Retrieval of chlorophyll concentration from leaf reflectance spectra using wavelet analysis. *Remote Sensing of Environment* 112(4):1614–1632.

Bork, E.W., West, N.E., and Price, K.P. 1999. Calibration of broad-and narrow-band spectral variables for rangeland covers component quantification, *International Journal of Remote Sensing* 20: 3641–3662.

Broge, N.H., and Leblanc, E. 2001. Comparing prediction power and stability of broadband and hyperspectral vegetation indices for estimation of green leaf area index and canopy chlorophyll density, *Remote Sensing of Environment* 76(2): 156–172.

Carter, G. 1998. Reflectance wavebands and indices for remote estimation of photosynthesis and stomatal conductance in pine canopies, *Remote Sensing of Environment* 63: 61–72.

Chan, J.C., and Paelinckx, D. 2008. Evaluation of Random Forest and Adaboost tree-based ensemble classification and spectral band selection for ecotope mapping using airborne hyperspectral imagery, *Remote Sensing of Environment* 112(6): 2999–3011.

Chen, J., Wang, R., Wang, C. 2008. A multiresolution spectral angle-based hyperspectral classification method. *International Journal of Remote Sensing* 29(11): 3159–3169.

Cho, M.A. 2007. Hyperspectral remote sensing of biochemical and biophysical parameters: The derivative red-edge "double peak feature," a nuisance or an opportunity? Ph.D. dissertation. p. 206. International Institute for Geo-information science and earth observation (ITC). Enschede, The Netherlands.

Cho, M.A., Skidmore, A.K., and Sobhan, I. 2009. Mapping beech (Fagus sylvatica L.) forest structure with airborne hyperspectral imagery, *International Journal of Applied Earth Observation and Geoinformation* 11(3): 201–211.

Clark, M.L., Roberts, D.A., and Clark, D.B. 2005. Hyperspectral discrimination of tropical rain forest tree species at leaf to crown scales, *Remote Sensing of Environment* 96: 375–398.

Coops, N.C., and Stone, C. 2005. A comparison of field-based and modelled reflectance spectra from damaged *Pinus radiata* foliage, *Australian Journal of Botany* 53: 417–429.

Colombo, R., Meroni, M., Marchesi, A., Busetto, L., Rossini, M., Giardino, C., and Panigada, C. 2008. Estimation of leaf and canopy water content in poplar plantations by means of hyperspectral indices and inverse modeling. *Remote Sensing of Environment* 112(4): 1820–1834

Curran, P. 1989. Remote sensing of foliar chemistry, *Remote Sensing of Environment* 30: 271–278.

Curran, P.J., Foody, G.M., Lucas, R.M., Honzak, M., and Grace, J. 1997. The carbon balance of tropical forests: From the local to the regional scale. In: van Gardingen, P.R., Foody, G.M. and Curran, P.J., Editors. *Scaling-up from cell to landscape*, Cambridge University Press, Cambridge, pp. 201–227.

Dalponte, M., Bruzzone, L., Vescovo, L., and Gianelle, D. 2009. The role of spectral resolution and classifier complexity in the analysis of hyperspectral images of forest areas, *Remote Sensing of Environment* 113(11): 2345–2355.

Darvishzadeh, R., Skidmore, A., Atzberger, C., and van Wieren, S. 2008a. Estimation of vegetation LAI from hyperspectral reflectance data: Effects of soil type and plant architecture, *International Journal of Applied Earth Observation and Geoinformation* 10: 358–373.

Darvishzadeh, R., Skidmore, A., Schlerf, M., and Atzberger, C. 2008b. Inversion of a radiative transfer model for estimating vegetation LAI and chlorophyll in a heterogeneous grassland, *Remote Sensing of Environment* 112: 2592–2604.

Darvishzadeh, R., Skidmore, A., Schlerf, M., Atzberger, C., Corsi, F., and Cho, M. 2008c. LAI and chlorophyll estimation for a heterogeneous grassland using hyperspectral measurements, *ISPRS Journal of Photogrammetry and Remote Sensing* 63: 409–426.

Daughtry, C.S.T., and Walthall, C.L. 1998. Spectral discrimination of *Cannabis sativa L.* leaves and canopies, *Remote Sensing of Environment* 64(2): 192–201.

De Foresta, H., and Michon, G. 1994. Agroforestry in Sumatra—where ecology meets economy, *Agroforestry Today* 6(4): 12–13.

Delegido, J., Alonso, L., González, G., and Moreno, J. 2010. Estimating chlorophyll content of crops from hyperspectral data using a normalized area over reflectance curve (NAOC), *International Journal of Applied Earth Observation and Geoinformation* 12(3): 165–174.

Drake, J.B., Dubayah, R.O., Clark, D.B., Knox, R.G., Blair, J.B., Hofton, M.A., Chazdon, R.L., Weishampel, J.F., and Prince, S. 2002. Estimation of tropical forest structural characteristics using large-footprint lidar, *Remote Sensing of Environment* 79(2–3): 305–319.

Draper, N.R., and Smith, H. 1981. *Applied regression analysis*, Wiley, New York.

Elvidge, C.D., and Chen, Z. 1995. Comparison of broad-band and narrow-band red and near-infrared vegetation indices, *Remote Sensing of Environment* 54: 38–48.

Fava, F., Colombo, R., Bocchi, S., Meroni, M., Sitzia, M., Fois, N., and Zucca, C. 2009. Identification of hyperspectral vegetation indices for Mediterranean pasture characterization, *International Journal of Applied Earth Observation and Geoinformation* 11(4): 233–243.

Ferwerda, J.G., Skidmore, A.K., and Mutanga, O. 2005. Nitrogen detection with hyperspectral normalized ratio indices across multiple plant species, *International Journal of Remote Sensing* 26(18): 4083–4095.

Filippi, A.M., and Jensen, J.R. 2006. Fuzzy learning vector quantization for hyperspectral coastal vegetation classification, *Remote Sensing of Environment* 100(4): 512–530.

Galvão, L.S., Epiphanio, J.C.N., Breunig, F.M., and Formaggio, A.R. 2011. Crop type discrimination using hyperspectral data. In: Thenkabail, P., Lyon, J., and Huerte, A., Editors. *Hyperspectral Remote Sensing of Vegetation*, Taylor and Francis. Chapter 18.

Galvão, L. S., Roberts, D.A., Formaggio, A.R., Numata, I., and Breunig, F.M. 2009. View angle effects on the discrimination of soybean varieties and on the relationships between vegetation indices and yield using off-nadir Hyperion data, *Remote Sensing of Environment* 113(4): 846–856.

Gillespie, T.W., Foody, G.M., Rocchini, D., Giorgi, A.P., and Saatchi, S. 2008. Measuring and modelling biodiversity from space, *Progress in Physical Geography* 32: 203–221.

Gitelson, A.A., and Merzlyak, M.N. 1997. Remote estimation of chlorophyll content in higher plant leaves, *International Journal of Remote Sensing* 18(12): 2691–2697.

Gong, P., Pu, R., and Miller, J. 1995. Coniferous forest leaf area index estimation along the Oregon transect using compact airborne spectrographic imager data, *Photogrammetric Engineering and Remote Sensing* 61: 1107–1117.

Guisan, A., Edwards, T.C. Jr., and Hastie, T. 2002. Generalized linear and generalized additive models in studies of species distributions: Setting the scene, *Ecological Modelling* 157(2–3): 89–100.

Guyot, G., and Baret, F. 1988. Utilisation de la haute résolution spectrale pour suivre l'état des couverts végétaux, *Proceedings of the 4th International Colloquium on Spectral Signatures of Objects in Remote Sensing. ESA SP-287* Aussois, France, pp. 279–286.

Haboudane, D., Miller, J.R., Pattey, E., Zarco-Tejada, P.J., and Strachan, I.B. 2004. Hyperspectral vegetation indices and novel algorithms for predicting green LAI of crop canopies: Modeling and validation in the context of precision agriculture, *Remote Sensing of Environment* 90: 337–352.

Haboudane, D., Miller, J.R., Tremblay, N., Zarco-Tejada, P.J., and Dextraze, L. 2002. Integrated narrow-band vegetation indices for prediction of crop chlorophyll content for application to precision agriculture, *Remote Sensing of Environment* 81: 416–426.

Hamilton, A.J. 2005. Species diversity or biodiversity? *Journal of Environmental Management* 75: 89–92.

Hill, M.J. 2004. Grazing agriculture: Managed pasture, grassland, and rangeland. In: Ustin, S.L. Editor. *Manual of Remote Sensing, volume 4, Remote Sensing for Natural Resource Management and Environmental Monitoring*, John Wiley & Sons, Hoboken, NJ, pp. 449–530.

Hochberg, E.J., Roberts, D.A., Dennison, P.E., Hulley, G.C. 2015. Special issue on the Hyperspectral Infrared Imager (HyspIRI): Emerging science in terrestrial and aquatic ecology, radiation balance and hazards, *Remote Sensing of Environment* 167: 1–5.

Horler, D.N.H., Dockray, M., and Barber, J. 1983. The red edge of plant leaf reflectance, *International Journal of Remote Sensing* 4(2): 273–288.

Houborg, R., and E. Boegh, 2008. Mapping leaf chlorophyll and leaf area index using inverse and forward canopy reflectance modelling and SPOT reflectance data. *Remote Sensing of Environment* 112:186–202.

Houborg, R., Anderson, Norman, J., Wilson, T., and Meyers, T. 2009. Intercomparison of a "bottom-up" and "top-down" modeling paradigm for estimating carbon and energy fluxes over a variety of vegetative regimes across the U.S., *Agricultural and Forest Meteorology* 149: 1875–1895.

Hsu, P.-H. 2007. Feature extraction of hyperspectral images using wavelet and matching pursuit, *ISPRS Journal of Photogrammetry and Remote Sensing* 62(2): 78–92.

Huang, Y., Chen, Z.-X., Yu, T., Huang, X.-Z., and Gu, X.-F. 2018. Agricultural remote sensing big data: Management and applications, *Journal of Integrative Agriculture* 17(9): 1915–1931, ISSN: 2095-3119, https://doi.org/10.1016/S2095-3119(17)61859-8, http://www.sciencedirect.com/science/article/pii/S2095311917618598.

Huete, A.R. 1988. A soil-adjusted vegetation index (SAVI), *Remote Sensing of the Environment* 25: 295–309.

Huete, A.R. 1989. Soil influences in remotely sensed vegetation-canopy spectra. In: Asrar, G., Editor. *Theory and applications of optical remote sensing*, Wiley, New York, pp. 107–141.

Ingram, J.C., Dawson, T.P., and Whittaker, R.J. 2005. Mapping tropical forest structure in southeastern Madagascar using remote sensing and artificial neural networks, *Remote Sensing of Environment* 94(4): 491–507.

Jensen, J.R. 2000. *Remote sensing of the environment: An earth resource perspective (Prentice Hall Series in Geographic Information Science)*, ISBN: 0134897331.

Jiang, J., Huete, A.R., Didan, K., and Miura, T. 2008. Development of a two-band enhanced vegetation index without a blue band, *Remote Sensing of Environment* 112(10): 3833–3845.

Jiang, Z., and Huete, A. R. 2010. Linearization of NDVI based on its relationship with vegetation fraction, *Photogrammetric Engineering and Remote Sensing* 76(8): 965–975.

Jollineau, M.Y., and Howarth, P.J. 2008. Mapping an inland wetland complex using hyperspectral imagery. *International Journal of Remote Sensing* 29(12): 3609–3671.

Kalacska, M., Sanchez-Azofeifa, G.A., Rivard, B., Caeilli, T., White, H.P., and Calvo-Alvarado, J.C. 2007. Ecological fingerprinting of ecosystem succession: Estimating secondary tropical dry forest structure and diversity using imaging spectroscopy, *Remote Sensing of Environment* 108: 82–96.

Kaufman, Y.J., and Tanré, D. 1992. Atmospherically resistant vegetation index (ARVI) for EOS-MODIS, *IEEE Transactions on Geoscience and Remote Sensing* 30(2): 261–270.

Kaufman, Y.J., and Tanre, D. 1996. Strategy for direct and indirect methods for correcting the aerosol effect on remote sensing from AVHRR to EOS-MODIS, *Remote Sensing of Environment* 55: 65–79.

Klecka, W.R. 1980. *Discriminant analysis*. Quantitative applications in social sciences series, No. 19. SAGE Publications, Thousand Oaks, CA.

Koger, C.J., Bruce, L.M., Shaw, D.R., and Reddy, K.N. 2003. Wavelet analysis of hyperspectral reflectance data for detecting pitted morning glory (*Ipomoea lacunosa*) in soybean (Glycine max), *Remote Sensing of Environment* 86(1): 108–119.

Lee, K.S., Cohen, W.B., Kennedy, R.E., Maiersperger, T.K., and Gower, S.T. 2004. Hyperspectral versus multispectral data for estimating leaf area index in four different biomes, *Remote Sensing of Environment* 91(3–4): 508–520.

Liu, Z.-Y., Wu, H.-F., and Huang, J.-F. 2010. Application of neural networks to discriminate fungal infection levels in rice panicles using hyperspectral reflectance and principal components analysis, *Computers and Electronics in Agriculture* 72(2): 99–106.

Lucas, K.L., and Carter, G.A. 2008. The use of hyperspectral remote sensing to assess vascular plant species richness on Horn Island, Mississippi, *Remote Sensing of Environment* 112(10): 3908–3915.

Lunetta, R., and Lyon, J. 2008. *Accuracy assessment of remote sensor and GIS*. CRC Press, Boca Raton, FL.

Lyon, J.G., Yuan, D., Lunetta, R.S., and Elvidge, C.D. 1998. A change detection experiment using vegetation indices, *Photogrammetric Engineering and Remote Sensing* 64: 143–150.

le Maire, G., Francois, C., and Dufrene, E. 2004. Towards universal broad leaf chlorophyll indices using PROSPECT simulated database and hyperspectral reflectance measurements, *Remote Sensing of Environment* 89: 1–28.

le Maire, G., François, C., Soudani, K., Berveiller, D., Pontailler, J.Y., Bréda, N., Genet, H., Davi, H., and Dufrêne, E. 2008. Calibration and validation of hyperspectral indices for the estimation of broadleaved forest leaf chlorophyll content, leaf mass per area, leaf area index and leaf canopy biomass, *Remote Sensing of Environment* 112: 3846–3864.

Maire, G.L., François, C., Soudani, K., Berveiller, D., Pontailler, J.V., Bréda, N., Genet, H., Davi, H., and Dufrêne, E. 2008. Calibration and validation of hyperspectral indices for the estimation of broadleaved forest leaf chlorophyll content, leaf mass per area, leaf area index and leaf canopy biomass, *Remote Sensing of Environment* 112(10): 3846–3864.

Martin, M., Newman, S., Aber, J., and Congalton, R. 1998. Determining forest species composition using high resolution remote sensing data, *Remote Sensing of Environment* 65: 249–254.

Martínez, B., and Gilabert, M.A. 2009. Vegetation dynamics from NDVI time series analysis using the wavelet transform, *Remote Sensing of Environment* 113(9): 1823–1842.

Melendez-Pastor, I., Navarro-Pedreño, J., Koch, M., and Gómez, I. 2010. Applying imaging spectroscopy techniques to map saline soils with ASTER images, *Geoderma* 158(1–2): 55–65.

Melesse, A., Weng, Q., Thenkabail, P., and Senay, G. 2007. Remote sensing sensors and applications in environmental resources mapping and modelling, *Sensors* 7(12): 3209–3241.

Muller, K., Bottcher, U., Meyer-Schatz, F., and Kage, H. 2008. Analysis of vegetation indices derived from hyperspectral reflection measurements for estimating crop canopy parameters of oilseed rape (Brassica napus L.), *Biosystems Engineering* 101: 172–182.

Mutanga, O., and Skidmore, A.K. 2004. Narrow-band vegetation indices overcome the saturation problem in biomass estimation, *International Journal of Remote Sensing* 25(2004): 1–16.

Nichol, C., Huemmrich, K., Black, T., Jarvis, P., Walthall, C., Grace, J., and Hall, F. 2000. Remote sensing of photosynthetic light-use efficiency of boreal forest, *Agricultural and Forest Meteorology* 101: 131–142.

Nolin, A.W., and Dozier, J. 2000. A hyperspectral method for remotely sensing the grain size of snow, *Remote Sensing of Environment* 74(2): 207–216.

Numata, I., Roberts, D.A., Chadwick, O.A., Schimel, J.P., Galvão, L.S., and Soares, J.V. 2008. Evaluation of hyperspectral data for pasture estimate in the Brazilian Amazon using field and imaging spectrometers, *Remote Sensing of Environment* 112(4): 1569–1683.

Numata, I., Roberts, D.A., Chadwick, O.A., Schimel, J., Sampaio, F.R., Leonidas, F.C., and Soares, J.V. 2007. Characterization of pasture biophysical properties and the impact of grazing intensity using remotely sensed data, *Remote Sensing of Environment* 109(2007): 314–327.

Okin, G.S., Roberts, D.A., Murray, B., and Okin, W.J. 2001. Practical limits on hyperspectral vegetation discrimination in arid and semiarid environments, *Remote Sensing of Environment* 77: 212–225.

Oldeland, J., Wesuls, D., Rocchini, D., Schmidt, M., and Jürgens, N. 2010. Does using species abundance data improve estimates of species diversity from remotely sensed spectral heterogeneity? *Ecological Indicators* 10: 390–396.

Ozdogan, M. 2010. The spatial distribution of crop types from MODIS data: Temporal unmixing using independent component analysis, *Remote Sensing of Environment* 114(6): 1190–1204.

Pacheco, A., and McNairn, H. 2010. Evaluating multispectral remote sensing and spectral unmixing analysis for crop residue mapping, *Remote Sensing of Environment* 114(10): 2219–2228.

Papeş, M., Tupayachi, R., Martínez, P., Peterson, A.T., and Powell, G.V.N. 2010. Using hyperspectral satellite imagery for regional inventories: A test with tropical emergent trees in the Amazon Basin, *Journal of Vegetation Science* 21(2): 342–354.

Pearlman, J.S., Barry, P.S., Segal, C.C., Shepanski, J., Beiso, D., and Carman, S.L. 2003. Hyperion, a space-based imaging spectrometer, *IEEE Transactions on Geoscience and Remote Sensing* 41: 1160–1173.

Penuelas, J., Filella, I., Lloret, P., Munoz, F., and Vilajeliu, M. 1995. Reflectance assessment of mite effects on apple trees, *International Journal of Remote Sensing* 16: 2727–2733.

Rao, N.R., Garg, P.K., and Ghosh, S.K. 2007. Estimation of plant chlorophyll and nitrogen concentration of agricultural crops using EO-1 Hyperion hyperspectral imagery, *Journal of Agricultural Science* 146: 1–11.

Sakamoto, T., Yokozawa, M., Toritani, H., Shibayama, M., Ishitsuka, N., and Ohno, H. 2005. A crop phenology detection method using time-series MODIS data, *Remote Sensing of Environment* 96: 366–374.

SAS Institute Inc. 2010. *SAS/STAT user's guide*, Version 6. 4th Ed., Volume 1, Cary, North Carolina.

Schlerf, M., Atzberger, C., and Hill, J. 2005. Remote sensing of forest biophysical variables using HyMap imaging spectrometer data, *Remote Sensing of Environment* 95(2): 177–194.

Schmidt, K.S., and Skidmore, A.K. 2003. Spectral discrimination of vegetation types in a coastal wetland, *Remote Sensing of Environment* 85(1): 92–108.

Schull, M.A., Knyazikhin, Y., Xu, L., Samanta, A., Carmona, P.L., Lepine, L., Jenkins, J.P., Ganguly, S., and Myneni, R.B. 2010. Canopy spectral invariants, Part 2: Application to classification of forest types from hyperspectral data, *Journal of Quantitative Spectroscopy and Radiative Transfer* 112(4): 736–750.

Shen, M., Tang, Y., Klein, J., Zhang, P., Gu, S., Shimono, A., and Chen, J. 2009. Estimation of aboveground biomass using *in situ* hyperspectral measurements in five major grassland ecosystems on the Tibetan Plateau, *Journal of Plant Ecology* 2(1): 43.

Shippert, P. 2001. Spectral and hyperspectral analysis with ENVI, *ENVI User's Group Notes*, April 22–28, *Annual meeting of the American Society of Photogrammetry and Remote Sensing*, St. Louis, MO.

Stagakis, S., Markos, N., Sykioti, O., and Kyparissis, A. 2010. Monitoring canopy biophysical and biochemical parameters in ecosystem scale using satellite hyperspectral imagery: An application on a *Phlomis fruticosa* Mediterranean ecosystem using multiangular CHRIS/PROBA observations, *Remote Sensing of Environment* 114(5): 977–994.

Thenkabail, P.S. 2003. Biophysical and yield information for precision farming from near-real time and historical Landsat TM images, *International Journal of Remote Sensing* 24(14): 2879–2904.

Thenkabail, P.S., Smith, R.B., and De-Pauw, E. 1999. *Hyperspectral vegetation indices for determining agricultural crop characteristics*, CEO research publication series No. 1, Center for earth Observation, Yale University, New Haven, pp. 47. ISBN: 0-9671303-0-1.

Thenkabail, P.S., Ward, A.D., and Lyon, J.G. 1995. Impacts of agricultural management practices on soybean and corn crops evident in ground-truth data and thematic Mapper vegetation indices, *Transactions of the American Society of Agricultural Engineers* 37(3): 989–995.

Thenkabail, P., Lyon, G.J., Turral, H., and Biradar, C.M. 2009a. *Remote Sensing of Global Croplands for Food Security*, CRC Press-Taylor and Francis Group.

Thenkabail, P.S., Biradar C.M., Noojipady, P., Dheeravath, V., Li, Y.J., Velpuri, M., Gumma, M., Reddy, G.P.O., Turral, H., Cai, X.L., Vithanage, J., Schull, M., and Dutta, R. 2009b. Global irrigated area map (GIAM), derived from remote sensing, for the end of the last millennium, *International Journal of Remote Sensing* 30(14): 3679–3733. July, 20, 2009.

Thenkabail, P.S., Enclona, E.A., Ashton, M.S., Legg, C., and Jean De Dieu, M. 2004a. Hyperion, IKONOS, ALI, and ETM+ sensors in the study of African rainforests, *Remote Sensing of Environment* 90: 23–43.

Thenkabail, P.S., Enclona, E.A., Ashton, M.S., and Van Der Meer, V. 2004b. Accuracy assessments of hyperspectral waveband performance for vegetation analysis applications, *Remote Sensing of Environment* 91(2–3): 354–376.

Thenkabail, P.S., GangadharaRao, P., Biggs, T., Krishna, M., and Turral, H. 2007. Spectral matching techniques to determine historical land use/land cover (LULC) and irrigated areas using time-series AVHRR pathfinder datasets in the Krishna river basin, India, *Photogrammetric Engineering and Remote Sensing* 73(9): 1029–1040.

Thenkabail, P.S., Hanjra, M.A., Dheeravath, V., and Gumma, M.A. 2010. A holistic view of global croplands and their water use for ensuring global food security in the 21st century through advanced remote sensing and non-remote sensing approaches, *Journal Remote Sensing* 2(1): 211–261.

Thenkabail, P.S., Smith, R.B., and De-Pauw, E. 2000. Hyperspectral vegetation indices for determining agricultural crop characteristics, *Remote Sensing of Environment* 71: 158–182.

Thenkabail, P.S., Smith, R.B., and De-Pauw, E. 2002. Evaluation of narrowband and broadband vegetation indices for determining optimal hyperspectral wavebands for agricultural crop characterization, *Photogrammetric Engineering and Remote Sensing* 68(6): 607–621.

Thenkabail, P.S., Ward, A.D., and Lyon, J.G. 1994a. LANDSAT-5 Thematic Mapper models of soybean and corn crop characteristics, *International Journal of Remote Sensing* 15(1): 49–61.

Thenkabail, P.S., Ward, A.D., Lyon, J.G., and Merry, C.J. 1994b. Thematic Mapper vegetation indices for determining soybean and corn crop growth parameters, *Photogrammetric Engineering and Remote Sensing* 60(4): 437–442.

Trombetti, M., Riaño, D., Rubio, M.A., Cheng, Y.B., and Ustin, S.L. 2008. Multi-temporal vegetation canopy water content retrieval and interpretation using artificial neural networks for the continental USA, *Remote Sensing of Environment* 112(1): 203–215.

Tsai, F., Lin, E., and Yoshino, K. 2007. Spectrally segmented principal component analysis of hyperspectral imagery for mapping invasive plant species, *International Journal of Remote Sensing* 28: 1023–1039.

Tucker, C.J. 1977. Spectral estimation of grass canopy variables, *Remote Sensing of Environment* 6: 11–26.

Ungar, S.G., Pearlman, J.S., Mendenhall, J.A., and Reuter, D. 2003. Overview of the earth observing one (EO-1) mission, *IEEE Transactions on Geoscience and Remote Sensing* 41: 1149–1159.

Ustin, S.L., Roberts, D.A., Gamon, J.A., Asner, G.P., and Green, R.O. 2004. Using imaging spectroscopy to study ecosystem processes and properties, *Bio Science* 54(2004): 523–534.

Vaiphasa, C.K., Skidmore, K.A., de Boer, W.F., and Vaiphasa, T. 2007. A hyperspectral band selector for plant species discrimination, *ISPRS Journal of Photogrammetry and Remote Sensing* 62(3): 225–235.

Wang, F.M., Huang, J.F., and Wang, X.Z. 2008. Identification of optimal hyperspectral bands for estimation of rice biophysical parameters, *Journal of Integrative Plant Biology* 50(3): 291–299.

White, J.C., Gómez, C., Wulder, M.A., and Coops, N.C. 2010. Characterizing temperate forest structural and spectral diversity with Hyperion EO-1 data, *Remote Sensing of Environment* 114(7): 1576–1589.

Yang, F., Li, J., Gan, X., Qian, Y., Wu, X., and Yang, Q. 2009. Assessing nutritional status of *Festuca arundinacea* by monitoring photosynthetic pigments from hyperspectral data, *Computers Electronics in Agriculture* 70(1): 52–59.

Yao, X., Zhu, Y., Tian, Y.C., Feng, W., and Cao, W.X. 2010. Exploring hyperspectral bands and estimation indices for leaf nitrogen accumulation in wheat, *International Journal of Applied Earth Observation and Geoinformation* 12(2): 89–100.

Zarco-Tejada, P.J., Berjón, A., López-Lozano, R., Miller, J.R., Martín, P., Cachorro, V., González, M.R., and de Frutos, A. 2005. Assessing vineyard condition with hyperspectral indices: Leaf and canopy reflectance simulation in a row-structured discontinuous canopy, *Remote Sensing of Environment* 99: 271–287.

Zhao, D., Huang, L., Li, J., and Qi, J. 2007. A comparative analysis of broadband and narrowband derived vegetation indices in predicting LAI and CCD of a cotton canopy, *ISPRS Journal of Photogrammetry and Remote Sensing* 62(1): 25–33.

Zhao, D.H., Li, J.L., and Qi, J.H. 2005. Identification of red and NIR spectral regions and vegetative indices for discrimination of cotton nitrogen stress and growth stage, *Computers and Electronics in Agriculture* 48(2): 155–169.

Zhao, G., and Maclean, L. 2000. A comparison of canonical discriminant analysis and principal component analysis for spectral transformation, *Photogrammetric Engineering and Remote Sensing* 66: 841–847.

Zhou, D., Zhu, Y., Tian, Y., Yao, X., and Cao, W. 2006. Monitoring leaf nitrogen accumulation with canopy spectral reflectance in rice, *Acta Agronomica Sinica* 32(9): 1316–1322.

Zhu, Y., Li, Y., Feng, W., Yao, X., and Cao, W. 2006. Monitoring leaf nitrogen in wheat using canopy reflectance spectra, *Canadian Journal of Plant Science* 86: 1037–1046.

Zhu, Y., Zhou, D., Yao, X., Tian, Y., and Cao, W. 2007. Quantitative relationships of leaf nitrogen status to canopy spectral reflectance in rice, *Australian Journal of Agricultural Research* 58(11): 1077–1085.

Section II

Hyperspectral Sensor Systems

Section II

2 Hyperspectral Sensor Characteristics

Airborne, Spaceborne, Hand-Held, and Truck-Mounted; Integration of Hyperspectral Data with LiDAR

Fred Ortenberg

CONTENTS

2.1 INTRODUCTION

This chapter sets forth the fundamental concepts of hyperspectral imaging—a powerful tool to collect precise remote sensing data involved in detection, recognition, and examination of the different objects of scientific, economic, and defense character. The development of the relevant equipment and systems is also discussed, as well as a few generic applications.

A hyperspectral sensor is one of the devices in use in remote sensing activities aimed at observing the Earth's surface with ground-based, airborne, or spaceborne imaging gear [1–7]. In outline definition, remote sensing consists in photo or electronic recording of the spatial and spectral distributions of electromagnetic radiation, as emitted by any object under monitoring either on the Earth's surface, atmosphere or underground. Present-day technologies enable such a recording throughout the entire spectrum, from the radio wavebands to the x-ray and gamma regions. The respective receivers differ widely, for each of them is based on specific physical principles optimized for collection in a given spectral region, zone or band.

Basically, three distinct spectral regions are used in the remote sensing community, namely: visible light with very-near infrared (IR), thermal, and radio wavelengths. In each of them, an image of any object is unique; it is defined by the object's specifics as to emission, reflection, absorption, and scattering of the electromagnetic waves. In the visible and very-near IR regions, it is the object's capacity to reflect solar energy in the first place, determined by the chemical composition of its surface. The thermal IR region is characterized by radiated energy, the latter being directly dependent on the object temperature. In the UHF radio region—that of the radars—the reflection is determined by the surface smoothness and texture.

The sensor which operates in a visible waveband, can either cover the entire spectral region in a so-called panchromatic imaging, or be confined to some of the spectrum zones or bands (e.g., green or red). In the first case, the basic sensor's characteristic is its resolution. At present, the sensors are developed of a very-high-resolution (VHR) class. A camera or instrument of such a class onboard a Low Earth Orbit (LEO) satellite allows Earth's surface imaging with a resolution of less than 1 m. In the second case, a multi-zonal or multiband image yields a set of distinct zone-specific images, which can be directly synthesized into color variants to bring out the details of interest, such as roads, installations, water surfaces, and vegetation.

The development of such equipment does not consist any more in enhancement of the images' spatial resolution, but in the increase of the number of the images taken in different spectral zones, that is, of the number of spectral channels, at a reasonable spatial resolution of each image. The more zones or bands, the more useful information. With several zones, the sensor becomes multispectral. With many narrow spectral zones, the sensor becomes hyperspectral.

It should be borne in mind, however, that the spectral region accessible to the human eye is but a minor source of such information. The ability of bees to single out melliferous herbs in a multi-hued summer meadow relates to the fact their vision extends into the ultra violet (UV) region. With such a "super-vision" directed at the Earth's surface, it is possible to distinguish healthy vegetation from degraded vegetation and human-made objects from natural.

Thermal IR radiation also carries particular information unattainable with the human eye. The natural radiation of any object indicates its surface temperature: the higher the latter, the lighter the shade of grey in the image. The eye being more sensitive to variations in hue than to those in brightness, the image contrasts are often deepened by means of a "scale" of the "false" colors commonly associated with temperatures—from violet, or blue ("cold") to red, or brown ("hot"). Thermal photography or videography is a widely-used tool for detection of the heat leaks in both industrial installations and built-up residential areas. They are also useful in reconnaissance: an underground defense plant can be invisible in the ordinary photograph, but since its functioning entails heat generation, it would shine brightly in the thermal region. Finally, such a picture of a person can provide valuable information on their state of health; much like thermal IR thermometers used in clinics that do not "touch" the patient but rather remotely sense their skin temperature.

Regarding the very-near IR region, it is possible to take any of its zones and substitute it for one of the primary colors in an ordinary color image. Such an operation, known as color synthesis, is performed by the "false-color" films used in remote sensing. The inexperienced eye may find it hard to recognize familiar objects in synthesized images but, on the other hand, they help overcome the limitations of human vision.

The most promising method of remote sensing is hyperspectral or HS imaging, which produces tens or even hundreds of images in narrow zones. The absorption spectra of substances and materials being specific, this technique makes it possible to identify vegetation, minerals, geological formations, soils, and structural materials in buildings and pavements via the physical/chemical compositions of the objects. Thanks to the extra-high resolution (of the order of that of a lab spectrometer) the volume of obtainable information can be increased by a four-figure factor or more.

Unlike the visible- and IR-region imaging devices, the radars belong to the active-sensor class. Whereas the former just passively capture either reflected and scattered solar radiation or that of the Earth's surface, the latter instrument emits their own electromagnetic waves and records their echo off the object. Radar imaging has features absent in other remote sensing techniques: it can't be hampered by blanket cloud cover; its images have their particular geometric distortions; and its coherence makes it possible to obtain detailed relief patterns accurate to tens of centimeters and it can operate both in daytime and at night.

Hyperspectral imaging is an evolutionary product of the multispectral (MS) systems, whereby—by means of new technologies—the number of information collection channels can be increased from 3–10 to 100–1000 with a high spectral resolution of 1–10 nm. The result is a multidimensional spatial-spectral image, in which each picture element (pixel) is characterized by its individual spectrum. Such an image is called an information cube, with two of its dimensions representing the projection of the imaged area on the plane, and the third—the frequency of the received radiation. A hyperspectral imaging device is an opto-electronic multichannel system designed for simultaneous independent generation of an image and corresponding video signal in a discrete or continuous sequence of spectrum intervals, supplemented by radiometric, spectral, and spatial image parameters pertaining to those ranges.

In other words, present-day multispectral and hyperspectral scanning devices are radiometrically calibrated multichannel video-spectrometers. The values of brightness, registered by the imaging system for some object in the different spectral zones, together with their graphical mappings serve as spectral curves allowing one to clearly discern the object and mark it off on the image.

The term "hyperspectral" has been repeatedly criticized, for such a collocation seems to be incongruous, without any physical sense. More justified would be using the prefix "hyper-" with a quantitative characterization to imply "too many," as in the word "hypersonic," that is, highly above the speed of sound. But in our case "spectrum" has a physico-mathematical, rather than a quantitative meaning. Similarly, this is why the words like "hyperoptical," "hyperoily," "hypermechanical," "hyperaerial," and so on, are devoid of sense. Consequently, the word "hyperspectral" might represent a metaphor, to express the excitement of the researchers as regards to the new technological means able to obtain a multitude of images of the same object, each of them in a very narrow spectral band. Spectroradiometer imaging technology emerged at the time when the panchromatic imagery was sufficiently advanced, and the picture quality was defined by the achieved spatial resolution. The images taken from the satellite orbit were said to be either of "high resolution" (about 1 m), or of "very high resolution" (less than 1 m). By the term "resolution," the spatial one was meant by default. Since the image sets have been provided by the spectroradiometer, the technique of obtaining these data cubes began to be referred to as the "hyper-resolution imaging." Until now, some researchers keep using this term, even without specifying what resolution they mean—spatial or spectral. The expression "spectral hyper-resolution imaging" could have a meaning of the visual spectroradiometer data, but the specialists have chosen instead this derivation—"hyperspectral." In any case, the term has taken root. Still, it should be made clear that what matters is not the discovery of a new phenomenon, but the development of the advanced video-spectrometers with high spectral resolution.

The interest in such opto-electronic devices is caused not only by their spectacular applications, but also by the fact that these technical means in many respects imitate the elements of the visual apparatus of higher animals and humans. It's known that all the living beings acquire 80%–90% of the information about their environment through vision. Therefore, the paths of developing the automatic systems of the technical vision for detection, recognition, and classification of the different objects have received close attention.

Hyperspectral imaging, being able to extract more precise and detailed information as compared with other techniques, is one of the most efficient and fast-developing directions of Earth remote sensing. The data on the energy reflected from objects provides extensive material for detailed analysis. Still, it is worth mentioning that development of the needed hyperspectral equipment proved to be a very complex task and hard to implement. This is why for a long enough time, designing hyperspectral sensors, in particular for spaceborne applications, had been of purely theoretical interest for the following reasons:

- The quality of the equipment precluded realization of hyperspectral sensors with high spatial and spectral resolution simultaneously.
- Reception of the vast volumes of hyperspectral data, their transmission through the communication channels, and their ground-based processing presented serious difficulties.
- The complexity of the data-flow containing the hyperspectral measurements required specialized and highly sophisticated processing software.

Nowadays, these difficulties are largely surmountable. At the time of writing, ways are opened for development of hyperspectral equipment in the optical and IR regions with high informational and operational indices, which in turn would mean a higher solution level for the problems to be solved, and spaceborne sensors are being developed with an eye towards operational and global data collection.

During the last three decades, the hi-tech industries in many countries were considerably involved in hyperspectral subjects. The dedicated element base was created for manufacturing the hyperspectral equipment. Various hyperspectral sensors, initially for the ground and airborne applications, and later on for the spaceborne ones, were designed, tested, and put to use. The modern spectrometer concept is based on the latest optical-system solutions involving large-format charge-coupled device (CCD) matrices. Currently available are imaging systems operating in hundreds of spectral bands, with signal-to-noise ratios providing information in 12 bits per count or byte. The advances in hyperspectral equipment and data processing software in recent years has significantly extended the range of possible applications, to be provided by the airborne and spaceborne imaging community.

The specifics just mentioned of putting hyperspectral sensors into space resulted in smaller availability of hyperspectral imagery as compared with other Earth remote sensing data. One of the reasons was that of a small number of the spacecraft with hyperspectral imaging systems, like Hyperion onboard the Earth Observing-1 (EO-1) spacecraft (NASA, USA), or CHRIS onboard PROBA (ESA). The increased hyperspectral-connected activity in space can be seen as of late, to overcome the backlog. Spaceborne hyperspectral sensors are expected to find use as general purpose instruments, providing data for a broad range of end-users.

Present-day airborne and spaceborne hyperspectral instrumentation operates in the optical and IR spectral bands and combines high spatial, spectral, and radiometric resolutions, and hence are now indispensable in recording distinctive features of the Earth's surface such as vegetation cover, landscape status, and anthropogenic impacts. Its broad, functional potential derives from the fact that the object's spectral parameters, as well as their surface distributions, are the most prolific sources of information on the object's state and the changes it is undergoing in the course of exploitation or its vital activity. For the hyperspectral imaging techniques evolving and getting introduced into new fields, their primary contribution will be in the exploration and development of the new applications via the selection of optimal spectral band parameters (bands position and widths). Practical operational considerations (sensor cost, data volume, data processing costs, etc.) for most current and future applications are favorable for the use of hyperspectral systems, taking into account their economic efficiency. Aerial and spaceborne hyperspectral sensors potentially provide an economic alternative to traditional ground-based observations alone, permitting data to be acquired for large areas in a minimum of time while enabling faster delivery and application turnaround.

By virtue of its ability to provide information beyond the capacity of traditional panchromatic and multispectral photography and imaging, hyperspectral sensor data is highly advantageous in solving economic, scientific and engineering problems. Some experts estimate [8] that hyperspectral sensors are capable of solving up to 70% of all Earth-observation problems, while visual information with high spatial resolution—only 30%.

As will be shown later in this and other chapters, additional information about the reflecting surfaces can be delivered by the return signals measured by LiDAR systems. This is why some new hyperspectral applications are based on the fusion of two imaging data types, as obtained by joint airborne or spaceborne hyperspectral and LiDAR systems.

2.2 HYPERSPECTRAL SENSOR CONCEPT

The hyperspectral sensor combines the following photonic technologies: conventional imaging, spectroscopy, and radiometry—to produce images together with spectral signatures associated with any spatial-resolution element (pixel). The position of spectral imaging relative to related technologies is shown in Figure 2.1.

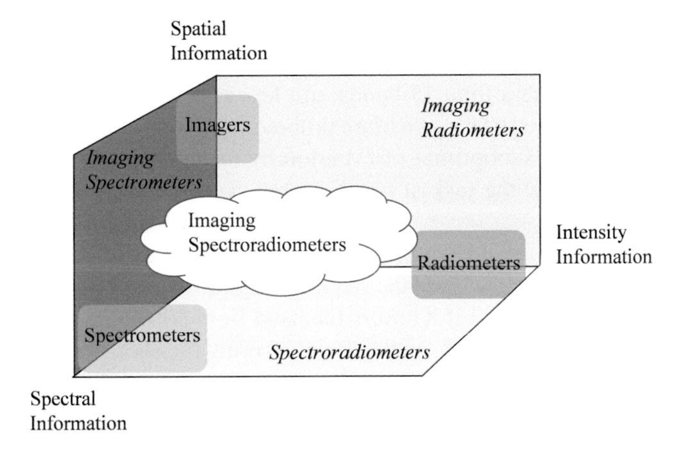

FIGURE 2.1 Relationship among radiometric, spectrometric, and imaging techniques.

Data produced by a spectral imager creates a cube, with position along two axes and wavelength along the third, as depicted in Figure 2.2. By proper calibration, the recorded values of the data cube can be converted to radiometric quantities that are related to the scene phenomenology (e.g., radiance, reflectance, emissivity, etc.). The latter provides a link to spatial and spectral analytical models, spectral libraries, and so on, to support various applications. The motion of the spacecraft is used to move the image line across the surface of the Earth, and the CCD for example is read out continuously to provide a complete hyperspectral dataset per line. The imagery for the whole area can be represented in the form of data cubes or a set of images, each image containing information from one wavelength.

To illustrate the strength of this technique, the combined spectral/spatial analysis allows detection in an image of optically unresolved (sub-pixel size) objects. Its applications range from Earth remote sensing to early cancer detection. It is obvious why the combination of imaging and spectroscopy is so attractive: conjoining of the two analyses, which can be considered as a data fusion, makes for enhanced image perception. Traditional color film-like imagery touches upon the idea, except that it is based on broad spectral bands and varying spatial resolution, so the details in colors are often achieved at the expense of spatial resolution as exemplified by mixing of red, green, and blue bands into true-color imagery from CCDs.

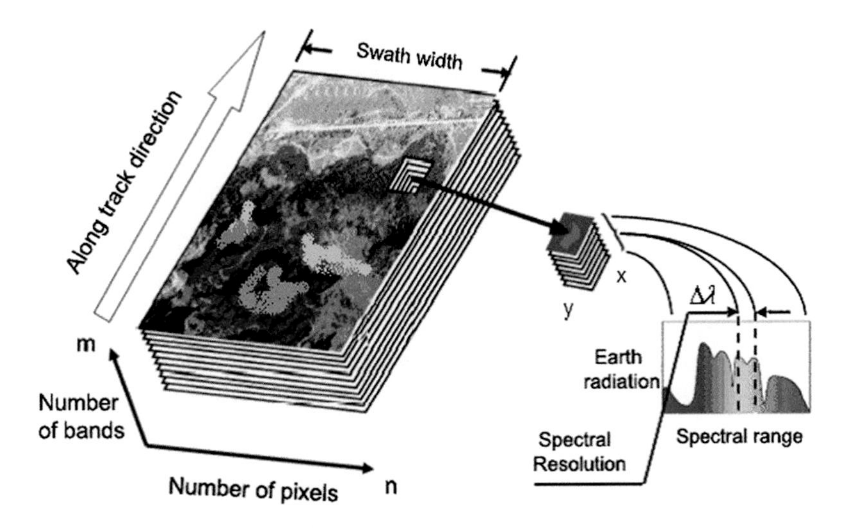

FIGURE 2.2 Imaging spectroradiometer concept—Data cube composed of individual images recorded over m spectral bands.

Hyperspectral imaging systems have the advantage of providing data at high resolution over a large number of bands, hence are suitable for a wide variety of applications. Their multispectral counterparts, operating over less than 15 bands, are less expensive, yield smaller datasets, and have higher signal-to-noise ratios. They are often tailored to a specific application, so that for other purposes their bands may be suboptimal or even totally unsuitable. However, once the optimal bands have been identified for the task at hand, such a system can provide an overall superior solution.

In hyperspectral imaging, the decisive factor is not the number of bands (channels), but the bandwidth—the narrower, the better—and the sequence of acquisition. In other words, a 15-channel system would count as hyperspectral if it covers the band from 500 to 700 nm, with each zone not exceeding 10 nm, while a similar 15-channel system covering the shortwave and visible range and the three IR ranges would count as multispectral.

The hyperspectral imaging system breaks down the incoming optical radiation into hundreds of bands and captures specific signatures unique to a scene. Those are processed in a way that permits reliable identification of the objects standing out against their backgrounds, which generate a visual representation of the "hidden" resource. By this means, it is possible to "penetrate" dense foliage and locate shallow underground installations, tunnels, pipelines, and so on. Incidentally, this has led to the term "hyperspectral" being interpreted as hypersensitive to buried or underground objects, which is obviously incorrect.

According to the terms of reference approved by the United Nations (UN), monitoring of the environment consists in regular observation of its constituents in space and time, for specific purposes and within the framework of programs worked out in advance. In the context of the vegetation cover, this means observation of its status, control of its current dynamics, forecasts for the future, advance warning of catastrophic upheavals, and recommendations on mitigation or prevention of damage. In the same context, hyperspectral measurements can serve the purposes of understanding: meteorological and climatic factors, pollution of the atmosphere, water bodies and soil, extraordinary human-made and natural situations, and information on land use and reforms.

Hyperspectral data have a high commercial potential. Through them, significant savings are expected in the development and operation of monitoring systems, which is a task both complex and expensive, demanding a collaborative effort by the scientific and manufacturing sectors at the national levels. At the same time, it should be noted that the terms of reference of such hyperspectral systems give rise to the following conflicts:

- Between the required coverage and the particularity of the image;
- Between the required spatial and spectral resolutions;
- Between the required and real quantization orders of the spectral channel signals;
- Between the video data volume and the limited traffic capacity of the current satellite communication channels.

Given below are specific features inherent in hyperspectral airborne and spaceborne applications:

- Through them, the detection and recognition steps yield accompanying spectral features. These admit lower spatial-resolution levels, which in turn allows recourse to smaller and lighter sensors. The latter can be accommodated on mini-vehicles which, operating in clusters, permit reduction of the observation periodicity for a tract to some hours or less. Another result of the reduced spatial-resolution level is a larger grid unit for the same receiver format.
- Hyperspectral data are readily amenable to computerized processing with involvement of geo-informational technologies. This in turn contributes to time-saving and facilitates employment of technician-level personnel. Advanced software has been developed, incorporating geo-informational technologies.

2.3 HYPERSPECTRAL SENSOR PHYSICS, PRINCIPLES, AND DESIGN

Imaging spectrometers typically use a 2D matrix array (e.g., a CCD), and produce progressive data cubes through successive recording—either of full spatial images, each at a different wavelength, or of narrow image swaths (1-pixel wide, multiple pixels long) with the corresponding spectral signature for each pixel in the swath. Remote imagers are designed to focus and measure the light reflected from contiguous areas on the Earth's surface. In many digital imagers, sequential measurements of small areas are taken in a consistent geometric pattern as the sensor platform moves, and subsequent processing is required to assemble them into a single image.

An optical dispersing element in the spectrometer splits light into narrow, adjoining wavelength bands, with the energy in each band being measured by a separate detector. Using hundreds, or even thousands of detectors, spectrometers can accommodate bands as narrow as 1 nm over a wide wavelength range, typically at least 400–2400 nm (from visible to middle-IR). Hyperspectral sensors cover bands narrower than their multispectral counterparts. Image data from several hundred bands are recorded concurrently, offering much higher spectral resolution than that provided by sensors covering broader bands. Sensors under development, designed to cover thousands of bands with even narrower bandwidth than hyperspectral examples, have a special name—ultraspectral.

The most common modes of image acquisition by means of hyperspectral sensors are known as "pushbroom scanning" (electronical) and "whiskbroom scanning" (electro-mechanical).

Pushbroom scanners (Figure 2.3) use a line of detectors over a 2D scene. The number of pixels equals that of ground cells for a given swath. The motion of the carrier aircraft or satellite realizes the scan in the along-track direction, thus the inverse of the line frequency equals the pixel dwell time. In a 2D detector, one dimension can represent the swath width (spatial dimension, y) and the other, the spectral range. Pushbroom scanners are lighter, smaller, and less complex than their whiskbroom counterparts because of fewer moving parts. They also have better radiometric and spatial resolution. Their major disadvantage is a large number of detectors amenable to calibration. These imaging spectrometers can be subdivided in Wide Field Imagers (MERIS, ROSIS) and Narrow Field Imagers (HSI, PRISM). These imager-types are conducive to highly repetitive or frequent global coverage when hyperspectral sensors are installed on board LEO satellites.

Whiskbroom scanners (Figure 2.4) are on-axis optics or telescopes with scan mirrors sweeping from one edge of the swath to the other. The field of view (FOV) of the scanner can be covered

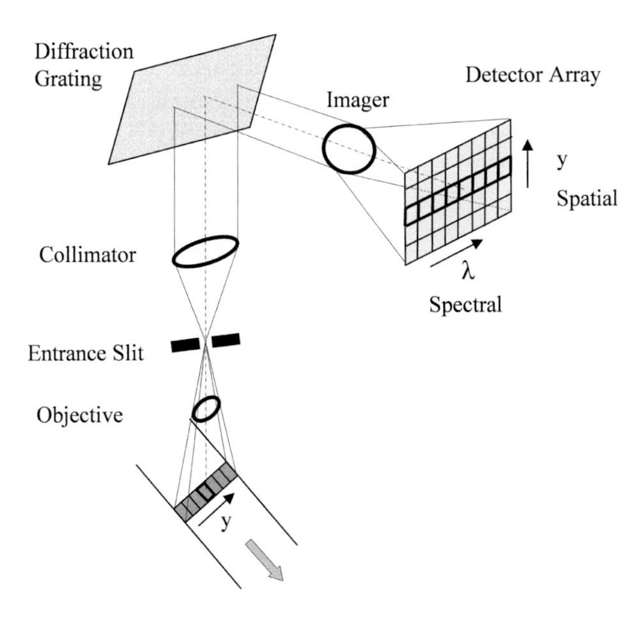

FIGURE 2.3 Principle of pushbroom scanning.

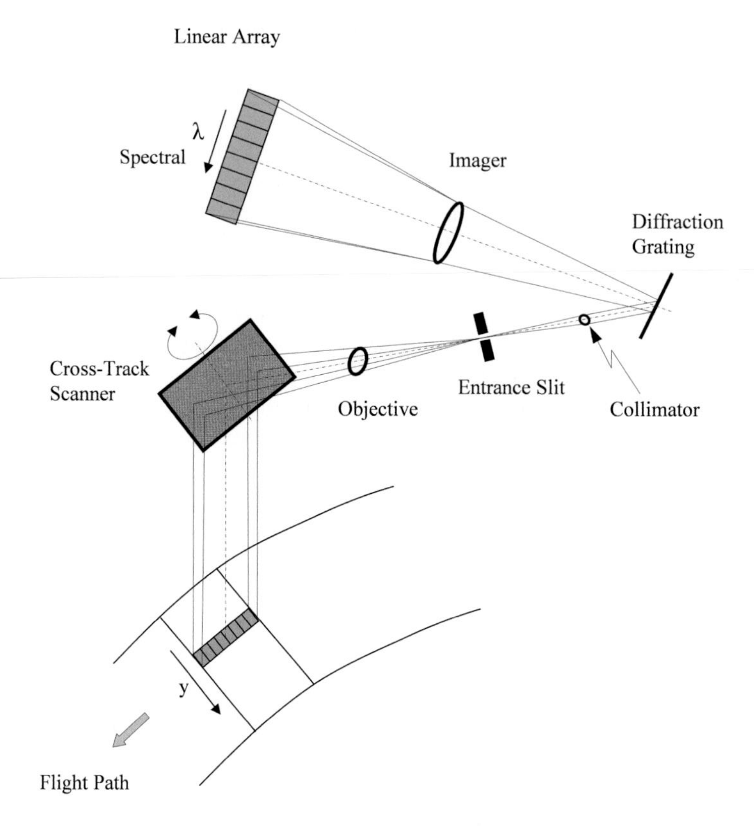

FIGURE 2.4 Principle of whiskbroom scanning.

by a single detector or a single-line detector, the carrier's motion implementing the sweep. This means that the dwell time for each ground cell must be very short at a given instantaneous field of view (IFOV), because each scan line consists of multiple ground cells to be covered. Whiskbroom scanners tend to be large and complex. The moving mirrors bring in spatial distortions to be corrected by preprocessing before delivering the information to the user. An advantage of the whiskbroom scanners is that they have fewer detectors subject to calibration, as compared to other types of sensors. Well-known examples are AVHRR, Landsat, and SeaWiFS.

Two approaches to hyperspectral sensors were launched and operational for many years in the new millennium: the dispersion approach (e.g., EO-1) and the Fourier Transform approach (e.g., MightySAT).

- Dispersion. By this technique, the spectral images are collected by means of a grating or a prism. The incoming electromagnetic radiation is separated under different angles. The spectrum of each ground pixel is dispersed and focused at different locations of the 1D detector array. This technique is used for both image acquisition modes (pushbroom and whiskbroom scanners). Hyperspectral imagers mainly use gratings as the dispersive element (HSI, SPIM).
- Fourier Transform Spectrometers (FTS). Spatial Domain Fourier Transform Spectrometers like the well-known sensors SMIFTS or FTHSI on the MightySat II spacecraft use the principle of the monolithic Sagnac interferometer. Unlike conventional FTS, the Earth Observation spectrometers in LEO operate with fixed mirrors. The optical scheme distributes the interferogram (spectrum) along one dimension of the detector, the other dimension representing the swath width.

Classical hyperspectral imaging is performed by one of two basic principles: scanning by a pushbroom approach or by a tuneable filter approach, for example using an Acousto-Optic Tuneable Filter (AOTF). In a pushbroom scanner, a line instrument is used where the light is diffracted into different color/wavelength components by means of a spectrograph onto a focal plane array. Unlike it, in the tuneable filter approach, instead of scanning the image stepwise in one spatial dimension, the entire spatial images are taken, as the camera scans through the different wavelengths. This is achieved by placing the AOTF between the image and the camera [9]. AOTF is solid-state device comprising a crystal filter whose transmission wavelength is controlled by attached piezoelectric transducers. The transducers create pressure waves in the crystal which changes its refractive index. These transducers are controlled by careful manipulation of the frequency of an applied electric field to vary the transmission wavelength of the filter in a stepwise manner. An image captured at each wavelength step is read into memory to form a slab of the hypercube. The data are then processed in order to produce the desired output image. The spectrograph in a pushbroom system provides better spectral resolution than the tuneable filter system, but at a cost of speed, while the latter device, with no moving parts, builds up the hypercube more swiftly.

For detailed studies of the Earth's surface in the reflection spectral band of 400–2400 nm, the onboard hyperspectral equipment should simultaneously provide high spatial resolution of 1–5 m, high spectral resolution of 5–10 nm in tens and even hundreds of channels, a swath of up to 100 km, and a signal-to-noise ratio $\geq$100. For devices to approach these requirements approximations have been made. One remedy consists in replacing the fixed-parameter sensors by adaptive systems. A major consideration in developing the latter systems is that they would operate under widely-varying conditions, including parameters of the atmosphere, characteristics and types of the Earth's surface, the space platform motion, and attitude. These variations are stochastic, which tends to impair both the efficacy of the observation process and the quality of the obtained images. Under the adaptive regimes, the information parameters of the system are selected and adjusted in real time for the problem in question. Among the adjustable information parameters are the spatial and spectral resolutions as well as the signal quantization width. The performance level of the observation missions can be raised to a qualitatively new stage by implementing the advanced operation modes for the onboard equipment.

One more prerequisite to this end is to reduce the volume of the transmitted imagery, combined with prevention of information losses. Redundancy should be curtailed by a three- or even four-figure factor, which maybe unfeasible by the traditional means alone. The proposed solution consists in real-time quality assessment of the observation data and selection of the informative onboard channels, to screen out the data with low informational content. This approach not only enables redundancy reduction, but also enhances the information output.

At present, the spectral intervals, their number, and their layout are chosen at the design stage and remain practically unchanged afterward. New methods of interval selection, currently under development, are based on analysis of the spectral reflection characteristics. Unlike the conventional procedure, the intervals and their limits are selected through a compromise between the required spectral and spatial resolutions. This is implemented, for example, in Full Spectral Imaging (FSI)—a method for acquiring, preprocessing, transmitting, and extracting information from full-spectrum remote sensing data by the "spectral curve" approach, instead of the current "bytes-per-band" one [10]. By virtue of the bandwidths' optimization, FSI admits simplified instrument characteristics and calibration as well as reduced data transmission and storage requirements. These improvements may be accomplished without loss of remote sensing information. FSI is neither "hyperspectral," nor "superspectral" or "ultraspectral" imaging; it is an end-to-end system which involves the whole technological sequence, from the observation technique to the data processing. FSI belongs to the class of hyperspectral-type imagers based on a dispersive element and a pushbroom image acquisition system. Instrument throughput, spatial coverage, and spectral response can be enhanced using multiple focal planes. The multiple entrance slits would be optically connected to several spectrometer-and-detector combinations. Each spectrometer would have identical optical elements except for the dispersion characteristics of the grating and the detector.

An important task is the screening of received hyperspectral data to obtain a true representation of the Earth's surface. The information involves random, systematic, as well as system-defined distortions caused by the atmosphere, curvature of the Earth, displacement of the hyperspectral equipment relative the Earth's surface at the moments of imaging, and physical characteristics of the sensors and channels. To remove these multiple distortions, a number of corrections should be applied, based on radiation, radiometric, geometrical, and calibration techniques. Radiation correction is related to adjustment of the EM energy amount as received by each sensor, thereby compensating for diverse atmosphere transparency to the different frequency bands. Radiometric correction represents elimination of the imaging system-defined distortions inserted both by the sensors and the transceivers. Geometrical correction, or image transformation, is intended to remove the distortions due to the Earth's curvature and rotation, inclination of the satellite's orbit with respect to the equatorial plane and, for high-resolution images, the terrain relief. This correction can be done automatically, with the hyperspectral sensors position known. More precise transformation and image linking to a specific coordinate frame is usually achieved by interactively setting the reference points. In the transformation process, the pixel coordinates conversion to a new raster frame may cause some changes in objects shape. In editing and joint processing of the different imaging data types, as well as the images of the same terrain taken at different times, it is internationally accepted to use, as an exchange standard, a so-called *orthoplan* projection. Calibration consists in converting the dimensionless data, as obtained by the sensors of various spectral bands, into true normalized values of the reflected or emitted energy.

2.4 HYPERSPECTRAL SENSOR OPERATIONAL MODES

The recent decades have been marked by the fast advance of the hyperspectral techniques. Development of the technical means for hyperspectral imaging went all the way from simple lab devices with limited capabilities to sophisticated space complexes, which enable excellent imaging of the Earth's surface simultaneously in hundreds of the narrow spectral bands, and successful downloading of this information to the Earth. Today, there are three basic methods of acquiring the hyperspectral images of the Earth: ground-based, airborne- and spaceborne imaging. Each method has peculiarities, advantages, shortcomings, and a range of problems of their own. Imaging method depends on the sensors' structural features imposed by their operating conditions—on the ground, in the atmosphere, or in space. The maximum efficiency can be attained when all the tools are available for working in all three spheres, as well as the opportunity of choosing the optimum way to achieve the goal. None of the cited methods duck challenges offered by the path of progress, and their simultaneous improvement brought an impressive rise in hyperspectral techniques.

Some general descriptions and results obtained by the hyperspectral techniques in the different hand-held, truck-mounted, airborne, and spaceborne applications are presented here by way of illustrating some of the points made, but not a substitution for the wealth of details provided by attending chapters (Figure 2.5).

2.4.1 Ground-Based Hyperspectral Imaging

Hyperspectral imagery is generated by the instruments called imaging spectroradiometers. The development of these sophisticated sensors involves the convergence and synergy of two related but distinct technologies: spectroscopy and remote imaging of the Earth. Spectroscopy is a study of emission, absorption, and reflection of light by the matter as a function of radiation wavelength. As applied to the field of optical remote sensing, spectroscopy deals with the spectrum of sunlight that is diffusely reflected (scattered) by the different materials on the Earth's surface. Ground-based or lab measurements of the light reflected from the test material became the first successful application of the spectrometers or spectroradiometers.

FIGURE 2.5 Landscape fantastic HS control.

On-ground hyperspectral imagers are ideal tools for the identification of obscured or resolution-limited targets, defeating camouflage, and recognition of both solid and gaseous chemicals. There are manufacturing processes that can make use of MS or hyperspectral data, which include inspection of color and paint quality, detection of rust, or the flaw detection in thin film coatings, and so on. Hyperspectral measurements made it possible to derive a continuous spectrum for each pixel. Once the adjustments for the sensor, atmospheric, and terrain effects are applied, these image spectra can be compared with field or lab reflectance spectra, in order to recognize and map surface materials, for example, specific types of vegetation. Collections of spectral curves for both the materials and the objects, routinely measured under lab or field conditions, are stored in the spectral libraries. These curves can be used for classification or automatic identification of objects and materials. Until recently, the imagers' resolution capability was restricted due to the limitations, imposed both by the detector design and the requirements to the data storage, transmission, and processing. Recent advances in these areas have contributed to the imagers' design, so their spectral ranges and resolution are now comparable with those of the ground-based spectrometers.

Surface Optics Corporation (SOC), which is specializing in the characterization and exploitation of the optical properties of surfaces, has developed a wide range of commercial products including hyperspectral video imagers and processors. Its SOC-700 family of hyperspectral imaging systems is devised to address the real-time processing requirements of various applications, such as machine vision, biological screening, or target detection. The hyperspectral imager (Figure 2.6) is a high-quality, portable, and easy-to-use spectral imaging instrument, radiometrically calibrated, with a software intended for both the analysis and viewing. The imager captures a 640 × 640-pixel image that has 120 spectral bands deep in as fast as four seconds. Hyperspectral sensors can acquire and process up to 100 lines of 12-bit data for all 120 bands and return an answer, to be used in controlling the process, screening the cell samples, or telling apart friend from foe. The hyperspectral analysis package provides the tools for serious spectral work, including complete radiometric calibration, reflectance calculations, matched filtering, and atmospheric correction. The sensor is set up within minutes and can withstand rugged outdoor environments. Since its inception, it has been used under rainforest and desert conditions and has been drop-tested from a 4-ft height.

Design and tests of hand-held hyperspectral devices has improved real-time detection and quantification for a wide spectrum of applications. An impressive hyperspectral imager, produced

FIGURE 2.6 HS target detection system on a field trial.

by Bodkin Design, can capture both the spectral and the spatial information of the scene instantly, without scanning [11]. It is designed for high-speed hyperspectral imaging of transient events and observation from moving platforms. The device has no moving components, making the system immune to mechanical failure. Its unique design simplifies the detector read-out and the optical system, and provides a fully registered, hyperspectral data cube on every video frame. It is ideally suited for airborne, hand-held, and vehicle-mounted systems.

The number of hyperspectral sensor applications in various areas of science and technology constantly improves. Thus, Terrax Inc. yields the Theia—a low-power, lightweight hyperspectral imaging system with automated image analysis and data optimization for real-time target detection and hyperspectral data comparison. It consists of cameras and a hyperspectral real-time image-capture and decoding subsystem. The Theia is designed to fit within the size, weight, and power envelope of a portable environment as well as fixed locations. It is robust and includes several on-board real-time target detection and data optimization algorithms based on a calibration library of 100 substance cubes.

The system called foveal hyperspectral imaging [12] employs a hyperspectral fovea with a panchromatic periphery. This approach results in a low-cost and compact hand-held hyperspectral imager suitable for applications such as surveillance and biological imaging. In these applications, the panchromatic peripheral image may be used for situational awareness and screening of the potential threats of the hyperspectral fovea to that region of interest for recognition or detailed characterization. As was demonstrated in the applications, the real-time capability of the foveal hyperspectral imaging shows particular promises including surveillance, retinal imaging, and medicine.

The hyperspectral imager is a compact asymmetric anamorphic imaging spectrometer covering broad spectral regions available from a specialty manufacturer. The complete imaging system is contained in a variety of packages suitable for deployment in a mobile vehicle environment or with the imaging head remote from the processing stack, as in building's perimeter security systems. The detection capability and corresponding display varies with the model and the database of cubes representing calibration for specific chemical or material signatures. Detection and determination can reliably work over distances from a few feet to 300 ft.

The Norwegians have built an interesting technology demonstrator for hyperspectral target detection [13]. The ground-to-ground demonstrator system was also developed, which is currently being used in tests. The package contains a HySpex VNIR hyperspectral camera, a line scanner monochrome camera, and a motorized turntable mounted on a tripod. The rotation axis and the linear FOV are vertical, so that the cameras scan around the horizon. Shown in Figure 2.6, the hyperspectral system was success tested on a field trial in northern Norway. The system is designed for detection of the small targets, as well as for search and rescue operations. Concerning the last option, it should be noted that in many cases there is no need to identify a specific target—what is needed is to detect the presence of a target with unknown a priori characteristics. Search and rescue operations provide good examples of such an approach. For instance, consider a hiker lost in the mountains or a boat on the high seas to be promptly detected. Rather than to conduct an extensive analysis of hyperspectral data cubes, which is quite a time-consuming task, a cursory technique based on "anomaly detection" can be applied.

Some researchers working with ground remote sensing systems have been carried out using sensors attached to long hydraulic booms hoisted above the crop canopy from the ground (the University of Illinois). The images collected from such a close distance have resolutions much higher than those taken from aircraft or satellites. Other ground-based systems use vehicle-mounted sensors that control variable-rate applicators in real time. For example, the remote sensors distinguishing the weeds from the crop are mounted on sprayers that change the application rate of herbicides, to be applied on the go. This form of remote sensing technology, called machine vision, is now in wide use.

Most of the hyperspectral imagers operate in the visible to shortwave IR bands. Hyperspectral imaging in the IR spectrum range has multiple applications, and it is an urgent problem to develop the relevant instrumentation. Many industrial firms develop and turn such a production out. One similar product that can operate both on the ground, and onboard a plane, is presented below. The Aerospace Corporation has designed and built the state-of-the-art narrowband hyperspectral Fourier Transform IR (FTIR) scanner—Spatially Enhanced Broadband Array Spectrograph System (SEBASS [49]), which offers some unique capabilities within the hyperspectral remote sensing arena. SEBASS can collect data either from the transportable ground-based sensor, or from a low-flying aircraft, depending on the application at hand. The SEBASS capability in the mid-wave and longwave IR is intended to remotely identify materials in the 3 to 13 μm "chemical fingerprint" spectral region. The hyperspectral sensor also provides high-resolution temperature data. SEBASS captures 128 spectral bands in the IR spectral range with its IFOV of 1 mrad, and FOV of 128 mrad (7.3°). Its typical remote IR products include spectrally and radiometrically calibrated, atmospherically corrected hyperspectral data cubes. SEBASS has three operational modes [49]:

- Vehicle Mounted Measurements—four mobile 2D scanning longwave IR FTIR vehicle mounted sensors. The vehicles can be driven to the site of interest to collect close-range hyperspectral data from 8 to 12 microns. The scanning FTIR sensors have been successfully deployed on over 100 ground-based data collections.
- Hand-held Measurements—provide in situ ground spectral measurements using hand-held instruments to verify material identifications derived from airborne measurements.
- Laboratory Measurements—to assist in material characterization of samples taken from regions of interest, establish ground truth, and improve interpretation of remote sensing data.

Incidentally, as was noticed by many researchers, the best results are those obtained by sharing the data as obtained by various hyperspectral sensors at different positions with respect to the target [14]. In this way, for example, sugar beet disease was detected by combining multitemporal hyperspectral remote sensing data, as provided by the different airborne, tractor and hand-held spectroradiometers.

Imaging spectrometry data are well established for detailed mineral mapping from airborne and spaceborne systems. Overhead data, however, have substantial additional potential when used together with ground-based hyperspectral measurements. United States researchers used the hyperspectral scanner system to acquire airborne data, outcrop scans, and to image boxed drill core and rock chips at approximately 6 nm nominal spectral resolution in 360 channels from 400 to 2450 nm. Analysis results using standardized hyperspectral methodologies demonstrate rapid extraction of representative mineral spectra and mapping of mineral distributions and abundances. A case history highlights the capabilities of these integrated datasets for developing improved understanding of relations between geology, alteration, and spectral signatures in three dimensions.

2.4.2 Airborne Hyperspectral Sensors

Up until 2000, a fleet of over 40 airborne sensor systems gave the mature form of the future of data acquisition opportunities. During most of the 1990s, hyperspectral imaging had been an area of active research and development, so the images were available only to researchers. With the recent appearance of commercial airborne systems and with applications in resource management, agriculture, mineral exploration, and environmental monitoring, hyperspectral imagery is poised to enter the mainstream of remote sensing [15–17]. But its effective use requires an understanding of the nature and limitations of both this data type and various strategies involved in its processing and interpretation.

Like the ground equipment, the airborne hyperspectral scanners are also designed to measure the radiance scattered by the Earth's surface, in the given spectral band—any pixel of the acquired image simultaneously contains both spatial and spectral information on the terrain. The input radiation flux is split into components, according to a wavelength. For each waveband, certain lines of the matrix are reserved. These data are digitized and written to storage of the unified control, data storage, and power supply module.

For spatial linking of the collected imagery, the inertial navigation system (INS) is involved in the post-flight processing of its own data together with those of the imaging scanner, to attribute geographic coordinates to each pixel. More advanced survey complexes are supplied with navigation systems which have the relevant software to control the airborne imagery process. Thus, along with imagery planning and recording, the system also controls both the aircraft and the imaging sensors during the flight. The estimation of the navigation and imagery precision is carried out in real time. To the credit of such imagery control systems and their high quality and reliability is the fact that about 150 of them are operational throughout the world today.

Companies-producers of such equipment in many countries have mastered manufacturing of the guide-beams for the hyperspectral scanners, including those to operate in visual and near-IR (VNIR), shortwave IR (SWIR), medium-wave IR (MWIR), and thermal IR (TIR) bands. The airborne scanners, which are currently becoming operational, are notable for their perfection and variety, the number of their different types supposedly exceeding one hundred. The scanning hyperspectral sensors of the leading companies have reliable, stable, and functional components; a software handy for operation; high capacity of digitizing and readout; high-precision and efficient optics; stable and reproducible parameters of the sensor calibration, and so on. The customers are often supplied by the companies not just with the imaging hardware, but with the complex imaging technologies. Such integrated digital airborne systems represent the off-the-shelf solutions for obtaining the final product.

Generally, each hyperspectral scanner consists of a sensor (optico-electronic unit); power supply; data accumulation and control unit; vibration damping platform; operator display; GPS receiver; inertial measurement unit; and data processing device. A matter of priority is also to maintain functional completeness of the hyperspectral scanners' spectrum, that is, a possibility to use them in all the applications somehow connected with the spectral analysis.

It is only in this case that the quantitative assessments can be made of the spectral characteristics, pertaining both to the underlying surface and specific ground objects, which, in turn, are extremely important for obtaining trustworthy results at the stage of applied analysis. This is why the companies are heavily investing in the development of their own optico-electronic tract, including CCD-receivers with high signal-to-noise ratio, stable radiometric calibration, and so on. The equipment design and the operational software make it mostly possible to make measurements at the moment of imaging and to account for it later on.

Airborne hyperspectral imaging ensures object recognition is made by their physicochemical makeup and enables one to identify plant species and vegetation status; while keeping under control the natural habitat of the different plants including weeds and unique plants; as well as spot vegetation and soil disturbances and find wetlands and salt disturbed soils.

Aircraft, drone, helicopter, and airships—each of them has some useful advantages as a platform for remote sensing systems. Aircraft can fly at relatively low altitudes, thus admitting a sub-meter sensor spatial resolution. The airborne sensors can be used more flexibly because of the variation of height (flight level) and the forward scan velocity. Therefore, spatial resolution and swath width can be easily adapted to the task requirements in the course of flight.

Aircraft can easily change their imagery schedule to avoid weather problems such as clouds, which may block a passive sensor's FOV, and to allow for the sun illumination. Sensor maintenance, repair, and configuration changes are easily applicable to aircraft platforms. There are no bounds for an aircraft but the state borders. At the same time, for a low-flying aircraft with a narrow sensor's FOV, a number of flybys are needed to cover a large ground area. The turnaround time it takes to get the data to the user is delayed, due to necessity of landing the aircraft before transferring the raw imagery to the data processing facility. Besides, the hyperspectral data can be easily and efficiently integrated with those obtained by the airborne LiDAR system, or the sensors operating in other spectral bands.

The hyperspectral scanners are the most advanced and sophisticated airborne optical devices of the Earth remote sensing. Here we name but a few of the state-of-the art hyperspectral airborne imagers with excellent characteristics, such as AVIRIS, HYDICE, AISA, HyMAP, ARES, CASI 1500, and AisaEAGLET and give basic specifications of the most interesting among them.

The Avionic Multisensors hyperspectral System, manufactured by the SELEX Galileo company, includes four hyperspectral/MS optical heads (e.g., VNIR, SWIR, MWIR, and TIR) which provide, in different configurations, a spectral coverage from the visible (400 nm) to the TIR (12,000 nm) band; the inertial platform with integrated GPS (INS/GPS) to take records of the aircraft position and attitude; the Instrument Control Unit/Pre-Processing Computer to control the optical head and store data in internal HD memory. The "modular" approach allows, with just a change of mechanical interface, a flexible arrangement of instrument accommodation, making it suitable for use on the different platforms, including unmanned aerial vehicles (UAVs) and ultra-light aircrafts.

The hyperspectral system can be applied both in airborne and ground operations, although its basic configuration is designed for airborne platforms, with pushbroom scanning to build up the spectral data cubes. To derive geo-registered images, the flight data coming from a dedicated GPS/INS unit are continuously logged in sync with the hyperspectral data. In the ground system, the same optical head is used as in the airborne system, as well as the scanning platform synchronized with the image acquisition. In this mode, the instruments can be used as "static" cameras for applications, where the linear platform movement needed for pushbroom mode is not available. The development follows a modularity principle aimed at creation of a flexible system to be utilized on manned or unmanned platforms for different applications, such as UAV, light aircraft, medium-range aircraft for maritime and coastal surveillance and patrol, and multi-mission maritime aircraft for high-altitude surveillance.

SPECIM airborne hyperspectral sensors provide market leading solutions for remote sensing, from small UAV systems to full-featured commercial, research, and defense tools. SPECIM brought to the market the AisaEAGLET hyperspectral sensor, particularly designed to meet space and weight limitations of UAVs and small piloted aircraft (Aisa stands for Airborne Imaging Spectrometer for Applications). This hyperspectral sensor is the most compact and complete airborne imaging system to acquire full, contiguous VNIR data with a high spatial resolution of 1600 pixels and excellent sensitivity even under low-light conditions, with the imager's spectral range of 400–1000 nm and spectral resolution of 3.3 nm (up to 200 spectral bands). Its FOV is $\sim$30°, IFOV—0.02°; resolution on the Earth's surface $\sim$0.35 m from 1000-m altitude. The system has a total mass of 10 kg, including the hyperspectral head (3.5 kg), compact data acquisition computer, and GPS/INS unit. The sensor

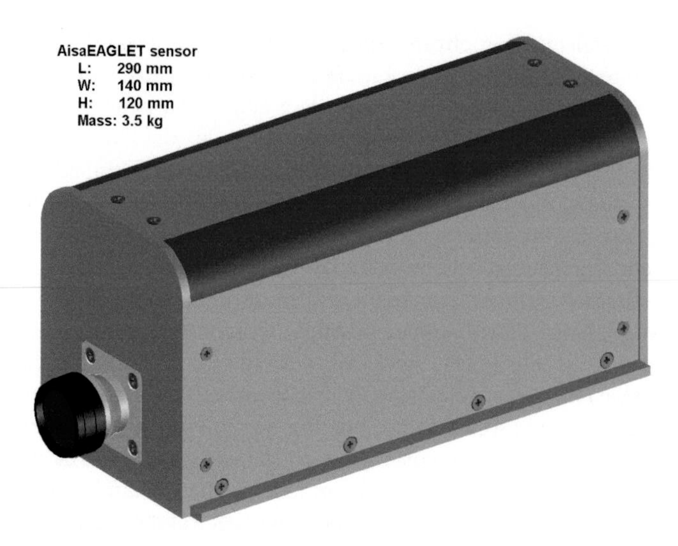

FIGURE 2.7 Optical head of AisaEAGLET hyperspectral sensors destined for operation on small aircrafts.

head, shown in Figure 2.7, contains a wavelength and radiometric calibration file. System output—12 bits digital, SNR in range of 130:1–300:1 (depending on the band configuration), power consumption <100 W, DC 10–30 V. The throughput of the high-efficiency imaging spectrograph is practically independent of polarization. The operator can create a hyperspectral and multispectral mode; apply specific band configurations; and quickly change from one mode or configuration to another in flight operation. The system's modular design facilitates its integration with different payloads. The company provides support for the implementation of the sensor system control from the ground through a telemetry link to the UAV. In addition to its high performance and compactness, its cost effectiveness makes it an exceptional and versatile remote sensing tool for environmental, forestry, agricultural, security, and defense applications. The AisaEAGLET hyperspectral sensor can be easily used in ground-based applications thanks to company proprietary scanning devices, which allow scanning of the target from a static platform, like a tripod or security monitoring facility.

The SpecTIR Company developed a pushbroom-imaging spectral instrument ProSpecTIR-VS with components for high dynamic range imaging devices, integrated GPS/INS sensor, durable housing, all integrated with flight operations and recording hardware. The hyperspectral instrument has dual sensors individually covering VNIR wavelengths of 400–1000 nm and SWIR in the 1000–2500 nm wavelength range. The dual sensors are co-boresighted and include all hardware and acquisition and processing software for flight operations and spectral mapping. The hyperspectral sensors described can be installed in almost any light aircraft with aerial camera capability. The imagery is navigated with the integrated DGPS/IMU and after processed provides geo-referenced, radiance, and reflectance files readily imported into spectral analysis programs. The instrument's main characteristics are: spectral resolution 2.9 nm, number of spectral bands 250, number of spatial pixels 320, FOV 24°, IFOV 1.3 mrad, SNR—350–500 VNIR, 800 SWIR.

An airborne hyperspectral sensor HyMap, developed in Australia and deployed for commercial operations around the world, provides 128 bands across the reflective solar wavelength region of 0.45–2.5 nm with contiguous spectral coverage (except for the atmospheric water vapor bands) and bandwidths between 15 and 20 nm. The sensor operates in a 3-axis gyro stabilized platform to minimize image distortions due to the aircraft motion. The system, to be transported between international survey sites by air freight, can be rapidly adapted into any aircraft with a standard aerial camera port. The HyMap provides a signal-to-noise ratio (>500:1) and image quality setting the industry standard. Lab calibration and operational system monitoring ensures the imagery to be calibrated as required for demanding spectral mapping tasks. Geo-location and image geocoding are based on differential GPS and INS. Hyperspectral sensors IFOV—2.5 mrad along track, 2.0 mrad cross track; FOV—61.3° (512 pixels); ground IFOV is 3–10 m for typical operational range.

The new AISA hyperspectral Imager in dual mode is capable of acquiring 492 bands of 2–6 nm bandwidths across the spectral range of 395–2503 nm, at pixel sizes between 75 cm and 4 m. It is mounted on a twin-engine Navajo along with a LiDAR system with 10-cm vertical accuracy and a digital camera. A known HYDICE sensor collects data of 210 bands over the range of 400–2500 nm with FOV of 320 pixels wide at the IFOV, projection of which on the ground (pixel size) of 1–4 m, depending on the aircraft altitude and ground speed.

Let's give some more examples of the current-day hyperspectral sensors which contributed in raising the technological level in the area. Successfully tested in the United States was the hyperspectral sensors onboard the reconnaissance UAV designed for spotting the tanks, missile launchers, and other camouflaged defense equipment located on the colorful background. The system developed by the TRW Company provides imaging in 384 working spectral bands. Another example is the Earth Remote Sensing (ERS) lab, with high spatial and spectral resolution, operated worldwide and lately, as an example, successfully commissioned in Kazakhstan. Its equipment consists of a widescreen 136-Mpixel UltraCAM X camera—one of the most powerful airborne devices of its kind in the world market—and a high-resolution CASI 1500 hyperspectral scanner, with 288 programmable channels in the visible and IR bands. By its technological level, the equipment meets standards of the satellite-borne ERS, the only difference being the carrier involved. The advantage of placing the equipment onboard the aircraft lies in the high resolution on the terrain (3–50 cm) and in highly accurate 5 cm linking, both in plane and in height, as ensured by the onboard navigation system data. The coverage resolution achieved by the airborne hyperspectral imaging is within 25–400 cm. At the maximum flight altitude, an 8 km swath can be covered, as with the high-resolution OrbView-3 satellite. To cope with huge data streams, the onboard equipment also includes the 13-processor supercomputer with 4 Tb memory.

2.4.3 Spaceborne Imaging

Big hopes are pinned on the imaging from space. For a long enough time, the MS systems performed onboard the spacecraft, generating several images in the separate wide spectral bands, from visual to IR. Until now, MS imaging from the new generation spacecraft (GeoEye: 1–4 spectral zones, RapidEye: 5 zones, and WorldView: 2–8 zones) was of a great practical interest. Nowadays, there are opportunities of developing spaceborne hyperspectral equipment for observations in the optical range, with high informational and operational characteristics, which would enhance a solution level of the space monitoring tasks [18–22].

Satellite platforms enable placement of various sensors for observation and provide coverage of the Earth's surface and periodic monitoring of the areas of interest during the satellite revisits. The image resolution is defined by the satellite orbital parameters, and its altitude in the first place. For satellites, there are no political borders, so they can sweep any part of the globe, regardless of the attitude of the states the onboard imaging equipment flies over. The basic weakness of the spaceborne imagery is its multi-hundred million dollars costs to develop the system as well as ground support facilities, while keeping in mind the spacecraft's relatively short operating life of about 5 years, or less. Decrease of the satellite mass can result in substantial cost saving, primarily due to less powerful and low-cost launchers to be involved in putting the satellite into orbit. This is why the concept of small satellites-based hyperspectral imaging was developed and implemented by means of mini-satellites (100–500 kg).

Basic characteristics of the space-based hyperspectral instruments are presented in Table 2.1. The list includes both flight-proven devices and those under development for future operations in space [23–35]. Most of them are devised for LEO small satellite missions, although placing possible hyperspectral sensors onboard the large satellites and satellites on geostationary orbit is also considered [36]. The first fourteen rows in Table 2.1 pertain to hyperspectral satellite with a swath on the Earth surface of 5–20 km, number of spectral bands ~200, and spatial and spectral resolution

TABLE 2.1
Space-Based Hyperspectral Mission Overview: Operational and Planned Instruments

Instrument (Satellite)	Altitude (km)	Pixel Size (m)	Number Bands	Spectral Range (nm)	Spectral Resolution (nm)	IFOV (μrad)	Swath (km)
HSI (SIMSA)	523	25	220	430–2400	20	47.8	7.7
FTHSI (MightySatII)	565	30	256	450–1050	10–50	50	13
Hyperion (EO-1)	705	30	220	400–2500	10	42.5	7.5
CHRIS (PROBA)	580	25	19	400–1050	1.25–11.0	43.1	17.5
COIS (NEMO)	605	30	210	400–2500	10	49.5	30
ARIES-1 (ARIES-1)	500	30	32	400–1100	22	60	15
			32	2000–2500	16		
			32	1000–2000	31		
UKON-B	400	20	256	400–800	4–8	50	15
Warfighter-1 (OrbView-4)	470	8	200	450–2500	11	20	5
			80	3000–5000			
EnMAP	675	30	92	420–1030	5–10	30	30
			108	950–2450	10–20		
HypSEO (MITA)	620	20	~210	400–2500	10	40	20
MSMI (SUNSAT)	660	15	~200	400–2350	10	22	15
PRISMA	695	30	250	400–2500	<10	40	30
ARTEMIS (TacSat-3)	425	4	400	400–2500	5	70	~10
HyspIRI	~700	60	>200	380–2500	10	80	145
SUPERSPEC (MYRIADE)	720	20	8	430–910	20	30	120
VENμS	720	5.3	12	415–910	16–40	8	27.5
Global Imager (ADEOS-2)	802	250–1000	36	380–1195	10–1000	310–1250	1600
WFIS (like MODIS)	705	1400	630	400–1000	1–5	2000	2400

of 10–30 m and 10–20 nm, respectively. The SUPERSPEC and Vegetation and Environment New Micro Satellite (VENμS) sensors represent multispectral imagers of the next generation; it was put in the table for comparison with previous hyperspectral instruments. The focal plane of multispectral cameras consists of a multi-line array, one for each spectral band with relative narrowband spectral filters. Along with the narrow FOV hyperspectral instruments, operational interests would also include some specimens of the wide FOV hyperspectral imaging systems. Examples of such hyperspectral missions are shown in the last two rows of Table 2.1. It is obvious that space applications of hyperspectral remote sensors are under constant development.

An indicative example of remote sensor advancement is the above mentioned joint Israeli/French scientific and technological project VENμS, launched in 2017. It is designed for high-definition photography of agricultural tracts as a means of ecological control. VENμS launch weight was 265 kg, and its expected lifetime in space was more than 4 years. It has superior parameters in terms of the number and width of its bands, spatial resolution, imaged strip width, mass-energy, and information capacities. The 90-W camera weighs 45 kg and discerns the objects down to 5.3 m in size from 720-km altitude, scanning across 27.5-km swaths. Photographing will be carried out in 12 narrow spectral channels in the 415–910 nm range, with each band varying from 16 to 40 nm. In this mission, most satellite and sensor design parameters are optimized.

Let us briefly review the worldwide state-of-the-art in the field. There are a small number of experimental operational spacecraft equipped with hyperspectral sensor, viz.: EO-1 (under the NASA program) and MightySat II.1, aka Sindri (under the US Air Force program); two other defense projects—ORBView-4/Warefighter and NEMO—are at various stages of readiness. The information capacity for different spectral bands was assessed by processing the measurement data as obtained in simultaneous imaging of the test objects by hyperspectral sensors and traditional optical technique (EO-1 with Landsat 7 and Terra; MightySat with the Keyhole visual reconnaissance satellite).

Investing in commercial satellites can be a risky business. TRW's Lewis satellite with hyperspectral sensors was lost shortly after launch in 1997. In the same year, EarthWatch lost its EarlyBird satellite 4 days after launch. In spite of the risk, every firm feels obligated to incorporate such systems in future spacecraft meant for the Earth remote sensing.

In Russia, for example, there is a metrological basis for the subject-related processing of the spectrometer data. Specific prototypes of the hyperspectral equipment were developed by the AFAR scientific production enterprise and the Reagent scientific-technical center [37]. The operation of the Yukon-B video-spectrometric system, with mass of 1.5 kg, is based on the optoacoustic effect in the visible band (500–800 nm). The overall number of zones is 256, from which 10 to 100, with width of 4–8 nm, are program-selected. Spatial resolution at 400-km altitude will be 20 m, and the frame size of the surface area—15 $\times$ 12 km, with one zonal image size amounting to 4–5 Mbit. Yukon-B is a component of the projected Yukon-UVIT complex designed to cover the UV, visible, and thermal spectral bands. As a platform for Yukon-B, the service module of the ISS Russian segment can be used, as well as small D33 spacecraft of the Monitor-E model manufactured by the Khrunichev Plant, and Kondor-E of the Scientific Production Enterprise of the Machine Building.

Another Russian project is the Astrogon mini-satellite, under development at the Research Institute of the Electric Machinery at Istra, near Moscow, as a part of the Gazprom Trust aerospace monitoring program. The Reagent's hyperspectral video-camera has the 700-channel capacity (with margin for increase up to 1000), with a spatial resolution of 5 m and adaptive control along both the spectral and spatial coordinates. The programmed control system permits the selection of the resolution level from 0.5 to 100 nm in the spectral bands of 200–900 nm and 1000–1400 nm, as well as one of two possible signal polarization modes. For the signal space-frequency transformation, both rearrangeable optoacoustic filters and micro-channel photon detectors are used. Total mass of the imaging equipment is 4 kg, with power consumption of 15 W. Designed precision of the geometric matching of the images is 5 m. The camera control system makes it possible to carry out both instrumental and virtual stereo-imaging.

Two more developments, pertaining to the hyperspectral field, are at different stages of implementation in the Russian space industry, viz.:

- On the basis of hyperspectral technology of the mentioned Astrogon project, tested in the experiments onboard planes and helicopters, under consideration is a joint development with Germany of the mini-satellite formation for space monitoring of the wildfires, volcano eruptions, and other calamities. Involved in the project is the IR technology, applied by the German party on the small BIRD satellite. The monitoring would make it possible to discern and locate the sources of the excess energy flux of both natural and anthropogenic origins.
- The Vavilov State Optical Institute (St. Petersburg, Russia) has developed a compact onboard hyperspectral camera for the small spacecraft, to receive simultaneously in $\sim$100 spectral channels of the visual and near-IR bands (200–1000 nm) a 2D distribution of the brightness field with angular resolution $\sim 20''$, which corresponds to 100 m spatial resolution on the Earth's surface for 1000-km orbit. The instrument provides detailed spectral information on the small gaseous constituents of the atmosphere, status of the inland waters, and vegetation and soil specifics.

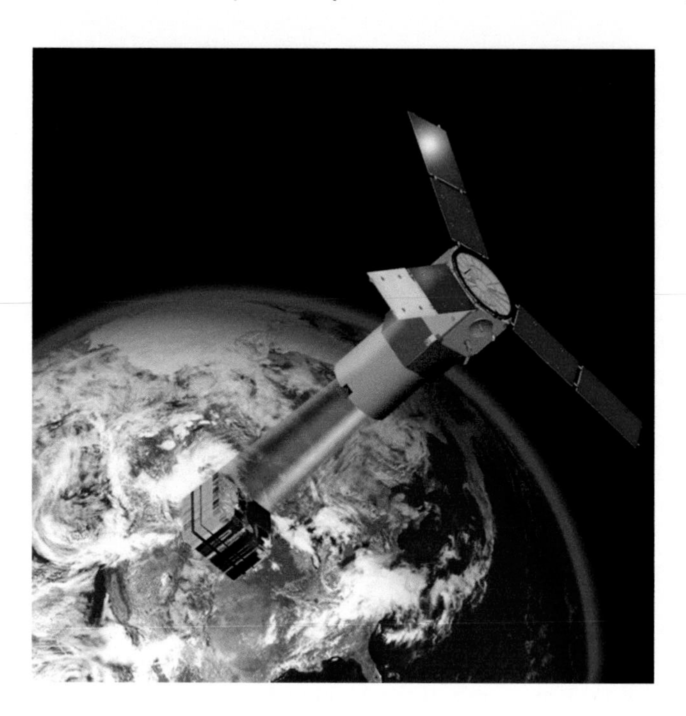

FIGURE 2.8 Artist's impression of TacSat-3 satellite in orbit.

A remarkable achievement in the hyperspectral-assisted space exploration was the successful launch in May 2009 of the multipurpose TacSat-3 US Air Force satellite from the NASA Wallops Flight Facility on the Virginia coast [38]. The satellite, shown in Figure 2.8, orbiting at a 425 km altitude, is required to deliver hyperspectral images within 10 minutes after an image acquisition. The main component of its specialized equipment, the ARTEMIS (Advanced Responsive Tactically-Effective Military Imaging Spectrometer), designed to operate in 400 spectral intervals in the 400–2500 nm band (the entire visible and very-near IR), is an advanced version of the Hyperion scanning spectrometer (NASA, 2000), which performs in 220 intervals with spatial resolution of 30 m/pixel. Space photos with such a resolution form the basis for scanning coverage of the popular geo-interfaces, like Google Earth. Besides the spectrometer, the ARTEMIS system includes a telescope and a signal processor with 16-Gb onboard memory. The secondary mirror of the telescope is integrated into the focal plane light-receiving matrix, common for all the bands, unlike its separate counterparts in the prior generation of hyperspectral detectors. This improvement enabled a simpler design of the telescope optical setup, easier data processing, and a cost saving. The duration of the TacSat-3 satellite demonstration mission was just under 3 years.

Let's return to the hyperspectral missions and look at some problems connected with the development of the hyperspectral satellites and their hyperspectral sensors. Energy resolution, defined as a number of resolved levels of the object's brightness, or of its image illumination, is chosen in accordance with required signal-to-noise ratio. It is well known that detection and recognition probability rise together with the ratio increase. Besides, consideration should be taken regarding the interconnection between spatial, spectral, and energy resolutions, which takes place in live hyperspectral sensors. For example, if higher spatial resolution is achieved by reducing the size of the image element, then the element would get a lesser share of the energy, needed for splitting by operating spectral ranges and obtaining the required signal-to-noise ratio in each of them. This is why various combinations of the basic characteristics can be found in the real devices, depending on the task at hand.

The true parameters—bandwidths and instantaneous fields of view (IFOV)—of existing hyperspectral imagery systems in the visible spectral range are presented in Figure 2.9. The interrelation between spectral and spatial resolutions, as calculated for a hyperspectral instrument

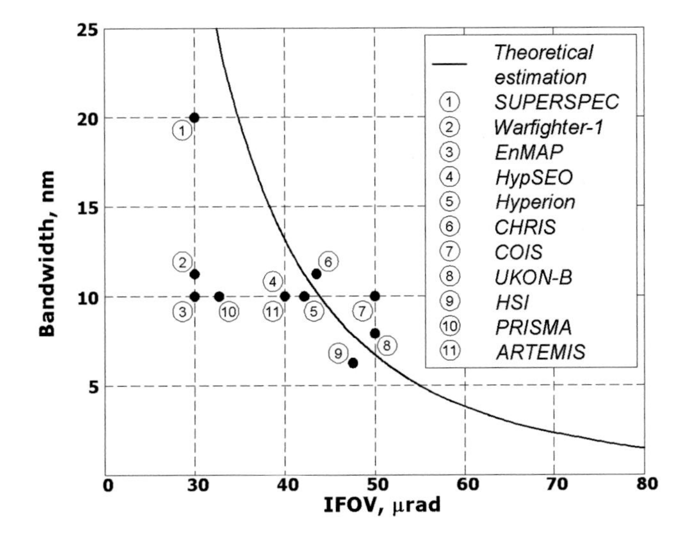

FIGURE 2.9 Bandwidth/IFOV relation for visible range.

with a 20-cm aperture, suitable for operation on the small satellites, is drawn by a solid line. As seen, there is a good conformity between the theory and practice.

The major obstacle in developing hyperspectral sensors is a well-known difficulty with designing the relatively cheap high-sensitive photo-detectors, operating in a wide spectral range, with high-spatial, spectral, and time resolution. This is also relevant with regard to constructing the efficient, durable, compact cooling systems for the radiation receivers, as well as to some more scientific and technical problems. Not all the attempts made thus far to design hyperspectral sensors as a Fourier-spectrometer were successful, since the requirement of the real-time performance failed to be met.

Another field of application for the small satellites may be the installation of a compact coarse-resolution imager to correct imagery, obtained by other sensors, for atmospheric variability caused mostly by water vapor and aerosols. Such an Atmospheric Corrector (LEISA AC) was used in the EO-1 land imaging mission. Using wedged filter technology, the AC provided the spectral coverage of 890–1580 nm with moderate resolutions (spectral ~5 nm and spatial ~250 m at nadir).

Conceiving the onboard communication system with a high-rate download of as many simultaneous images as there are channels, powered by additional solar panels, it should be taken into consideration that enhancing spatial resolution would imply the weight increase. Likewise, adding more channels would also result in more weight as well as in more power consumption. Moreover, with the orbit altitude increase (e.g., from 400 to 1200 km), or with wavelength increase (from visible to IR) the imager aperture diameter should also be enlarged. For higher ground resolution, the satellite dimensions are defined not only by dimensions of the communication and power units, but also by those of the imager. Thereby, the requirements to the transmitting antenna, optical system, total power consumption, and so on, would impose operational constraints on the mission design, which should be eventually translated into restrictions on the spacecraft dimensions and mass.

Given the existing physical and technological constraints, the resulting ambiguity as to the satellite dimensions urged us to determine the estimation criteria for the satellite's mass versus spatial resolution and channel array. Depicted in Figure 2.10 are feasible combinations of the hyperspectral imagers with LEO observation satellites of various size and weight, depending on the devices' spatial and spectral resolutions. Three different domains represent the conventional satellite classes, viz. micro (up to 100 kg), mini (100–500 kg) and large (over 500 kg). The completed analysis clearly shows that it is impossible to carry out all conceivable LEO hyperspectral missions with microsatellites alone. Meanwhile, most of scientific, commercial, and defense hyperspectral tasks can be implemented just with mini-satellites on the basis of conventional technology.

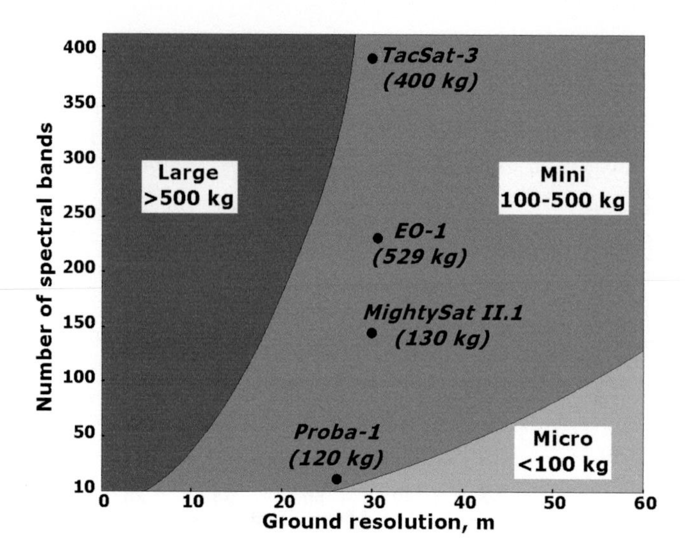

FIGURE 2.10 Hyperspectral sensors accommodation on satellites of different subclasses as function of imager's ground resolution and number of bands.

Still, all spatial, spectral, and radiation requirements growing higher, the mass value would exceed the upper limit for mini-satellites, therefore ensuring 200–300 spectral bands with spatial resolution of some meters (especially in the IR region at high-LEO altitudes) would be feasible only with large satellites. Microsatellites would match well only in the visible region, with large limitations on the resolution (the ground resolutions of 30–60 m and the number of spectral bands 30–100, respectively). The spatial and spectral resolutions of the well-known operated hyperspectral satellites EO-1, MightySat II, Proba-1, and TacSat-3 are also drawn in Figure 2.10 with bullet points, their respective weights being put in parentheses. As can be seen, the point's positions match well with the estimations.

However used onboard, the small satellites were special-purpose instruments, with high spectral resolution in a limited number of wavelengths. As an example, the atmospheric ozone mapping by the small satellites in a short spectral range of ozone bands in UV using a simple filter wheel photometer was a successful application of this principle. Similar experiments with poor spatial resolution are available in small satellites [39].

Instrumentation mass and size, affordable by the present technology level, having such an impact on the future mission's feasibility, big efforts are made to reduce the power consumption, mass, volume, and the unit costs of the hyperspectral sensors under development, as well as to lower demands to the data downlink. Whether these efforts will lead to the emergence of miniature hyperspectral instruments suitable for the small satellites' missions, only time can show.

2.5 LiDAR AND HYPERSPECTRAL DATA INTEGRATION

Virtually all remote sensing techniques, including those presented in the previous paragraphs, rely upon passive sensing of the amount of solar radiation as reflected toward the sensor by clouds, ocean, or solid land, and IR-radiation emitted from the natural or artificial thermal sources on the Earth's surface. Unlike them, the LiDAR system is an active sensor. Since LiDAR carries a source of radiation of its own, it can determine where and when, in the daytime or at night, to take the measurements. LiDAR is similar to radar in the sense that it also can track any target of interest, from airplanes in flight to thunderstorms. But unlike radar, LiDAR uses short pulses of coherent laser light, with very short wavelengths and high instant capacity, so its laser beam does not disperse while traveling away from the source, as ordinary light does. Besides, lasers offer great advantages over conventional light sources in terms of peak power and narrow spectral bandwidth.

A primary functional part of LiDAR is the laser, its other components being the scanner, collimator, lens, receiver, amplifiers, samplers, and other optical and electronic elements. The main function of the laser is to generate emission which, having been reflected by the Earth's surface or the ground objects, returns to the source, to be detected by the highly sensitive receiver. The response time being directly proportional to the target range, it is equivalent to measuring the distance between the emitting source and the reflecting object. The wavelength choice depends on the laser function, as well as on the safety and security requirements. The most commonly used are Nd:YAG lasers, which operate at the wavelengths such as 1550 nm; 1064 nm—near-IR; 532 nm—green light; 355 nm—near-UV.

While ground-based LiDAR instruments profile a single viewing site, the airborne LiDAR systems offer one of the most accurate, expedient, and cost-effective ways to capture upward- or downward-looking data over a wider area. Vertical accuracies of less than 5 cm are possible at 500 m above ground, amounting to less than 20 cm at 3-km altitude. This enables LiDAR to be involved in various applications, such as surveys of corridors like pipeline routes, roads, power lines, highways, and railways; urban environments and mapping; flood plain surveys; forestry (mapping of tree canopies); archaeology; seismic exploration; coastal zone surveys; and oil and gas exploration.

In the airborne LiDAR systems, the laser range-finder is mounted over an opening in the aircraft floor and scans beneath the aircraft, producing a wide swath over which the distance to the ground is measured. The scanning device controls the direction of the laser pulses' propagation, to provide coverage of some predetermined scan swath. In most cases, the cross-track scanning is implemented by using the oscillating mirror, while the along-track scanning is by the motion of the carrier aircraft itself along its operational path.

All the aircraft movements are recorded by its inertial navigation system (INS), to be used in the data post-processing. The LiDAR can also be installed on a stabilized platform. By merging laser ranging, GPS positioning, and inertial attitude technologies, LiDAR can directly measure the shape of the Earth's surface beneath the aircraft's flight path. Data collection rates, accuracy, and other characteristics of some excising and operating airborne LiDARs correspond to the user's requirements and allow them to carry out various research efforts.

A fundamental limitation of airborne laser scanners and imagery consists in their generally straight-down perspective, which makes impossible the accurate and detailed mapping of vertical structures such as cliff faces, coastal ridges, or any vertical side of a natural or man-made structure. To reveal the vertical faces of structures from the air, imaging from the different perspectives should be introduced in the LiDAR data acquisition process by means of some specialized mounting.

The airborne applications being limited to comparatively small regions, it is the spaceborne LiDAR systems that can provide continuous geospatial data and offer a truly global view of the Earth's surface. Furthermore, unlike an airborne laser, a spaceborne one poses no hazard to the general public, even when viewing with the naked eye, binoculars, or small telescopes.

The first LiDAR system in space was the LITE (Lidar In-Space Technology Experiment) instrument, designed and built by NASA. From a 260-km Space Shuttle orbit, its pencil-wide laser beam spread to approximately 300 meters wide at the surface—about the size of three football fields. The data obtained by LITE were used in the Earth atmosphere studies, to measure vegetative cover and distinguish various types of surfaces. LiDAR is a very reliable tool for active remote sensing from space. A future spaceborne high-performance imaging LiDAR for global coverage is now under development.

While early LiDAR scanning systems were capable of registering a single return pulse for each transmitted pulse, most systems in operation today are able to record multiple discrete returns, which occurs when an emitted LiDAR beam encounters in its path an object like a forest canopy. The data may be classified as first, second, third, or last returns, and bare-Earth. A return might be generated from a reflection caused by the top of the tree, but a sufficient amount of laser light energy is able to reflect from lower portions of the tree, or finally, from the ground. These multiple returns can be analyzed and classified to produce information about the objects as well as the bare ground surface. Buildings, trees, and power lines are individually discernible features and may be classified separately. Most often used are the first and last returns, the first one showing the highest features such as the tree canopy, buildings, and so on, and the last generally assumed to refer to the ground level. Being digital, it can be directly processed into a detailed bare-Earth digital elevation model at vertical accuracies within 0.15–1 m. Derived products include contour maps, slope aspect, 3D topographic images, virtual reality visualizations, and more.

The LiDAR instruments have an option to measure the intensity of the returned light. This captured information can be displayed as an orthorectified image—a 256-gray level mapping of the light pulse reflectance amplitude. Such an image, comparable with a coarse photograph of the area under survey, can be painted according to the elevation levels.

The amplitudes of the laser scan return signals, as measured by LiDAR, do not allow proper reconstruction of the ground reflectance, mainly because of spreading and absorption terms, which depend on the laser-target distance. This prevents not only accurate imaging of the ground, but also the integration of the intensity information with the segmentation/classification process. This difficulty can be overcome by a suitable calibration of the laser scan return values. A number of signal calibration procedures are known, insufficiently precise as they are. However, the radiometric information of the laser scan is relatively poor in terms of information content being the laser light is monochromatic.

Some researchers utilize the return intensities to extract more information about the reflecting surfaces than it can be done by using LiDAR solely as a ranging tool. Using the intensity measurements combined with height data for classification of vegetation types, it was found, for example, that the reflectance values, as shown by pure broadleaf forests, are significantly higher than those for pure conifer forests.

The exploratory data analysis was carried out to assess the potential of the laser return type and intensity as variables for classification of individual trees or forest stands according to species. The evaluation of the irregular behavior of some major ground indices, fulfilled on the basis of integrating the LiDAR and hyperspectral data, made it possible to refine the technique of discovering new archeological sites. In the land cover classification, it proved effective to process LiDAR intensity data clouds together with the high quality, large-coverage images provided by the airborne cameras.

Besides the intensity, the primary products derived from the raw LiDAR data also include the waveform returns. The new full-waveform laser scanning technique offers advanced data analysis options not available in traditional LiDAR sensors. This technique, implemented in the latest generation of scanners, will promote evolution in such areas, as continuous multilayer vegetation modeling; waveform-based feature classification; single-pulse slope determination, and so on.

The LiDAR images, including orthophotos, can be seamlessly integrated with MS, hyperspectral, and panchromatic data sets. Combined with Geographic Information System (GIS) data and other surveying information, LiDAR imagery enables the generation of complex geomorphic-structure mapping products, building renderings, advanced 3D Earth modeling, and many more high-quality mapping products. Galileo is one of a few companies providing highly accurate fused image data products. Using high spatial-resolution LiDAR and hyperspectral imaging data, the maps were improved by adding a 3D perspective.

As was discovered, the airborne LiDAR can be used for monitoring bio-fluorescence when coupled with a hyperspectral tunable filter [40]. The promising technique of hyperspectral fluorescence imaging, as induced by a 632-nm laser, was applied for assessing the soluble solids content of fruit. The laser-induced hyperspectral bio-fluorescence imaging of different objects is now under study in lab experiments, airborne applications, and fieldwork.

The efficiency of the advanced geospatial technology, based on the fusion of the hyperspectral imagery and LiDAR data, was assessed in Canada and the United States, as applied to the mapping of the location and condition of different tree species, for both the effective management of the forested ecosystems and counteraction to invasive tree pests. The hyperspectral imaging was obtained by the AISA Dual system with the Eagle (395–970 nm) and Hawk (970–2503 nm) sensors, enabling simultaneous acquisition in 492 narrow spectral channels. LiDAR data were collected using a TRSI

Mark II two-return sensor. Also used in the data processing was the collection of ground spectral signatures.

German and Israeli scientists are implementing the project on fusion of hyperspectral images and LiDAR data for engineering structure monitoring [41]. They confirm that the assessed degradation of urban materials in artificial structures by exploring possible chemical and physical changes using spectral information across the VIS-NIR-SWIR spectral region (400–2500 nm). This technique provides the ability for easy, rapid, and accurate in situ assessment of many materials on a spatial domain under near real-time condition and high temporal resolution. LiDAR technology, on the other hand, offers precise information about the geometrical properties of the surfaces within the areas under study and can reflect different shapes and formations of the complex urban environment. Generating a monitoring system based on the integrative fusion between hyperspectral sensors and LiDAR data may enlarge the application envelop of each technology separately and contribute valuable information on urban runoff and planning.

The LiDAR and hyperspectral data were also investigated from the viewpoint of their potential to predict the canopy chlorophyll and carotenoid concentrations in a spatially complex boreal mixed wood. Using canopy scale application of hyperspectral reflectance and derivative indices, LiDAR data analysis was conducted to identify structural metrics related to chlorophyll concentration. Then the LiDAR metrics and hyperspectral indices were combined to determine whether concentration estimates could be further improved. Integrating mean LiDAR first-return heights for the 25th percentile with the hyperspectral derivative chlorophyll index enabled further strengthening of the relationship to canopy chlorophyll concentration. Maps of the total chlorophyll concentration for the study site revealed distinct spatial patterns, indicative of the spatial distribution of species at the site.

The project [42] applied advanced geospatial technology, including high-resolution airborne AISA hyperspectral sensors and LiDAR data collection, in conjunction with analytical applications of GIS, to develop new tools needed for improved species mapping, risk assessment, forest health monitoring, rapid early detection, and management of invasive species. Similar technologies play an increasingly important role in offering accurate, timely, and cost-effective solutions to such problems.

Airborne LiDAR has become a fully operational tool for hydrographic surveying. Present-day airborne laser bathymetry systems can simultaneously measure both water depth and adjacent surface topography. hyperspectral imagery from abovementioned Compact Airborne Spectrographic Imager (CASI) (see Section 2.4.2) has proven to be a valuable instrument for coastal measurements and analysis. CASI spectral resolution of 288 bands for each spatial pixel allows for the extraction of a vast amount of information such as water clarity and temperature, bottom type, bathymetry, as well as water quality (chlorophyll, dissolved organic carbon, and suspended minerals), soil types, and plant species [50]. In order to achieve a comprehensive hydrographic capability, the LiDAR and hyperspectral sensors CASI sensors were integrated to provide the different agencies with the relevant information. The Naval Oceanographic Office, for example, uses the airborne laser bathymetry system to collect hydrographic information about the littoral zone for the warfighter and, by adding the hyperspectral capability, enhance its efficiency in rendering a quick and more adequate environmental picture.

An interesting research issue pertains to the integration of LiDAR and hyperspectral data to improve the discovery of new archeological sites [43]. Under investigation was the possibility of using such an integrated dataset in evaluation of the irregular behavior of some major ground indices. While hyperspectral data allow identifying specific humidity, vegetation, and thermal conditions in the target area, LiDAR data provide the accurate geometric information. In fact, accurate filtering of the laser scanning data allows the computation of the Digital Terrain Model, while Lambertian-based calibration of LiDAR intensity enhances the automatic data segmentation and thus the detection of possible sites of interest. In order to fully extract hyperspectral information from MIVIS and AISA sensor data, some specific procedures have been implemented together with the adequate computer analysis. Such data processing has been applied to both already discovered archeological sites of the ancient city of Aquileia, and the new areas in its northern part. When integrated, the resulting datasets revealed with sensible accuracy the presence of surface/below surface archaeological heritage. As was confirmed by the archaeologists, the present-day available airborne imaging, namely LiDAR-assisted and the hyperspectral, is the most technologically advanced approach. In some cases, these instruments were put together as an integrated unique sensor mounted on the helicopter.

The new remote sensing technologies on a basis of LiDAR data integration with optical hyperspectral and MS imaging data were developed for natural hazard detection and decision support systems. The innovative imagery fusion refines the identification and mapping of past geologic events, such as landslides and faults, while also providing quicker and simpler processes for forecasting and mitigating future environmental hazards.

United States researchers used the remote sensing data to improve mapping and characterizing mechanical properties of rocks. In early 2007, simultaneous hyperspectral and LiDAR imagery was obtained from a plane over Cuprite, Nevada [51]. The imaging instrumentation included the Optech LiDAR, with an operational wavelength of 1.064μm, next-generation hyperspectral sensors—Mapping Reflected-energy Spectrometer, working in the 0.4–2.5 μm spectral range, and the Nikon D2X digital camera to capture high-spatial-resolution true-color images. Results of the data analysis suggested, for some surfaces, a correlation between mineral content and surface roughness, although the LiDAR resolution (~1 m ground sampling distance) turned out to be too coarse to extract surface texture properties of clay minerals in some of the alluvial fans captured in the imagery. Such experiments may provide valuable information about the mechanical properties of the surface cover in addition to generating another variable of use for material characterization, image classification, and scene segmentation.

A short description of the successful research efforts, based on joint hyperspectral sensors and LiDAR data, does not pretend to be complete; it just demonstrates high efficiency of this imaging technique in various applications. Future mission planning should include consideration of determining optimal ground sampling to be used by LiDAR and hyperspectral systems [44,45]. The fusion of LiDAR elevation data and MS and hyperspectral classification results is a valuable tool for imagery analysis and should be explored more extensively.

2.6 SUMMARY AND OUTLOOK

The hyperspectral imaging sensors are designed to provide a detailed analysis of the entire light spectrum from visible light up to far IR. Present-day sensors can break this wavelength range into hundreds, or even thousands, of sections for individual analysis and examine such extremely narrow sections, which makes it possible to look for specific chemical compositions that reflect light only in the chosen wavebands. Hyperspectral sensors simultaneously capture both spectral and spatial information of the scene and provide a fully registered spatial and spectral data cube for every video frame, and are equally well suited for ground, airborne, and space operations. Due to their unique capabilities, they can collect data from hand-held and vehicle-mounted ground systems, low- and high-altitude aircraft, as well as LEO or even GEO satellites, according to the application in question. One shortcoming of flight operations with hyperspectral sensors, also with aerial- and space-based LiDAR, is the need for relatively cloud-free conditions over the target area.

To verify the identifications derived from the airborne hyperspectral imagery, ground spectral measurements are often carried out at the sites of interest by hand-held or truck-mounted hyperspectral instruments, providing the ground truth, that is, the material characterization of the natural land samples in UV-VNIR-SWIR-MWIR-LWIR spectral bands.

The spaceborne hyperspectral imaging has huge potential with respect to the duration of the observations, Earth's surface coverage, and abundance of the valuable information. Still there have been few successful hyperspectral space missions due to the complexities mentioned earlier. Described in this chapter were civilian, dual-use, and defense hyperspectral sensors that have been put into orbit, as well as the sensors currently under development or design and some systems planned but never implemented, such as NEMO and Warfighter.

At present, most of the hyperspectral servicing firms are ready to assist clients in their remote sensing data demands. They can support the entire end-to-end data collection effort, starting with the site identification, observation planning, imaging operations, and data analysis. Typical remote hyperspectral products include spectrally and radiometrically calibrated hyperspectral data cubes; geo-reference files for all data cubes; atmospherically corrected data cubes; field spectra from hand-held spectrometers; laboratory spectra of samples; and apparent emissivity data cubes for IR sensors.

Without doubt, the hyperspectral data from operating sensors provide an important contribution for the research community [46–48]. The current state-of-the-art hyperspectral data are far ahead of the first tentative steps taken about 30 years ago. New hyperspectral technology, including instruments performance, processing schemes, and calibration, improved significantly. Well-established spheres of hyperspectral application include agriculture, forestry, water resources, atmosphere, geology, mineralogy, wetlands, environment, management of coastal waters, defense and security, and urban areas.

Hyperspectral mission and sensor performance parameters are derived from the requirements for specific applications. For the spaceborne hyperspectral imaging, along with cost considerations, this leads to the choice of the orbit; re-visit capability; SNR; spatial resolution; data volume per time period; spectral resolution; spectral sampling interval; swath width; radiometric accuracy and stability; spectral range of interest; and required time for the data delivery to the users and processing steps.

An individual user's hyperspectral imagery and related data requirements, which are essential to the event under examination, differ from each other. The most important for all users proved to be not hyperspectral sensors and imagery qualities, but better service and reduced equipment costs.

Hyperspectral spaceborne imaging is getting mature and more satellites with hyperspectral sensors are expected to be put into orbit in the next decade. These may include the German EnMAP; Indian TWSat; South African MSMI hyperspectral sensor; satellites with sensors like Hyperon or CHRIS; wide-area synoptic sensors similar to MODIS and MERIS with finer spectral resolution; and NASA HyspIRI and ESA FLEX, in the case of available funding. Future Landsat and NPOESS satellites might have the hyperspectral capability, as well as UK Disaster Monitoring and German RapidEye constellations. Italy, China, Israel, Canada, and other countries have also announced their hyperspectral space projects. Along with this, the development of the airborne and drone hyperspectral imaging for civilian and public-good applications is also anticipated.

REFERENCES

1. Glackin D.L. Peltzer G.R. Civil: Commercial, and international remote sensing systems and geoprocessing, *The Aerospace Corporation, AIAA*, p. 89, 1999.
2. Kramer H.J. *Observation of the Earth and Its Environment, Survey of Missions and Sensors*, 4th edition, Springer, Berlin, 2002.
3. Nieke J., Schwarzer H., Neumann A., Zimmermann G. Imaging spaceborne and airborne sensor systems in the beginning of the next century. *European Symposium on Aerospace Remote Sensing, IEEE Conference on Sensors Systems and Next Generation Satellites III*, 1997. https://ieeexplore.ieee.org/document/1356079/references.
4. Puschell J.J. Hyperspectral imagers for current and future missions, *Proceedings of SPIE*, 4041, 121–132, 2000.
5. Lillesand T.M., Kiefer R.W., Chipman J.W. *Remote Sensing and Image Interpretation*, 5th edition, Wiley, NY, p. 763, 2004.
6. Chang C.-I. *Hyperspectral Imaging. Technigues for Spectral Detection and Classification*, Kluwer Academic/Plenum Publishers, NY, 370, 2003.
7. Borengaser M. *Hyperspectral Remote Sensing*, Lewis, Boca Raton, FL, 2004.
8. Кучейко А. Российские перспективы в гиперспектре, *Новости космонавтики*, 222, p. 24, 2001 (Kuchako A., Russian Hyperspectral Perspective, J. Novosti Kosmonavtiki, N 222, p. 24, 2001).
9. Bjorn A. Hyperspectral Chemical and Property Imaging, Patent 20090015686, USPC Class 3482221, 01-15-2009.

10. Bolton J.F. Full spectral imaging: A revisited approach to remote sensing, *SPIE Conference on Remote Sensing*, September, Barcelona, Spain, 2003.
11. Bodkin A. Development of a miniature, hyperspectral imaging digital camera, Navy SBIR FY2005.1, Proposal N051-971-0772, 2005.
12. Fletcher-Holmes D.W., Harvey A.R. Real-time imaging with a hyperspectral fovea, *Journal of Optics A: Pure and Applied Optics*, 7, 298–302, 2005.
13. Skauli T., Kåsen I., Haavardsholm T., Kavara A., Tarabalka Y., Farsund O. Status of the Norwegian hyperspectral technology demonstrator, NATO OTAN, RTO-MP-SET-130, Norwegian defence research establishment.
14. Yang C., Everitt J.H., Davis M.R., Mao, C. A CCD camera-based hyperspectral imaging system for stationary and airborne applications, *Geocarto International Journal*, 18(2), 71–80, 2003.
15. Anger C.D. Airborne hyperspectral remote sensing in the future, *Proceedings of the 4th International Airborne Remote Sensing Conference and Exhibition/21st Canadian Symposium on Remote Sensing*, Vol. 1, ERIM International Inc., Ann Arbor, Michigan, USA, 1–5, 1999.
16. Buckingham R., Staenz K., Hollinger A. A review of Canadian airborne and space activities in hyperspectral remote sensing, *Canadian Aeronautics and Space Journal*, 48(1), 115–121, 2002.
17. Jianyu W., Rong S., Yongqi X. The development of Chinese hyperspectral remote sensing technology, *SPIE*, 5640, 358, 2005.
18. Briottet X. et al. Military applications of hyperspectral imagery, *Proceedings of SPIE Defense & Security Symposium*, paper No. 62390B, 2004.
19. Jason S., Cutter M., Meerman M., Curiel A.S. Low cost hyperspectral imaging from a microsatellite, *15-th Annual AIAA/USU Conference on Small Satellites*, SSC01-II-1, Utah, 2001.
20. Curiel A.S., Cawthorne A., Sweeting M. Progress in small satellite technology for earth observation missions, *5-th IAA Symposium on Small Satellites for Earth Observation*, IAA-B5-0301, Berlin, 2005.
21. Cutter M. Review of a small satellite hyper-spectral mission, *19-th Annual AIAA/USU Conference on Small Satellites*, SSC05-IV-2, Utah, 2005.
22. Ortenberg F., Guelman M. Small satellite's role in future hyperspectral earth observation missions, *Acta Astronautica*, 64(11–12), 1251–1262, 2009.
23. Pearlman J., Barry P., Segal C., Shepanski J., Beiso D., Carman S. Hyperon, a space-based imaging spectrometer, *IEEE Transactions on Geoscience and Remote Sensing*, 41(6), 1160–1173, 2003.
24. Vidi R., Chiaratini L., Bini A. Hyperresolution: An hyperspectral and high resolution imager for Earth observation, *Proceedings of the 5th International Conference on Space Optics (ICSO 2994)*, Toulouse, France, 105–111, 2004.
25. Davis C. et al. Ocean PHILLS hyperspectral imager: Design, characterization and calibration, *Optics Express*, 10(4), 210–221, 2002.
26. Hollinder A. et al. Resent developments in hyperspectral environment and resource observer (HERO) mission, *Proceedings of the International Geoscience and Remote Sensing Symposium and 27th Canadian Symposium on Remote Sensing*, July, Denver, Colo, 1620–1623, 2006.
27. Pretil G., Cisbanil A., De Cosmo V., Galeazzi C., Labate D., Melozzi M. Hyperspectral instruments for earth observation, *International Conference on Space Optics (ICSO)*, Toulouse, France, 2008.
28. Cutter M., Sweeting M. The CHRIS hyperspectral mission—Five years in-orbit experience, IAC-07-B1.3.02, *58th International Astronautical Congress*, September 24–28, Hyderabad, India, 2007.
29. Stuffer T. et al. The EnMAP hyperspectral imager—An advanced optical payload for future applications in earth observation programmes, *Acta Astronautica*, 61, 115–120, 2007.
30. Sigernes F. et al. Proposal for a new hyper spectral imaging micro satellite: SVALBIRD, *5-th IAA Symposium on Small Satellites for Earth Observation*, IAA-B5-0503, Berlin, 2005.
31. Morea G.D., Sabatini P. Perspectives and advanced projects for small satellite missions at Carlo Gavazzi Space, *Proceedings of the 4S Symposium Small Satellites Systems and Services*, La Rochelle, France, 2004.
32. Poinsignon V., Duthil P., Poilve H. A superspectral micro satellite system for GMES land cover applications, *Proceedings of the 4S Symposium Small Satellites Systems and Services*, La Rochelle, France, 2004.
33. Haring R.E., Pollock R., Cross R.M., Greenlee T. Wide-field-of-view imaging spectrometer (WFIS): From a laboratory demonstration to a fully functional engineering model, *Proceedings of SPIE*, 4486, 403–410, 2002.
34. Schoonwinkel A., Burger H., Mostert S. Integrate hyperspectral, multispectral and video imager for microsatellites, *19-th Annual AIAA/USU Conference on Small Satellites*, SSC05-IX-6, Utah, 2005.

35. Cooley T., Lockwood R., Gardner J., Nadile R., Payton A. ARTEMIS: A rapid response hyperspectral imaging payload, *Paper No. RS4-2006-5002, 4th Responsive Space Conference*, April 24–27, Los Angeles, CA, 2006.
36. Li J., Sun F., Schmit T., Venzel W., Gurka J. Study of the hyperspectral environmental suit (HES) on GOES-R, *Proceedings of the20th International Conference on Interactive Information and Processing System forMeteorolgy, Oceanography, and Hydrology*, January, Seattle, Wash., Paper p2.21.6, p. 6, 2004.
37. Astapenko V.M., Ivanov V.I., Khorolsky P.P. Review of current status and prospects of hyperspectral satellite imaging, *Space Science and Technology*, Kyiv, Ukraine, 8(4), 73, 2002.
38. Davis T., Straight S. Development of the tactical satellite 3 for responsive space missions, *Proceedings of the 4th Responsive Space Conference*, April 24–27, Los Angeles, CA, p. 10, 2006.
39. Guelman M., Ortenberg F., Shiryaev A., Waler R. Gurwin-Techsat: Still alive and operational after nine years in orbit, *Acta Astronautica*, 65(1–2), 157–164, 2009.
40. Liu M., Zhang L., Guo E. Hyperspectral laser-induced fluorescence imaging for nondestructive assessing soluble solids content of orange, *Computer and Computing Technologies in Agriculture*, I, 51–59, 2008.
41. Brook A., Ben-Dor E., Richter R. Fusion of hyperspectral images and LIDAR data for civil engineering structure monitoring, *Commission VI, WG VI/4, Proceedings ISPRS XXXVIII-1-4-7,* W5, paper 127, 2009.
42. Souci J., Hanou I., Puchalski D. High-resolution remote sensing image analysis for early detection and response planning for Emerald Ash Borer, *Photogrammetric Engineering and Remote Sensing*, August, 905–907, 2009.
43. Johnson J.K. Sensor and data fusion technologies in archaeology, *ISPRS Workshop on Remote Sensing Methods for Change Detection and Process Modeling*, Germany, 2010.
44. Fujii T., Fukuchi T. *Laser Remote Sensing*, Taylor & Francis, Abingdon, UK, p. 888, 2005.
45. Zhang J. Multi-source remote sensing data fusion: Status and trends, *International Journal of Image and Data Fusion*, 1(1), 5–24, March 2010.
46. Nieke J., Seiler B., Itten K., Ils Reusen I., Adriaensen S. Evaluation of user-oriented attractiveness of imaging spectroscopy data using the Value-Benefit Analysis (VBA), *5th EARSeL Workshop on Imaging Spectroscopy*, April 23–25, Bruges, Belgium, 2007.
47. Puschell J. Hyperspectral imagers for current and future missions, *Proceedings of SPIE*, 4041, 121–132, 2000.
48. Buckingham R., Staenz K. Review of current and planned civilian space hyperspectral sensors for EO, *Canadian Journal of Remote Sensing*, 34(Suppl. 1), S187–S197, 2008.
49. Schulenburg N., Riley D., Kupferman P., Bryant G. On using SEBASS as a Mineral Exploration Tool, *GRSG AGM—Exploration Remote Sensing*, Geologic Society of America, 2009. https://www.grsg.org.uk/abstracts/on-using-sebass-as-a-mineral-exploration-tool/.
50. Lopez R., Frohn R. *Remote Sensing for Landscape Ecology: New Metric Indicators—Monitoring, Modeling, and Assessment of Ecosystems*, Taylor & Francis-CRC Press, Boca Raton, FL, 2017.
51. Riley D., Hecker C. Mineral mapping with airborne hyperspectral thermal infrared remote sensing at Cuprite, Nevada, USA, Chapter 24, In: *Thermal Infrared Remote Sensing*, Springer, 2013.

3 Hyperspectral Remote Sensing in Global Change Studies

Jiaguo Qi, Yoshio Inoue, and Narumon Wiangwang

CONTENTS

3.1 INTRODUCTION

A few decades ago, hyperspectral imagery data and processing software were available to only spectral remote sensing experts. Nowadays, one of the fastest growing technologies in the field of remote sensing has been investing on the research and development of hyperspectral sensors for data acquisition and software for data analysis [1]. Unlike multispectral imaging systems (e.g., Landsat, SPOT, IKONOS) that capture reflected or emitted incoming radiation from the Earth's surface in a few broad wavelength bands across the electromagnetic spectrum, hyperspectral imagers measure reflected radiation at numerous narrow, contiguous wavelength channels. The substantially finer spectral resolution data from hyperspectral sensors enhance the capability to characterize the Earth's surface more effectively than do the broadband multispectral data [2].

The distinction between hyperspectral and multispectral is based on the narrowness and contiguous nature of the measurements, not the "number of bands" [3]. For example, a sensor that measures only 20 spectral bands can be considered hyperspectral if those bands are narrow (e.g., 10 nm) and contiguous where there is useful content to be collected. On the other hand, if a sensor measures 20 wider spectral bands (e.g., 100 nm), or is separated by non-measured wavelength ranges, the sensor is no longer considered hyperspectral [1]. The detailed contiguous range of spectral bands of a hyperspectral sensor provides an ability to produce a contiguous "spectrum," which is one of the characteristics that distinguishes it from multispectral sensors.

Radiances measured by multispectral sensors are generally adequate for rough discrimination of surface cover into categories; however, they are rather limited in the amount of quantitative information that can be inferred from the spectral content of the data. The spectra, or *spectral reflectance curves*, from hyperspectral remote sensors provide much more detailed information about absorption regions of the surface of interest, very much like the spectra that would be measured in a spectroscopy laboratory. This unique characteristic of hyperspectral data is useful for a wide range of applications, such as mining, geology, forestry, agriculture, and environmental assessment.

This chapter is to focus on existing hyperspectral remote sensing systems, global change requirements, application examples, and challenges ahead.

3.2 HYPERSPECTRAL SENSORS AND CHARACTERISTICS

In the 1970s, space-based multispectral remote sensors were launched and produced images of the Earth's surface. Even with only a few broad wavelength bands, the images greatly improved our understanding of our planet's surface. The idea of developing hyperspectral, imaging sensors, also known as *imaging spectroscopy*, emerged in the early 1980s to improve our ability to better characterize the Earth's surface. The Airborne Visible/Infrared Imaging Spectrometer (AVIRIS) developed at the NASA Jet Propulsion Laboratory in California was the first spectrometer being used on moving platforms such as aircraft.

In the early imaging spectroscopy era, most of the hyperspectral sensors were mounted on aircraft (e.g., AVIRIS). After decades of research and development, hyperspectral technology was expanded to space-based remote sensing systems and several satellite hyperspectral sensors were proposed and subsequently launched. The very first spaceborne hyperspectral sensors were NASA's Hyperion sensor on the Earth Observing-1 (EO-1) satellite, and the U.S. Air Force Research Lab's Fourier Transform Hyperspectral Imager (FTHSI) on the MightySat II satellite. With more satellite-based sensors being planned, more hyperspectral imagery will be available to provide near global coverage at regular repeated cycles [1] suitable for global change studies.

3.2.1 Spaceborne Systems

Spaceborne hyperspectral sensors aboard satellites may provide continuous acquisition of the Earth's surface images at lower relative cost. However, wide spatial coverage by a hyperspectral sensor is often compromised with its spatial resolution or ground sampling interval and other challenges. Consequently, repeated hyperspectral images are not widely available. Table 3.1 lists currently available spaceborne hyperspectral instruments, with a wide range in the number of spectral bands, spectral range, and swath width.

The FTHSI program initiated in 1995, was successfully launched in July 2000, maybe the only Department of Defense (DoD) space-based hyperspectral imager to discern spectrally unique objects with the Fourier transform technique. The width of the image footprint was 13 km with a spatial resolution 30 m, covering a spectral range from 450 to 1050 nm by 256 spectral bands (Table 3.1).

In November 2000, NASA launched the Hyperion sensor onboard the EO-1 satellite as part of a 1-year technology validation/demonstration mission. The Hyperion imaging spectrometer had a 30 m spatial resolution, 7.7 km swath width, and 10 nm contiguous spectral resolution (Table 3.1). With its high radiometric accuracy of 220 spectral bands, complex landscapes of the Earth's surface could be imaged and spectrally characterized.

The Compact High Resolution Imaging Spectrometer (CHRIS) was launched in October 2001 on board the PROBA platform (Table 3.1). The sensor was developed by in the United Kingdom with support from the British National Space Centre. The sensor was designed for the study of atmospheric aerosols, land surfaces, and coastal and inland waters with its 62 spectral bands ranging from 400 to 1050 nm and a spatial resolution of 17 m. Despite the fact that the mission was designed for a 1-year life span, the sensor was in operation for more than ten years [4].

The Global Imager (GLI) was part of the ADEOS II mission, an international satellite mission led by the Japanese Aerospace Exploration Agency (JAXA) with participations from the United States (NASA) and France (CNES—Centre Nationale d'Etudes Spatiales; Table 3.1). Its spectral range was from 250 to 1000 nm and its image size ws very large (1600 km). The GLI mission was to collect data to aid better understandings of water, energy, and carbon circulations in order to contribute to global environmental change studies. ADEOS II was launched on December 14, 2002, but unfortunately, the mission ended 10 months later, due to a failure of the solar panel on October 24, 2003.

The Ozone Monitoring Instrument (OMI) flown on the EOS Aura spacecraft was designed to measure atmospheric composition (Table 3.1). The sensor was launched on July 15, 2004, into an ascending node 705 km sun-synchronous polar orbit. The OMI was a nadir pushbroom hyperspectral imaging sensor that observed solar backscatter radiation in the UV and visible wavelengths (264–504 nm). It had 780 spectral bands with a swath large enough to provide global coverage in one day (14 orbits) at a spatial resolution of 13 × 24 km at nadir. The key air quality components include NO_2, SO_2, and aerosol characteristics, as well as ozone profiles.

The Hyperspectral Imager for the Coastal Ocean (HICO), the Navy's "Sea Strike" was designed and built by the Naval Research Laboratory (NRL) to be the first spaceborne imaging spectrometer optimized for scientific investigation of the coastal ocean and nearby land regions with high signal-to-noise ratios in the blue spectral region and full coverage of water-penetrating wavelengths (Table 3.1). Due to the fact that water absorbs most of the light in the electromagnetic spectrum, visible light is the only part of the spectrum that sufficiently penetrates the water column to sense the water and the bottom surface properties. The sensor was launched on the H-2 Transfer Vehicle (HTV) and was rendezvoused with the International Space Station (ISS) in September 2009. The sensor has been serving as a spaceborne hyperspectral method to detect submerged objects, to provide environmental data products to Naval forces, and to develop the coupled physical and bio-optical models of coastal ocean regions globally.

A few more sensors have been proposed for the near future launches, including PRecursore IperSpettrale della Missione Applicativa (PRISMA) under development by the Italian Space Agency (ASI). Listed in Table 3.1, PRISMA is a pushbroom hyperspectral sensor and is optimized to derive information about land cover, soil moisture and agricultural land uses, quality of inland waters, status of coastal zones, pollution, and the carbon cycle [5]. The hyperspectral instrument is to acquire images in 250 spectral bands at 30 m spatial resolution. When combined with a panchromatic camera, a higher spatial resolution (5 m) could be produced.

TABLE 3.1
Operational and Planned Satellite Hyperspectral Instruments

Instrument (Satellite)	Altitude (km)	Pixel Size (m)	Number of Bands	Spectral Range (nm)	Spectral Resolution (nm)	IFOV (μrad)	Swath (km)
Hyperion (EO-1)	705	30	220	400–2500	10	42.5	7.5
FTHSI (MightySat II)	575	30	256	450–1050	10–50	50	13
CHRIS (PROBA)	580	25	19	400–1050	1.25–11.0	43.1	17.5
OMI (AURA)	705	13,000	780	270–500	0.45–1.0	115	2600
HICO	~390	92	102	380–900	5.7	<20	42 × 192
COIS (NEMO)	605	30	210	400–2500	10	49.5	30
HIS (SIMSA)	523	25	220	430–2400	20	47.8	7.7
Warfighter-1 (OrbView-4)	470	8	200	450–2400	11	20	5
	470	8	80	3000–5000	25	20	5
EnMAP (Scheduled 2014)	650	30	94	420–1000	5–10	30	30
	650	30	155	900–2450	10–20	30	30
HypSEO (MITA)	450	20	210	400–2500	10	40	20
MSMI (SUNSAT)	660	15	200	400–2350	10	22	15
PRISMA	695	30	250	400–2500	10	43	30
Global Imager (ADEOS-2)	803	250–1000	36	380–1195	10–1000	310–1250	1600
WFIS	705	1400	630	360–1000	1–5	2000	2400

Environmental Mapping and Analysis Program (EnMAP) was a German pushbroom hyperspectral satellite (Table 3.1) with a pointing feature for fast target revisits (4 days), providing high-quality hyperspectral image data on a timely and frequent basis. Aboard EnMAP were Hyperspectral Imagers (HSI) sensors designed to derive surface physical parameters on a global scale, with accuracy not achievable by currently available spaceborne sensors. Data from the 249 spectral channels, with a 30 m pixel size, were assimilated in physically based ecosystem models, and ultimately provided information products on the status of various terrestrial ecosystems [6].

The wide field-of-view imaging spectrometer (WFIS) was a sensor that represented another aspect of imaging spectroscopy (Table 3.1), by providing a possibility for entire global surface study with frequent revisits like the Moderate Resolution Imaging Spectroradiometer (MODIS) aboard Terra and Aqua satellites. WFIS was a pushbroom sensor designed to operate in the visible and near-infrared spectral region (360–1000 nm) with an approximately 1-nm sampling interval [7].

3.2.2 Airborne Systems

Airborne hyperspectral imagery has become more accessible due to the increasing number of companies operating hyperspectral spectrometers. Similar to the scanning mechanism of a spaceborne hyperspectral sensor, an airborne sensor generates hundreds (often image columns) of individual pixels along the scan line direction (often perpendicular to the flight direction). As the aircraft moves forward, a new array of pixels is generated along the flight direction. As such, geometric quality of airborne images can be affected by environmental conditions such as wind and/or by flight operations such as aircraft speed and alignment. Therefore, airborne hyperspectral image processing/calibration may be complex and occasionally bring in errors to the analysis [8].

Airborne hyperspectral sensors are more flexible than those satellite-based sensors in terms of acquisition schedules adjustable to weather conditions, spectral and spatial resolution requirements, and flight line arrangements. However, airborne hyperspectral images can be costly for large area coverage, due to limited swath width and slower speed of the carrier as compared to spaceborne examples.

In further comparison with spaceborne hyperspectral sensors, the configuration of airborne systems varies widely in terms of spectral range, number of spectral bands, manufacturers, and spatial and temporal coverage. A survey of existing airborne hyperspectral sensors and their spectral characteristics is presented in Table 3.2. This is not an exhausted list but represents what is available for airborne systems at the time of publication.

3.2.3 Ground-Based Systems

Ground-based hyperspectral systems are currently available from a few commercial companies and their spectral characteristics are listed in Table 3.3. These systems can be mounted on low-elevation platforms such as trucks or be handheld, due to their light weight and small size. In general, the spectral range of these systems is from ultraviolet to middle infrared (200–2500 nm) with varying bandwidth or spectral resolution. The ground sampling interval or the footprint of these sensors varies depending on the height of the sensor and their total field-of-view (FOV). One of the advantages of the ground-based sensing systems is that they are flexible in deployment and can often be used for both field-based and in-lab spectral measurements. Specifications on the operating system requirements, software support, as well as data storage, also vary greatly from sensor to sensor, and from manufacturer to manufacturer.

TABLE 3.2
Current Airborne Hyperspectral Sensors and Data Providers

Airborne Sensors	Manufacturer	Number of Bands	Spectral Range (μm)
AISA EAGLE (Airborne Imaging Spectrometer)	Spectral Imaging	up to 488	0.40–0.97
AISA EAGLET (Airborne Imaging Spectrometer)	Spectral Imaging	up to 410	0.40–1.00
AISA HAWK (Airborne Imaging Spectrometer)	Spectral Imaging	254	0.97–2.50
AISA DUAL (Airborne Imaging Spectrometer)	Spectral Imaging	up to 500	0.40–2.50
AISA OWL (Airborne Imaging Spectrometer)	Spectral Imaging	up to 84	8.00–12.00
AVIRIS (Airborne Visible/Infrared Imaging Spectrometer)	NASA Jet Propulsion Lab	224	0.40–2.50
CASI-550 (Compact Airborne Spectrographic Imager)	ITRES Research	288	0.40–1.00
CASI-1500 Wide-Array (Compact Airborne Spectrographic Imager)	ITRES Research	288	0.38–1.05
SASI-600 (Compact Airborne Spectrographic Imager)	ITRES Research	100	0.95–2.45
MASI-600 (Compact Airborne Spectrographic Imager)	ITRES Research	64	3.00–5.00
TASI-600 (Compact Airborne Spectrographic Imager)	ITRES Research	32	8.00–11.5
DAIS 7915 (Digital Airborne Imaging Spectrometer)	GER Corporation	32	0.43–1.05
		8	1.50–1.80
		32	2.00–2.50
		1	3.00–5.00
		6	8.70–12.3
DAIS 21115 (Digital Airborne Imaging Spectrometer)	GER Corporation	76	0.40–1.00
		64	1.00–1.80
		64	2.00–2.50
		1	3.00–5.00
		6	8.00–12.0
EPS-H (Environmental Protection System)	GER Corporation	76	0.43–1.05
		32	1.50–1.80
		32	2.00–2.50
		12	8.00–12.5
HYDICE (Hyperspectral Digital Imagery Collection Experiment)	Naval Research Lab	210	0.40–2.50
HyMap	Analytical Imaging and Geophysics	32	0.45–0.89
		32	0.89–1.35
		32	1.40–1.80
		32	1.95–2.48
HySpex	Norsk Elektro Optikk	128 (VIS/NIR1)	0.40–1.00
		160 (VIS/NIR2)	0.40–1.00
		160 (SWIR1)	0.90–1.70
		256 (SWIR2)	1.30–2.50
PROBE-1	Earth Search Sciences Inc.	128	0.40–2.50

3.3 HYPERSPECTRAL REMOTE SENSING METHODS

Many of the methods developed for multispectral imagery analysis and processing can be adopted for hyperspectral images. However, the following are more specifically developed and optimized for hyperspectral analysis.

3.3.1 Support Vector Machines

Support Vector Machines (SVMs) are new methods that have been successfully used for hyperspectral data classification [9–13], and can efficiently work with large inputs, handle noise-attached data in

TABLE 3.3
Current Handheld Hyperspectral Sensors and Data Providers

Handheld Sensors	Manufacturer	Spectral Resolution	Spectral Range (μm)
FieldSpec 3 Hi-Res Portable Spectroradiometer	ASD Inc. (Analytical Spectral Devices)	3 nm @700 nm 8.5 nm @1400 nm 6.5 nm @2100 nm	0.35–2.50
FieldSpec 3 Max Portable Spectroradiometer	ASD Inc. (Analytical Spectral Devices)	3 nm @700 nm 10 nm @1400/2100 nm	0.35–2.50
Handheld 2 Portable Spectroradiometer	ASD Inc. (Analytical Spectral Devices)	<3 nm @700 nm	0.325–1.075
UV-VIS Spectrometers (USB4000-UV-VIS)	Ocean Optics Inc.	~1.5	0.20–0.85
VIS-NIR Spectrometers (USB4000-VIS-NIR)	Ocean Optics Inc.	~1.5	0.35–1.00
UV-NIR Spectrometers (HR4000CG)	Ocean Optics Inc.	0.75	0.20–1.10
UV-VIS Hyperspectral USB Spectrometer	Edmund Optics Inc.	1.5	0.20–0.72
VIS-NIR Hyperspectral USB Spectrometer	Edmund Optics Inc.	2	–1.05

a robust way, and produce sparse *or efficient* solutions [14,15]. SVMs are based on *kernel methods* that map data from the original input feature space to a kernel feature space of higher dimensionality and then solve for a linear problem solution in that space [15]. These methods allow an interpretation of learning algorithms geometrically in the kernel space (which is nonlinearly related to the input space), and thus combining statistics and geometry in an effective way [15] to take advantages of hyperspectral imagery.

3.3.2 Kernel Fisher Discriminant Analysis

The Kernel Fisher Discriminant (KFD) analysis is another new and effective method for hyperspectral data classification [16]. The KFD method adopts the same concept of kernel used in SVMs to obtain nonlinear solutions, however, KFD minimizes a different function than the SVMs do, and thus the solution is expressed in a different way for potentially more accurate classification.

3.3.3 Matched Filtering

Some hyperspectral applications are only focused on searching for the existence or fractional abundance of one or a few single target materials. Matched filtering (MF) is a type of unmixing procedure that identifies only targets of interest [17]. It is sometimes called a *Partial Unmixing* because spectra of all endmembers in an image are not required. Matched filtering was originally developed to compute abundances of targets of interest that were relatively rare in the image [1].

Matched filtering algorithms perform a mathematical transformation to maximize the contribution of the target spectrum while minimizing the background [1,18]. Therefore, the approaches perform best when target material is rare and does not contribute significantly to the background signature [1]. A modified version of matched filtering uses derivatives of the spectra to enhance the differentiation ability [18]. The output image presents the fraction of the pixel that contains the target material [1].

3.3.4 Libraries Matching Techniques

Spectral libraries are collections of reflectance spectra measured from materials of known composition, usually in the field or laboratory [1] that are highly desirable for hyperspectral image analysis. Laboratory-derived spectra may be found at the ASTER Spectral Library, which contains the NASA Jet Propulsion Laboratory Spectral Library and the U.S. Geological Survey Spectral

Library. Other publicly accessible reference spectral libraries are also available, such as those in digital image processing software (e.g., ENVI, ERDAS, PCI Geomatica) and other sources [19–25].

Absorption features of specific materials within a given IFOV are typically present in the spectra. These absorption features provide the information needed for the Earth's surface characterization based on the surface spectral absorption locations, relative depths, and widths [26,27]. Characterization and automatic detection of such absorption features on the basis of the spectral similarity between the pixel and target spectra is the fundamental principle of library matching techniques [26]. A measured spectrum may be divided into several spectral regions before absorption features in each of the regions are detected and matched with those from the spectral library. The algorithm assigns the pixel to the class that its spectrum most closely resembles [26]. Library matching techniques perform best when the scene includes extensive areas of pure materials that have corresponding reflectance spectra in the reference library.

Comparing the spectral properties in a hyperspectral image with the one stored in libraries can take a significant amount of time and computing resources. Therefore, coding techniques (e.g., *Binary Spectral Encoding*) are developed to represent a pixel spectrum, which has high degree of redundancy, in a simple and effective manner [27]. Library matching techniques may not perform well when mixtures of targets are present in one pixel, or some materials have very similar spectral characteristics. Sample mixed spectra can be included in the library to improve the accuracy; however, it is not likely that all possible mixtures (and all mixture proportions) can be included in the reference library [18].

3.3.5 Derivative Spectroscopy

Among the techniques developed in remote sensing analysis, derivative spectroscopy is particularly promising for use with hyperspectral image data. Differentiation of a spectral curve estimates the slope at each wavelength over the entire spectral range. A first-order derivative is the rate of change of the absorbance with respect to wavelength. Although differentiation of the spectra does not provide more information than the original spectra, however, it can emphasize the potentially unique target features while suppressing other unwanted information [26]. Second-order derivative spectra, which are insensitive to substrate reflectance, have been used to mitigate soil background influences in vegetation studies [28,29].

Second or higher derivatives are relatively insensitive to variations in illumination (due to cloud cover), solar angle variance, or topographic effects [26]. Although some of the researchers have used high-order derivatives, first- and second-order derivatives have been the most common [30,31]. Talsky [32] suggested that the signal-to-noise ratio (SNR) decreases as derivative order increases. Spectral derivatives have successfully been used in remote sensing applications for decades [28,33,34]. In addition, several studies used this method directly toward specific applications, such as water quality assessment [35–37].

3.3.6 Narrow Band Spectral Indices

Hyperspectral indices have been developed for quantification of biophysical parameters, based on specific absorption features that best describe the biophysical indicators. Examples of such indices are provided below [26].

3.3.6.1 Normalized Difference Vegetation Index: NDVI

The traditional normalized difference vegetation index or NDVI has been modified or computed with narrow spectral bands such as the one below to emphasize the sensitivity to green vegetation density.

$$\mathrm{NDVI}_{narrow\,band} = \frac{\rho(860\mathrm{nm}) - \rho(660\mathrm{nm})}{\rho(860\mathrm{nm}) + \rho(660\mathrm{nm})} \tag{3.1}$$

3.3.6.2 Yellowness Index: YI

The "yellowness" index (YI) is sensitive to decreased chlorophyll content or leaf chlorosis and, therefore, is an indicator of stresses in plant leaves. The YI measures the change in shape of reflectance spectra between the 550 nm (maximum green reflectance band) and the 650 nm (maximum red absorption band). The YI uses only wavelengths in the visible spectrum, a region that is relatively insensitive to change in leaf water content and structure [26,33,38].

3.3.6.3 Normalized Difference Water Index: NDWI

NDWI is used to determine vegetation liquid water content, and can be derived from narrow spectral bands that are sensitive to water content.

$$NDWI_{narrow\,band} = \frac{\rho(860\text{nm}) - \rho(1240\text{nm})}{\rho(860\text{nm}) + \rho(1240\text{nm})} \tag{3.2}$$

3.3.6.4 Red-Edge Position Determination: REP

The REP is defined as the point of maximum slope on a vegetation reflectance spectrum between the red and near-infrared wavelengths. The REP is strongly correlated with foliar chlorophyll content and, therefore, can be a sensitive indicator of vegetation stress [26]. A linear method, based on narrow spectral bands features, was proposed by Clevers [39] to highlight the red-edge changes in a given spectrum:

$$REP = 700 + 40\left[\frac{\rho(860\,\text{nm}) - \rho(1240\,\text{nm})}{\rho(860\,\text{nm}) + \rho(1240\,\text{nm})}\right] \tag{3.3}$$

where

$$\rho_{(red\ edge)} = \frac{\rho(670\,\text{nm}) + \rho(780\,\text{nm})}{2}$$

3.3.6.5 Crop Chlorophyll Content Prediction

This narrowband vegetation index, developed by Haboudane et al. [40], integrates the capabilities of indices that minimize soil background affects and indices that are sensitive to chlorophyll concentration. The commonly used indices include the Transformed Chlorophyll Absorption in Reflectance Index (TCARI) [41] and the Optimized Soil-Adjust Vegetation Index (OSAVI) [42].

$$\text{Crop Chlorophyll Content} = \frac{TCARI}{OSAVI}$$

$$TCARI = 3\left[(\rho_{700} - \rho_{670}) - 0.2(\rho_{700} - \rho_{550})\left(\frac{\rho_{700}}{\rho_{670}}\right)\right] \tag{3.4}$$

$$OSAVI = \frac{(1 + 0.16)(\rho_{800} - \rho_{670})}{(\rho_{800} + \rho_{670} + 0.16)}$$

3.3.7 Neural Network

The Neural Network (NN) is one of the promising feature selection methods. NNs are mathematical models that simulate brain dynamics [43] that are supposed to be quite powerful in remote sensing imagery analysis, especially in image classification, due to their nonlinear properties. It should be noted that NN is highly sensitive to the Hughes phenomenon (the curse of dimensionality), which is particularly a problem for hyperspectral images, and may not work effectively when dealing with a high number of spectral bands [15]. Moreover, the use of NN for hyperspectral image classification has been limited primarily because of the lengthy computational time required for the training process. Nonetheless, several researches have successfully used NN algorithms to estimate vegetation types and biophysical parameters, such as in coastal and ocean waters [44–46].

3.4 GLOBAL CHANGE REQUIREMENTS AND APPLICATIONS

Hyperspectral imagery has been used to assess, analyze, detect, and monitor several key global environmental change variables. For example, hyperspectral data have been used to estimate sediments, chlorophyll *a*, and algal type information in oceans and inland waters [37,47–50], identify vegetation species [51], study plant canopy chemistry [52–56], detect vegetation stress [57,58], monitor biogeochemical and greenhouse gas cycles [59–63], improve land cover classification accuracy, and more details in plant species recognition [64]. Geologists also use imaging spectroscopy for soil organic matter estimation, salinity and moisture content detection, and mineral mapping [51,65].

3.4.1 Global Change Requirements

Because the Earth is a dynamic system, sufficient understanding of the complex interactions among physical and ecological processes is needed for global change studies. To achieve this, both long- and short-term observations are required to quantify, analyze, and subsequently understand the spatial and temporal variability, trend, and magnitudes of changes in ecosystems dynamics. Multispectral remote sensing systems such as NASA's Landsat and NOAA's AVHRR sensors have been instrumental and inspirational in providing global coverage for systematic analysis of the Earth's dynamics [61,63]. However, the spectral characteristics and the sensor design of these systems limit their applications in the areas that require specific and more accurate assessment of, for example, nutrient deficiencies in plants, algal information of lakes and streams, invasive species identification and detection [66], soil composition, and specific atmospheric gas concentrations. For example, reflectance spectra of agro-ecosystems and the seasonal changes of the spectra in a paddy rice field, shown in Figure 3.1, are much better spectrally characterized by hyperspectral signatures than by that of multispectral data [67]. Broadband spectral signatures would not be able to detect such subtle changes, but hyperspectral measurements enable the detection and quantification of plant, soil, and ecosystem variables due to their high spectral resolution and continuity. These requirements lead to the exploration of hyperspectral sensing systems where spectral information is much richer for enhanced and new applications in global change studies than multispectral data.

3.4.2 Global Change Applications

Hyperspectral data have been used in numerous applications to specifically address global environmental issues. The following are not meant to be an exhausted application list; rather they are examples demonstrating the type of issues one can address with hyperspectral data.

3.4.2.1 Water Quantity and Quality

Imaging spectrometry is a cost-effective technique for water quality studies over large areas of aquatic systems, such as lakes, coastal areas, bays, estuaries, and even oceans. For example, spaceborne and airborne hyperspectral images have been used to assess the trophic status of lakes and to map the spatial distribution of water quality parameters, such as temperature, chlorophyll *a*, turbidity, and total suspended solids, over large areas [37,47,68–70]. These studies are possible because of narrow, unique spectral absorption features of water bodies that are only detectable by hyperspectral data [37]. Owing to the unique spectral signatures of algal pigments, compositions of algal populations in

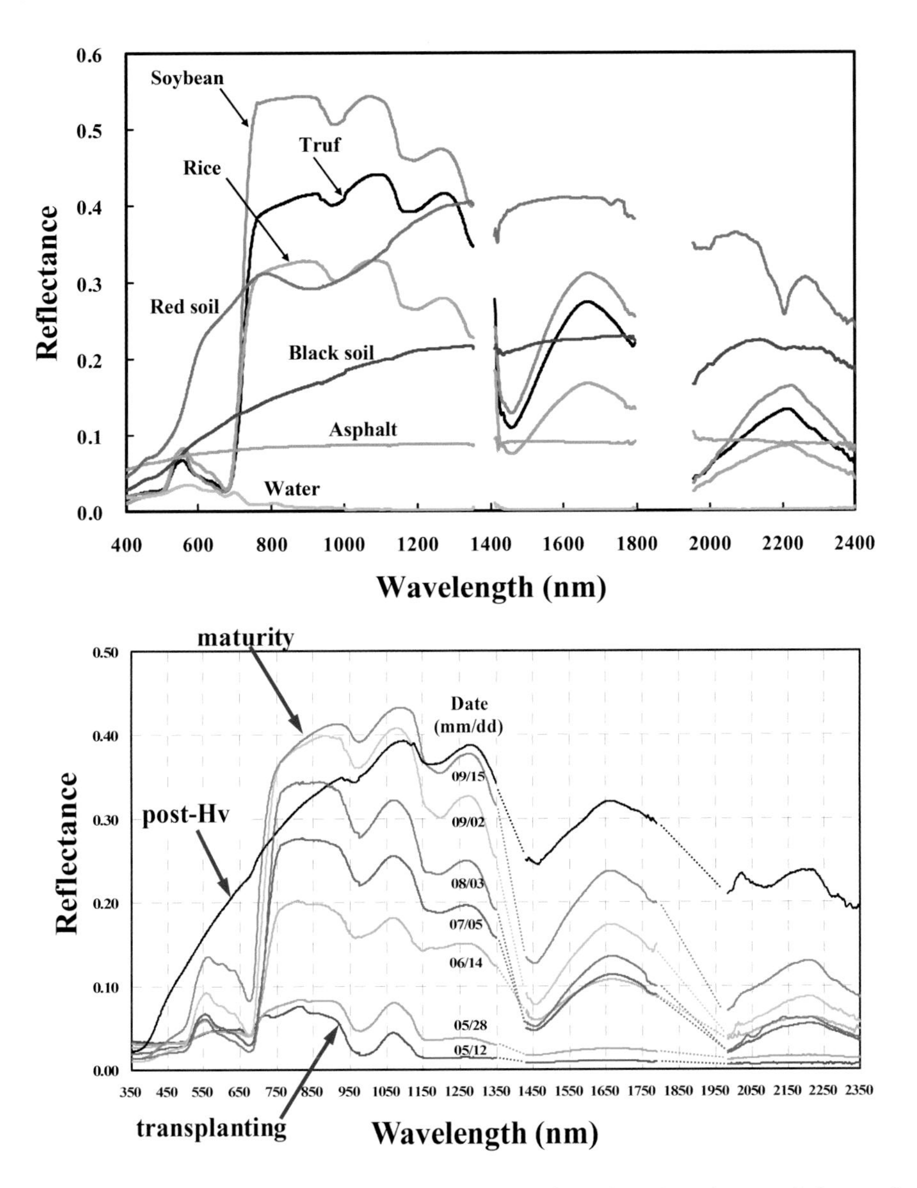

FIGURE 3.1 Typical reflectance spectra in agro-ecosystem surfaces (upper), and seasonal change of spectra in a paddy rice field (lower).

aquatic systems can be detected by analyzing absorption properties in the region between 400 and 700 nm [49]. Various methods are effective in mapping water quality, including pigment-specific absorption algorithms, spectral angle mapping algorithms, spectral libraries comparison, principle component analysis, derivative spectroscopy, regression techniques, and other spectral indicators like band ratios [30,31,37,71,72].

3.4.2.2 Carbon Sequestration and Fluxes

Carbon sequestration is a process of removing carbon from the atmosphere and depositing it in a reservoir or carbon sink. The main natural sinks are photosynthesis by terrestrial plants, and physicochemical and biological absorption of carbon dioxide by the oceans. The oceans are the largest active carbon sinks on Earth, absorbing more than a quarter of the carbon dioxide released from human activities [4].

Hyperspectral data proved to be useful to estimate and map primary production in the oceans and other open surface waters [48–50] while multispectral data are equally suitable for terrestrial primary production estimation. The use of hyperspectral remote sensing to measure chlorophyll *a* from space has been a highly successful technique for mapping phytoplankton distribution on a global basis, which could be used to estimate the amount of carbon sequestration in the oceans.

3.4.2.3 Greenhouse Gas Emissions

Greenhouse gases, as defined by UNFCCC, are "the atmospheric gases responsible for causing global warming and climate change. The major greenhouse gases are carbon dioxide (CO_2), methane (CH_4) and nitrous oxide (N_2O). Less prevalent—but very powerful—greenhouse gases are hydrofluorocarbons (HFCs), perfluorocarbons (PFCs) and sulphur hexafluoride (SF_6)." Hyperspectral remote sensing has shown to be successful in detecting methane gas concentrations [59,60,62], despite some technical challenges. Although methane is transparent in the visible part of the electromagnetic spectrum, it does contain a number of spectral features at longer wavelengths that can be detected. Three significant absorption features, at 3.31, 3.28, and 3.21 μm, were detected in the methane spectrum [59]. Some of the spectral features may appear to be obscured by regions of water vapor absorption; however, the obscured spectral features are still detectable at high methane concentrations [73]. The methane absorption features at 0.88 and 7.7 μm have also been proposed for atmospheric studies in the presence of significant water vapor content [73]. Once the spectral bands associated with absorption features are determined, band ratio techniques could be used develop indicators of methane concentration.

3.4.2.4 Atmospheric Chemistry

Although multispectral remote sensing has been used for atmospheric monitoring, hyperspectral imagery proved to be more suitable due to its fine spectral resolution and sampling intervals. Space measurements of the ozone column have been conducted since the 1970s with the series of Solar Backscatter Ultraviolet (SBUV) and Total Ozone Mapping Spectrometer (TOMS) sensors [74]. These spectrometers have been used to monitor the ozone layer, measure stratospheric dynamics, and detect tropospheric ozone pollution on a regional scale [75]. The SBUV and TOMS, with 1-nm spectral resolution, measure backscattered radiance at two wavelengths and estimate the ozone column [76,77]. In 1995, the Global Ozone Monitoring Experiment (GOME) was launched and operated as the first space instruments that measured the ultraviolet and visible part of the spectrum with a high spectral resolution [75]. In 2002, the Scanning Imaging Absorption Spectrometer for Atmospheric Cartography (SCIAMACHY) was launched onboard the Environmental Satellite (ENVISAT) to facilitate the measurement of atmospheric absorptions from the ultraviolet to the near-infrared spectral range (240–2380 nm), providing knowledge about the composition, dynamics, and radiation balance of the atmosphere.

The OMI hyperspectral sensor aboard the EOS Aura satellite was successfully launched in 2004, as a continuation of the long-term global total ozone monitoring from satellite measurements that began in 1970 with SBUV and TOMS. Data from OMI were used to derive ozone columns using Differential Optical Absorption Spectroscopy (DOAS) [75,78–80]. In comparison, TOMS estimated the spatial distribution of the total ozone by observing changes of solar radiation in several near UV spectral bands. DOAS derived the ozone column by fitting a reference ozone absorption cross section to the measured sun-normalized radiance. The main advantages of DOAS compared to the original SBUV/TOMS techniques are that DOAS was less sensitive to changes in variability of the periodic radiometric calibration of the instrument and less sensitive to disturbing factors like absorbing aerosols [75].

OMI can distinguish between aerosol types, such as smoke, dust, and sulfates, and can measure cloud pressure and coverage. Other instruments on the Aura satellite, such as the Tropospheric Emission Spectrometer (TES) may provide global measurements of tropospheric ozone and its photochemical precursors. With the entire spectrum from 3.2 to 15.4 μm at high spectral resolution, many other gases such as carbon monoxide, ammonia, and organics can be retrieved [81]. Aura satellite is providing the next level of atmospheric measurements in the stratospheric and tropospheric layers in order to understand the recovering process of the stratospheric ozone layer, the composition of chemistry in the troposphere, and the roles of upper tropospheric aerosols, water vapor, and ozone in climate change. The four instruments on Aura provide valuable data to global change studies as well as continuing important atmospheric composition monitoring that began earlier with other satellites such as TOMS [81,82,110].

3.4.2.5 Vegetation Ecology

Vegetation is a key attribute of land-use and land cover change dynamics not only for its role in food production but also its role in land-atmosphere interactions. The exchange of biochemicals between terrestrial ecosystems and the atmosphere could be determined by the properties of vegetation. Therefore, an accurate assessment of vegetation properties and temporal dynamics is very important in Earth system science [83]. Due to the extended spectral dimension, hyperspectral remote sensing has improved modeling, monitoring, and understanding of vegetation canopies [52,67] and enhanced quantitative mapping of key vegetation properties [54–56]. For example, data from hyperspectral sensors have been used to derive plant species composition, and biological and biochemical properties of the forests [54,111], such as chlorophyll and nitrogen concentrations. These studies suggest that hyperspectral remote sensing brings new capabilities to estimate vegetation properties that otherwise would not be possible with traditional multispectral sensors. Still, interest has increased in using hyperspectral remote sensing for biodiversity monitoring [84,85] and invasive species mapping [66].

3.4.2.6 Vegetation Biochemical Properties

Foliar chemistry measurements of plant canopies allows a better understanding of ecosystem function and service, since many biochemical processes, such as photosynthesis, respiration, and litter decomposition, are related to the foliar chemistry of plants. Important chemical components of vegetation foliage that could be used as bio-indicators are the concentration of nitrogen and carbon, and the content of water [86]. For example, hyperspectral remote sensing combined with canopy radiative transfer models provided consistent and accurate information of these chemical compositions [54]. Forest biomass and aboveground carbon stocks have been estimated using hyperspectral imagery from AVIRIS using partial least squares (PLS) regressions[87].

Plant pigments, such as chlorophyll and carotenoids, have specific absorption spectra, which play essential roles in the photochemical cycle in plant leaves. These specific absorption features are only detectable remotely with hyperspectral sensors, as demonstrated by various studies. For example, Inada [88] showed that the narrow spectral band ratio of 800 nm/550 nm was the most effective index for estimating the leaf chlorophyll content of rice. The reflectance at 675 and 550 nm has been used to determine chlorophyll content as well as for nitrogen stress detection.

The total nitrogen content of a canopy can be estimated using narrowband width measurements in visible and near-infrared wavelengths such as 480, 620, and 840 nm [89], and the estimate accuracy can be further improved by using sharp absorption features in the shortwave infrared wavelengths, 1650 and 2200 nm [90].

3.4.2.7 Invasive Plant Species Detection

Species invasion is recognized as a significant threat to global biodiversity and ecosystem health. In some cases, invasive species could irreversibly change the structure and functioning of entire ecosystems and result in biological diversity loss [66,91]. Ecosystem research suggests that invasive and aggressive plant species may be the result of general ecosystem stress related to changes in the frequency of landscape disturbance, such as road construction, deforestation, land-use conversion to agricultural or urban development, or other hydrologic alterations [91,92]. Identification of the extent of landscapes being stressed by invasive plant species using spectral signatures can help target vulnerable areas in need of restoration or protection [91,111]. Contrary to multispectral remote sensing, which can only detect invasions when the effects have spread out in a wide area, hyperspectral imagery offers a unique potential for analyzing the signals of ecological changes at an early stage and, therefore, an indication of biological invasion and biogeochemical change [66,92]. For example, non-native species ice plant, jubata grass, fennel, and giant reed grass (*Phragmites* ssp.) from a range of habitats in California were mapped with AVIRIS imagery at relatively high accuracy [93] and using CASI for reed grasses in the Laurentian Great Lakes [91,101], and pure pixels of Brazilian pepper was detected with hyperspectral imagery [94]. It was also demonstrated that an early detection of invasive weeds (spotted knapweed and baby's breath) was also possible [95] using hyperspectral sensors.

3.4.2.8 Vegetation Health

Information about vegetation health, such as disease, fire disturbance, and insect attack, is crucial in ecosystem protection and management and hyperspectral remote sensing can provide diagnostic indicators for early detection. For example, Lawrence and Labus [96] successfully used high spatial resolution imagery to identify different levels of tree stress resulting from Douglas-fir beetle attack. Koetz et al. [97] also mapped spatially distributed fuel moisture content and fuel properties with inversion of radiative transfer models to serve as input for forest fire spread and mitigation models. Diagnostic analysis of specific disease, however, has been a challenging issue with hyperspectral remote sensing, as sensors only "see" plant symptoms rather than the causes. Discrimination of diseases may be possible with knowledge of the physiological effect of the disease on leaf and canopy elements. For example, necrotic diseases can cause a darkening of leaves in the visible spectrum and a cell collapse that would decrease near-infrared reflectance. Chlorosis induced diseases (mildews and some viruses), for example, cause marked changes in the visible reflectance (similar to nitrogen deficiency). Other diseases, however, may be detected by their effects on canopy geometry.

3.5 HYPERSPECTRAL REMOTE SENSING CHALLENGES

There are numerous challenges facing hyperspectral remote sensing, ranging from system design and data processing to methodological developments. Because little is available from literature, the following encompass general issues that can be addressed for broader and improved applications of hyperspectral remote sensing imagery.

3.5.1 System Design Challenges

From global change perspective, the design of a hyperspectral sensor entails the configuration of the following key parameters and requirements: spectral regions, number of spectral bands, spectral bandwidth, spatial resolution, swath width, revisit cycle, signal-to-noise ratios, onboard storage, and data downlinks to ground stations. These system parameters will determine the appropriateness of a specific global change application.

There is a challenge in balancing the spectral bandwidth and signal-to-noise ratio. Existing hyperspectral sensors operate in the spectral region from approximately 200–2500 nm. The spectral bandwidth widens from ultraviolet to shortwave infrared, in order to maintain an acceptable signal-to-noise ratio in the longer wavelength region. This widening presents a challenge in methodological development of hyperspectral remote sensing as significant sharp absorption features will likely not be detectable with wider spectral bandwidth.

The signal-to-noise ratio needs to be balanced with spatial resolution requirements as well. As global change studies increasingly require quantitative information about the Earth's surface properties, there is a need to acquire hyperspectral images at high spatial resolution and over large coverage areas. High spatial resolution requires small instantaneous field of view (IFOV) of the sensor, but smaller IFOV results in lower signal-to-noise ratio and compromises the sensor's ability to have large geographic area coverage.

There is also a conflict in the number of spectral bands, spatial resolution, swath width, and onboard data storage. The requirements for a large number of contiguous spectral bands and high spatial resolution of large geographic coverage undoubtedly increases the data volume, which presents challenges for both onboard storage capacity and the time required for downlinking to ground stations. For global change studies, the priority ought to be on the geographic extent of coverage as this allows a broader and diverse ecosystem analysis and at the same time increases global access to hyperspectral imagery.

3.5.2 Processing and Visualization Challenges

For a given geographical area imaged, the data can be viewed as a two-dimensional image that represents spatial location and spectral information. Displaying hyperspectral data is more challenging than it is for multispectral data. Hyperspectral images contain far more spectral bands than can be displayed with a standard red-green-blue (RGB) display. A convenient visualization approach is to reduce the dimensionality of the image (from tens to hundreds of dimensions) to three dimensions at the expense of information losses [98,99]. The optimal hyperspectral display methods for quantitative and qualitative analysis of the data should enhance natural colors, preserve natural edges or contours of the features, highlight target features of interest, and enable simple and quick computational processing [99].

Several different techniques have been proposed and implemented for useful dimensionality reduction of hyperspectral images. Color matching functions (CMF) are one of these methods that specify how much of each of three primary colors must be mixed to create the "color sensation" in the form of monochromatic light at a particular wavelength in support of human image interpretation [98]. The technique linearly projects hyperspectral data in the visible range onto the color matching functions to determine the amount of the three primary colors that would create the same color sensation as viewing the original spectrum. It creates consistent images where hue, brightness, and saturation have interpretable and relevant meaning [98]. A disadvantage of the CMF is that there might be a decrease in sensitivity of human vision at the edges of the visible spectrum [98].

Principal component analysis (PCA) is also used to reduce hyperspectral data dimensionality by assigning the first three principal components to RGB [100,101]. Recent work found that the use of wavelets reduced noise in spectra before applying PCA could improve visualization [102]. Disadvantages of PCA include the difficulty to interpret the displayed image because the displayed colors represent principal components that do not typically represent natural colors of the features. The colors change drastically, depending on the data, and they do not correlate strongly with data variation. The standard saturation used in PCA display leads to simultaneous contrast problems and the computational complexity is high [98].

A number of linear methods have been used to optimize the hyperspectral imagery display [72,99,101,103–105]. Jacobson and Gupta [105] used fixed linear spectral weighting envelopes to create natural looking palettes while other information could still easily be added using highlight colors. The method maximized usefulness for human analysis while maintaining the natural look of the imagery [98]. Another data dimensionality reduction method is artificial neural networks (ANN). After the neural network training process, images can be processed very quickly, making it reasonable to use this approach for real-time analysis. However, a disadvantage of ANN is that it is unclear how the neurons handle new spectral inputs that were not in the training dataset.

3.5.3 Data Volumes and Redundancy

Hyperspectral images are composed of a large number of spectral bands in order to generate fine enough spectral resolution needed to characterize the spectral properties of surface materials. As a result, the volume of data in a single scene can be overwhelming. Although hyperspectral imagery provides the potential for more accurate and detailed information extraction than possible with other types of remotely sensed data, it can be spectrally over-determined or over-calculated. A tremendous amount of the data in a scene are redundant and much of the additional data do not add to the inherent information content for a particular application [27]. Spectral redundancy means that the information content of one spectral band can be fully or partly predicted from other bands within the scene [27]. The adjacent spectral bands are often found to be highly correlated to one another and their reflectance values, therefore, appear nearly identical. One way to identify spectral redundancy is by computing the correlation matrix for the image where high correlation values between bands indicate high degrees of redundancy or dimensionality [27]. This is another manifestation of inter-band redundancy known as the Hughes Phenomenon.

The greater the number of bands in an image, the more storage and processing time is required for the analysis. Therefore, developing effective tools and approaches to reduce the dimensionality of hyperspectral data, while retaining the information content in the imagery, remains a challenge. When analyzing a hyperspectral image, the focus has been on extracting spectral information within individual pixels, rather than spatial variations within each band. The traditional statistical classification methods that have been developed and used for multispectral image analysis may not be suitable for hyperspectral images unless they are modified to account for the high dimensionality nature of the hyperspectral data.

3.5.4 Radiometric Calibration

One of the most critical steps in hyperspectral data analysis is to convert the measured radiance data to surface reflectance so that individual spectra can be compared directly with laboratory or field data for appropriate interpretation [18,26,106]. A comprehensive conversion method must account for the solar irradiance spectrum, lighting effects due to solar angle and topography, atmospheric transmission, sensor gain and offset, and path radiance due to atmospheric scattering.

Because hyperspectral sensors acquire data at near continuous wavelengths, atmospheric correction should be taken into account the atmospheric absorption properties as shown in Figure 3.2. These absorption regions are dominated by water vapor (1.4 and 1.9 μm) with smaller contributions from carbon dioxide (CO_2), ozone (O_3), and other gases [18,107]. For example, narrow atmospheric water absorption bands in the visible and near-infrared spectrum are used and include bands at 0.69, 0.72, and 0.76 μm, an oxygen (O_2) absorption band at 0.76 μm, and carbon dioxide (CO_2) absorption bands in the shortwave infrared region at 2.005 and 2.055 μm and all have been used in atmospheric correction algorithms [106,108].

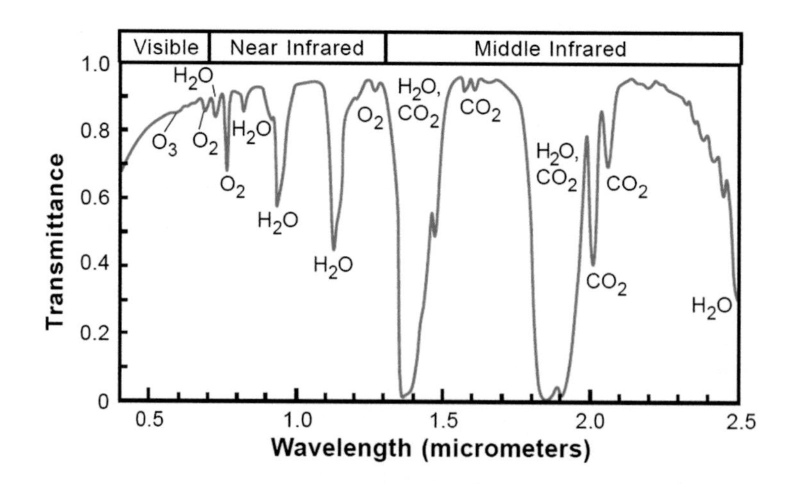

FIGURE 3.2 Plot of atmospheric transmittance versus wavelength for typical atmospheric conditions.

3.5.5 Methodological Challenges

Hyperspectral images provide rich information about the Earth's surface and therefore are desirable for global change studies. However, several issues should be considered in analysis and interpretation of such data. The large number of spectral bands in hyperspectral imagery and the small number of known target spectra in most image scenes create the problem known as the curse of dimensionality. The use of traditional image classification methods developed for multispectral analysis, such as the Maximum Likelihood Classifier (MLC) and Multiple Linear Regression (MLR or ordinary least squares OLS), without a modification to account for the high dimensionality of the hyperspectral data usually result in low efficiency and accuracy of the classification process. The MLR method assumes no intercorrelation between the independent variables, and the number of samples (endmembers) should be larger than the number of independent variables (spectral bands). Therefore, if independent variables (spectral bands) have significant correlations among each other, which are common for hyperspectral data, the MLR technique will be subjected to a multicollinearity issue [109]. For hyperspectral data that normally have a tremendous number of bands, it would be very difficult to have an adequate number of endmembers to make MLR work effectively. A solution is to engage in feature selection to reduce the dimensionality of the dataset and remove redundant spectral bands and arrived at a dataset with enough bands to address the application but not overwhelm the system with redundancy as discussed in Section 3.5.3.

3.6 DISCUSSION AND FUTURE DIRECTIONS

Accurate and timely information about land cover dynamics is essential for global change studies and hyperspectral data, either from ground-based, airborne, and/or spaceborne systems, has proven to have a greater potential for detailed information extraction than can be achieved from multispectral imagery. Significant progress has been made in the development of new sensors, new technologies for data processing, new methods for analysis, and new models for enhanced information extraction from hyperspectral data over the past decade. However, challenges exist in data access, data storage, data visualization, and analytical methodologies. These challenges are to be addressed by continued research effort and new technology inventions in such areas as sensor design and data compression or data storage or data compression.

Future research is likely to continue to be in the area of methodological developments with a focus on information extraction algorithms from hyperspectral images. At the same time, one would see an increase in hyperspectral image availability for research development, as more and more agencies are planning to launch hyperspectral sensors. Continued progress in sensor design, data availability, and new analytical methods will further promote broader hyperspectral applications in global change studies.

REFERENCES

1. Shippert, P., Introduction to hyperspectral image analysis, *Online Journal of Space Communication, Issue No. 3: Remote Sensing of Earth via Satellite*, 2003.
2. Birk, R.J. and McCord, T.B., Airborne hyperspectral sensor systems, *Aerospace and Electronic Systems Magazine*, 9(10):26–33, 1994.
3. Goetz, A.F.H. et al., Imaging spectrometry for earth remote sensing, *Science*, 228:1147–1153, 1985.
4. Earth Institute News, Oceans' uptake of manmade carbon may be slowing, *Earth Institute News*, The Earth Institute at Columbia University, Columbia University, http://www.earth.columbia.edu/articles/view/2586, 2009.
5. Labate, D. et al., The PRISMA payload optomechanical design, a high performance instrument for a new hyperspectral mission, *Acta Astronautica*, 65(9–10):1429–1436, 2009.
6. Stuffler, T. et al., The EnMAP hyperspectral imager—An advanced optical payload for future applications in Earth observation programmes, *Acta Astronautica*, 61(1–6):115–120, 2007.

7. Haring, R.E. et al., WFIS: A wide field-of-view imaging spectrometer, *Proceeding SPIE*, Vol. 3759, 1999.
8. Cetin, H., Comparison of spaceborne and airborne hyperspectral imaging systems for environmental mapping, *Proceeding of ISPRS Congress Istanbul*, 2004.
9. Gualtieri, J.A. and Cromp, R.F., Support vector machines for hyperspectral remote sensing classification, *Proceedings SPIE of 27th AIPR Workshop Advance in Computer Assisted Recognition*, Vol. 3584:221–232, 1998.
10. Gualtieri, J.A. et al., Support vector machine classifiers as applied to AVIRIS data, *Summaries of the Eighth JPL Airborne Earth Science Workshop*, JPL Publication, 99–17:217–227, 1999.
11. Huang, C., Davis, L.S., and Townshend, J.R.G., An assessment of support vector machines for land cover classification, *International Journal of Remote Sensing*, 23(4):725–749, 2002.
12. Camps-Valls, G. et al., Robust support vector method for hyperspectral data classification and knowledge discovery, *IEEE Transactions on Geoscience and Remote Sensing*, 42(7):1530–1542, 2004.
13. Melgani, F. and Bruzzone, L., Classification of hyperspectral remote sensing images with support vector machines, *IEEE Transactions on Geoscience and Remote Sensing*, 42(8):1778–1790, 2004.
14. Cristianini, N. and Shawe-Taylor, J., *An Introduction to Support Vector Machines*, Cambridge University Press, UK, 2000.
15. Camps-Valls, G., Kernel-based methods for hyperspectral image classification, *IEEE Transactions on Geoscience and Remote Sensing*, 43(6):1351–1362, 2005.
16. Mika, S. et al., Invariant feature extraction and classification in kernel spaces, *Advances in Neural Information Processing Systems*, 12, MIT Press, Cambridge, MA, 1999.
17. Boardman, J.W., Kruse, F.A., and Green, R.O., Mapping target signatures via partial unmixing of AVIRIS data, *Summaries of the Fifth JPL Airborne Earth Science Workshop*, 95(1):23–26, 1995.
18. Smith, R.B., *Introduction to Hyperspectral Imaging*, MicroImages, Inc., 1–24, 2006.
19. Elvidge, C.D., Visible and infrared reflectance characteristics of dry plant materials, *International Journal of Remote Sensing*, 11(10):1775–1795, 1990.
20. Salisbury, J.W., D'Aria, D.M., and Jarosevich, E., Midinfrared (2.5–13.5 micrometers) reflectance spectra of powdered stony meteorites, *International Journal of Solar System Studies (Icarus)*, 92:280–297, 1991.
21. Salisbury, J.W. et al., *Infrared (2.1–25 micrometers) Spectra of Minerals*, John Hopkins University Mineral Library, Johns Hopkins University Press, p. 294, 1991.
22. Grove, C.I., Hook, S.J., and Paylor, E.D., *Laboratory Reflectance Spectra for 160 Minerals 0.4–2.5 Micrometers*, JPL Publication, 1992.
23. Clark, R.N. et al., The U. S. Geological Survey, digital spectral library, version 1: 0.2 to 3.0 microns, *U.S. Geological Survey Open File Report* 93-592, p. 1340, 1993.
24. Salisbury, J.W., Wald, A., and D'Aria, D.M., Thermal-infrared remote sensing and Kirchhoff's law 1. Laboratory measurements, *Journal of Geophysical Research*, 99:11897–11911, 1994.
25. Korb, A.R. et al., Portable FTIR spectrometer for field measurements of radiance and emissivity, *Applied Optics*, 35:1679–1692, 1996.
26. Jensen, J.R., *Introductory Digital Image Processing: A Remote Sensing Perspective*, Pearson Prentice Hall, Upper Saddle River, New Jersey, p. 526, 2005.
27. Richards, J.A. and Jia, X., *Remote Sensing Digital Image Analysis: An Introduction*, Fourth Edition, Springer, p. 439, 2006.
28. Demetriades-Shah, T.H., Steven, M.D., and Clark, J.A., High resolution derivative spectra in remote sensing, *Remote Sensing of Environment*, 33:55–64, 1990.
29. Li, Y. et al., Use of second derivatives of canopy reflectance for monitoring prairie vegetation over different soil backgrounds, *Remote Sensing of Environment*, 44:81–87, 1993.
30. Butler, W.L. and Hopkins, D.W., Higher derivative analysis of complex absorption spectra, *Photochemistry and Photobiology*, 12:439–450, 1970.
31. Fell, A.F. and Smith, G., Higher derivative methods in ultraviolet, visible and infrared spectrophotometry, *The Analytical Proceedings*, 19:28–32, 1982.
32. Talsky, G., *Derivative Spectrophotometry: Low and Higher Order*, VCH Publishers, New York, p. 228, 1994.
33. Philpot, W.D., The derivative ratio algorithm: Avoiding atmospheric effects in remote sensing, *IEEE Transactions on Geoscience and Remote Sensing*, 29(3):350–357, 1991.
34. Penuelas, J. et al., Reflectance indices associated with physiological changes in nitrogen- and water-limited sunflower leaves, *Remote Sensing of Environment*, 48:135–146, 1994.
35. Dick, K. and Miller, J.R., Derivative analysis applied to high resolution optical spectra of freshwater lakes, *Proceedings of 14th Canadian Symposium on Remote Sensing*, Calgary, Alberta, California, 1991.

36. Chen, Z., Curran, P.J., and Hansom, J.D., Derivative reflectance spectroscopy to estimate suspended sediment concentration, *Remote Sensing of Environment*, 40:67–77, 1992.
37. Wiangwang, N., Hyperspectral data modeling for water quality studies in Michigan's inland lakes, *PhD dissertation*, Department of Geography, Michigan State University, p. 243, 2006.
38. Adams, M.L., Philpot, W.D., and Norvell, W.A., Yellowness index: An application of the spectral second derivative to estimate chlorosis of leaves in stresses vegetation, *International Journal of Remote Sensing*, 20(18):3663–3675, 1999.
39. Clevers, J.G., Imaging spectrometry in agriculture: Plant vitality and yield indicators, *Imaging Spectrometry: A Tool for Environmental Observations*, Kluwer Academic, Netherlands, 193–219, 1994.
40. Haboudane, D. et al., Integrated narrow-band vegetation indices for prediction of crop chlorophyll content for application to precision agriculture, *Remote Sensing of Environment*, 81(2–3):416–426, 2002.
41. Daughtry, C.S.T. et al., Estimating corn leaf chlorophyll concentration from leaf and canopy reflectance, *Remote Sensing of Environment*, 74(2):229–239, 2000.
42. Rondeaux, G., Steven, M., and Baret, F., Optimization of soil-adjusted vegetation indices, *Remote Sensing of Environment*, 55:95–107, 1996.
43. Demouth, H. and Beale, M., *Neural Network Toolbox User's Guide Version 4*, Natick, Massachusetts, 840 p. 2003.
44. Cipollini, P. et al., Retrieval of sea water optically active parameters from hyperspectral data by means of generalized radial basis function neural networks, *IEEE Transactions on Geoscience and Remote Sensing*, 39(7):1508–1524, 2001.
45. Zhang, Y. et al., Application of an empirical neural network to surface water quality estimation in the Gulf of Finland using combined optical data and microwave data, *Remote Sensing of Environment*, 81(2–3):327–336, 2002.
46. Zhang, Y. et al., Application of empirical neural networks to chlorophyll-*a* estimation in coastal waters using remote optosensors, *IEEE Sensors Journal*, 3(4):376–382, 2003.
47. Malthus, T.J. et al., An evaluation of the airborne thematic mapper sensor for monitoring inland water quality, *Proceeding of the 22nd Annual Conference of the Remote Sensing Society*, University of Durham, 317–342, 1996.
48. Mumby, P.J. et al., A bird's-eye view of the health of coral reefs, *Nature*, 413:36–37, 2001.
49. Richardson, L.L., Hyperspectral imaging sensors and the marine coastal zone, *Hyperspectral Remote Sensing of the Ocean (Proceedings Volume)*, 4154:115–123, 2001.
50. Davis, C. et al., Ocean PHILLS hyperspectral imager: Design, characterization, and calibration, *Optics Express*, 10(4):210–221, 2002.
51. Clark, R.N. and Swayze, G.A., Mapping minerals, amorphous materials, environmental materials, vegetation, water, ice, and snow, and other materials: The USGS tricorder algorithm, *Summaries of the Fifth Annual JPL Airborne Earth Science Workshop*, 95-1(1):39–40, 1995.
52. Curran, P.J., Remote sensing of foliar chemistry, *Remote Sensing of Environment*, 30:271–278, 1989.
53. Ustin, S., *Remote Sensing for Natural Resource Management and Environmental Monitoring (Manual of Remote Sensing – Third Edition) Volume 4*, John Wiley & Sons, 679–729, 2004.
54. Goodenough, D.G., Li, J.Y., and Dyk, A., Combining hyperspectral remote sensing and physical modeling for applications in land ecosystems, *IEEE International Conference on Geoscience and Remote Sensing Symposium*, IGARSS'06, 2000–2004, Denver, Colorado, 2006.
55. Koetz, B., Estimating biophysical and biochemical properties over heterogeneous vegetation canopies—Radiative transfer modeling in forest canopies based on imaging spectrometry and lidar, *PhD thesis*, Remote Sensing Laboratories, Department of Geography, University of Zurich, Zurich, 2006.
56. Huber, S. et al., The potential of spectrodirectional CHRIS/PROBA data for biochemistry estimation, *Envisat Symposium 2007, Montreux, Switzerland*, the Netherlands, p. 6, 2007.
57. Merton, R.N., Multi-temporal analysis of community scale vegetation stress with imaging spectroscopy, *PhD thesis*, Geography Department, University of Auckland, New Zealand, p. 492, 1999.
58. Apan, A. et al., Detecting sugarcane "orange rust" disease using EO-1 Hyperion hyperspectral imagery, *International Journal of Remote Sensing*, 25:489–498, 2004.
59. Krier, A. and Sherstnev, V.V., Powerful interface light emitting diodes for methane gas detection, *Journal of Physics D: Applied Physics*, 33(2):101–106, 2000.
60. Wei, H., The seasonal variation of column abundance of atmospheric CH_4 and precipitable water derived from ground-based IR solar spectra, *Infrared Physics and Technology*, 41:313–319, 2000.
61. Roy, J., Saugier, B., and Mooney, H.A., *Terrestrial Global Productivity*, Academic Press, p. 573, 2001.
62. Barnhouse, W.D., Methane plume detection using passive hyper-spectral remote sensing, *MS thesis*, Bowling Green State University, p. 141, 2005.

63. Grace, J., Role of forest biomes in the global carbon balance, *The Carbon Balance of Forest Biomes*, Taylor and Francis Group, 19–46, 2005.
64. Goodenough, D.G. et al., Processing Hyperion and ALI for forest classification, *IEEE Transactions on Geoscience and Remote Sensing*, 41:1321–1331, 2003.
65. Clark, R.N., Swayze, G.A., and Gallagher, A., Mapping the mineralogy and lithology of Canyonlands, Utah with imaging spectrometer data and the multiple spectral feature mapping algorithm, *Summaries of the Third Annual JPL Airborne Geoscience Workshop*, 92-14(1):11–13, 1992.
66. Asner, G.P. and Vitousek, P.M., Remote analysis of biological invasion and biogeochemical change, *Proceedings of the National Academy of Sciences*, 102:4383–4386, 2005.
67. Ustin, S.L. et al., Remote sensing of environment: State of the science and new directions, *Remote Sensing of Natural Resources Management and Environmental Monitoring*, 679–729, 2004.
68. Jupp, D.L.B., Kirk, J.T.O., and Harris, G.P., Detection, identification and mapping of cyanobacteria—Using remote sensing to measure the optical quality of turbid inland waters, *Australian Journal of Marine and Freshwater Research*, 45:801–828, 1994.
69. Roelfsema, C. et al., Remote sensing of a cyanobacterial bloom (lyngbya majuscule) in Moreton Bay, Australia, *IEEE International Transactions on Geoscience and Remote Sensing Symposium*, IGARSS'01 Proceedings, 613–615, 2001.
70. Wiangwang, N., Water clarity/trophic condition monitoring using satellite remote sensing data, *Master's thesis*, Department of Geography, Michigan State University, p. 152 , 2003.
71. Tsai, F. and Philpot, W., Derivative analysis of hyperspectral data, *Remote Sensing of Environment*, 66:41–51, 1998.
72. Richards, J.A. and Jia, X., *Remote Sensing Digital Image Analysis: An Introduction*, Third Edition, Springer, p. 363, 1999.
73. Des Marais, D.J. et al., Remote sensing of planetary properties and biosignatures on extrasolar terrestrial planets, *Astrobiology*, 2(2):153–181, 2002.
74. Heath, D.F. and Park, H., The solar backscatter ultraviolet (SBUV) and Total Ozone Mapping Spectrometer (TOMS) experiment, *The Nimbus-7 Users Guide*, NASA Goddard Space Flight Center, 175–211, 1978.
75. Veefkind, J.P. et al., Total ozone from the Ozone Monitoring Instrument (OMI) using the DOAS technique, *IEEE Transactions on Geoscience and Remote Sensing*, 44(5):1239–1244, 2006.
76. Bhartia, P.K., OMI ozone product, *NASA Goddard Space Flight Center, OMI Algorithm Theoretical Basis Document Volume II*, p. 91, 2002.
77. Bhartia, P.K. et al., Highlights of the version 8 SBUV and TOMS datasets released at this symposium, *Proceedings of the XX Quadrennial Ozone Symposium*, Athens, Greece, p. 294, 2004.
78. Burrows, J.P. et al., The global monitoring experiment (GOME): Mission concept and first scientific results, *Journal of Atmospheric Sciences*, 56(2):151–175, 1999.
79. Piters, A.J.M. et al., *GOME ozone fast delivery and value-added products, version 3.0*, KNMI Report, GOFAP-KNMI-ASD-01, De Bilt, the Netherlands, 2000.
80. Spurr, R., Thomas, W., and Loyola, D., GOME level 1 to 2 algorithms description, *DLR Technical note ER-TN-DLR-GO- 0025*, Oberpfaffenhofen, Germany, 2002.
81. Hilsenrath, E. et al., Early data from Aura and continuity from UARS and TOMS, *Space Science Reviews*, 125(1-4):417–430, 2006.
82. Levelt, P.F. et al., The ozone monitoring instrument, *IEEE Transactions on Geoscience and Remote Sensing*, 44(5):1093–1101, 2006.
83. Itten, K.I. et al., APEX—The hyperspectral ESA airborne prism experiment, *Sensors*, 8(10):6235–6259, 2008.
84. Turner, W. and Spector, S., Remote sensing for biodiversity science and conservation trends, *Ecology and Evolution*, 18:306–314, 2003.
85. Schaepman, M. and Malenovsky, Z., Bridging scaling gaps for the assessment of biodiversity from space, *The Full Picture*, Group on Earth Observations (GEO), Geneva, Switzerland, 258–261, 2007.
86. Huber, S., Estimating foliar biochemistry from hyperspectral data in mixed forest canopy, *Forest Ecology and Management*, 256:491–501, 2008.
87. Goodenough, D.G. et al., Mapping forest biomass with AVIRIS and evaluating SNR impact on biomass prediction, *Natural Resources Canada*, internal report (presented at NASA JPL AVIRIS Workshop), 2005.
88. Inada, K., Spectral ratio of reflectance for estimating chlorophyll content of leaf, *Japanese Journal of Crop Science*, 154:261–265, 1985.
89. Shibayama, M. and Akiyama, T.A., Spectroradiometer for field use. VII. Radiometric estimation of nitrogen levels in filed rice canopies, *Japanese Journal of Crop Science*, 55:433–438, 1986.

90. Inoue, Y., Moran, M.S., and Horie, T., Analysis of spectral measurements in rice paddies for predicting rice growth and yield based on a simple crop simulation model, *Plant Production Science*, 1:269–279, 1998.
91. Lopez, R.D., *An Ecological Assessment of Invasive and Aggressive Plant Species in Coastal Wetlands of the Laurentian Great Lakes: A Combined Field-based and Remote Sensing Approach*, The United States Environmental Protection Agency, Environmental Sciences Division, 2001.
92. Lopez, R.D. et al., *Using Landscape Metrics to Develop Indicators of Great Lakes Coastal Wetland Condition*. The United States Environmental Protection Agency, EPA/600/X-06/002, Washington, DC, p. 31 , 2006.
93. Ustin, S.L. et al., Hyperspectral remote sensing for invasive species detection and mapping, *Geoscience and Remote Sensing Symposium*, 3:1658–1660, 2002.
94. Lass, L.W. and Prather, T.S., Detecting the locations of Brazilian pepper trees in the everglades with a hyperspectral sensor, *Weed Technology*, 18(2):437–442, 2004.
95. Lass L. W. et al., A review of remote sensing of invasive weeds and example of the early detection of spotted knapweed (*Centaurea maculosa*) and babysbreath (*Gypsophila paniculata*) with a hyperspectral sensor, *Weed Science*, 53(2):242–251, 2005.
96. Lawrence, R. and Labus, M., Early detection of Douglas-fir beetle infestation with subcanopy resolution hyperspectral imagery, *Western Journal of Applied Forestry*, 18:202–206, 2003.
97. Koetz, B. et al., Radiative transfer modeling within a heterogeneous canopy for estimation of forest fire fuel properties, *Remote Sensing of Environment*, 92:332–344, 2004.
98. Wyszecki, G. and Stiles, W.S., *Color Science: Concepts and Methods, Quantitative Data and Formulae*, Second Edition, John Wiley and Sons, p. 968, 2000.
99. Jacobson, N.P., Gupta, M.R., and Code, J.B., Linear fusion of image sets for display, *IEEE Transactions on Geoscience and Remote Sensing*, 45(10):3277–3288, 2007.
100. Ready, P.J. and Wintz, P.A., Information extraction, SNR improvement, and data compression in multispectral imagery, *IEEE Transactions on Communications*, 21(10):1123–1131, 1973.
101. Tyo, J.S. et al., Principal components-based display strategy for spectral imagery, *IEEE Transactions on Geoscience and Remote Sensing*, 41(3):708–718, 2003.
102. Kaewpijit, S., Moigne, J.L., and El-Ghazawi, T., Automatic reduction of hyperspectral imagery using wavelet spectral analysis, *IEEE Transactions on Geoscience and Remote Sensing*, 41:863–871, 2003.
103. Harsanyi, J.C. and Chang, C.I., Hyperspectral image classification and dimensionality reduction: An orthogonal subspace projection approach, *IEEE Transactions on Geoscience and Remote Sensing*, 32:779–785, 1994.
104. Tyo, J.S. and Olsen, R.C., Principal-components-based display strategy for spectral imagery, *IEEE Workshop on Advances in Techniques for Analysis of Remotely Sensed Data*, 276–281, 2003.
105. Jacobson, N.P. and Gupta, M.R., Design goals and solutions for display of hyperspectral images, *IEEE Transactions on Geoscience and Remote Sensing*, 43(11):2684–2692, 2005.
106. Gao, B.C. et al., Atmospheric correction algorithms for hyperspectral remote sensing data of land and ocean, *Remote Sensing of Environment*, 113(S1):S17–S24, 2009.
107. Gao, B.C. and Goetz, A.F.H., Column atmospheric water vapor and vegetation liquid water retrievals from airborne imaging spectrometer data, *Journal of Geophysical Research*, 95(D4):3549–3564, 1990.
108. Kruse, F.A., Imaging spectrometer data analysis—A tutorial, *Proceedings of the International Symposium on Spectral Sensing Research (ISSSR)*, San Diego, California, 44–54, 1994.
109. Centner, V. et al., Comparison of multivariate calibration techniques applied to experimental NIR data sets, *Applied Spectroscopy*, 54:608–623, 2000.
110. Pierce, R.B. et al., Impacts of background ozone production on Houston and Dallas, Texas, air quality during the Second Texas Air Quality Study field mission, *Journal of Geophysical Research* 114, 2009.
111. Lopez, R.D. and Frohn, R.C., *Remote Sensing for Landscape Ecology: New Metric Indicators*, 2nd Edition, CRC Press, 269, 2017.

Section III

Hyperspectral Libraries of Agricultural Crops and Vegetation

4 Monitoring Vegetation Diversity and Health through Spectral Traits and Trait Variations Based on Hyperspectral Remote Sensing

Angela Lausch and Pedro J. Leitão

CONTENTS

4.1 INTRODUCTION

In order to understand the significance of hyperspectral remote sensing (HRS) techniques and thus the spectral traits (ST) and spectral trait variations (STV) approach for recording, monitoring, and assessing the status, stress, disturbances, or resource limitations of vegetation, it is first necessary to define both vegetation diversity and vegetation health.

Vegetation diversity refers to the diversity of living vegetation, which includes the phylogenetic, taxonomic, structural, and functional properties of vegetation at different organization levels. These include (but are not limited to) molecular, genetic, individual, species, populations, communities, biomes, ecosystems, and landscapes characteristics. These properties emerge as a result of evolutionary history, epigenetic disturbances, changing environmental conditions, and anthropogenically-modified ecological processes (Lausch et al., 2016b). The essential characteristics of vegetation diversity are: (i) the phylo diversity, which measures the length of evolutionary pathways that are linked to a given set of plant taxa; (ii) the taxonomic diversity, which is the composition and configuration of different plant taxa; (iii) the structural diversity, which refers to the composition and configuration of 2D and 3D vegetation entities, (iv) the functional diversity, which refers to the diversity realized in ecological functions and processes in the ecosystem, and (v) the trait diversity, which refers the distribution of biochemical, physiological, morphological, structural, phenological, or functional characteristics (Figure 4.1) (see also Lausch et al., 2018b).

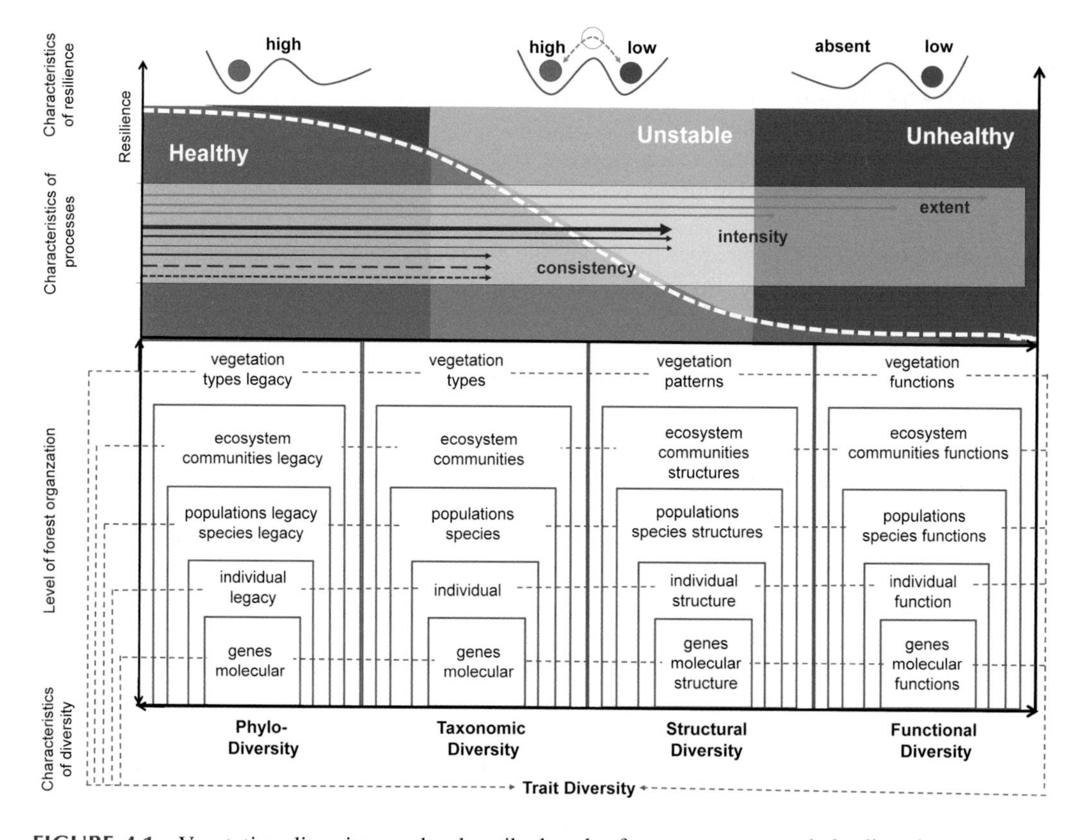

FIGURE 4.1 Vegetation diversity can be described under four components: phylo diversity, taxonomic diversity, structural diversity, functional and trait diversity. All of the characteristics involve different levels of vegetation organization from molecular, genetic, individual, species, populations, communities, landscapes, and ecosystems up to biomes. Different characteristics of the processes (the extent, process intensity, process consistency, resilience, and their characteristics) influence the resilience of vegetation health. (From Lausch A. et al., 2018b. *Remote Sens.*, 10, 1120, doi:10.3390/rs10071120.)

Vegetation health, on the other hand, refers to changes in taxonomic, phylogenetic, structural, functional, or trait vegetation diversity, at different levels of biological organization (as defined for vegetation diversity), which are not attributed to evolutionary changes or adjustment mechanisms, and which can disrupt the resilience of ecosystems (Lausch et al., 2018a,b). While natural and anthropogenic drivers can lead to disturbances in vegetation health, healthy vegetation is characterized by a high degree of stability, elasticity, and resistance to stress, disturbances, and resource limitations when vegetation diversity is maintained or at least restored after the impacts of natural or anthropogenic drivers (after Lausch et al., 2016b).

HRS has great potential for monitoring vegetation diversity and health, in its different components and levels of organization. Several studies have used HRS for mapping vegetation phylogenetic diversity, for example, by linking it with an integrative spectral diversity indicator (Schweiger et al., 2017). Furthermore, the "spectranomic approach" (Asner and Martin, 2016) makes use of HRS to establish links between the phylogenetic, taxonomic, and functional characteristics of vegetation. The taxonomic diversity of vegetation has been described with HRS in different environments, such as forests (Fassnacht et al., 2016) savannas (Leitão et al., 2017), and grasslands (Schmidtlein et al., 2007), as well as invasive species mapping (Somers and Asner, 2013; Santos et al., 2016). HRS is also widely used for measuring the structural diversity of vegetation at different levels of organization, such as biochemical (Asner and Martin, 2008) organism (Asner, 2015), communities (Lopatin et al., 2017), as well as ecosystems (Leitão et al., 2015; Santos et al., 2016). HRS techniques are able to record status and shifts in both functional traits at both the species level (functional traits) and the ecosystem level (ecosystem functions). Ustin and Gamon (2010) have demonstrated the use of HRS for mapping plant functional types, based on their functional traits. Moreover, several studies have used HRS to map functional diversity in different environments such as the Amazon (Asner et al., 2014a,b,c) or temperate forests (Schneider et al., 2017). Recently, HRS has been identified as being able to provide unprecedented opportunities for characterizing several ecosystem functions related to vegetation, such as disturbance regulation, production of biomass, or supporting habitats (Pettorelli et al., 2017).

The knowledge of many factors is required to understand how HRS techniques can be used to record and evaluate the status, stress, and disturbances of vegetation diversity as well as its stability, resilience, and health. Therefore, the following points are discussed in this chapter:

i. Which different approaches exist for monitoring vegetation diversity and health?
ii. The concept of ST and STV using HRS techniques: which ST/STV can be recorded using the HRS technique?
iii. How can HRS techniques be used for monitoring the four characteristics of vegetation diversity (phylogenetic, taxonomic, structural, and functional diversity)?
iv. Why can HRS be used for monitoring stress, disturbances, processes, or resource limitations in vegetation diversity and health?
v. What are the limitations, the constraints, and the potential of HRS for monitoring vegetation diversity and its health?
vi. How does HRS compare to multispectral sensors as well as in situ processes for recording vegetation diversity and health?

4.2 APPROACHES FOR MONITORING VEGETATION DIVERSITY AND HEALTH

There are two different methods to monitor vegetation diversity, its characteristics, influences, and adaptations as a result of stress, disturbances, and resource limitations. One is in situ or field-based monitoring, and the other is the remote sensing (RS) approach. In situ observation refers to a direct identification, recording, and monitoring of plant species or communities and even landscapes by taxonomists. By contrast, RS observation is based entirely on spectral reflectance values. RS sensors do not come into contact with the object and are situated anywhere from a few millimeters away up to thousands of kilometers away. The sensors are mounted or integrated on platforms such as Analytical Spectral Devices (ASD), cameras and sensors in a laboratory, wireless sensor networks (WSN), tower installations (close-range RS techniques), drones, airplanes (airborne RS), or satellites (spaceborne RS).

4.2.1 In Situ Approaches

In situ approaches conducted by taxonomists for monitoring vegetation diversity and its pressures and processes are the basic concept of biogeography and macroecology (Violle et al., 2014, Lausch et al., 2018a). The characterization of vegetation diversity is based on different species concepts (Wheeler and Meier, 2000). The most important are the phylogenetic species concept (PSC) (Eldredge and Cracraft, 1980), the biological species concept (BSC) (Mayr, 1942), and the morphological species concept (MSC) (Mayr, 1969).

The Phylogenetic Species Concept (PSC): Vegetation diversity has evolved over long processes of evolution and explains why everything from plant species to ecosystems differ in their evolutionary relationships (Eldredge and Cracraft, 1980). When comparing the diversity of species assemblages (each with the same number of species), their species richness is the same, but their phylogenetic diversity or the diversity of evolutionary relationships represented in an assemblage might differ. It follows that taxonomic diversity and phylogenetic diversity are not necessarily correlated to each other in a positive way. Rather, an increase in taxonomic diversity can just as well be accompanied by a decrease in phylogenetic diversity (Knapp et al., 2017). Phylogenetic diversity is not distributed randomly across the globe (Sechrest et al., 2002) and is affected by land use (Knapp et al., 2008; Frishkoff et al., 2014), the climate (Willis et al., 2008), biological invasions, and extinctions (Knapp et al., 2017). Phylogenetic diversity thus provides insight into the status and trends of biodiversity beyond taxonomic diversity.

The Biological Species Concept (BSC): This provides answers to fundamental questions at the organism level and in populations and communities about species richness, abundances, diversity, distributions, thresholds, and fitness (Mayr, 1942). Based on taxonomic units, the BSC is, however, limited in terms of understanding how plant species change with regard to taxonomy, structure and function based on species adaptations, interactions and responses to stress, disturbances or resource availability (Boulangeat and Thuiller, 2012). To understand and explain the effects of natural and anthropogenic stressors using BSC approaches, gradients in monitoring are important such as latitude, biomass, experimental diversity gradients, and climate gradients (Purvis and Hector, 2000).

The Morphological Species Concept (MSC): This concept (Mayr, 1969) allows us to adopt a completely new approach to understanding species diversity and vegetation health. Traits are anatomical, morphological, biochemical, physiological, structural, or phenological features and characteristics of organisms (Kattge et al., 2011). Plant traits are measured, described, and standardized across species depending on their response to environmental factors and constraints, trophic levels, and survival or dispersal strategies (Pérez-Harguindeguy et al., 2013). Traits can be measured at all hierarchical organization levels of the vegetation from leaves (Kühn et al., 2004) to individual plants (Violle et al., 2007), to plant populations and communities (Cadotte et al., 2015), and thus at the landscape and biogeographical levels (Violle et al., 2014). Therefore, mapping species traits captures vegetation features and diversity at all levels, from a local to a global scale (Abelleira Martínez et al., 2016). Trait and trait variations help us to understand, explain, and predict why organisms live where they do, how they will respond to environmental change (Green et al., 2008), and how they interact with different stressors, disturbances, or resource limitations. Furthermore, traits and their variations filter for status, environmental changes, land-use intensity (LUI), stress, or disturbances (Garnier et al., 2007; Gámez-Virués et al., 2015). Traits can indicate resource depletion (Kühn et al., 2006), define the state, presence (Kühn et al., 2006), and dispersal strategies (Díaz et al., 2015), or indicate functional disturbances and shifts in ecosystems (Lavorel and Garnier, 2002; Reichstein et al., 2014).

4.2.2 Remote Sensing Approach: Spectral Traits and Spectral Trait Variations

RS commonly records biochemical, biophysical, physiognomic, morphological, structural, phenological, and functional traits at all organization levels, from the molecular and the individual level right up to communities and the whole ecosystem (Abelleira Martínez et al., 2016). This is based on the principles of image spectroscopy across the electromagnetic spectrum from visible to microwave bands (Ustin and Gamon, 2010). Compared to the traits approach in the morphological species concept by taxonomists, RS approaches are not able to record all traits and trait variations (Homolová et al., 2013; Lausch et al., 2016b). The traits and trait variations that can be monitored by HRS techniques are therefore called "Spectral Traits" (ST), the changes to the spectral traits are called "Spectral Trait Variations" (STV), and the RS approach is thus called the "remote sensing-spectral trait/spectral trait variations-concept"—RS-ST/STV-C (Lausch et al., 2016a,b). Traits are crucial in bridging gaps between in situ and remote sensing approaches to monitor, understand, and assess status, stress, disturbances, or resource limitations of vegetation diversity and vegetation health (Figure 4.2).

HRS techniques record the characteristics (spectral traits) as well as the changes in the characteristics (spectral trait variations) as a result of processes, stress, and disturbances from the molecular level to ecosystems and landscapes based on the principles of spectroscopy in the spectral range of 400–2500 nm (Ustin and Gamon, 2010). Therefore, the spectral traits for plants are anatomical, morphological, biochemical, biophysical, physiological, structural, phenological, or functional characteristics of plants, populations, and communities that are influenced by phylogenetic, taxonomic, populations', and communities' characteristics, which can be directly or indirectly recorded using RS techniques in space (Lausch et al., 2016b).

Given the high spectral resolution of HRS in the range of 400–2500 nm with bandwidths of 6–10 nm, the ability to record spectral traits compared to broadband RS techniques with <20 (broad) spectral bands is much higher. The spectral traits of vegetation that can be recorded with HRS techniques (Figure 4.3) can be grouped into classes depending on the focus of the recording such as chemical, biochemical, biophysical phenotypical, physiognomic, or morphological properties, among others.

ST/STV are recorded either as single traits such as plant phenology, leaf carbon, or nitrogen content, or they emerge from a combination of different spectral traits (trait combinations), for example, ecological strategy types, biomass, vegetation structure, or leaf area index. ST/STV can be recorded either directly (content of chlorophyll, xanthophyll, or water) or/and also indirectly through processes triggered by the driver (see Table 4.1). For instance, heavy metals (Götze et al., 2010), plant disease (Mahlein, 2016), or the application of pesticides to crops lead to biochemical-biophysical molecular changes in plant traits, which leads to trait variations that can be recorded with HRS. The ST that can be recorded using HRS techniques can be seen in Figure 4.3. The factors that are responsible for the recordability of ST/STV are described in Section 4.4 under the constraints of HRS techniques for recording traits and trait variations.

4.3 MAPPING STATUS, STRESS, DISTURBANCES, AND RESOURCE LIMITATIONS IN VEGETATION WITH HYPERSPECTRAL (HRS)

Vegetation diversity is complex (Hillebrand and Matthiessen, 2009), multidimensional in space and time, and is poorly understood when it comes to its various interactions, drivers, and disturbances. This is why there is a focus on recording disturbances and changes to the resilience of ecosystems. As a result of high spectral resolution, hyperspectral technology is particularly suited to record shifts, changes, and disturbances of numerous traits and trait variations on all scales of species organization. Therefore, spectral trait variations (STV) can be defined as "changes to Spectral Traits (ST)" in terms of physiology, senescence, or phenology. However, "changes are also caused by stress, disturbances, and the resource limitations of plants, populations, and communities, which can be directly or indirectly recorded by remote sensing techniques in space and over time" (Lausch et al., 2016b).

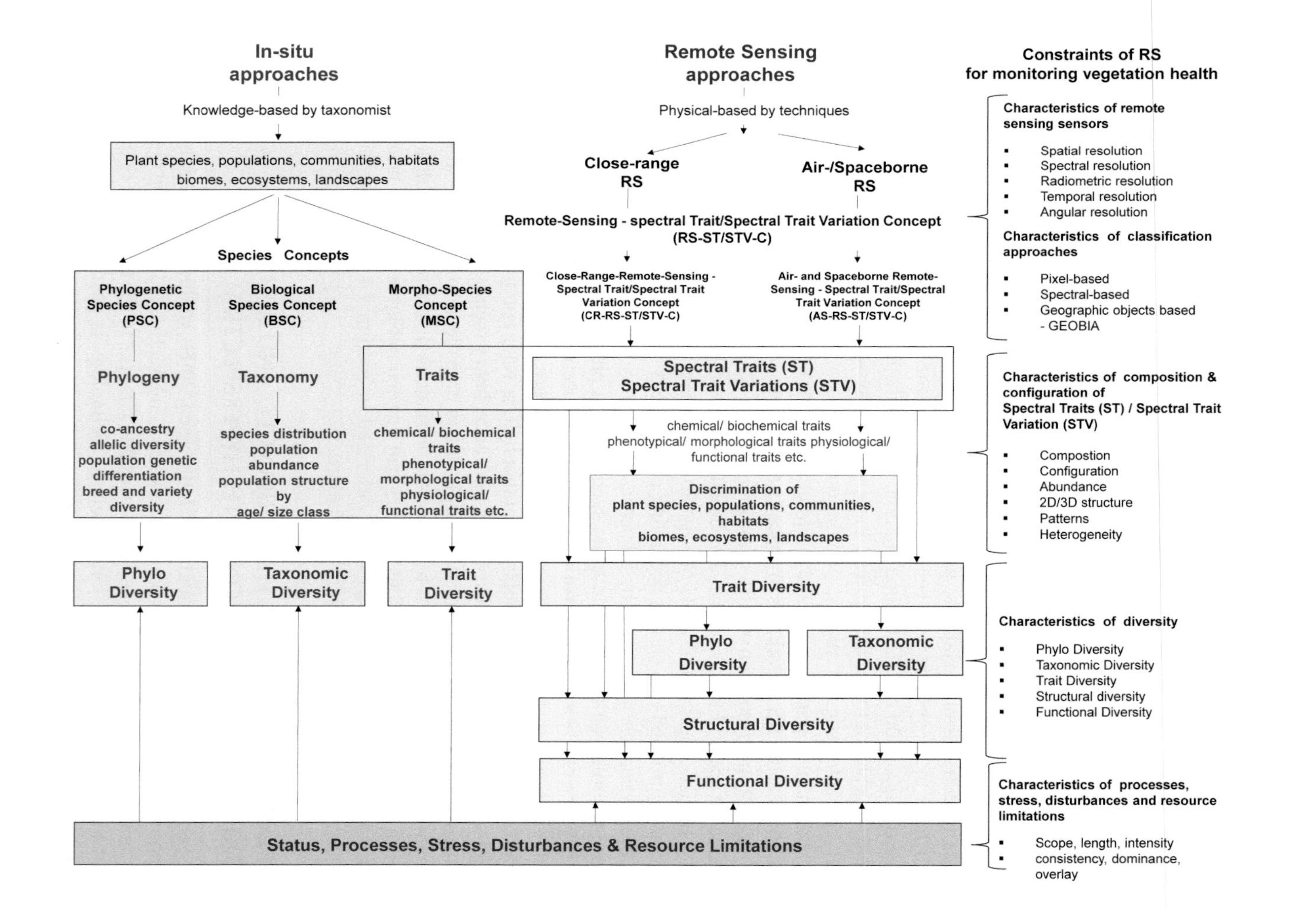

FIGURE 4.2 Illustration of in situ and remote sensing approaches to monitor status, stress, disturbances, and resource limitations, when incorporating the four characteristics of vegetation diversity and assessing vegetation health and resilience in ecosystems. (From Lausch A. et al., 2018b. *Remote Sens.*, 10, 1120, doi:10.3390/rs10071120.)

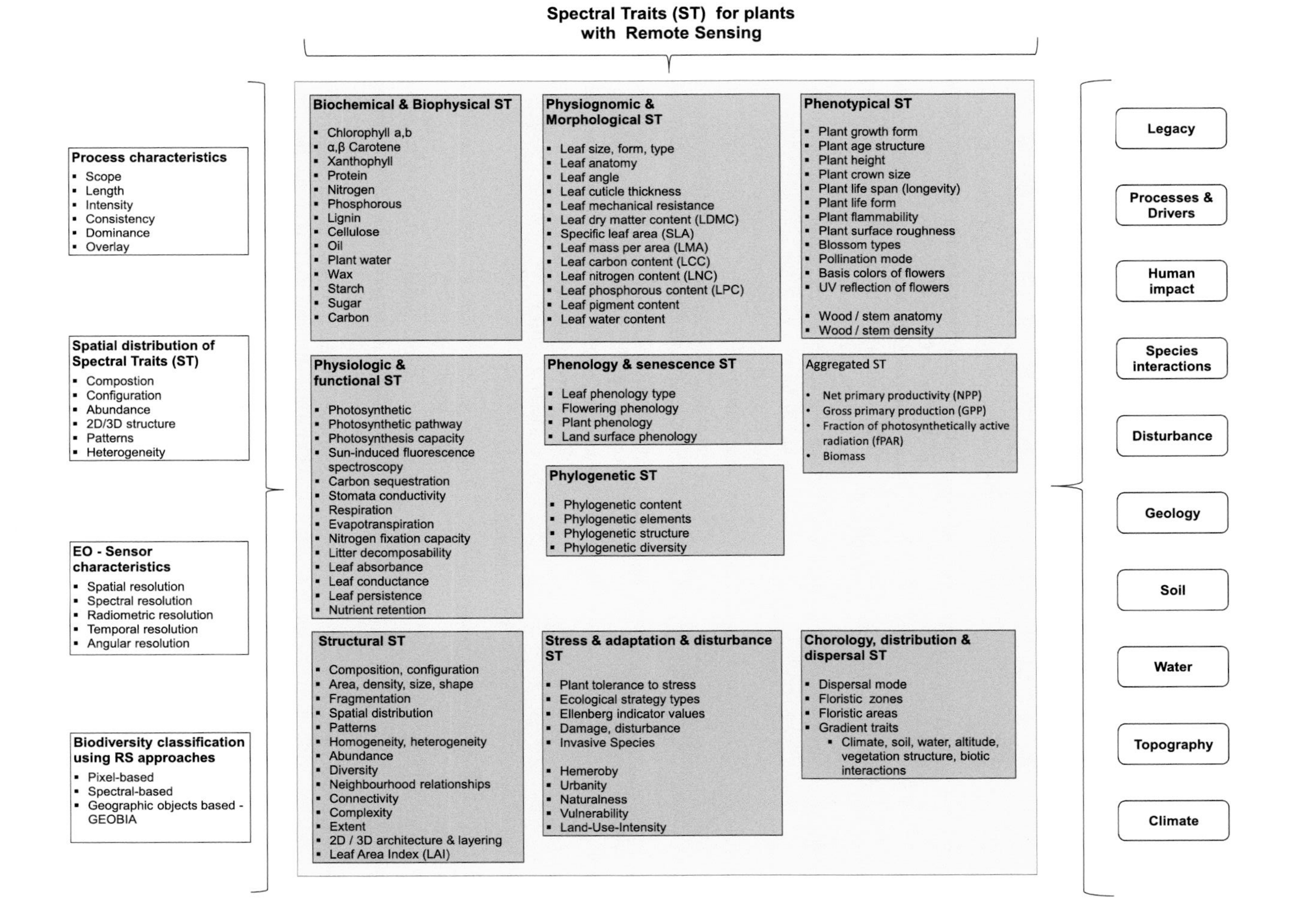

FIGURE 4.3 Spectral traits (ST) for observing and assessing phylogenetic, taxonomic, structural, functional and trait diversity using hyperspectral remote sensing techniques. (Modified after Lausch A. et al., 2016b. *Remote Sens.* 8, 1029. doi: 10.3390/RS8121029)

TABLE 4.1

Examples of Spectral Traits (ST) and Spectral Trait Variations (STV) for Monitoring Vegetation Diversity as Well as the Response to Adaptation, Stress, Disturbances, and Resource Limitations[a]

Spectral Traits (ST) of Plants	References
Biochemical-Biophysical ST	
Pigment content (chlorophyll a, b, α, β carotene, xanthophyll)	Asner et al. (2011a), Malenovský et al. (2015), Zarco-Tejada et al. (2016), Ustin et al. (2009), Hilker et al. (2011), Pandey et al. (2017), Gitelson et al. (2003), Blackburn (2007)
Nitrogen	Asner et al. (2016b), Asner and Martin (2008), Balzotti et al. (2016), Carlson et al. (2007), Knyazikhin et al. (2013), Schlerf et al. (2010), Pandey et al. (2017), McManus et al. (2016)
Phosphorus	Asner et al. (2016a,b), Asner et al. (2015a,b), Asner and Martin (2016), Mutanga and Kumar (2007), Porder et al. (2005), Pandey et al. (2017), McManus et al. (2016)
Lignin	Asner et al. (2015a,b), McManus et al. (2016) Serrano et al. (2002), Smith et al. (2002), Thenkabail et al. (2016), Ustin et al. (2009), McManus et al. (2016)
Cellulose	Asner et al. (2015a,b), Thenkabail et al. (2016), McManus et al. (2016)
Phenol	Asner et al. (2015a), Klosterman et al. (2014), McManus et al. (2016)
Plant water content	Asner et al. (2015b), Assal et al. (2016), Bergen et al. (2009), Brosinsky et al. (2013), Lausch et al. (2015b), Zarco-Tejada et al. (2003), Zhu et al. (2017)
Wax, starch, sugar	Asner et al. (2015b), Asner et al. (2014a,b,c), Ustin et al. (2004)
Carbon content	Asner et al. (2013), Asner et al. (2012c), Asner and Mascaro (2014), Marvin and Asner (2016)
Tanins	McManus et al. (2016)
Physiognomic-Morphological ST	
Leaf size, form, type, leaf anatomy, leaf optical properties, leaf wettability traits	Lukeš et al. (2013), Roth et al. (2016), Sims and Gamon (2002), Féret et al. (2017)
Leaf dry matter content (LDMC)	Asner et al. (2015b), Kokaly et al. (2009), Wang et al. (2011)
Specific leaf area (SLA)	Ali et al. (2016), Asner and Martin (2009, 2008), Ball et al. (2015), Czerwinski et al. (2014), Lukeš et al. (2013)
Leaf mass per area (LMA)	Asner et al. (2016a,b), Asner et al. (2015a, 2015b, 2011b), McManus et al. (2016), Asner et al. (2011b)
Leaf carbon content (LCC)	Asner et al. (2013), Asner and Mascaro (2014), Marvin and Asner (2016)
Leaf nitrogen content (LNC)	Dechant et al. (2017), Pandey et al. (2017), Oppelt and Mauser (2004), Wang et al. (2017)
Leaf phosphorus content (LPC)	Asner et al. (2015a,b), Mutanga and Kumar (2007), Porder et al. (2005)
Leaf pigment content	Zhang et al. (2008), Asner et al. (2011a), Malenovský et al. (2015), Zarco-Tejada et al. (2016), Ustin et al. (2009), Hilker et al. (2011), Pandey et al. (2017), Gitelson et al. (2003), Matsuda et al. (2012)
Leaf water content	Asner et al. (2015b), Assal et al. (2016), Brosinsky et al. (2013), Lausch et al. (2015b), Zarco-Tejada et al. (2003), Asner et al. (2011a), Goodenough et al. (2004)

(*Continued*)

TABLE 4.1 (*Continued*)
Examples of Spectral Traits (ST) and Spectral Trait Variations (STV) for Monitoring Vegetation Diversity as Well as the Response to Adaptation, Stress, Disturbances, and Resource Limitations[a]

Spectral Traits (ST) of Plants	References
Plant leaf chemical properties	Pandey et al. (2017), McManus et al. (2016), Asner and Martin (2008), Paz-Kagan et al. (2017)
Phenotypical ST	
Plant growth properties	El-Hendawy et al. (2017)
Tree age, forest age structure, forest stand age	Buddenbaum et al. (2005), Wu et al. (2017)
Wood, stem density, timber volume	Jusoff (2008), Jusoff and Yusoff (2009), Nink et al. (2015), Schlerf and Atzberger (2006)
Tree crown size, crown characteristics	Kandare et al. (2017), Clasen et al. (2015)
Physiological and Functional ST	
Photosynthesis, photosynthesis pathway, chlorophyll fluorescence	Damm et al. (2015), Kraft et al. (2012), Meroni et al. (2009), Moreno et al. (2014), Rascher (2007), Rascher et al. (2015), Zhao et al. (2005), Inoue et al. (2008), El-Hendawy et al. (2017), Rascher et al. (2015), Townsend et al. (2008), Malenovský et al. (2007)
Sun-induced fluorescence spectroscopy	Garzonio et al. (2017), Damm et al. (2015)
Carbon sequestration	Male et al. (2010), Asner et al. (2012a, 2014a,b,c), Lee et al. (2015), Yu et al. (2014)
Phenology and Senescence ST	
Leaf phenology type, leaf age, leaf development, flower status, plant and canopy phenology	Chen et al. (2009), Lausch et al. (2015b), de Moura et al. (2017), Rajah et al. (2017), Götze et al. (2017), Burkart et al. (2017)
Flower mapping, pollination types	Feilhauer et al. (2016), Abdel-Rahman et al. (2015), Landmann et al. (2015)
Stress, Adaptation, and Disturbance ST	
Damage, disturbances by species, defoliating and tree mortality insects, parasites, forest insect outbreaks and pest damage, plant disease, plant stress	Rajah et al. (2017), Behmann et al. (2014), Lausch et al. (2013a), Couture et al. (2013), Mišurec et al. (2016), Malenovský et al. (2007), Fassnacht et al. (2014), Behmann et al. (2014), Mahlein (2016), Dash et al. (2017)
Damage, disturbances (fire, water, storm, fallen tree, dead wood, deforestation, degradation)	Smit et al. (2016), Behmann et al. (2014)
Ecological strategy types, plant functional types (PFT), Ellenberg indicator values, CSV strategy types, C3 and C4 grasses functional types	Kattenaorn et al. (2017), Ustin and Gamon (2010), Schmidtlein et al. (2012), Schmidt et al. (2017a), Asner (2015), Schmidtlein (2005), Shoko and Mutanga (2017), Schweiger et al. (2017)
Naturalness, intact landscapes, monitoring of protected areas, conservation and landscapes, habitat quality, vegetation health	Schmidtlein and Sassin (2004), Feilhauer et al. (2014), Degerickx et al. (2017), Leitão et al. (2015), Schmidt et al. (2017b), Feilhauer et al. (2014), Neumann et al. (2015), Schwieder et al. (2014), Suess et al. (2015), Doktor et al. (2014)
Invasive species	Somers and Asner (2013), Skowronek et al. (2016), He et al. (2011), Skowronek et al. (2016), Underwood et al. (2006), Somers and Asner (2012)
Land-use intensity, land-use management (soil, water, crop, pesticide management)	Smit et al. (2016), Nansen et al. (2010), Holasek et al. (2017)
Plant traits as sensor for soil characteristics and soil moisture	Lausch et al. (2013c)

(*Continued*)

TABLE 4.1 (*Continued*)
Examples of Spectral Traits (ST) and Spectral Trait Variations (STV) for Monitoring Vegetation Diversity as Well as the Response to Adaptation, Stress, Disturbances, and Resource Limitations[a]

Spectral Traits (ST) of Plants	References
Chorology, Distribution and Dispersal ST	
Gradient traits (climate, soil, water, altitude, biotic, biochemical, floristic continuum)	Harris et al. (2015), Schmidtlein et al. (2007), Feilhauer et al. (2014), Neumann et al. (2015), Leitão et al. (2015), Asner et al. (2016a,b), Schmidtlein and Sassin (2004)
Structural ST	
Biogeochemical heterogeneity, spatial distribution of biochemical ST, patterns, structure, heterogeneity, homogeneity, diversity (alpha, beta, gamma diversity), abundance, connectivity, neighbourhood relationship, area, density, size, shape, extent of areas of plant species, communities, vegetation types	Schneider et al. (2017), Hakkenberg et al. (2017), Leutner et al. (2012), Pottier et al. (2014), Asner et al. (2012b), Asner (2013), Townsend et al. (2008), Feilhauer et al. (2010), Aneece et al. (2017), Asner et al. (2017), Lausch et al. (2013b)
Functional plant diversity	Schneider et al. (2017)
Leaf area index (LAI)	Wang et al. (2011), Tanaka et al. (2015), Danner (2017), Yuan et al. (2017), Haboudane et al. (2004)
Aggregated ST	
Net primary production (NPP), gross primary production (GPP), productivity, light use efficiency estimation, photosynthesis capacity	Yu et al. (2014), Zhang and Zhou (2017), Alton (2017), Mariotto et al. (2013), Peng and Gitelson (2012), Hill et al. (2014)
Biomass	Bratsch et al. (2017), Marshall and Thenkabail (2015), Luo et al. (2017), Meyer et al. (2017)
Fraction of photosynthetically active radiation FAPAR	Zhang and Zhou (2017)
Phylogenetic ST	
Phylogenetic information of traits, phylogenetic structure of foliar spectral traits	McManus et al. (2016), Cavender-Bares et al. (2016), Asner and Martin (2011), Schweiger et al. (2018)
Additional Indicators of Vegetation Health	
Shifting biochemical traits (photosynthesis respiration, plant productivity, phenology, growing season length, variation in carbon dioxide exchange and carbon balance, greening response)	Asner et al. (2016a,b), Asner et al. (2015a)
Vegetation health and conditions	Dash et al. (2017), Banerjee et al. (2017), Piro et al. (2017), Lowe et al. (2017), Götze et al. (2010), Lausch et al. (2013a)
Discrimination of plant species, communities, vegetation types, dominant species, and mapping of functional guilds, vegetation types	Lopatin et al. (2017), Roth et al. (2015), Buddenbaum et al. (2005), Laurin et al. (2016), Möckel et al. (2014), Feret and Asner (2013), Baldeck and Asner (2014), Richter et al. (2016), Bratsch et al. (2016), Liu et al. (2017)
Vegetation monitoring, vegetation change, land-use and land cover change, crop-classification	Lamine et al. (2017), Zhong et al. (2017), Deshpande et al. (2017), Guidici and Clark (2017), Priem and Canters (2016), Bareth et al. (2015)
Mapping indicators of ecosystem services	Braun et al. (2017)

[a] Which can be recorded with hyperspectral remote sensing techniques (HRS) on different platforms—laboratory, tower, drones, airborne, spaceborne. Many other applications for mapping the status and shifts in phylogenetic, taxonomic, trait, structural and functional vegetation diversity are discussed throughout this book.

The reasons why HRS is suitable to record stress, disturbances, or resource limitations from molecular to landscape scales are manifold and include the following (after Lausch et al., 2016b):

- Processes, stress, disturbances, and resource limitations cause changes in traits and lead to trait variations. These changes can be found at all levels in vegetation organization from molecular, genetic, and plant species right up to landscapes (see Figure 4.4).
- Traits and trait variations are filters and indicators for phylogenetic and epigenetic adaptations and disturbance processes which were triggered by different drivers and stress indicators (see Figure 4.5).
- The nature and strength of the trait variations, trait adaptations, or trait modifications which influence the status, stability, and resilience of ecosystems, depend on the characteristics of the processes (scope, length, intensity, consistency, dominance, overlay of processes, or history).
- Plant traits and trait variations are therefore a proxy and a filter for the status of vegetation diversity (phylogeny, taxonomy, structurefunction and traits) and the processes and disturbances which have an impact on them, and hence have utility (see Figures 4.4 and 4.5).
- HRS approaches can potentially record direct and/or indirect traits and trait variations on all spatio-temporal scales of vegetation hierarchy.
- Spectral HRS responses, their patterns and heterogeneity are therefore a proxy for plant trait status, diversity, composition, and configuration and the result of processes, stresses, disturbances, or resource limitations on traits at various spatio-temporal scales of the vegetation hierarchy (see Figure 4.6).

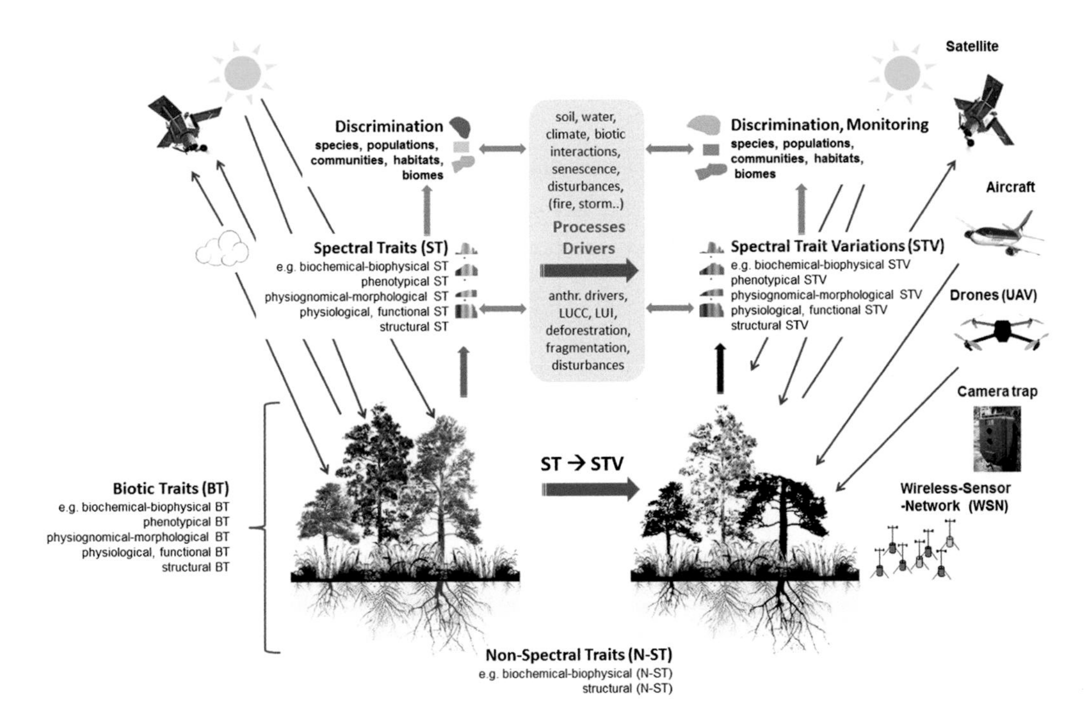

FIGURE 4.4 Mapping spectral traits (ST) and spectral trait variations (STV) and their interactions with drivers and processes to monitor status, stress, and disturbances of vegetation diversity with hyperspectral remote sensing techniques. (From Lausch A. et al., 2016b. *Remote Sens.* 8, 1029. doi: 10.3390/RS8121029)

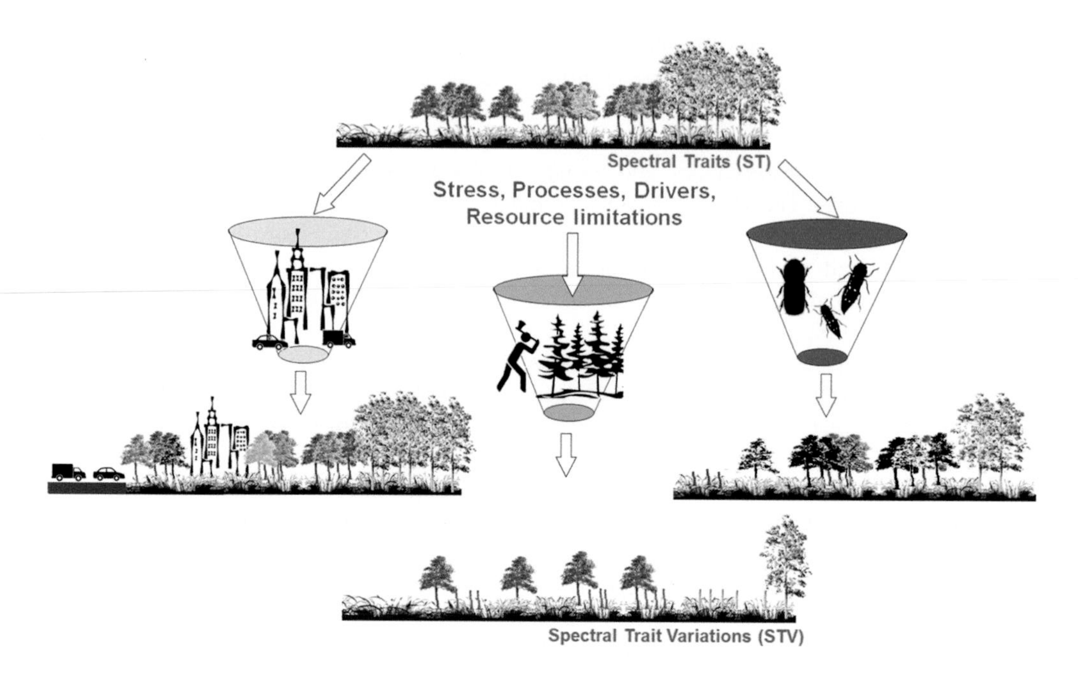

FIGURE 4.5 Plant traits and trait variations are a proxy and a filter for the status, stress, disturbances, or resource limitations of vegetation diversity and their characteristics. (From Lausch A. et al., 2016b. *Remote Sens.* 8, 1029. doi: 10.3390/RS8121029)

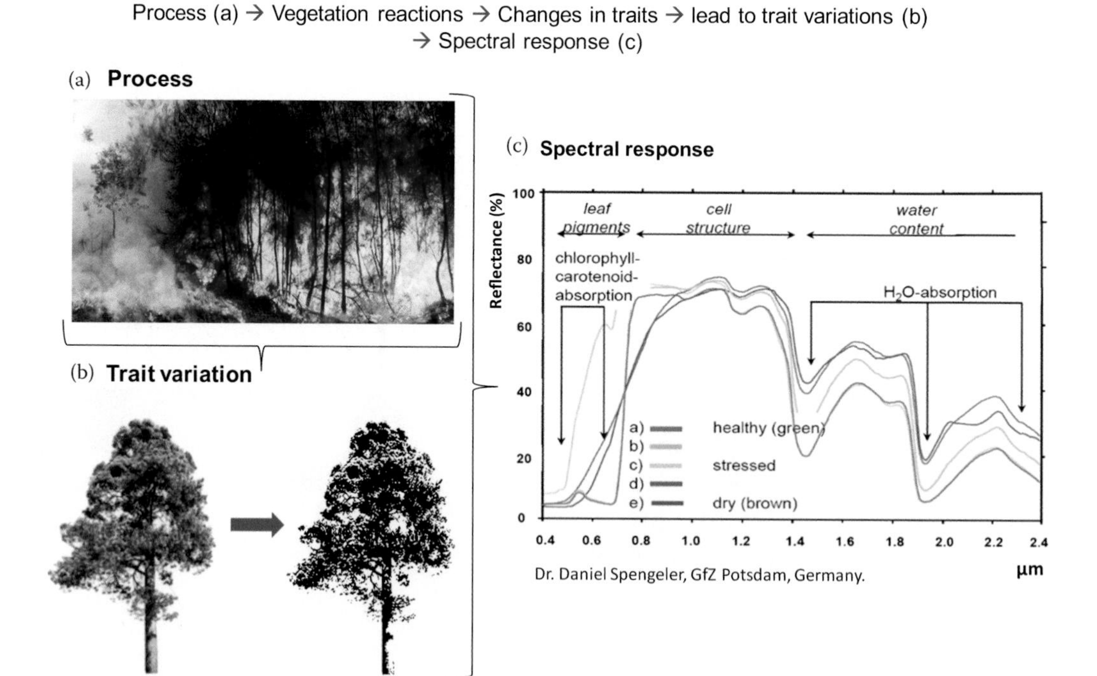

FIGURE 4.6 (a) The process of a forest fire (b) leads to spectral changes of traits and thus also to trait variations, which (c) leads to a change in the spectral response of the hyperspectral remote sensing data. The change in the spectral response is thus a proxy for the status, processes, stress, disturbances, or resource limitations in vegetation diversity and health, Image: Dr. Daniel Spengeler, GfZ Potsdam, Germany.

4.4 CONSTRAINTS OF HRS TECHNIQUES FOR RECORDING TRAITS AND TRAIT VARIATIONS

It has been demonstrated that by using the ST and STV concept, HRS is capable of recording status as well as stress and disturbances in phylogenetic, taxonomic, structural, and functional vegetation diversity. Nonetheless, HRS also exhibits some constraints and limitations, which have an influence on the recording of ST/STV and as a result on the status, changes and shifts in vegetation, which are summarized in the following (modified after Lausch et al., 2016a):

1. The characteristics of ST are important, as are the STV of plants, populations, communities, or landscape types. Traits of various taxa or landscape types are characterized by their phylogenetic development and are changed in an epigenetic way by stress, disturbances, or resource limitations. As a result of this, taxa-specific traits develop specific reactions, interactions ,and ecological potencies which can lead to taxa or landscape-type specific trait variations given various stressors and disturbances. Not all characteristics of ST/STV can be recorded by HRS, although its high spectral resolution is often an advantage (Figures 4.2 and 4.3).
2. The shape, density, and distribution of ST/STV are crucial. The spatial and temporal shape, density, and distribution of the composition and configuration (Lausch et al., 2015a) of traits and trait variations can potentially influence their ability to be recorded, and potentially correlate directly with the spatial, spectral, radiometric, angular, and temporal characteristics of HRS. For instance, ST/STV with unfavorable trait forms (long linear traits like small red bells) and distributions such as elongated ones, or a low density and unfavorable distribution in space can limit their recordability with HRS.
3. Additional limitations are the spatial, spectral, radiometric, angular, and temporal characteristics of the HRS sensors. These limitations are mostly dependent on the type of HRS techniques and installed RS platforms such as Analytical Spectral Devices (ASD) including cameras and sensors in a laboratory, sensor networks in tower installations (close-range RS techniques), drones, airplanes (airborne RS), or satellites (spaceborne RS) (Lausch et al., 2017). The characteristics of HRS sensors influence the ability to record ST/STV and hence the recording and monitoring of status and changes in phylogenetic, taxonomic, structural, and functional vegetation diversity at different organization levels.
4. The choices of the classification method are also important (pixel-based or geographic/object-based approaches).
5. How well the applied algorithm and its assumptions fit the RS data and the ST of the species will determine the measurability, discrimination, and thus the derivation and assessment of the entities of biodiversity with RS.

All five constraints for recording and monitoring the status and changes in phylogenetic, taxonomic, structural, and functional vegetation diversity and their shifts and disturbances to assess vegetation health as well as their resilience on different organization levels are inseparable and must be considered in conjunction with one another.

4.5 MONITORING PHYLOGENETIC, TAXONOMIC, STRUCTURAL, AND FUNCTIONAL VEGETATION DIVERSITY AND HEALTH USING HRS

The characteristics of phylogenetic vegetation diversity, taxonomy, traits, structure and function overlap and interact in different ways and to varying degrees. However, in order to gain a basic understanding of the potential as well as the limitations of HRS techniques, the ability to record and monitor the status, stress and resource limitations of vegetation diversity using HRS will be considered individually (Figure 4.1).

4.5.1 Phylogenetic Vegetation Diversity

Phylogenetic vegetation diversity promotes ecosystem resilience and stability (Cadotte et al., 2012). Therefore, measuring phylogenetic diversity is crucial. HRS and future developments in imaging spectroscopy using sensors like the Environmental Mapping and Analysis Program—EnMAP (Guanter et al., 2015), Hyperspectral Infrared Imager—HyspIRI (Lee et al., 2015), or the Hyperspectral Precursor of the Application Mission—PRISMA hyperspectral mission (Stefano et al., 2013), with its high spectral resolution is, to date, the most successful and only approach to record phylogenetic vegetation diversity using RS techniques. Based on continuous imaging spectroscopy, the "spectranomics approach" was developed (Asner and Martin, 2009, 2016), which links phylogenetic, taxonomic, and functional vegetation characteristics based on the chemical makeup or chemical phylogeny of plant species and canopy properties (McManus et al., 2016). Following this approach, 21 biochemical elements and their molecular properties, for example, the content of various photosynthetic pigments, nitrogen, phosphorous, polyphenols, cellulose, lignin, or water in leaves were recorded with HRS (Asner and Martin, 2009; see also Figure 4.7). Furthermore,

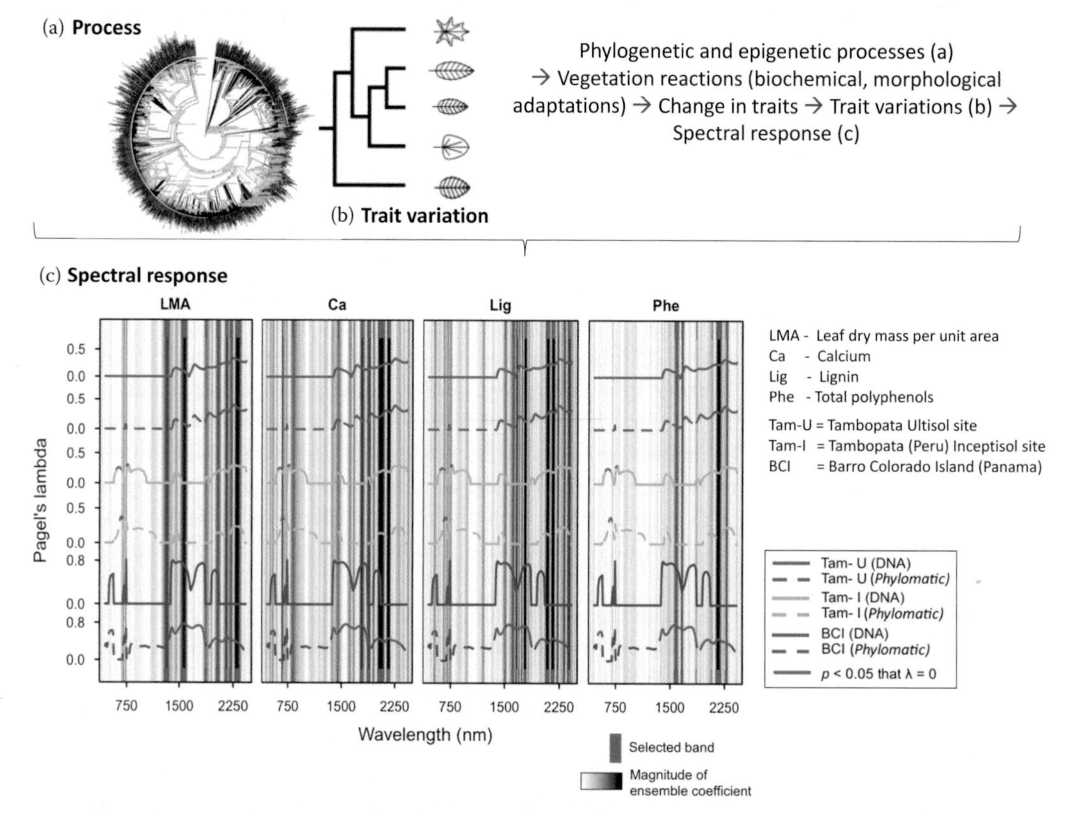

FIGURE 4.7 (a) Phylogenetic and epigenetic processes lead to vegetation responses in the form of biochemical or morphological adaptations in vegetation, (b) lead to changes in traits and trait variations, which can be measured as a (c) result in the spectral response of hyperspectral RS data. This enables phylogenetic diversity to be monitored by HRS. Mean reflectance spectra for each site (colored lines) and range of all reflectance spectra for all sites included in the study (grey area); Phylogenetic signal, as Pagel's lambda, of species reflectance coefficients at each site for the phylogenetic signal, as Pagel's lambda, of species reflectance coefficients at each site for four representative functional traits, overlaid on the multimodel ensemble regression coefficients for representative functional traits, overlaid on the multimodel ensemble regression coefficients (grey bars) as measures of band importance to each trait model. Site-specific lambda spectra are (grey bars) as measures of band importance to each trait model. Copyright: Licence number: 4390510942829. (From McManus K.M. et al., 2016. *Remote Sens.* 8, 1–16. doi: 10.3390/rs8030196.)

McManus et al. (2016) analyzed the phylogenetic structure of foliar spectral traits in tropical forest canopies using HRS techniques. The spectral fingerprint of each plant species and canopy were generated from their chemical phylogeny characteristics, based on the similarity and uniqueness of the chemical makeup of plant taxa and communities (Asner and Martin, 2016). Schweiger et al. (2018) predict that 97% of phylogenetic diversity in plant species is based on an integrative spectral diversity indicator on the basis of the leaf economic spectrum (Díaz et al., 2015). Cavender-Bares et al. (2016) investigated the linkage of leaf spectral using HRS techniques with genetic and phylogenetic variation in oaks and showed that the spectral similarity is significantly linked to the phylogenetic similarity among oak species. Many more examples are provided in Table 4.1.

4.5.2 Taxonomic Vegetation Diversity Using HRS

Plant species taxonomic diversity, heterogeneity, and richness and their shifts are key parameters for describing the status, stability, and resilience of ecosystems (Richter et al., 2016). Due to its high spectral resolution, HRS is suitable for recording a variety of different plant traits and trait variations. Compared to broadband RS techniques, HRS is also very well suited for mapping and distinguishing tree species (Roth et al., 2015), tree communities (Fassnacht et al., 2016; Leitão et al., 2017), floristic compositions (Schmidtlein and Sassin, 2004; Leutner et al., 2012), dominant species, mapping functional guilds, invasive plant species (Somers and Asner, 2013; Santos et al., 2016), and forest or other vegetation types (Laurin et al., 2016).

In spite of being able to record different traits and trait variations well, the mapping of taxonomic vegetation diversity does have its restrictions. Indeed, different taxa in plant species, communities, guilds, vegetation types, and biomes can only be distinguished with HRS if the plant taxa or plant communities can be differentiated from one another by their trait and/or trait variations. Various taxa can therefore only be distinguished if their species-dependent (phylogenetic and epigenetic-determined) traits as well as their development processes (senescence or phenology traits) can be distinguished.

In this way, different taxa in plant communities right up to landscape types can be distinguished from each other and monitored using HRS under the following prerequisites (see also Figure 4.8):

a. The chemical, biochemical-biophysical, morphometric, geometric, or physiology ST/STV of taxa can be discriminated using HRS.
b. The life cycle and the development of the taxa, such as senescence, phenology, flowering characteristics, and flowering time or growth characteristics, also allow for an additional temporal discrimination on trait variations.
c. Specific regional resource limitations are present which determine the geographic presence of plant species.
d. Different taxa differ from each other in the development of stress, adaptations, and disturbance traits, for example, leaf indumentum, strength of the cuticle, leaf morphology, change to the intercellular storage tissue, adaptions and specific habitats, and stress factors (e.g., ecological strategy types, Ellenberg indicator plant species, hemeroby types, plants for urbanity, or naturalness).

Furthermore, there are some important limitations when it comes to the discrimination of plant taxa using HRS. These relate to the characteristics, combinations, and variations of taxa traits and the spatial composition and configuration of taxa traits such as shape, density, vertical layering, and distribution in space and/or over time, which lead to a critical level of spectroscopic measurability using HRS. This critical level is in turn dependent upon the spatial and directional resolution of HRS data. Furthermore, multitemporal HRS approaches, which enable high temporal HRS data, improve the recording and discrimination of further traits for various taxa. The choice of the classification method (pixel-based or object-based approach) and how well the RS algorithm and its assumptions

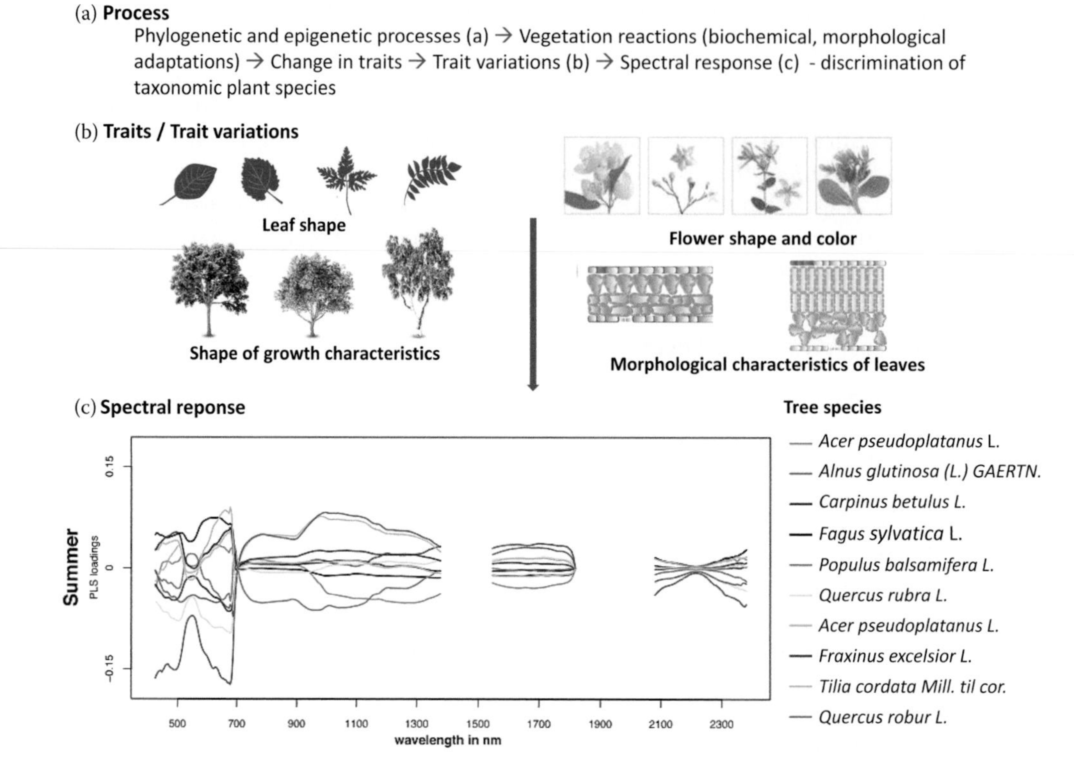

FIGURE 4.8 (a) Phylogenetic and epigenetic processes lead to vegetation responses in the form of biochemical or morphological adaptations in vegetation, which (b) lead to changes in traits and trait variations. A discrimination of different taxonomic plant species is possible if the spectral responses of the individual species are different from one another. (c) Discrimination of different forest tree species based on airborne hyperspectral remote sensing data (AISA-DUAL) with partial least square discriminant analysis (PLS-DA), mean PLS loadings for each species as obtained for seven latent variables (99% explained variance). (Modified after Richter R. et al., 2016. *Int. J.* 52, 464–474. doi: 10.1016/j.jag.2016.07.018.)

fit with HRS data, the characteristics, and the spatial distribution of traits from various taxa will influence the discrimination, derivation, as well as the assessment of taxa using HRS (Lausch et al., 2016a). Finally, the integration and coupling of hyperspectral databases for taxa (SPECCHIO; Hueni et al., 2009) and close-range HRS techniques mounted on different platforms (e.g., ASD, WSN, stationary and mobile towers, image spectroscopy in the laboratory, airborne and spaceborne-HRS) provide crucial information on the calibration, the validation, and thus on the discrimination and monitoring of different taxa.

If various taxa can be discriminated using RS time series, the HRS time series are, compared to broadband RS techniques, very well suited for monitoring the status and changes to taxonomic diversity. However, mapping taxonomic vegetation diversity using HRS techniques, when compared to in situ techniques, will always remain a spectral recording of the taxonomic diversity, which only partially corresponds to the taxonomic vegetation diversity by in situ processes.

4.5.3 Structural Vegetation Diversity

Natural and anthropogenic disturbances lead to changes in structural vegetation diversity and thus to changes in structural traits. There is a direct correlation between the structural characteristics of plant traits and trait variations in response to the natural and human influences on vegetation and landscapes. Since traits can exist at all levels of biological organization, it is possible to record

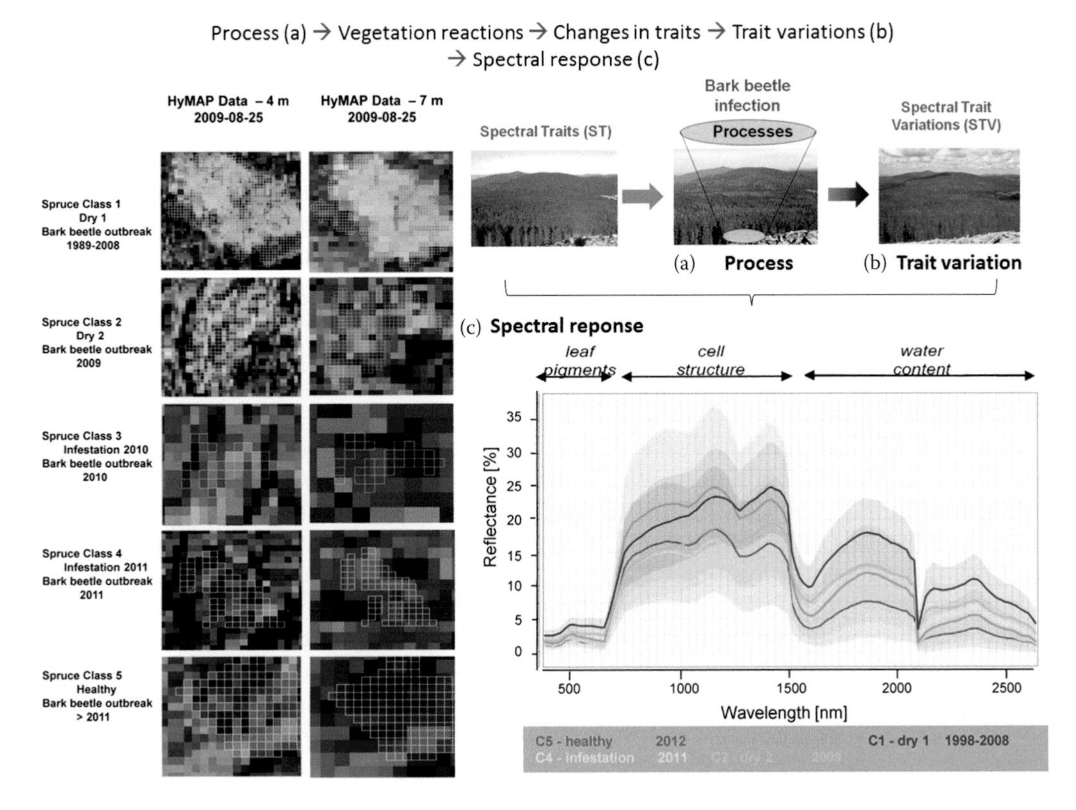

FIGURE 4.9 (a) Processes of bark beetle infestations lead to vegetation reactions, (b) which lead to changes in traits and trait variations, and to a (c) change in the spectral response. Trees infested with bark beetle gradually die, resulting in a change in the structural diversity of forest ecosystems. (Modified after Lausch A. et al., 2013a. *For. Ecol. Manage.* 308, 76–89. doi: 10.1016/j.foreco.2013.07.043.)

the structural vegetation diversity (composition and configuration) at the phylogenetic (McManus et al., 2016), molecular, chemical, and biochemical (Asner and Martin, 2008; Ustin et al., 2009) organismic levels (Asner, 2015), as well as species and communities (Lopatin et al., 2017), population and community structure levels (Baldeck et al., 2016, 2015; Laurin et al., 2016) up to biomes and ecosystem and landscape types (Leitão et al., 2015). Thanks to the high spectral resolution as well as improved temporal resolution, HRS techniques are especially suitable for monitoring structural traits, for example, floristic continuums (Schmidtlein et al., 2007), plant species heterogeneity, diversity, or richness (Leitão et al., 2015, 2017), land-use/land cover mapping (Schwieder et al., 2014), fragmentation and connectivity, as well as bark beetle defoliator infestations (see Figure 4.9). They are also potentially suitable for monitoring and describing urbanization, neighborhood relationships, area, density, size, and shape characteristics related to habitats and harvesting, and land-use intensity and grassland management strategies as well as assessing the naturalness, habitat quality, homogenization in vegetation, monitoring of protected areas, conservation, and landscapes. Here, HRS technologies record hyperspectral vegetation structures not only in 2D, but also and increasingly in 3D (Aasen et al., 2015).

4.5.4 Functional Vegetation Diversity Using HRS

Recording the status and shifts in functional vegetation diversity is a comprehensive application area of HRS techniques. What is important here is the question of why HRS can record the status and shifts in ecosystem functions. Indeed trait-based approaches in ecology help to improve our understanding of why organisms live where they do and how they will respond to environmental change (Green et al., 2008). Trait-based approaches therefore follow the principle of the functionality of the trait concept (Violle et al., 2007). Since HRS is highly suitable for recording traits and trait variations, HRS techniques are able to record the status and shifts in ecosystem functions on different platforms and at all levels of organization. If the intention is to record ecosystem functions using HRS, the traits which regulate and influence the ecosystem functions have to be defined and the trait variations, triggered by processes, have to be identified. Figure 4.3 outlines specific "physiological and functional spectral traits" such as ST for photosynthesis, photosynthetic pathway, carbon sequestration, or evapotranspiration, which can be recorded by HRS. Some traits and trait combinations are involved that can have an impact on functional diversity. In order to be able to record and discriminate the status and shifts of functional trait diversity, the same criteria, limitations, and constraints apply as those for discriminating among different taxa to record taxonomic diversity using HRS.

Braun et al. (2017) present a method to map regional heterogeneity and patterns of tree functional diversity using a combination of imaging spectroscopy and laser scanning. They generated a continuous map of functional diversity that considered various traits without the need to identify species or individuals using taxonomists. This approach is a fundamental step forward in mapping functional vegetation diversity using HRS.

Plant functional types (PFT) are functional convergences based on environmental resources and stress constraints and measurements conducted with remote sensing are based on plant traits (Ustin and Gamon, 2010). The most important example of plant functional types are the CSR strategy types by Grime (1974), where plant traits change according to their stressors and to disturbance regimes. According to this system, community composition is established from ruderal plants, competitive plants, and stress-tolerant plants. Plant traits such as the dry matter content, the canopy height, the onset of flowering, the flowering period, lateral spread, specific leaf area, and leaf dry weight are crucial for the position and mapping of plant species in the CSR trait space (Schmidtlein et al., 2012). Furthermore, PFTs also play a decisive role in deriving plant functions. The establishment of plant functional types, and complex plant strategy types as a form of adaptation of plants and communities, can be excellently demonstrated with high-resolution hyperspectral sensoring (Schmidtlein et al., 2012; Schmidt et al., 2017a; Figure 4.10).

A more comprehensive overview of the existing references that use HRS techniques to monitor the status, stress, and disturbances in vegetation diversity and health can be found in Table 4.1.

4.6 SUMMARIZING IMPORTANT QUESTIONS WHEN USING HRS TECHNIQUES FOR MONITORING VEGETATION DIVERSITY AND HEALTH

4.6.1 Which Traits of Vegetation and Vegetation Health Are Best Established Using HRS?

- HRS techniques can record traits and trait variations in vegetation in a direct way (e.g., chlorophyll content) or in an indirect way, meaning that processes are triggered by stress in the plant (e.g., fungicide application in a wheat field). In the same way, traits and trait variations can be recorded as single traits or as trait combinations.
- The ST and STV which can be recorded using the HRS technique are shown in Figure 4.3 and Table 4.1.

4.6.2 What Are the Factors Determining the Recordability of Traits and Trait Variations Using the HRS Technique?

The recordability of traits and trait variations in vegetation depends on the following criteria, (modified after Lausch et al., 2016a):

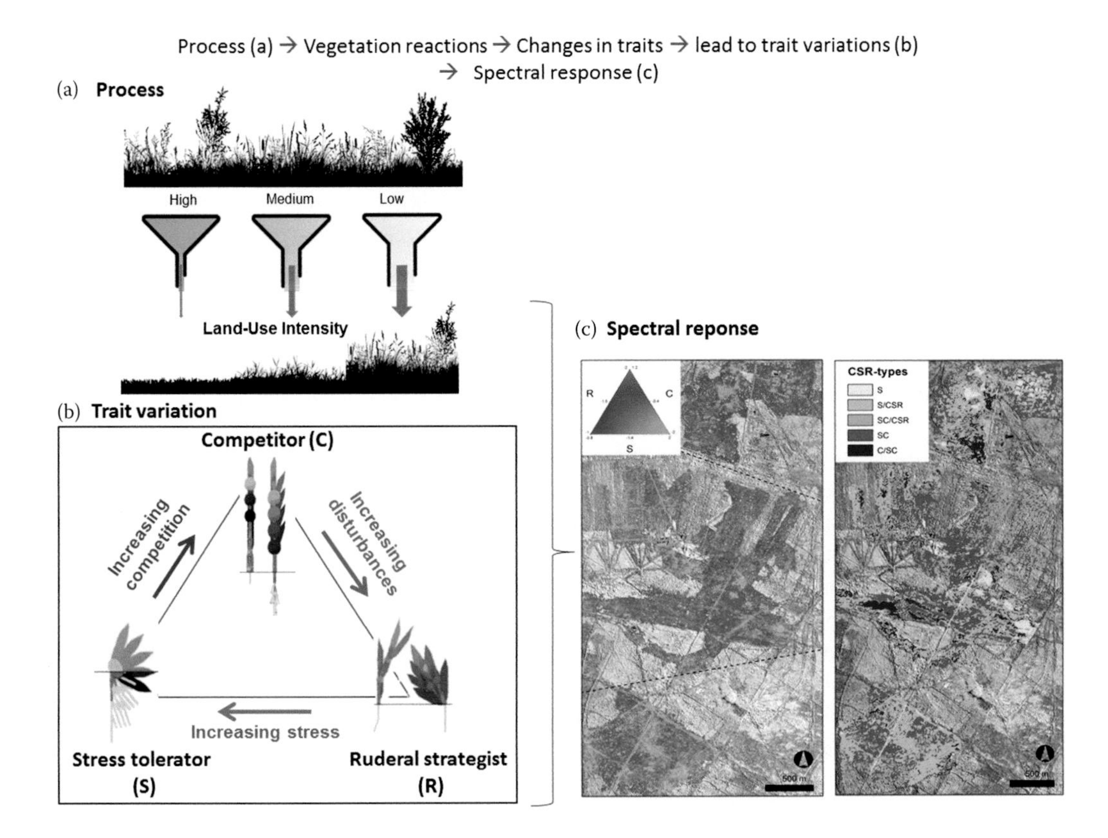

FIGURE 4.10 (a) Processes of land-use intensity (LUI), (b) lead to vegetation reactions, (c) resulting in an adaptation of plant strategy types (competitor-C, stress tolerator-S, ruderal-R) with altered traits and trait variations and consequently to a new spectral response because of the adaptation to the LUI. (Modified after Wellmann T. et al., 2018. *Ecol. Indic.* 85, 190–203; Schmidt, J. et al., 2017a. *Ecol. Indic.* 73, 505–512.)

- The characteristics of traits and trait variations at the different levels of vegetation organization (from genetic, molecular, and individual to the landscape level).
- The shape, density, and distribution of traits and trait variations (from genetic, molecular, and individual to the landscape level) in space and over time.
- The spatial, spectral, radiometric, angular, and temporal characteristics of the hyperspectral sensor.
- The hyperspectral sensor platform to be used (close-range, airborne, or spaceborne).
- The choice of the classification method (pixel-based or [geographic] object-based approach).
- How well the hyperspectral RS algorithm and its conclusions fit the hyperspectral data and the ST and STV of the species.

4.6.3 What Is the Potential and What Are the Limitations of HRS When Studying Vegetation Diversity and Health?

Potential:

- Traits and trait variations exist at all levels of the vegetation hierarchy: molecular, genetic, individual, species, populations, communities, biomes, ecosystems, and landscapes. Therefore, HRS that is based on different platforms (close-range, airborne, spaceborne) is crucial and, indeed, the only approach to record a large variety of ST and STV on all of the scales.

- Additionally, due to its high spectral resolution, HRS is the only approach capable of monitoring status, adaptations, shifts, stress, and disturbances in all of the four characteristics of vegetation diversity, namely: phylogenetic, taxonomic, structural, and functional vegetation diversity.
- In terms of the spectrometric approach, HRS is to date also the only RS technique which makes it possible to draw conclusions on the status and shifts in phylogenetic diversity.
- Recording and distinguishing different plant taxa and communities as far as landscape types with HRS is much more feasible than with broadband (multispectral) RS, due to its high spectral resolution and, thus, its ability to distinguish spectral traits.
- As a result of recent developments in 3D hyperspectral sensors, traits and trait variations can increasingly be recorded in 3D.
- The potential to record traits and trait variations at all levels of the vegetation hierarchy will be improved immensely when multitemporal and multisensor RS techniques and data are used. As a result of the diversity of various multisensors with their differentiated spatial, spectral, radiometric, angular, and temporal characteristics, a range of traits and trait variations can be recorded and monitored.
- In the same way, traits and trait variations define the location, proliferation strategies, and occurrence and indicate disturbances and processes in plant species from local to global levels. Since HRS is the only type of sensor which can depict a variety of traits and trait variations extensively, objectively, comparatively, and repeatedly, HRS is suited to assess areas quantitatively and qualitatively, to evaluate protection status such as the Natura 2000 areas, and to define vegetation types on the basis of traits.
- Moreover, by using HRS techniques, marked improvement is expected in the recording and evaluation of human pressures on landscapes such as land-use intensity (LUI). In this way, important indicators of LUI can be recorded using HR such as the application of pesticides, fungicides, and fertilizers, the infestation of insects and fungi, or management indicators for agriculture that are intensive or extensive in nature.
- HRS also makes it possible to record status, shifts, and stress in structural and functional vegetation diversity. The tremendous ability of HRS to distinguish ST and STV makes it the most successful technique for recording indicators to assess and quantify ecosystem services.

Limitations

- The HRS approach is, however, also limited in comparison to other RS techniques when it comes to recording taxonomic diversity such as in the in situ approaches used by taxonomists.
- Depicting trait information in 3D is at the moment very limited with HRS.
- HRS can record status, stress, disturbances, resource limitations, and processes from local to global levels in an extensive, objective, comparative, and repeated way. Without coupling with high-frequency, close-range RS data like Wireless Sensor Networks (WSN) or information from spectral laboratory and plant phenomics facilities (fully automated operation), or Ecotrons (controlled environmental facility; Lausch et al., 2017), the recorded processes (due to the overlapping of different drivers for processes) cannot be sufficiently explained using HRS techniques. In the same way, important calibration and validation information for HRS data is missing.
- The anticipated hyperspectral spaceborne sensors like EnMAP and HyspIRI have to cope with big RS data. If HRS data is to be used in the future, requirements in data management, archiving, and evaluation for HRS techniques will be required. Topics like data science, big data, machine learning methods, as well as linked open data (LOD), semantification, or ontologization will then play a huge role (Lausch et al., 2015c, 2018).

4.6.4 A Comparison of HRS with Broadband RS Techniques

- The potential of HRS discussed in Section 4.6.3 at the same time constitutes the advantage of HRS compared to the broadband RS technique.
- To date, a tremendous advantage of multispectral (MS) data lies in the presence of long time series with free access to MS data like the Landsat-Mission.
- Moreover, for monitoring vegetation diversity and health, RS data with a high temporal repetition rate is required, as is the case for sensors like Modis, RapidEye, or Sentinel-2. Due to the tremendous amount of data, which for instance is expected with EnMAP, the repetition rate of spaceborne HRS compared to MS data is limited. Since not many papers have been written about the comparative assessment of spectral, spatial, or temporal resolution of remote sensing data, there is no definitive clarification as to whether a high temporal resolution compared to a high spectral resolution is more important for recording (Lausch et al., 2013b).
- As a result of the spatial resolution of EnMAP data with 30 m, future HRS data compared to MS data is only partially suitable for examining certain topics like urban structures and processes. But in the near future, HRS sensors are expected to feature improved spatial resolution.

4.6.5 What Are the Advantages and Disadvantages of Studying Vegetation Diversity and Health Using Hyperspectral Remote Sensing as Opposed to the Conventional In Situ Approaches Used by Biologists and Taxonomists?

When all monitoring approaches of the four vegetation characteristics are compared with one another, then it is possible to summarize the following advantages and disadvantages.

4.6.5.1 In Situ Approach

Advantages

- High accuracy of the mapping and taxonomic allocation of species, populations, and communities up to biotopes.
- Accurate knowledge about phylogeny, ecological power, the occupation of niches, as well as reactions and resilience to stress and disturbances and their health.

Disadvantages

- Monitoring the environment, stress, and process interactions in plant species is only possible with the assistance of gradients (spatial or temporal gradient).
- Limited monitoring of short-term stress or disturbances.
- Only a partial systematic monitoring from local to global levels is possible.
- Recording of often selective information.
- High costs, time, and data management.
- Objectivity, repeatability, and comparability of the monitoring is given, but dependent upon the availability of a plant species specialist.

4.6.5.2 Hyperspectral Remote Sensing Approach

Advantages

- Monitoring the environment, stress, and process interactions in plant species.
- Very good possibility to monitor short- to long-term stress or disturbances.
- Very good possibility for systematic monitoring from local to global levels.
- Recording of primarily surface information.
- Low costs, time, and data management.
- Objectivity, repeatability, and comparability of the monitoring are given, but dependent upon the availability and the know-how of a plant species specialist.

Disadvantages

- Low rate of thematic exactness of the mapping of species, populations, and communities.
- Limited monitoring of the phylogenetic and taxonomic vegetation diversity.
- Objectivity, repeatability, and comparability of the monitoring are given, but dependent upon the availability and the know-how of a plant species specialist.

4.7 CONCLUSIONS

The trait-based approach in RS is the basis for recording, monitoring, and assessing status, shifts, and changes caused by processes, stress, disturbances, and resource limitations in vegetation diversity. Trait and trait variations serve as a filter and a proxy for status, adaptations, shifts, and changes by processes, stress, disturbances, and resource limitations.

HRS is a crucial technique when it comes to recording spectral-based trait and trait variations, but compared to the morphological species concept (MSC) used by taxonomists, not all the traits and trait variations can be recorded using HRS. Therefore, we refer to traits and trait variations recorded and monitored using HRS techniques as being "spectral traits" (ST), the changes in spectral traits as "spectral trait variations" (STV) and the RS approach as the "remote sensing–spectral trait/spectral trait variations concept" (Lausch et al., 2016b).

The four characteristics of vegetation diversity not only overlap but also interact directly as well as indirectly with one another. Therefore, assigning ST/STV to specific groups or to the four characteristics of vegetation diversity only makes sense for a better classification. It is however imperative to adopt an approach whereby ST/STV support different characteristics of vegetation diversity and health in different ways, intensity, and characteristics.

Consequently, investigating the biological significance, interactions, and complementarity of the ST/STV for assessing vegetation diversity and health will be the greatest challenge in future research to assess the significance of characteristics and the type of spectral information for vegetation diversity and health.

Vegetation diversity is multidimensional and enormously complex in space and time. Since HRS techniques can be integrated into all platforms, it is very easy to monitor status, shifts, and disturbances in vegetation diversity from a local to a global scale, over the short to long term and from a low to a high frequency.

It is not possible to record the complexity of and the disturbances in ecological systems with one single scale or platform. Furthermore, given the fact that data-driven information is needed, which is geared toward calibrating and validating different platforms, future approaches will have to merge the different platforms with one another.

REFERENCES

Aasen, H., Burkart, A., Bolten, A., Bareth, G., 2015. Generating 3D hyperspectral information with lightweight UAV snapshot cameras for vegetation monitoring: From camera calibration to quality assurance. *ISPRS J. Photogramm. Remote Sens.* 108, 245–259. doi: 10.1016/j.isprsjprs.2015.08.002

Abdel-Rahman, E.M., Makori, D.M., Landmann, T., Piiroinen, R., Gasim, S., Pellikka, P., Raina, S.K., 2015. The utility of AISA eagle hyperspectral data and random forest classifier for flower mapping. *Remote Sens.* 7, 13298–13318. doi: 10.3390/rs71013298

Abelleira Martínez, O.J., Fremier, A.K., Günter, S., Ramos Bendaña, Z., Vierling, L., Galbraith, S.M., Bosque-Pérez, N.A., Ordoñez, J.C., 2016. Scaling up functional traits for ecosystem services with remote sensing: Concepts and methods. *Ecol. Evol.* doi: 10.1002/ece3.2201

Ali, A.M., Darvishzadeh, R., Skidmore, A.K., van Duren, I., Heiden, U., Heurich, M., 2016. Estimating leaf functional traits by inversion of PROSPECT: Assessing leaf dry matter content and specific leaf area in mixed mountainous forest. *Int. J. Appl. Earth Obs. Geoinf.* 45, 66–76. doi: 10.1016/j.jag.2015.11.004

Alton, P.B., 2017. Retrieval of seasonal Rubisco-limited photosynthetic capacity at global FLUXNET sites from hyperspectral satellite remote sensing: Impact on carbon modelling. *Agric. For. Meteorol.* 232, 74–88. doi: 10.1016/j.agrformet.2016.08.001

Aneece, I.P., Epstein, H., Lerdau, M., 2017. Correlating species and spectral diversities using hyperspectral remote sensing in early-successional fields. *Ecol. Evol.* 7, 3475–3488. doi: 10.1002/ece3.2876

Asner, G.P., 2013. Biological diversity mapping comes of age. *Remote Sens.* 5, 374–376. doi: 10.3390/rs5010374

Asner, G.P., 2015. Organismic remote sensing for tropical forest ecology and conservation. *Ann. Missouri Bot. Gard.* 100, 127–140. doi: 10.3417/2012016

Asner, G.P., Anderson, C.B., Martin, R.E., Tupayachi, R., Knapp, D.E., Sinca, F., 2015a. Landscape biogeochemistry reflected in shifting distributions of chemical traits in the Amazon forest canopy. *Nat. Geosci.* 8, 567–573. doi: 10.1038/ngeo2443

Asner, G.P., Clark, J.K., Mascaro, J., Galindo García, G.A., Chadwick, K.D., Navarrete Encinales, D.A., Paez-Acosta, G. et al., 2012a. High-resolution mapping of forest carbon stocks in the Colombian Amazon. *Biogeosciences Discuss.* 9, 2445–2479. doi: 10.5194/bgd-9-2445-2012

Asner, G.P., Knapp, D.E., Anderson, C.B., Martin, R.E., Vaughn, N., 2016a. Large-scale climatic and geophysical controls on the leaf economics spectrum. *Proc. Natl. Acad. Sci. U. S. A.* 113, E4043–E4051. doi: 10.1073/pnas.1604863113

Asner, G.P., Knapp, D.E., Martin, R.E., Tupayachi, R., Anderson, C.B., Mascaro, J., Sinca, F. et al., 2014a. Targeted carbon conservation at national scales with high-resolution monitoring. *Proc. Natl. Acad. Sci.* 111, E5016–5022. doi: 10.1073/pnas.1419550111

Asner, G.P., Martin, R.E., 2008. Spectral and chemical analysis of tropical forests: Scaling from leaf to canopy levels. *Remote Sens. Environ.* 112, 3958–3970. doi: 10.1016/j.rse.2008.07.003

Asner, G.P., Martin, R.E., 2009. Airborne spectranomics: Mapping canopy chemical and taxonomic diversity in tropical forests. *Front. Ecol. Environ.* 7, 269–276. doi: 10.1890/070152

Asner, G.P., Martin, R.E., 2011. Canopy phylogenetic, chemical and spectral assembly in a lowland Amazonian forest. *New Phytol.* 189, 999–1012. doi: 10.1111/j.1469-8137.2010.03549.x

Asner, G.P., Martin, R.E., 2016. Spectranomics: Emerging science and conservation opportunities at the interface of biodiversity and remote sensing. *Glob. Ecol. Conserv.* 8, 212–219. doi: 10.1016/j.gecco.2016.09.010

Asner, G.P., Martin, R.E., Anderson, C.B., Knapp, D.E., 2015b. Quantifying forest canopy traits: Imaging spectroscopy versus field survey. *Remote Sens. Environ.* 158, 15–27. doi: 10.1016/j.rse.2014.11.011

Asner, G.P., Martin, R.E., Anderson, C.B., Kryston, K., Vaughn, N., Knapp, D.E., Bentley, L.P. et al., 2016b. Scale dependence of canopy trait distributions along a tropical forest elevation gradient. *New Phytol.* doi: 10.1111/nph.14068

Asner, G.P., Martin, R.E., Carranza-Jimenéz, L., Sinca, F., Tupayachi, R., Anderson, C.B., Martinez, P., 2014b. Functional and biological diversity of foliar spectra in tree canopies throughout the Andes to Amazon region. *New Phytol.* 204, 127–139. doi: 10.1111/nph.12895

Asner, G.P., Martin, R.E., Knapp, D.E., Tupayachi, R., Anderson, C., Carranza, L., Martinez, P., Houcheime, M., Sinca, F., Weiss, P., 2011a. Spectroscopy of canopy chemicals in humid tropical forests. *Remote Sens. Environ.* 115, 3587–3598. doi: 10.1016/j.rse.2011.08.020

Asner, G.P., Martin, R.E., Knapp, D.E., Tupayachi, R., Anderson, C.B., Sinca, F., Vaughn, N.R., Llactayo, W., 2017. Airborne laser-guided imaging spectroscopy to map forest trait diversity and guide conservaton. *Science* 355, 385–389.

Asner, G.P., Martin, R.E., Suhaili, A.B., 2012b. Sources of canopy chemical and spectral diversity in Lowland Bornean Forest. *Ecosystems* 15, 504–517. doi: 10.1007/s10021-012-9526-2

Asner, G.P., Martin, R.E., Tupayachi, R., Anderson, C.B., Sinca, F., Carranza-Jiménez, L., Martinez, P., 2014c. Amazonian functional diversity from forest canopy chemical assembly. *Proc. Natl. Acad. Sci. U. S. A.* 111, 5604–5609. doi: 10.1073/pnas.1401181111

Asner, G.P., Martin, R.E., Tupayachi, R., Emerson, R., Martinez, P., Sinca, F., Powell, G.V.N., Wright, S.J., Lugo, A.E., 2011b. Taxonomy and remote sensing of leaf mass per area (LMA) in humid tropical forests. *Ecol. Appl.* 21, 85–98. doi: 10.1890/09-1999.1

Asner, G.P., Mascaro, J., 2014. Mapping tropical forest carbon: Calibrating plot estimates to a simple LiDAR metric. *Remote Sens. Environ.* 140, 614–624. doi: 10.1016/j.rse.2013.09.023

Asner, G.P., Mascaro, J., Anderson, C., Knapp, D.E., Martin, R.E., Kennedy-Bowdoin, T., van Breugel, M. et al., 2013. High-fidelity national carbon mapping for resource management and REDD+. *Carbon Balance Manag.* 8, 7. doi: 10.1186/1750-0680-8-7

Asner, G.P., Mascaro, J., Muller-Landau, H.C., Vieilledent, G., Vaudry, R., Rasamoelina, M., Hall, J.S., van Breugel, M., 2012c. A universal airborne LiDAR approach for tropical forest carbon mapping. *Oecologia* 168, 1147–1160. doi: 10.1007/s00442-011-2165-z

Assal, T.J., Anderson, P.J., Sibold, J., 2016. Spatial and temporal trends of drought effects in a heterogeneous semi-arid forest ecosystem. *For. Ecol. Manage.* 365, 137–151. doi: 10.1016/j.foreco.2016.01.017

Baldeck, C., Asner, G., 2014. Single-species detection with airborne imaging spectroscopy data: A comparison of support vector techniques. 1–12.

Baldeck, C.A., Asner, G.P., Martin, R.E., Anderson, C.B., Knapp, D.E., Kellner, J.R., Wright, S.J., 2015. Operational tree species mapping in a diverse tropical forest with airborne imaging spectroscopy. *PLoS One* 10. doi: 10.1371/journal.pone.0118403

Baldeck, C.A., Tupayachi, R., Sinca, F., Jaramillo, N., Asner, G.P., 2016. Environmental drivers of tree community turnover in western Amazonian forests. *Ecography (Cop.)*. doi: 10.1111/ecog.01575

Ball, A., Sanchez-Azofeifa, A., Portillo-Quintero, C., Rivard, B., Castro-Contreras, S., Fernandes, G., 2015. Patterns of leaf biochemical and structural properties of Cerrado life forms: Implications for remote sensing. *PLoS One* 10, 1–15. doi: 10.1371/journal.pone.0117659

Balzotti, C.S., Asner, G.P., Taylor, P.G., Cleveland, C.C., Cole, R., Martin, R.E., Nasto, M., Osborne, B.B., Porder, S., Townsend, A.R., 2016. Environmental controls on canopy foliar N distributions in a neotropical lowland forest. *Ecol. Appl.* 26, 2449–2462. doi: 10.1002/eap.1408

Banerjee, B.P., Raval, S., Zhai, H., Cullen, P.J., 2017. Health condition assessment for vegetation exposed to heavy metal pollution through airborne hyperspectral data. *Environ. Monit. Assess.* 189, 604. doi: 10.1007/s10661-017-6333-4

Bareth, G., Aasen, H., Bendig, J., Gnyp, M.L., Bolten, A., Jung, A., Michels, R., Soukkamäki, J., 2015. Low-weight and UAV-based hyperspectral full-frame cameras for monitoring crops: Spectral comparison with portable spectroradiometer measurements. *Photogramm. Fernerkundung Geoinf.* 2015, 69–79. doi: 10.1127/pfg/2015/0256

Behmann, J., Steinrücken, J., Plümer, L., 2014. Detection of early plant stress responses in hyperspectral images. *ISPRS J. Photogramm. Remote Sens.* 93, 98–111. doi: 10.1016/j.isprsjprs.2014.03.016

Bergen, K.M., Goetz, S.J., Dubayah, R.O., Henebry, G.M., Hunsaker, C.T., Imhoff, M.L., Nelson, R.F., Parker, G.G., Radeloff, V.C., 2009. Remote sensing of vegetation 3-D structure for biodiversity and habitat: Review and implications for lidar and radar spaceborne missions. *J. Geophys. Res.* 114, G00E06. doi: 10.1029/2008JG000883

Blackburn, G.A., 2007. Hyperspectral remote sensing of plant pigments. *J. Exp. Bot.* 58, 855–867. doi: 10.1093/jxb/erl123

Boulangeat, I., Thuiller, W., 2012. Accounting for dispersal and biotic interactions to disentangle the drivers of species distributions and their abundances. *Ecol. Lett.* 15, 584–593. doi: 10.1111/j.1461-0248.2012.01772.x

Bratsch, S.N., Epstein, H.E., Buchhorn, M., Walker, D.A., 2016. Differentiating among four Arctic tundra plant communities at Ivotuk, Alaska using field spectroscopy. *Remote Sens.* 8, 51. doi: 10.3390/rs8010051

Bratsch, S.N., Epstein, H.E., Buchhorn, M., Walker, D.A., Landes, H., 2017. Relationships between hyperspectral data and components of vegetation biomass in Low Arctic tundra communities at Ivotuk, Alaska. *Environ. Res. Lett.* 12, 025003. doi: 10.1088/1748-9326/aa572e

Braun, D., Damm, A., Paul-Limoges, E., Revill, A., Buchmann, N., Petchey, O.L., Hein, L., Schaepman, M.E., 2017. From instantaneous to continuous: Using imaging spectroscopy and in situ data to map two productivity-related ecosystem services. *Ecol. Indic.* 82, 409–419. doi: 10.1016/j.ecolind.2017.06.045

Brosinsky, A., Lausch, A., Doktor, D., Salbach, C., Merbach, I., Gwillym-Margianto, S., Pause, M., 2013. Analysis of spectral vegetation signal characteristics as a function of soil moisture conditions using hyperspectral remote sensing. *J. Indian Soc. Remote Sens.* 42, 311–324. doi: 10.1007/s12524-013-0298-8

Buddenbaum, H., Schlerf, M., Hill, J., 2005. Classification of coniferous tree species and age classes using hyperspectral data and geostatistical methods. *Int. J. Remote Sens.* 26, 5453–5465. doi: 10.1080/01431160500285076

Burkart, A., Hecht, V.L., Kraska, T., Rascher, U., 2017. Phenological analysis of unmanned aerial vehicle based time series of barley imagery with high temporal resolution. *Precis. Agric.* doi: 10.1007/s11119-017-9504-y

Cadotte, M.W., Arnillas, C.A., Livingstone, S.W., Yasui, S.L.E., 2015. Predicting communities from functional traits. *Trends Ecol. Evol.* 30, 510–511. doi: 10.1016/j.tree.2015.07.001

Cadotte, M.W., Dinnage, R., Tilman, D., 2012. Phylogenetic diversity promotes ecosytem stability. *Ecology* 93, S223–S233. doi: 10.1890/11-0426.1

Carlson, K.M., Asner, G.P., Hughes, R.F., Ostertag, R., Martin, R.E., 2007. Hyperspectral remote sensing of canopy biodiversity in Hawaiian lowland rainforests. *Ecosystems* 10, 536–549. doi: 10.1007/s10021-007-9041-z

Cavender-Bares, J., Meireles, J.E., Couture, J.J., Kaproth, M.A., Kingdon, C.C., Singh, A., Serbin, S.P. et al., 2016. Associations of leaf spectra with genetic and phylogenetic variation in oaks: Prospects for remote detection of biodiversity. *Remote Sens.* 8. doi: 10.3390/rs8030221

Chen, J., Shen, M., Zhu, X., Tang, Y., 2009. Indicator of flower status derived from in situ hyperspectral measurement in an alpine meadow on the Tibetan Plateau. *Ecol. Indic.* 9, 818–823. doi: 10.1016/j.ecolind.2008.09.009

Clasen, A., Somers, B., Pipkins, K., Tits, L., Segl, K., Brell, M., Kleinschmit, B., Spengler, D., Lausch, A., Förster, M., 2015. Spectral unmixing of forest crown components at close range, airborne and simulated Sentinel-2 and EnMAP spectral imaging scale. *Remote Sens.* 7, 15361–15387. doi: 10.3390/rs71115361

Couture, J.J., Serbin, S.P., Townsend, P.A., 2013. Spectroscopic sensitivity of real-time, rapidly induced phytochemical change in response to damage. *New Phytol.* 198, 311–319. doi: 10.1111/nph.12159

Czerwinski, C.J., King, D.J., Mitchell, S.W., 2014. Mapping forest growth and decline in a temperate mixed forest using temporal trend analysis of Landsat imagery, 1987–2010. *Remote Sens. Environ.* 141, 188–200. doi: 10.1016/j.rse.2013.11.006

Damm, A., Guanter, L., Paul-Limoges, E., van der Tol, C., Hueni, A., Buchmann, N., Eugster, W., Ammann, C., Schaepman, M.E., 2015. Far-red sun-induced chlorophyll fluorescence shows ecosystem-specific relationships to gross primary production: An assessment based on observational and modeling approaches. *Remote Sens. Environ.* 166, 91–105. doi: 10.1016/j.rse.2015.06.004

Danner, M., 2017. Retrieval of biophysical crop variables from multi-angular canopy spectroscopy. *Remote Sens.* 9, 726. doi: 10.3390/rs9070726

Dash, J.P., Watt, M.S., Pearse, G.D., Heaphy, M., Dungey, H.S., 2017. Assessing very high resolution UAV imagery for monitoring forest health during a simulated disease outbreak. *ISPRS J. Photogramm. Remote Sens.* 131, 1–14. doi: 10.1016/j.isprsjprs.2017.07.007

Dechant, B., Cuntz, M., Vohland, M., Schulz, E., Doktor, D., 2017. Estimation of photosynthesis traits from leaf reflectance spectra: Correlation to nitrogen content as the dominant mechanism. *Remote Sens. Environ.* 196, 279–292. doi: 10.1016/j.rse.2017.05.019

Degerickx, J., Hermy, M., Somers, B., 2017. Mapping functional urban green types using hyperspectral remote sensing. *2017 Jt. Urban Remote Sens. Event*, 1–4. doi: 10.1109/JURSE.2017.7924553

de Moura, Y.M., Galvão, L.S., Hilker, T., Wu, J., Saleska, S., do Amaral, C.H., Nelson, B.W. et al., 2017. Spectral analysis of amazon canopy phenology during the dry season using a tower hyperspectral camera and modis observations. *ISPRS J. Photogramm. Remote Sens.* 131, 52–64. doi: 10.1016/j.isprsjprs.2017.07.006

Deshpande, S., Inamdar, A., Vin, H., 2017. Urban land use/land cover discrimination using image-based reflectance calibration methods for hyperspectral data. *Photogramm. Eng. Remote Sens.* 83, 365–376. doi: 10.14358/PERS.83.5.365

Díaz, S., Kattge, J., Cornelissen, J.H.C., Wright, I.J., Lavorel, S., Dray, S., Reu, B. et al., 2015. The global spectrum of plant form and function. *Nature* 529, 1–17. doi: 10.1038/nature16489

Doktor, D., Lausch, A., Spengler, D., Thurner, M., 2014. Extraction of plant physiological status from hyperspectral signatures using machine learning methods. *Remote Sens.* 6, 12247–12274. doi: 10.3390/rs61212247

Eldredge, N., Cracraft, J., 1980. *Phylogenetic Patterns and the Evolutionary Process.* University Press, New York, Columbia.

El-Hendawy, S., Al-Suhaibani, N., Hassan, W., Tahir, M., Schmidhalter, U., 2017. Hyperspectral reflectance sensing to assess the growth and photosynthetic properties of wheat cultivars exposed to different irrigation rates in an irrigated arid region. *PLoS One* 12, 1–22. doi: 10.1371/journal.pone.0183262

Fassnacht, F.E., Latifi, H., Ghosh, A., Joshi, P.K., Koch, B., 2014. Assessing the potential of hyperspectral imagery to map bark beetle-induced tree mortality. *Remote Sens. Environ.* 140, 533–548. doi: 10.1016/j.rse.2013.09.014

Fassnacht, F.E., Latifi, H., Stereńczak, K., Modzelewska, A., Lefsky, M., Waser, L.T., Straub, C., Ghosh, A., 2016. Review of studies on tree species classification from remotely sensed data. *Remote Sens. Environ.* 186, 64–87. doi: 10.1016/j.rse.2016.08.013

Feilhauer, H., Dahlke, C., Doktor, D., Lausch, A., Schmidtlein, S., Schulz, G., Stenzel, S., 2014. Mapping the local variability of Natura 2000 habitats with remote sensing. *Appl. Veg. Sci.* 17, 765–779. doi: 10.1111/avsc.12115

Feilhauer, H., Doktor, D., Schmidtlein, S., Skidmore, A.K., 2016. Mapping pollination types with remote sensing. *J. Veg. Sci.* 27, 999–1011. doi: 10.1111/jvs.12421

Feilhauer, H., Oerke, E., Schmidtlein, S., 2010. Remote sensing of environment quantifying empirical relations between planted species mixtures and canopy reflectance with PROTEST. *Remote Sens. Environ.* 114, 1513–1521. doi: 10.1016/j.rse.2010.02.006

Feret, J.-B., Asner, P.G., 2013. Tree species discrimination in tropical forests using airborne imaging spectroscopy. *IEEE Trans. Geosci. Remote Sens.* 51, 73–84. doi: 10.1109/TGRS.2012.2199323

Féret, J.B., Gitelson, A.A., Noble, S.D., Jacquemoud, S., 2017. PROSPECT-D: Towards modeling leaf optical properties through a complete lifecycle. *Remote Sens. Environ.* 193, 204–215. doi: 10.1016/j.rse.2017.03.004

Frishkoff, L.O., Karp, D.S., M'Gonigle, L.K., Hadly, E.A., Daily, G.C., 2014. Loss of avian phylogenetic diversity in neotropical agricutural systems. *Science* 345, 1343–1346. doi: 10.7910/DVN/26910

Gámez-Virués, S., Perović, D.J., Gossner, M.M., Börschig, C., Blüthgen, N., de Jong, H., Simons, N.K. et al., 2015. Landscape simplification filters species traits and drives biotic homogenization. *Nat. Commun.* 6, 8568. doi: 10.1038/ncomms9568

Garnier, E., Lavorel, S., Ansquer, P., Castro, H., Cruz, P., Dolezal, J., Eriksson, O. et al., 2007. Assessing the effects of land-use change on plant traits, communities and ecosystem functioning in grasslands: A standardized methodology and lessons from an application to 11 European sites. *Ann. Bot.* 99, 967–985. doi: 10.1093/aob/mcl215

Garzonio, R., di Mauro, B., Colombo, R., Cogliati, S., 2017. Surface reflectance and sun-induced fluorescence spectroscopy measurements using a small hyperspectral UAS. *Remote Sens.* 9, 1–24. doi: 10.3390/rs9050472

Gitelson, A.A., Gritz, Y., Merzlyak, M.N., 2003. Relationships between leaf chlorophyll content and spectral reflectance and algorithms for non-destructive chlorophyll assessment in higher plant leaves reflectance and algorithms for non-destructive chlorophyll assessment. *J. Plant Physiol.* 282. doi: 10.1078/0176-1617-00887

Goodenough, D.G., Pearlman, J., Chen, H., Dyk, A., Han, T., Li, J., Miller, J., Niemann, K.O., 2004. Forest information from hyperspectral sensing. In: *IEEE International IEEE International IEEE International Geoscience and Remote Sensing Symposium, 2004. IGARSS '04. Proceedings. 2004.* IEEE, pp. 2585–2589. doi: 10.1109/IGARSS.2004.1369826

Götze, C., Gerstmann, H., Gläßer, C., Jung, A., 2017. An approach for the classification of pioneer vegetation based on species-specific phenological patterns using laboratory spectrometric measurements. *Phys. Geogr.* 38, 524–540. doi: 10.1080/02723646.2017.1306672

Götze, C., Jung, A., Merbach, I., Wennrich, R., Gläßer, C., 2010. Spectrometric analyses in comparison to the physiological condition of heavy metal stressed floodplain vegetation in a standardised experiment. *Cent. Eur. J. Geosci.* 2, 132–137. doi: 10.2478/v10085-010-0002-y

Green, J.L., Bohannan, J.M., Whitaker, R.J., 2008. Microbial biogeography: From taxonomy to traits. *Science* 320, 1039–1043. doi: 10.1126/science.1153475

Grime, J.P., 1974. Vegetation classification by reference to strategies. *Nature* 250, 26–31. doi: 10.1038/250026a0

Guanter, L., Kaufmann, H., Segl, K., Foerster, S., Rogass, C., Chabrillat, S., Kuester, T. et al., 2015. The EnMAP spaceborne imaging spectroscopy mission for earth observation. *Remote Sens.* 7, 8830–8857. doi: 10.3390/rs70708830

Guidici, D., Clark, M., 2017. One-dimensional convolutional neural network land-cover classification of multi-seasonal hyperspectral imagery in the San Francisco Bay Area, California. *Remote Sens.* 9, 629. doi: 10.3390/rs9060629

Haboudane, D., Miller, J.R., Pattey, E., Zarco-Tejada, P.J., Strachan, I.B., 2004. Hyperspectral vegetation indices and novel algorithms for predicting green LAI of crop canopies: Modeling and validation in the context of precision agriculture. *Remote Sens. Environ.* 90, 337–352. doi: 10.1016/j.rse.2003.12.013

Hakkenberg, C.R., Peet, R.K., Urban, D.L., Song, C., 2017. Modeling plant composition as community-continua in a forest landscape with LiDAR and hyperspectral remote sensing. *Ecol. Appl.* 28, 177–190. doi: 10.1002/eap.1638

Harris, A., Charnock, R., Lucas, R.M., 2015. Hyperspectral remote sensing of peatland floristic gradients. *Remote Sens. Environ.* 162, 99–111. doi: 10.1016/j.rse.2015.01.029

He, K.S., Rocchini, D., Neteler, M., Nagendra, H., 2011. Benefits of hyperspectral remote sensing for tracking plant invasions. *Divers. Distrib.* 17, 381–392. doi: 10.1111/j.1472-4642.2011.00761.x

Hilker, T., Gitelson, A., Coops, N.C., Hall, F.G., Black, T.A., 2011. Tracking plant physiological properties from multi-angular tower-based remote sensing. *Oecologia* 165, 865–876. doi: 10.1007/s00442-010-1901-0

Hill, V.J., Zimmerman, R.C., Bissett, W.P., Dierssen, H., Kohler, D.D.R., 2014. Evaluating light availability, seagrass biomass, and productivity using hyperspectral airborne remote sensing in Saint Joseph's Bay, Florida. *Estuaries and Coasts* 37, 1467–1489. doi: 10.1007/s12237-013-9764-3

Hillebrand, H., Matthiessen, B., 2009. Biodiversity in a complex world: Consolidation and progress in functional biodiversity research. *Ecol. Lett.* 12, 1405–1419. doi: 10.1111/j.1461-0248.2009.01388.x

Holasek, R., Nakanishi, K., Ziph-Schatzberg, L., Santman, J., Woodman, P., Zacaroli, R., Wiggins, R., 2017. The selectable hyperspectral airborne remote sensing kit (SHARK) as an enabler for precision agriculture. In: Bannon, D.P. (Ed.), *Proc. SPIE 10213, Hyperspectral Imaging Sensors: Innovative Applications and Sensor Standards 2017.* 22 May 2017. Vol. 1021304. doi: 10.1117/12.2267856

Homolová, L., Malenovský, Z., Clevers, J.G.P.W., García-Santos, G., Schaepman, M.E., 2013. Review of optical-based remote sensing for plant trait mapping. *Ecol. Complex.* 15, 1–16. doi: 10.1016/j.ecocom.2013.06.003

Hueni, A., Nieke, J., Schopfer, J., Kneubühler, M., Itten, K.I., 2009. The spectral database SPECCHIO for improved long-term usability and data sharing. *Comput. Geosci.* 35, 557–565. doi: 10.1016/j.cageo.2008.03.015

Inoue, Y., Peñuelas, J., Miyata, A., Mano, M., 2008. Normalized difference spectral indices for estimating photosynthetic efficiency and capacity at a canopy scale derived from hyperspectral and CO2flux measurements in rice. *Remote Sens. Environ.* 112, 156–172. doi: 10.1016/j.rse.2007.04.011

Jusoff, H.K., 2008. Estimating acacia mangium plantation's standing timber volume using an airborne hyperspectral imaging system. 61–67.

Jusoff, K., Yusoff, M.H.M., 2009. New Approaches in Estimating Rubberwood Standing Volume Using Airborne Hyperspectral Sensing. *Mod. Appl. Sci.* 3. doi: 10.5539/mas.v3n4p62

Kandare, K., Ørka, H.O., Dalponte, M., Næsset, E., Gobakken, T., 2017. Individual tree crown approach for predicting site index in boreal forests using airborne laser scanning and hyperspectral data. *Int. J. Appl. Earth Obs. Geoinf.* 60, 72–82. doi: 10.1016/j.jag.2017.04.008

Kattenaorn, T., Kattenborn, T., Fassnacht, F.E., Pierce, S., Lopatin, J., Grime, J.P., Schmidtlein, S., 2017. Linking plant strategies and plant traits derived by radiative transfer modelling radiative transfer modelling. *J. Veg. Sci.* doi: 10.1111/jvs.12525

Kattge, J., Díaz, S., Lavorel, S., Prentice, I.C., Leadley, P., Bönisch, G., Garnier, E. et al., 2011. TRY—a global database of plant traits. *Glob. Chang. Biol.* 17, 2905–2935. doi: 10.1111/j.1365-2486.2011.02451.x

Klosterman, S.T., Hufkens, K., Gray, J.M., Melaas, E., Sonnentag, O., Lavine, I., Mitchell, L., Norman, R., Friedl, M.A., Richardson, A.D., 2014. Evaluating remote sensing of deciduous forest phenology at multiple spatial scales using PhenoCam imagery. *Biogeosciences* 11, 4305–4320. doi: 10.5194/bg-11-4305-2014

Knapp, S., Kühn, I., Schweiger, O., Klotz, S., 2008. Challenging urban species diversity: Contrasting phylogenetic patterns across plant functional groups in Germany. *Ecol. Lett.* 11, 1054–1064. doi: 10.1111/j.1461-0248.2008.01217.x

Knapp, S., Winter, M., Klotz, S., 2017. Increasing species richness but decreasing phylogenetic richness and divergence over a 320-year period of urbanization. *J. Appl. Ecol.* 54, 1152–1160. doi: 10.1111/1365-2664.12826

Knyazikhin, Y., Schull, M.A., Stenberg, P., Mõttus, M., Rautiainen, M., Yang, Y., Marshak, A. et al., 2013. Hyperspectral remote sensing of foliar nitrogen content. *Proc. Natl. Acad. Sci. U. S. A.* 110, E185–92. doi: 10.1073/pnas.1210196109

Kokaly, R.F., Asner, G.P., Ollinger, S.V., Martin, M.E., Wessman, C.A., 2009. Characterizing canopy biochemistry from imaging spectroscopy and its application to ecosystem studies. *Remote Sens. Environ.* 113. doi: 10.1016/j.rse.2008.10.018

Kraft, S., Del Bello, U., Bouvet, M., Drusch, M., Moreno, J., 2012. FLEX: ESA's Earth Explorer 8 candidate mission. *Int. Geosci. Remote Sens. Symp.* 7125–7128. doi: 10.1109/IGARSS.2012.6352020

Kühn, I., Bierman, S.M., Durka, W., Klotz, S., 2006. Relating geographical variation in pollination types to environmental and spatial factors using novel statistical methods. *New Phytol.* 172, 127–39. doi: 10.1111/j.1469-8137.2006.01811.x

Kühn, I., Durka, W., Klotz, S., Diversity, S., Issue, S., Invasion, P., Sep, E., 2004. BiolFlor: A new plant-trait database as a tool for plant invasion ecology. *Diversity Distrib.* 10, 363–365. Linked references are available on JSTOR for this article: Database as a tool for plant invasion ecology.

Lamine, S., Petropoulos, G.P., Singh, S.K., Szabó, S., Bachari, N.E.I., Srivastava, P.K., Suman, S., 2017. Quantifying land use/land cover spatio-temporal landscape pattern dynamics from Hyperion using SVMs classifier and FRAGSTATS ®. *Geocarto Int.* 1–17. doi: 10.1080/10106049.2017.1307460

Landmann, T., Piiroinen, R., Makori, D.M., Abdel-Rahman, E.M., Makau, S., Pellikka, P., Raina, S.K., 2015. Application of hyperspectral remote sensing for flower mapping in African savannas. *Remote Sens. Environ.* 166, 50–60. doi: 10.1016/j.rse.2015.06.006

Laurin, G.V., Puletti, N., Hawthorne, W., Liesenberg, V., Corona, P., Papale, D., Chen, Q., Valentini, R., 2016. Discrimination of tropical forest types, dominant species, and mapping of functional guilds by hyperspectral and simulated multispectral Sentinel-2 data. *Remote Sens. Environ.* 176, 163–176. doi: 10.1016/j.rse.2016.01.017

Lausch, A., Bannehr, L., Beckmann, M., Boehm, C., Feilhauer, H., Hacker, J.M., Heurich, M. et al., 2016a. Linking earth observation and taxonomic, structural and functional biodiversity: Local to ecosystem perspectives. *Ecol. Indic.* 70, 317–339. doi: 10.1016/j.ecolind.2016.06.022

Lausch, A., Bastian O., Klotz, S., Leitão, P. J., Jung, A., Rocchini, D., Schaepman, M.E., Skidmore, A.K., Tischendorf, L., Knapp, S. 2018a. Understanding and assessing vegetation health by in-situ species and remote sensing approaches. *Methods in Ecology and Evolution*, 00: 1–11. https://doi.org/10.1111/2041-210X.13025.

Lausch, A., Blaschke, T., Haase, D., Herzog, F., Syrbe, R.-U., Tischendorf, L., Walz, U., 2015a. Understanding and quantifying landscape structure—A review on relevant process characteristics, data models and landscape metrics. *Ecol. Modell.* 295, 31–41. doi: 10.1016/j.ecolmodel.2014.08.018

Lausch, A., Borg, E., Bumberger, J., Dietrich, P., Heurich, M., Huth, A., Jung, A., Klenke R., Knapp, S., Mollenhauer, H., Paasche, H., Paulheim, H., Pause, P., Schweitzer, C., Schmulius, C., Settele, J., Skidmore, A.K., Wegmann, M., Zacharias, S., Kirsten, T., Schaepman, M.E., 2018b. Understanding Forest Health with Remote Sensing, *Part III: Requirements for a Scalable Multi-Source Forest Health Monitoring Network Based on Data Science ApproachesRemote Sensing*, 10, 1120, doi:10.3390/rs10071120.

Lausch, A., Erasmi, S., King, D.J., Magdon, P., Heurich, M., 2016b. Understanding forest health with remote sensing-Part I—A review of spectral traits, processes and remote-sensing characteristics. *Remote Sens.* 8, 1029. doi: 10.3390/RS8121029

Lausch, A., Erasmi, S., King, D.J., Magdon, P., Heurich, M., 2017. Understanding forest health with remote sensing-Part II—A review of approaches and data models. *Remote Sens.* 9, 129. doi: 10.3390/rs9020129

Lausch, A., Heurich, M., Gordalla, D., Dobner, H.-J., Gwillym-Margianto, S., Salbach, C., 2013a. Forecasting potential bark beetle outbreaks based on spruce forest vitality using hyperspectral remote-sensing techniques at different scales. *For. Ecol. Manage.* 308, 76–89. doi: 10.1016/j.foreco.2013.07.043

Lausch, A., Pause, M., Doktor, D., Preidl, S., Schulz, K., 2013b. Monitoring and assessing of landscape heterogeneity at different scales. *Environ. Monit. Assess.* 185, 9419–9434. doi: 10.1007/s10661-013-3262-8

Lausch, A., Salbach, C., Schmidt, A., Doktor, D., Merbach, I., Pause, M., 2015b. Deriving phenology of barley with imaging hyperspectral remote sensing. *Ecol. Modell.* 295, 123–135. doi: 10.1016/j.ecolmodel.2014.10.001

Lausch, A., Schmidt, A., Tischendorf, L., 2015c. Data mining and linked open data—New perspectives for data analysis in environmental research. *Ecol. Modell.* 295, 5–17. doi: 10.1016/j.ecolmodel.2014.09.018

Lausch, A., Zacharias, S., Dierke, C., Pause, M., Kühn, I., Doktor, D., Dietrich, P., Werban, U., 2013c. Analysis of vegetation and soil patterns using hyperspectral remote sensing, EMI, and gamma-ray measurements. *Vadose Zo. J.* 12. doi: 10.2136/vzj2012.0217

Lavorel, S., Garnier, E., 2002. Predicting changes in community composition and ecosystem functioning from plant traits: Revisting the Holy Grail. *Funct. Ecol.* 16, 545–556. doi: 10.1046/J.1365-2435.2002.00664.X

Lee, C.M., Cable, M.L., Hook, S.J., Green, R.O., Ustin, S.L., Mandl, D.J., Middleton, E.M., 2015. An introduction to the NASA Hyperspectral InfraRed Imager (HyspIRI) mission and preparatory activities. *Remote Sens. Environ.* 167, 6–19. doi: 10.1016/j.rse.2015.06.012

Leitão, P.J., Schwieder, M., Senf, C., 2017. sgdm: An R Package for performing sparse generalized dissimilarity modelling with tools for gdm. *ISPRS Int. J. Geo-Information* 6, 23. doi: 10.3390/ijgi6010023

Leitão, P.J., Schwieder, M., Suess, S., Okujeni, A., Galvão, L.S., van der Linden, S., Hostert, P., 2015. Monitoring natural ecosystem and ecological gradients: Perspectives with EnMAP. *Remote Sens.* 7, 13098–13119. doi: 10.3390/rs71013098

Leutner, B.F., Reineking, B., Müller, J., Bachmann, M., Beierkuhnlein, C., Dech, S., Wegmann, M., 2012. Modelling forest α-diversity and floristic composition—On the added value of LiDAR plus hyperspectral remote sensing. *Remote Sens.* 4, 2818–2845. doi: 10.3390/rs4092818

Liu, L., Coops, N.C., Aven, N.W., Pang, Y., 2017. Mapping urban tree species using integrated airborne hyperspectral and LiDAR remote sensing data. *Remote Sens. Environ.* 200, 170–182. doi: 10.1016/j.rse.2017.08.010

Lopatin, J., Fassnacht, F.E., Kattenborn, T., Schmidtlein, S., 2017. Mapping plant species in mixed grassland communities using close range imaging spectroscopy. *Remote Sens. Environ.* 201, 12–23. doi: 10.1016/j.rse.2017.08.031

Lowe, A., Harrison, N., French, A.P., 2017. Hyperspectral image analysis techniques for the detection and classification of the early onset of plant disease and stress. *Plant Methods* 13, 1–12. doi: 10.1186/s13007-017-0233-z

Lukeš, P., Stenberg, P., Rautiainen, M., Mõttus, M., Vanhatalo, K.M., 2013. Optical properties of leaves and needles for boreal tree species in Europe. *Remote Sens. Lett.* 4, 667–676. doi: 10.1080/2150704X.2013.782112

Luo, S., Wang, C., Xi, X., Pan, F., Peng, D., Zou, J., Nie, S., Qin, H., 2017. Fusion of airborne LiDAR data and hyperspectral imagery for aboveground and belowground forest biomass estimation. *Ecol. Indic.* 73, 378–387. doi: 10.1016/j.ecolind.2016.10.001

Mahlein, A.-K., 2016. Present and Future Trends in Plant Disease Detection 1–11.

Male, E.J., Pickles, W.L., Silver, E.A., Hoffmann, G.D., Lewicki, J., Apple, M., Repasky, K., Burton, E.A., 2010. Using hyperspectral plant signatures for CO2 leak detection during the 2008 ZERT CO2 sequestration field experiment in Bozeman, Montana. *Environ. Earth Sci.* 60, 251–261. doi: 10.1007/s12665-009-0372-2

Malenovský, Z., Bartholomeus, H.M., Acerbi-Junior, F.W., Schopfer, J.T., Painter, T.H., Epema, G.F., Bregt, A.K., 2007. Scaling dimensions in spectroscopy of soil and vegetation. *Int. J. Appl. Earth Obs. Geoinf.* 9, 137–164. doi: 10.1016/j.jag.2006.08.003

Malenovský, Z., Turnbull, J.D., Lucieer, A., Robinson, S.A., 2015. Antarctic moss stress assessment based on chlorophyll content and leaf density retrieved from imaging spectroscopy data. *New Phytol.* 208, 608–624. doi: 10.1111/nph.13524

Mariotto, I., Thenkabail, P.S., Huete, A., Slonecker, E.T., Platonov, A., 2013. Hyperspectral versus multispectral crop-productivity modeling and type discrimination for the HyspIRI mission. *Remote Sens. Environ.* 139, 291–305. doi: 10.1016/j.rse.2013.08.002

Marshall, M., Thenkabail, P., 2015. Advantage of hyperspectral EO-1 Hyperion over multispectral IKONOS, GeoEye-1, WorldView-2, Landsat ETM+, and MODIS vegetation indices in crop biomass estimation. *ISPRS J. Photogramm. Remote Sens.* 108, 205–218. doi: 10.1016/j.isprsjprs.2015.08.001

Marvin, D.C., Asner, G.P., 2016. Spatially explicit analysis of field inventories for national forest carbon monitoring. *Carbon Balance Manag.* 11, 9. doi: 10.1186/s13021-016-0050-0

Matsuda, O., Tanaka, A., Fujita, T., Iba, K., 2012. Hyperspectral imaging techniques for rapid identification of arabidopsis mutants with altered leaf pigment status. *Plant Cell Physiol.* 53, 1154–1170. doi: 10.1093/pcp/pcs043

Mayr, E., 1942. Systematics and the origin of Species from the Viewpoint of a Zoologist. *Nature.* doi: 10.1038/151347a0

Mayr, E., 1969. The biological meaning of species. *Biol. J. Linn. Soc.* 1, 311–320. doi: 10.1111/j.1095-8312.1969.tb00123.x

McManus, K.M., Asner, G.P., Martin, R.E., Dexter, K.G., Kress, W.J., Field, C.B., 2016. Phylogenetic structure of foliar spectral traits in tropical forest canopies. *Remote Sens.* 8, 1–16. doi: 10.3390/rs8030196

Meroni, M., Rossini, M., Guanter, L., Alonso, L., Rascher, U., Colombo, R., Moreno, J., 2009. Remote sensing of solar-induced chlorophyll fluorescence: Review of methods and applications. *Remote Sens. Environ.* 113, 2037–2051. doi: 10.1016/j.rse.2009.05.003

Meyer, H., Lehnert, L.W., Wang, Y., Reudenbach, C., Nauss, T., Bendix, J., 2017. From local spectral measurements to maps of vegetation cover and biomass on the Qinghai-Tibet-Plateau: Do we need hyperspectral information? *Int. J. Appl. Earth Obs. Geoinf.* 55, 21–31. doi: 10.1016/j.jag.2016.10.001

Mišurec, J., Kopačková, V., Lhotáková, Z., Campbell, P., Albrechtová, J., 2016. Detection of spatio-temporal changes of Norway spruce forest stands in ore mountains using landsat time series and airborne hyperspectral imagery. *Remote Sens.* 8, 92. doi: 10.3390/rs8020092

Möckel, T., Dalmayne, J., Prentice, H., Eklundh, L., Purschke, O., Schmidtlein, S., Hall, K., 2014. Classification of grassland successional stages using airborne hyperspectral imagery. *Remote Sens.* 6, 7732–7761. doi: 10.3390/rs6087732

Moreno, J., Alonso, L., Delegido, J., Rivera, J.P., Ruiz-Verdú, A., Sabater, N., Tenjo, C., Verrelst, J., Vicent, J., 2014. Misión Flex (Fluorescence Explorer): Observación de la fluorescencia por teledetección como nueva técnica de estudio del estado de la vegetación terrestre a escala global. *Rev. Teledetec.* (41): 111–119. doi: 10.4995/raet.2014.2296

Mutanga, O., Kumar, L., 2007. Estimating and mapping grass phosphorus concentration in an African savanna using hyperspectral image data. *Int. J. Remote Sens.* 28, 4897–4911. doi: 10.1080/01431160701253253

Nansen, C., Abidi, N., Sidumo, A.J., Gharalari, A.H., 2010. Using spatial structure analysis of hyperspectral imaging data and fourier transformed infrared analysis to determine bioactivity of surface pesticide treatment. *Remote Sens.* 2, 908–925. doi: 10.3390/rs2040908

Neumann, C., Weiss, G., Schmidtlein, S., Itzerott, S., Lausch, A., Doktor, D., Brell, M., 2015. Gradient-based assessment of habitat quality for spectral ecosystem monitoring. *Remote Sens.* 7, 2871–2898. doi: 10.3390/rs70302871

Nink, S., Hill, J., Buddenbaum, H., Stoffels, J., Sachtleber, T., Langshausen, J., 2015. Assessing the suitability of future multi- and hyperspectral satellite systems for mapping the spatial distribution of norway spruce timber volume. *Remote Sens.* 7, 12009–12040. doi: 10.3390/rs70912009

Oppelt, N., Mauser, W., 2004. Hyperspectral monitoring of physiological parameters of wheat during a vegetation period using AVIS data. *Int. J. Remote Sens.* 25, 145–159. doi: 10.1080/0143116031000115300

Pandey, P., Ge, Y., Stoerger, V., Schnable, J.C., 2017. High throughput in vivo analysis of plant leaf chemical properties using hyperspectral imaging. *Front. Plant Sci.* 8, 1–12. doi: 10.3389/fpls.2017.01348

Paz-Kagan, T., Vaughn, N.R., Martin, R.E., Brodrick, P.G., Stephenson, N.L., Das, A.J., Nydick, K.R., Asner, G.P., 2017. Landscape-scale variation in canopy water content of giant sequoias during drought. *For. Ecol. Manage.* 0–1. doi: 10.1016/j.foreco.2017.11.018

Peng, Y., Gitelson, A.A., 2012. Remote estimation of gross primary productivity in soybean and maize based on total crop chlorophyll content. *Remote Sens. Environ.* 117, 440–448. doi: 10.1016/j.rse.2011.10.021

Pérez-Harguindeguy, N., Díaz, S., Lavorel, S., Poorter, H., Jaureguiberry, P., Bret-Harte, M.S., Cornwell, W.K. et al., 2013. New Handbook for standardized measurment of plant functional traits worldwide. *Aust. J. Bot.* 23, 167–234. doi: http://dx.doi.org/10.1071/BT12225

Pettorelli, N., Schulte to Bühne, H., Tulloch, A., Dubois, G., Macinnis-Ng, C., Queirós, A.M., Keith, D.A. et al., 2017. Satellite remote sensing of ecosystem functions: Opportunities, challenges and way forward. *Remote Sens. Ecol. Conserv.* 1–23. doi: 10.1002/rse2.59

Piro, P., Porti, M., Veltri, S., Lupo, E., Moroni, M., 2017. Hyperspectral monitoring of green roof vegetation health state in sub-Mediterranean climate: Preliminary results. *Sensors (Switzerland)* 17, 1–17. doi: 10.3390/s17040662

Porder, S., Asner, G.P., Vitousek, P.M., 2005. Ground-based and remotely sensed nutrient availability across a tropical landscape. *Proc. Natl. Acad. Sci. U. S. A.* 102, 10909–12. doi: 10.1073/pnas.0504929102

Pottier, J., Malenovsky, Z., Psomas, A., Homolova, L., Schaepman, M.E., Choler, P., Thuiller, W., Guisan, A., Zimmermann, N.E., 2014. Modelling plant species distribution and diversity in alpine grasslands using airborne imaging spectroscopy. *Biol. Lett.* 10, 4. doi: 10.1098/rsbl.2014.0347

Priem, F., Canters, F., 2016. Synergistic use of LiDAR and APEX hyperspectral data for high-resolution urban land cover mapping. *Remote Sens.* 8, 1–22. doi: 10.3390/rs8100787

Purvis, A., Hector, A., 2000. Getting the measure of biodiversity. *Nature* 405, 212–219.

Rajah, P., Odindi, J., Abdel-Rahman, E., Mutanga, O., 2017. Determining the optimal phenological stage for predicting common dry bean (*Phaseolus vulgaris*) yield using field spectroscopy. *South African J. Plant Soil* 34, 379–388. doi: 10.1080/02571862.2017.1317854

Rascher, U., 2007. FLEX—Fluorescence Explorer: A remote sensing approach to quatify spatio-temporal variations of photosynthetic efficiency from space. *Photosynth. Res* 91.

Rascher, U., Alonso, L., Burkart, A., Cilia, C., Cogliati, S., Colombo, R., Damm, A. et al., 2015. Sun-induced fluorescence—a new probe of photosynthesis: First maps from the imaging spectrometer HyPlant. *Glob. Chang. Biol.* 21, 4673–4684. doi: 10.1111/gcb.13017

Reichstein, M., Bahn, M., Mahecha, M.D., Kattge, J., Baldocchi, D.D., 2014. Linking plant and ecosystem functional biogeography. *Proc. Natl. Acad. Sci.* 111, 201216065. doi: 10.1073/pnas.1216065111

Richter, R., Reu, B., Wirth, C., Doktor, D., 2016. The use of airborne hyperspectral data for tree species classification in a species-rich Central European forest area. *Int. J.* 52, 464–474. doi: 10.1016/j.jag.2016.07.018

Roth, K.L., Casas, A., Huesca, M., Ustin, S.L., Alsina, M.M., Mathews, S.A., Whiting, M.L., 2016. Leaf spectral clusters as potential optical leaf functional types within California ecosystems. *Remote Sens. Environ.* 184, 229–246. doi: 10.1016/j.rse.2016.07.014

Roth, K.L., Roberts, D.A., Dennison, P.E., Alonzo, M., Peterson, S.H., Beland, M., 2015. Differentiating plant species within and across diverse ecosystems with imaging spectroscopy. *Remote Sens. Environ.* 167, 135–151. doi: 10.1016/j.rse.2015.05.007

Santos, M.J., Khanna, S., Hestir, E.L., Greenberg, J.A., Ustin, S.L., 2016. Measuring landscape-scale spread and persistence of an invaded submerged plant community from airborne remote sensing. *Ecol. Appl.* 26, 1733–1744. doi: 10.1890/15-0615

Schlerf, M., Atzberger, C., 2006. Inversion of a forest reflectance model to estimate structural canopy variables from hyperspectral remote sensing data. *Remote Sens. Environ.* 100, 281–294. doi: 10.1016/j.rse.2005.10.006

Schlerf, M., Atzberger, C., Hill, J., Buddenbaum, H., Werner, W., Schüler, G., 2010. Retrieval of chlorophyll and nitrogen in Norway spruce (*Picea abies* L. Karst.) using imaging spectroscopy. *Int. J. Appl. Earth Obs. Geoinf.* 12, 17–26. doi: 10.1016/j.jag.2009.08.006

Schmidt, J., Fassnacht, F.E., Lausch, A., Schmidtlein, S., 2017a. Assessing the functional signature of heathland landscapes via hyperspectral remote sensing. *Ecol. Indic.* 73, 505–512. doi: 10.1016/j.ecolind.2016.10.017

Schmidt, J., Fassnacht, F.E., Neff, C., Lausch, A., Kleinschmit, B., Förster, M., Schmidtlein, S., 2017b. Adapting a Natura 2000 field guideline for a remote sensing-based assessment of heathland conservation status. *Int. J. Appl. Earth Obs. Geoinf.* 60, 61–71. doi: 10.1016/j.jag.2017.04.005

Schmidtlein, S., 2005. Imaging spectroscopy as a tool for mapping Ellenberg indicator values. *J. Appl. Ecol.* 42, 966–974. doi: 10.1111/j.1365-2664.2005.01064.x

Schmidtlein, S., Feilhauer, H., Bruelheide, H., 2012. Mapping plant strategy types using remote sensing. *J. Veg. Sci.* 23, 395–405. doi: 10.1111/j.1654-1103.2011.01370.x

Schmidtlein, S., Sassin, J., 2004. Mapping of continuous floristic gradients in grasslands using hyperspectral imagery. *Remote Sens. Environ.* 92, 126–138. doi: 10.1016/j.rse.2004.05.004

Schmidtlein, S., Zimmermann, P., Schupferling, R., Weiss, C., 2007. Mapping the floristic continuum: Ordination space position estimated from imaging spectroscopy. *J. Veg. Sci.* 18, 131–140. doi: 10.1111/j.1654-1103.2007.tb02523.x

Schneider, F.D., Morsdorf, F., Schmid, B., Petchey, O.L., Hueni, A., Schimel, D.S., Schaepman, M.E., 2017. Mapping functional diversity from remotely sensed morphological and physiological forest traits. *Nat. Commun.* 8(1), 1441. doi: 10.1038/s41467-017-01530-3

Schweiger, A.K., Cavender-Bares, J., Townsend, P.A., Hobby, S.E., Madritch, M.D., Wang, R., Tilman, D., Gamon, J.A., 2018. Plant spectra integrate components of biodiversity and predict ecosystem function. *Nat. Ecol. Evol.* . doi:10.1038/s41559-018-0551-1

Schweiger, A.K., Schütz, M., Risch, A.C., Kneubühler, M., Haller, R., Schaepman, M.E., 2017. How to predict plant functional types using imaging spectroscopy: Linking vegetation community traits, plant functional types and spectral response. *Methods Ecol. Evol.* 8, 86–95. doi: 10.1111/2041-210X.12642

Schwieder, M., Leitão, P.J., Suess, S., Senf, C., Hostert, P., 2014. Estimating fractional shrub cover using simulated enmap data: A comparison of three machine learning regression techniques. *Remote Sens.* 6, 3427–3445. doi: 10.3390/rs6043427

Sechrest, W., Brooks, T.M., da Fonseca, G.A.B., Konstant, W.R., Mittermeier, R.A., Purvis, A., Rylands, A.B., Gittleman, J.L., 2002. Hotspots and the conservation of evolutionary history. *Proc. Natl. Acad. Sci.* 99, 2067–2071. doi: 10.1073/pnas.251680798

Serrano, L., Penuelas, J., Ustin, S.L., 2002. Remote sensing of nitrogen and lignin in Mediterranean vegetation\rfrom AVIRIS data:\rDecomposing biochemical from structural signals. *Remote Sens. Environ.* 81, 355–364.

Shoko, C., Mutanga, O., 2017. Seasonal discrimination of C3 and C4 grasses functional types: An evaluation of the prospects of varying spectral configurations of new generation sensors. *Int. J. Appl. Earth Obs. Geoinf.* 62, 47–55. doi: 10.1016/j.jag.2017.05.015

Sims, D.A., Gamon, J.A., 2002. Relationships between leaf pigment content and spectral reflectance across a wide range of species, leaf structures and developmental stages. *Remote Sens. Environ.* 81, 337–354. doi: 10.1016/S0034-4257(02)00010-X

Skowronek, S., Ewald, M., Isermann, M., Van De Kerchove, R., Lenoir, J., Aerts, R., Warrie, J. et al., 2016. Evaluating the potential of hyperspectral remote sensing to map an invasive bryophyte species. *Biol. Invasions.* doi: 10.1007/s10530-016-1276-1

Smit, I.P.J., Asner, G.P., Govender, N., Vaughn, N.R., van Wilgen, B.W., 2016. An examination of the potential efficacy of high-intensity fires for reversing woody encroachment in savannas. *J. Appl. Ecol.* 53, 1623–1633. doi: 10.1111/1365-2664.12738

Smith, M.L., Ollinger, S.V., Martin, M.E., Aber, J.D., Hallett, R.A., Goodale, C.L., 2002. Direct estimation of aboveground forest productivity through hyperspectral remote sensing of canopy nitrogen. *Ecol. Appl.* 12, 1286–1302. doi: 10.1890/1051-0761(2002)012[1286:DEOAFP]2.0.CO;2

Somers, B., Asner, G.P., 2012. Hyperspectral time series analysis of native and invasive species in Hawaiian rainforests. *Remote Sens.* 4, 2510–2529. doi: 10.3390/rs4092510

Somers, B., Asner, G.P., 2013. Multi-temporal hyperspectral mixture analysis and feature selection for invasive species mapping in rainforests. *Remote Sens. Environ.* 136, 14–27. doi: 10.1016/j.rse.2013.04.006

Stefano, P., Angelo, P., Simone, P., Filomena, R., Federico, S., Tiziana, S., Umberto, A. et al., 2013. The PRISMA hyperspectral mission: Science activities and opportunities for agriculture and land monitoring. *Int. Geosci. Remote Sens. Symp.* pp. 4558–4561. doi: 10.1109/IGARSS.2013.6723850

Suess, S., Van Der Linden, S., Okujeni, A., Leitão, P.J., Schwieder, M., Hostert, P., 2015. Using class probabilities to map gradual transitions in shrub vegetation from simulated EnMAP data. *Remote Sens.* 7, 10668–10688. doi: 10.3390/rs70810668

Tanaka, S., Kawamura, K., Maki, M., Muramoto, Y., Yoshida, K., Akiyama, T., 2015. Spectral index for quantifying leaf area index of winter wheat by field hyperspectral measurements: A case study in Gifu Prefecture, Central Japan. *Remote Sens.* 7, 5329–5346. doi: 10.3390/rs70505329

Thenkabail, P.S., Lyon, J.G., Huete, A., 2016. *Hyperspectral Remote Sensing of Vegetation.* CRC Press, New York.

Townsend, A.R., Asner, G.P., Cleveland, C.C., 2008. The biogeochemical heterogeneity of tropical forests. *Trends Ecol. Evol.* 23, 424–431. doi: 10.1016/j.tree.2008.04.009

Underwood, E.C., Mulitsch, M.J., Greenberg, J.A., Whiting, M.L., Ustin, S.L., Kefauver, S.C., 2006. Mapping invasive aquatic vegetation in the Sacramento-San Joaquin Delta using hyperspectral imagery. *Environ. Monit. Assess.* 121, 47–64. doi: 10.1007/s10661-005-9106-4

Ustin, S.L., Gamon, J.A., 2010. Remote sensing of plant functional types. *New Phytol.* 186, 795–816. doi: 10.1111/j.1469-8137.2010.03284.x

Ustin, S.L., Gitelson, A.A., Jacquemoud, S., Schaepman, M., Asner, G.P., Gamon, J.A., Zarco-Tejada, P., 2009. Retrieval of foliar information about plant pigment systems from high resolution spectroscopy. *Remote Sens. Environ.* 113. doi: 10.1016/j.rse.2008.10.019

Ustin, S.L., Roberts, D.A.R.A., Gamon, J.A., Green, R.O., 2004. Using imaging spectroscopy to study ecosystem processes and properties. *Bioscience* 54, 523–534. doi: 10.1641/0006-3568(2004)054[0523:UISTSE]2.0.CO;2

Violle, C., Navas, M.L., Vile, D., Kazakou, E., Fortunel, C., Hummel, I., Garnier, E., 2007. Let the concept of trait be functional! *Oikos* 116, 882–892. doi: 10.1111/j.2007.0030-1299.15559.x

Violle, C., Reich, P.B., Pacala, S.W., Enquist, B.J., Kattge, J., 2014. The emergence and promise of functional biogeography. *Proc. Natl. Acad. Sci.* 111, 13690–13696. doi: 10.1073/pnas.1415442111

Wang, L., Qu, J.J., Hao, X., Hunt, E.R., 2011. Estimating dry matter content from spectral reflectance for green leaves of different species. *Int. J. Remote Sens.* 32, 7097–7109. doi: 10.1080/01431161.2010.494641

Wang, Z., Skidmore, A.K., Wang, T., Darvishzadeh, R., Heiden, U., Heurich, M., Latifi, H., Hearne, J., 2017. Canopy foliar nitrogen retrieved from airborne hyperspectral imagery by correcting for canopy structure effects. *Int. J. Appl. Earth Obs. Geoinf.* 54, 84–94. doi: 10.1016/j.jag.2016.09.008

Wellmann, T., Haase, D., Knapp, S., Salbach, C., Selsam, P., Lausch, A., 2018. Urban land use intensity assessment: The potential of spatio-temporal spectral traits with remote sensing. *Ecol. Indic.* 85, 190–203.

Wheeler, Q.D., Meier, R., 2000. *Species Concepts and Phylogenetic Theory.* Columbia University Press, New York.

Willis, C.G., Ruhfel, B., Primack, R.B., Miller-Rushing, A.J., Davis, C.C., 2008. Phylogenetic patterns of species loss in Thoreau's woods are driven by climate change. *Proc. Natl. Acad. Sci.* 105, 17029–17033. doi: 10.1073/pnas.0806446105

Wu, J., Chavana-Bryant, C., Prohaska, N., Serbin, S.P., Guan, K., Albert, L.P., Yang, X. et al., 2017. Convergence in relationships between leaf traits, spectra and age across diverse canopy environments and two contrasting tropical forests. *New Phytol.* 214, 1033–1048. doi: 10.1111/nph.14051

Yu, Q., Wang, S., Mickler, R.A., Huang, K., Zhou, L., Yan, H., Chen, D., Han, S., 2014. Narrowband bio-indicator monitoring of temperate forest carbon fluxes in northeastern China. *Remote Sens.* 6, 8986–9013. doi: 10.3390/rs6098986

Yuan, H., Yang, G., Li, C., Wang, Y., Liu, J., Yu, H., Feng, H., Xu, B., Zhao, X., Yang, X., 2017. Retrieving soybean leaf area index from unmanned aerial vehicle hyperspectral remote sensing: Analysis of RF, ANN, and SVM regression models. *Remote Sens.* 9. doi: 10.3390/rs9040309

Zarco-Tejada, P.J., González-Dugo, M.V., Fereres, E., 2016. Seasonal stability of chlorophyll fluorescence quantified from airborne hyperspectral imagery as an indicator of net photosynthesis in the context of precision agriculture. *Remote Sens. Environ.* 179, 89–103. doi: 10.1016/j.rse.2016.03.024

Zarco-Tejada, P.J., Rueda, C.A., Ustin, S.L., 2003. Water content estimation in vegetation with MODIS reflectance data and model inversion methods. *Remote Sens. Environ.* 85, 109–124. doi: 10.1016/S0034-4257(02)00197-9

Zhang, F., Zhou, G., 2017. Deriving a light use efficiency estimation algorithm using in situ hyperspectral and eddy covariance measurements for a maize canopy in Northeast China. *Ecol. Evol.* 7, 4735–4744. doi: 10.1002/ece3.3051

Zhang, Y., Chen, J.M., Miller, J.R., Noland, T.L., 2008. Leaf chlorophyll content retrieval from airborne hyperspectral remote sensing imagery. *Remote Sens. Environ.* 112, 3234–3247. doi: 10.1016/j.rse.2008.04.005

Zhao, D., Reddy, K.R., Kakani, V.G., Reddy, V.R., 2005. Nitrogen deficiency effects on plant growth, leaf photosynthesis, and hyperspectral reflectance properties of sorghum. *Eur. J. Agron.* 22, 391–403. doi: 10.1016/j.eja.2004.06.005

Zhong, Y., Cao, Q., Zhao, J., Ma, A., Zhao, B., Zhang, L., 2017. Optimal decision fusion for urban land-use/land-cover classification based on adaptive differential evolution using hyperspectral and LiDAR data. *Remote Sens.* 9, 868. doi: 10.3390/rs9080868

Zhu, X., Wang, T., Skidmore, A.K., Darvishzadeh, R., Niemann, K.O., Liu, J., 2017. Canopy leaf water content estimated using terrestrial LiDAR. *Agric. For. Meteorol.* 232, 152–162. doi: 10.1016/j.agrformet.2016.08.016

5 The Use of Hyperspectral Proximal Sensing for Phenotyping of Plant Breeding Trials

Andries B. Potgieter, James Watson, Barbara George-Jaeggli, Gregory McLean, Mark Eldridge, Scott C. Chapman, Kenneth Laws, Jack Christopher, Karine Chenu, Andrew Borrell, Graeme L. Hammer, and David R. Jordan

CONTENTS

> Crop phenotyping based on spectral reflectance information relies on the properties of the light emerging from the canopy after multiple interactions (such as reflections, transmissions, and absorptions) with the tissues of the plant.
>
> **Li et al., 2014**

5.1 INTRODUCTION

The use of remotely-sensed reflectance spectra to assess vegetation biomass and plant physiological status goes back to the analysis of data acquired in the 1970s by Landsat 1 (Rouse et al., 1974). While the resolution then was at the scale of entire vegetation systems or regions, subsequent improvements in the accuracy of Global Positioning Systems (GPS) and the reduction in costs of high-resolution hyperspectral sensors enable the proximal use of these sensors so we can zoom in to the scale of crops, individual trial plots, and even plant organs. This opens up new opportunities for crop breeders to non-destructively phenotype a large number of progeny or potential parental lines. While the term "phenotype" commonly refers to the combination of all, a subset, or a single trait of an organism (Mahner and Kary, 1997), "phenotyping" typically refers to the measurement of these traits. As a phenotype is not only a consequence of the organism's genes—or its genotype—but also of the environment the organism is grown in, it is important for breeders to evaluate the phenotypic variation of their traits of interest in various environments (Sadras et al., 2013; Ghanem et al., 2015). Morphological traits—for example, leaf size or plant height—may be directly observed or measured with simple equipment, while others such as physiological responses to biotic or abiotic stress—for example, stomatal conductance—are invisible and difficult to measure. As a consequence, phenotyping has been either generally limited to simple morphological traits, or to only a small number of genotypes.

Rapid advances in various DNA analysis techniques, on the other hand, have made comprehensive genetic characterizations (or genotyping) of large numbers of individual breeding lines efficient and cost-effective. To fully capitalize on this new possibility to understand the genetic architecture of complex or quantitative traits of interest, methods to accurately and efficiently phenotype large numbers of lines for morphological as well as physiological characteristics are needed. High-throughput plant phenotyping (HTPP) has thus been described as the new frontier in plant breeding (Araus and Cairns, 2014).

For many applications, such as agronomic decision-making in precision agriculture (Haboudane et al., 2004; Verger et al., 2014), studying plant responses to biotic and abiotic stress (Mahlein, 2016; Rebetzke et al., 2016) and screening for general phenotypic variation (Lelong et al., 2008; Hazratkulova et al., 2012; Chapman et al., 2014; Potgieter et al., 2017), the use of well-established vegetation indices calculated from a few distinct reflectance spectra is entirely sufficient.

However, to detect genotypic differences in phenology, chemical composition, photosynthetic capacity, water relations, responses to environmental cues, two- and three-dimensional canopy characteristics, and interactions between these traits at very high spatial and temporal resolution, HTPP platforms integrating narrow-band hyperspectrometry with a suite of other sensors are required (Furbank and Tester, 2011; White et al., 2012). Such platforms create large datasets from disparate types of hardware. Integrated analysis pipelines are needed to turn this plethora of raw data into plot-level data which input into data analysis programs for interpretation by plant breeders and researchers (Li et al., 2014).

Living plants absorb photons from the sun in the visible spectrum (400–710 nm), while photons are mainly reflected in the infrared region (710 nm to 1 mm) across the continuum of the electromagnetic radiation spectrum (hubblesite.org). Traditionally, multispectral analysis of broad regions of these wavelength ranges have been used to investigate crop responses relating to canopy structure, vegetation health, nitrogen content, and chlorophyll status. However, with the advent of more cost-effective narrow-band (<10 nm) hyperspectral cameras, imaging in the hyperspectral domain has become an ideal tool for investigating the interactions between plant organs and solar radiation. More specifically, hyperspectral approaches are proposed to measure crop-physiological traits relating to crop cover percentage (Cover%), canopy structure, light use efficiency (LUE), and photosynthesis (Deery et al., 2014; Rahaman et al., 2015).

While hyperspectral sensing provides similar information to that from multispectral sensing, the larger number of wavelengths captured by hyperspectral sensing confers two key advantages. The first is greater flexibility in terms of analysis. Data collection in the field is expensive, and it is desirable to incorporate new indices into a study without having to buy new hardware or repeat a trial. The second is that hyperspectral hardware often provides narrower spectral bandwidths, allowing greater fidelity in spectral ratio calculations. For example, the NANO-hyperspectral camera described below has a full width half maximum (FWHM) bandwidth of approximately 6 nm (2.6 nm nominal steps), while the multispectral Micasense RedEdge camera has a FWHM of 10–20 nm around specific wavelengths (www.micasense.com.au). In addition, for spectral fidelity, high spatial resolution is required when analyzing phenotypic traits that are related to the plant organ level aggregated across a canopy. Furthermore, the ability to identify and extract reflectance from sorghum heads and leaves separately requires a high pixel resolution of ~5 mm. However, the advantages of high-resolution spectral and spatial measurements come at the cost of significantly more data captured per unit area. This in turn increases the infrastructural and algorithmic complexity required to convert the data into useful information.

We describe a cost-effective high-throughput plant phenotyping (HTPP) system, developed at the University of Queensland (UQ). This mobile proximal-sensing platform and analysis pipeline integrates outputs from sun-induced hyperspectral sensors (point and spatial) with pseudo red-green-blue (RGB), thermal infrared (TIR) images and light detection and ranging (LiDAR) and delivers plot-level information on genotypic variation in plant traits that are of interest to plant breeders. Here, we use the examples of three preselected narrow-band hyperspectral indices relating to canopy structure, and light use measured in breeding trials in northeastern Australia for spring habit wheat (*Triticum aestivum* L.) grown in winter and summer sorghum (*Sorghum bicolor* [L.] Moench). Lastly, we outline how to use this HTPP in a framework that links the different islands of knowledge including *Genomics*, *Phenomics*, and *Crop Modeling* to deliver targeted outcomes across scales that will enhance scientists' understanding and knowledge of the relationships between traits, crop-physiological processes, and genotypes. The Food and Agriculture Organization (FAO) estimates that an increase in agricultural production of more than 60% is required between 2005 and 2050 (Alexandratos and Bruinsma, 2012). If this should be achieved without significantly increasing the area of land under agricultural production, crop yields per area need to be increased. However, advances in global yield potential trends have been slowing significantly during the last two decades (Ray et al., 2012; Grassini et al., 2013). We believe this framework will facilitate the step change and paradigm shift in cereal breeding that is needed to meet this production increase.

5.2 MATERIAL AND METHODS

5.2.1 Hyperspectral Sensors

Two types of hyperspectral sensors were used in this study: a NANO-Hyperspec camera, and a set of USB2000+ point spectrometers (OO). These sensors are summarized in Table 5.1. The OO sensors provide higher spectral fidelity but relatively poor spatial resolution, while the NANO provides high-resolution spatial information with slightly less spectral fidelity than the OO sensors. The NANO was selected as the primary sensor for measuring sorghum and wheat reflectance (Figure 5.1). Each NANO scan consists of 640 pixels perpendicular to the lens, with 270 spectral bands per pixel.

Reflectance measurements need to be corrected for local atmospheric and light conditions using white reference (WR) measurements. For proximal-sensing applications, data quality is directly related to how close a WR is to its corresponding measurement, both spatially and temporally. In

TABLE 5.1
Comparison of the Two Types of Hyperspectral Sensors Used in This Study

Sensor	Headwall NANO-Hyperspec	Ocean Optics USB2000+
Abbreviation	NANO	OO
Spatial values per scan	640	1
Spectral values per scan	270	2028
Spectral range	400–1000 nm	200–1100 nm
FWHM (bandwidth)	6 nm	~0.1–10 nm (configurable)
Interface	Ethernet	USB
Manufacturer	Headwall Photonics (www.headwallphotonics.com)	Ocean Optics (www.oceanoptics.com)

order to minimize the effect of changes in local conditions (such as passing cloud cover), one of the OO sensors was fitted with a CC-3-DA cosine corrector, and was used to continuously collect incoming solar irradiance at each wavelength constituting a WR. Each sensor's reflectance value was calculated according to its own calibration values, and then these values were scaled by the relative change in conditions detected by the WR.

5.2.2 Sensing Platform

To achieve the spatial resolution needed to differentiate between plant organs such as heads and leaves, the NANO and OO sensors needed to remain close to the canopy, at a consistent height. Also, since the NANO is a line scanner, it should also be moved at a constant speed, so that its scan rate could be set appropriately. Due to the size of typical trials (hundreds of plots) and these two requirements, a tractor-based sensing platform was chosen as the most suitable method to convey

FIGURE 5.1 UQ HTPP platform "GECKO." (1) Infrared Thermal Camera (Gobi 640 GigE). (2) Hyperspectral Line Scanner (NANO Headwall). (3) LiDAR (SICK LMS400). (4) Point Spectrometer (3×Ocean Optics USB2000). (5) Infrared thermometers (3×Apogee S100). (6) Point Spectrometer hub. (7) Thermometer hub. (8) Power. (9) Global Positioning System (GPS) and computer. (10) Point Spectrometer (Ocean Optics) for detecting direct solar radiance. (11) GPS antenna. (12) Tractor. (13) Weather Station (on-site).

the sensors. We designated this system the GECKO (Genotype and Environment Characterisation through Kinetic Observation, Figure 5.1).

This system is designed to enable simultaneous crop canopy proximal sensing with a large number of sensor types, and to coordinate the initialization and data capture of these sensors. The GECKO positions the NANO approximately three meters above the ground, and the fiber-optic inputs for the canopy-facing OO sensors just above the crop canopy (within about 0.5 m). The GECKO including the sensors moved at a constant 1.1 m per second. The fiber-optic input of the OO white reference sensor was installed on the roof of the GECKO's cabin and pointing skyward (Figure 5.1).

The GECKO was integrated with a GPS RTK (Real Time Kinematic) system with 2-cm accuracy. The latitude and longitude position of each sensor measurement was derived by integrating the GPS RTK positions with the GECKO's trajectory and the relative position of each sensor with respect to the GPS antenna.

5.2.3 Narrow-Band Hyperspectral Vegetation Indices (HVI)

Apart from providing reflectance information across a wide range of the electromagnetic spectrum, hyperspectral sensors have the advantage of high spectral fidelity and high spatial resolution. This allows the calculation of indices that relate closely to particular plant and crop characteristics. For exemplification of sensing outputs from this framework, we only discuss three indices derived from the hyperspectral camera. Furthermore, these three indices relate to specific physiological and structural canopy attributes, which are of interest to breeders. Firstly, the traditional Normalized Difference Vegetation Index (NDVI, Table 5.2), which is a normalized ratio of the NIR and red bands (Rouse et al., 1974), has been shown to be a good predictor of sorghum leaf area index (Potgieter et al., 2017). Secondly, the Transformed Chlorophyll Absorption in Reflectance Index (TCARI, Table 5.2) uses bands corresponding to the minimum absorption of the photosynthetic pigments, centered at 550 and 700 nm in conjunction with the chlorophyll-*a* maximum absorption band (around 670 nm), while minimizing effects of soil reflectance and non-photosynthetic canopy materials and is sensitive to variability in leaf chlorophyll content (Haboudane et al., 2002). Lastly, the Photochemical Reflectance Index (PRI; Table 5.2) was selected because of its relationship to photosynthetic activity (Gamon et al., 1997; Gamon, 2010). We exemplified the utility of the HTPP framework through the application of these preselected vegetation indices (Table 5.2) to capture hyperspectral measurements of experimental plots in a field trial on a weekly basis.

TABLE 5.2
The Specific Vegetation Indices Used Identify Canopy Architecture, Light Use Efficiency, and Chlorophyll Abundance

Name	Calculation	Phenotypic Trait	Reference
Normalized difference vegetation index (NDVI)	$\frac{R_{800} - R_{670}}{R_{800} + R_{670}}$	Plant/canopy structure	Rouse et al. (1974)
Transformed chlorophyll absorption reflectance index (TCARI)	$3[(R_{700} - R_{670}) - 0.2(R_{700} - R_{550})(R_{700}/R_{670})]$	Chlorophyll concentration	Haboudane et al. (2002)
Photochemical reflectance index (PRI)	$\frac{R_{570} - R_{531}}{R_{570} + R_{531}}$	Photosynthetic light use efficiency	Gamon et al. (1997)

Note: R_n denotes reflectance R at wavelength n.

5.2.4 Hyperspectral-Sensing Pipeline

To manage the phenotyping process, a software pipeline was developed to interpret the raw sensor data. This pipeline also allows for the extraction, download, upload, and conversion of data into plot-level information (Figure 5.2). With respect to the NANO, the Ocean Optics white reference, and the GPS RTK hardware, this software pipeline performs the following functions:

- Control of the hardware (initialization, start/stop of data capture, and data transfer).
- Reliable transfer and storage of data.
- Hardware-specific interpretation of sensor outputs.
- Geo-positioning of the NANO and Ocean Optics white reference data according to the positions and time-stamps of the GPS RTK output, taking into account the position of each sensor relative to the GPS antenna, and the GECKO's speed and direction of travel.
- Calculation of reflectance values from the hardware-specific NANO and Ocean Optics digital data.
- Correction of the NANO reflectance values using the geo-positioned Ocean Optics white reference reflectance.
- Assignment of geo-positioned reflectance values to trial plots.
- Identification of measurements taken over vegetation, and calculation of plot-level indices such as NDVI, PRI, and TCARI.

The scale of the data involved (the NANO collects approximately 10,752,000 values per square meter), the variable size of individual trials, and the need to share raw and post-processed data across multiple sites means that "cloud" (remotely hosted servers) computing infrastructure is an ideal

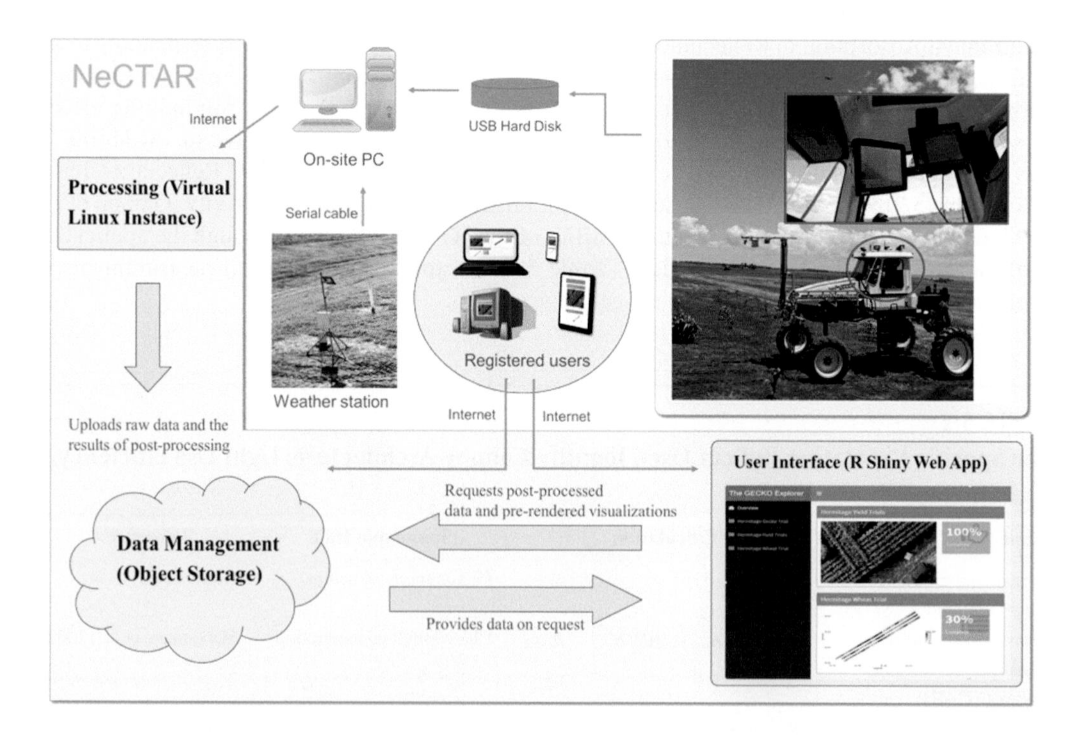

FIGURE 5.2 Schematic overview of the proximal-sensing pipeline. Data collected in the field via GECKO are uploaded to a cloud-hosted processing server for analysis. The raw data and results of processing are then stored in "Object Storage," which provides reliable cloud storage. Users are then able to access both the raw data files and a custom web interface via the Internet.

platform to host the pipeline. The GECKO project uses the Australian NeCTAR (National eResearch Collaboration Tools and Resources project, www.nectar.org.au.research-cloud) research cloud infrastructure, which allows computing hardware and disk storage to be dynamically expanded as needed. After a field run, data were uploaded to the cloud-hosted processing server for analysis. Both the raw data and results of processing are then stored on the NeCTAR object storage infrastructure, which provides data redundancy across sites, and fast access to both the raw data and a custom web server. Figure 5.2 summarizes these steps.

5.2.5 Experimental Trials

5.2.5.1 Wheat

A field trial was established during the normal growing season for spring habit wheat in Australia over the winter of 2016 at a site near Warwick (28°12′ S, 152°05′ E; 480 m above sea level) in southern Queensland. Crops were sown in 2 m × 6 m plots with row spacing of 25 cm and a target population density of 100 plants m^{-2}. Non-limiting levels of nutrients were applied while weeds and diseases were controlled as necessary. In total, 10 wheat genotypes were chosen that are known to differ in adaptation to water-limited environments having variation for root architecture traits and stay-green (Manschadi et al., 2006, 2010; Christopher et al., 2013, 2014, 2016).

The trial was sown on July 22, 2016, using a complete replicated, latinized design with 4 replicate plots of each genotype. NDVI measurements were taken manually at approximately weekly intervals from early development to maturity using a model 505 GreenSeeker hand-held™, model 505 (NTech Industries Inc., www.ntechindustries.com) as previously described by Christopher et al. (2013). Measurements were also taken with the GECKO on 7 dates from before August 22, which was before crop flowering up to November 15, which was near the time of crop maturity. GreenSeeker measurements and those made using the GECKO were compared using linear regression in R (R Core Team, 2017).

5.2.5.2 Sorghum

To establish algorithms between various sorghum crop traits and hyperspectral data from the proximal-sensing platform (GECKO), we conducted a large field trial during the 2016/2017 summer season in the sorghum-growing region of northeastern Australia. Seven hundred sorghum genotypes from the Sorghum Conversion Program (SCP) were selected. The SCP is a backcross breeding scheme in which genomic regions conferring early maturity and dwarfing from an elite donor were introgressed (bred) into approximately 800 exotic sorghum accessions representing the breadth of genetic diversity in sorghum (Stephens et al., 1967). Using genotypes from this program allowed us to test the suitability of proximally-sensed hyperspectral vegetation indices (HVI) to assess traits to do with leaf, canopy architecture, and growth parameters across a wide range of phenotypes. However, to allow for enough clearance under the GECKO during the later data capturing missions, we excluded genotypes that were more than 1.3 m tall in previous trials. The remaining 693 genotypes were grown in a 920-plot trial. The experimental design for the trial was a partially replicated design (Cullis et al., 2006) where 30% of the genotypes were planted in two plots and the remaining genotypes were only planted once. The replicated entries can be resolved into two equal blocks in the row direction. Included in the 920-plot main trial were 8 plots each of 4 standard sorghum hybrids as check plots. Immediately adjacent to the main trial we planted a further 60 plots with a subset of the genotypes to be used as a training set to establish relationships between sensed data and measured plot data.

The study site was located at Hermitage Research Facility (28°12′ S, 152°06′ E; 480 m above sea level) in Warwick in southeastern Queensland. Each plot was 4.5 m long and consisted of 4 rows planted to the same genotype. Only the middle two rows of each plot were used for data capture. Row widths were 0.6 m between the middle two rows and 0.75 m between the middle rows and the outside rows. The trial was conducted on a near-level site on a self-mulching alluvial clay with a high montmorillonite clay content (McKeown, 1978). Volumetric soil moisture content to a depth of

1.25 m was determined by drilling 4 soil cores across the trial site. Non-limiting levels of nutrients were applied using 280 kg ha^{-1} urea containing 46% nitrogen prior to sowing. The trial was sown on December 6, 2016, and the plots were kept weed and pest free with a pre-emergent herbicide and in-trial applications of insecticide as needed.

5.3 VISUALIZATION AND ANALYSIS OF SPECTRAL METRICS

5.3.1 Case Study in Wheat

5.3.1.1 Validation of the Plot-Level Hyperspectral Reflectance Signatures Obtained with the GECKO

For the wheat experiment, plot NDVI values were aggregated from the point-scale Ocean Optics (OO) sensor and compared with measurements from the hand-held GreenSeeker. GreenSeeker data averaged to plot level has been previously used by practitioners to analyze environmental and genotypic effects on the expression of the stay-green trait, which leads to longer green leaf area retention and prolonged photosynthesis (Christopher et al., 2013, 2016).

A linear regression was performed to determine the relationship for NDVI between aggregated plot-level GreenSeeker and OO measurements across the different dates and genotypes. For index validation, we contrasted the NDVI from a hand-held GreenSeeker crop sensor (NDVI_GS), which has an active light source, against the NDVI derived from the OceanOptics spectrometer (NDVI_OO) on the GECKO platform which uses passive reflectance with adjustment for WR. Strong correlation ($r^2 = 0.98$, root mean square error ($RMSE$) = 0.085) was observed between NDVI_GS and NDVI_OO (Figure 5.3).

Measurements from both the GECKO (OO) and the GreenSeeker (GS) allowed differentiation of genotypic variations. Figure 5.4 depicts the comparison of the temporal profiles of the NDVI_GS (Figure 5.4a) and NDVI_OO (Figure 5.4b) for an elite cultivar (Suntop) and a breeding line (HIL63)

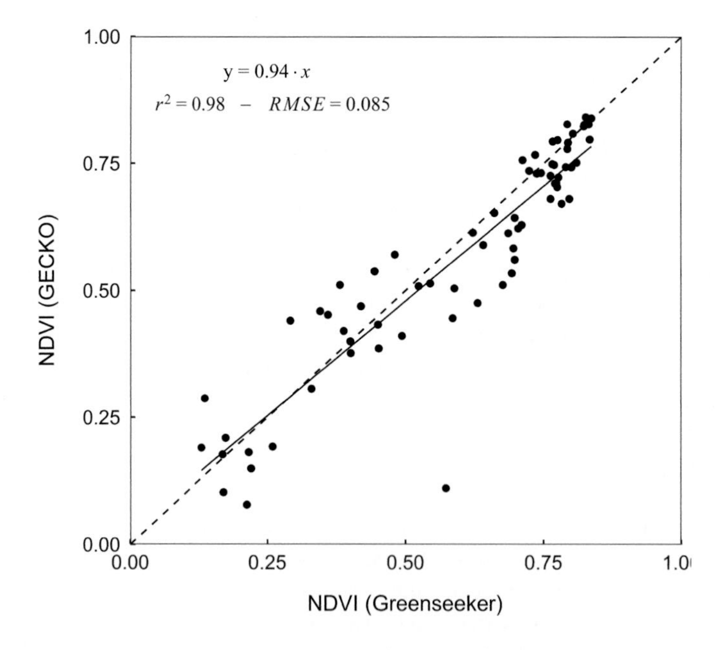

FIGURE 5.3 NDVI measured from GECKO (point spectrometer [Ocean Optics] detecting direct solar radiance) and GreenSeeker for all wheat plots over the growing season for all dates. Linear regression through the origin (y intercept = 0) is given with high R square and small root mean square error (rmse).

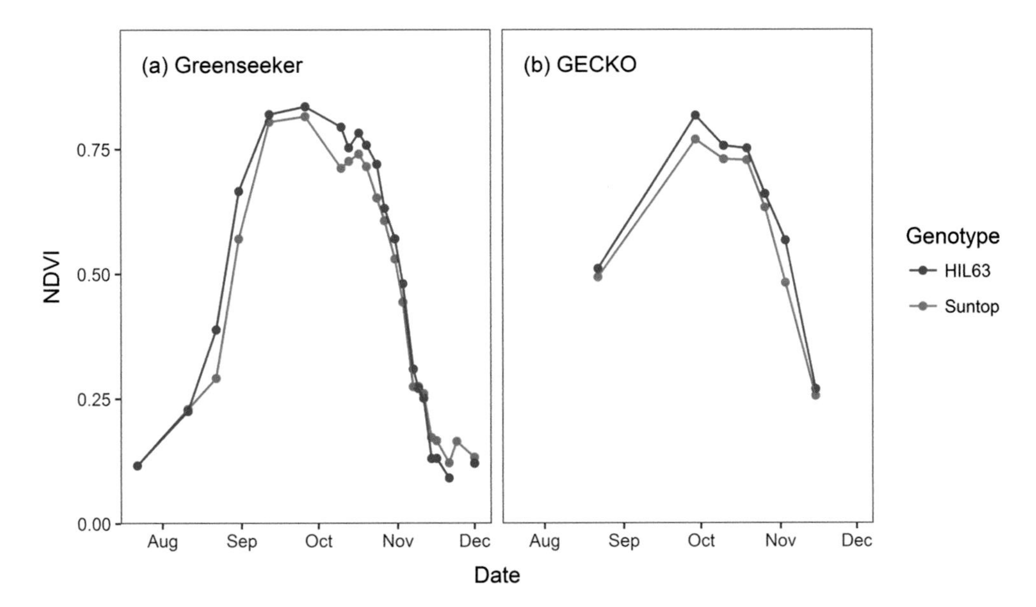

FIGURE 5.4 Dynamics of NDVI measured by (a) GreenSeeker (using and active light source) and (b) the GECKO (point spectrometer [Ocean Optics] detecting direct solar radiance) for two wheat genotypes (HIL63 and Suntop).

over the growing season. The NDVI_OO measurements from GECKO exhibited similar temporal and genotypic patterns to those of the NDVI_GS. Both systems provided estimates of NDVI that for genotype HIL63 were equal to or greater than those for Suntop for most of the crop cycle. This suggests that the NDVI_OO, like the NDVI_GS, can detect genotypic differences. Thus, NDVI_OO could be used in large-scale trials to investigate the genetics underlying crop growth and/or crop senescence (Christopher et al., 2018) and/or assist breeders in selecting for elite material.

5.3.1.2 Spectral Signatures of Different Plant Organs in Wheat

One advantage of a hyperspectral-sensing platform, such as the GECKO, is the ability to both (1) extract narrow-band spectral signatures of leaves, heads, dead material, and soil, and (2) generate such reflectance values at high resolution (4.5 mm × 4.5 mm) for each pixel within a plot. Figure 5.5 shows the aggregated reflectance for each wavelength for the soil and specific parts within a wheat canopy. Aggregated reflectance values were extracted for wheat heads (orange, 267 pixels), leaves (green, 201 pixels), dead material (yellow, 240 pixels), and soil (dark red, 1123 pixels) from a sample wheat plot between flowering and grain filling. Dead material showed the highest reflectance across most wavelengths compared to the other classes. The reflectance of both living leaves and the green wheat heads had similar patterns with peaks around the green and NIR bands typical of chlorophyll. However, the reflectance of wheat heads was greater than that of living leaves throughout the spectrum. The reflectance patterns for living and non-living leaf tissue versus soil align well with previous studies (Asner, 1998).

5.3.1.3 Cross-Sectional (Single Date) Comparisons of Breeding Plots

Figure 5.6 depicts the images for the visible RGB (Figure 5.6a) and the three indices described in Table 5.1 captured on November 3, 2016, near flowering. These indices relate to canopy structure and greenness (NDVI; Figure 5.6b), chlorophyll abundance (TCARI; Figure 5.6c) and light use efficiency (PRI; Figure 5.6d) for a sample wheat plot. Although the NDVI (Figure 5.6b) discriminates reasonably well between dead material (dark red) and living material (blue colors), the effect of shadow and background reflectance on the edge of plants (yellow) are noticeable. In the sample wheat plot considered, mean NDVI values for soil, leaves, heads, and dead material were 0.28, 0.82, 0.58,

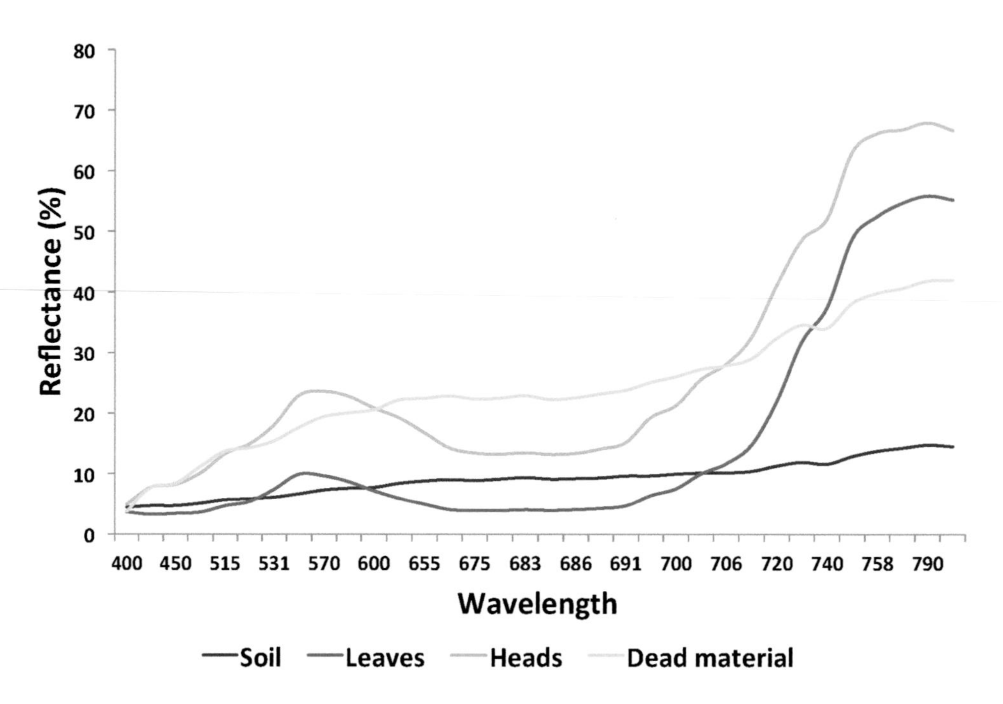

FIGURE 5.5 Aggregated spectral profiles across wavelengths of living tissues (green and orange) and non-living parts (red and yellow) in a sample wheat plot. Date captured on the November 3, 2016, at the flowering stage when wheat heads were fully green. Classes had the following pixel sample sizes: 267 on wheat heads, 201 on leaves, 240 on dead material, and 1123 on soil.

and 0.3, respectively. This concurs with previous research suggesting that NDVI is more sensitive to background reflectance and in addition is likely to saturate at dense canopy levels (Huete et al., 1994, 2002). TCARI (Figure 5.6c) allows a more crisp image especially at the edges of the canopy since it relates more closely to chlorophyll absorbance (Haboudane et al., 2002). PRI (Figure 5.6d) relates more closely to photosynthetic activity at the canopy level (Gamon et al., 1997) and distinguishes well between heads and green leaves, as well as sunlit (light green) and shaded regions within the canopy.

5.3.2 Case Study in Sorghum

5.3.2.1 Validation of the Plot-Level Hyperspectral-Reflectance Signatures Obtained with the GECKO

In the sorghum trial, we compared the spectral reflectance signatures from the NANO Headwall Hyperspectral camera on the GECKO with the spectral reflectance signature obtained with a hand-held leaf spectrometer (ASD FieldSpec4, Analytical Spectral Devices Inc. Longmont, Colorado, USA, https://www.asdi.com/products-and-services/fieldspec-spectroradiometers). This sensor's wide-range detection capacity (350–2500 nm) provides uniform VIS/NIR/SWIR data collection across a broad range of the solar irradiance spectrum. However, for comparison against hyperspectral plot reflectance data from the NANO Headwall on the Gecko, the range was limited between 400 and 800 nm. Furthermore, while the ASD FieldSpec4 has a constant light source, the hyperspectral reflectance captured by the NANO Headwall camera on the GECKO, captures reflectance from the crop canopy induced through solar irradiance. Despite this, the spectral signatures across 40 selected wavelengths from both spectrometers were quite similar (Figure 5.7) and a significant and strong correlation ($R^2 = 0.98$) between ASD FieldSpec and NANO Headwall spectral reflectance data was observed under the sunny conditions in which the NANO data were collected. The two spectral profiles showed closest correlation in the VIS wavelengths (<710 nm), while the largest

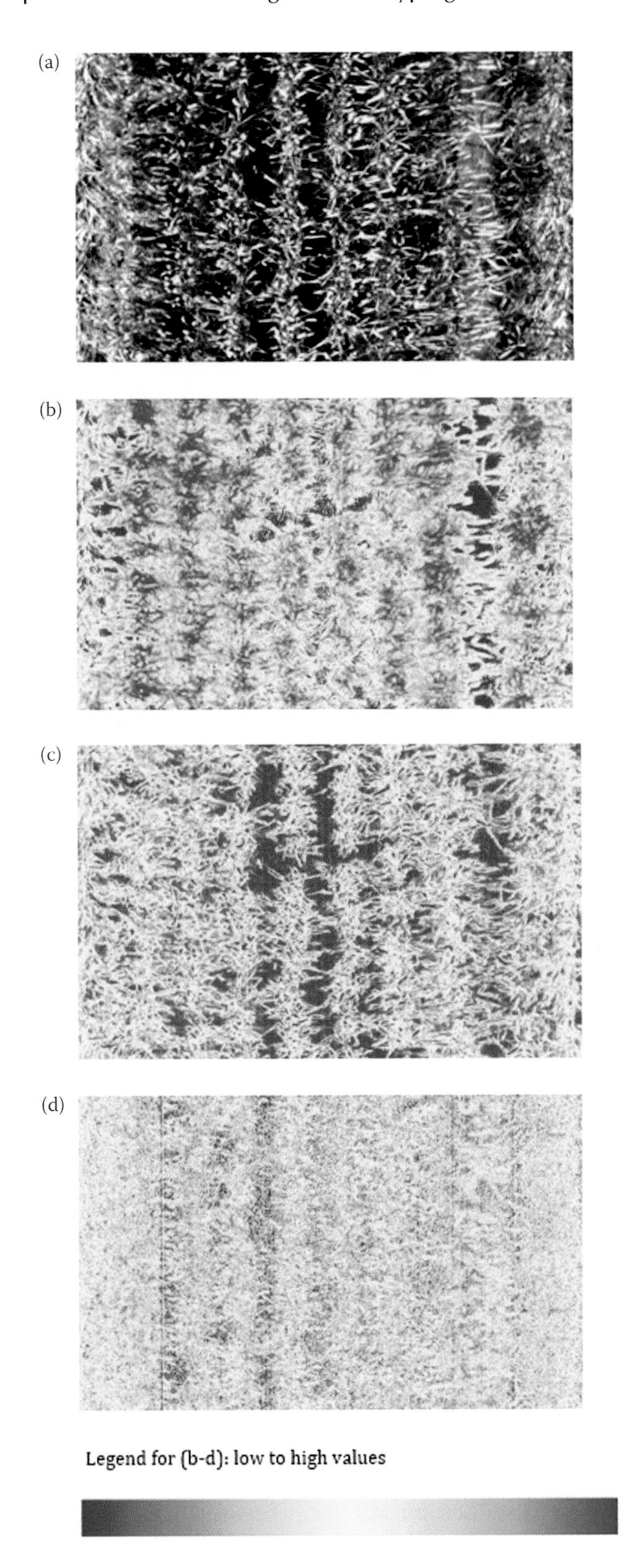

FIGURE 5.6 Example wheat plot captured using the NANO camera on the GECKO on the November 3, 2016. From top to bottom, (a) RGB, (b) NDVI, (c) TCARI, and (d) PRI.

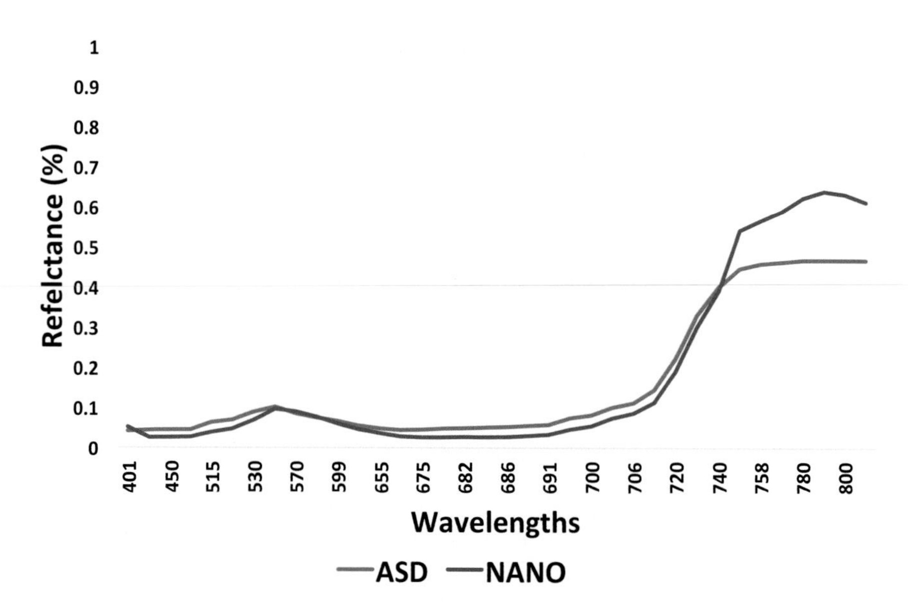

FIGURE 5.7 Reflectance signature from a spot measurement on a *single leaf* at the top of the canopy obtained from an ASD FieldSpec4 (ASD) and reflectance for 400 pixels near the top of the canopy from the NANO Headwall hyperspectral camera on the GECKO (NANO). For comparison, only wavelengths up to 800 nm were plotted.

divergence was noticed in the NIR wavelength spectrum (710–850 nm). These slight differences might be anticipated since the ASD FieldSpec measurement came from a spot on the second last fully expanded leaf, while the NANO hyperspectral reflectance was calculated as the average per wavelength from ~400 selected pixels on different parts of the canopy that correspond to the second last fully expanded leaf. Furthermore, measured canopy reflectance is greatly impacted by leaf angle, LAI, and bidirectional reflectance from soil (Asner, 1998), which largely explains discrepancies observed here, that is, reflectance from a single leaf (or part leaf) and canopy reflectance. Similar relationships were found for other plots (data not shown).

5.3.2.2 Spectral Signatures of Different Plant Organs in Sorghum

Figure 5.8 shows the spectral differences across wavelengths (400–800 nm) captured by the hyperspectral camera at plant level within a sample sorghum plot captured 91 days after sowing when heads were clearly visible. Here, we extracted spectral profiles for sorghum heads (orange, 488 pixels), leaves (green, 1069 pixels), dead material (yellow, 319 pixels), and soil (dark red, 5197 pixels). Clear differences in spectral reflectance values were observed for green leaves and other main parts of the canopy and soil within a plot. Reflectance for green leaves had a small peak in the green visible region and high in the NIR region as was expected. Dead material showed the highest reflectance in the visible region, while reflectance for sorghum heads was low in the green visible region and between that of living and dead material in the red visible region. Soil reflectance was near zero across all wavelengths.

5.3.2.3 Cross-Sectional (Single Date) Comparisons of Breeding Plots

A comparison of an RGB image and these three indices for one time point and one sorghum breeding trial plot is given in Figure 5.9.

The NDVI index (Figure 5.9b) discriminates clearly between living plant parts (blue), dead leaves (yellow), and soil (dark red). Aggregated NDVI values for soil, leaves, heads, and dead material were 0.32, 0.78, 0.41, and 0.24 respectively. Chlorophyll abundance activity is depicted in Figure 5.9c and shows clear differences in the distribution of chlorophyll in values at soil (red), heads (dark

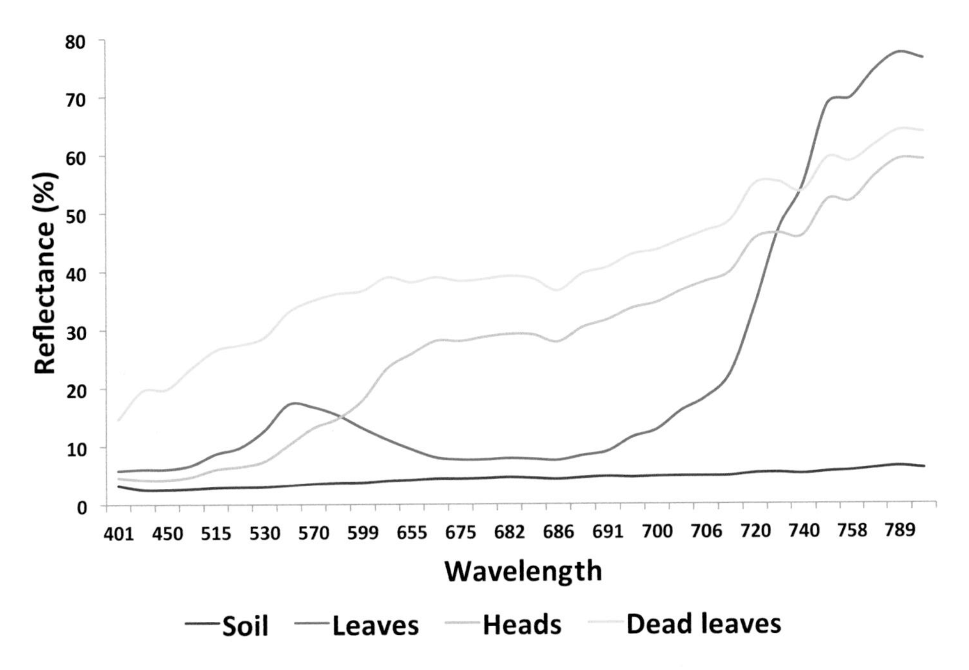

FIGURE 5.8 Aggregated spectral profiles across wavelengths of living plants (green and orange) and non-living parts (red and yellow) in a sorghum-breeding plot. Sample pixel sizes were as follows: 488 on sorghum heads, 1069 on leaves, 319 on dead material, and 5197 on soil.

red), and leaf (green to blue) levels within the canopy. There was also a clear discrepancy between shadowed leaves and sunlit leaves where sunlit leaves had the highest TCARI values. The PRI (Figure 5.9d), which relates to photosynthetic activity, shows significant differences between heads (green) and leaves (yellow). The PRI also distinguished well between sunlit (yellow) and shaded leaves (light orange). However, it is important to note that quantifying PRI values from sensing with

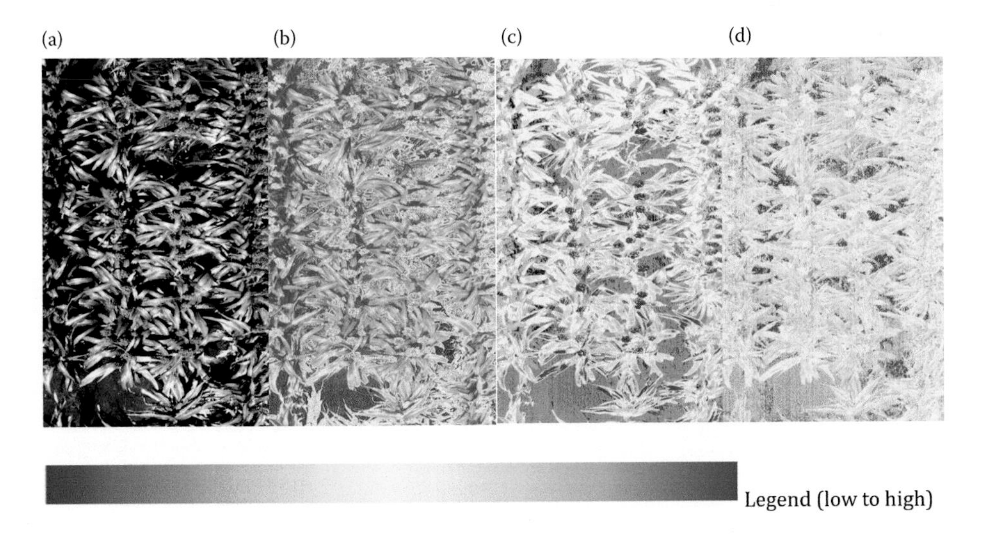

FIGURE 5.9 Example of (a) visible Red Green Blue (RGB), (b) Normalized Vegetation Index (NDVI), (c) Transformed Chlorophyll Absorption in Reflectance Index (TCARI), and (d) Photochemical Reflectance Index (PRI) images generated from the NANO Headwall hyperspectral camera on the GECKO for one time point and sorghum breeding trial plot. Legend: Dark red represents low values, while dark blue to white represents high values of the specific index.

actual measurements is unlikely since no actual in situ robust approach exists currently for actual determining of PRI in situ in canopies (Mõttus et al., 2017).

5.3.2.4 Crop Growth Dynamic Metrics Derived from Hyperspectral-Sensing

Retrieval of HVI over time allows for the investigation of a spectral crop growth profile of each plot (genotype) through the growing season. Figure 5.10 shows the RGB images for several plots (of the 980 plots), at 70 days after sowing. Most genotypes in the image have reached flowering by this date. From the zoomed-in image of the combined mosaic, it is clearly noticeable that sorghum genotypes in pre-breeding trials are diverse in terms of phenotypic traits like leaf size, leaf angle, flowering, and Cover%. Such data are critical in determining the amount of light that is captured by the crop canopy during the different crop growth stages. In the early stages of the season, plant breeders use Cover% as a measure of crop establishment per genotype.

Figure 5.11a and b show the dynamic metrics generated for the Hermitage GECKO trial for the 2016/2017 summer season. Here, the values per plot represent the aggregated Maximum NDVI (a) and rate of senescence (b). Such crop growth dynamic metrics are known to be useful in assessing leaf area duration of canopies and, in particular, the onset and rate of leaf senescence (Borrell et al., 2001, 2014; Jordan et al., 2012). Delayed leaf senescence (stay-green) is a drought adaptation mechanism in sorghum (Jordan et al., 2012; Borrell et al., 2014) and wheat (Christopher et al., 2008, 2016) that is positively correlated with grain yield under post-flowering drought.

Apart from cross-sectional screens, that is, comparing plant and canopy characteristics of different genotypes at one time point, the capturing of hyperspectral information and derived indices at regular intervals allows the assessment of dynamic canopy characteristics. This includes the tracking of changes in green leaf area and differentials in leaf greenness and photosynthetic parameters due to water availability.

Temporal metrics can be derived from various hyperspectral narrow-band indices. Figure 5.12 shows such an example for several indices relating to structure (NDVI), chlorophyll abundance (TCARI), and photosynthesis activity (PRI) for the GECKO sorghum breeding trial. Further validation of these indices against measured biomass and canopy light use will be undertaken in the near future to determine likely drivers and predictability of morphological and physiological crop traits at the canopy scale.

The maximum NDVI value was reached around the start of February, which is close to flowering (on average plots flowered on February 15). While NDVI values slightly decreased after this maximum, the TCARI remained reasonably similar or increased slightly at later dates, which was to be expected since the indices were extracted from living plant organs. Pixels were aggregated from living cover based on those pixels that had NDVI values greater than 0.5. This threshold was mainly to account (i.e., remove soil reflectance) for the impact of soil reflectance values on the canopy at the aggregated plot level (Huete and Tucker, 1991, Potgieter et al., 2017). The PRI recovered slightly after

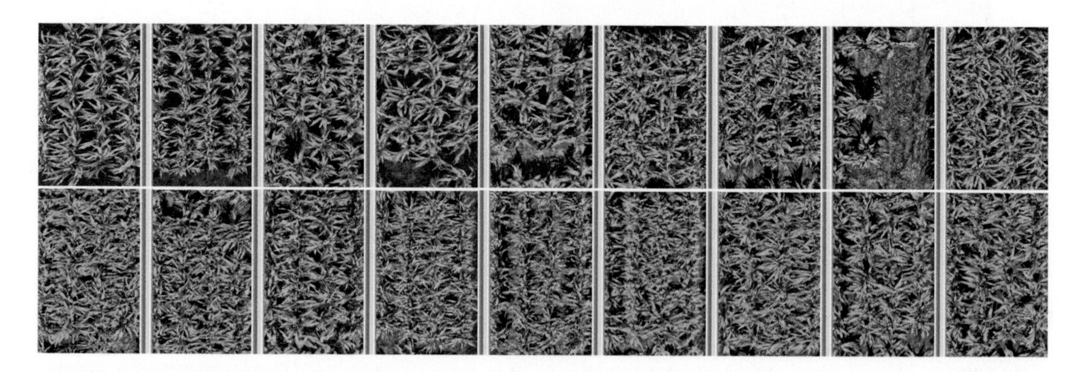

FIGURE 5.10 A visible (RGB) image of a few sample plots from the mosaic created from the GECKO field run for February 14, 2017, across 980 plots. Each plot represents a different genotype.

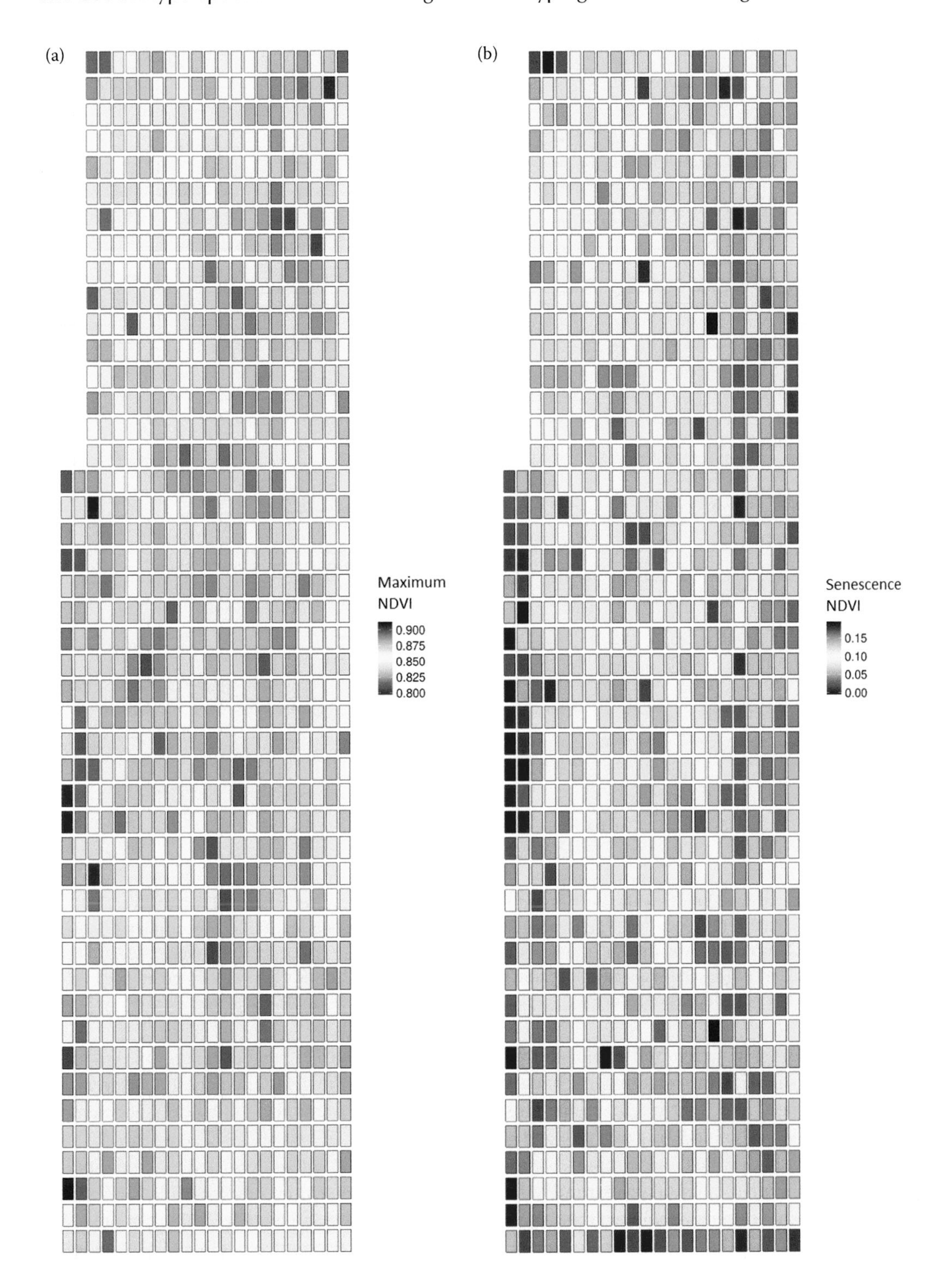

FIGURE 5.11 (a) Maximum NDVI and (b) rate of senescence (red = slow senescence, with higher retention of greenness, and green fast senescence) for the 980 sorghum plots at Hermitage during the 2016/2017 summer cropping season.

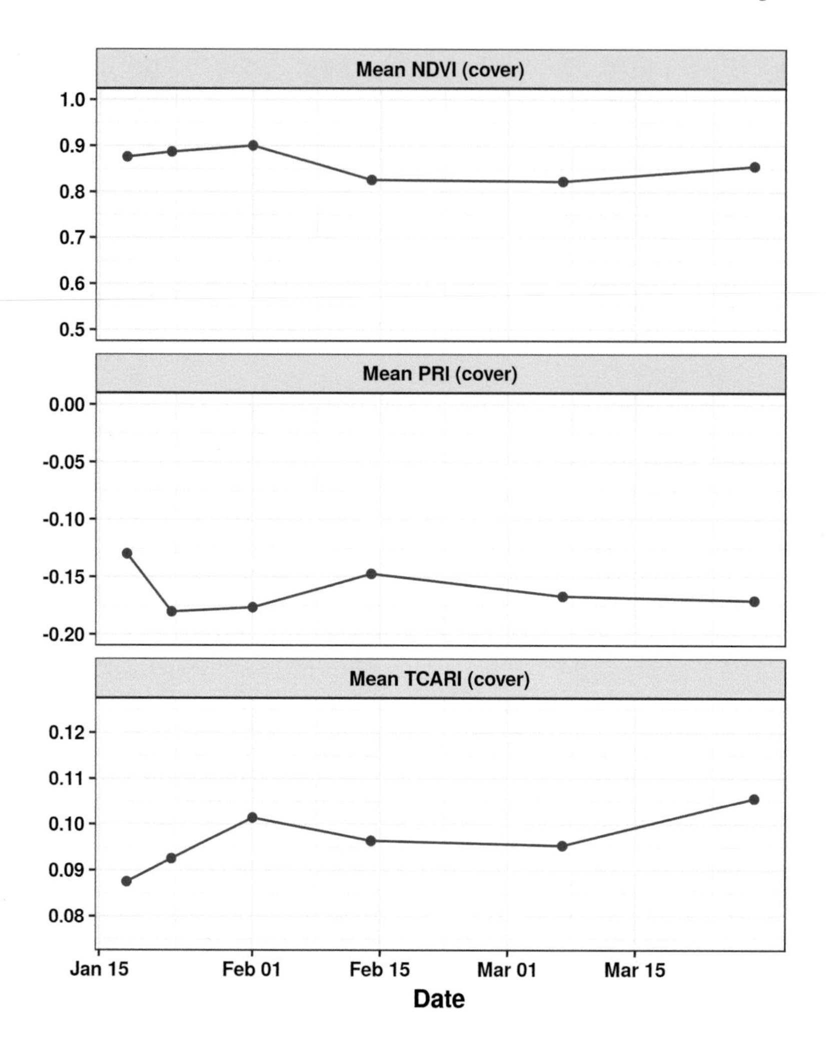

FIGURE 5.12 Averaged index values estimated for Normalized Vegetation Index (NDVI; masked between 0.5 and 1 to exclude reflectance from non-vegetation material), Photochemical Reflectance Index (PRI) and Transformed Chlorophyll Absorption in Reflectance Index (TCARI) for six dates between the middle of January and the end of March for a single sorghum breeding trial plot. Large variability existed in index values over time between plots (data not shown).

February 1, following irrigation of plots between February 15 and 22. Further research is currently in progress to analyze the relationships between targeted sensing metrics and actual measurements of LAI, biomass, and light interception. This is envisaged to enable the linkage of such sensing approaches with crop modeling at a plot and genotype level.

Overall, for both wheat and sorghum breeding trials, the HTPP pipeline showed high fidelity and accuracy in the manipulation of extremely large data sets. Furthermore, it exemplified the ability to transform data into morphological and physiological traits that can be utilized by plant breeders, crop modelers, and other researchers to enhance their knowledge and understanding of physiological traits operating at canopy levels.

5.4 OPPORTUNITY FOR A STEP CHANGE IN CROP BREEDING

While technology advancements have facilitated high-throughput genotyping, high-throughput phenotyping still lags greatly. HTPP platforms and data analysis pipelines such as the one presented here have, when used in a fully integrated research framework, the potential to lead to improved understanding of complex plant traits and ultimately provide a step change that is needed to further increase crop yields.

An integrated research framework, as depicted in Figure 5.13, links accurate HTPP with sophisticated tools, such as dynamic crop growth modeling and high-throughput high-resolution

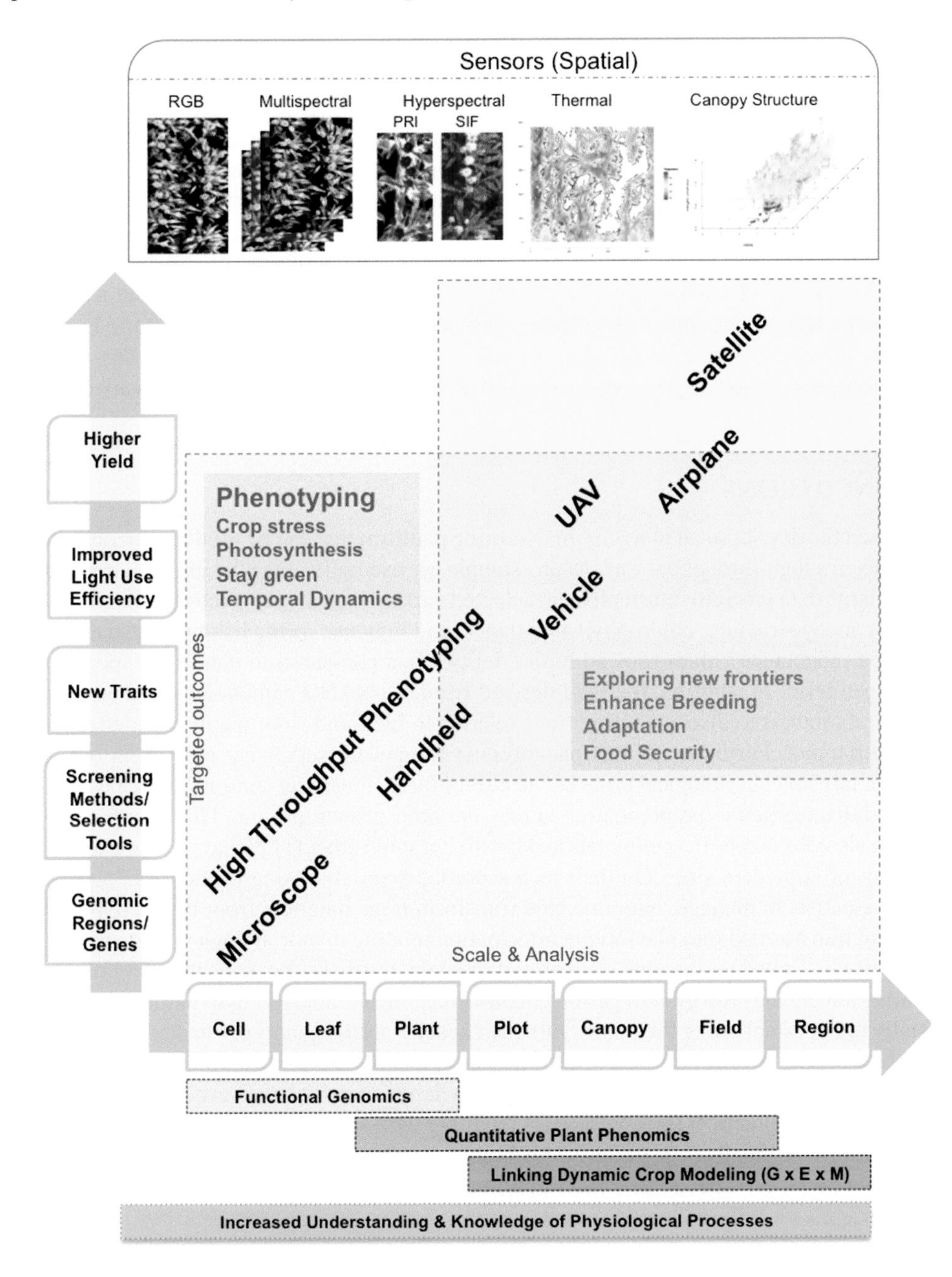

FIGURE 5.13 Schematic diagram showing linkages between platforms, sensors, and data analysis pipelines operating across scales to deliver targeted outcomes to crop breeders. (Adapted from E. Oerke et al. 2014. Proximal sensing of plant diseases. In: M. Gullino and P. Bonants, editors, *Detection and Diagnostics of Plant Pathogens. Plant Pathology in the 21st Century (Contributions to the 9th International Congress).* Springer, Dordrecht; A.-K. Mahlein. 2016. *Plant Disease* 100: 241–251. doi: 10.1094/PDIS-03-15-0340-FE.)

genotyping. Dynamic crop models enable us to predict genotype performance in various environments and under different agronomic management (GxExM simulations; Chapman, 2008; Hammer et al., 2010; Chenu et al., 2017), which reduces the need for expensive yield trials. In this way, they can also be employed to enhance the value of genomic selection by providing an avenue to account for such interactions (Hammer et al., 2016; Messina et al., 2017). Crop models are also useful in identifying and evaluating target traits (Ghanem et al., 2015; Casadebaig et al., 2016; Hammer et al., 2016). Accurate phenotyping in turn enables the continuous updating of crop models in a near real-time feedback approach for prediction of crop growth in breeding field experiments. Such an integrated framework allows for scaling from genes to improved yield, or from cellular to regional scales. However, the most critical aspect in a successful framework is the improved understanding of, and linkage to, the physiological traits, or the underlying mechanisms of the target trait (Hammer et al., 2016). A hyperspectral framework that simply correlates spectra with yield is unlikely to achieve more than simply analyzing yield differences (White et al., 2012). Therefore, the most value is likely to come from integrating HTPP with modeling platforms. This will not only allow the prediction of morphological and physiological traits outside the observed genotypic data set, but it will serve as a diagnostic tool to determine specific trait responses of new genotypes at canopy level under different environments across large scales. Such developments will enable crop breeders to explore and discover new frontiers in genotypic traits linked to generating higher yielding varieties across different environments.

5.5 CONCLUSIONS

We discussed the development of a proximal-sensing platform that can be used to generate accurate phenotyping in a high-throughput way. As an example, we exemplified the flexibility in transforming extremely large data sets into informative preselected narrow-band hyperspectral vegetation indices relating to canopy structure, chlorophyll abundance, and light use spring habit wheat and sorghum. The derived reflectance values showed high fidelity when compared to hand-held spectrometers at plant organ level. In addition, the HVI derived from the NANO camera exemplified its spatial and temporal prowess to discriminating traits relating to LUE and chlorophyll abundance at organ levels within a plot. Furthermore, this phenotyping framework serves not only as an out-scaling approach of targeted physiological traits but also illuminates the direct contrast of such traits within a large and diverse genotype population across the crop growth period. This proximal-sensing framework also overcomes the computational limitation some other HTPP have in the manipulation and analysis of large data sizes. On their own accord, proximal-sensing frameworks augment the ability of scientists to manage, integrate, and transform large data sets from in situ hyperspectral data that are transformed into phenotypic information relating to morphological and physiological traits that can be linked to dynamic crop growth models. Furthermore, concatenating of retrieved information during the crop growth period enables phenometric analysis of dynamic traits that will potentially aid in establishing the photosynthetic capacity across genotypes and environments. As demonstrated here, for a HTPP platform to be operationally successful requires a scientific and biological sound foundation as well as an extremely large computational capacity that allows for efficacy in deriving practical sensing metrics from big data sets. Finally, it is envisaged that such a HTPP platform will deliver quantitative phenomic metrics that can serve as input into dynamic crop modeling with linkages to genomics and thus enhancing a breeder's ability to select higher yielding varieties for different environments.

ACKNOWLEDGMENTS

We would like to acknowledge the contributions to the sensor assembly and programming made by Ed Holland. We would also like to thank Dr. Alexandre Ivakov and Sean Williamson for collecting the FieldSpec4 reflectance data, Drs Robert Furbank and Jose Antonio Jiménez-Berni for advice and comments, Peter George for driving the GECKO, and staff from the Queensland Department of Agriculture sorghum breeding program for providing the seed and helping with data collection.

This research was funded partially by the Australian Government through the Australian Research Council Centre of Excellence for Translational Photosynthesis (grant CE140100015), the Bill & Melinda Gates Foundation (grant OPPGD1197 iMashilla "A targeted approach to sorghum improvement in moisture stress areas of Ethiopia"), and a Major Equipment and Infrastructure Grant "Phenotype Sensing Platform to Enhance Plant Breeding" by the University of Queensland.

REFERENCES

Alexandratos, N. and J. Bruinsma. 2012. World agriculture towards 2030/2050: The 2012 revision. ESA Working paper No. 12-03. Rome, FAO.

Araus, J.L. and J.E. Cairns. 2014. Field high-throughput phenotyping: The new crop breeding frontier. *Trends Plant Sci* 19: 52–61. doi: 10.1016/j.tplants.2013.09.008

Asner, G.P. 1998. Biophysical and biochemical sources of variability in canopy reflectance. *Remote Sensing of Environment* 64: 234–253.

Borrell, A.K., G. Hammer and E. Van Oosterom. 2001. Stay-green: A consequence of the balance between supply and demand fornitrogen during grain filling? *Annals of Applied Biology* 138: 91–95.

Borrell, A.K., J.E. Mullet, B. George-Jaeggli, E.J. van Oosterom, G.L. Hammer, P.E. Klein and D.R. Jordan. 2014. Drought adaptation of stay-green sorghum is associated with canopy development, leaf anatomy, root growth, and water uptake. *Journal of Experimental Botany* 65: 1–13. doi: 10.1093/jxb/eru232

Casadebaig, P., B. Zheng, S. Chapman, N. Huth, R. Faivre and K. Chenu. 2016. Assessment of the potential impacts of wheat plant traits across environments by combining crop modeling and global sensitivity analysis (Report). *PLoS ONE* 11: e0146385. doi: 10.1371/journal.pone.0146385

Chapman, C.S., T. Merz, A. Chan, P. Jackway, S. Hrabar, F.M. Dreccer, E. Holland, B. Zheng, J.T. Ling and J. Jimenez-Berni. 2014. Pheno-Copter: A low-altitude, autonomous remote-sensing robotic helicopter for high-throughput field-based phenotyping. *Agronomy* 4: 279–301. doi: 10.3390/agronomy4020279

Chapman, S. 2008. Use of crop models to understand genotype by environment interactions for drought in real-world and simulated plant breeding trials. *International Journal of Plant Breeding* 161: 195–208. doi: 10.1007/s10681-007-9623-z

Chenu, K., J.R. Porter, P. Martre, B. Basso, S.C. Chapman, F. Ewert, M. Bindi and S. Asseng. 2017. Contribution of crop models to adaptation in wheat. *Trends in Plant Science* 22: 472–490. https://doi.org/10.1016/j.tplants.2017.02.003

Christopher, J., M.J. Christopher, A.K. Borrell, S. Fletcher and K. Chenu. 2016. Stay-green traits to improve wheat adaptation in well-watered and water-limited environments. *Journal of Experimental Botany* 67: 5159–5172.

Christopher, J., M. Christopher, R. Jennings, S. Jones, S. Fletcher, A. Borrell, A.M. Manschadi, D. Jordan, E. Mace and G. Hammer. 2013. QTL for root angle and number in a population developed from bread wheats (*Triticum aestivum*) with contrasting adaptation to water-limited environments. *Theoretical and Applied Genetics* 126: 1563–1574.

Christopher, J.T., A.M. Manschadi, G.L. Hammer and A.K. Borrell. 2008. Developmental and physiological traits associated with high yield and stay-green phenotype in wheat. *Australian Journal of Agricultural Research* 59: 354–364. doi: 10.1071/AR07193

Christopher, J.T., M. Veyradier, A.K. Borrell, G. Harvey, S. Fletcher and K. Chenu. 2014. Phenotyping novel stay-green traits to capture genetic variation in senescence dynamics. *Functional Plant Biology* 41: 1035–1048.

Christopher, M., K. Chenu, R. Jennings, S. Fletcher, D. Butler, A. Borrell and J. Christopher 2018. QTL for stay-green traits in wheat in well-watered and water-limited environments. *Field Crops Research* 217: 32–44.

Cullis, B.R., A.B. Smith and N.E. Coombes. 2006. On the design of early generation variety trials with correlated data. *Journal of Agricultural, Biological, and Environmental Statistics* 11: 381. doi: 10.1198/108571106x154443

Deery, D., J. Jimenez-Berni, H. Jones, X. Sirault and R. Furbank. 2014. Proximal remote sensing buggies and potential applications for field-based phenotyping. *Agronomy* 5: 349–379.

Furbank, R.T. and M. Tester. 2011. Phenomics—Technologies to relieve the phenotyping bottleneck. *Trends in Plant Science*. 16: 635–644.

Gamon, J.A. 2010. The Photochemical Reflectance Index (PRI)—A measure of photosynthetic light-use efficiency. *HyspIRI Meeting*.

Gamon, J.A., L. Serrano and J.S. Surfus. 1997. The photochemical reflectance index: An optical indicator of photosynthetic radiation use efficiency across species, functional types, and nutrient levels. *Oecologia* 112: 492–501. doi: 10.1007/s004420050337

Ghanem, M.E., H. Marrou and T.R. Sinclair. 2015. Physiological phenotyping of plants for crop improvement. *Trends in Plant Science* 20: 139–144. https://doi.org/10.1016/j.tplants.2014.11.006

Grassini, P., K.M. Eskridge and K.G. Cassman. 2013. Distinguishing between yield advances and yield plateaus in historical crop production trends. *Nature Communications* 4: 1–11.

Haboudane, D., J.R. Miller, E. Pattey, P.J. Zarco-Tejada and I.B. Strachan. 2004. Hyperspectral vegetation indices and novel algorithms for predicting green LAI of crop canopies: Modeling and validation in the context of precision agriculture. *Remote Sensing of Environment* 90: 337–352. http://dx.doi.org/10.1016/j.rse.2003.12.013

Haboudane, D., J.R. Miller, N. Tremblay, P.J. Zarco-Tejada and L. Dextraze. 2002. Integrated narrow-band vegetation indices for prediction of crop chlorophyll content for application to precision agriculture. *Remote Sensing of Environment* 81: 416–426. http://dx.doi.org/10.1016/S0034-4257(02)00018-4

Hammer, G., C. Messina, E. van Oosterom, S. Chapman, V. Singh, A. Borrell, D. Jordan and M. Cooper. 2016. Molecular breeding for complex adaptive traits: How integrating crop ecophysiology and modelling can enhance efficiency. In: X. Yin and P. C. Struik, editors, *Crop Systems Biology: Narrowing the Gaps between Crop Modelling and Genetics*. Springer International Publishing, Cham. pp. 147–162.

Hammer, G.L., E. van Oosterom, G. McLean, S.C. Chapman, I. Broad, P. Harland and R.C. Muchow. 2010. Adapting APSIM to model the physiology and genetics of complex adaptive traits in field crops. *Journal of Experimental Botany* 61: 2185–2202.

Hazratkulova, S., R.C. Sharma, S. Alikulov, S. Islomov, T. Yuldashev, Z. Ziyaev, Z. Khalikulov, Z. Ziyadullaev and J. Turok. 2012. Analysis of genotypic variation for normalized difference vegetation index and its relationship with grain yield in winter wheat under terminal heat stress. *Plant Breeding* 131: 716–721. doi: 10.1111/pbr.12003

Huete, A., K. Didan, T. Miura, E.P. Rodriguez, X. Gao and L.G. Ferreira. 2002. Overview of the radiometric and biophysical performance of the MODIS vegetation indices. *Remote Sensing of Environment* 83: 195–213. http://dx.doi.org/10.1016/S0034-4257(02)00096-2

Huete, A., C. Justice and H. Liu. 1994. Development of vegetation and soil indices for MODIS-EOS. *Remote Sensing of Environment* 49: 224–234.

Huete, A.R. and C.J. Tucker. 1991. Investigation of soil influences in AVHRR red and near-infrared vegetation index imagery. *International Journal of Remote Sensing* 12: 1223–1242.

Jordan, D.R., C.H. Hunt, A.W. Cruickshank, A.K. Borrell and R.G. Henzell. 2012. The relationship between the stay-green trait and grain yield in elite sorghum hybrids grown in a range of environments. *Crop Science* 52: 1153–1161. doi: 0.2135/cropsci2011.06.0326

Lelong, C.C.D., P. Burger, G. Jubelin, B. Roux, S. Labbe and F. Baret. 2008. Assessment of unmanned aerial vehicles imagery for quantitative monitoring of wheat crop in small plots. *Sensors* 8: 3557–3585. doi: 10.3390/s8053557

Li, L., Q. Zhang and D. Huang. 2014. *A Review of Imaging Techniques for Plant Phenotyping. Sensors* 14: 20078–20111. MDPI AG, Basel, Switzerland. doi: 10.3390/s141120078

Mahlein, A.-K. 2016. Plant disease detection by imaging sensors—Parallels and specific demands for precision agriculture and plant phenotyping. *Plant Disease* 100: 241–251. doi: 10.1094/PDIS-03-15-0340-FE

Mahner, M. and M. Kary. 1997. What exactly are genomes, genotypes and phenotypes? and what about phenomes? *Journal of Theoretical Biology* 186: 55–63. https://doi.org/10.1006/jtbi.1996.0335

Manschadi, A.M., J. Christopher, P. Devoil and G.L. Hammer. 2006. The role of root architectural traits in adaptation of wheat to water-limited environments. *Functional Plant Biology* 33: 823–837. doi: 10.1071/fp06055

Manschadi, A.M., J.T. Christopher, G.L. Hammer and P. Devoil. 2010. Experimental and modelling studies of drought-adaptive root architectural traits in wheat (*Triticum aestivum* L.). *Plant Biosystems—An International Journal Dealing with all Aspects of Plant Biology* 144: 458–462. doi: 10.1080/11263501003731805

McKeown, F.R. 1978. *A Land Classification of the Hermitage Research Station*. Division of Land Utilisation. Queensland Department of Primary Industries, Brisbane, Australia.

Messina, C.D., F. Technow, T. Tang, R.L. Totir, C. Gho and M. Cooper. 2017. Leveraging biological insight and environmental variation to improve phenotypic prediction: Integrating crop growth models (CGM) with whole genome prediction (WGP). *bioRxiv*. doi: 10.1101/100057

Mõttus, M., R. Hernández-Clemente, V. Perheentupa and V. Markiet. 2017. In situ measurement of Scots pine needle PRI. *Plant Methods* 13: 35. doi: 10.1186/s13007-017-0184-4.

Oerke, E., A. Mahlein and U. Steiner. 2014. Proximal sensing of plant diseases. In: M. Gullino and P. Bonants, editors, *Detection and Diagnostics of Plant Pathogens. Plant Pathology in the 21st Century (Contributions to the 9th International Congress)*. Springer, Dordrecht.

Potgieter, A.B., B. George-Jaeggli, S.C. Chapman, K. Laws, L.A. Suárez Cadavid, J. Wixted, J. Watson, M. Eldridge, D.R. Jordan and G.L. Hammer. 2017. Multi-spectral imaging from an unmanned aerial vehicle enables the assessment of seasonal leaf area dynamics of sorghum breeding lines. *Frontiers in Plant Science* 8: 1532.

Rahaman, M.M., D. Chen, Z. Gillani, C. Klukas and M. Chen. 2015. Advanced phenotyping and phenotypedataanalysis for the study of plant growth and development. *Frontiers in Plant Science* 6: 619. doi: 10.3389/fpls.2015.00619

Ray, D.K., N. Ramankutty, N.D. Mueller, P.C. West and J.A. Foley. 2012. Recent patterns of crop yield growth and stagnation. *Nature Communications* 3: 1293. doi: 10.1038/ncomms2296

R Core Team. 2017. *A Language and Environment for Statistical Computing*. R Foundation for Statistical Computing, Vienna, Austria.

Rebetzke, G.J., J.A. Jimenez-Berni, W.D. Bovill, D.M. Deery and R.A. James. 2016. High-throughput phenotyping technologies allow accurate selection of stay-green. *Journal of Experimental Botany* 67: 4919–4924. doi: 10.1093/jxb/erw301

Rouse Jr, J., R. Haas, J. Schell and D. Deering. 1974. Monitoring vegetation systems in the great plains with ERTS. *Proceedings, 3rd Earth Resource Technology Satellite-1 (ERTS-1) Symposium*. NASA, Goddard Space Flight Center.

Sadras, V.O., G.J. Rebetzke and G.O. Edmeades. 2013. The phenotype and the components of phenotypic variance of crop traits. *Field Crops Research* 154: 255–259. https://doi.org/10.1016/j.fcr.2013.10.001

Stephens, J.C., F.R. Miller and D.T. Rosenow. 1967. Conversion of alien sorghums to early combine genotypes1. *Crop Science* 7: 396. doi: 10.2135/cropsci1967.0011183X000700040036x.

Verger, A., N. Vigneau, C. Chéron, J.-M. Gilliot, A. Comar and F. Baret. 2014. Green area index from an unmanned aerial system over wheat and rapeseed crops. *Remote Sensing of Environment* 152: 654–664. http://dx.doi.org/10.1016/j.rse.2014.06.006

White, J.W., P. Andrade-Sanchez, M.A. Gore, K.F. Bronson, T.A. Coffelt, M.M. Conley, K.A. Feldmann et al. 2012. Field-based phenomics for plant genetics research. *Field Crops Research* 133: 101–112. http://dx.doi.org/10.1016/j.fcr.2012.04.003

6 Linking Online Spectral Libraries with Hyperspectral Test Data through Library Building Tools and Code

Muhammad Al-Amin Hoque and Stuart Phinn

CONTENTS

6.1 INTRODUCTION

Here we demonstrate how libraries or databases of consistently collected, attributed, and stored data can be valuable to develop and assess remote sensing applications. These data are collected from various platforms including field, airborne, and spaceborne, and the data types include reflectance, absorbance, and transmission.

Although the literature on field-based, airborne, and spaceborne hyperspectral applications are extensive, covering atmospheric, vegetation, soils, freshwater, and oceanic water, and benthic environments, there is limited published material explaining the systematic collection, storage, and sharing of these data. By "systematic collection," we refer to: data collection protocols; data format/storage/metadata/license standards, and publication of data in a format and location where it can be found, assessed, and downloaded for use.

These systematically collected, stored, and accessible databases of spectral information are becoming more important due to:

- Requirements to publish and make accessible, publicly funded research datasets; and
- A shift to large-scale, semi-automated processing and machine learning by using extensive training data systems.

Here, we assess the literature to determine the extent to which standardised and curated collections of laboratory, field, airborne, and satellite hyperspectral datasets are available. These "spectral libraries" are evaluated for: (1) established or known targets, (2) their form and contents, (3) common applications, (4) approaches used to link them to processing other laboratory, field, airborne, and satellite measurements, (5) known limitations, and (6) specified future directions for development. The data considered includes all forms of measurements, from raw digital numbers, to radiance, irradiance, reflectance, absorption, and transmission measures. In some cases, the libraries also contain modelled spectral properties using radiative transfer models.

6.2 SPECTRAL PROPERTY LIBRARIES: COMMON ELEMENTS AND STANDARDS

Libraries of spectral properties, such as reflectance, have several common elements and usually include data that have been collected, processed, and curated following set procedures. Three forms of libraries are evident: (1) a collection of measurements associated with a specific project, for example, Advance Spaceborne Thermal Emission and Reflection Radiometer (ASTER); (2) a collection of measurements made by one or more instruments used by one laboratory or institution, across multiple projects, for example, Vegetation Spectral Library (VSL); and (3) a collection of measurements made by a range of instruments, either for a single application, or multiple applications, stored in a common format, for example, SPECCHIO (http://specchio.ch/), ECOSIS (https://ecosis.org/), and SPECTATION (http://www-app2.gfz-potsdam.de/spectation/?file=main). The common properties of each of these libraries are: (1) use of an indexed database of one form or other; (2) standard metadata and licensing, but with varying amounts of detail; (3) standard file formats; and (4) varying abilities to search, evaluate, and download data for re-use. Each of these points are discussed and the libraries are summarised in Table 6.1 and their applications in Table 6.2.

6.2.1 Use of an Indexed Database of One Form or Another

In this context, data are stored as a spectral library, that is, an indexed collection of measurements, either in very simple forms (a website listing), or more advanced forms (e.g., a database, such as SPECCHIO, http://specchio.ch/) allowing search, query, and retrieval of specific records. Simple websites, allowing listing, download, and viewing of datasets are used by the United States Geological Survey (USGS; Kokaly et al., 2017), and other users including ASTER (Baldridge et al., 2009), Arizona State University (ASU; Christensen et al., 2000), and the Santa Barbara urban spectral library (Herold et al., 2004). Database interfaces are used for storing the VSL data (Goswami and Matharasi, 2015). The SPECCHIO spectral library was created using a MySQL database server where spectral data are stored within a database system based on the entity-relationship model (Hueni et al., 2009). Excel worksheets are also used for storing spectral libraries in a flat format data style. Examples of this include spectral reflectance libraries stored on data portals such as Remote Sensing Research UQ-Pangea (Roelfsema and Phinn, 2012).

6.2.2 Standard Metadata and Licensing

The completeness and quality of metadata are vital factors to consider in the development of a standard spectral library intended for exchanging and sharing spectra datasets globally (Rasaiah et al., 2015).

TABLE 6.1
Overview of Spectral Libraries

	Spectra Sources								
Library Name	Field	Laboratory	Airborne Sensor	Spaceborne Sensor	Compilation	Nature of Samples	Database Type	Web Address	References
USGS	–	√	–	–	–	Earth materials	Web-based archive	https://speclab.cr.usgs.gov/spectral-lib.html	Kokaly et al. (2017)
ASTER	–	–	–	–	√	Earth materials	Web-based archive	https://speclib.jpl.nasa.gov/	Baldridge et al. (2009)
ASU	–	√	–	–	√	Earth/planetary materials	Web-based archive	http://speclib.asu.edu/	Christensen et al. (2000)
SPECCHIO	√	√	√	√	√	Earth materials	MySQL	http://v473.vanager.de:8080/SPECCHIO_Web_Interface/	Hueni et al. (2009)
EcoSIS	√	√	√	√	√	Vegetation functional traits	Web-based archive	https://ecosis.org/	Serbin et al. (2013)
SPECTATION	√		√			Vegetational parameters	Postgre SQL	http://www-app2.gfz-potsdam.de/spectation/?file=main	–
Auscope	√				√	Earth materials	Web-based archive	www.auscope.org.au/)	Woodcock et al. (2010)
VSL		√			√	Vegetation and related land covers	Postgre SQL	http://spectrallibrary.utep.edu/	Goswami and Matharasi (2015)
Santa Barbara Urban Spectral Library	√			√		Urban materials	Web-based archive	http://www.geogr.uni-jena.de/~c5hema/spec/sburbspec_main.htm	Herold et al. (2004)
Remote Sensing Research UQ-Pangea	√					Coral reef, sea grass	Web-based archive	https://doi.pangaea.de/10.1594/PANGAEA.864310	Roelfsema and Phinn (2013)

TABLE 6.2
Summary of Relevant Spectral Library Applications

Spectral Library and form of Library	Data, Metadata, and Measurement Types Included in Library	Standards Applied to Data Prior to Inclusion in Library, Collection Protocol, or Method	Known Applications of Library
Spectral library developed from: i. Imagery -AVIRS, Hyperion, Landsat ETM+ ii. Field spectrometer data iii. Laboratory Measurement iv. Simulated with radiative transfer model Online spectral libraries and web addresses: i. USGS spectral library (https://speclab.cr.usgs.gov/spectral-lib.html) ii. ASTER spectral Library(https://speclib.jpl.nasa.gov/) iii. VSL spectral library(http://spectrallibrary.utep.edu/) iv. ASU spectral library(http://speclib.asu.edu/) v. SPECCHIO spectral library(http://v473.vanager.de:8080/SPECCHIO_Web_Interface/) vi. Santa Barbara(http://www.geogr.uni-jena.de/~c5hema/spec/sburbspec_main.htm) File format: ASCII File Database type: i. Web-based archive ii. MySQL iii. Postgre SQL	Meta data: i. Location information: Location description, Referencing Datum, Latitude and Longitude ii. Instrument: Make and model, Manufacturer, dark signal correction iii. Reference Standard: Reference material, serial number iv. Hyperspectral signal properties: Data type (Reflectance, Radiance), Wavelength Interval v. Illumination information: Source of illumination (e.g., sun, lamp) vi. Viewing geometry: Distance from target, Illumination angle, Sensor angle vii. Atmospheric condition: Cloud cover (%), Humidity, wind speed viii. Target sampling information: target/sample type, target/sample ID, target digital Photographs ix. General project name: Project participants (owners), name of experiment/project	Collection protocol/Method: A. Sampling protocol 1. Sampling strategy • Data collection date • Site number and distribution (site identification and location) • Species identification and name 2. Spectra acquisition • Spectrometer configuration (field of view, number files) • Observation geometry (solar and observation angle) • Target sample (Measurement time : Solar time 8:0018:00, local noon time 11:00–13:00 and target selection) • Distance from target (1.5 m above the top of target) • Web length range (350–2500 nm) • Resolution (10 nm) B. Data processing 1. Spectra pre-processing • Reflectance (reference panel) • Preparation (number of spectra) 2. Separability estimation Standards applied to data prior for inclusion in library: • Standards were measured multiple times during the acquisition of sample spectra to ensure that there were no major deviations in instrument performance. • Intra species variability was not considered in most of the online spectral library. Spectra collection from satellite image: • Geometric, radiometric and atmospheric corrections conducted in the image before extracting the endmembers. • Spectra were extracted from image using the reference polygons • Spectra were extracted from each hyperspectral image using the verified targeted species patches. Pixels within patches were selected to extract species-specific spectral information.	• Vegetation cover/fractions mapping • Vegetation species mapping • Tree species mapping • Land cover mapping • Mapping canopy foliar dry biomass • Mapping land cover fractions • Monitoring the spread of invasive plant species • Mapping crops/cover and weed • Mapping wetland composition • Soil contamination mapping • Mapping changing distribution of dominant species • Mapping plant functional type

Transformations Applied to Link Spectral Properties Library to Satellite Data or Satellite Data Product	Known Limitations of Methods Used	Directions for Future Work	References
A. Spectral unmixing approaches: 1. sSMA 2. MESMA 3. Endmember bundle unmixing 4. Monte Carlo unmixing 5. BSMA 6. VMESMA 7. Sparse unmixing 8. WASMA B. Classification approaches: 1. SAM 2. SID 3. Spectral similarity value 4. SFF 5. LDA 6. CDA C. Inverse models: 1. CWA 2. PLS 3. SVR 4. Marko Chain Monte Carlo	A. Spectral Unmixing: approaches: • sSMA assumes a constant spectral signature for each endmember. Actually, endmembers do not vary on per pixel basis • Presence of nonlinear mixing and noise • Endmember bundle vectors are uniformly or symmetrically distributed within each endmember bundle domain, which may not be true. • BSMA requires more than l spectra to represent each material. • VMESMA needs expert monitoring for guided injection of prior spatial and spectral knowledge from multiple data sources • Many endmember extraction algorithm do not consider for spatial or temporal variability in spectral endmembers • The high computational complexity of sparse unmixing algorithm • MESMA is CPU-intensive and unable to properly account for interclass variability • The a priori knowledge of pixel is required for WASMA B. Classification approaches: • SAM cannot distinguish between negative and positive correlations between image and target spectra. Spectral divergence is also not accounted. • SID is computationally extensive and works mainly for multi-pixel target spectra. • SFF has limited range of material types for application since its focus on the unique spectral features. • A sufficient number of training data are required to confirm a positive definite class covariance matrix for LDA. • CD transformation is human guided and class dependent. C. Inverse models: • Need consistent data collection • More a priori information is required.	A. Spectral Unmixing: approaches: • Need to improve the computational efficiency of unmixing approaches • An approach which will count the interclass variability. • Need robust methodologies to properly select endmembers models. • An approach which can properly account for spatial and temporal endmember variability B. Classification approaches: • Need an approach which can account the spectral variability more effectively. • An approach which can deal with the very large dimensionality and complexity of hyperspectral data • Need automated and scene dependent process C. Inverse models: • Need models that can adjust to multitemporal domains • Need hybrid models to deal training data issues.	Bateson et al. (2000) Asner and Lobell (2000) Song (2005) García-Haro et al. (2005) Powell et al. (2007) Zomer et al. (2009) Dobigeon et al. (2008) Asner and Martin (2009) Byambakhuu et al. (2010) Iordache et al. (2011) Youngentob et al. (2011) Nidamanuri and Zbell (2011) Schaaf et al. (2011) Somers et al. (2012) Tits et al. (2012) Cheng et al. (2014b) Nidamanuri and Ramiya (2014) Somers and Asner (2014) Roth et al. (2015) Beland et al. (2016) Dudley et al. (2015) Amaral et al. (2015)

However, the amounts of metadata provided and its quality varies widely in published spectral libraries. Common metadata elements included in spectral libraries are: (1) location information including location description, referencing datum, latitude, and longitude; (2) instrument name with make, model, calibration status, and manufacturer; (3) reference standard with reference material and serial number; (4) hyperspectral measurement parameters, including measurement type(s), and wavelength interval; (5) illumination information; (6) viewing geometry including distance from target, illumination angle, and sensor angle; (7) atmospheric conditions; (8) target sampling information covering target/sample type, target/sample ID, target digital photographs; and (9) project name. A large number of online spectral libraries include licensing information, linked to permission forms for downloading as well as conditions for re-using data and appropriate citation.

6.2.3 Standard File Formats

Different file formats are used to store reflectance, absorbance, and transmission spectra data in spectral libraries. One of the most common file formats is the American Standard Code for Information Interchange (ASCII). Other commonly used file formats are spectrometer and image analysis software, including ENVI's ".sli" format. Another native file format for storing spectral data is SPECtrum Processing Routines (SPECPR), which is used in the SPECPR software (Clark et al., 2007).

6.2.4 Abilities to Search, Evaluate, and Download Data for Re-Use

The capabilities of spectral libraries vary in terms of searching, evaluating and downloading data for re-use. While most spectral libraries have the capacity to provide access for downloading data for re-use, some of the spectral libraries only have web-interface capabilities to search and view datasets online, and not allowing analysis. The searching procedure includes checking the metadata on location, sample type, measurement process, instrument, data processing, and format of data. Measurements of spectral reflectance published as data records on Pangea (e.g., Roelfsema and Phinn, 2012, 2013, 2017), Auscope, ASTER, VSL, and ASU spectral libraries all have web interfaces to facilitate these processes to users.

6.3 EXISTING SPECTRAL PROPERTY LIBRARIES

In this section, we outline the three common types of spectral property libraries: (1) a collection of measurements associated with specific project (e.g., ASTER); (2) a collection of measurements made by one or more instruments used by one laboratory or institution, across multiple projects (e.g., ASU); and (3) a collection of measurements made by a range of instruments, either for a single application, or multiple, stored in a common format (e.g., SPECCHIO).

1. *A collection of measurements associated with a specific project*: The ASTER spectral library (https://speclib.jpl.nasa.gov/) contains data from the ASTER imaging instrument on NASA's Terra satellite and contributes to a wide array of global change related applications. The library is a compilation of over 2400 spectra of natural and man-made materials ranging the wavelength from 0.4–15.4 μm to support ASTER applications (Baldridge et al., 2009). The NASA Jet Propulsion Laboratory (JPL), Johns Hopkins University (JHU), and the USGS are all contributors for this library. All of the compiled data were converted into common standards and metadata.
2. *A collection of measurements made by one or more instruments used by one laboratory or institution, across multiple projects*: The VSL, Pangea data records, UC Santa Barbara, and ASU spectral libraries can be included in this category. The VSL (http://spectrallibrary.utep.edu/) was created using reflectance of vegetation and land cover collected from different

research projects and is maintained by the University of Texas. The storage, query, and retrieval of data are provided via a PostgresSQL database interface. Currently, the library is comprised of 235 datasets acquired from different parts of the world (Goswami and Matharasi, 2015). Pangea is an online data publishing website, where discrete datasets or collections can be published along with detailed metadata, Remote Sensing Research Centre (RSRC) at the University of Queensland has completed extensive coral reef and seagrass spectrometry and these data were published as individual collections at (Roelfsema and Phinn 2012; Roelfsema and Phinn 2013, 2017; Roelfsema et al., 2016) (Figure 6.1). Spectra of algal, seagrass, biotic and abiotic features of coral reef collected under different projects from various Australian marine environments and published in Pangea. Pangea is an open access data publisher for the Earth and environmental science. The Santa Barbara (www.geogr.uni-jena.de/~c5hema/spec/sburbspec.htm) is an urban spectral library at the University of California with 4500 spectra ranging from 350–2400 nm collected from the Santa Barbara region (Herold et al., 2004). The ASU (http://speclib.asu.edu/) is a thermal emission spectral library of Arizona State University which includes the spectra of pure minerals as well rocks.

3. *A collection of measurements made by a range of instruments, either for a single application, or multiple, stored in a common format*: The USGS spectral library (https://speclab.cr.usgs.gov/spectral-lib.html) is a comprehensive database of spectra collected from laboratory, field, and airborne spectrometer measurements using a range of instruments (Kokaly et al., 2017). The current version (splib07) includes 1300 spectra covering mineral, vegetation, and related microorganism, rocks, coatings, soil and mixtures, volatiles, chemicals, and plants. Each spectrum is converted into a common format and stored as an image plot with required metadata. As a static library, the users can only download the data without the access for uploading new spectra. In contrast, SPECCHIO (http://specchio.ch/) is a dynamic spectral library where users can upload spectra and metadata from their own projects. For storing reference spectra and campaign data, the SPECCHIO was developed and maintained by remote sensing laboratories at the University of Zurich. At present, this library contains 111,023 spectra of various targets acquired from 71 campaigns (Hueni et al., 2009). Each spectrum is stored with detailed metadata at the spectrum and campaign level. The users can download their required spectra easily using SPECCHIO web interface (Figure 6.2). Auscope (www.auscope.org.au/) is the Australian national comprehensive spectral library where Australian geoscience and geospatial research data from various projects are stored to provide free access to researchers. Another two web-based spectral libraries are EcoSIS and SPECTATION. EcoSIS is a comprehensive ecology-themed spectral library which currently contains 54,260 spectra collected from different parts of the world through different research projects. On the other hand, SPECTATION is a spectral library of vegetational parameters designed for collecting, archiving, and sharing spectral datasets for future applications. This library was developed by Humboldt Universität zu Berlin, Universität Potsdam, and GFZ Potsdam.

6.4 COMMON APPLICATIONS OF SPECTRAL PROPERTY LIBRARIES

Applications for spectral libraries cover a very broad range, depending on the scale of applications, type of environment, and sensors. They also include algorithm development and testing in a modelling context; as well as implementation of algorithms for target detection, broad-scale composition or feature type mapping; and mapping of a biophysical properties.

6.4.1 Algorithm Development and Testing in a Modelling Context

Spectral libraries are often used for algorithm development and testing for improving hyperspectral data analysis. Dobigeon et al. (2008) have proposed a hierarchical Bayesian model for hyperspectral

Roelfsema, Christiaan M; Phinn, Stuart R (2012): Spectral reflectance library of selected biotic and abiotic coral reef features in Heron Reef. *Centre for Remote Sensing & Spatial Information Science, School of Geography, Planning & Environmental Management, University of Queensland, Brisbane, Australia, PANGAEA,* https://doi.org/10.1594/PANGAEA.804589

Always quote above citation when using data! You can download the citation in several formats below.

RIS Citation | BIBTEX Citation | Text Citation | Facebook | Twitter | Google+ | Show Map | Google Earth

Abstract:

Underwater spectral reflectance was measured for selected biotic and abiotic coral reef features of Heron Reef from June 25-30, 2006. Spectral reflectance's of 105 different benthic types were obtained in-situ. An Ocean Optics USB2000 spectrometer was deployed in an custom made underwater housing with a 0.5 m fiber-optic probe mounted next to an artificial light source. Spectral readings were collected with the probe(bear fibre) about 5 cm from the target to ensure that the target would fill the field of view of the fiber optic (FOV diameter ~4.4 cm), as well as to reduce the attenuating effect of the intermediate water (Roelfsema et al., 2006). Spectral readings included for one target included: 1 reading of the covered spectral fibre to correct for instrument noise, 1 reading of spectralon panel mounted on divers wrist to measure incident ambient light, and 8 readings of the target. Spectral reflectance was calculated for each target by first subtracting the instrument noise reading from each other reading. The corrected target readings were then divided by the corrected spectralon reading resulting in spectral reflectance of each target reading. An average target spectral reflectance was calculated by averaging the eight individual spectral reflectance's of the target. If an individual target spectral reflectance was visual considered an outlier, it was not included in the average spectral reflectance calculation. See Roelfsema at al. (2006) for additional info on the methodology of underwater spectra collection.

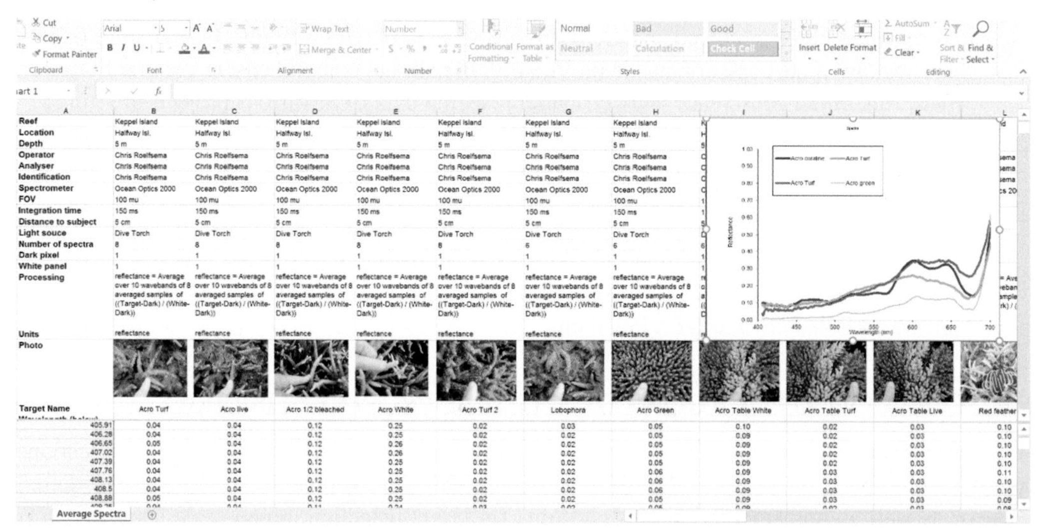

FIGURE 6.1 Example of spectral reflectance library stored in the Pangea database.

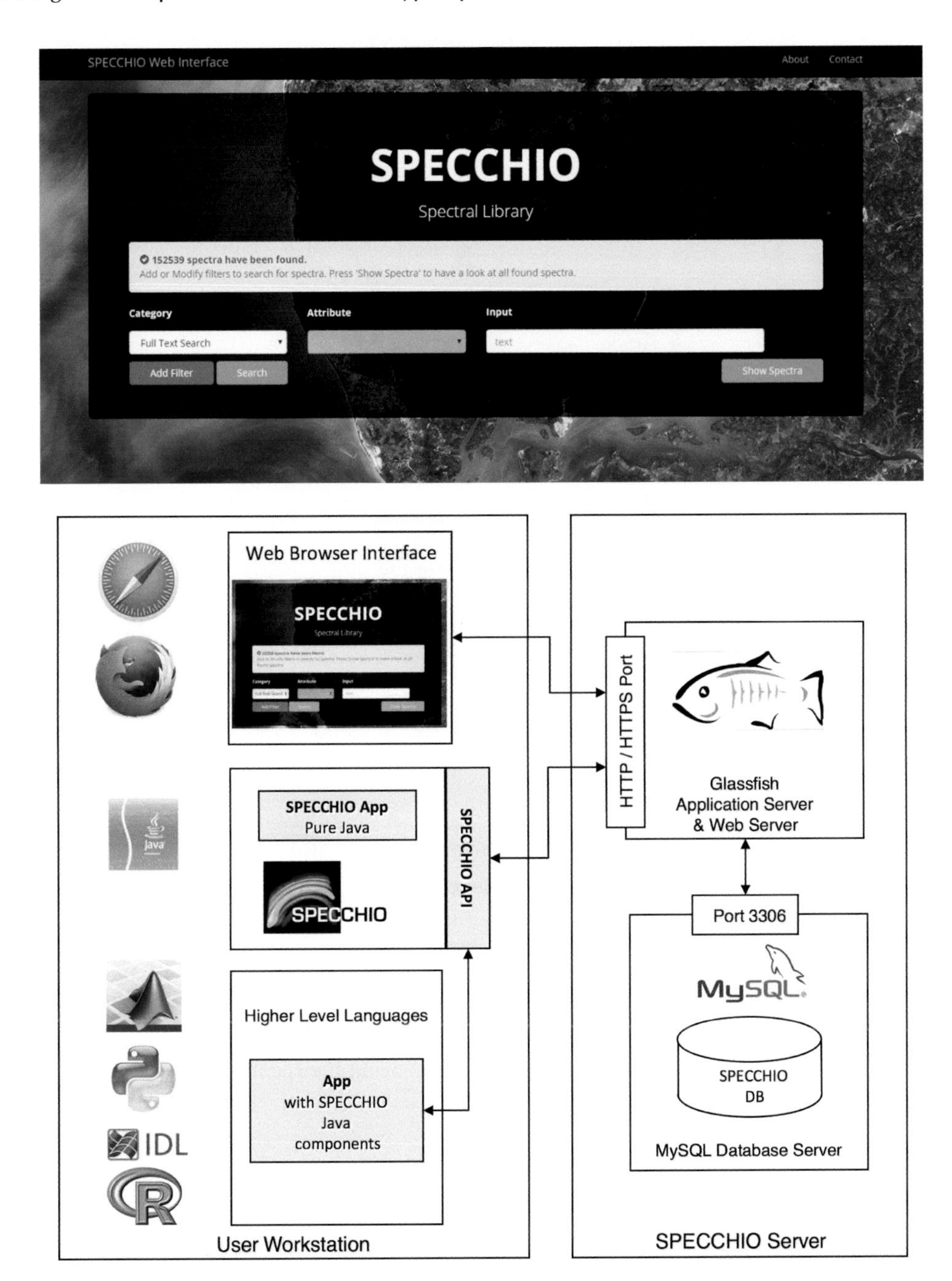

FIGURE 6.2 SPECCHIO web-browser interface and system architecture. (From http://specchio.ch/index.php.)

image unmixing. The performance of this unmixing approach was evaluated via simulations using the laboratory spectra from the NASA JPL, JHU, and the USGS. Similarly, Iordache et al. (2011) have developed a semi-supervised approach to linear spectral unmixing which called the sparse regression unmixing approach for hyperspectral data interpretation. They used the spectra database from the USGS and ASTER spectral libraries for conducting experiments with the developed approach. In another study, Nidamanuri and Ramiya (2014) evaluated the applicability and quality of some hyperspectral data classifier algorithms, for example, spectral angle mapper (SAM), spectral information divergence (SID), and spectral feature fitting (SFF). They generated the single-reference

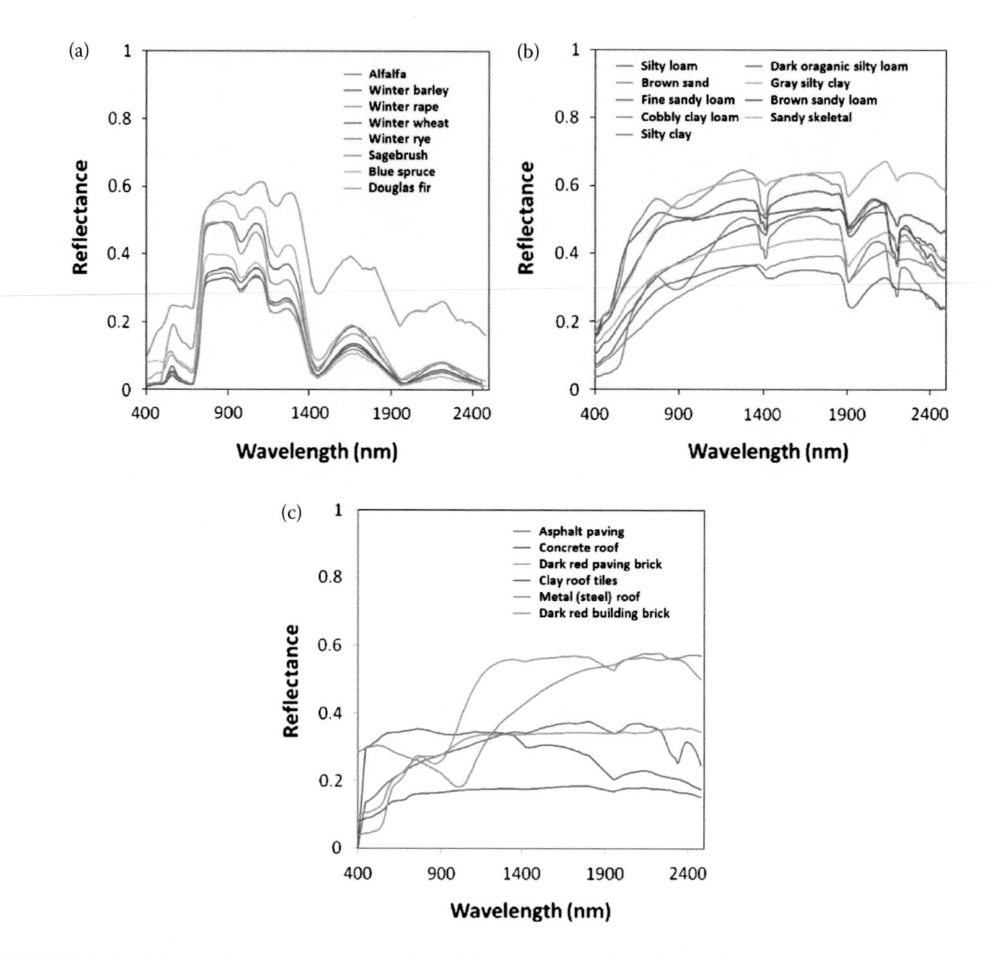

FIGURE 6.3 Mean reflectance spectra of the (a) vegetation, (b) soils, and (c) urban materials used from spectral libraries to evaluate the applicability and quality of some hyperspectral data classifier algorithms. (From Nidamanuri RR, Ramiya A 2014. *Geocarto International* 29:609–624.)

spectral library of vegetation, soils, and urban materials using the spectra database from ASTER, USGS, and VSL for their experiments (Figure 6.3).

6.4.2 Implementation of Algorithms for Target Detection, Broad-Scale Composition, or Feature Type Mapping

Spectral libraries are widely used as reference data to implement algorithms for target detection, broad-scale composition, or feature type mapping. This includes vegetation species and plant functional type mapping. Dudley et al. (2015) used the single multitemporal spectral library to map the vegetation species across spatial and temporal phenological gradients in Santa Barbara, California. The spatial variation in the dates of endmembers was selected from the multi-temporal endmember library to map each pixel in images (Figure 6.4). The same image date was considered in the endmember selection to classify each pixel. Likewise, vegetation species were also mapped by Youngentob et al. (2011) in Australia and Roth et al. (2015) in different North American ecosystems using the developed spectral libraries. In the same way, some other studies successfully mapped the vegetation cover, plant species, and its functional type (Bateson et al., 2000; Asner and Martin, 2009; Schaaf et al., 2011; Somers et al., 2012; Amaral et al., 2015). Spectral libraries are also used for mapping land cover and their fractions. Byambakhuu et al. (2010) assessed the land cover fractions in

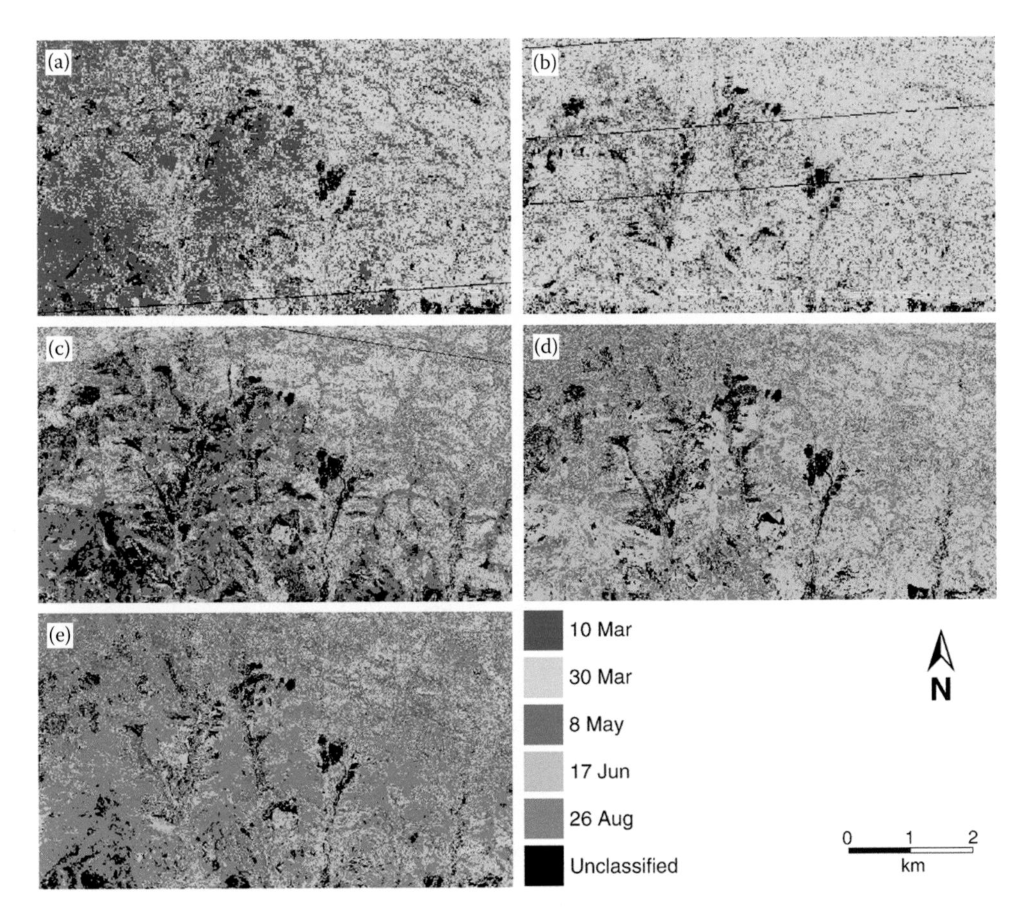

FIGURE 6.4 Classification result showing the date of endmembers from the multitemporal endmember library used to classify pixels in images for (a) Mar 10; (b) Mar 30; (c) May 8; (d) Jun 17; and (e) Aug 26 in the process of vegetation species mapping. (From Dudley KL et al. 2015. *Remote Sensing of Environment* 167:121–134. doi:10.1016/j.rse.2015.05.004.)

Mongolian steppe regions where they have used the spectral library, which was created from the field spectra. Similarly, Powell et al. (2007) mapped urban land covers, and Zomer et al. (2009) classified wetland land covers through the different algorithms using a spectral library. In addition, detailed crops were also mapped by Nidamanuri and Zbell (2011) using library spectra as a reference data

6.4.3 Mapping of a Biophysical Property

Spectral libraries also drive algorithms for mapping biophysical properties, such as leaf mass per area (LMA), photosynthetic and non-photosynthetic pigment concentrations in vegetation, concentrations of organics and in-organic suspended and dissolved materials in water, and the moisture and mineral content soil. Kokaly et al. (2003) estimated the chlorophyll and leaf water absorption in Yellowstone National Park using the developed spectral library. Similarly, Cheng et al. (2014b) estimated the LMA across a variety of plant species. In another study, Kutser (2004) used the spectral library to produce the chlorophyll concentration map in cyanobacterial blooms. On the other hand, García-Haro et al. (2005) mapped the sludge abundance in the oil-contaminated soil and Kokaly et al. (2013) assessed the concentration of suspended oil in marshes using spectral library databases. A reference spectral library of oiled and non-oiled materials presented in Figure 6.5 was developed for delineating suspended oil in marshes.

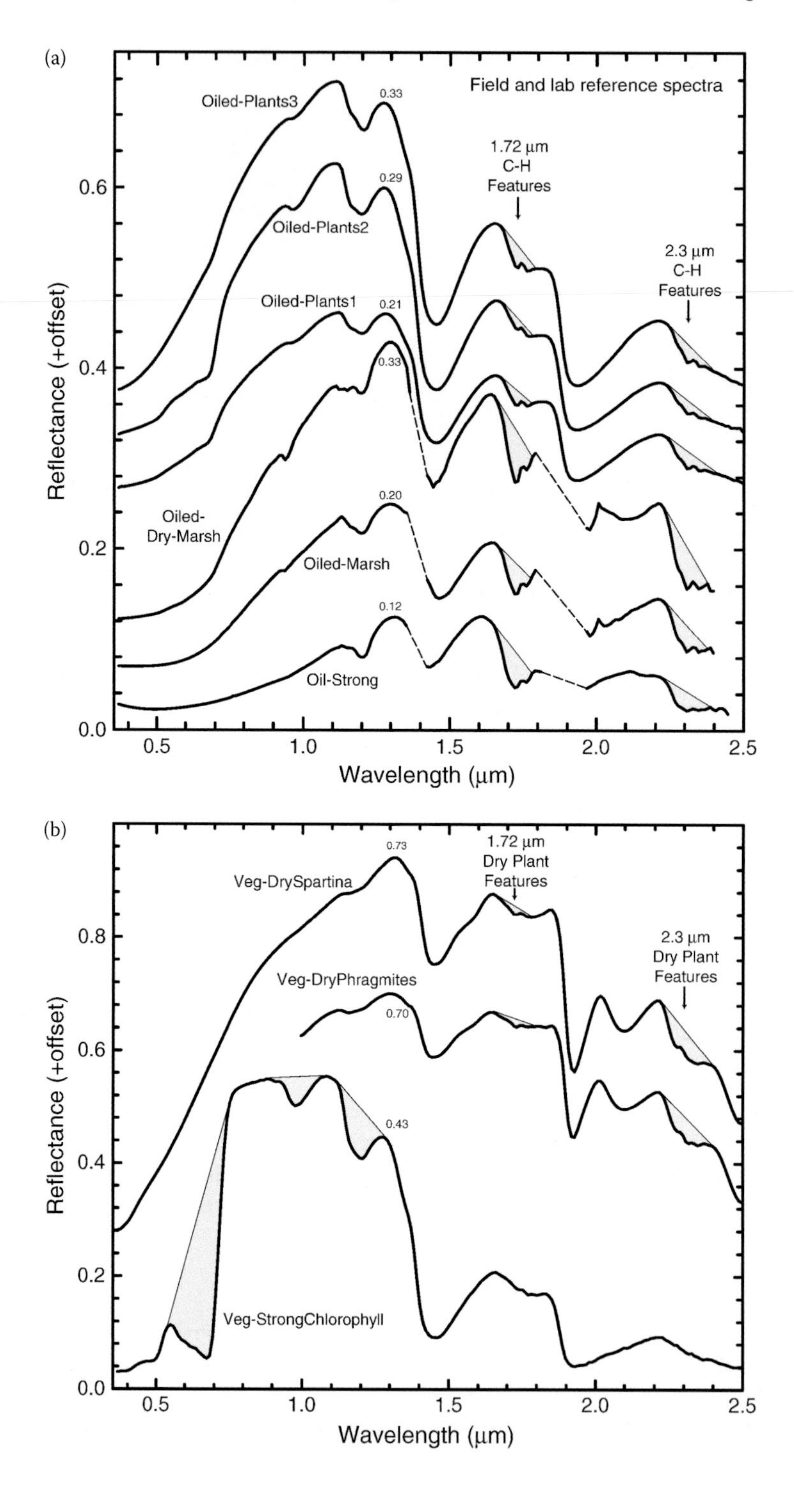

FIGURE 6.5 Reference spectral library to delineate the distribution of oil-damaged canopies in marshes. (a) Oiled plants, oiled marsh, and oiled poled, (b) Non-oiled vegetation. (From Kokaly RF et al. 2013. *Remote Sensing of Environment* 129:210–230. doi:10.1016/j.rse.2012.10.028.)

6.5 METHODS USED TO LINK SPECTRAL PROPERTY LIBRARIES TO AIRBORNE AND SATELLITE HYPERSPECTRAL DATA

Using spectral library data to process other laboratory, field, and airborne and satellite images data requires the library data to meet specific standards and contain certain metadata, and then uses several different algorithms to ingest the spectral library data and use it with the other data.

The data and metadata requirements include: data collected in the appropriate physical unit and using appropriate methods that it is physically feasible for it to represent the reflectance/absorbancetransmittance of the feature to be identified or mapped in the laboratory, field, and airborne and satellite image data. Part of this assessment can only be completed if the metadata in the spectral reflectance library is sufficient to describe: data file format, collection methods and scales, environmental conditions at the time of capture, and collection instrument specifications in detail, especially spectral bandwidths and units used.

Once these conditions are met, the following are examples of common algorithms used to map targets or a range of features, or estimate biophysical properties:

a. *Spectral unmixing approaches*: Also known as spectral mixture analysis (SMA), is one of the widely used approaches for hyperspectral data analysis (Somers et al., 2011). Mixed pixel signals are generally modelled within this approach as a linear or nonlinear mixture model. In a simple SMA, endmembers do not vary on a per pixel basis (Youngentob et al., 2011). To overcome this limitation, several algorithms have been developed including Multiple Endmember Spectral Mixture Analysis (MESMA) by Roberts et al. (1998), endmember bundle unmixing by Bateson et al. (2000), Monte Carlo unmixing by Asner and Lobell (2000), and Bayesian Spectral Mixture Analysis (BSMA) by Song (2005). Other algorithms under spectral unmixing approaches are Variable Multiple Endmember Spectral Mixture Analysis (VMESMA; García-Haro et al., 2005); Sparse Unmixing (Iordache et al., 2011); and Wavelength Adaptive Spectral Mixture Analysis (WASMA; Somers and Asner, 2014).
b. *Classification approaches*: Applications of classification approaches using online libraries for hyperspectral data analysis are increasing due to changes in image processing approaches associated with the shift to online storage and processing (Bioucas-Dias et al., 2013). Commonly used classification approaches are SAM (Zomer et al., 2009); SID (Nidamanuri and Ramiya, 2014); spectral similarity value; SFF (Nidamanuri and Zbell, 2011); Fisher's linear discriminant analysis (LDA; Roth et al., 2015); and canonical discriminant analysis (CDA; Beland et al., 2016).
c. *Inverse models*: This approach is used to estimate physical parameters from hyperspectral data, such as fractional coverage of vegetation. Example approaches include continuous wavelet analysis (CWA; Cheng et al., 2014a), Partial Least Square Regression (PLSR; Chadwick and Asner, 2016), Support Vector Regression (SVR; Malenovský et al., 2017), and Marko Chain Monte Carlo.

Several challenges are associated with the approaches outlined above for hyperspectral data analysis. Most of the spectral unmixing approaches are unable to properly account for spatial and temporal variability between and among endmembers. In particular, the BSMA algorithm requires more than one spectrum to represent each material, and VESMA needs expert monitoring for guided injection of prior spatial and spectral knowledge from multiple data sources. A priori knowledge of pixels is required for WASMA. On the other hand, unbalance between high dimensionality and limited availability of training data, the presence of mixed pixels as well as complexity in the geometry of hyperspectral data are some common challenges of hyperspectral data classification approaches. More a priori information and consistent data collection are required for inverse model applications.

6.6 CONCLUSIONS AND FUTURE DIRECTIONS TO IMPROVE SPECTRAL PROPERTY LIBRARIES AND THEIR LINKS TO AIRBORNE AND SATELLITE HYPERSPECTRAL DATA

Based on this partial evaluation, more effective use can be made of hyperspectral laboratory, field, and airborne and satellite image datasets collected globally if they are:

- Collected using standard protocols as described by Rasaiah et al. (2015);
- Stored in a commonly used file format with internationally accepted metadata, containing sufficient details on the data collection methods and formats that enables the "fit for use" purpose of the data to be assessed;
- Published as part of an online database (e.g., SPECCHIO), or internationally recognized data publishing site (e.g., PANGEA).

Reviews of the collection and use of online databases presented in this chapter showed that there are significant benefits to the producers of the data and broader Earth observation applications communities (Rasaiah et al., 2015). These benefits include improved data quality and greater ability to collaborate and build applications using spectral libraries. More and more recent studies and large-scale projects are now collecting, storing, and publishing data systematically, due to the global shift in research practices to open data and software (Tenopir et al., 2011).

There are significant limitations with some aspects of online data collections and sharing, and these could be addressed by:

- Greater publication and discussion and standardised field and laboratory spectral measurements across multiple disciplines;
- International Earth observation programs, such as Committee on Earth Observation Satellite (CEOS), working with space agencies to develop and build spectral library standards that complement their "Analysis Ready Data" satellite image data format specifications and tools for delivering this;
- Development of a suitable international community around this arena that is visible, accessible, and provides useable and practical guidance on protocols for collecting spectral data, storing it and its metadata, suitable spectral property libraries, and instructions on how to build a library if required;
- Algorithm-specific improvements including the need to improve the computational efficiency of unmixing approaches and their capability to properly account for spatial and temporal endmember variability. Appropriate solutions are also required to select the optimal number and type of endmembers for a specific scene. Development is required for effective and efficient classification approaches that can deal with the very large dimensionality and complexity of hyperspectral data. Moreover, there is also a need to reduce the computation complexity and time of hyperspectral data classification approaches. The availability of large training data and specification models that can adjust to multitemporal domains are required to foster improvement needed for inversion approaches.

REFERENCES

Amaral CH, Roberts DA, Almeida TIR, Souza CR 2015. Mapping invasive species and spectral mixture relationships with neotropical woody formations in southeastern Brazil. *ISPRS Journal of Photogrammetry and Remote Sensing* 108:80–93. doi:10.1016/j.isprsjprs.2015.06.009.

Asner GP, Lobell DB 2000. A biogeophysical approach for automated SWIR unmixing of soils and vegetation. *Remote Sensing of Environment* 74:99–112. doi:10.1016/s0034-4257(00)00126-7.

Asner GP, Martin RE 2009. Airborne spectranomics: Mapping canopy chemical and taxonomic diversity in tropical forests. *Frontiers in Ecology and the Environment* 7:269–276.

Baldridge A, Hook S, Grove C, Rivera G 2009. The ASTER spectral library version 2.0. *Remote Sensing of Environment* 113:711–715.

Bateson CA, Asner GP, Wessman CA 2000. Endmember bundles: A new approach to incorporating endmember variability into spectral mixture analysis. *IEEE Transactions on Geoscience and Remote Sensing* 38:1083–1094. doi:10.1109/36.841987.

Beland M et al. 2016. Mapping changing distributions of dominant species in oil-contaminated salt marshes of Louisiana using imaging spectroscopy. *Remote Sensing of Environment* 182:192–207. doi:10.1016/j.rse.2016.04.024.

Bioucas-Dias JM, Plaza A, Camps-Valls G, Scheunders P, Nasrabadi N, Chanussot J 2013. Hyperspectral remote sensing data analysis and future challenges. *IEEE Geoscience and Remote Sensing Magazine* 1:6–36.

Byambakhuu I, Sugita M, Matsushima D 2010. Spectral unmixing model to assess land cover fractions in Mongolian steppe regions. *Remote Sensing of Environment* 114:2361–2372.

Chadwick KD, Asner GP 2016. Organismic-scale remote sensing of canopy foliar traits in lowland tropical forests. *Remote Sensing* 8:87. doi:10.3390/rs8020087.

Cheng T, Riano D, Ustin SL 2014a. Detecting diurnal and seasonal variation in canopy water content of nut tree orchards from airborne imaging spectroscopy data using continuous wavelet analysis. *Remote Sensing of Environment* 143:39–53. doi:10.1016/j.rse.2013.11.018.

Cheng T, Rivard B, Sanchez-Azofeifa AG, Feret JB, Jacquemoud S, Ustin SL 2014b. Deriving leaf mass per area (LMA) from foliar reflectance across a variety of plant species using continuous wavelet analysis. *ISPRS Journal of Photogrammetry and Remote Sensing* 87:28–38. doi:10.1016/j.isprsjprs.2013.10.009.

Christensen PR et al. 2000. A thermal emission spectral library of rock-forming minerals. *Journal of Geophysical Research: Planets* 105:9735–9739.

Clark RN, Swayze GA, Wise R, Livo KE, Hoefen T, Kokaly RF, Sutley SJ 2007. USGS digital spectral library splib06a. US geological survey, digital data series 231.

Dobigeon N, Tourneret J-Y, Chang C-I 2008. Semi-supervised linear spectral unmixing using a hierarchical Bayesian model for hyperspectral imagery. *IEEE Transactions on Signal Processing* 56:2684–2695.

Dudley KL, Dennison PE, Roth KL, Roberts DA, Coates AR 2015. A multi-temporal spectral library approach for mapping vegetation species across spatial and temporal phenological gradients. *Remote Sensing of Environment* 167:121–134. doi:10.1016/j.rse.2015.05.004.

García-Haro F, Sommer S, Kemper T 2005. A new tool for variable multiple endmember spectral mixture analysis (VMESMA). *International Journal of Remote Sensing* 26:2135–2162.

Goswami S, Matharasi K 2015. Development of a web-based vegetation spectral library (VSL) for remote sensing research and applications. *Peer Journal of Pre Prints.*

Herold M, Roberts DA, Gardner ME, Dennison PE 2004. Spectrometry for urban area remote sensing—Development and analysis of a spectral library from 350 to 2400 nm. *Remote Sensing of Environment* 91:304–319.

Hueni A, Nieke J, Schopfer J, Kneubuhler M, Itten KI 2009. The spectral database SPECCHIO for improved long-term usability and data sharing. *Computers & Geosciences* 35:557–565. doi:10.1016/j.cageo.2008.03.015.

Iordache M-D, Bioucas-Dias JM, Plaza A 2011. Sparse unmixing of hyperspectral data. *IEEE Transactions on Geoscience and Remote Sensing* 49:2014–2039.

Kokaly RF, Despain DG, Clark RN, Livo KE 2003. Mapping vegetation in Yellowstone National Park using spectral feature analysis of AVIRIS data. *Remote Sensing of Environment* 84:437–456.

Kokaly RF et al. 2013. Spectroscopic remote sensing of the distribution and persistence of oil from the Deepwater Horizon spill in Barataria Bay marshes. *Remote Sensing of Environment* 129:210–230. doi:10.1016/j.rse.2012.10.028.

Kokaly RF et al. 2017. USGS Spectral Library Version 7. US Geological Survey.

Kutser T 2004. Quantitative detection of chlorophyll in cyanobacterial blooms by satellite remote sensing. *Limnology and Oceanography* 49:2179–2189.

Malenovský Z, Lucieer A, King DH, Turnbull JD, Robinson SA 2017. Unmanned aircraft system advances health mapping of fragile polar vegetation. *Methods in Ecology and Evolution.*

Nidamanuri RR, Ramiya A 2014. Spectral identification of materials by reflectance spectral library search. *Geocarto International* 29:609–624.

Nidamanuri RR, Zbell B 2011. Use of field reflectance data for crop mapping using airborne hyperspectral image. *ISPRS Journal of Photogrammetry and Remote Sensing* 66:683–691.

Powell RL, Roberts DA, Dennison PE, Hess LL 2007. Sub-pixel mapping of urban land cover using multiple endmember spectral mixture analysis: Manaus, Brazil. *Remote Sensing of Environment* 106:253–267. doi:10.1016/j.rse.2006.09.005.

Rasaiah BA, Jones SD, Bellman C, Malthus TJ, Hueni A 2015. Assessing field spectroscopy metadata quality. *Remote Sensing* 7:4499–4526.

Roberts DA, Gardner M, Church R, Ustin S, Scheer G, Green R 1998. Mapping chaparral in the Santa Monica Mountains using multiple endmember spectral mixture models. *Remote Sensing of Environment* 65:267–279.

Roelfsema CM, Phinn SR 2012. Spectral reflectance library of selected biotic and abiotic coral reef features in Heron Reef. *PANGAEA*. doi:10.1594/PANGAEA.804589.

Roelfsema CM, Phinn SR 2013. Spectral reflectance library of selected biotic and abiotic coral reef features in Glovers Reef, Belize. *PANGAEA*. doi:10.1594/PANGAEA.824861.

Roelfsema CM, Phinn SR 2017. Spectral reflectance library of corals and benthic features in the Cook Islands (Aitutaki). *PANGAEA*. doi:org/10.1594/PANGAEA.872505.

Roelfsema CM, Phinn SR, Joyce K 2016. Spectral reflectance library of algal, seagrass and substrate types in Moreton Bay, Australia. *PANGAEA*. doi:10.1594/PANGAEA.864310.

Roth KL, Roberts DA, Dennison PE, Alonzo M, Peterson SH, Beland M 2015. Differentiating plant species within and across diverse ecosystems with imaging spectroscopy. *Remote Sensing of Environment* 167:135–151. doi:10.1016/j.rse.2015.05.007.

Schaaf AN, Dennison PE, Fryer GK, Roth KL, Roberts DA 2011. Mapping Plant functional types at multiple spatial resolutions using imaging spectrometer data. *Giscience & Remote Sensing* 48:324–344. doi:10.2747/1548-1603.48.3.324.

Serbin S et al. Assimilation of leaf and canopy spectroscopic data to improve the representation of vegetation dynamics in terrestrial ecosystem models. In: *AGU Fall Meeting Abstracts*, 2013.

Somers B, Asner GP 2014. Tree species mapping in tropical forests using multi-temporal imaging spectroscopy: Wavelength adaptive spectral mixture analysis. *International Journal of Applied Earth Observation and Geoinformation* 31:57–66. doi:10.1016/j.jag.2014.02.006.

Somers B, Asner GP, Tits L, Coppin P 2011. Endmember variability in spectral mixture analysis: A review. *Remote Sensing of Environment* 115:1603–1616. doi:10.1016/j.rse.2011.03.003.

Somers B, Zortea M, Plaza A, Asner GP 2012. Automated extraction of image-based endmember bundles for improved spectral unmixing. *IEEE Journal of Selected Topics in Applied Earth Observations and Remote Sensing* 5:396–408. doi:10.1109/jstars.2011.2181340.

Song C 2005. Spectral mixture analysis for subpixel vegetation fractions in the urban environment: How to incorporate endmember variability? *Remote Sensing of Environment* 95:248–263.

Tits L, De Keersmaecker W, Somers B, Asner GP, Farifteh J, Coppin P 2012. Hyperspectral shape-based unmixing to improve intra- and interclass variability for forest and agro-ecosystem monitoring. *ISPRS Journal of Photogrammetry and Remote Sensing* 74:163–174. doi:10.1016/j.isprsjprs.2012.09.013.

Tenopir C, Allard S, Douglass K, Aydinoglu AU, Wu L, Read E, Manoff M, Frame M 2011. Data sharing by scientists: Practices and perceptions. *PloS One* 6(6):e21101.

Woodcock R, Simons B, Duclaux G, Cox S 2010. AuScope's use of standards to deliver earth resource data. In: *EGU General Assembly Conference Abstracts*, p. 1556.

Youngentob KN, Roberts DA, Held AA, Dennison PE, Jia XP, Lindenmayer DB 2011. Mapping two Eucalyptus subgenera using multiple endmember spectral mixture analysis and continuum-removed imaging spectrometry data. *Remote Sensing of Environment* 115:1115–e21128. doi:10.1016/j.rse.2010.12.012.

Zomer RJ, Trabucco A, Ustin S 2009. Building spectral libraries for wetlands land cover classification and hyperspectral remote sensing. *Journal of Environmental Management* 90:2170–2177.

7 The Use of Spectral Databases for Remote Sensing of Agricultural Crops

Andreas Hueni, Lola Suarez, Laurie A. Chisholm, and Alex Held

CONTENTS

7.1 INTRODUCTION

The evaluation, assessment, and monitoring of agricultural production is of vital importance in an age of changing climate, environmental perturbations, and genetic modifications of food crops already susceptible to pests and disease (Craik et al., 2017). Further, precision management of agriculture is critical in order to meet customer demands and adjust farming practices according to varying environmental conditions. Combined with the increasing price of resources (i.e., labour, water and fertilisers), off-farm environmental impacts, and dietary narrowing, the need to manage virtually any aspect of an agricultural crop is essential and contributes to more sustainable agricultural production (Teluguntla et al., 2017). Data obtained from remote sensing continues to play a steadfast role for many aspects of agricultural production, ranging from estimation of crop growth and productivity, to detection of more subtle reflectance features associated with crop health, nutrition, and plant stress, to the evaluation of environmental factors which directly contribute to productivity, such as soil type and soil moisture content. In situ spectral measurements facilitate the establishment of regional soil spectral libraries from which properties such as soil texture and classification can contribute to management decisions related to agricultural production (Lacerda et al., 2016).

The decline of agricultural production, whether through disease, soil, and fertilizer variability or climatic fluctuations is a threat to biosecurity (Mohanty et al., 2016), which necessitates a short reaction time to alleviate any potential of crop loss. Technological advances allow for the development of smart networks for timely data collection and subsequent analytical processing using automated and batch-mode techniques, combined with advances in algorithmic approaches to processing remote sensing data indicative of plant disease diagnosis (Mohanty et al., 2016).

Portable, proximal sensors, such as spectroradiometers and fluorescence instruments, provide the laboratory equivalence of measurements directly related to productivity measures (Duda et al., 2017). Data derived from such instruments allows for interrogation using a range of novel algorithms such as spectral similarity measures which draw from collections of in situ spectra collection as input into classifications which use aggregated data to finely discriminate between crop classes (Chauhan and Mohan, 2014).

Further, such measurements at the leaf scale provide information on variability and are associated with differences in leaf age, the degree of tissue damage, and the adverse presence of biotic stressors that can affect leaf pigment content (Stone et al., 2001), produce necrosis (Stone et al., 2005), or affect leaf structure. Within landscapes, proximal remote sensing provides the degree of contribution from the underlying soil surface and non-photosynthetic components, such as the woody components of bark and wood which also vary (Coops et al., 2003). Spectra can assist species identification through the development of spectral libraries (Lucas et al., 2008; Garcia-Ruiz et al., 2015; Asner and Martin, 2016).

Multitemporal acquisition of remote sensing data facilitates monitoring the health status of crops and evaluating the impacts of any perturbations which affects production ranging from broad climatic influences to local solar radiation (Ashcroft et al., 2011). Ranging from in situ point measurements to drone-, airborne-, and satellite-based measurements of an entire crop or regional area, remotely-sensed data have been widely used to inform spatially implicit management decisions concerning agricultural production.

Yet, the interpretation of in situ data to address very specific agricultural crop conditions is still restricted to operators with a very high level of expertise, in most cases in the scientific domain. The translation of remote sensing data into digested, ready-to-use management information is an area where remote sensing specialists still need to invest effort. In order to derive accurate products, the user needs to create a model considering all the parameters involved. In most cases, results are highly location-, instrument-, or species-specific and hypothesis testing is highly demanding of resources. Unfortunately, this tends to restrict the extension of research outcomes for different species or at larger spatial scales.

Scientific groups continuously collect data for specific studies that are only used for that purpose and stored on local media on local computers. This data collection is time consuming and usually involves gathering ancillary measurements that properly describe the measuring environment. When this information stays in local computers within a research group, the fullest potential of use is not reached. If, on the other hand, these disparate data collections are combined and compiled into well-designed databases, data re-use could grow exponentially, reach other dimensions, and more fully support research initiatives and contribute to a collective effort toward the assessment of a problem.

The use of remote sensing for the assessment of physiological processes (e.g., Cogliati et al., 2015) and responses related to the dynamic growing conditions of any type of vegetation requires testing a set of hypotheses that may or may not work for all functional types or species. While there are ecological models and a good basic knowledge of vegetation internal functioning processes, how those can be tracked from a distance is an evolving area of research and techniques are subject to continuous improvement.

Recent computational advances provide techniques that benefit from large data volumes and countless observations to build robust algorithms and estimations. Machine learning, for example, has been applied to numerous scientific fields demonstrating how large datasets provide realistic, generalised models absorbing outliers or individual point errors (Mucherino et al., 2009). Open-source, "BigData" solutions that provide computationally efficient image analysis (e.g., Lewis et al., 2017), are also emerging as significant opportunities for integration of spectral databases into time-series analysis and detailed monitoring of crop condition over time.

Funding new studies that require high levels of resources often requires demonstrating a valid hypothesis. If there were a shared repository of data collected for other similar studies that could be used to test new ideas, this would bring a complete new source of hypothesis testing before investing on further data collection.

A spectral database is a repository for the long-term storage of spectral data and their metadata, enabling data sharing (Hueni et al., 2009). Spectral data stored within spectral databases generally consists of single spectral signatures acquired by point spectrometers under field or laboratory conditions, irrespective of their wavelength range. Spectral libraries are a term for collections of spectral data. Well-known examples include the USGS (Kokaly et al., 2017) and ASTER (Baldridge et al., 2009) spectral libraries. Any such library can be loaded as a collection into a spectral database.

Compiling all the relevant metadata and sharing existing data would enhance the long-term usability of data, especially in cases where data collection and analysis are resource and time consuming. This practice also helps standardising data collection protocols in individual research groups and enables the transfer of data collections between generations of scientists.

7.1.1 Spectral Databases and Information Systems: State of the Art

Collections of in situ or laboratory spectroscopy measurements of known materials are compiled as spectral libraries and stored in an organised spectral database. Optimally, spectral databases should be designed to store data from a variety of spectral sensor sources and accommodate diverse metadata sets to describe the spectral data. The usability of spectral data is largely driven by metadata. Ideally, the metadata are detailed enough to raise a contextual awareness, describing the reasons for and conditions under which the spectral data were acquired (Milton et al., 2006). In this sense, the purpose of use for spectral data can be assessed by the user.

The large number of spectroradiometer models on the market requires generic storage and data handling in the spectral database. Spectral signatures acquired by different instruments will typically differ in the number of spectral bands and their associated centre wavelengths and spectral bandwidths. Data files produced by spectroradiometers often include detailed metadata pertaining to the configuration of the instrument. Data ingestion processes must be written to glean such information automatically (Hueni et al., 2007).

Metadata related to the description of sampling design, sampling environment, and target are often application dependent. Associated measurement values acquired by non-spectral methods such as chemical analysis in laboratories may also be linked to the spectral data by entering them in the spectral database. Hence, the required metadata are application domain specific (Rasaiah et al., 2014) and the database must be designed to handle such diverse metadata in a generic fashion.

Often, spectral databases are designed for one specific application (e.g., Bhojaraja et al., 2015) or are region-specific and include spectra from a single spectroradiometer (e.g., Nidamanuri and Zbell, 2011; Manjunath et al., 2014). Only a few truly generic spectral databases exist, such as SPECCHIO (SPECCHIO, 2017). A recent development, EcoSIS, is an example of a comprehensive spectral library. It specifically houses ecological data, but can handle data in a generic way (EcoSIS, 2017). Currently, SPECCHIO is the only spectral information system developed to house generic spectroscopy data for multiple applications, consistently available for over a decade (Zhao et al., 2017). Table 7.1 gives an overview of existing spectral libraries and spectral information systems and their implementation and application details.

The current trend in spectral database development leads toward the concept of comprehensive spectral information systems (SIS) by adding software components to aid information building and retrieval. This can be achieved in flexible ways by providing an application programmer interface (API), allowing direct spectral database interaction from higher-level programming languages like MATLAB® or R (Hueni et al., 2012). Enabling these types of functionality—data through to analysis—produces a system which can directly underpin the researcher efficiently and with quality outcomes.

TABLE 7.1
List of Current Spectral Libraries (SL) and Spectral Information Systems (SIS)

Name	Type	Main Use and Details	Access	Reference
SPECCHIO	SIS	Generic with comprehensive API and flexible metadata handling	Open Source	Hueni et al. (2009), www.specchio.ch
SSD's Spectral Library Database	SIS	Vegetation and minerals, mineral assemblages, and soil samples	Private	Pfitzner et al. (2005); Pfitzner et al. (2008)
DLR Spectral Archive	SIS	Ground reference library for airborne campaigns. Migrated to SPECCHIO system in 2016.	Private	
EcoSIS	SIS	Generic, mainly used for vegetation. Designed as data portal for sharing. Web based.	Open	EcoSIS (2017), www.ecosis.org
Ahvaz Spectral Geodatabase Platform	SIS	Demonstrator system for combining a spectral database with a GIS system.	Private	Karami et al. (2013)
IREA-CNR Database	SIS	Ground reference library featuring rocks, crops, and metals. Storage is file system based.	Private	Pompilio et al. (2013)
NARSS Spectral Database	SIS	Ground reference library for different land covers in Egypt.	Private	Arafat et al. (2013)
KCM's Multispectral Database	SIS	Imaging spectroscopy data acquired by a laboratory setup	Read-only access	Irgenfried and Hock (2015) http://msidb.iosb.fraunhofer.de
Spectral library for outcrop characterisation	SIS	Ground reference library, mainly for non-vegetated land	Private	Colini et al. (2017)
SPECTATION	SIS	Designed for a vegetation study in Germany. Microsoft Windows only.	Open	http://www-app2.gfz-potsdam.de/spectation
USGS Spectral Library	SL	Reference library of mainly laboratory-based samples	Open	Kokaly et al. (2017), https://speclab.cr.usgs.gov/spectral-lib.html
ASTER Spectral Library	SL	Natural and man-made materials	Open	Baldridge et al. (2009), https://speclib.jpl.nasa.gov
Santa Barbara Urban Spectral Library	SL	Ground reference library, mainly of urban materials. Also available within EcoSIS.	Open	Herold et al. (2004)

7.2 SELECTED AGRICULTURAL APPLICATIONS THAT MAY BENEFIT FROM SPECTRAL DATABASES

Almost a decade ago, Schaminée et al. (2009) were calling for an urgent wide-scale scientific and applied vegetation common research effort to gather data in the context of a rapidly changing world and ensuing loss of biodiversity. The so-called data banking was considered a source of data to study regional vegetation trends under all sorts of environmental and anthropogenic pressures over time. This presents a larger challenge today, with access to remotely-sensed data that can be acquired at virtually any scale—from local to regional to continental or global scale.

The next sections present a number of applications that would benefit from spectral dataset collections. Table 7.2 shows some of these studies where specific agricultural needs are assessed. Spectral databases can be used to support physical modelling, radiative transfer modelling parametrisation and validation, data interpolation models for different spatial and time scales,

TABLE 7.2
Summary of Agricultural Applications That Would Benefit from Dedicated Spectral Databases

Application	Crop	Method	Example of Previous Studies
Water stress	Fruit orchards & tree plantations	RTM—Vegetation indices	Suárez et al. (2008); Suárez et al. (2009)
		Fluorescence	Zarco-Tejada et al. (2009)
		TIR/CWSI	Berni et al. (2009a)
	Vineyards	TIR	Bellvert et al. (2016)
	Herbaceous crops & grasslands	RTM—Vegetation indices	Suárez et al. (2009)
		Fluorescence	Lichtenthaler and Babani (2000); Rossini et al. (2015)
	Various crops types	TIR	Hsiao et al. (1976); Hsiao and Bradford (1983)
Nutrient stress/ Pigment estimation	Fruit orchards & tree plantations	Vegetation indices	O'Connell et al. (2016); Perry et al. (2016)
		RTM	Kempeneers et al. (2008)
	Vineyards	RTM inversion—Vegetation indices	Zarco-Tejada et al. (2005)
		Fluorescence	Cendrero-Mateo et al. (2016)
	Herbaceous crops & grasslands	RTM inversion—Vegetation indices	Haboudane et al. (2002)
		Spectral signatures	Filella and Penuelas (1994); Thulin et al. (2014)
		RTM Inversion	Darvishzadeh et al. (2008)
		Vegetation indices	Fitzgerald et al. (2010); Gabriel et al. (2017); Perry et al. (2017)
	Various crops types	Vegetation indices	Gitelson and Merzlyak (1996, 1997)
		RTM Inversion	Bacour et al. (2002); Berni et al. (2009b)
Disease detection	Fruit orchards & tree plantations	Vegetation indices	Luedeling et al. (2009); Calderón et al. (2013); López-López et al. (2016); Wang et al. (2017)
		Fluorescence	Calderón et al. (2013); Calderón et al. (2015)
	Vineyards	Vegetation indices	Meggio et al. (2010)
	Herbaceous crops & grasslands	Vegetation indices	Apan et al. (2004); Perry et al. (2017)
		Pattern classification	Camargo and Smith (2009)
		Machine learning	Fuentes et al. (2017)
	Various crops types	Machine learning	Mohanty et al. (2016)
Vigour/Leaf area index	Fruit orchards & tree plantations	RTM Inversion	Meroni et al. (2004); le Maire et al. (2011)
	Vineyards	Vegetation indices	Johnson (2003)
	Herbaceous crops & grasslands	Vegetation indices	Delegido et al. (2015)
		Spectral signatures	Filella and Penuelas (1994)
		RTM Inversion	Darvishzadeh et al. (2008)
	Various crops types	RTM Inversion	Bacour et al. (2002); Berni et al. (2009b)

vegetation index design, to build empirical models, machine learning techniques, or simply used as lookup tables for parameter retrieval.

7.2.1 Water Stress Detection

Water stress occurs when the demand of water exceeds the available supply during a certain period or when poor water quality restricts its use. It is well known that severe water deficits affect many physiological processes and have a strong impact on agricultural yield (Hsiao et al., 1976). However, even moderate water deficits, which are not easy to detect, can also have important negative effects on yield (Hsiao and Bradford, 1983). It is important to be able to assess the level of stress through pertinent indicators. The early detection of water stress is a key issue to avoid yield loss, which can be affected even by short-term water deficits (Hsiao et al., 1976).

Water stress affects plant spectral signal through three main ways: changes in the photosynthetic processes, pigments, and heat dissipation (e.g., xanthophyll cycle and chlorophyll fluorescence); direct changes in reflectance features associated with liquid water in leaf tissues, and changes in the canopy temperature consequence of a decrease of the evaporative cooling associated with stomata closure. Signal changes under stress conditions are located in specific spectral areas and in some cases do not surpasses the 3% of the signal under non-stress conditions (Figure 7.1, from Zarco-Tejada et al., 2000).

Up to date, radiative transfer models are not fully integrating these plant protection mechanisms. The processes and pigments affected by pre-visual water stress vary quickly throughout the day, therefore, complete collection of data describing the photoprotection processes is complex as all measurements need to be done almost concurrently (i.e., leaf spectra, stomatal conductance, water potential, photosynthetic rate, fluorescence, and leaf destructive sampling for pigment quantification). As the resources involved in field data collection are many, the benefit of sharing existing data is enormous.

The measurement of the fluorescence component of the plant functional apparatus is gaining increased attention with the development of the fluorescence explorer (FLEX) satellite mission (Vicent et al., 2016). An important prerequisite of the success of the mission is the establishment of

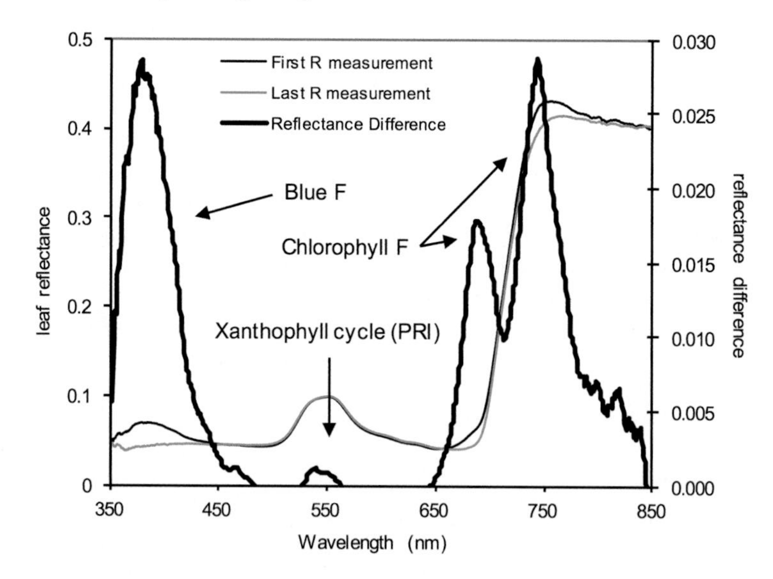

FIGURE 7.1 Leaf reflectance of dark adapted leaves (dotted line, first measurement) and under steady-state condition (continuous line, last measurement), showing the reflectance difference associated with blue fluorescence, chlorophyll fluorescence, and xanthophylls pigment cycle. (From Zarco-Tejada, P. J. et al. 2000. *Remote Sensing of Environment* 74(3): 582–595.)

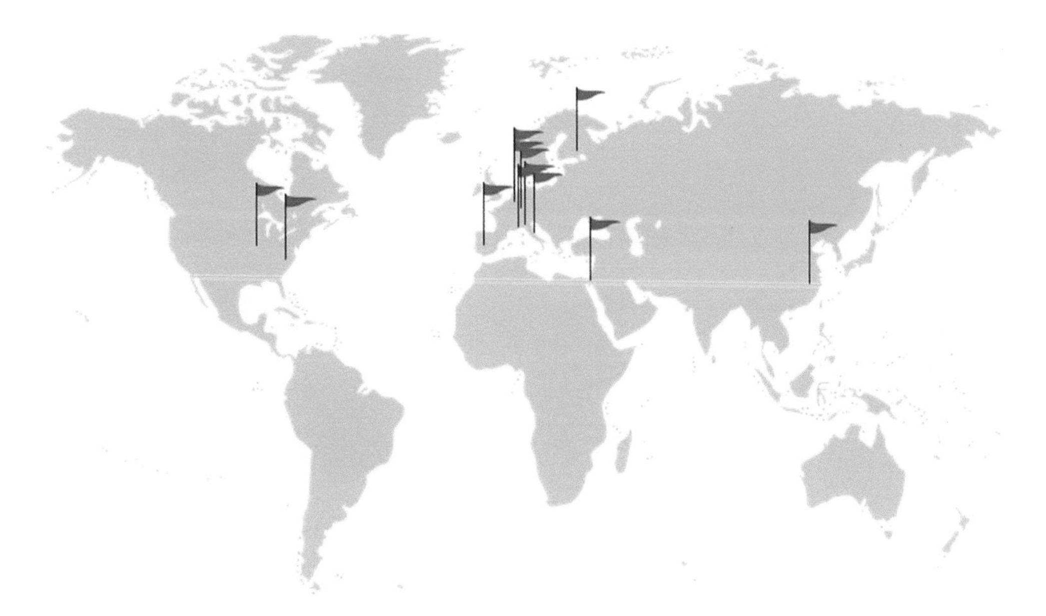

FIGURE 7.2 Spatial locations of FloX spectrometers mounted on towers by the end of 2017, measuring plant fluorescence. (From Burkart, A. 2017. Measuring sun-induced-fluorescence with the FloX—From the scientific problem towards a market ready product. *GHG FLUX Workshop*. Potsdam.)

a spectral database to provide experimental evidence of the reliability of the atmospheric correction schemes developed for FLEX. Data from selected specialized point spectrometers (FloX; JB Hyperspectral Devices) mounted on towers (Figure 7.2) will be ingested into a spectral database and linked with HyPLANT (Rossini et al., 2015) and Sentinel-2 satellite data within the AtmoFLEX project in the framework of the ESA Earth Observation Envelope Programme EOEP-5.

7.2.2 Nutrient Stress and Pigment Content Detection

Indices for nutrient content are generally derived at the leaf, or patch scales in the case of pastures, based on the specific spectral absorption features (Thulin et al., 2014). The difficulty of diagnosing specific nutrient stress lies in the overlapping absorption spectral features for many primary nutrients (Suárez and Berni, 2012). Another difficulty associated to nutrient stress detection is the use of indices derived from leaf spectral data for canopy spectra. The structure of the canopy and the background characteristics play a very important role in the overall canopy signal (Suárez et al., 2008). For that reason, some indices derived at the leaf scale cannot be used at the canopy scale. Some authors apply index combinations to account for the background and structural effects (Haboudane et al., 2002), but this technique is not as straightforward for every case. Coupling leaf and canopy spectral databases with the pertinent metadata would bridge this gap and provide robust spectral indices for nutrient content detection at all scales.

The estimation of pigment content has been helped by spectral databases since radiative transfer models include chlorophyll concentration as input parameter (PROSPECT, Jacquemoud and Baret, 1990). Once the simulation of unlimited leaf spectra covering all ranges of input parameters was possible, the search for optimal vegetation indices could start. Lookup tables allowed the design of specific experiments and new generation sensors based on their capability to represent vegetation spectral responses (Richter et al., 2012).

Radiative transfer models have been inverted to derive parameters (Meroni et al., 2004 and others; Darvishzadeh et al., 2008; Jacquemoud et al., 2009) but this method can only be applied to existing model inputs, Figure 7.3 shows how lookup tables built from leaf-canopy coupled model

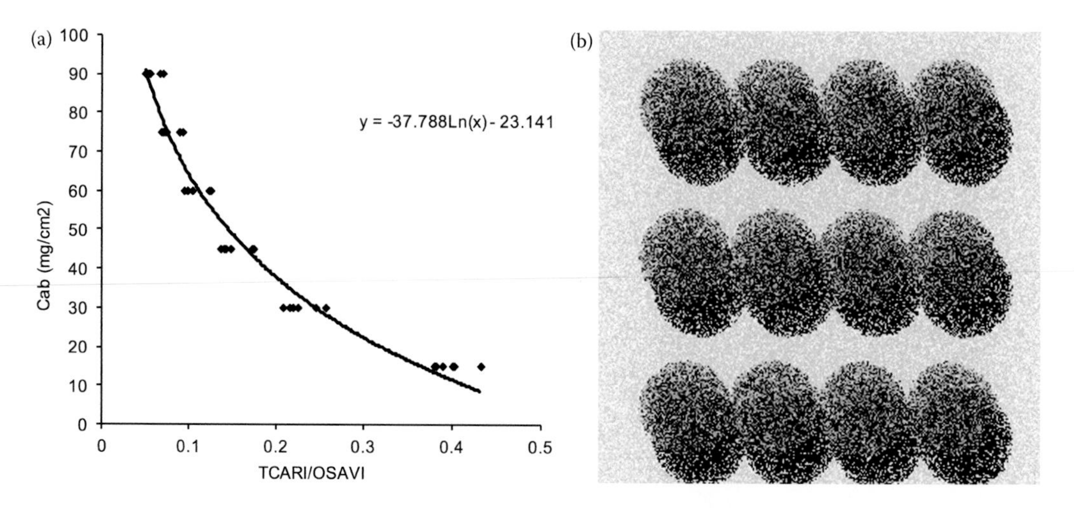

FIGURE 7.3 (a) Algorithm to estimate chlorophyll a + b from TCARI/OSAVI index developed with the FLIGHT 3D radiative transfer model. Input parameters for simulating the peach orchard canopy reflectance ranged between 0.5 and 7 (LAI) and 15 and 90 g/cm^2 (Cab) (From Berni, J. A. J. et al. 2009a. *Remote Sensing of Environment* 113(11): 2380–2388.); (b) Sample 3D scene simulated with FLIGHT model for developing the scaling-up algorithm to estimate chlorophyll concentration. (*IEEE Transactions on Geoscience and Remote Sensing* 47(3): 722–738.)

simulations can be used to derive algorithms to compute chlorophyll content from a vegetation index (from Berni et al., 2009a).

There are many pigments that are still not integrated in leaf models and are of interest for specific physiological processes. Another limitation of the use of radiative transfer modelling is the inability to simulate realistic scenes or constrain the model inputs properly avoiding ill-posed solutions (Zurita-Milla et al., 2015). Apart from general models derived from extensive lookup tables built from radiative transfer model simulations, locally derived empirical models are a common approach. The latter is very accurate within a small area but lacks of applicability in others. More generalised empirical models can be built integrating existing data collected over multiple experimental sites with inputs of all vegetation types if existing data collections were shared. If that was the case, the parameterisation of radiative transfer models with leaf optical properties could heavily benefit as well from rich data collections of spectral data with associated pigment concentrations to allow for a stratified trait allocation dependent on leaf position within the canopy.

7.2.3 Disease Detection

Disease detection is one of the most critical elements of plant management, especially for agricultural crops. Plants can be affected by abiotic stresses like shortage of water or nutrients; or by biotic stresses like insects, fungi, bacteria, or viruses. In many cases, the exact cause of the disease can only be discerned through destructive sampling, but knowing the symptoms in an environmental, spatial, and historical context can help pinpoint the cause.

Regional and federal organisations are fostering collaborations to keep spread of disease as limited as possible and avoid large-area infestations. The collaborative initiatives stem mostly from producers safeguarding the industrial sector and consumer safety. However, there is little knowledge interchange between the scientific community and agricultural crop growers. Hence, there is a great potential in data collected in affected fields that could help the early detection of affected areas. Examples of how spectral changes can lead to disease detection are shown in Figure 7.4.

Studies focusing on the detection of specific diseases in vineyards (Blanchfield et al., 2006; Renzullo et al., 2007; Meggio et al., 2010), fruit orchards (Calderón et al., 2013), grains (Mewes et al., 2011),

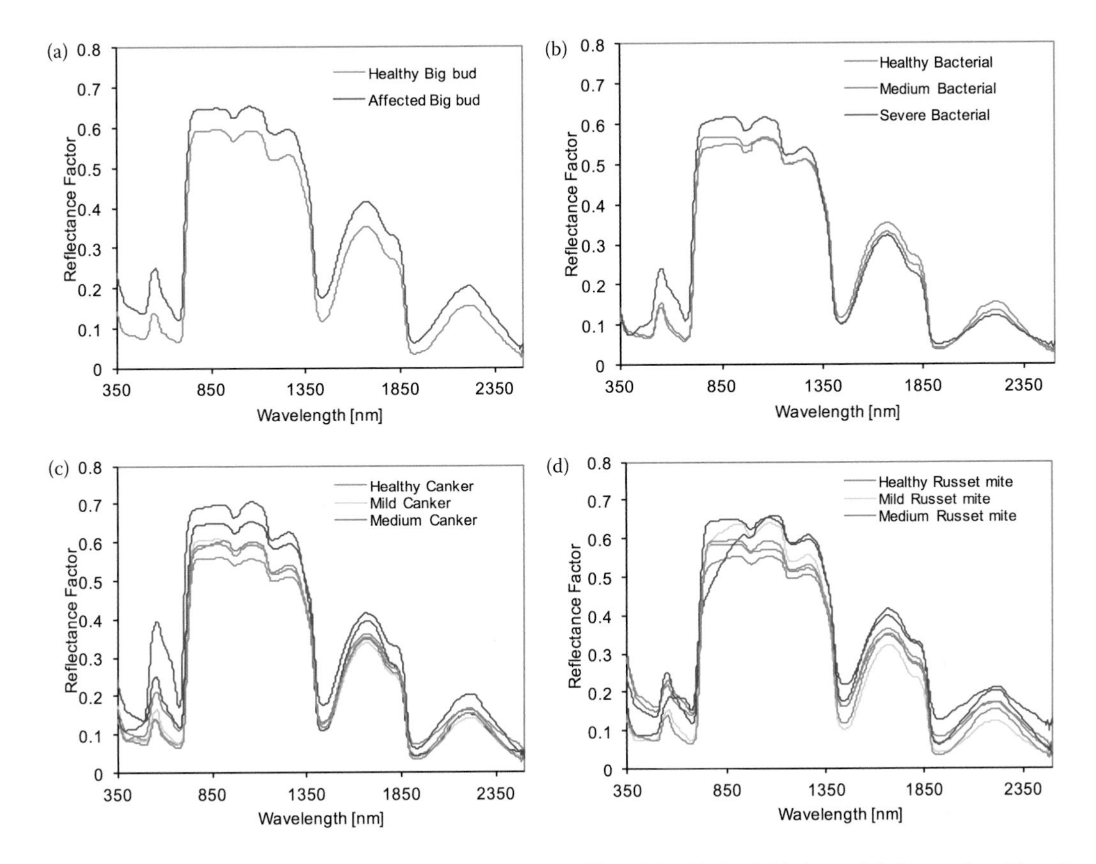

FIGURE 7.4 Leaf spectra collected on tomato plants affected by big bud (a), bacterial (b), canker (c) and russet mite (d), diseases at different levels. (Adapted from Ryu, D. et al. 2017. *Australian Processing Tomato Grower* 38: 35–38.)

and herbaceous crops like cotton (Camargo and Smith, 2009) exist. The further application of the results of these studies is spatially limited (local studies) and in many cases the detection is sensor and species specific. The scientific community focuses on the results and apparently does not yet realise the potential of raw data collected across species and territories. Mohanty et al. (2016) presented a study where over 54,000 crowdsourced red-green-blue (RGB) pictures helped identify 26 diseases in 14 crop species. This study encouraged other authors and Wang et al. (2017) and Fuentes et al. (2017) repeated the technique for disease severity in apples and tomatoes, respectively, the following year. These results show how powerful databases can be for detecting crop damage. These techniques do not, however, allow early detection as they are based on visible symptoms and use limited spectral data. Studies have demonstrated that hyperspectral data can be used for early detection of diseases (Rumpf et al., 2010; Serranti et al., 2017) and, therefore, the assimilation of spectral data with disease-specific metadata in spectral databases could inform detection hypothesis testing and algorithm development.

Some diseases are systemic, affecting the whole plant system. In these cases, plants present similar symptoms to the ones of water stress, closing stomata and reducing evaporation and photosynthetic efficiency in the short term and vegetation growth and wilting in the long term (Beaumont, 1995). Other diseases present symptoms in the leaf surface in the form of spots or discolouration. In those cases, remote sensing can assist in detecting the pigmentation differences. Lastly, there are diseases that present both systemic effects and discolouration symptoms.

Since determining the cause of plant disease can be very difficult, spectral libraries of leaf and canopy measurements for different species and disease levels accompanied with complementary metadata may establish a good starting point for investigating the best remote detection techniques.

7.2.4 Food Security: Food Grading, Quality, and Source of Origin

In a time of mass agricultural production dominated by large corporations supported by political processes, food security is of global concern. Impacts of climate change, compounded by environmental shocks such as floods, and a decreasing amount of arable land, all serve to emphasise the need to ensure a more resilient agricultural production system (Turner et al., 2017). In Australia, the need to evaluate and monitor the impact of pests and diseases on agricultural productivity, seen to be of high risk to food biosecurity, has been identified as a high priority area to help manage risk (Craik et al., 2017).

The evaluation and assessment of food quality control and safety is a highly topical area of research related to food security and strategies which attempt to correct flaws in the global food system by making agricultural production more secure and sustainable for all. While well established, food chemistry studies are dominated by the use of FTIR microscopy centred on spectra of infrared absorption and emission. The development of spectral libraries across the full-range optical spectroscopy range (350–2500 nm) offers complementary and expanded absorption and reflectance data yielding information on water status, cellulose, and lignin for studies in food quality. Detailed spectra collected across a range of pure to contaminated agricultural crop products could contribute to consistent analytical methods to further our understanding of introduced contaminants and spoilage to food and food products. Grains and meals used for livestock feed are susceptible to contamination and spoilage, with potential to affect human consumption (Shen et al., 2016). Benefits would include the development of comprehensive detection screening methods to detect additives. Further, grain-based spectra, whether sub-products for biofuel production or livestock feed, can be used to determine food origins and traceability (Tena et al., 2015), an important component within strategies such as food sovereignty which emphasise food security and sustainability. Investigations which shift research into food grading, quality control, and origin traceability into comprehensive SIS accompanied by full-range optical spectral libraries is a clear area of potential research as humans enter into an increasingly complex global food system.

If spectral libraries are built from consistently acquired, full-range spectra with sufficient metadata to determine fitness for purpose, it has been shown that spectral matching algorithms, better known as successful for semi-automated mineral mapping, have potential to be effective for agricultural crop classification of hyperspectral imagery, particularly if data are collected to include seasonal and phenological state (Nidamanuri and Zbell, 2011). Teluguntla et al. (2017) provided an example of a carefully constructed field-based data collection which encompassed a range of crops and conditions over extensive areas. Class-based spectra could be generated using a quantitative spectral matching technique (QSMT) based on spectral correlation similarity to sort spectra into classes subsequently compared to an ideal spectrum for each class. While part of a satellite-based, spatially extensive automated cropland classification process, the conceptual framework and use of techniques rules could be a potential model for how to better address spectral matching across varying vegetation type and condition level. Rich ground-based spectra collection across crop types, condition, phenology state, and season could allow researchers improved use of machine learning algorithms and move semi-automated mapping of agricultural crop status to a new level.

7.2.5 Soil-Vegetation-Atmosphere-Transfer Models

Soil-Vegetation-Atmosphere-Transfer (SVAT) models compile the energy and water transfer processes on Earth into numerical equations based on the physical processes. These models are difficult to parameterise due to their complexity, especially when it comes to the vertical fluxes (Van Loon and Troch, 2001). Remote sensing data, as acquired systematically at the global scale, presents itself as convenient input to these models. Satellite information has been widely introduced as training and input data in the form of derived Leaf Area Index (LAI), or the fraction of Absorbed Photosynthetic Active Radiation (fAPAR), evapotranspiration, surface albedo, or temperature (Caselles et al., 1992; Olioso et al., 1999; de Wit et al., 2012).

SVAT models have been adapted to incorporate the whole energy balance and simulate resulting reflectance for all spectral regions (Wigneron et al., 1993; Olioso et al., 1999, 2002; Sobrino et al., 2011; Widlowski et al., 2011; Qiu et al., 2016). The full implementation of ecological models resolving the energy balance together with satellite sensor evolution puts us in the best position for global vegetation monitoring (Betts et al., 1996). There is still pending work in this sector and compiling the existing data would be of help. Other models like PROSPECT (Jacquemoud and Baret, 1990) were built upon spectral databases and have been widely used in direct and forward mode to inform vegetation assessment and monitoring. The inversion of radiative transfer models (RTMs) is very powerful once the simulation parameters have been properly defined. This step can be extremely complex for some vegetation types and the inversion is essentially ill-posed. This problem can be overcome by having spectral databases covering the full range of input trait parameters. In the past, databases based on model simulations have been used to derive vegetation indices (Haboudane et al., 2002). More recently, statistical models or emulators, based on databases, have been used to approximate the functioning of RTM inversions (Rivera et al., 2015).

Latest advances include the SCOPE model that incorporates the water and energy balance effects on vegetation physiology and the resulting signals including photoprotection mechanisms and resulting fluorescence signal (van der Tol et al., 2009). The performance of this model has been tested for many species and only failed to describe spectral variations related to xanthophyll pigment concentration (van der Tol et al., 2104). The next model generations would benefit from datasets covering all mechanisms and the model could be used to inform on global vegetation performance using fluorescence provided by the FLEX mission.

7.3 DATABASE IMPLEMENTATION AND SPECTROSCOPY DATA LIFE CYCLE

The actual implementation of a spectral database or spectral information system may vary to a certain degree, depending on the choice of software system components. We illustrate the common structure with the example of the SPECCHIO database (Figure 7.5).

The backend of the system is a relational database, storing spectral data and their metadata, as well as handling user accounts and user groups. Access to the database is strictly provided through a web service with a defined application programme interface (API) with most interfaces requiring a user authentication. Clients can access the SPECCHIO server by a variety of software tools, but always utilise the same API, written in Java. This allows communication with the server using either the SPECCHIO Java client or any programming language that provides a Java bridging technology. All lower level calls are handled by the Java components, presenting abstract data entry and access methods without requiring intrinsic knowledge of the database schema. This abstraction allows the running of optimisation routines to reduce metadata redundancy. Typically, spectroradiometer measurements are often replicated to establish the target variation. In these cases, most of the metaparameters are highly redundant and the system can reduce the stored metadata volume typically by around 30–60%, depending on the metaparameter set. This feature in turn increases the speed of data queries due to significantly reduced entries in the database.

Spectral information systems support the spectroscopy data life cycle to a large extent from: data acquisition where required metaparameters may already be pre-informed and aligned with available metadata attributes offered by the spectral information system; to loading, augmenting, and processing the data; to finally retrieving information to build and test new hypotheses, leading to potentially new experiments (Figure 7.6).

Metadata augmentation and data processing must be transparent to the user through the storage data provenance information. The system thus permits the storage of several different processing levels, for example, raw digital numbers, radiances, and reflectances, with clear dependencies and information about the processing algorithms. The handling of provenance adds to the replicability of scientific findings, constituting an important pillar of transparent science.

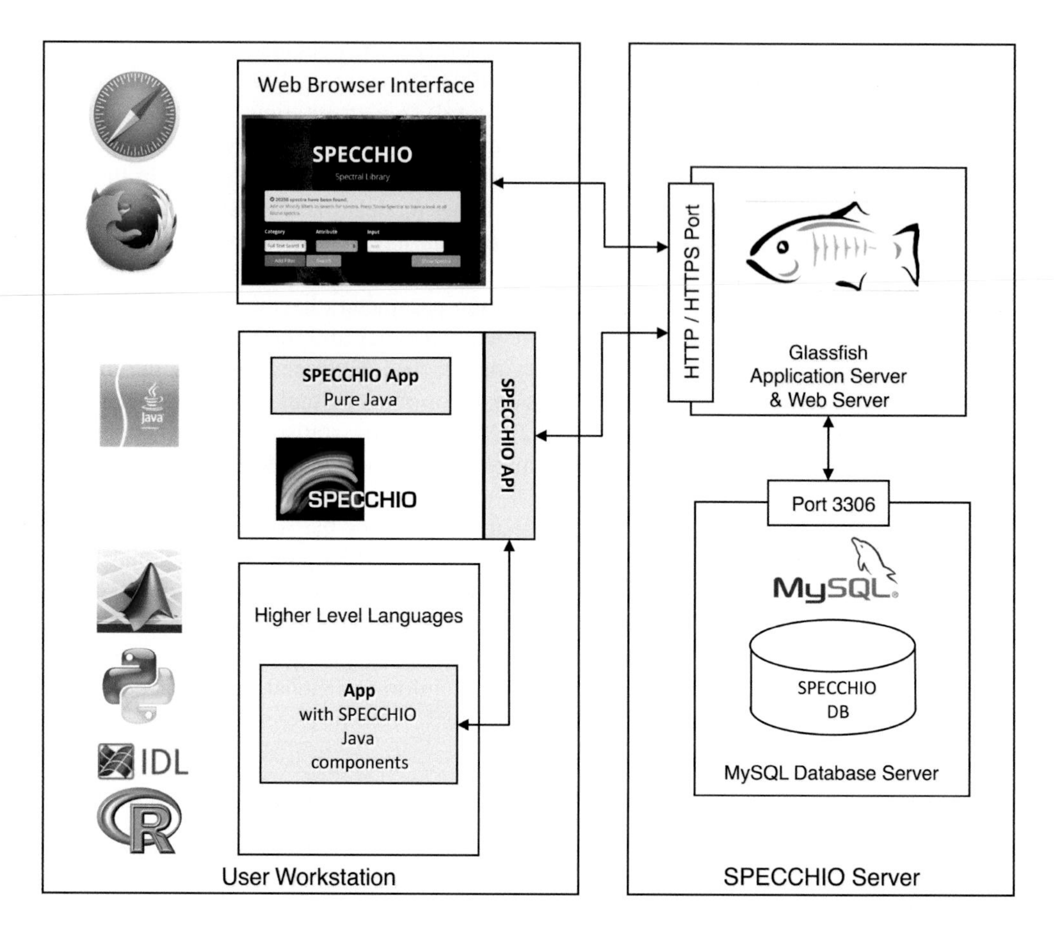

FIGURE 7.5 System architecture of the SPECCHIO Spectral Information System. (From SPECCHIO 2017. "SPECCHIO Spectral Information System." Retrieved 05.04.2017, 2017, from www.specchio.ch.)

7.4 CASE STUDY: SPECTRAL DATABASE CONTAINING PIGMENT CONTENT IN A WATER STRESS EXPERIMENT

The spectral data of this case study were loaded into a SPECCHIO database through the standard data ingestion routine. The initial number of metaparameters inserted automatically was 12, including the filename, date and time of acquisition, and instrument settings like integration time and field of view. A spreadsheet containing the results of chemical leaf analysis in the laboratory was used during data augmentation, connecting the records in the database with the spreadsheet rows based on the tree identifier encoded in the spectral filename. This led to a mean number of 24 metaparameters per spectrum with some deviations as not all biophysical parameters were available for all samples (Figure 7.7).

The distribution of available biophysical metaparameters of this dataset was then queried from MATLAB through the SPECCHIO Java API, giving histograms per attribute (Figure 7.8). Based on these histograms, spectral investigations could be started, for example, querying which spectra have a Lutein content above 10 ugrams/cm^2. Such queries can be formulated in the SPECCHIO graphical user interface, but also implemented in a query condition that is then sent to the API.

```
cond = EAVQueryConditionObject('eav', 'spectrum_x_eav', 'Lutein', 'double_val');
cond.setValue('10.0');
cond.setOperator('>=');
query.add_condition(cond);
```

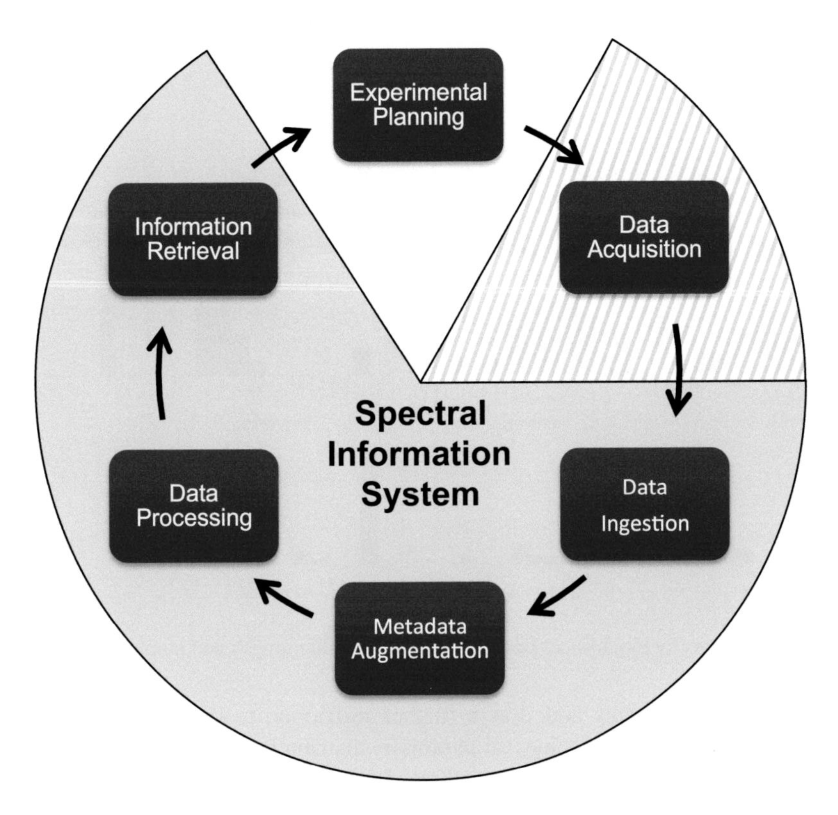

FIGURE 7.6 Spectroscopy data life cycle, largely supported by the spectral information system. (From Hueni, A. 2017. *SPECCHIO Spectral Information System—Leaflet*, U. o. Zurich.)

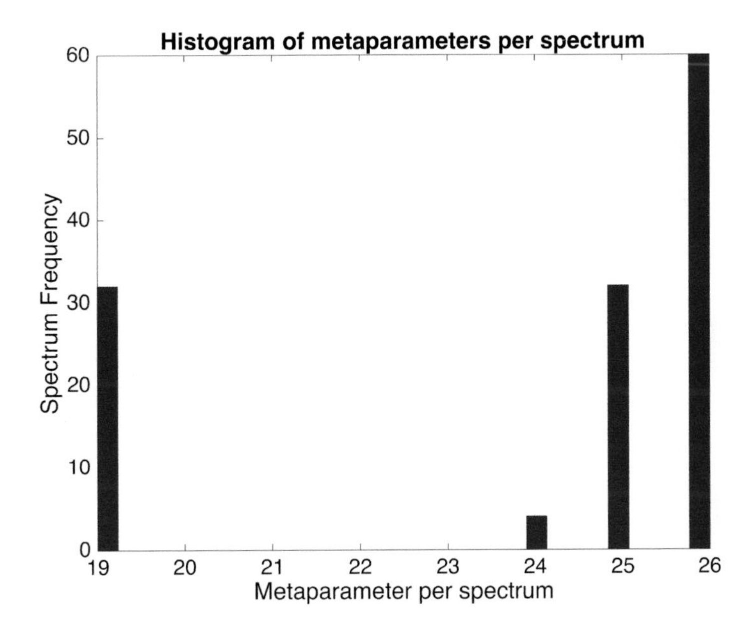

FIGURE 7.7 Histogram of the metadata space density after biophysical metadata augmentation.

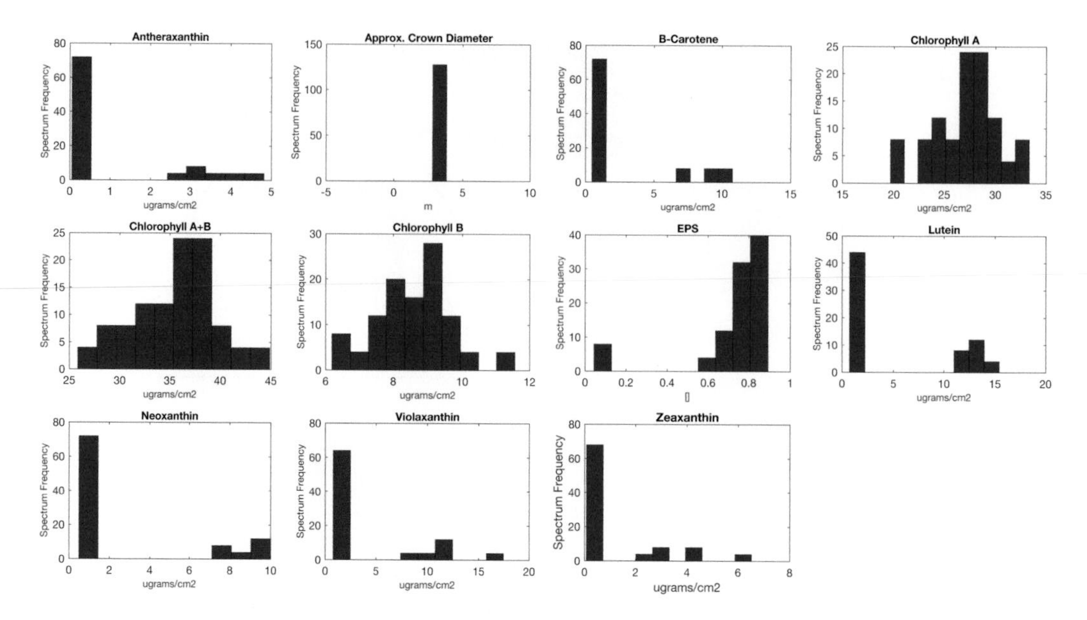

FIGURE 7.8 Histograms by biophysical parameter showing value ranges and number of spectra per value.

This condition was then simply added as a further restriction to the histogram-based analysis, producing a new plot where the biophysical parameter distributions of the spectra with high Lutein could easily be compared to the complete dataset (Figure 7.9). High variation of pigment content was found between samples with varying water stress. Pigment content histograms show differences in samples collected from plots where water stress was induced or on days where the evaporative demand was higher.

The specific case of Lutein concentration was further investigated by first averaging the spectral data and pigment data per sampled leaf and day of sampling. This was achieved by first splitting the dataset based on timestamps into daily blocks, then grouping the data within the blocks by the tree identifier stored in the database.

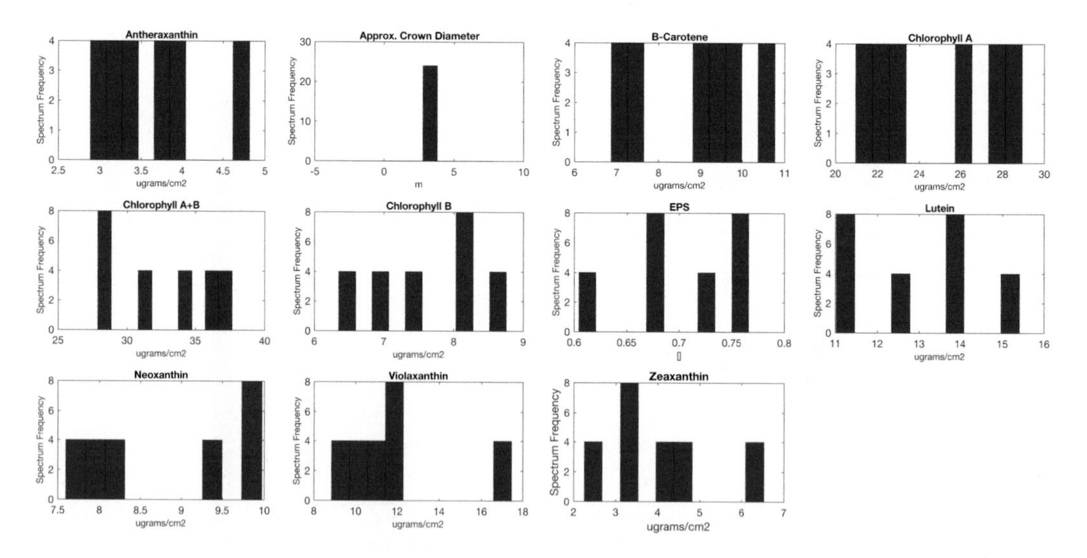

FIGURE 7.9 Histograms by biophysical parameter showing value ranges and number of spectra per value for spectra with high associated Lutein content.

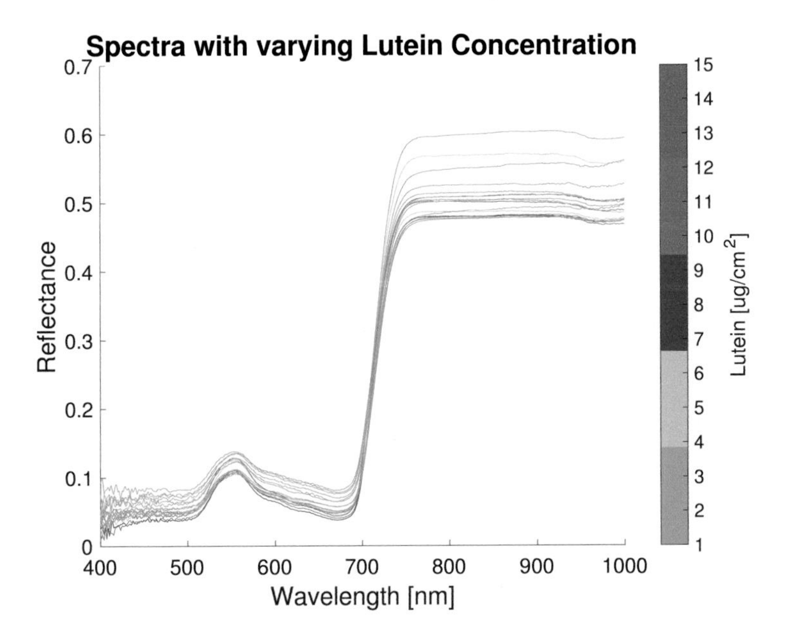

FIGURE 7.10 Averaged spectra colour coded by their Lutein concentration, demonstrating a generally low reflectance in the 400–700 nm range for high Lutein contents.

A simple spectral plot of the averaged dataset and coloured by the mean Lutein content shows that the absorption in the visible wavelength range is indeed at the lower end of the spectral reflectance range (Figure 7.10).

A correlation analysis between the Lutein content and two spectral indices (a simple ratio and an NDVI-type index) illustrates the potential to estimate Lutein content from leaf spectra (Figure 7.11).

Spectral database users interested in developing new index formulations can benefit from information on spectral regions affected by specific pigment concentration. For example, if Lutein is the pigment of interest, it is important to know which wavelengths are highly related to Lutein concentration (Figure 7.12). At the same time, other pigments with overlapping absorption features and structural effects on the spectra may decrease the usability of the index. Normally, reference spectral bands are required to account for those effects, being simple ratio and normalise difference formulations commonly used for this purpose. Figure 7.11 shows how these formulations can be easily computed and compared using SPECCHIO in combination with MATLAB.

7.5 CONCLUSIONS AND RECOMMENDATIONS

Agriculture is long associated with the development of societies. Projected increases in world population levels clearly point towards increased customer demand which presents challenges to the management of agricultural crop production. Over-exploitation of natural resources combined with pressures related to climate change are of high concern, which foster strategies to move towards more sustainable use of resources and more resilience in the global food system. Demands for increased agricultural production push producers towards higher levels of mechanisation, chemicals, and resource use. This is countered by local movements to return to a fairer trading system and more sustainable practices.

Tools for monitoring the multifaceted aspects of agricultural crop production are becoming more sophisticated, with instruments previously restricted to research laboratories becoming more accessible to the farmer. Instead of disjointed efforts of investment in creating disparate datasets that are localised or are of single-purpose, combined efforts amongst scientific groups should be supported. Examples in this chapter have illustrated the exciting potential of well-designed spectral

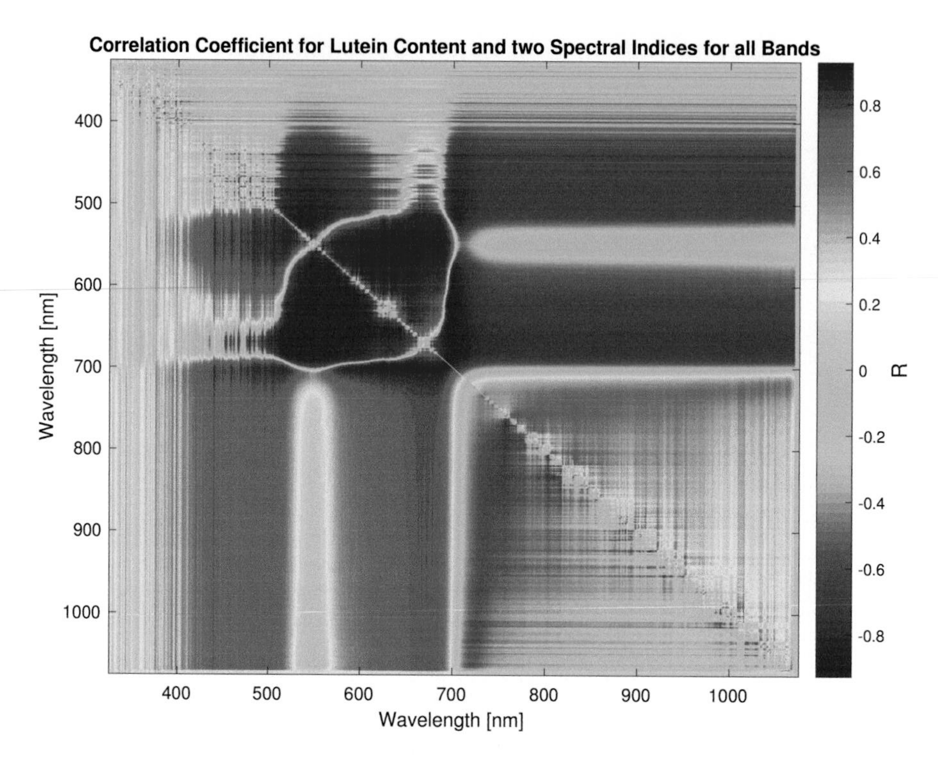

FIGURE 7.11 Correlation coefficients of Lutein concentration for two spectral indices calculated for all bands: upper triangle: simple band ratio (Rx/Ry), lower triangle: normalised difference index (Rx − Ry)/(Rx + Ry).

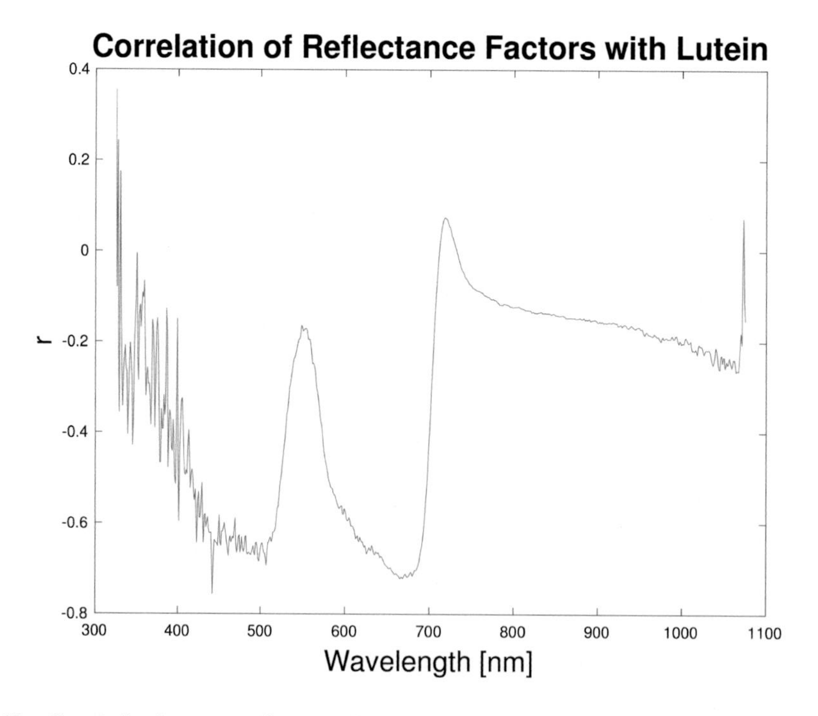

FIGURE 7.12 Correlation between reflectance factors and Lutein content.

database systems which allow the compilation of datasets collected for a wide range of agricultural-related measurements over varying conditions, which, if established, in itself could become a tool as a source of data for hypothesis testing and validation. Such systems would further open the potential to apply machine learning techniques for prediction using large, quality-controlled datasets. This chapter presents an overview and case study which demonstrates the potential of compiled spectral datasets to address specific needs for agricultural system management. Such systems must be based upon data collection protocols and metadata descriptions that enhance data transferability within and among working groups.

REFERENCES

Apan, A., Held, A., Phinn, S. and Markley, J. 2004. "Detecting sugarcane 'orange rust' disease using EO-1 Hyperion hyperspectral imagery." *International Journal of Remote Sensing* 25(2): 489–498.

Arafat, S., Farg, E., Shokr, M. and Al-Kzaz, G. 2013. "Internet-based spectral database for different land covers in Egypt." *Advances in Remote Sensing* 2(2): 85–92.

Ashcroft, M. B., French, K. O. and Chisholm, L. A. 2011. "An evaluation of environmental factors affecting species distributions." *Ecological Modelling* 222(3): 524–531.

Asner, G. P. and Martin, R. E. 2016. "Spectranomics: Emerging science and conservation opportunities at the interface of biodiversity and remote sensing." *Global Ecology and Conservation* 8: 212–219.

Bacour, C., Jacquemoud, S., Leroy, M., Hautecoeur, O., Weiss, M., Prévot, L., Bruguier, L. and Chauki, H. 2002. "Reliability of the estimation of vegetation characteristics by inversion of three canopy reflectance models on airborne POLDER data." *Agriculture and Environment* 22(6): 555–565.

Baldridge, A. M., Hook, S. J., Grove, C. I. and Rivera, G. 2009. "The ASTER spectral library version 2.0." *Remote Sensing of Environment* 113(4): 711–715.

Beaumont, P. 1995. Reflectance foliaire et acclimatation à un deficit hydrique: cas des feuilles de tournesol. Ecophysiologie végétale et Télédétection. Toulouse, France, Université Paul Sabatier. *PhD Thesis*: 129.

Bellvert, J., Zarco-Tejada, P. J., Marsal, J., Girona, J., González-Dugo, V. and Fereres, E. 2016. "Vineyard irrigation scheduling based on airborne thermal imagery and water potential thresholds." *Australian Journal of Grape and Wine Research* 22(2): 307–315.

Berni, J. A. J., Zarco-Tejada, P. J., Sepulcre-Cantó, G., Fereres, E. and Villalobos, F. 2009a. "Mapping canopy conductance and CWSI in olive orchards using high resolution thermal remote sensing imagery." *Remote Sensing of Environment* 113(11): 2380–2388.

Berni, J. A. J., Zarco-Tejada, P. J., Suarez, L. and Fereres, E. 2009b. "Thermal and narrowband multispectral remote sensing for vegetation monitoring from an unmanned aerial vehicle." *IEEE Transactions on Geoscience and Remote Sensing* 47(3): 722–738.

Betts, A. K., Ball, J. H., Beljaars, A. C. M., Miller, M. J. and Viterbo, P. A. 1996. "The land surface-atmosphere interaction: A review based on observational and global modeling perspectives." *Journal of Geophysical Research: Atmospheres* 101(D3): 7209–7225.

Bhojaraja, B. E., Hegde, G., Pruthviraj, U., Shetty, A. and Nagaraj, M. K. 2015. "Mapping agewise discrimination of arecanut crop water requirement using hyperspectral remote sensing." *Aquatic Procedia* 4(Supplement C): 1437–1444.

Blanchfield, A. L., Robinson, S. A., Renzullo, L. and Powell, K. S. 2006. "Can leaf pigment composition help us identify grapevines infested with phylloxera?" *Functional Plant Biology* 33: 507–517.

Burkart, A. 2017. Measuring sun-induced-fluorescence with the FloX—From the scientific problem towards a market ready product. *GHG FLUX Workshop*. Potsdam.

Calderón, R., Navas-Cortés, J. A., Lucena, C. and Zarco-Tejada, P. J. 2013. "High-resolution airborne hyperspectral and thermal imagery for early detection of Verticillium wilt of olive using fluorescence, temperature and narrow-band spectral indices." *Remote Sensing of Environment* 139: 231–245.

Calderón, R., Navas-Cortés, J. and Zarco-Tejada, P. 2015. "Early detection and quantification of verticillium wilt in olive using hyperspectral and thermal imagery over large areas." *Remote Sensing* 7(5): 5584.

Camargo, A. and Smith, J. S. 2009. "Image pattern classification for the identification of disease causing agents in plants." *Computers and Electronics in Agriculture* 66(2): 121–125.

Caselles, V., Sobrino, J. A. and Coll, C. 1992. "On the use of satellite thermal data for determining evapotranspiration in partially vegetated areas." *International Journal of Remote Sensing* 13(14): 2669–2682.

Cendrero-Mateo, M. P., Moran, M. S., Papuga, S. A., Thorp, K. R., Alonso, L., Moreno, J., Ponce-Campos, G., Rascher, U. and Wang, G. 2016. "Plant chlorophyll fluorescence: Active and passive measurements at canopy and leaf scales with different nitrogen treatments." *Journal of Experimental Botany* 67(1): 275–286.

Chauhan, H. and Mohan, B. K. 2014. Effectiveness of spectral similarity measures to develop precise crop spectra for hyperspectral data analysis. *ISPRS Annals of the Photogrammetry, Remote Sensing and Spatial Information Sciences, ISPRS Technical Commission VIII Symposium*. Hyderabad, India. II-8, 2014: 83–90.

Cogliati, S., Rossini, M., Julitta, T., Meroni, M., Schickling, A., Burkart, A., Pinto, F., Rascher, U. and Colombo, R. 2015. "Continuous and long-term measurements of reflectance and sun-induced chlorophyll fluorescence by using novel automated field spectroscopy systems." *Remote Sensing of Environment* 164: 270–281.

Colini, L., Doumaz, F., Silvestri, M., Musacchio, M., Spinetti, C., Buongiorno, M. F. and Lombardo, V. 2017. Mt Etna (Italy) and Sahara desert (Algeria) sites: CAL/VAL activities for hyperspectral data and development of spectral libraries for outcropping surfaces characterization. *EARSeL SIG Imaging Spectroscopy Workshop*. Zurich, Switzerland.

Coops, N. C., Stone, C., Culvenor, D. S., Chisholm, L. A. and Merton, R. N. 2003. "Chlorophyll content in eucalypt vegetation at the leaf and canopy scales as derived from high resolution spectral data." *Tree Physiology* 23(1): 23–31.

Craik, W., Palmer, D. and Sheldrake, R. 2017. Priorities for Australia's biosecurity system, An independent review of the capacity of the national biosecurity system and its underpinning Intergovernmental Agreement: 188.

Darvishzadeh, R., Skidmore, A., Schlerf, M. and Atzberger, C. 2008. "Inversion of a radiative transfer model for estimating vegetation LAI and chlorophyll in a heterogeneous grassland." *Remote Sensing of Environment* 112(5): 2592–2604.

Delegido, J., Verrelst, J., Rivera, J. P., Ruiz-Verdú, A. and Moreno, J. 2015. "Brown and green LAI mapping through spectral indices." *International Journal of Applied Earth Observation and Geoinformation* 35: 350–358.

de Wit, A., Duveiller, G. and Defourny, P. 2012. "Estimating regional winter wheat yield with WOFOST through the assimilation of green area index retrieved from MODIS observations." *Agricultural and Forest Meteorology* 164: 39–52.

Duda, B. M., Weindorf, D. C., Chakraborty, S., Li, B., Man, T., Paulette, L. and Deb, S. 2017. "Soil characterization across catenas via advanced proximal sensors." *Geoderma* 298: 78–91.

EcoSIS 2017. "EcoSIS." from https://ecosis.org/.

Filella, I. and Penuelas, J. 1994. "The red edge position and shape as indicators of plant chlorophyll content, biomass and hydric status." *International Journal of Remote Sensing* 15(7): 1459–1470.

Fitzgerald, G., Rodriguez, D. and O'Leary, G. 2010. "Measuring and predicting canopy nitrogen nutrition in wheat using a spectral index—The canopy chlorophyll content index (CCCI)." *Field Crops Research* 116(3): 318–324.

Fuentes, A., Yoon, S., Kim, C. S. and Park, S. D. 2017. "A Robust deep-learning-based detector for real-time tomato plant diseases and pests recognition." *Sensors* 17(9).

Gabriel, J. L., Zarco-Tejada, P. J., López-Herrera, P. J., Pérez-Martín, E., Alonso-Ayuso, M. and Quemada, M. 2017. "Airborne and ground level sensors for monitoring nitrogen status in a maize crop." *Biosystems Engineering* 160: 124–133.

Garcia-Ruiz, F. J., Wulfsohn, D. and Rasmussen, J. 2015. "Sugar beet (Beta vulgaris L.) and thistle (Cirsium arvensis L.) discrimination based on field spectral data." *Biosystems Engineering* 139: 1–15.

Gitelson, A. A. and Merzlyak, M. N. 1996. "Signature analysis of leaf reflectance spectra: Algorithm development for remote sensing of chlorophyll." *Journal of Plant Physiology* 148: 495–500.

Gitelson, A. A. and Merzlyak, M. N. 1997. "Remote estimation of chlorophyll content in higher plant leaves." *International Journal of Remote Sensing* 18(12): 2691–2697.

Haboudane, D., Miller, J. R., Tremblay, N., Zarco-Tejada, P. J. and Dextraze, L. 2002. "Integrated narrow-band vegetation indices for prediction of crop chlorophyll content for application to precision agriculture." *Remote Sensing of Environment* 81(2): 416–426.

Herold, M., Roberts, D. A., Gardner, M. E. and Dennison, P. E., 2004. "Spectrometery for urban remote sensing—Development and analysis of a spectral library from 350 to 2400 nm." *Remote Sensing of Environment* 91: 304–319.

Hsiao, T. C. and Bradford, K. J. 1983. Physiological consequences of cellular water deficits. *Limitations to Efficient Water Use in Crop Production*. H. M. Taylor, W. R. Jordan and T. R. Sinclair, Madison, WI: 227–265.

Hsiao, T. C., Acevedo, E., Fereres, E., and Henderson, D. W. 1976. "Water stress, growth, and osmotic adjustment." *Philosophical Transactions of the Royal Society B* 273: 479–500.

Hueni, A. 2017. *SPECCHIO Spectral Information System—Leaflet*, U. o. Zurich.

Hueni, A., Chisholm, L., Suarez, L., Ong, C. and Wyatt, M. 2012. Spectral information system development for Australia. *Geospatial Science Research Symposium*. Melbourne, Australia. 1328: 1–11.

Hueni, A., Nieke, J., Schopfer, J., Kneubühler, M. and Itten, K. I. 2007. Metadata of spectral data collections. *5th Workshop on Imaging Spectroscopy*. Bruges (B), EARSeL.

Hueni, A., Nieke, J., Schopfer, J., Kneubühler, M. and Itten, K. I. 2009. "The spectral database SPECCHIO for improved long-term usability and data sharing." *Computers & Geosciences* 35(3): 557–565.

Irgenfried, S. and Hock, J. 2015. *Acquisition and Storage of Multispectral Material Signatures–Workflow Design and Implementation*, KIT Scientific Publishing.

Jacquemoud, S. and Baret, F. 1990. "Prospect: A model of leaf optical properties spectra." *Remote Sensing of Environment* 34: 75–91.

Jacquemoud, S., Verhoef, W., Baret, F., Bacour, C., Zarco-Tejada, P. J., Asner, G. P., François, C. and Ustin, S. L. 2009. "PROSPECT + SAIL models: A review of use for vegetation characterization." *Remote Sensing of Environment* 113: S56–S66.

Johnson, L. F. 2003. "Temporal stability of an NDVI-LAI relationship in a Napa Valley vineyard." *Australian Journal of Grape and Wine Research* 9(2): 96–101.

Karami, M., Rangzan, K. and Saberi, A. 2013. "Using GIS servers and interactive maps in spectral data sharing and administration: Case study of Ahvaz Spectral Geodatabase Platform (ASGP)." *Computers & Geosciences* 60(0): 23–33.

Kempeneers, P., Zarco-Tejada, P. J., North, P. R. J., de Backer, S., Delalieux, S., Sepulcre-Cantó, G., Morales, F. et al. 2008. "Model inversion for chlorophyll estimation in open canopies from hyperspectral imagery." *International Journal of Remote Sensing* 29(17–18): 5093–5111.

Kokaly, R. F., Clark, R. N., Swayze, G. A., Livo, K. E., Hoefen, T. M., Pearson, N. C., Wise, R. A. et al. 2017. USGS Spectral Library Version 7: 61.

Lacerda, P. M., Demattê, A. J., Sato, V. M., Fongaro, T. C., Gallo, C. B. and Souza, B. A. 2016. "Tropical texture determination by proximal sensing using a regional spectral library and its relationship with soil classification." *Remote Sensing* 8(9).

le Maire, G., Marsden, C., Verhoef, W., Ponzoni, F. J., Lo Seen, D., Bégué, A., Stape, J.-L. and Nouvellon, Y. 2011. "Leaf area index estimation with MODIS reflectance time series and model inversion during full rotations of Eucalyptus plantations." *Remote Sensing of Environment* 115(2): 586–599.

Lewis, A., Oliver, S., Lymburner, L., Evans, B., Wyborn, L., Mueller, N., Raevksi, G. et al. 2017. "The Australian geoscience data cube—Foundations and lessons learned." *Remote Sensing of Environment* 202: 276–292.

Lichtenthaler, H. K. and Babani, F. 2000. "Detection of photosynthetic activity and water stress by imaging the red chlorophyll fluorescence." *Plant Physiology and Biochemistry* 38(11): 889–895.

López-López, M., Calderón, R., González-Dugo, V., Zarco-Tejada, J. P. and Fereres, E. 2016. "Early detection and quantification of almond red leaf blotch using high-resolution hyperspectral and thermal imagery." *Remote Sensing* 8(4).

Lucas, R., Bunting, P., Paterson, M. and Chisholm, L. 2008. "Classification of Australian forest communities using aerial photography, CASI and HyMap data." *Remote Sensing of Environment* 112(5): 2088–2103.

Luedeling, E., Hale, A., Zhang, M., Bentley, W. J. and Dharmasri, L. C. 2009. "Remote sensing of spider mite damage in California peach orchards." *International Journal of Applied Earth Observation and Geoinformation* 11(4): 244–255.

Manjunath, K. R., Kumar, A., Meenakshi, M., Renu, R., Uniyal, S. K., Singh, R. D., Ahuja, P. S., Ray, S. S. and Panigrahy, S. 2014. "Developing spectral library of major plant species of Western Himalayas using ground observations." *Journal of the Indian Society of Remote Sensing* 42(1): 201–216.

Meggio, F., Zarco-Tejada, P. J., Núñez, L. C., Sepulcre-Cantó, G., González, M. R. and Martín, P. 2010. "Grape quality assessment in vineyards affected by iron deficiency chlorosis using narrow-band physiological remote sensing indices." *Remote Sensing of Environment* 114(9): 1968–1986.

Meroni, M., Colombo, R. and Panigada, C. 2004. "Inversion of a radiative transfer model with hyperspectral observations for LAI mapping in poplar plantations." *Remote Sensing of Environment* 92(2): 195–206.

Mewes, T., Franke, J. and Menz, G. 2011. "Spectral requirements on airborne hyperspectral remote sensing data for wheat disease detection." *Precision Agriculture* 12(6): 795.

Milton, E. J., Fox, N. P. and Schaepman, M. 2006. Progress in field spectroscopy. *Geoscience and Remote Sensing Symposium*. Denver, CO, US, IEEE International.

Mohanty, S. P., Hughes, D. P. and Salathé, M. 2016. "Using deep learning for image-based plant disease detection." *Frontiers in Plant Science* 7: 1419.

Mucherino, A., Papajorgji, P. and Pardalos, P. M. 2009. "A survey of data mining techniques applied to agriculture." *Operational Research* 9(2): 121–140.

Nidamanuri, R. R. and Zbell, B. 2011. "Transferring spectral libraries of canopy reflectance for crop classification using hyperspectral remote sensing data." *Biosystems Engineering* 110(3): 231–246.

O'Connell, M., Whitfield, D. and Abuzar, M. 2016. *Satellite Remote Sensing of Vegetation Cover and Nitrogen Status in Almond*, International Society for Horticultural Science (ISHS), Leuven, Belgium.

Olioso, A., Braud, I., Chanzy, A., Courault, D., Demarty, J., Kergoat, L., Lewan, E. et al. 2002. "SVAT modeling over the Alpilles-ReSeDA experiment: Comparing SVAT models over wheat fields." *Agronomie* 22(6): 651–668.

Olioso, A., Chauki, H., Courault, D. and Wigneron, J.-P. 1999. "Estimation of evapotranspiration and photosynthesis by assimilation of remote sensing data into SVAT models." *Remote Sensing of Environment* 68(3): 341–356.

Perry, E. M., Bluml, M., Goodwin, I., Cornwall, D. and Swarts, N. D. 2016. *Remote Sensing of N Deficiencies in Apple and Pear Orchards*, International Society for Horticultural Science (ISHS), Leuven, Belgium.

Perry, E. M., Nuttall, J. G., Wallace, A. J. and Fitzgerald, G. J. 2017. "In-field methods for rapid detection of frost damage in Australian dryland wheat during the reproductive and grain-filling phase." *Crop and Pasture Science* 68(6): 516–526.

Pfitzner, K., Bartolo, R. E., Ryan, B. and Bollhöfer, A. 2005. Issues to consider when designing a spectral library database. *SSC 2005 Spatial Intelligence, Innovation and Praxis: The national biennial Conference of the Spatial Sciences Institute*. Melbourne, Spatial Sciences Institute.

Pfitzner, K., Esparon, A. and Bollhoefer, A. 2008. SSD's spectral library database. *14th Australasian Remote Sensing and Photogrammetry Conference*. Darwin, AU.

Pompilio, L., Villa, P., Boschetti, M. and Pepe, M. 2013. "Spectroradiometric field surveys in remote sensing practice: A workflow proposal, from planning to analysis." *IEEE Geoscience and Remote Sensing Magazine* 1(2): 37–51.

Qiu, B., Guo, W., Xue, Y. and Dai, Q. 2016. "Implementation and evaluation of a generalized radiative transfer scheme within canopy in the soil-vegetation-atmosphere transfer (SVAT) model." *Journal of Geophysical Research: Atmospheres* 121(20): 12,145–12,163.

Rasaiah, B., Jones, S., Bellman, C. and Malthus, T. 2014. "Critical metadata for spectroscopy field campaigns." *Remote Sensing* 6(5): 3662–3680.

Renzullo, L. J., Blanchfield, A. L. and Powell, K. S. 2007. *Insights into the Early Detection of Grapevine Phylloxera from in situ Hyperspectral Data*, International Society for Horticultural Science (ISHS), Leuven, Belgium.

Richter, K., Hank, T. B., Vuolo, F., Mauser, W. and D'Urso, G. 2012. "Optimal exploitation of the sentinel-2 spectral capabilities for crop leaf area index mapping." *Remote Sensing* 4(3).

Rivera, P. J., Verrelst, J., Gómez-Dans, J., Muñoz-Marí, J., Moreno, J. and Camps-Valls, G. 2015. "An emulator toolbox to approximate radiative transfer models with statistical learning." *Remote Sensing* 7(7).

Rossini, M., Nedbal, L., Guanter, L., Ač, A., Alonso, L., Burkart, A., Cogliati, S. et al. 2015. "Red and far red sun-induced chlorophyll fluorescence as a measure of plant photosynthesis." *Geophysical Research Letters* 42(6): 1632–1639.

Rumpf, T., Mahlein, A. K., Steiner, U., Oerke, E. C., Dehne, H. W. and Plümer, L. 2010. "Early detection and classification of plant diseases with support vector machines based on hyperspectral reflectance." *Computers and Electronics in Agriculture* 74(1): 91–99.

Ryu, D., Mann, L., Wang, Y., Suarez, L., Gupta, D. and Fuentes, S. 2017. "Application of ground-based and unmanned aerial vehicle (UAV)-borne spectroscopy to detect tomato diseases." *Australian Processing Tomato Grower* 38: 35–38.

Schaminée, S. M., Hennekens, M. C. and Rodwell, J. S. 2009. "Vegetation-plot data and databases in Europe: An overview." *Preslia* 81: 173–185.

Serranti, S., Bonifazi, G., Luciani, V. and D'Aniello, L. 2017. Classification of Peronospora infected grapevine leaves with the use of hyperspectral imaging analysis. *Proc. SPIE 10217.*

Shen, G., Fan, X., Yand, Z. and Han, L. 2016. "A feasibility study of non-targeted adulterant screening based on NIRM spectral library of soybean meal to guarantee quality: The example of non-protein nitrogen." *Food Chemistry* 210: 35–42.

Sobrino, J. A., Mattar, C., Gastellu-Etchegorry, J. P., Jiménez-Muñoz, J. C. and Grau, E. 2011. "Evaluation of the DART 3D model in the thermal domain using satellite/airborne imagery and ground-based measurements." *International Journal of Remote Sensing* 32(22): 7453–7477.

SPECCHIO 2017. "SPECCHIO Spectral Information System." Retrieved 05.04.2017, 2017, from www.specchio.ch.

Stone, C., Chisholm, L. and Coops, N. 2001. "Spectral reflectance characteristics of Eucalyptus foliage damaged by insects." *Australian Journal of Botany* 49(6): 687–698.

Stone, C., Chisholm, L. and McDonald, S. 2005. "Effects of leaf age and psyllid damage on the spectral reflectance properties of Eucalyptus saligna foliage." *Australian Journal of Botany* 53(1): 45–54.

Suárez, L. and Berni, J. A. J. 2012. Spectral response of citrus and their application to nutrient and water constraints diagnosis. *Advances in Citrus Nutrition*. A. K. Srivastava, Dordrecht, Springer Netherlands: 125–141.

Suárez, L., Zarco-Tejada, P. J., Berni, J. A. J., González-Dugo, V. and Fereres, E. 2009. "Modelling PRI for water stress detection using radiative transfer models." *Remote Sensing of Environment* 113(4): 730–744.

Suárez, L., Zarco-Tejada, P. J., Sepulcre-Cantó, G., Pérez-Priego, O., Miller, J. R., Jiménez-Muñoz, J. C. and Sobrino, J. 2008. "Assessing canopy PRI for water stress detection with diurnal airborne imagery." *Remote Sensing of Environment* 112(2): 560–575.

Teluguntla, P., Thenkabail, P. S., Xiong, J., Gumma, M. K., Congalton, R. G., Oliphant, A., Poehnelt, J., Yadav, K., Rao, M. and Massey, R. 2017. "Spectral matching techniques (SMTs) and automated cropland classification algorithms (ACCAs) for mapping croplands of Australia using MODIS 250-m time-series (2000–2015) data." *International Journal of Digital Earth* 10(9): 944–977.

Tena, N., Boix, A. and von Holst, C. 2015. "Identification of botanical and geographical origin of distillers dried grains with solubles by near infrared microscopy." *Food Control* 54: 103–110.

Thulin, S., Hill, M., Held, A., Jones, S. and Woodgate, P. 2014. "Predicting levels of crude protein, digestibility, lignin and cellulose in temperate pastures using hyperspectral image data." *American Journal of Plant Sciences* 5: 997–1019.

Turner, G. M., Larsen, K. A., Candy, S., Ogilvy, S., Ananthapavan, J., Moodie, M., James, S. W., Friel, S., Ryan, C. J. and Lawrence, M. A. 2017. "Squandering Australia's food security—The environmental and economic costs of our unhealthy diet and the policy Path We're On." *Journal of Cleaner Production*.

van der Tol, C., Berry, J. A., Campbell, P. K. E. and Rascher, U. 2014. "Models of fluorescence and photosynthesis for interpreting measurements of solar-induced chlorophyll fluorescence." *Journal of Geophysical Research: Biogeosciences* 119(12): 2312–2327.

van der Tol, C., Verhoef, W., Timmermans, J., Verhoef, A. and Su, Z. 2009. "An integrated model of soil-canopy spectral radiances, photosynthesis, fluorescence, temperature and energy balance." *Biogeosciences* 6(12): 3109–3129.

Van Loon, E. E. and Troch, P. A. 2001. "Tikhonov regularization as a tool for assimilating soil moisture data in distributed hydrological models." *Hydrological Processes* 16(2): 531–556.

Vicent, J., Sabater, N., Tenjo, C., Acarreta, J. R., Manzano, M., Rivera, J. P., Jurado, P., Franco, R., Alonso, L. and Verrelst, J. 2016. "FLEX end-to-end mission performance simulator." *IEEE Transactions on Geoscience and Remote Sensing* 54(7): 4215–4223.

Wang, G., Sun, Y. and Wang, J. 2017. "Automatic image-based plant disease severity estimation using deep learning." *Computational Intelligence and Neuroscience* 2017: 8.

Widlowski, J. L., Pinty, B., Clerici, M., Dai, Y., De Kauwe, M., de Ridder, K., Kallel, A. et al. 2011. "RAMI4PILPS: An intercomparison of formulations for the partitioning of solar radiation in land surface models." *Journal of Geophysical Research: Biogeosciences* 116(G2): 1–25.

Wigneron, J.-P., Kerr, Y., Chanzy, A. and Jin, Y.-Q. 1993. "Inversion of surface parameters from passive microwave measurements over a soybean field." *Remote Sensing of Environment* 46(1): 61–72.

Zarco-Tejada, P. J., Berjón, A., López-Lozano, R., Miller, J. R., Martín, P., Cachorro, V., González, M. R. and de Frutos, A. 2005. "Assessing vineyard condition with hyperspectral indices: Leaf and canopy reflectance simulation in a row-structured discontinuous canopy." *Remote Sensing of Environment* 99(3): 271–287.

Zarco-Tejada, P. J., Berni, J. A. J., Suárez, L., Sepulcre-Cantó, G., Morales, F. and Miller, J. R. 2009. "Imaging chlorophyll fluorescence with an airborne narrow-band multispectral camera for vegetation stress detection." *Remote Sensing of Environment* 113(6): 1262–1275.

Zarco-Tejada, P. J., Miller, J. R., Mohammed, G. H. and Noland, T. L. 2000. "Chlorophyll fluorescence effects on vegetation apparent reflectance: I. leaf-level measurements and model simulation." *Remote Sensing of Environment* 74(3): 582–595.

Zhao, L., Hu, S., Zeng, X., Wu, Y., Lin, Y., Liu, J. and Fan, S. 2017. An integrated software system for supporting real-time near-infrared spectral big data analysis and management. *IEEE International Congress on Big Data*. Honolulu, HI, USA. 97–104.

Zurita-Milla, R., Laurent, V. C. E. and van Gijsel, J. A. E. 2015. "Visualizing the ill-posedness of the inversion of a canopy radiative transfer model: A case study for Sentinel-2." *International Journal of Applied Earth Observation and Geoinformation* 43: 7–18.

8 Characterization of Soil Properties Using Reflectance Spectroscopy

E. Ben-Dor, S. Chabrillat, and José A. M. Demattê

CONTENTS

8.1 INTRODUCTION

Reflectance spectroscopy has become a very useful tool for the soil sciences over the past 25 years. This technique enables the extraction of quantitative and qualitative information on many soil attributes in real time and can shed light on the soil's composition without the need for labor-intensive wet-chemistry analyses (e.g., Dunn et al., 2002; Brown et al., 2006; Nanni and Dematte, 2006). Reflectance spectroscopy information is acquired in both point and image arrangements in the laboratory, field, and from air and space domains. Whereas in the laboratory, soil reflectance measurements are performed under controlled conditions, preferably using standard protocols (and hence with minimum interference), in the field, reflectance measurements are fraught with a variety of problems, such as variations in viewing angle, changes in illumination, soil roughness, management, and soil sealing (Ben-Dor et al., 1998). Acquiring soil reflectance data from air and space involves additional difficulties, resulting from, for example: relatively low signal-to-noise ratio sensors, lack of optimal illumination time, and atmospheric attenuation. Laboratory-based measurements enable an understanding of the chemical and physical principles of soil reflectance, and are widely used for practical applications requiring a quantitative approach. As the sensitivity of portable field spectrometers increases, field soil spectroscopy is becoming a promising tool for rapid point-by-point monitoring of the soil environment. Recently, considerable effort has been invested in commercializing field spectroscopy-based sensors for agricultural applications (e.g., VERIS Technologies http://www.veristech.com/index.aspx). Several papers have described the possibility of using soil reflectance in the field for precision agriculture applications (e.g., Colin et al., 2004; Mouazen et al., 2007; Demattê et al., 2014; Franceschini et al., 2017). In this regard, the future looks bright for the development of a new spectral technology, termed imaging spectroscopy (IS), which combines the spectral and spatial domains (Ben-Dor et al., 2009; Ustin and Schaepman, 2009). Due to the large number of airborne IS sensors operating today for many terrestrial applications, this technology is slowly but surely entering the field of soil science, where its use will rely heavily on the spectral foundation generated over the past two decades in soil analysis laboratories (Ben-Dor et al., 2007). Understanding the principles and limitations of soil spectra is crucial for use of the forthcoming soil-IS technology. Information about soils from reflectance spectra in the visible near-infrared (VIS-NIR: 0.4–1.1 μm) and shortwave-infrared (SWIR, 1.1–2.5 μm) spectral regions constitutes almost all of the data that passive solar sensors can provide and, therefore, this chapter will only cover these regions. We provide a historical overview of soil reflectance spectroscopy and a general overview of the chemical–physical principles of the soil reflectance spectrum in this spectral region. We also discuss the basic interactive processes between soils and electromagnetic radiation, and shed light on the principle of quantitative soil spectral approaches. An important section will discuss the emergence of the Soil Spectral Library (SSL) activity as well as new standards and protocols for both laboratory and field spectral measurements. In addition, this chapter will provide recent examples on how soil reflectance spectroscopy is used in the modern remote-sensing arena, with both point and imaging sensors, as well as future notes on the potential of this methodology.

8.2 SOIL

8.2.1 General Background

Soil is a complex material that is extremely variable in physical and chemical composition. It is formed from exposed masses of partially weathered rocks and minerals of the Earth's crust. Soil formation, or genesis, is strongly dependent upon the environmental conditions of both the atmosphere and lithosphere. Soils are a product of five factors: climate, vegetation, fauna, topography, and parent materials. The great variability in soils is the result of interactions among these factors and their influence on the formation of different soil profiles (Buol et al., 1973). Whereas soil taxonomy and soil mapping require knowledge of the entire soil profile, remote sensing measures only the thin (about 50 μm) upper surface layer (Ben-Dor et al., 1998). Thus, reflectance remote sensing cannot be an effective tool for classifying soils from a pedological standpoint where examination of the entire soil profile is required. This limitation becomes more salient as the natural surface is altered, for example, by agricultural activities and, therefore, less of the natural soil body exists (Aase and Tanaka, 1983; Baumgardner et al., 1985). Another limiting factor in the use of remote sensing as a pedological tool is the masking effect of vegetation and snow, both of which affect radiant flux at the surface. Nonetheless, although soil spectral information is complex, it is less complex than the soil itself and hence can reduce its uncertain dimensionality to a practical utilization.

8.2.2 Soil Compositions

Any given soil mixture is made up of all three phases of matter: solid, liquid, and gas. A typical soil may consist of about 50% pore space, containing spatially and temporally varying proportions of gas and liquid.

8.2.2.1 The Solid Phase

This phase contains organic and inorganic matter in a complicated and generic mixture of primary and secondary minerals, organic components, and salts. The solid phase consists of three main particle size fractions—sand (2–0.2 mm), silt (0.2–0.002 mm), and clay (<0.002 mm), which together govern two major soil properties: texture and structure. Soil texture is a function of these three main components distribution and is generally described in terms of quantities of gravel, sand, silt, and clay. Soil structure (a function of adhesive forces between generally fine particles) describes the aggregation characteristics of a soil. These two properties play a major role in soil behavior and influence some major soil characteristics, such as drainage, fertility, moisture, and erosion. The inorganic portion of the solid phase consists of soil minerals, which are generally categorized as either primary or secondary. Primary minerals are derived directly from the weathering of parent materials that were formed under much higher temperatures and pressures than those found at the Earth's surface. Secondary minerals are formed by geochemical weathering of the primary minerals. An extensive description of minerals in the soil environment is given by Dixon and Weed (1989), and readers who wish to expand their knowledge in this area are referred to that text. In general, the dominant primary minerals are quartz, feldspar, orthoclase, and plagioclase. Some layer silicate minerals, such as mica and chlorite, and ferromagnesian silicates, such as amphibole, peroxide, and olivine also exist. The secondary minerals in soils—often termed clay minerals—are aluminosilicates such as smectite, illite, vermiculite, sepiolite, kaolinite, and gibbsite. The type of clay mineral present is strongly dependent on the soil's weathering stage and can be a significant indicator of the environmental conditions under which it was formed. Other secondary minerals in soils are aluminum and iron oxides and hydroxides, carbonates (calcite and dolomite), sulfates (gypsum), and phosphates (apatite). Most of these minerals are relatively insoluble in water and maintain equilibrium with the water solution. Soluble salts such as halite may also be found in soil but they are mobile in water. Clay minerals are most likely found in the fine-sized soil particles (<2 mm; clay fraction) and are characterized by relatively high specific surface areas (50–800 m^2/g). The primary minerals and other non-clay minerals are usually found in both the sand and silt portions, and consist of relatively small specific surface areas (<1 m^2/g). In addition to the inorganic components in the solid phase, organic components also exist. Although the organic matter content in mineral soils does not exceed 15% (and is usually less), it plays a major role in the soil's chemical and physical behavior (Schnitzer and Khan, 1978). Organic matter is composed of decaying tissues from vegetation and the bodies of microfauna and macrofauna. Soil organic matter can be found in various stages of degradation, from coarse dead to complex fine components called humus (Stevenson, 1982). Its content is naturally higher in the upper soil horizon, making consideration of organic matter essential for remote-sensing applications, where only the upper thin layer is detected.

8.2.2.2 The Liquid and Gas Phases

These phases in soils are complementary to the solid phase and occupy about 50% of the soil's total volume. The liquid consists of components of water and dissolved anions in various amounts and positions. The water molecules either fill the entire pore volume in the soil ("saturated"), occupy a portion of its pore volume ("wet"), or are adsorbed on the surface areas ("air dry"). The composition of the soil's gaseous phase is normally very similar to that of the atmosphere, except for the concentrations of oxygen and carbon dioxide, which vary according to the biochemical activity in the root zone.

8.2.3 Soil Minerals

8.2.3.1 Clay Minerals

Clay minerals (also referred to as phyllosilicate minerals) are crystalline aluminosilicates organized in a layered structure. The crystal structure consists of two basic units: the Si tetrahedron, which is formed by a Si^{3+} ion surrounded by four O^{2-} ions in a tetrahedral configuration, and the Al octahedron formed by an Al^{4+} ion surrounded by four O^{2-} and two OH^- ions in an octahedral configuration. These structural units are joined together into tetrahedral and octahedral sheets, respectively, by adjacent Si tetrahedrons sharing all three basal corners and by Al octahedrons sharing edges. These sheets, in turn, form the clay mineral layer by sharing the optical O of the tetrahedral sheet. Layer silicates are classified into eight groups according to layer type, layer charge, and type of interlayer cations. The layer type designated 1:1 is organized with one octahedral and one tetrahedral sheet, whereas the 2:1 layer type is organized with two octahedral and one tetrahedral sheet. A one-layer octahedral sheet (1) is also found in highly leached acid soils. The layer silicate charge is a function of isomorphic substitution which occurs in both the tetrahedral and octahedral positions during the weathering process. Charge density is one of the major factors governing soil behavior and, therefore, mineral species and composition are considered to be key to understanding soil behavior. Clay minerals are spectrally active mostly in the SWIR region (see later).

8.2.3.1.1 Origin of Layer Minerals in Soils

All clay minerals are derived from the weathering of primary minerals. The occurrence of smectite, vermiculite, illite, or kaolinite is related to the degree of weathering and the chemical nature of the soil environment. Muscovite tends to produce illite, whereas biotite tends to produce vermiculite. Both illite and vermiculite are associated with slightly weathered materials. Vermiculite requires large amounts of magnesium during clay formation, which is most likely to occur in neutral to slightly alkaline soils. Illite occurs to a greater extent than vermiculite in soils of moderate acidity. Illite tends to form smectite as surface potassium ions are removed by weathering processes and new cation substitution occurs. Smectite minerals (2:1 configuration) are an important component of slightly to moderately weathered soils, which are formed under relatively high pH values and specific Si and Al concentrations in the soil solution. Kaolinite minerals (1:1 configuration) are predominant in highly weathered, leached soils that turn, under stronger weathering and more acid conditions, into gibbsite (1 octahedral configuration). Whereas gibbsite is quite rare, smectite and kaolinite are more commonly found in soils. Kaolinite may be formed from a 2:1 mineral during the weathering process and requires an environment in which both silica and alumina are accumulated in a ratio that favors its formation. Illite and vermiculite are associated with youthful materials. Smectite is formed in the middle stages of the weathering process and, therefore, is most likely to be found in many soils as the major or secondary mineral. Similarly, the kaolinite component in soils tends to increase with increasing stages of weathering. Soils of warm temperate regions have a high percentage of kaolinite in the clay fraction, whereas cold areas tend to form more illitic- and smectitic-type minerals. Of all of the soil minerals discussed above, smectite is thought to be the most active, because of its high specific surface area and electrochemical reactivity. These characteristics are known to affect many of the soil's properties, as reported by Banin and Amiel (1970) and others.

8.2.3.2 Non-Clay Minerals

The most common of these are divided into five groups: silicates, phosphates, oxides and hydroxides, carbonates, and sulfides and sulfates. The fraction of each mineral in soils depends upon the environmental conditions and the parent materials. Primary minerals will most likely be found in young soils, where the weathering process is weak. Whereas silicates such as feldspars are rarely found in mature soils, quartz may also be found in some developed soils, depending on their environmental conditions and parent material. In general, the quartz mineral is spectrally inactive in the VIS–NIR–SWIR region and, therefore, diminishes other spectral features in the soil mixture. Other non-clay silicate minerals such as feldspars may have some diagnostic absorption features that make the soil spectrum less monotonous.

Oxide-group minerals occur in highly weathered areas such as those associated with slopes, highly leached profiles, or "mature" soils. Phosphate and sulfate minerals can be found in such soils as apatite and gypsum, respectively. Although both minerals have unique spectral features, their occurrence in soils may be relatively rare and even undetectable. Other oxides, such as iron, are strongly spectrally active, mostly in the VIS region, because of their crystal field and charge-transfer mechanism. The content of free oxides (both iron and aluminum) is low in young soils but increases gradually as the soil ages, similar to organic matter.

8.3 SOIL SPECTROSCOPY

8.3.1 Definitions and Limitations

A (soil) spectrum is a collection of discrete energies, covering a wide spectral range, of photons that travel along the sun (or other equivalent source)–surface (soil)–sensor pathways after removing the atmospheric and solar (source) effects. The soil reflectance spectrum is a collection of values obtained from the ratio of radiance I to irradiance (L) fluxes across most of the spectral region of the solar emittance function. The values are traditionally described, from a practical standpoint, by a relative ratio against a perfect reflector spectrum measured for the same soil geometry and position (Palmer, 1982; Baumgardner et al., 1985; Jackson et al., 1987). To illustrate the reflectance product of the above calculation, Figure 8.1 provides spectra of five representative soil types in a semiarid environment (taken from the Tel Aviv Soil Spectral Library, SSL, collection).

The electromagnetic energy in question covers the VIS (0.4–0.7 μm), NIR (0.7–1.1 μm), and SWIR (1.1–2.5 μm) spectra. Soil reflectance is an inherent soil property that should not be affected

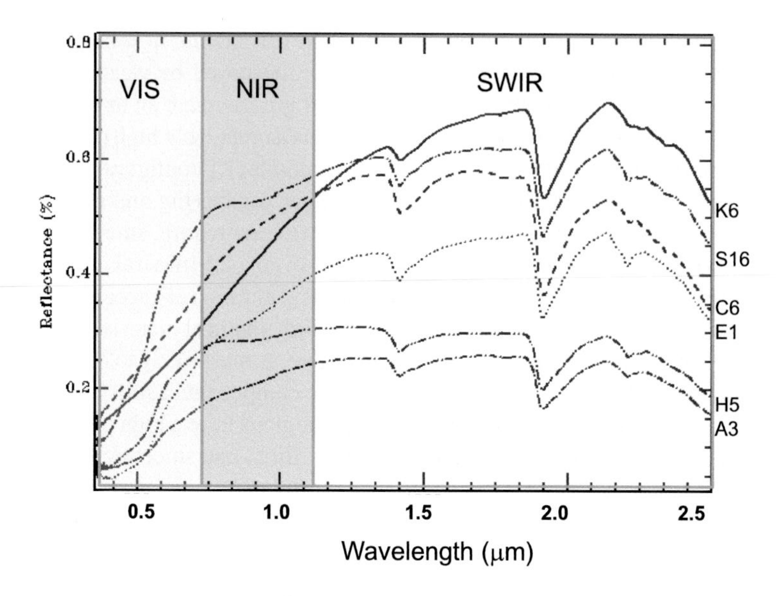

FIGURE 8.1 Reflectance spectra of five different mineral soils from arid and semi-arid area from Israel. (USDA definition: K6-Caliorthid, S16-Torriorthrnt, C6-Xerothent, El-Rhodoxeralf, H5-Xeret, A3-Rhodoxeralf.) (TAU spectral library.)

by external conditions such as radiation intensity or the instrument used. Observational capabilities can be extended by the use of spectrometers and radiometers that can quantify the characteristics of radiation scattered from the soil. Soil reflectance data have been acquired with such instruments in a substantial number of remote-sensing studies performed in both the field and the laboratory (Baumgardner et al., 1985). Although most of these studies have focused on the spectral distribution of the scattered radiation, some data on the directional distribution and polarization state of radiation scattered by soils are also available in the literature. The studies generally demonstrate relationships between spectral reflectance data and certain soil properties which correspond to the well-known relationships with soil color. Recent studies have also made use of new sensitive sensors from either point or image domains that can provide an immediate spatial view of the spectral information. To provide a better understanding of soil spectra, this section will provide some background information on the electromagnetic spectrum, radiation interactions with soil, and soil attributes that affect the spectral response. This discussion is mainly limited to the 0.4–2.5-μm region of the electromagnetic spectrum.

8.3.2 Spectral Measurements

Soil reflectance measurements can be acquired in the laboratory or in the field, as well as from both the air and orbit. Whereas in the laboratory, soil reflectance is measured under controlled conditions of geometry and illumination, no atmospheric interference, and using standard protocols, in the field, reflectance measurements face a number of interfering factors, such as variations in viewing angle, changes in illumination, soil roughness, soil sealing, and differences in measured areas (Milton, 1987). The acquisition of soil reflectance data from the air or orbit involves additional difficulties, such as low signal-to-noise ratio, atmospheric interference, and the need to acquire data under high sun elevation angles. In this regard, laboratory-based measurements enable an understanding of the chemical and physical principles of soil reflectance, but do not guarantee that all of the laboratory applications can be adopted for the remote sensing of soils. An extensive body of work has been applied to soil-laboratory analyses (see Section 8.7.2.2) whereas only a few (but growing) number of studies have been conducted for field soil (i.e., using a portable spectrometer—Odlare et al., 2005) or soil–air or space (see Section 8.10.2). In general, field measurements by point spectrometer are characterized by better spectral performance than that provided by IS sensors, and thus, they are preferable for extracting soil properties in the NIR–SWIR region. However, using a field (point) spectrometer rather than an image (point-by-point) spectrometer reduces the accuracy of the first mapping process. This is because interpolation of selected points to a thematic soil map is less accurate than information obtained from a mosaic of hundreds of individual pixels having chemical–physical information. Some common portable point spectrometers are available for remote sensing, such as ASD (http://www.asdi.com/prod/ps2.html), Spectral Evaluation (http://www.spectralevolution.com/spectrometers_oreXpress.html), GER (http://www.ger.com), LICOR (http://licor.alcavia.net/), and PIMA (http://www.terraplus.com/products/properties/pima.htm). A comprehensive technical review of field spectrometers can be found at http://www.themap.com.au/overview_spectrometer.htm. Point spectrometers are characterized by a high signal-to-noise ratio and good stability and better spectral resolution such as the new ultra high resolution SR-6500 spectromter from Spectral Evolution. Their quick response time can make them operable from aircraft, where they can spectrally track kinetic processes (e.g., soil and vegetation drying processes). Based on this ability, Karnieli et al. (1999) mounted an ASD spectrometer on a light aircraft equipped with a video camera and acquired measurements along a climate cross section in Israel (personal communication). They showed for the first time that a laboratory/field point spectrometer can be used as an airborne tool for assessing soil and vegetation status with very high signal-to-noise ratio standards. In 2010, Pimstein et al. observed that even in a single well-calibrated spectrometer, instabilities can occur; therefore, they suggested a standard protocol and correction factor to normalize internal and external variations of the spectrometer being used. Several papers followed up on this idea. Ben-Dor et al. (2015) developed a standard and protocol approach for measuring soil reflectance such that system effects (spectrometer make, illumination characterization, geometry) would not affect the results. They suggested an internal soil standard (ISS) approach that uses the same samples in all laboratories and aligns all measurements to a "mother" spectrometer. This would enable gathering and comparing spectral information regardless of the protocol and spectrometer used. Later, Kopačková and Ben-Dor (2016) demonstrated this approach with several soils from Australia and Israel using two different spectrometer makes (ASD and Spectral Evolution) under laboratory conditions. This method was recently adopted by Romero et al. (2018) who showed significant progress using the SSL project with Brazilian soil and by Gholizadeh et al. (2017) who demonstrated that by using the ISS method, data mining of Czech Republic soils was improved. It should be pointed out that measuring soil reflectance in the field is different from the laboratory, due to changing sun illumination, atmospheric attenuation effects (for example, at 1.4 and 1.9 μm, there is strong absorption of water vapor), and soil sealing in the former. To solve the first problem, an artificial light source is used to directly illuminate the target. For example, in the ASD spectrometer, an external device with a small closed chamber (contact probe) is used, whereas the PIMA spectrometer has a built-in illumination source. To solve the second problem, a prototype device was developed (named 3S-HEAD) to acquire reflectance measurements from soil boreholes, and it provides excellent information on the soil profile without the need to open trenches (Ben-Dor et al., 2008a,b). A new version of this device, named SpectralTool©, is furnished with a video camera and a unique interface for field work. This device can also acquire soil reflectance from a soil surface that has not been mixed in the field as well as several soil attributes based on the spectroscopy. (see later discussion at Section 8.10.3). Whereas the above assemblies still capturing small area of the soil surface, Ben-Dor et al. (2017) recently developed an assembly (SoilPRO®) that deals with the problems preventing reliable field measurements. The apparatus can be connected to any portable spectrometer and measure the (large) soil surface under natural (undisturbed) conditions at laboratory quality. It consists of a large and lightweight closed chamber covering a wide surface area furnished with controllable and adjustable illumination at a constant geometry. The chamber, isolated from solar illumination and atmospheric attenuation, provides consistent, high-quality soil spectra, enabling a high volume of measurements in a short time by an unskilled operator. Figure 8.2 shows the SoilPRO® device and working scheme in the field.

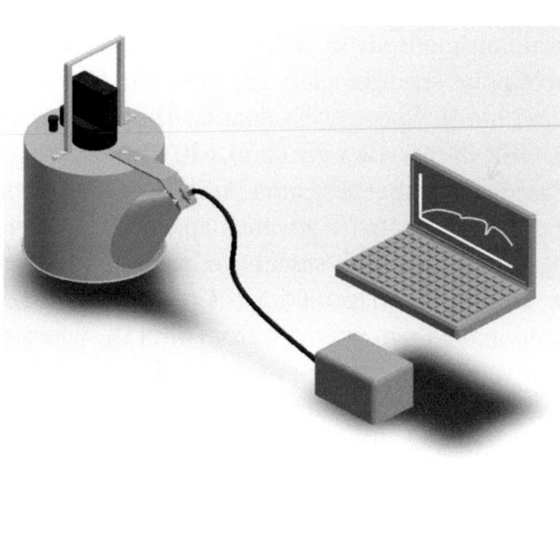

FIGURE 8.2 The SoilPRO® assembly to measure soil reflectance of undisturbed soils in the field with laboratory quality. (After Ben-Dor E., Granot A. and Notesco G. 2017. *Geoderma* 306:73–80.)

This is an important issue that prevents the effects of external parameters such as sun illumination and atmospheric conditions on the final results. Whereas in the laboratory, a standard protocol is used and the samples are properly prepared for measurement (sieved to 2 mm and well mixed), while the illumination and fore optics geometry remain constant, in the field, these factors vary and produce uncertain variations in the spectral response. The geometries of both irradiance and radiance play a major role in deriving the soil spectrum while the SoilPRO® assembly tries to standardize it in the field. The bidirectional reflectance distribution function (BRDF, see Section 8.6.2) assumes that the radiation source, the target, and the sensor are all points in the measurement space and that the ratio calculated between absolute values of radiance and irradiance is strongly dependent upon the geometry of their positions.

8.4 MECHANISMS OF SOIL-RADIATION INTERACTIONS

A comprehensive description of the physical mechanisms governing the electromagnetic radiation of diverse minerals and rocks is provided by Clark (1998). This section focuses on the most common soil-radiation processes across the VIS–NIR–SWIR spectral region.

8.4.1 Radiation Interactions with a Volume of Soil

The process of radiation scattering by soils results from a multitude of quantum mechanical interactions with the enormous number and variety of atoms, molecules, and crystals in a macroscopic volume of soil. In contrast to certain absorption features, most characteristics of the scattered radiation are not attributable to a specific quantum mechanical interaction. The effects of a particular mechanism often become obscure in the composite effect of all of the interactions. The difficulty in accounting for the effects of a large number of complex quantum mechanical interactions often leads to the use of non-quantum mechanical models of electromagnetic radiation. Physicists frequently resort to the classical wave theory or even to geometrical optics to elucidate the effects of a macroscopic volume of matter on radiation.

8.4.1.1 Refractive Indices

When light passes through a medium, some part of it will always be absorbed. This can be conveniently taken into account by defining a complex index of refraction:

$$\tilde{n} = n + i\kappa. \tag{8.1}$$

Here, the real part of the refractive index n indicates the phase speed, while the imaginary part κ indicates the amount of absorption loss when the electromagnetic wave propagates through the material. That κ corresponds to absorption can be seen by inserting this refractive index into the expression for the electric field of a plane electromagnetic wave traveling in the z-direction. The relation of wave number to refractive index is given by:

$$k = \frac{2\pi n}{\lambda_0} \tag{8.2}$$

where λ_0 is the vacuum wavelength. With complex wave number $\tilde{k}$ and refractive index $n + i\kappa$, this can be inserted into the plane wave expression as:

$$\mathbf{E}(z,t) = \mathrm{Re}(\mathbf{E}_0 e^{i(\tilde{k}z-\omega t)}) = \mathrm{Re}(\mathbf{E}_0 e^{i(2\pi(n+i\kappa)z/\lambda_0-\omega t)}) = e^{-2\pi\kappa z/\lambda_0}\,\mathrm{Re}(\mathbf{E}_0 e^{i(kz-\omega t)}). \tag{8.3}$$

Here, we see that κ gives an exponential decay, as expected from the Beer–Lambert law. Since intensity is proportional to the square of the electric field, the absorption coefficient becomes:

$$\frac{4\pi\kappa}{\lambda_0} \tag{8.4}$$

κ is often called the extinction coefficient in physics, although this has a different definition in chemistry.

Both n and κ are dependent on the frequency. In most circumstances $\kappa > 0$ (light is absorbed) or $\kappa = 0$ (light travels forever without loss). In special situations, especially in the gain medium of lasers, $\kappa < 0$ can also correspond to an amplification of the light. In the soil matrix, the radiation travels through a thin layer of particles, is reflected back to the sensor, and provides a spectrum whose shape and nature is affected by the abovementioned process (consisting of both the real and the imaginary part of the complex refraction index). Any substance in the soil matrix that affects these indices is termed a "chromophore." Knowing the chromophore's behavior can shed light on the physical and chemical constituents of the soil matrix in question. It can be either be derived with the naked eye as "color vision" (across the VIS region) or by careful analysis of the spectral responses (across the VIS–NIR–SWIR regions) according to the above theory. In general, due to its complexity, a given soil sample consists of a variety of chromophores, which vary with environmental conditions. In many cases, the spectral signals related to a given chromophore will overlap with other chromophore signals and hinder the assessment of the direct effect of the chromophore in question. Accordingly, it is important to understand the chromophores' physical processes as well as their origin and nature. Another point to mention is that in soil, there are often interactions between chromophoric and non-chromophoric properties (see Section 8.4.2).

The factors affecting soil spectral shape are defined as "physical" if the real part of the refractive index is associated and "chemical" if the spectral changes are associated with the imaginary part of the refraction index. This terminology is adopted from the weathering processes in soil where "physical" weathering refers to "size" changes in the soil matrix with no chemical alterations, and "chemical" weathering refers to chemical "alteration" of the soil's materials.

8.4.2 Spectral Processes and Chromophores

The chromophore can also be defined as a parameter or substance (chemical or physical) that significantly affects the shape and nature of a soil spectrum via its attenuation of incident radiation. There are direct chromophores that are associated with direct chemical or physical properties of a given material (e.g., absorption feature of water molecules), and indirect chromophores that are correlated to the direct chromophores (e.g., specific surface area, i.e., correlated to clay minerals). The discussion in this section refers to direct chromophores in soils, either chemical or physical. A given soil sample consists of a variety of chromophores, which vary with environmental conditions. In many cases, the spectral signals related to a given chromophore overlap with the signals of other chromophores, hindering the assessment of a given chromophore's effect. Because of the complexity of the chromophores in soil, it is important to understand their physical activity, as well as their origin and nature. The spectra of pure minerals are extensively discussed elsewhere and readers are referred to those sources (e.g., Hunt and Salisbury, 1970, 1971, 1976; Hunt et al., 1971a,b,c; Clark, 1998). In the following section, our discussion focuses primarily on factors affecting soil spectra, directly or indirectly, of both chemical and physical chromophores.

8.5 CHEMICAL PROCESS AND CHROMOPHORES

Chemical chromophores are those materials that absorb incident radiation at discrete energy levels. The absorption process usually appears on the reflectance spectrum as troughs whose positions are attributed to specific chemical groups in various structural configurations. All features in the VIS–NIR–SWIR spectral regions have a clearly identifiable physical basis. In soils, three major chemical chromophores can be roughly categorized as follows: minerals (mostly clay and iron oxides), organic matter (living and decomposing), and water (solid, liquid, and gas phases).

8.5.1 Clay Minerals

Basically, the spectral features of clay minerals in the NIR–SWIR region are associated with overtone and combination modes of fundamental vibrations of functional groups in the IR region. Of all of the clay mineral elements, only the hydroxide group is spectrally active in the VIS–NIR–SWIR region. The OH group can be found either as part of the mineral structure (mostly in the octahedral position, termed lattice water) or as a thin water molecule directly or indirectly attached to the mineral surfaces (termed adsorbed water). Three major spectral regions are active for clay minerals in general and for smectite minerals in particular: around 1.3–1.4, 1.8–1.9, and 2.2–2.5 μm. For Ca-montmorillonite (SCa-2), a common clay mineral in the soil environment, the lattice OH features are found at 1.410 μm (assigned 2υOH, where υOH symbolizes the stretching vibration at around 3630 cm^{-1}) and at 2.206 μm (assigned υOH + δOH where δOH symbolizes the bending vibration at around 915 cm^{-1}). In comparison, OH features of free water (W) are found at 1.456 μm (assigned υW + 2δW, where υW symbolizes the stretching vibration at around 3420 cm^{-1}, and δW the bending vibration at around 1635 cm^{-1}), 1.910 μm (assigned υ'W + δW where υ'W symbolizes the high-frequency stretching vibration at around 3630 cm^{-1}) and 1.978 μm (assigned υW + δW). Note that these assigned positions can change slightly from one smectite to the next, depending upon their chemical composition and surface activity. The spectra of three smectite endmembers are given in Figure 8.3, as follows: montmorillonite (dioctahedral, aluminous), nontronite (dioctahedral, ferruginous), and hectorite (trioctahedral, manganese).

The OH absorption feature of the νOH + δOH in combination mode at around 2.2 μm is slightly, but significantly shifted for each endmember. In highly enriched Al smectite

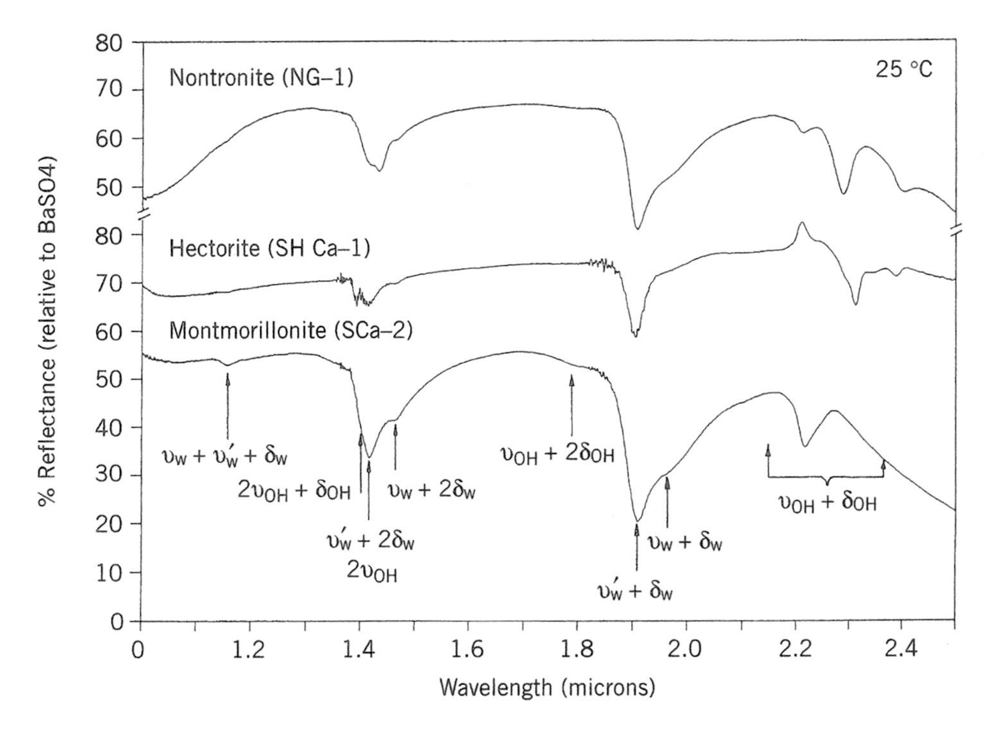

FIGURE 8.3 Reflectance spectra of three pure smectite endmembers in the NIR-SWIR region (nontronite = Fe smectite; hectorite = Mg smectite; montmorillonite Al-smectite). Also given are possible combination and overtone modes for explaining each of the spectral feature. (TAU spectral library.)

(montmorillonite), the Al-OH bond is spectrally active at 2.16–2.17 μm. In highly enriched iron smectite (nontronite), the Fe-OH bond is spectrally active at 2.21–2.24 μm and in highly enriched magnesium smectite (hectorite), the Mg-OH bond is spectrally active at 2.3 μm. Based on these wavelengths, Ben-Dor and Banin (1990) found a significant correlation between the absorbance values derived from the reflectance spectra and the total content of Al_2O_3, MgO, and Fe_2O_3. Except for a significant lattice OH absorption feature at around 2.2 μm in smectite, invaluable information about OH in free water molecules can be culled at around 1.4 and 1.9 μm. Because smectite minerals contribute relatively high specific surface areas to the soils, and these are covered by free and hydrated water molecules, these absorption features can be significant indicators of soil water content.

Kaolinite and illite minerals are also spectrally active in the SWIR region as they both consist of octahedral OH sheets. From Figure 8.4, which presents pure spectra of non-smectite layer clay minerals (kaolinite, chlorite, vermiculite, and illite), one can see that different positions and spectral shapes of the lattice OH in the layer minerals affect soil spectra across the SWIR region.

These changes are a result of the different structures and chemical compositions of the minerals. In the case of kaolinite, a 1:1 mineral (one octahedron and one tetrahedron), the fraction of the OH group is higher than in 1:2 minerals (one octahedron and two tetrahedrons), and hence, the lattice OH signals at around 1.4 and 2.2 μm are relatively strong, whereas the signal at 1.9 μm is very weak (because of relatively low surface areas and adsorbed water molecules). In the case of gibbsite, an octahedral aluminum structure (1), the 1.4 μm signal is even stronger, but the signal at 2.2 μm is shifted significantly to the IR region relative to kaolinite. An important feature is presented at 2265 nm (Madeira Netto, 1996) also related with gibbsite. Note that under relatively high signal-to-noise ratio conditions, a second overtone feature of the structural OH ($3\upsilon_{OH}$) can be observed at around 0.95 μm in OH-bearing layer minerals as well (Goetz et al., 1991).

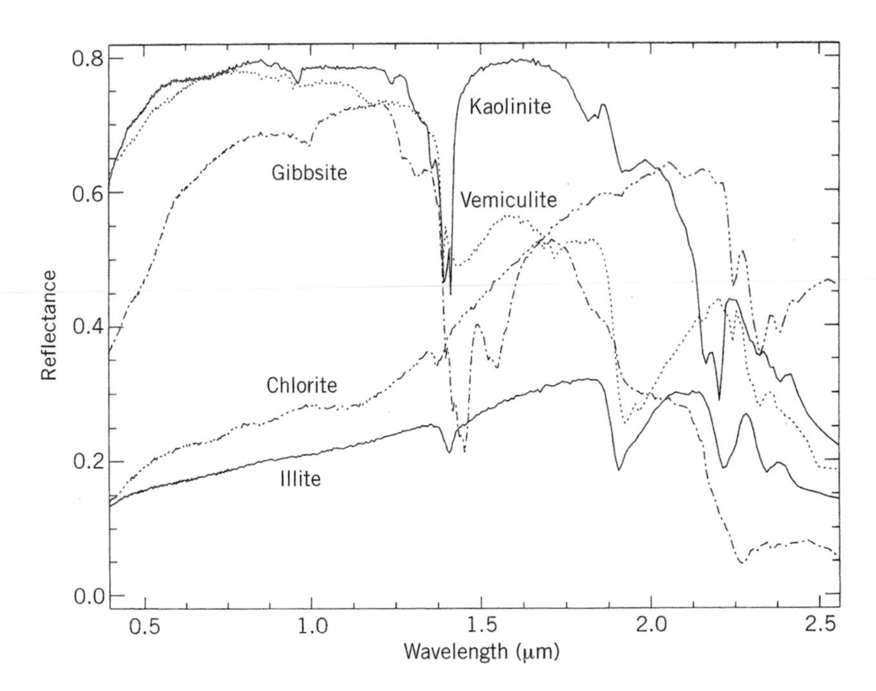

FIGURE 8.4 Reflectance spectra of representative pure non-smectite clay minerals. (USGS spectral library.)

In Figure 8.5, the direct spectral chromophores of OH in soils' clay minerals (along with other chromophores) with the exact absorbance mechanisms and intensities are presented in the SWIR (1.1–2.5 μm) spectral regions.

Based on the above spectral features, Chabrillat and Goetz (1999) and Chabrillat et al. (2002) used AVIRIS sensor data to assess and map expansive clay soils in Colorado for urban planning and environmental applications.

The affinity of water molecules to clay mineral surfaces is correlated to the latter's specific surface area. The sequence of specific surface areas of the above minerals is: smectite > vermiculite > illite > kaolinite > chlorite > gibbsite, which usually provide a similar spectral sequence at the water absorption feature near 1.8 μm (area and intensity). As smectite and kaolinite are often found in soils, they can also appear in a mixed-layer formation that overlaps spectrally. Kruse et al. (1991) described a specific case in Paris Basin, France, where interstratifications of smectite/kaolinite (a result of the alkaline weathering process of the flint-bearing chalk) was identified. Figure 8.6 presents the spectra of smectite, kaolinite, and halloysite (hydrated kaolinite) endmembers with the two representative spectra from the basin area soils examined by Kruse et al. (1991).

The noticeable asymmetrical OH absorption feature at 2.2 μm was further examined by those authors to yield a graph that predicts the relative amount of kaolinite in the mixture (Figure 8.7).

8.5.2 Carbonates

Carbonates, and especially calcite and dolomite, are found in soils that are formed from carbonic parent materials, or in a chemical environment that permits calcite and dolomite precipitation. Carbonates, especially those of fine particle size, play a major role in many of the soil chemical processes that are most likely to occur in the root zone. A relatively high concentration of fine carbonate particles may cause fixation of iron ions in the soil and consequently, inhibition of chlorophyll production. On the other hand, the absence of carbonate in soils may affect the soil's buffering capacity, and hence negatively affect biochemical and physicochemical processes. The

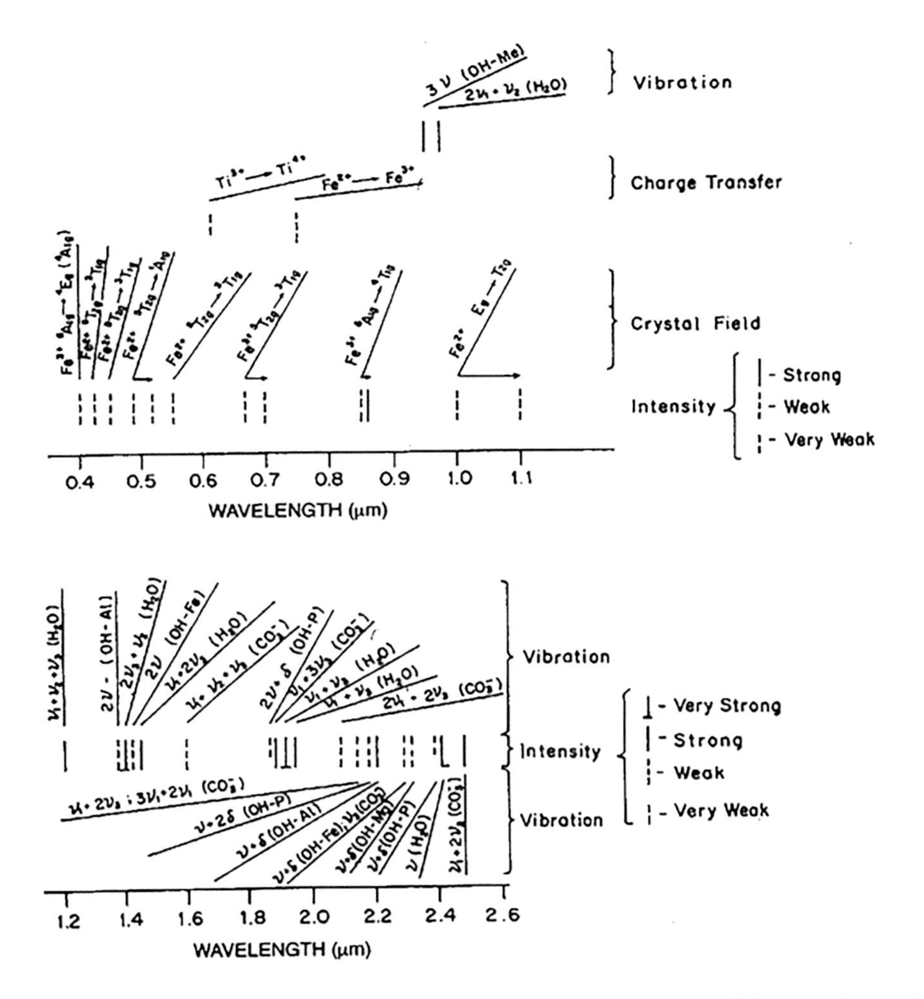

FIGURE 8.5 Active groups and mechanism in the soil chromophors. For each possible group the wavelength range and absorption feature intensity is given. (After Ben-Dor E., Irons J.R. and Epema G.F./Rencz A.N., Ryerson R.A. (Eds): *Manual of Remote Sensing: Remote Sensing for Earth Science*. pp. 111–187. 1999. Copyright Wiley-VCH Verlag GmbH & Co. KGaA.)

C–O bond, part of the $-CO_3$ radical in carbonate, is the spectrally active chromophore. Hunt and Salisbury (1970, 1971) indicated the availability of five major overtones and combination modes to describe the C–O bond in the SWIR region. In their table, υ_1 accounts for the symmetric C–O stretching mode, υ_2 for the out-of-plane bending mode, υ_3 for the asymmetric stretching mode, and υ_4 for the in-plane bending mode in the IR region. Gaffey (1986) added two additional significant bands centered at 2.23–2.27 μm (moderate) and at 1.75–1.80 μm (very weak), whereas Van-der-Meer (1995) summarized the seven possible calcite and dolomite absorption features with their spectral widths. It is evident that significant differences occur between the two minerals. This enabled Kruse et al. (1990), Ben-Dor and Kruse (1995), and others to differentiate between calcite and dolomite formations using airborne spectrometer data with bandwidths of 10 nm. Aside from the seven major C–O bands, Gaffey and Reed (1987) were able to detect copper impurities in the calcite minerals, as indicated by the broad band between 0.903 and 0.979 μm. However, such impurities are difficult to detect in soils, because overlap with other strong chromophores may occur in this region. Gaffey (1985) showed that Fe impurities in dolomite shift the carbonate's absorption band toward longer wavelengths, whereas Mg in calcite shifts the band toward shorter wavelengths. As carbonates in

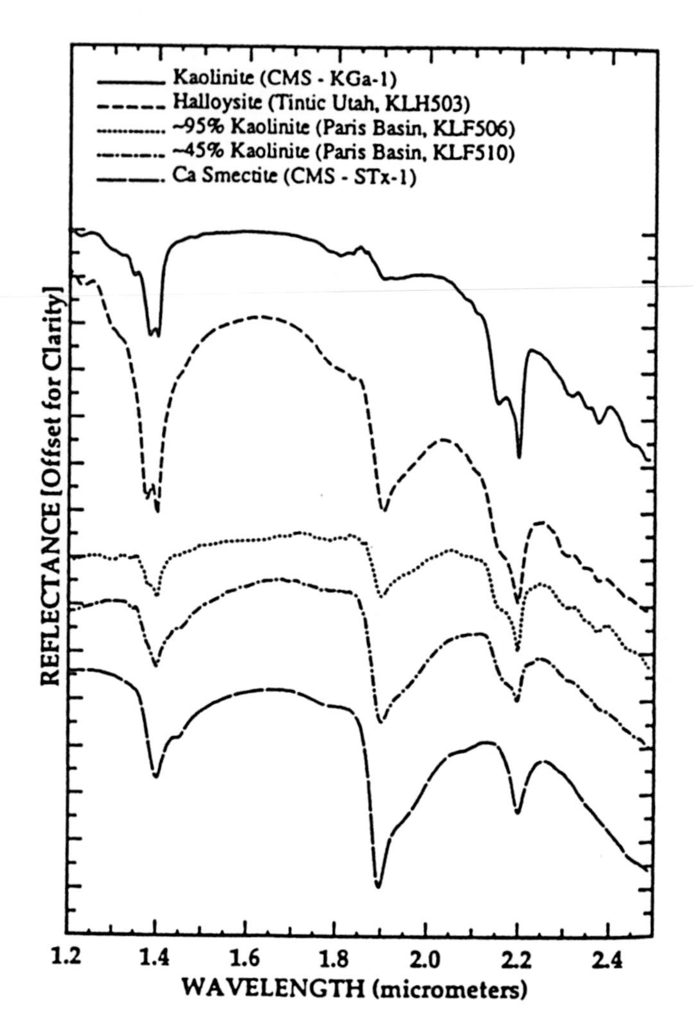

FIGURE 8.6 Reflectance spectra of kaolinite mixed-layer kaolinite/smectite from Paris Basis, halloysite from Tintic Utha and Ca-smectite. (After Kruse F.A., Thiry M. and Hauff P.L. 1991. Spectral identification (1.2–2.5 mm) and characterization of Paris basin kaolinite/smectite clays using a field spectrometer. *Proceedings of the 5th International Colloquium, Physical Measurements and Signatures in Remote Sensing*, Courchevel, France. I: pp. 181–184.)

soils are quite likely to be impure, it is only reasonable to expect that the carbonates' absorption feature positions will differ slightly from one soil to the next. A correlation between reflectance spectra and soil carbonate concentration was found by Ben-Dor and Banin (1990). Those authors used a calibration set of soil spectra and their chemical data to find three wavelengths that best predict the calcite content in arid soil samples (1.8, 2.35, and 2.36 μm). They concluded that the strong and sharp absorption features of the C–O bands in the examined soils provide an ideal tool for studying soil carbonate content solely from reflectance spectra. The best performance obtained for quantifying soil carbonate content ranged between 10% and 60%. Figure 8.8 provide the spectra of pure carbonate and other non-clay minerals in soils.

In Figure 8.5, the direct spectral chromophores of CO_3 in carbonates (along with other chemical attributes) with the exact absorbance mechanisms and intensities as describe in this section are presented in the SWIR (1.1–2.5 μm) spectral regions.

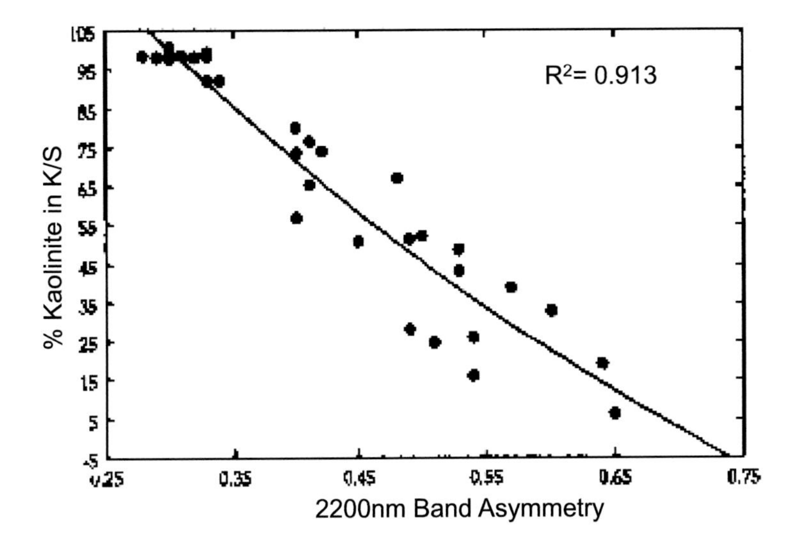

FIGURE 8.7 A correlation between the asymmetry of the 2.2 μm absorption band and percentage kaolinite form Paris Basin soil samples consisting of interstratification of kaolinite/smectite. (After Kruse F.A., Thiry M. and Hauff P.L. 1991. Spectral identification (1.2–2.5 mm) and characterization of Paris basin kaolinite/smectite clays using a field spectrometer. *Proceedings of the 5th International Colloquium, Physical Measurements and Signatures in Remote Sensing*, Courchevel, France. I: pp. 181–184.)

8.5.3 Organic Matter

Organic matter plays a major role in many chemical and physical processes in the soil environment, and has a strong influence on soil reflectance characteristics. Soil organic matter is a mixture of decomposing tissues of plants, animals, and secreted substances. The sequence of organic matter decomposition in soils is strongly determined by the soil microorganism activity. In the initial stages of the decomposition process, only marginal changes occur in the chemistry of the parent organic material. The mature stage refers to the final stage of microorganism activity, in which new, complex

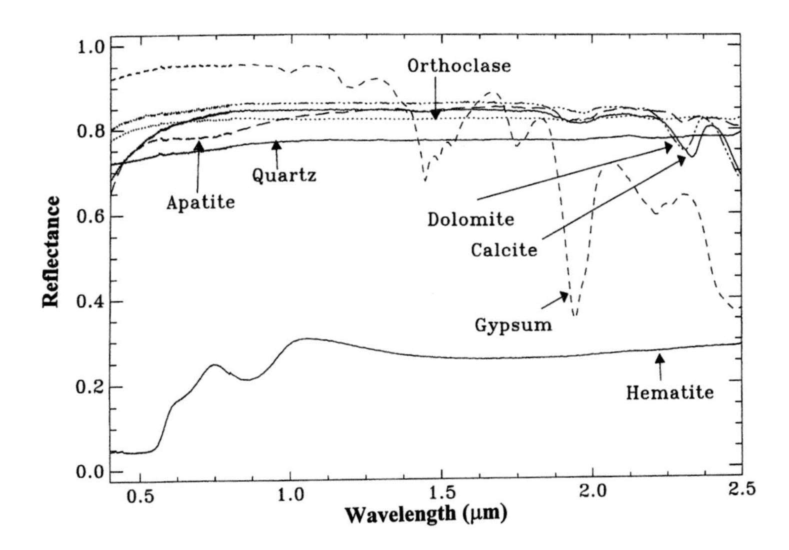

FIGURE 8.8 Reflectance spectra of representative pure carbonate and other non-clay minerals in soils.

compounds, often called humus, are formed. The most important factors affecting the amount of soil organic matter are those involved in soil formation, that is, topography, climate, time, type of vegetation, and oxidation state. Organic matter, particularly humus, plays an important role in many soil properties, such as aggregation, fertility, water retention, ion transformation, and color.

Because organic matter exhibits spectral activity throughout the entire VIS–NIR–SWIR region, but especially in the VIS region, workers have extensively studied organic matter from a remote-sensing standpoint (e.g., Kristof et al., 1971). Baumgardner et al. (1970) noted that if the organic matter in soils drops below 2%, there is only a minimal effect on the reflectance property, which was recently ratified by Demattê et al. (2017) in Brazilian wetland soils. Montgomery (1976) indicated that organic matter content as high as 9% does not appear to mask the contribution of other soil parameters to soil reflectance. In another study, Schreier (1977) indicated that the relation of organic matter content to soil reflectance follows a curvilinear exponential function. Mathews et al. (1973) found that organic matter correlates with reflectance values in the 0.5–1.2 μm range, whereas Beck et al. (1976) suggested that the 0.90–1.22-μm region is suited for mapping organic matter in soils. Krishnan et al. (1980) used a slope parameter at around 0.8 μm to predict organic matter content and Da-Costa (1979) found that simulated Landsat channels (bands 4, 5, and 6) yield reflectance readings that are significantly correlated with organic carbon content in soils. Downey and Byrne (1986) showed that it is possible to predict both moisture and bulk density of milled peat using spectral information.

The wide spectral range found by different workers for assessing organic matter content suggests that organic matter has important chromophores across the entire spectral region. Figure 8.9 shows the reflectance spectra of coarse organic matter (in the NIR–SWIR region), isolated from an Alfisol, and of the humus compounds (humic acid) extracted from this organic matter.

Numerous absorption features exist that relate to the high number of functional groups in the organic matter. These can all be spectrally explained by combination and vibration modes of organic functional groups (Chen and Inbar, 1994). Vinogradov (1981) developed an exponential model to predict the humus content in the upper horizon of plowed forest soils by using reflectance parameters between 0.6 and 0.7 μm for two extreme endmembers (humus-free parent material and humus-enriched soil). Schreier (1977) found an exponential function that accounts for soil organic matter content in reflectance spectra. Al-Abbas et al. (1972) used a multispectral scanner, with 12 spectral bands covering the 0.4–2.6 μm range, from an altitude of 1200 m and showed that a polynomial

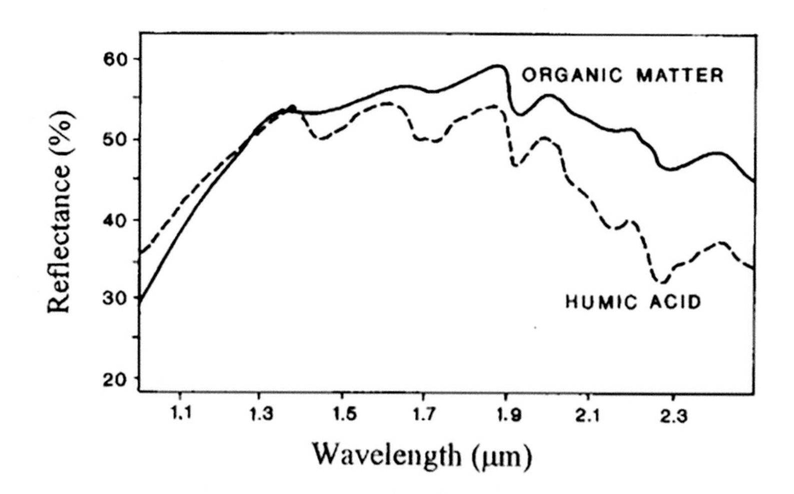

FIGURE 8.9 The spectral reflectance curves of pure organic matter isolated from Alfisol and its extracted humic acid. (After Ben-Dor E., Irons J.R. and Epema G.F./Rencz A.N., Ryerson R.A. (Eds): *Manual of Remote Sensing: Remote Sensing for Earth Science*. pp. 111–187. 1999. Copyright Wiley-VCH Verlag GmbH & Co. KGaA.)

equation will predict the organic matter content from only five channels. They implemented the equation on a pixel-by-pixel basis to generate an organic content map of a 25-ha field. Dalal and Henry (1986) were able to predict the organic matter and total organic nitrogen content in Australian soils using wavelengths in the SWIR region (1.702–2.052 μm) combined with chemical parameters derived from the soils. Using a similar methodology, Morra et al. (1991) showed that the SWIR region is suitable for identification of organic matter composition between 1.726 and 2.426 μm. Evidence that organic matter assessment from soil reflectance properties is related to soil texture, and more likely to the soil's clay, was provided by Leger et al. (1979) and Al-Abbas (1972). Aber et al. (1990) noted that the organic matter, including its decomposition stage, affects the reflectance properties of mineral soil.

Baumgardner et al. (1985) showed that three organic soils at different levels of decomposition yielded different spectral patterns. A study by Ben-Dor et al. (1997), using a controlled decomposition process over more than a year, revealed significant spectral changes across the entire VIS–NIR–SWIR region as the organic matter aged. Figure 8.10 shows a typical spectrum of grape (CGM) organic matter during 392 days of decomposition.

Significant changes can be seen in the slope values across the VIS–NIR region and in the spectral features across the entire spectrum. Ben-Dor et al. (1997) postulated that some of the analyses traditionally used to assess organic matter content in soils from reflectance spectra may be biased by the age factor. As many soils consist of dry vegetation in various different stages of degradation, assessment of organic matter using reflectance spectra should consider the vegetation's aging status. Although mineral soil has relatively low organic matter content (around 0%–4%), its accurate assessment requires high spectral resolution data across the entire VIS–NIR–SWIR region. Gomez et al. (2008) demonstrated the retrieval of organic carbon content using and data acquired both in the field and from Hyperion data and found a favorable similarity between the two methods. Recently, comprehensive attention has been given to developing a spectral assessment tool to map organic matter in soil remotely using IS devices which are based on laboratory and field models (Bartholomeus et al., 2008, 2010; Stevens et al., 2008; Terra et al.,

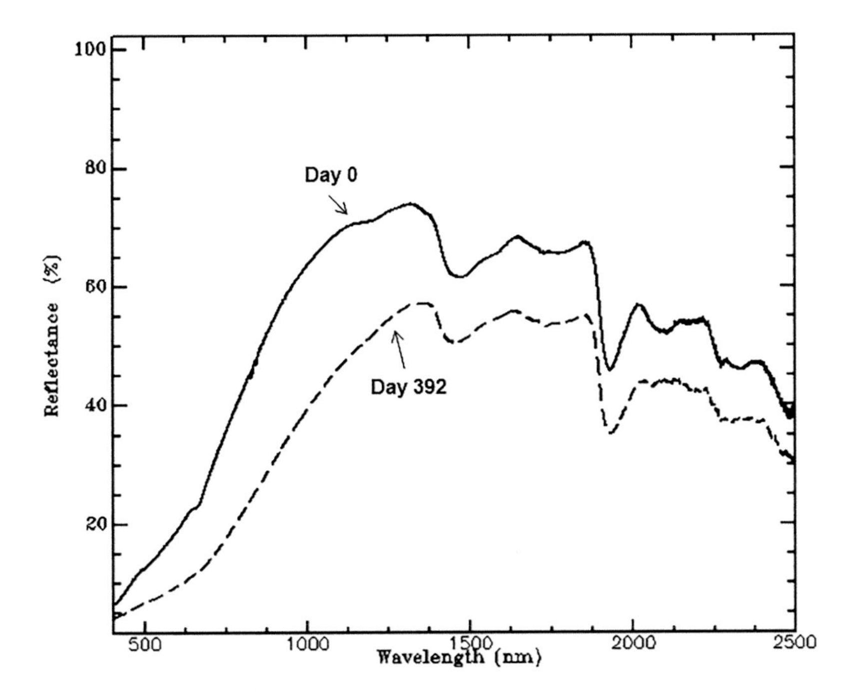

FIGURE 8.10 The reflectance spectra of two endmembers that represent two extreme composting stage t0=0 days and t8=378 days for grape marc material (CGM).

2015). A proximal approach for monitoring soil organic stock using soil spectroscopy has been recently demonstrated by Viscarra Rossel et al. (2017). Nonetheless, a new study demonstrated that organic matter assessment is highly affected by the soil type as well as by the initial organic source (organic matter specious). These results hinder a robust spectral analysis of organic matter content in soils and call for precaution while several soil types and organic matter specious are involved in the spectral analysis.

8.5.4 Water

The various forms of water in soils are all active in the VIS–NIR–SWIR region (based on the vibration activity of the OH group) and can be classified into three major categories: (1) *hydration water* which is incorporated into the lattice of the mineral, for example, limonite ($Fe_2O_3 \cdot 3H_2O$) and gypsum ($CaSO_4 \cdot 4H_2O$), (2) *hygroscopic water* which is adsorbed on the soil surface as a thin layer, and (3) *free water* which occupies the soil pores. Each of these categories influences soil spectra differently, providing the ability to identify the soil's water status (see Section 8.8.3.3). Three basic fundamentals in the IR regions exist for water molecules, particularly the OH group: υw_1—asymmetric stretching, δw—bending and υw_3—symmetric stretching vibrations. Theoretically, in a mixed system of water and minerals, combination modes of these vibrations can yield OH absorption features at around 0.95 μm (very weak), 1.2 μm (weak), 1.4 μm (strong), and 1.9 μm (very strong) related to $2\omega_1 + \upsilon w_3$, $\upsilon w_1 + \upsilon w_3 + \upsilon w$, $\upsilon w_3 + 2\upsilon w$, and $\upsilon w_3 + \upsilon w$, respectively.

1. *Hydration water* can be seen in minerals such as gypsum as strong OH absorption features at around 1.4 and 1.9 μm (Hunt et al., 1971b).
2. *Hygroscopic (adsorbed) water* is adsorbed on the surface areas of clay minerals (especially smectite) and organic matter (especially humus). Early results by Obukhov and Orlov (1964) in the VIS region showed that the slope of the spectral curve for soils is not affected by wetting and that the ratio of the reflectance of moist soil to that of dry soil remains practically constant. Shields et al. (1968) also pointed out that "moisture has no significant effect on the hue or chroma of several soils." Peterson et al. (1979) observed linear relationships between bidirectional reflectance factors at 0.71 μm in oven-dried soil samples that consisted of water tensions between 15 and 0.33 bars. These findings actually suggest that soil albedo is the first factor in the soil spectrum that is altered upon soil wetting (Idso et al., 1975). The primary reason for this is the change in the medium surrounding the particles from air to water, which decreases their relative refractive index (Twomey et al., 1986; Ishida et al., 1991). Based on this idea, Ishida et al. (1991) developed a quantitative theoretical model to estimate the effect of soil moisture on soil reflection. The shape of soil reflectance curves is strongly affected by the presence of water absorption bands at around 1.4 and 1.9 μm, and occasionally weaker absorption bands at around 0.95 and 1.2 μm. Because the amount of hygroscopic water in soil is governed by atmospheric conditions (i.e., relative humidity), the significant spectral changes are related to changes in the adsorbed water molecules on the mineral surfaces. It is interesting to note that a similar observation was already made years ago by Bowers and Hanks (1965) with soils that consisted of different moisture values (ranging from 0.8% to 20.2%). This observation demonstrates that the gas phase (water vapor in this case) in the soil environment plays a major role in the quantitative assessment of both structural and free water OH. Further insight into this problem was provided by Montgomery and Baumgardner (1974) and Montgomery (1976), who indicated that it is not possible to quantitatively assess water content in soils because of the different states of dryness under which the soils were measured. Peterson (1980) demonstrated the capability to derive soil moisture tension of soil using spectral information. Using reflectance spectra of several treated smectite minerals, Cariati et al. (1983) examined shifts in the OH absorption features at 1.4, 1.9, and 2.2 μm. They found that vibration properties of the

adsorbed water strongly depend upon the composition of the smectite structure. In another study, Cariati et al. (1981) indicated that several kinds of interactions are responsible for the vibration properties of the hygroscopic molecules, and that these may even change with water content. Because smectite is the most effective clay mineral in the soil environment at affecting the reflectance spectrum in the major water absorption features, Cariati et al.'s (1981) observations may help us understand the spectral activity of hygroscopic moisture in soils. However, further work is still required to implement the results obtained for pure smectite in the complex soil system.

3. *Free pore water (wet condition)* is water that is not in either the hygroscopic phase or filling the entire pore volume size (saturated condition). The rate of movement of this water into the plant is governed by water tension or water potential gradients in the plant–soil system. Water potential is a measure of the water's ability to do work compared to pure free water, which has zero energy. In soils, water potential is less than that of pure free water due in part to the presence of dissolved salts and the attraction between soil particles and water. Water will flow from areas of high potential to lower potential and hence flow from the soil to the root and up the plant occurs along potential gradients. In agricultural systems, plant growth occurs at soil water potentials between 15 and 0.3 bar (note these are actually negative water potentials); however, water tensions in desert environments are far greater. Baumgardner et al. (1985) studied the reflectance spectra of a representative soil (Typic Hapludalf by the USDA) with various water tensions (Figure 8.11).

As expected, when water tension decreased (and hence, water content increased), the general albedo decreased and the area under the strong 1.4 and 1.9 μm water absorption peaks also decreased. Clark (1981) examined the reflectance of montmorillonite at room temperature for two different water conditions (Figure 8.12), and showed a dramatic decrease in albedo from dry to wet material.

Other changes related to water and lattice OH can be observed across the entire spectrum as well. Some of these changes are directly related to the total amount of free and adsorbed water and some, to the increase in the spectral reflectance fraction of the soil (wet) surface. In kaolinite minerals, a similar trend was observed under two moisture conditions; however, the changes around the water OH absorption features were less pronounced than in montmorillonite. In the latter, adding water to the sample enhanced the water OH features at 0.94, 1.2, 1.4, and 1.9 μm, because of the relatively high surface area and the correspondingly high content of adsorbed water. In kaolinite, the relatively

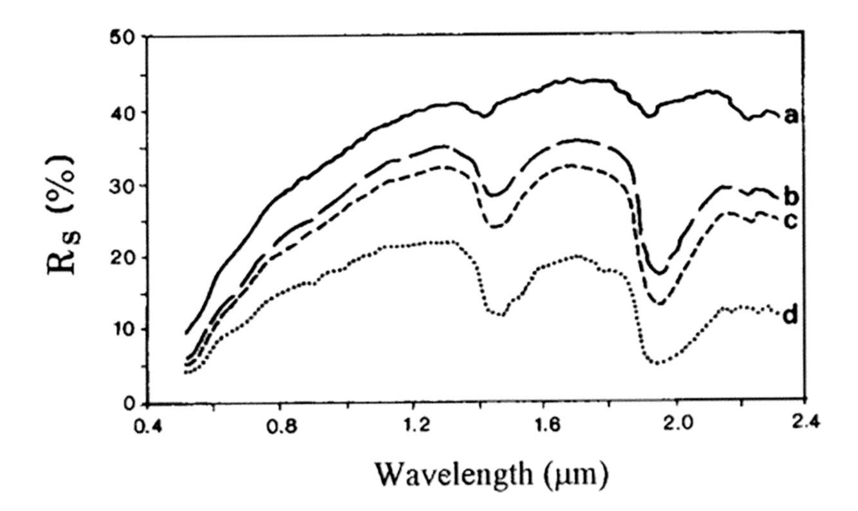

FIGURE 8.11 Spectra curve of Typic Haplludalf soil at four different moisture tensions: oven dry (a), 15 bar (b), 0.3 bar (c), 0.1 bar (d). (After Baumgardner M.F. et al. 1985. *Advances in Agronomy* 38:1–44.)

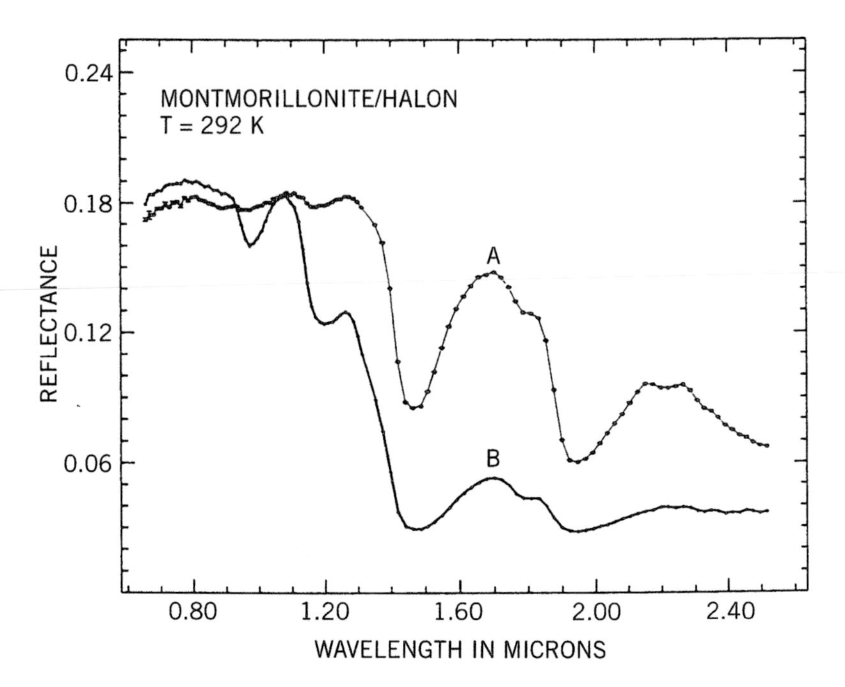

FIGURE 8.12 Reflectance spectra of montmorillonite with 50% (A) and 90% (B) water mixed in the sample (by weigh) at room temperature. (After Clark R.N. 1981. *Journal of Geophysical Research* 86:3074–3086.)

low specific surface area obscured a similar response and hence only small changes were noticeable. In the montmorillonite, the lattice-OH features at 2.2 μm decreased, suggesting that hygroscopic moisture is a major factor affecting the clay minerals' (and soil's) spectra. In soils in which the entire pore volume size (or more) is filled with water (under saturated or flooded conditions, respectively), the soil reflectance is more likely to consist of more secular than Lambertian components. It should be noted that under remote-sensing conditions, the water vapor absorptions overlap with the soil water signals, putting use of the above relationship into question. In Figure 8.5, the direct spectral chromophores of water with the exact absorbance mechanisms and intensities describe in here are presented in the SWIR (1.1–2.5 mm) spectral regions.

8.5.5 Iron

Iron is the most abundant element on the Earth's surface and the fourth-most abundant element in the Earth's crust. The average Fe concentration in the Earth's crust is 5.09 mass %, and the average Fe^{3+}/Fe^{2+} ratio is 0.53 (Ronov and Yaroshevsky, 1971). The geochemical behavior of iron in the weathering environment is largely determined by its significantly higher mobility in the divalent vs. trivalent state. Changes in its oxidation state, and consequently in its mobility, tend to occur under different soil conditions. The major Fe-bearing minerals in the Earth's crust are the mafic silicates, Fe-sulfides, carbonates, oxides, and smectite clay minerals. All Fe^{3+} oxides have striking colors, ranging from red and yellow to brown, due to selective light absorption in the VIS range caused by transitions in the electron shell. It is well known that even a small amount of iron oxides can change the soil's color significantly. The red, brown, and yellow "hue" values, all caused by iron, are widely used in soil-classification systems in almost all countries and languages. Representative soil spectra with various amounts of total Fe_2O_3 are presented in Figure 8.13. Also given in Figure 8.8 is a pure spectrum of hematite, a common iron oxide mineral in arid and semiarid soils and in Figure 8.14 pure spectrum of more iron oxide minerals.

The iron's feature assignments in the VIS–NIR region result from the electronic transition of iron cations (3+, 2+), either as the main constituent (as in iron oxides) or as impurities (as in iron

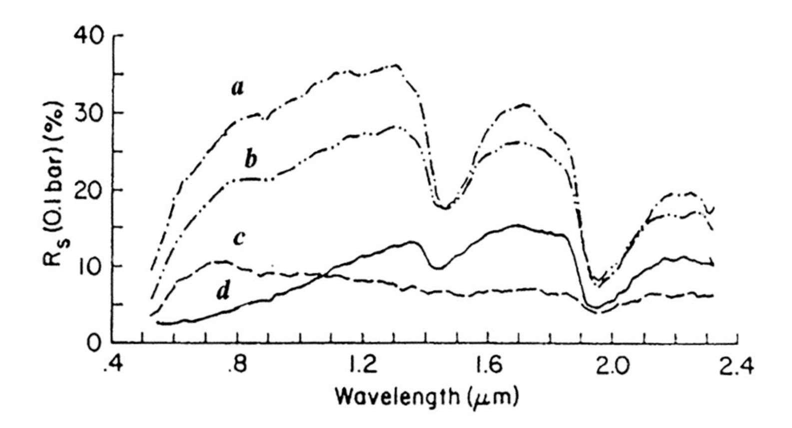

FIGURE 8.13 Reflectance spectra of soils consisting of different textures but exhibiting iron absorption bands: Fine sand, 0.20% Fe_2O_3 (a); sandy loam, 0.64% Fe_2O_3 (b); clay, 25.6% Fe_2O_3 (c); silty loam, 0.76% Fe_2O_3 (d). (After Baumgardner M.F. et al. 1985. *Advances in Agronomy* 38:1–44.)

smectite). Hunt et al. (1971a) summarized the physical mechanisms responsible for Fe^{2+} (ferrous) and Fe^{3+} (ferric) spectral activity in the VIS–NIR region as follows: the ferrous ion typically produces a common band at around 1 μm due to the spin allowed during transition between the E_g and T_{2g} quintet levels into which the D ground state splits into an octahedral crystal field. Other ferrous bands are produced by transitions from the $5T_{2g}$ to $3T_{1g}$ states at 0.55 μm, to $1A_{1g}$ at around 0.51 μm, to $3T_{2g}$ at 0.45 μm and to $3T_{1g}$ at 0.43 μm. For the ferric ion, the major bands produced in the spectrum are the result of the transition from the $6A_{1g}$ ground state to $4T_{1g}$ at 0.87 μm, $4T_{2g}$ at 0.7 μm, and either $4A_{1g}$ or $4E_g$ at 0.4 μm.

Just as organic matter is an important indicator for soils, iron oxides provide significant evidence that soil is being formed in a given area of the Earth's crust (Schwertmann, 1988). Iron oxide content and species are strongly correlated with short- and long-term soil-weathering processes. Iron oxide transformation in the soil often occurs under natural soil conditions. Hematite and goethite are common iron oxides in soils and their relative content is strongly controlled by soil temperature, water, organic matter, and annual precipitation. Hematitic soils are reddish and goethitic soils are yellowish brown. Their reflectance spectra also differ, as can be seen in Figure 8.14.

Hematite (α-Fe_2O_3) has Fe^{3+} ions in octahedral coordination with oxygen. Goethite (α-FeOOH) also has Fe^{3+} in octahedral coordination, but different site distortions along with the oxygen ligand (OH) provide the main absorption features that appear near 0.9 μm. Lepidocrocite (α-FeOOH), which is associated with goethite but rarely with hematite, is another common unstable iron oxide found in soils. It appears mostly in subtropical regions and is often found in the upper subsoil position (Schwertmann, 1988). Maghemite (α-Fe_2O_3) is also found in soils, mostly in subtropical and tropical regions, and it has occasionally been identified in soils in humid temperate areas. Ferrihydrite is a highly disordered Fe^{3+} oxide mineral found in soils in cool or temperate, moist climates, characterized by young iron oxide formations and soil environments that are relatively rich in other compounds (e.g., organic matter, silica, and so on). Iron associated with clay mineral structures is also an active chromophore in both the VIS–NIR and SWIR spectral regions. This can be seen in the nontronite-type mineral presented in Figure 8.3. Based on the structural OH–Fe features of smectite in the SWIR region, Ben-Dor and Banin (1990) generated a prediction equation to account for the total iron content in a series of smectite minerals. The wavelengths selected automatically by their method were 2.2949, 2.2598, 2.2914, and 1.2661 μm. Stoner (1979) also observed a higher correlation between reflectance in the 1.55–2.32-μm region and iron content in soils, whereas Coyne et al. (1989) found a linear relationship between total iron content in montmorillonite and the absorbance measured in the 0.6–1.1 μm spectral region. Ben-Dor and Banin (1995a) used spectra of 91 arid soils to show that their total iron content (both free and structural iron) can be predicted by multiple linear regression analysis and wavelengths 1.075, 1.025, and 0.425 μm.

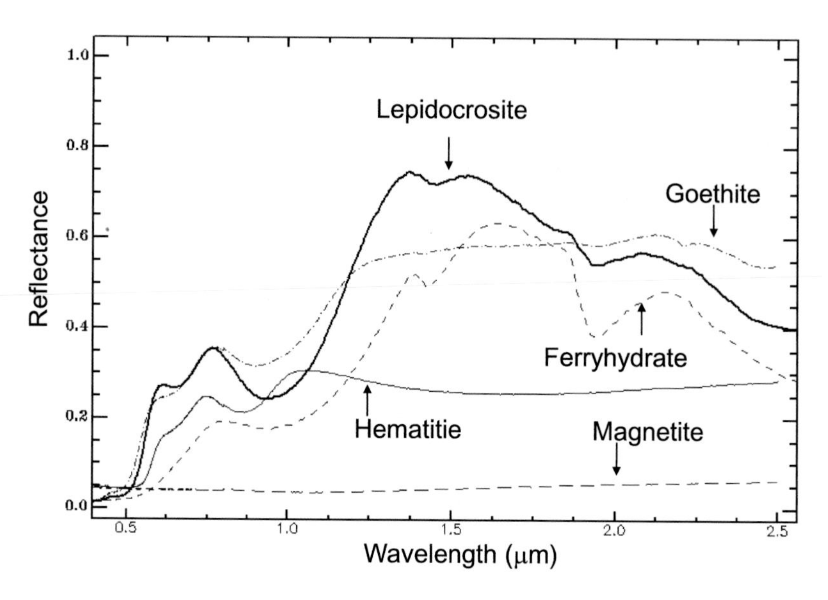

FIGURE 8.14 The spectra of pure iron oxides found in soils. (After Ben-Dor E., Irons J.R. and Epema G.F./ Rencz A.N., Ryerson R.A. (Eds): *Manual of Remote Sensing: Remote Sensing for Earth Science.* pp. 111–187. 1999. Copyright Wiley-VCH Verlag GmbH & Co. KGaA.)

Obukhov and Orlov (1964) generated a linear relationship between reflectance values at 0.64 μm and the total percentage of Fe_2O_3 in other soils. Taranik and Kruse (1990) showed that a binary encoding technique for the spectral slope values across the VIS–NIR spectral region is capable of differentiating a hematite mineral from a mixture of hematite–goethite–jarosite. The iron oxides species are strongly affected by the pH and thus the resulted color can be used as an indicator to estimate pH in regalities (e.g., Zabcic et al., 2014). In Figure 8.5, the direct spectral chromophores of iron are presented in both VIS-NIR (0.4–1.1 mm) and SWIR (1.1–2.5 mm) spectral regions.

It is important to mention that iron can often have an indirect influence on the overall spectral characteristics of soils. In the case of free iron oxides, it is well known that soil particle size is strongly related to absolute iron oxide content (Soileau and McCraken, 1967; Stoner and Baumgardner 1981; Ben-Dor and Singer 1987). As iron oxide content increases, the size fraction of the soil particles increases as well, because of the cementing effect of the free iron oxides. As a result, problems resulting from different scattering effects are introduced into the soil analysis. Moreover, free iron oxides, mostly in their amorphous state, may coat the soil particles with a film that prevents natural interaction between the soil particle (clay or non-clay minerals) and the sun's photons. Karmanova (1981) found that well-crystallized iron compounds have the strongest effect on the spectral reflectance of soil, and that removal of non-silicate iron (mostly iron oxides) helps enhance other chromophores in the soil. In this respect, Kosmas et al. (1984) demonstrated a second-derivative technique in the VIS region as a feasible approach for differentiating even small features of synthetic goethite from clays, and they suggested that such a method may be adopted to assess quantities of iron oxide in mixtures. Based on these spectral characteristics, Demattê et al. (1999) showed the possibility of spectrally assessing the alteration of soil properties, and Gerbermann and Neher (1979) showed that soil mixtures of clay and sand can be predicted from reflectance spectra. Ben-Dor et al. (2005) modeled iron oxides' absorption features in a sand dune and were able to account for the rubification process of the soil formed over the dune. Bartholomeus et al. (2007) demonstrated the retrieval of iron content in partially vegetated Mediterranean soil using both ground spectroscopy and IS data. Studies by Lugassi et al. (2010) have shown that iron oxide spectra of soils which have burned can be used as a quantitative indicator to assess the temperature of the fire. Iron oxide alteration during the fire events played a major role in the spectral domains which were used by the authors to study the fire days after it occurred. Other reflectance changes within the soil subjected to fire have been described by Kokaly et al. (2007).

It can thus be concluded that iron is a very strong chromophore in soil, and that a determination of its content in clay and soil from reflectance spectra in the entire VIS–NIR–SWIR region is feasible. Based on the complexity of the iron component in the soil environment, as well as on the intercorrelation between iron and other soil components, sophisticated methods and relatively high spectral resolution data are absolutely required to determine iron content from reflectance spectra.

8.5.6 Soil Salinity

Soil salinity is one of the major factors affecting biomass production and is the principal cause of soil degradation (Csillag et al., 1993). Salt-affected areas cover about 7% of the Earth's land surface (Toth et al., 1991) and are located mostly in arid and semiarid regions (Verma et al., 1994). However, salt-affected soils can also be found in sub-humid and coastal areas associated with hydrogeological structures. Soil salts have been reported to be in the form of Na_2CO_3, $NaHCO_3$, and NaCl, which are very soluble and mobile components of the soil environment. Typically, saline soils have a poor structure, are highly erosive, have low fertility, low microbial activity, and other attributes that are not conducive to plant growth.

The spectral signature of saline soils can be a result of the salt itself or, indirectly, of other chromophores related to the presence of the salt (e.g., organic matter, particle size distribution). Hunt et al. (1971c) reported an almost featureless spectrum for halite (NaCl 433B from Kansas). Although salt is spectrally featureless, Hick and Russell (1990) raised the hypothesis that there are certain wavelengths in the VIS–NIR–SWIR region that can provide more accurate information about saline-affected areas. Dwivedi and Sreenivas (1998) applied an image-manipulation tool to the study of soil salinity by remote-sensing means, whereas Rao et al. (1995) investigated the spectral reflectance of salt-affected soils and found some spectral variations.

Vegetation is an indirect factor that can facilitate the detection of saline soils from reflectance measurements (Hardisky et al., 1983; Wiegand et al., 1994). Gausman et al. (1970), for example, showed that cotton leaves growing in saline soils have a higher chlorophyll content than those growing in low-salt soil. Hardisky et al. (1983), using the spectral reflectance of a *Spartina alterniflora* canopy, showed a negative correlation between soil salinity and spectral vegetation indices. In the absence of vegetation, the major influence of salt is on the structure of the upper soil surface.

Because no direct significant spectral features are found in the VIS–NIR–SWIR region for sodic soil, indirect techniques are thought to be more suitable for classifying salt-affected areas (Sharma and Bhargava, 1988; Verma et al., 1994). Salt in water is most likely to affect the hydrogen bond in water molecules, causing subtle spectral changes; based on this, Hirschfeld (1985) suggested that high-spectral-resolution data are required. Support for this idea was provided by Szilagyi and Baumgardner (1991) who reported on the feasibility of characterizing soil-salinity status with high-resolution laboratory spectra. A relatively high number of spectral channels are also important in identifying indirect relationships between salinity and other soil properties that appear to involve chromophores in the VIS–NIR–SWIR regions. Csillag et al. (1993) analyzed high-resolution spectra taken from about 90 soils in the USA and Hungary for their chemical parameters, including clay and organic matter content, pH and salt. They claimed that because salinity is such a complex phenomenon, it cannot be attributed to a single soil property. While studying the capability of commercially available Earth-observing optical sensors, they were able to show that six broad bands in the VIS–NIR–SWIR region best discriminate soil salinity. These six channels were selected solely on the basis of their overall spectral distribution, which provided complete information about salinity status. In another study, Metternicht and Zink (1997) showed that by using six combined Landsat bands (1, 2, 4, 5, 6, 7), it is possible to discriminate salt and sodium-affected soil with varying confidence limits. They discussed the indirect salt effect on soil spectral responses and suggested the addition of more electromagnetic radiation bands to shed more light on this problem. Thus, it can be concluded that the entire spectral region needs to be considered in evaluating salinity levels in different environments and unknown soil systems. Mougenot et al. (1993) noted that in addition to an increase in reflectance with salt content, high salt content may mask ferric ion absorption in the VIS region. Those authors concluded that salts are not easily identified in proportions below 10% or 15%.

Another important factor in saline soils is the fact that in modern agriculture, farmers add gypsum to sodic soils for soil reclamation (Singh, 1994). The artificial increase in gypsum content in such soils may alter the soil reflectance spectra significantly and this, therefore, requires attention. It should, therefore, be remembered that although salt is not a strong or direct chromophore, its interaction with other soil components (water, structure, iron, and organic matter) makes its assessment possible but complicated. Farifteh et al. (2006) studied the reflectance spectra of soils affected by salt, both artificially and naturally, and established some interesting findings toward understanding the spectral features of salt-affected soils. Ben-Dor et al. (2008a,b) and Metternich et al. (2008) reviewed the spectral-based studies of remote sensing of soil salinity using passive and active data combined with other electromagnetic means. Studies by Livne et al. (2009) describe a robust spectral model to account for soil salinity which has been used with both Israeli and Uzbekistani soils. Further developments in soil spectroscopy and salinity interactions are anticipated in the near future. This is mainly based on the potential of soil reflectance to shed light on soil salinity and the increasing need to assess soil salinity in the field rapidly and frequently.

8.5.7 Chemical Chromophores: Summary

To provide an overview of direct chemical chromophore activity in soils, Figure 8.15 illustrates a spectrum of a selected soil from Israel (Haploxeralf) with the positions of all possible chromophores.

Figure 8.5 summarizes the chromophores associated with soil and geological matter as collected from the literature and summarized by Ben-Dor et al. (1998).

A summary of important direct soil chromophores were generated by Soriano-Disla et al. (2014) is given in Table 8.1. Note that this table also consists of spectral signatures of the spectral foundation associated with the thermal region TIR (Thermal Infrared; 3000–14,000 nm) that is, divided into the Mid-Wave Infrared (MWIR; 3000–5000 nm) and the Longwave Infrared (LWIR; 8000–14,000 nm). It also lists the intensities of each chromophore in the VIS–NIR–SWIR spectral regions as they

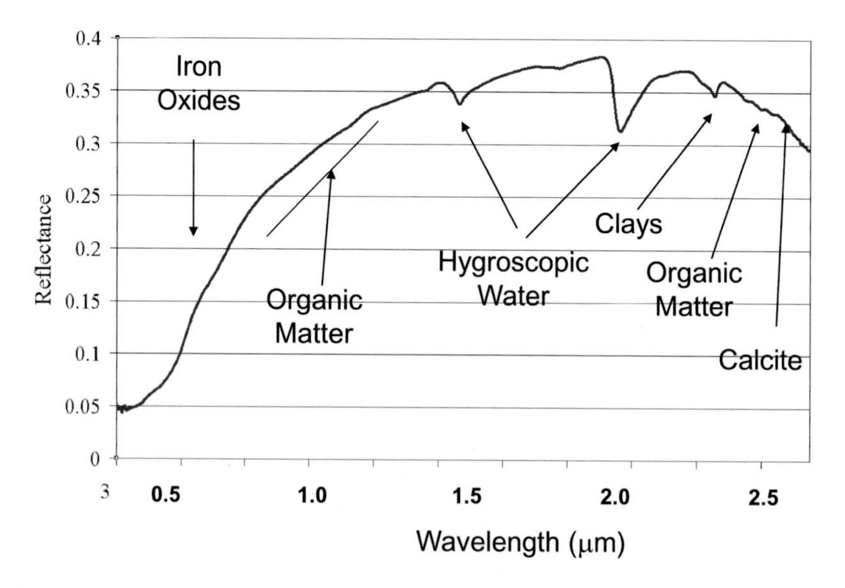

FIGURE 8.15 A soil spectra, (Haploxeralf) that represents the major chromophors in soils (see text for more details).

TABLE 8.1
Approximate Wavelength (nm) and Wavenumbers (cm-1) of Direct Spectral Absorptions in the VNIR-SWIR-TIR Spectral Regions of Major Soil Components and Those Described in Figure 8.5 for Its Exact Spectral Mechanisms

	Soil Component	Wavenumbers (cm-1)	Wavelengths (nm)	References
MIR	Quartz (sand)	1100–1000	9091–10,000	Nguyen et al. (1991), Van der Marel and Beutelspacher (1976)
	Clay minerals	3690–3620 Kaolinite	2710–2762	Nguyen et al. (1991), Van der Marel and Beutelspacher (1976), Viscarra-Rossel and Behrens (2010)
		3620–3630 and 400–3300 Smectite and Illite	2755–2762 and 2941–3030	Ibid
	Carbonates	1430 and 2520	6993 and 3968	Ibid
	Iron oxides	600–700	14,286–16,667	Van der Marel and Beutelspacher (1976)
	Iron Oxyhydroxides	3100, 900, 800	3226, 11,111 12,500	Ibid
	Organic matter	2930–2850 Alkyl ($-CH_2$)	3413–3509	Nguyen et al. (1991), Van der Marel and Beutelspacher (1976), Viscarra-Rossel and Behrens (2010)
		1670 and 1530 Protein amide (OC-NH)	5952 and 6535	Ibid
		1720 Carboxylic acid (COOH)	5814	Ibid
		1630 Water associated	6135	Ibid
		1600 and 1400 Carboxylate anion ($-COO^-$)	6250 and 7143	Nguyen et al. (1991), Van der Marel and Beutelspacher (1976)
		1600–1570 Aromatic group	6250–6369	Van der Marel and Beutelspacher (1976)
VIS–NIR	Water	7143 and 5263	1400 and 1900	Viscarra-Rossel and Behrens (2010), Ben-Dor and Banin (1995), Dalal and Henry (1986)
	Clay mineral	7143 and 4545 Kaolinite	1400 and 2200	Viscarra-Rossel and Behrens (2010)
		4545, 4274, and 4090 Illite	2200, 2340, and 2445	Viscarra-Rossel and Behrens (2010)
		4545 Smectite	2200	Viscarra-Rossel and Behrens (2010), Ben-Dor and Banin (1995)
	Carbonates	4283	2335	Ibid
	Iron oxides	25,000, 22,222, 20,000, 15,385, and 11,111	400, 450, 500, 650, and 900	Viscarra-Rossel and Behrens (2010)
	Organic matter	9091, 6250, 5882, 5556, 5000, and 4167–4545	1100, 1600, 1700, 1800, 2000, and 2200–2400	Viscarra-Rossel and Behrens (2010), Ben-Dor and Banin (1995), Dalal and Henry (1986), Murra et al. (1991)

Source: Taken from Soriano-Disla et al. 2014. *Applied Spectroscopy Reviews* 49(2):139–186.
Note: For the exact reference, please refer to the original paper of Soriano-Disla et al. 2014.

appear in those studies. The current review demonstrates that high-resolution spectral data can provide additional, sometimes quantitative, information about soil properties that are strongly correlated to those chromophores, that is, primary and secondary minerals, organic matter, iron oxides, water, and salt.

8.6 PHYSICAL PROCESSES AND CHROMOPHORES

8.6.1 General

In addition to chemical processes, the reflectance of light from the soil surface is dependent upon numerous physical processes. Reflection, or scattering, is clearly described by Fresnel's equation and depends upon the angle of the incident radiation and the refraction index of the materials in question. Generally, physical factors are those parameters which affect soil spectra with regard to Fresnel's equation, but which do not cause changes in the position of the specific chemical absorption. These parameters include particle size, sample geometry, viewing angle, radiation intensity, incident angle, and azimuth angle of the source. Changes in these parameters are most likely to affect the shape of the spectral curve through changes in baseline height and absorption feature intensities. In the laboratory, measurement conditions can be maintained constant. In the field, several of these parameters are unknown and may hinder an accurate assessment of soil spectra.

Many studies covering a wide range of materials have shown that differences in particle size alter the shape of soil spectra (Hunt and Salisbury, 1970; Pieters, 1983; Baumgardner et al., 1985). Specifically, Hunt and Salisbury (1970) quantified effects of about 5% in absolute reflectance due to particle size differences, and these changes occurred without altering the position of the diagnostic spectral features. Under field conditions, aggregate size rather than particle size distributions may be more important in altering soil spectra (Orlov, 1966; Baumgardner et al., 1985). In the field, aggregate size may change over a short time due to tillage, soil erosion, aeolian accumulation, or physical crust formation (e.g., Jackson et al., 1990; Chappell et al., 2005). Basically, the aggregate size, or more accurately roughness, plays a major role in the shape of field and airborne soil spectra (e.g., Cierniewski, 1987, 1989). Escadafal and Hute (1991) showed strong anisotropy reflectance properties in five soils with a rough surface.

A practical solution for evaluating the effects of physical parameters is to evaluate the reflectance of a given target relative to a perfect reflector measured at the same geometry and viewing angle as the target in question. In reality, such conditions are impossible to achieve in the field, and complex effects, such as those of particle size, cannot be completely eliminated by this method. It is postulated that more effort should be expended on obtaining a more precise accounting of physical effects under field conditions (from both spectroscopy and IS points of view), as Pinty et al. (1989) tried to do by simulating the bidirectional effect over bare soils.

8.6.2 Models of Radiation Scattering by Soils

The geometry of both irradiance and radiance plays a major role in deriving the soil spectrum. The BRDF assumes that the radiation source, the target, and the sensor are all points in the measurement space and that the ratio calculated between absolute values of radiance and irradiance is strongly dependent upon the geometry of their positions. Theories and models explaining the BRDF phenomenon in relation to soil components are widely discussed and covered in the literature (Hapke, 1981a,b, 1984, 1986, 1993; Pinty et al., 1989; Jacquemoud et al., 1992; Liang and Townshend, 1996). The following equation describes the basic BRDF value:

$$fr(q,q',f,f') = \frac{dL(q,q',f,f',E)}{dE(q,f)} \quad (8.5)$$

where E is the radiance, L is the irradiance and q, q', f, f', are source and sensor zenith angles, and source and target azimuth angles, respectively.

Whereas the BRDF is better suited to remote-sensing applications, hemispheric and bihemispheric reflectance factors are also used in the laboratory (Baumgardner et al., 1985). To reduce the effects of geometry and to eliminate systematic and non-systematic measurement interference, reflectance standards such as MgO, $BaSO_4$, and Halon are often used to correct the relative reflectance spectrum (Tkachuk and Law, 1978; Young et al., 1980; Weindner and Hsia, 1981). Another factor that affects soil spectra is the sensor's field of view (FOV) and sun target geometry. If the soil is homogeneous, a small FOV may be sufficient. However, where some variation occurs in the soil, the FOV should be adjusted to cover a representative portion of that soil. Studies by Feingersh et al. (2010) have shown BRDF measurements of selected soils in the laboratory using a controlled geometry setup. Use of this device enabled Feingersh et al. (2007) to develop a correction scheme for soil urban material and vegetation BRDF effects that appear on IS images. This correction has afterwards implemented in an operational software BREFCOR (Schläpfer et al., 2010). In an earlier study, Chappell et al. (2006) used a bidirectional model to detect induced in situ erodibility in a dust process playa in Australia.

8.7 RELATIONSHIP BETWEEN SOIL CHROMOPHORE AND PROPERTIES

8.7.1 Qualitative Aspects

Color is perceived by the human eye and brain; these serve as day-to-day spectrometers and analyzers. Pedologists have long used soil color to describe and classify soils and infer their characteristics (Buol et al., 1973; Escadafal, 1993). As stated by Baumgardner et al. (1985) "ever since soil science evolved into an important discipline for study and research, color has been one of the most useful variables in characterizing and describing a particular soil." Certain qualitative relationships between color and soil properties are well recognized by pedologists on the basis of their collective observations and of a conceptual understanding of the interaction of visible light with soil material. Today's instruments can convert soil reflectance curves into color parameters (Escadafal et al., 1989; Escadafal and Hute, 1991; Escadafal, 1993), and soil color continues to play a major role in modern soil classification. Models that formulate color as mathematical functions of soil properties, however, are not well established in the pedological community. Soil color is a spectral phenomenon and today, spectrometers and computers are replacing the human eye and brain, respectively, and the following section discusses various models describing soil reflectance properties in the VIS through SWIR regions of the electromagnetic spectrum.

8.7.2 Quantitative Aspects of Proximal Soil Spectroscopy

8.7.2.1 Historical Notes

Quantitative soil spectroscopy is a mature discipline which has come quite a long way since the mid-1960s, when Bowers and Hanks (1965) published their paper on the correlation between soil reflectance and soil moisture content. That pioneering study, followed by a series of papers by Hunt and Salisbury (1970–1976) and Hunt (1980), proved that water and minerals in the soil environment have unique spectral fingerprints that can be further used for specific recognition. In parallel to this development, Ben-Gera and Norris (1968) published a paper showing a correlation between reflectance readings and moisture contents in soybeans. Based on this finding, the new discipline of near-infrared spectral analysis (NIRS) emerged, which focused on extracting quantitative information from reflectance data, first in the food sciences (Davies and Grant, 1987) and then in other disciplines using the "chemometrics" approach also known as soil proximal sensing. The first conference to gather together scientists in this field was held in 1987 in Norwich, UK: The First *International Near-Infrared Spectroscopy Conference*. Today, many conferences, workshops, and scientific meetings are dedicated to the NIRS concept, along with a specialized scientific journal dedicated to the NIRS method which was established in 1993 (*The Journal of Near Infrared Spectroscopy*). In 1987, Davies published an article in *European Spectroscopy News* entitled "Near Infrared Spectroscopy Analysis: Time for the Giant to Wake Up." He appealed to potential users at the time to use reflectance spectroscopy across the NIR wavelength region for chemical analysis of powders. A decade later, Davies (1998) published another article entitled "The History of Near Infrared Spectroscopic Analysis: Past, Present and Future 'From Sleeping Technique to the Morning Star of Spectroscopy'" that showed that the NIRS technique had come a long way and that the quantitative optical approach (mostly for food products) was mature, successful, and applicable. The disciplines that made use of NIRS were the food sciences, pharmacology, the textile, tobacco, and oil industries, agriculture, art, the paper industry, and more. Learning from these sectors' successes, Dalal and Henry (1986) applied the NIRS approach to soils. This pioneering study captured the attention of many researchers who realized the potential of soil reflectance spectroscopy. Whereas today, a scientific community namely "Group of Proximal Sensing" is very active in all aspects of soil spectroscopy. The first scientists to systematically gather soil spectral information and publish it in the form of a soil spectral atlas were Stoner et al. (1980). Their Soil Spectral Library (SSL) very soon became a classical tool that soil scientists came to rely on. Later, when laboratory and portable field spectrometers were introduced into the market (around 1993), more scientists realized the potential of soil spectroscopy, and consequently more spectral libraries were assembled (e.g., Demattê and Garcia, 1999; Bellinaso et al., 2010). A comprehensive summary of the quantitative applications of soil reflectance spectroscopy was provided by Ben-Dor (2002) and in April 2008, a world soil spectroscopy group was established by Viscarra Rossel (http://groups.google.com/group/soil-spectroscopy), who gathered soil spectra and corresponding attributes from more than 80 countries worldwide to generate a global soil spectral/attribute database providing soil-NIRS capability to all. This initiative was based on the idea that the NIRS approach in soil sciences, which had become well established and applicable, should be more collaborative. This was the obvious step after understanding that only information sharing would help advance soil quantitative spectroscopy (e.g., Condit, 1970; Shepherd and Walsh, 2002; Brown et al., 2006). Comprehensive reviews on NIRS applications for soils can be found in Malley et al. (2004) and Viscarra-Rossel et al. (2006), and other important reviews focusing on soil reflectance theory and applications can be found in Clark and Roush (1984), Irons et al. (1989), Ben-Dor et al. (1999), and Ben-Dor (2002). The readers are referred to a comprehensive summary of many (spectral-based) models to predict soil properties (with their relevant citations) as generated by Soriano-Disla et al. (2014). They separated the NIRS ("chemometric") models into specific spectral regions that are based on the spectral mechanisms discussed in Section 8.5 (electronic—VIS-NIR, overtones and combinations modes—SWIR, and fundamental vibration—MWIR). Note, however, that the spectral definition used by Soriano-Disla et al. is the one which is adopted by the Soil Chemomatrix Community (SCC) that is slightly different from the one which is used by the Remote Sensing Community (RSC). The following translation enable comparison between the communities: $UV_{(SCC)} = UV_{(RSC)}$, $VIS_{(SCC)} = VIS_{(RSC)}$, $NIR_{(SCC)} = NIR\text{-}SWIR_{(RSC)}$, $MIR_{(SCC)} = LWIR\text{-}MWIR_{(RSC)}$. This chapter keeps the RSC notation.

8.7.2.2 Quantitative Applications in Soils

In general, soil reflectance spectra are directly affected by chemical and physical chromophores, as already discussed. The spectral response is also a product of the interaction between these parameters, calling for a precise understanding of all chemical and physical reactions in soils. For example, even in a simple mixture of iron oxide, clay, and organic matter, the spectral response cannot be judged simply by linear mixing models of the three endmembers. Strong chemical interactions between these components are, in most cases, nonlinear and rather complex. For instance, organic components, mostly humus, affect soil clay minerals in chemical and physical ways. Similarly, free iron oxides may coat soil particles and mask photons that interact with the real mineral components or the iron oxides themselves (and organic matter as well). In addition, the coating material may collate fine particles into coarse aggregates that may physically change the soil's spectral behavior from a physical standpoint. Karmanova (1981) selectively removed the iron oxides from soil samples and concluded that the effects of various iron compounds on the spectral reflectance and color of soils were not proportional to their relative contents. Another example of the strong relationship between chromophores is given by Bedidi et al. (1990, 1991), who showed that the normally accepted view of decreasing soil baseline height with increasing moisture content (VIS region) does not hold for lateritic (highly leached low-pH) soils. They concluded that the spectral behavior of such soils under various moisture conditions is more complex than originally thought. In this context, Galvao et al. (1995) showed spectra from laterite soils (VIS–NIR region) consisting of complex spectral features that appeared to deviate from other soils. Seeking for more detailed information, Demattê and Terra (2014) observed important relationship between soil spectral properties and pedogenetic alterations, showing its relevance on this matter. Al-Abbas et al. (1972) found a correlation between clay content and reflectance data in the VIS–NIR–SWIR region and suggested that this was not a direct but an indirect relationship, strongly controlled by the organic matter chromophore. Another anomaly that relates to the interactions between soil chromophores was identified by Gerbermann and Neher (1979). They carefully measured the reflectance properties in the VIS region of a clay–sand mixture extracted from the upper horizon of a montmorillonite soil and found that "adding of sand to a clay soil decreases the percent of soil reflectance." This observation stood in contrast to what was traditionally expected from adding coarse (sand) to fine (clay) particles in a mixture (soil), that is, that this would tend to increase soil reflectance. Similarly, Ben-Dor and Banin (1994, 1995a,b,c) concluded that intercorrelations between feature and featureless properties play a major role in assessing unexpected information about soil solely from their reflectance spectra in either the VIS–NIR or SWIR regions. Ben-Dor and Banin (1995b) examined arid and semiarid soils from Israel and showed that "featureless" soil properties (i.e., properties without direct chromophores such as K_2O, total SiO_2, and Al_2O_3) can be predicted from the reflectance curves due to their strong correlation with "feature" soil properties (i.e., properties with direct chromophores). Csillag et al. (1993) best described the effect of multiple factors indirectly affecting soil spectra in their discussion on soil salinity, which can be considered a featureless property. They stated that "salinity is a complex phenomenon and therefore variation in the (soil) reflectance spectra cannot be attributed to a single (chromophoric) soil property." To get the most out of soil spectra, they examined chromophoric properties of organic matter and clay content, among others, and ran a principal component analysis to fully account for the salinity status culled from the soil reflectance spectra. Whiting et al. (2005) demonstrated that water content in soil can be quantitatively assessed by using spectral features away from the central water peak in the thermal IR (TIR) region. Demattê and Terra (2014) demonstrated that soil spectroscopy can assess different weathering levels of soils and consequently highlighted soil formation processes. These examples show that soil chromophores do not stand alone in the soil matrix and that spectral anomalies are often found in the soil environment. Examining all available information on a soil's population (spectral and chemical) is key to understanding soil reflectance spectra and their relationship to soil properties. To some extent, this suggests that soil spectra should be judged and examined with caution to obtain quantitative information about the soil, despite the fact that the chemical and physical mechanisms in the VIS–NIR–SWIR region, taken separately, are well understood.

It is interesting to note that the soil spectral community is growing rapidly worldwide, with concomitant development and use of commercial applications and methods to assess soil attributes from reflectance spectroscopy. A review of soil attributes that are already available from reflectance spectroscopy can be found in Malley et al. (2004) and Viscarra-Rossel et al. (2006), and others in the literature, such as Schwartz et al. (2013), who demonstrated the possibility of detecting small amounts of carbohydrates in soil using spectroscopy. Recently, NIRS has also been incorporated with IS data and used to map soil surface properties (see Section 8.10.2). Over the past decade, many users have discovered the potential of soil spectroscopy and much work has, therefore, been published. Malley et al. (2004) summarized all of the quantitative applications of soil spectroscopy prior to this surge in activity, and later, Viscarra-Rossel (2006) published a review that covered the applications which had been added by the growing community of soil spectral users. McBratney et al. (2003) also shed light on this technology through their pioneering work over the years. Brown et al. (2006) concluded that the NIRS technique has the potential to replace or augment standard soil characterization techniques, and based their conclusion in 3768 soil samples from the United States. In view of the growing soil spectral community, Viscarra Rossel (2009) generated an initiative (Soil World Spectral Group, http://groups.google.com/group/soil-spectroscopy) in which all members of this community were asked to join together and contribute to their local spectral libraries to generate a worldwide spectral library that would be accessible to all. Using these SSL and spectral data mining approaches can provide an interesting view of soil attributes worldwide. To that end, using an updated version of the world SSL, Viscarra Rossel et al. (2016) generated several world visualizations of soil attributes based on this database (Figure 8.16).

It is important to note that the analysis of soil spectral information in either point or image domains is a "big data" issue. Accordingly, extracting a spectral-based model to assess a soil attribute solely from spectroscopy requires a tool that will challenge the analysis in a simple way while assisting non-skilled personnel to obtain an optimal and reliable model in a short time. To that end, Carmon and Ben-Dor (2017) developed the PARACUDA-II® tool in which any spectral database (point or image) can be automatically analyzed by its optimal model. The tool executes various pre-processing treatments for the spectral data and runs several models using neural network and partial least squares regression (PLSR). The tool uses parallel high-computing resources to provide the five best models, which are then judged by the users for physical assignments and explanations. The tool reduces labor time and cost and enables better exploitation of the "big data" provided by soil spectra. For example, Kopačková et al. (2017) examined the potential of PARACUDA-II® to analyze organic matter in soils using combined optical and thermal spectral regions, demonstrating remarkable results. Recently, Gholizadeh et al. (2018) demonstrated that PARACUDA II(R) performs better than other well know data mining routines traditionally used to model soil carbon content from spectroscopy.

8.8 FACTORS AFFECTING SOIL REFLECTANCE

8.8.1 General

In the laboratory, where soil spectra are recorded under controlled conditions, it was thought that the data could be simply analyzed. This assumption was reexamined by Pimstein et al. (2010) who realized that even under controlled conditions, it is important to maintain a strict protocol and a standard internal procedure to enable comparisons between users. In the laboratory, samples are mixed and homogenized prior to measurement. In the field, natural soil is affected by different factors, such as dust accumulation and soil crust (both biogenic and physical) that prevent sensing the "real" soil surface that is measured in the laboratory. In the field, the soil might also be only air-dried, which can significantly affect the soil spectral signatures. Vegetation also plays a major role in masking the "real" soil signals, and can be classified as biospheric interference. The FAO Production Yearbook (1994) states that about 56% of the land area is covered by green vegetation, such as forest, pasture, and crops, whereas the rest is bare or covered by dry vegetation, snow, or urban development. Within the non-vegetated areas, only a portion of the soils are characterized by an unaltered surface layer (e.g., not tilled or not having undergone natural soil sealing) and hence, even partial sensing of the natural soil surface is difficult. This effect can be termed surface-coverage interference. Another important problem in acquiring accurate soil reflectance spectra from air and space is atmospheric interference. Electromagnetic energy interacts with atmospheric gas molecules and aerosol particles that may cause misinterpretation of the "soil spectrum" derived from airborne sensors. The measurement geometry (referred to earlier as BRDF) also plays a major role in affecting soil reflectance. It is also important to mention other factors that can change the soil spectrum by natural incidence, such as fire. This latter occurrence can significantly change the mineralogy of the upper surface of the soil and bias exact identification of several important soil attributes (e.g., clay content).

8.8.2 Biosphere

8.8.2.1 Higher Vegetation

Soil is a growth environment for green plants (natural and agricultural) and a sink for decomposing tissues of vegetation and fauna. Large parts of the world's soils are vegetated (green or dry), and the

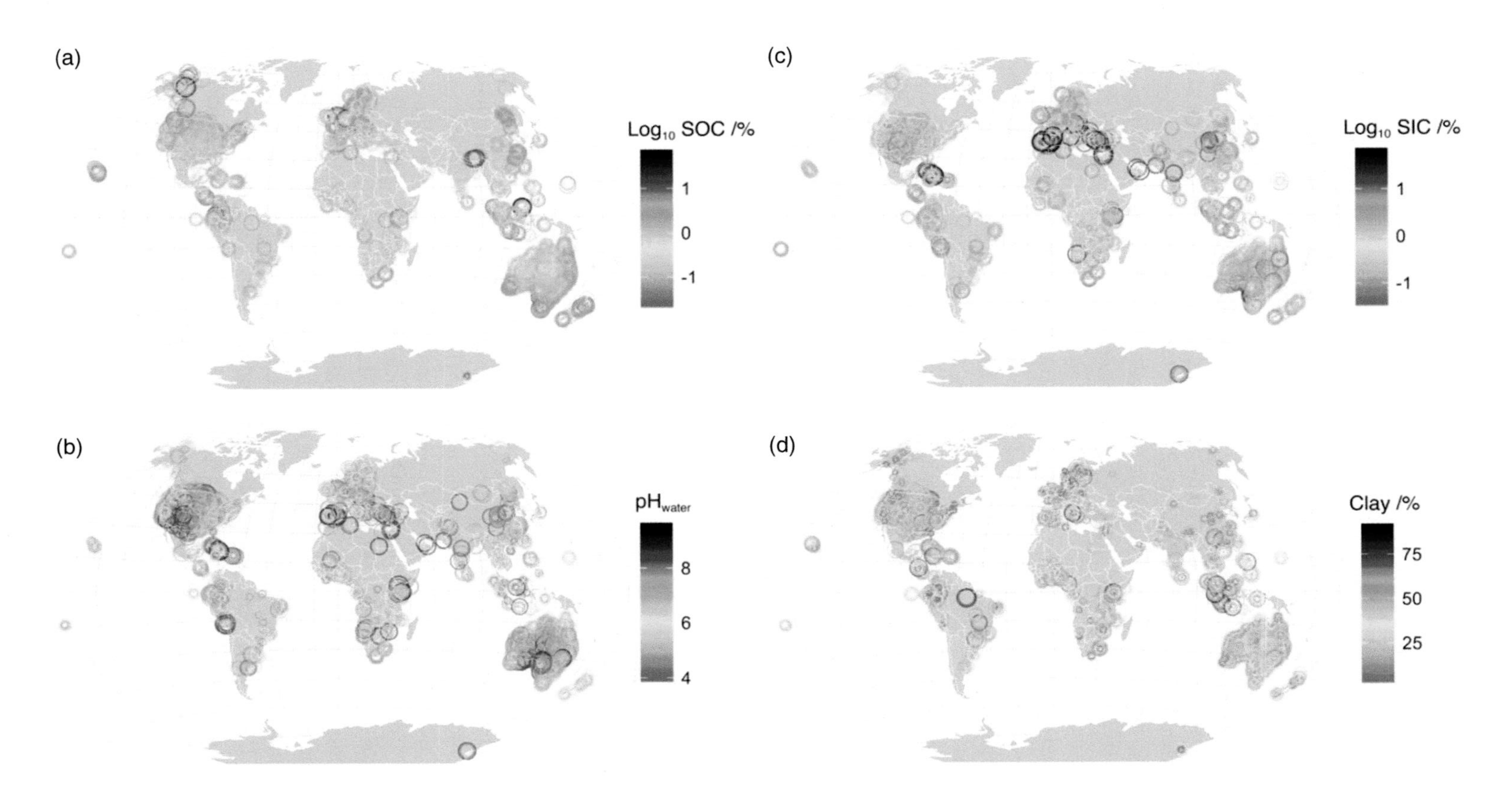

FIGURE 8.16 A global predictor maps as generated from the Global Soil Spectral Library. (After Viscarra Rossel R.A. et al. 2016. *Earth-Science Reviews* 155:198–230.)

problem of deriving soil spectra from the mixture of soil and vegetation signals is complex. Siegal and Goetz (1977) postulated that "the effect of naturally occurring vegetation on spectral reflectance of earth materials is a subject that deserves attention." At one extreme are situations in which the canopy cover is so dense that reflectance from soils is too difficult to interpret. Where the vegetation cover is only partial, a mixed signal from soil and vegetation is received and, to some extent, the chemical and physical components can be resolved (Murphy and Wadge, 1994). In a soil–vegetation mixture, nonlinear models are typically used to resolve the soil spectral components (Goetz, 1992; Ray and Murray, 1996). Otterman et al. (1995) noted that the relationship between the amount, type, and architecture of a vegetation cover and the reflectance properties of the underlying soil is an important issue (e.g., low-albedo soils are those most significantly affected by vegetation). The soil spectral region most strongly affected by green vegetation is 0.68–1.3 μm, due to the steep rise in reflectance that it causes (e.g., Ammer et al., 1991). Dry vegetation does not alter the spectrum in the VIS–NIR region, aside from changing the albedo, whereas in the SWIR region, significant vegetation affects are related to cellulose, lignin, and water. The low reflectance of green vegetation beyond 1.4 μm indicates that if a soil–vegetation mix exists, most of the spectral information will relate to rock and soil types (Siegal and Goetz, 1977). Two chromophores that exist in both plant and soil material—water and organic matter—can complicate interpretation of the spectra, particularly in the SWIR region. In the green vegetation–soil mix, liquid water of green and dry vegetation may overlap with the soil water forms. Signatures of lignin, cellulose, and protein can also significantly affect the soil components in the soil–vegetation mixture. Murphy and Wadge (1994) showed that in one case, although live vegetation had a greater impact on the SWIR region of the soil spectra, dead vegetation had a greater impact on the 2.2 μm absorption features (see, e.g., the reflectance spectra of pure organic matter in Figure 8.9). Murphy and Wadge (1994) concluded that dead vegetative tissue has a greater impact on soil spectra than live vegetation, and they suggested that workers consider this effect more seriously.

From a vegetation point of view, Tucker and Miller (1977) postulated that "remotely sensed data of vegetated surfaces could be analyzed more accurately if the contribution of the underlying soils spectra are known." Tueller (1987) and Smith et al. (1990) noted that it is difficult to extract vegetation information when its coverage consists of less than 30%–40%. The normalized differential vegetation index (NDVI) is a parameter that is commonly used to estimate the green vegetation cover in satellite and airborne data. The index, which is based on the normalized difference between the NIR and VIS reflectance values, is very sensitive to soil background, atmosphere, and sun-angle conditions (Qi et al., 1994). Based on that background, Huete (1988) developed a new index called soil-adjusted vegetation index (SAVI) which accounts for soil brightness and shadows, and Liu and Huete (1995) presented another index, the modified NDVI, which accounts for atmospheric attenuation as well. The SAVI has been shown to significantly minimize soil-related problems in nadir measurements over a variety of plant canopies and densities. Data from Huete (1988) and later from Huete et al. (1991) noted that that the optimal correction factor was achieved at the point at which dark and light SAVI values were the same. More precise models take into account the vegetation architecture (Otterman et al., 1995) or contain additional correction factors (Rondeaux et al., 1996). Richardson et al. (1975) developed three plant canopy models for extracting plant, soil, and shadow reflectance components of a cropped field. Using such models, Murphy and Wadge (1994) were able to separate soil and vegetation spectra using GER 63-channel IS data (Ben-Dor and Kruse, 1995). Roberts et al. (1993) also incorporated an unmixing procedure to discriminate vegetation, litter, and soils using AVIRIS 224-channel IS data (Vane et al., 1993) and were able to account for different soil types using a residual spectrum technique. It can be concluded that soil spectral signatures can be extracted from areas that are partially covered by decaying or live vegetation; however, caution should be exercised when assessing the "true" soil reflectance spectra in a soil—vegetation mixture. A recent paper by Ahmadian et al. (2016) demonstrated how soil line (SL) can be estimated from LANDSAT 8 data using different statistical approach and accordingly to separate between the vegetation and soil background spectral response. Guerschman et al. (2009) describes a method for resolving photosynthetic, non-photosynthetic, and soils across the ~2 million km^2 Australian tropical savanna zone with hyperspectral and multispectral imagery. A spectral library compiled from field campaigns in 2005 and 2006, together with three Earth Observing-1 (EO-1) Hyperion scenes acquired during the 2005 growing season were used to explore the spectral response space of these three endmembers. Accordingly, a linear unmixing approach was developed using the Normalized Difference Vegetation Index (NDVI) and the Cellulose Absorption Index (CAI).

8.8.2.2 Lower Vegetation

A major vegetation component in arid soil areas that is usually ignored by workers is the biogenic crust. However, this issue has been receiving more attention, and its importance to explaining anomalies in field soil spectra and satellite data has been demonstrated (Pinker and Karnieli, 1995). The biogenic crust consists mainly of lower, non-vascular (microphytic) plants covering the upper soil surface in a thin layer (Rogers and Langer, 1972; West, 1990). The microphytic community consists of mosses, lichens, algae, fungi, cyanobacteria, and bacteria. Each of these groups have pigments that are spectrally active in the VIS region under certain environmental conditions (Figure 8.17), and thus may mask soil features or, of bigger concern, may be interpreted as part of the soil signature (Karnieli et al., 1999).

O'Neill (1994) showed that some soil spectral features (between 2.08 and 2.10 μm) can be attributed to the microphytic crust and speculated that this was due to cellulose. Karnieli and Tsoar (1994) showed that the microphytic crust causes a decrease in the soil's overall albedo, leading to the false identification of anomalies in arid soils. The spectral response related to the biogenic crust permits linear mixing models, unlike the complex architecture of higher vegetation which requires nonlinear models to analyze mixed signals. During a long-term study, Zaady et al. (2007) demonstrated the validity of several indices for assessing the crust based on its recovery condition. It is, however, important to note that in addition to more basic research and consideration of the biogenic crust issues, more quantitative studies are still needed to fully account for its effects on soil spectra. Malec et al. (2015) have shown the spectral capability to assess soil, non-photosynthetic, and photosynthetic vegetation by spectral fractionating all endmember.

8.8.3 Lithosphere

8.8.3.1 Soil Cover and Crust

Soil seal and cover can be formed by different processes. The biogenic crust, as outlined above, is one example of such interference. Aeolian material and desert varnish are others. A lithosphere crust

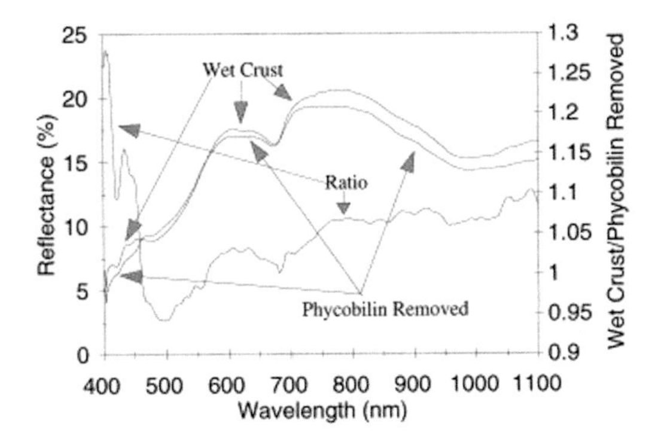

FIGURE 8.17 Spectra of wetted cyanobacteria crust, phycobilin extracted crust, and the ratio between them. It is shown that relative high reflectivity of the crust in the blue region is due to the spectral characteristics of the phycobilin pigments.

that is often found in soil is the "rain crust." This crust is formed by raindrops (Morin et al., 1981) which cause segregation of fine particle sizes at the surface of the soil. This can increase runoff and lead to soil erosion. The crusting effect is more pronounced in saline soils and well studied with relation to the mineralogical and chemical changes in the soil surface (Shainberg, 1992). The immediate observation after a rainstorm is enhanced "hue" and "value" of the soil color because of an increase in the fine fraction on the surface. One can assume that the reflectance spectrum of the "rain crust" will be totally different from that of the original soil, because it contains a greater clay fraction with a different textural component. In the literature, the issue of "rain crust" as it affects the spectral signature of soils has not received considerable attention; therefore, we encourage workers to consider this problem in their studies. The spectral changes observed at the soil surface are caused by changes in the soil's texture (clay-fraction enrichment), structure (from loose to compact), and roughness. Several innovative studies have shown a significant relationship between spectral information and the infiltration rate of water into the soil profile as measured in the laboratory (e.g., Ben-Dor et al., 2004a,b; Goldshlager et al., 2010). The next requisite step after the rain-simulation studies was to test the use of an airborne IS sensor to characterize a structural crust in the field. Ben-Dor et al. (2004a,b) used an IS sensor that covered the VIS–NIR region (AISA Eagle hyperspectral IS) over highly crusted areas in southern Israel. Using a spectral-based index (the normalized spectral area [NSA]), which is the area under the ratio curve generated by using a tested spectrum against a standard reference spectrum, they were able to generate a possible erosion hazard map of the soil area (Figure 8.18).

An important question based on that finding is whether a generic spectral model can describe the crust status rather than the kinetics of the formation process. A study by Goldshlager et al. (2010) showed that the spectral model used to predict crust status might be more robust than originally thought. By using four soils from Israel and three soils from the USA subjected to rain events in a rain simulator, promising results were obtained using a combined prediction equation for infiltration

FIGURE 8.18 The infiltration image of a Loess soil as generated on the basis of soil reflectance information and rain simulator measurements using AISA-Eagle sensor. (After Ben-Dor E. et al. 2005. *Geoderma* 131:1–21.)

rate with a root mean square error (RMSE) of cross-validation of 15.2% and a ratio of prediction to deviation (RPD) of 1.98. In another study, Chappell et al. (2005) also investigated the effect of soil structure changes due to rain and wind-tunnel events. Their results showed that the spectral information can shed more light on the soil composition and structure generated by these two factors (rain crust and aeolian abrasion).

8.8.3.2 Surface Affected by Fire

Fire affects a variety of physical and chemical soil properties, including loss of structure and soil organic matter, reduced soil porosity, increased pH, and alteration of the soil minerals. Soil degradation by fire plays a major role in the indirect cost of wildfires worldwide. Therefore, a large effort is being invested in the development of monitoring tools for fires (Veraverbeke et al., 2012). The interaction between heat and soil minerals has been extensively studied in the last decades by thermal-analysis methods such as differential thermal analysis, differential scanning calorimetric, and thermal gravimetric analysis during a controlled heating process. The integration of the knowledge gained from soil spectroscopy (both imaging and point) with that on changes in soil properties that might be induced by the heating process could serve as a powerful tool for studying post-fire consequences for the environment. In a study by Lugassi et al. (2010), a heterogeneous natural fire simulation that lasted 10 min at 190°C–450°C was applied to a uniform area of loess soil (Xeric Torriorthent). The burned soil (after cooling) exhibited a higher albedo along the entire solar illumination region relative to the original soil. Common spectral analysis, using continuum removal, showed that heating the soil caused a shift in the absorption bands of iron oxides and a gradual disappearance of illite absorption. The significant spectral and mineralogical changes with heat enabled these authors to relate the maximum soil surface temperatures measured with a net of thermocouples to the associated soil spectral reflectance (0.35–2.5 μm) measured after the soil had cooled down, and thus to understand the new mineralogy and degradation potential of the burned soil. Figure 8.19 shows the reconstructed soil temperature map as derived from the spectral information. An important finding was that the estimation of the maximum surface temperature using soil reflectance information was not affected by the thin layer of scorched organic matter present on the burned surface; however, the interference of larger amounts of ash on the soil surface in this analysis has not yet been tested and further investigations are, therefore, very important. We hypothesize that we can further use IS to assess and map fire areas and to generate a hazard map indicating the potential dangers and ways of preventing consequent soil degradation. Further study to this end is highly warranted.

8.8.3.3 Soil Moisture

Water is considered to be one of the most significant chromophores in the soil system (Idso et al., 1975; Stoner et al., 1980; Baumgardner et al., 1985; Hummel et al., 2001). Muller and Décamps (2000) determined that the impact of soil moisture on reflectance could be greater than the differences in reflectance between soil categories; hence, they stressed its importance in the previously discussed applications. Soil moisture affects the baseline height (albedo) as well as several spectral features across the entire spectral range (see e.g., Lekner and Dorf, 1988). Figure 8.20 (from Bishop et al., 1994) shows the features directly associated with the OH group in the water molecule (at 1.400 and 1.9 μm), and some that are indirectly associated with the strong OH group in the LWIR region (around 2.75–3.0 μm), which affect the lattice OH in clay (at 2.2 μm) and CO_3 in carbonates (at 2.33 μm).

Bowers and Hanks' (1965) frequently quoted work defined a loss of albedo and spread of absorptions in the 1.44 and 1.9 μm regions, whereas a clay-OH band at 2200 nm diminished with increasing water content. Figure 8.21 shows the loss in albedo from the VIS through SWIR regions (0.4–2.5 μm) with increasing moisture, increasing water band depths, and decreasing band depth of the 2.2-μm region (the OH-lattice band), described by Bowers and Hanks (1965) in a sample with high clay content (Whiting et al., 2004b).

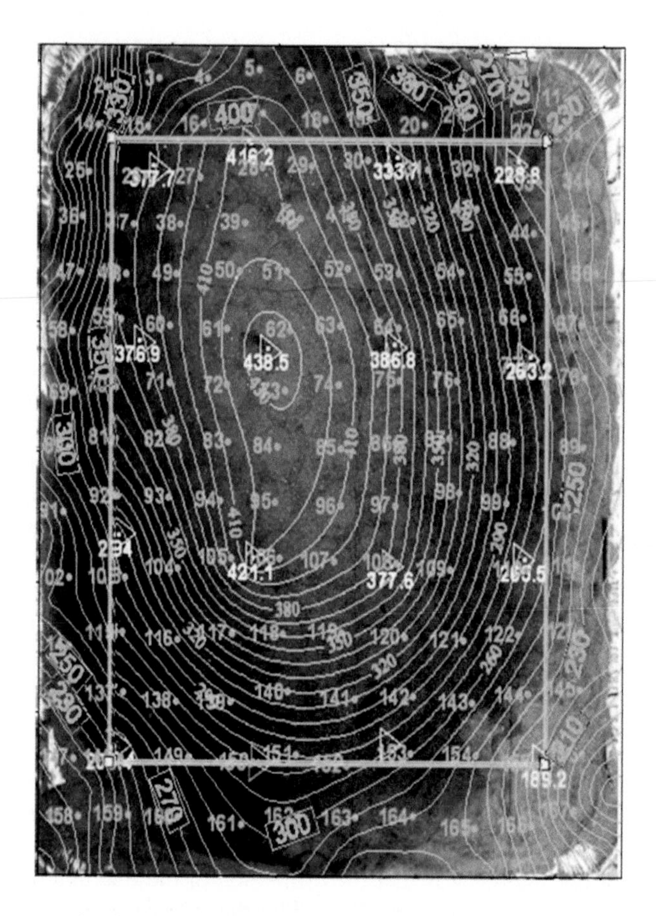

FIGURE 8.19 The extended isothermal maps of the burned soil based on spectral measurements and prediction models to reconstruct the surface temperature. (Taken from Lugassi R., Ben-Dor E. and Eshel G. 2010. *Remote Sensing of Environment* 114:322–331.).

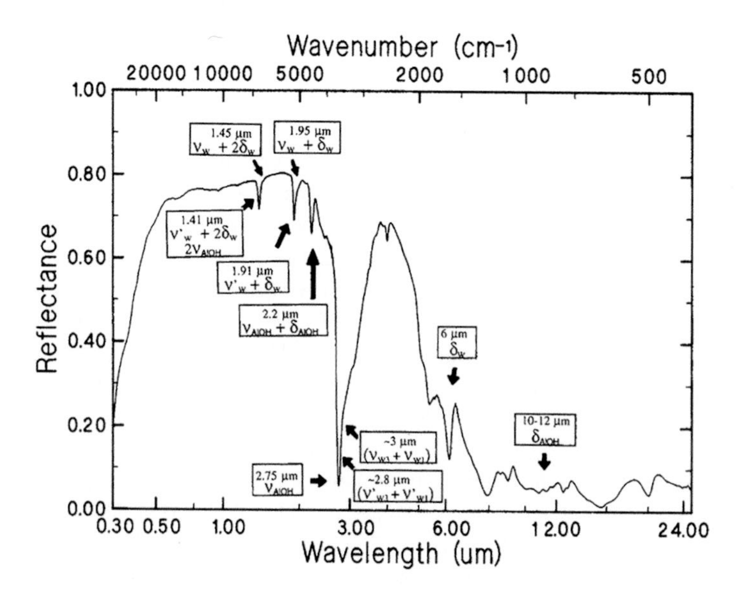

FIGURE 8.20 The bond stretching and bending vibrations in montmorillonite clay due water and aluminum hydroxyl. (After Bishop J.L., Pieters C.M. and Edwards J.O. 1994. *Clays and Clay Minerals* 42:702–716.).

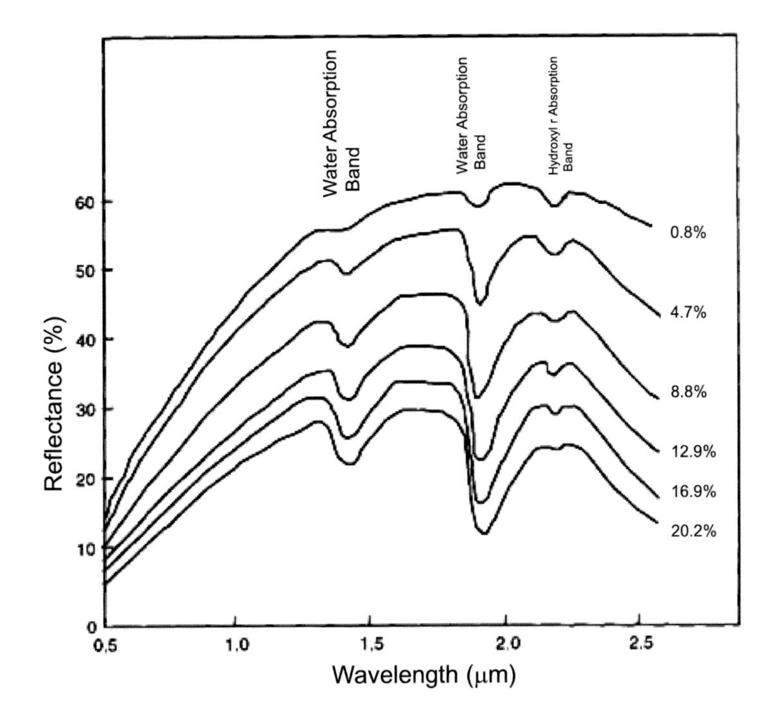

FIGURE 8.21 Spectral reflectance curves for Newton silt loam at various moisture contents. (After Bowers S. and Hanks R.J. 1965. *Soil Science* 100:130–138.)

Ben-Dor et al. (1998) noted a decrease in the 2.2 μm absorption feature in Ca-montmorillonite at various relative humidities (Figure 8.22). In the highly sensitive 1900-nm region, a water OH combination band showed excellent nonlinear fit to the increase in water content.

Demattê et al. (2006) applied this feature and others to practical field use. They found that the best interpretation of water content emerges when both dry and wet soil samples were spectrally measured. Dalal and Henry (1986) isolated the main differences in absorbance (log 1/reflectance), and found them to be related to the variation in moisture contents across the 1.1–2.50 μm SWIR spectral region. In this region, they determined that the correlation coefficient was greater than 0.92 when the bands at 1926, 1954, and 2.15 μm were used in a NIRS approach with gravimetric moisture ranging from air-dry (ca. 4%) to intermediate moisture (ca. 13%); the standard error of prediction was 0.58% water content with finely ground samples (<0.25 mm), and greater error for coarse ground samples (<2 mm). Lobell and Asner (2002) also showed that the SWIR region was much more sensitive than the VIS region for assessing soil moisture, and suggested that this region be used for practical purposes in the field. However, for the four soils that they examined, different exponential decay rates between the volumetric content and spectral parameters were noted. This suggests that their method cannot be considered generic, and that special attention must be given to every soil group examined. A robust spectral technique to estimate soil moisture content was developed by Whiting et al. (2004a) using a broad range of soils. In their approach, they isolated the influence of the fundamental water band from a sequence of gravimetric moisture contents in two distinctly different soils in the California Central Valley, USA (high clay content, low carbonate), and La Mancha, Spain (low clay content, high carbonate). They fitted an inverted Gaussian function centered on the assigned fundamental water absorption region at 2800 nm, beyond the limit of commonly used instruments, over the logarithmic soil spectrum continuum found with convex hull boundary points (Figure 8.23).

The area of the inverted function, the soil moisture Gaussian model (SMGM), accurately estimated the water content within a RMSE of 2.7% and coefficient of determination (R^2) of 0.94 among both soil regions, and a RMSE of 1.7%–2.5% with R^2 ranging from 0.94 to 0.98 when samples were

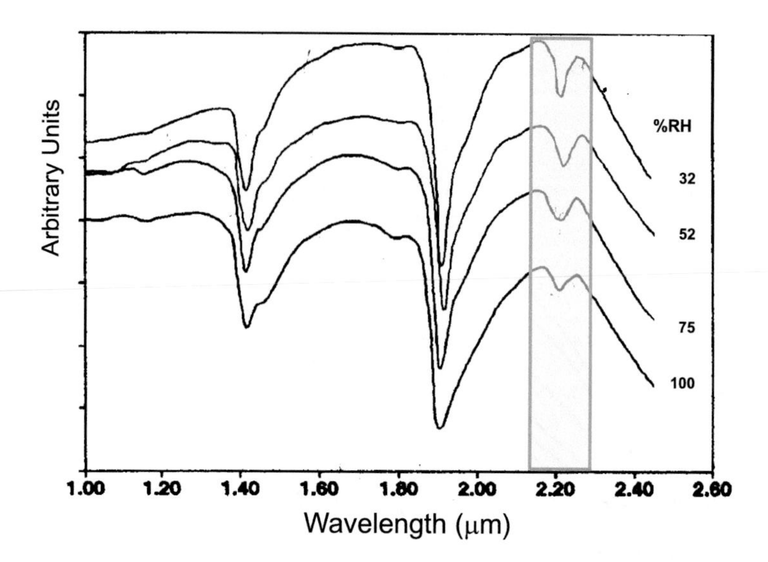

FIGURE 8.22 Reflectance spectra (in the NIR-SWIR region) of montrmorillonite, that were exposed to four different relative humidity atmosphere four more than twenty days. (After Ben-Dor E., Irons J.R. and Epema G.F. / Rencz A.N., Ryerson R.A. (Eds): *Manual of Remote Sensing: Remote Sensing for Earth Science*. pp. 111–187. 1999. Copyright Wiley-VCH Verlag GmbH & Co. KGaA.)

separated according to landform position (Spain) or salinity (USA). Using AVIRIS hyperspectral images of these soil regions in an air-dried state, they improved the abundance estimates by 10% of the regression mean by including the SMGM area as a parameter in the empirical determination of clay-OH and carbonate abundance based on the continuum-removed mineral-band depth (Whiting et al., 2005). This method is novel since it uses the entire SWIR region, it is not directly affected by the atmospheric water vapor, and it works in the real IS domain. Based on the above method,

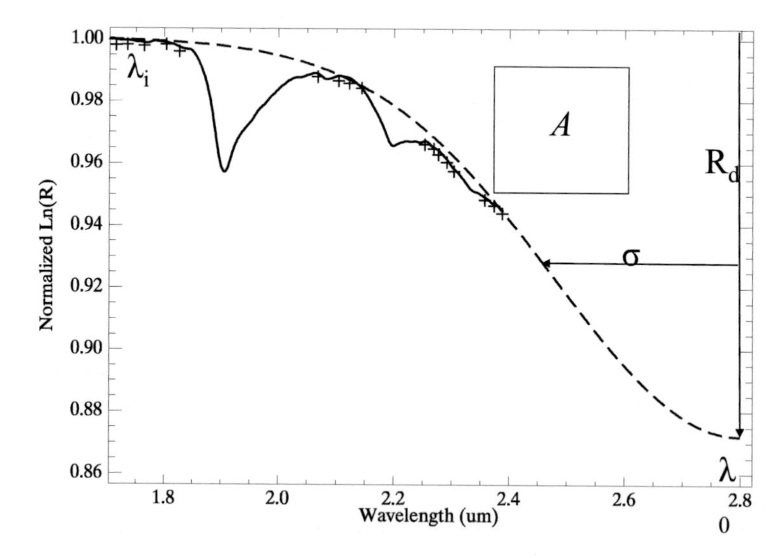

FIGURE 8.23 Inverted Gaussian function is fitted to the fundamental water absorption center at 2800 nm to the convex hull boundary points of a logarithmic transformed SWIR region, where λ_i, wavelength value at maximum reflectance; λ, wavelength value of 2800; σ, distance at inflexion; R_d, depth of Gaussian, and *A*, area above the continuum. (After Whiting M.L., Li L. and Ustin S.L. 2004a. *Remote Sensing of Environment* 89:535–552.)

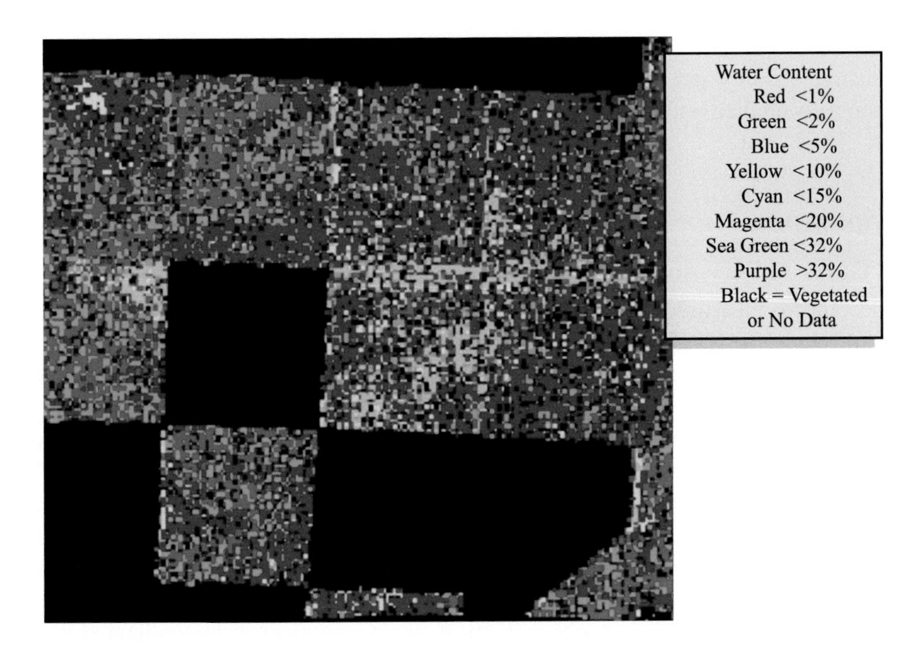

FIGURE 8.24 Surface water content (gravimetric) from AVIRIS data (3 May 2003, near Lemoore, California) as estimated with the SMGM. (After Whiting M.L., Li L. and Ustin S.L. 2004b. Correcting mineral abundance estimates for soil moisture. In: 13th *Annual JPL Airborne Earth Science Workshop*, Green R.O. (ed.), Pasadena, California, March 30–April 3, 2004, JPL Publication 05-3-1.).

the authors were also able to present a processed AVIRIS image that provides soil moisture content (Figure 8.24; Whiting et al., 2004b).

Working on the similar goal of developing a novel approach to estimating soil moisture content solely from spectral readings not affected by atmospheric attenuations, Haubrock et al. (2008a,b) developed and successfully tested a new model for determining soil moisture by means of remote-sensing techniques. This model was based on combining multitemporal high-spatial-resolution IS observations with field and laboratory spectral studies, along with hydrological measurements. This method, termed the normalized soil moisture index (NSMI), was tested for the best spectral prediction of soil moisture content in the field using the 400–2450-nm spectral region. R^2 was 0.61 (up to 0.71) for natural field samples, taking into account the influence of different environmental factors: heterogeneous soil types and related field moisture content, variable soil water profiles, and the presence of soil crust and vegetation cover. Moisture is an integral part of a soil's reflectance, and future modeling attempts may support the contribution of soil background to vegetation radiative transfer models. Jacquemoud et al. (1992), using a modification of Hapke's single-scattering albedo model (Hapke, 1981a, 1981b), separated the surface geometry component in a radiative transfer model for soil reflectance, SOILSPECT. They also noticed "quasihomothetic variations" in the VIS–NIR and SWIR regions with moisture content, though the moisture dataset was limited (Jacquemoud et al., 1992). As Pinty et al. (1998) noted, investigations to account for this decline in albedo may help resolve this modeling problem. This was further confirmed by Lesaignoux et al. (2009) who demonstrated the moisture impact on soil reflectance. Recently, some works have been applied to predict how the dry soil spectrum can be predicted from its wet spectrum. To that end for example, demonstrated that in several soil types from Israel it is possible to predict the dry spectrum from the wet one by using a dedicated SSL for the moisturized case. This success paved the road for using the existing SSLs (dry) databases for assessing soil attributes despite the strong masking effect of the water molecules on the chromophores' spectral response. In summary, it can be concluded that soil surface water content should be estimated with caution, and due to its effect on other soil components, its spectral absorptions require proper attention. Soil moisture is an important property, not only for assessing the water content available for plant utilization, but also for assessing the direct exchange of soil water with the atmosphere (i.e., evaporation). This innovative direction has not yet been fully studied and developed for use in IS, though it appears very promising and highly necessary.

8.8.4 Atmosphere

8.8.4.1 Gases and Aerosols

When reflectance is measured from air- or spaceborne sensors, the atmosphere's gases and aerosols play a major role in the VIS–NIR–SWIR spectral regions and thus may attenuate soil reflectance. Absorption and scattering of electromagnetic radiation takes place across these regions. Water vapor, oxygen, carbon dioxide, methane, ozone, nitrous oxides, and carbon monoxide are the spectrally active components across approximately half of the VIS–NIR–SWIR regions. Some good models for retrieving gas and aerosol interference exist and are widely used by many workers (e.g., LOWTRAN-7 [Kneizys et al., 1988], 5S & 6S codes [First and Second Simulation of Satellite Signal in the Solar Spectrum; Tanre et al., 1986], and ATOCR [Richter and Schläpfer, 2002]). Although a discussion of these models is beyond the scope of this chapter (see, e.g., Ben-Dor et al., 2004a,b) one should be aware that in many cases, the models do not perfectly remove all atmospheric attenuation and may alter the soil spectrum. This problem is most likely to appear in hyperchannel data, where discrete absorption features are more pronounced relative to the multichannel data, which practically averages small features into one wide value.

Figure 8.25 illustrates the spectral regions under which atmospheric attenuation can affect the soil spectrum. This figure shows the reflectance spectrum of an E-7 soil from Israel (Haploxeralf, taken from TAU spectral library) overlain on its simulated (soil) radiance as calculated by MODTRAN. The latter is normalized to the Plank sun function on top, at the atmosphere, to illustrate only atmospheric transmittance. As the atmospheric attenuations remain, the most affected spectral regions can be clearly seen. The VIS region is affected by aerosol scattering (monotonous decay from 0.4 to 0.8 μm) and absorption of ozone (around 0.6 μm), water vapor (0.73, 0.82 μm), and oxygen (0.76 μm). The

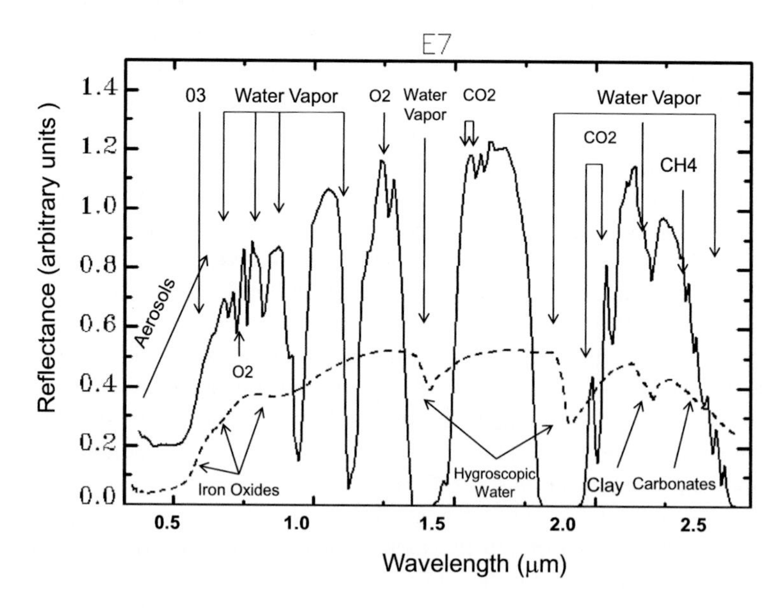

FIGURE 8.25 An simulated AVIRIS spectrum of a Haploxeralf soil from semi arid environment from Israel (E7 from TAU spectral library), after removing the solar effect. Across the spectrum the major gas absorption's absorption features are annotated to show area where atmospheric attenuation's might overlap with soil features.

NIR–SWIR regions are affected by absorption of water vapor (0.94, 1.14, 1.38, 1.88 μm), oxygen (around 1.3 μm), carbon dioxides (around 1.56, 2.01, 2.08 μm), and methane (2.35 μm).

Also seen are the absorption peaks of the soil chromophores at 2.33 μm (carbonates), 2.2 μm (clay), 1.9 and 1.4 μm (hygroscopic water), and 0.5, 0.6, and 0.9 μm (iron oxides) that are overlapped by the above atmospheric chromophores. As already mentioned, even weak spectral features in the soil spectrum can contain very useful information. Therefore, great caution must be taken before applying any quantitative models to soil reflectance spectra derived from air- or spaceborne hyperchannel sensors. Validation of the (atmospherically) corrected data is an essential step in ensuring that the reflectance spectrum consists of reliable soil information.

8.9 SOIL SPECTRAL LIBRARY (SSL)

8.9.1 The Importance of SSL

Terrestrial spectral libraries are important databases for the analysis of hyperspectral remote-sensing information under optimal conditions from all domains, as they provide the spectral features of rocks minerals and even manmade objects (e.g. Herold et al., 2004). These features can then be used as fingerprints to identify unknown (sometimes mixed) spectral information captured by either point or image spectroscopy. In general, spectral libraries are a collection of material and their spectral data, followed up with solid alternative identification analyses such as x-ray diffraction or wet-chemistry analyses. Several comprehensive spectral libraries are available today in the literature and on the web for rocks, minerals, and urban materials (e.g., USGS, JPL; Herold et al., 2002). Building soil spectral libraries is neither possible nor pedagogical to start from remote-sensing platforms. These platforms are relatively far from the target and are hindered by several issues, such as climate, atmospheric degradation, distance from the target, field issues, and others. The spectral measurement under laboratory condition minimizes those variations and promises that the obtained spectral signatures belongs only to the soil medium. The first soil spectral databanks consisted mostly of soils from the USA and some samples from Brazil (Baumgardner et al., 1985). In 1992, Brazil established a large databank (for one state), which was then published by Formaggio et al. (1996). However, its importance was only realized in 2002 by Shepherd and Walsh (2002). The point is that, for many years, the idea of soil spectral libraries was neglected, mostly due to technological issues, cost, and difficult access to sensors. On the other hand, in the first decade of the twenty-first century, the technology improved, and new sensors appeared with powerful optics and software, driving a soil-sensing revolution ahead (Ben-Dor et al., 2009).

Understanding the importance of spectral libraries for the soil proxy and for the high-spectral-resolution remote-sensing arenas that emerged in the early twenty-first century, the soil science community, which deals with both spectroscopy and remote-sensing data, has turned their attention to formulating an SSL archive. As previously discussed, the first SSL archive was brought to the attention of the community in the early 1980s by Baumgardner et al. (1985), who published the first "Soil Reflectance Atlas" consisting of hundreds of soils from the USA and Brazil. Thereafter, and based on the growing interest in the soil spectroscopy field, more SSLs were generated, first on a regional scale (e.g., Brazil—Bellinaso et al., 2010; Czech Republic—Brodský et al., 2011; France—Gogé et al., 2012; Denmark—Knadel et al., 2012; Mozambique—Cambule et al., 2012; Spain and Australia—Viscarra Rossel and Webster, 2012; China—Zhou Shi et al., 2014a,b) and then on a continent scale (LUCAS) and even global coverage (Viscarra Rossel et al., 2016).

In general, the SSL is different from other spectral libraries as it consists of a collection of soil material that has been sampled from different horizons with detailed metadata consisting of exact location, pedogenic characterization, and measurement protocol in both field and laboratory. In addition, the SSL must include detailed chemical and physical data from well-accepted wet laboratory methods. The common soil attributes in SSLs are: mechanical composition (clay, silt, and sand content), organic matter, carbonate, iron oxides, hygroscopic moisture, and specific surface area.

Some SSLs also present more soil attributes, such as heavy metals, x-ray diffraction descriptions, and micro- and macro-elements (e.g., Kemper and Sommer, 2002).

With the rise in availability of dual-use portable spectrometers, more users in the soil disciplines are entering into spectral-based remote sensing of soils. As a result, many soil spectral databases are being constructed today in the laboratory and shared and used to assist other domains (i.e., http://www.isric.org/data/icrafisric-spectral-library). In the last decade, there has been a growing effort to generate more local SSLs by professional users who then go on to contribute and exchange their data internationally to form a worldwide SSL archive (Viscarra Rossel, 2009). Generating SSLs in a large-scale domain is a growing and important mission that has been adopted by many organizations and entities, at both the national (e.g., Knadel et al., 2012) and continental (e.g., the LUCAS SSL in Europe; Tóth et al., 2013) levels, whereas other initiatives have understood its importance and are beginning to establish local and national SSLs (e.g., Israel). To that end, the first publication using a SSL with global coverage was presented by Brown et al. (2006). The world SSL was extended by Viscarra Rossel et al. (2016) after their first attempt in 2006. Many groups from different countries contributed to their local SSLs and these were merged to yield a global view of several soil attributes using proxy analysis. The world spectral library idea, more than simply an attempt to gather spectral information on the world's soils, was an important step toward establishing a standard protocol and quality indicators that would be accepted by all members of the growing soil spectral community. The new world SSL (Viscarra Rossel et al., 2016) provides standards and a protocol in its appendix. It is important to mention that SSL initiatives such as those being carried out in Africa (Shepherd and Walsh, 2002), Brazil (Romero et al., 2018), China (Shi et al., 2014a,b), and recently, under the GEO-CRADLE H202 project (http://geocradle.eu/en) for Mediterranean countries, still require standardization. As it is anticipated that there will be an acceleration in this direction, a better way to ensure accurate SSL sharing is strongly needed. In 2013, Ben-Dor et al. suggested the ISS idea which is based on using a common and agreed upon stable "soil sample" as the ISS and adjusting all measurements to a "mother" spectrometer at the CSIRO in Perth, Australia. This was based on Pimstein et al.'s (2010) idea to adopt the internal standards process from the wet chemistry discipline. Later, in 2013, Ben-Dor et al. found ideal ISS samples from Western Australia—namely Lucky Bay (LB) and Wylie Bay (WB)—both quartzic sand dunes that are being dispersed to the global spectral community as a common ISS, free of charge.

Today, there is a growing effort to use the ISS method for better exploitation of SSL data by enabling effective and accurate sharing of SSLs from different sources. Kopačková and Ben-Dor (2016) demonstrated the ability of the ISS procedure to align soils from Israel and Australia which had been measured by two different spectrometers and geometries (ASD and Spectral Evolution). In another recent study, Romero et al. (2018) demonstrated the practical utilization of the ISS approach to establish the Brazilian SSL, in which several countrywide laboratories are involved. Recently, Gholizadeh et al. (2017) demonstrated the LB—ISS improve chemometric analysis of Czech soils. The power of the global SSL is demonstrated by Viscarra Rossel et al. (2016) who applied chemometrics models to extract predictors for several soil attributes in a global sale as presented in Figure 8.16.

In summary, it can be said that the main objectives of a SSL are to: (1) be a databank for patterns (spectral signatures of the target and digital numbers); (2) be the basis for the generation of proxy models to quantify soils; (3) indicate patterns of different soils with specific properties; and (4) validate and shed light on the remote sensing of soils from any sensors (assuming that hyperspectral information can be resampled into super and multispectral domains). With a spectral library, the user can relate several properties with soil property quantification, classification, mapping, and monitoring, such as: soil degradation, moisture content, chemistry composition, pedogenesis, geochemistry, physics, microbiology, mineralogy, mining, and others. Technology, associated with years of research and a strong background, has brought us to this point. It is now a matter of grabbing the opportunity and moving forward.

8.10 SOIL REFLECTANCE AND REMOTE SENSING

8.10.1 General

A comprehensive description of soil spectral remote sensing can be found in Ben-Dor et al. (2007). In general, many studies have been conducted with the intention of classifying soils and their properties using optical sensors onboard orbital satellites, such as Landsat MSS & TM, SPOT, and NOAA-AVHRR (e.g., Cipra et al., 1980; Frazier and Cheng, 1989; Kierein-Young and Kruse, 1989; Agdu et al., 1990; Moran et al., 1992). Qualitative classification approaches have traditionally been used to analyze multichannel data in cases where the spectral information was relatively scarce. Nevertheless, it has also been possible to obtain useable sets of information on soil type, soil degradation, and soil conditions from "broad" channel sensors by applying sophisticated classification approaches (Price, 1990; Ben-Dor and Banin, 1995c). Over the years, soil spectra have been collected and analyzed in the laboratory both quantitatively and qualitatively, by many workers (e.g., Latz et al., 1981; Price, 1995). Although use of these libraries involves many limitations, it is understood that the spectral domain is very important for soil mapping. The last 25 years have seen the development of a new remote-sensing technique, termed Hyperspectral Remote Sensing (HSR, also termed IS in this chapter). This is an advanced tool that provides high-spectral-resolution data in an image, with the aim of providing near-laboratory quality reflectance or emittance for each individual picture element (pixel) from far or near distances (Vane et al., 1984). This information enables the identification of objects based on the spectral absorption features of their chromophores and has found many uses in terrestrial and marine applications (Clark and Roush, 1984; Vane et al., 1984; Dekker et al., 2001). Figure 8.26 illustrates the concept, In which the spectral information of a given pixel shows a new dimension that cannot be obtained by traditional point spectroscopy, air photography, or other multiband images. IS can thus be described as an "expert" geographical information system (GIS) in which layers are built on a pixel-by-pixel basis, rather than on a selected group of points (McBratney et al., 2003). This enables spatial recognition of the phenomenon in question with a precise spatial view and use of the traditional GIS-interpolation technique in precise thematic images. Since the spatial–spectral-based view may provide better information than viewing either

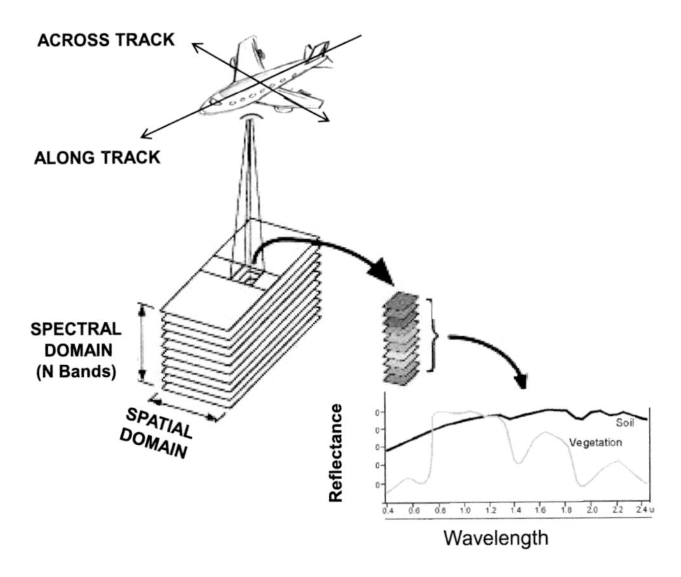

FIGURE 8.26 The Image Spectroscopy (IS) concept: An IS image is composed of N spectral bands that generates a spectral cube. For each pixel a spectrum can be extracted representing a spectral foot print of the object. (Taken from www.ccrs.nrcan.gc.ca/ccrs/misc/issues/hyperview_e.html.)

the spatial or spectral views separately, IS serves as a powerful and promising tool in the modern remote-sensing arena. Since 1983, when the first airborne IS sensor (AIS; Vane et al., 1984) ushered in the IS era, this technique has been used mostly for geology, water, and vegetation applications. It appears that its application does not yet extend to soils because these present a complex matrix: it is only recently, with the advent of better signal-to-noise sensors, the manufacture of less expensive IS sensors, and the development of many soil (point) spectroscopy applications, that soil-IS activity has progressed somewhat.

8.10.2 IS Applications in Soils: From the Ground to Airborne Platforms

Spaceborne IS sensors are currently being built, and soon to be hopefully launched (Staenz et al., 2013). For example: the EnMAP (Environmental Mapping and Analysis Program, Stuffler et al., 2009) from Germany with ~240 spectral bands covering the VIS–NIR–SWIR region at a pixel size of 30 m and a high signal-to-noise ratio (Guanter et al., 2015) launch 2020; the PRISMA (PRecursore IperSpettrale della Missione Applicativa, Hyperspectral Precursor of the Application Mission) from Italy with ~250 bands and a pixel size of 30 m (Pignatti et al., 2013) launch 2018; and the HISUI (Hyperspectral Imager Suite) (Matsunaga et al., 2017), Japan initiative, at 30 m, launch 2019 for a 3 years deployment on ISS (International Space Station); and the SHALOM (Spaceborne Hyperspectral Applicative Land and Ocean Mission) (Feingresh and Ben Dor 2015), an Italy-Israel initiative, at 10 m, launch 2022. Furthermore, more global missions are in the design phase such as the HyspIRI from the USA (pixel size of 60 m), (Lee et al., 2015), plus EMIT on the ISS, and the candidate Sentinel-10/CHIME mission (Copernicus Hyperspectral Imaging Mission for the Environment) (ESA 2018) which is in phase A/B. Also, the ECOSTRESS (LWIR) already deployed on the ISS and DESIS (VIS-NIR) sensors assumed to be deployed in summer 2018 where others have been emerged and are planed to be places in orbit soon (e.g. PRISMA).

The upcoming availability of these high signal-to-noise ratio spaceborne IS data is expected to provide a major step toward operational quantitative mapping and monitoring of soil surfaces on large scales. Indeed, these instruments will be able to provide global spectroscopic data for mapping quantitative soil properties at low cost, and could enable accurate assessment and monitoring of soil erosion such as, for example, carbon loss or increase under degradation processes or recultivation effects in soils, based on routine delivery on regional to global scales (spatial resolution from 10 to 30 m) of soil maps at a reasonable temporal resolution. Nonetheless, advances are still needed to fully develop IS products for soils that can support, in a credible manner, global digital mapping and monitoring of soils. As already discussed, soil spectroscopy based on laboratory, field, and airborne data has been shown to be an adequate method for mapping the spatial distribution of soil surface properties, moving into the quantitative domain based on multivariate statistical methodologies, as long as soil chromophores are present, the soils are well exposed and homogeneously distributed, and local ground data are available. In view of the current limitations for determining soil properties in terms of methods, spatial extent, and observational and environmental conditions, a main topic of investigation in the framework of preparing for future hyperspectral satellite programs is the potential of these sensors and their expected accuracy for the delivery of soil products. Indeed, the real accuracy that can be expected from the upcoming satellite sensors and the operability of the predictions linked with harmonious methodologies for applications at regional to global scales have yet to be demonstrated.

The first paper demonstrating the capability of adopting airborne IS for the soil proxy approach was by Ben-Dor et al. (2002). They demonstrated a significant potential to extract the following soil attributes in agricultural fields from the DAIS-7915 (Strobl et al., 1997) sensor: salinity, organic matter, and hygroscopic moisture. Figure 8.27 shows the quantitative maps generated from the spectral-based model of these attributes.

Several recent studies have examined the potential of upcoming hyperspectral missions for soil mapping using different extraction methods, such as PLSR, and spectral-feature techniques based

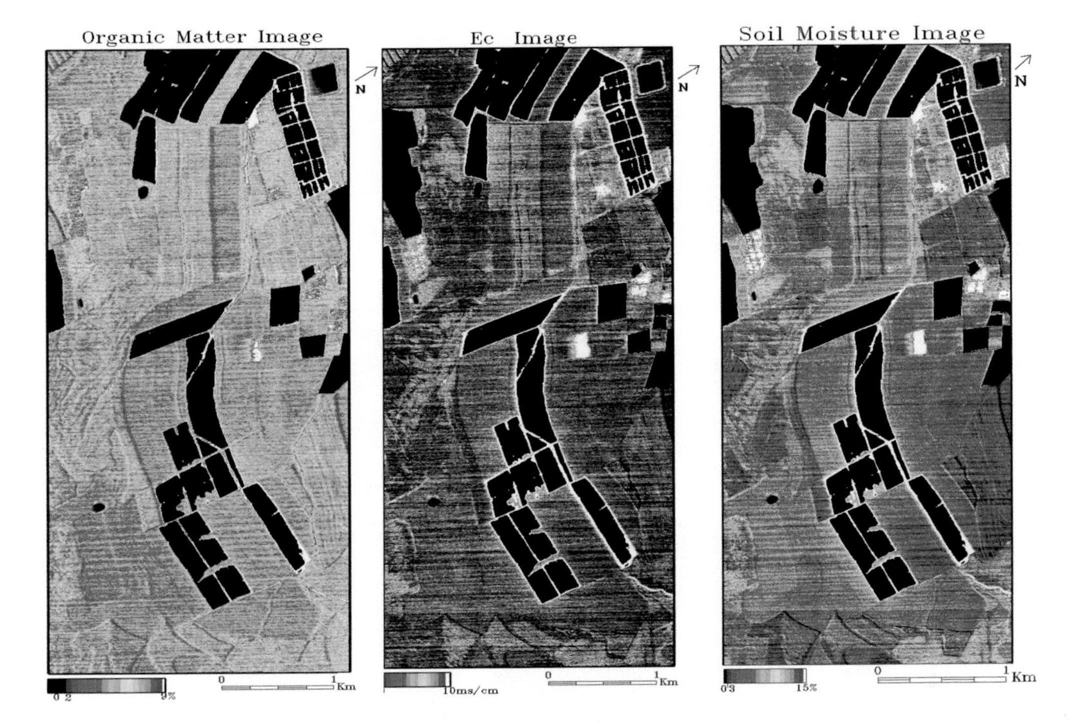

FIGURE 8.27 A map showing three soil attributes: Soil organic matter, Soil salinity (measured as Electrical Conductivity EC) and Soil hygroscopic moisture in agriculture field in Israel . The results are derived by applying chemometric models to the reflectance image for these three attributes. The dark areas are vegetation masks. (Taken from Ben Dor E. et al. 2002. *International Journal of Remote Sensing* 23:1043–1062.)

on the EnMAP end-to-end simulator (Chabrillat et al., 2014; Steinberg et al., 2016), the PRISMA simulation and spectral resampling added to noise degradation for HyspIRI and EnMAP simulations (Castaldi et al., 2016), or the different spatial resampling added to atmospheric noise degradation (Gomez et al., 2015). As an example, Figure 8.28 shows a simulation of upcoming EnMAP soil products compared to airborne mapping for the determination of clay, iron oxide, and soil organic carbon contents.

Steinberg et al. (2016) showed that EnMAP data-derived soil models using semi-operational methods (auto-PLSR) can predict iron oxide, clay, and Soil Organic Carbon (SOC) contents with an R^2 of 0.53–0.67 compared to airborne hyperspectral imagery with R^2 of 0.64–0.74. The spatial distribution of the soil properties was generally coherent between the simulated EnMAP and the airborne mapping, demonstrating the potential of the upcoming sensor system for the mapping of surface soil properties on a larger scale. Furthermore, Castaldi et al. (2016) clearly demonstrated an improvement in the accuracy of estimated soil variables from bare soil imagery using forthcoming hyperspectral imagers, as compared to current-generation sensors such as ALI, Landsat8, and Sentinel-2. Overall, further research will need to focus on estimation accuracies for the retrieval of soil variables using, once they are available, real data from next-generation hyperspectral satellite imagers.

8.10.3 Limitations of IS for Soil Mapping

It appears that the limited number of studies in soil-IS applications (relative to other disciplines such as geology, vegetation, and water) is due to the difficulties encountered on the journey from point spectroscopy to a cognitive (imaging) spectral view of soils. Whereas the number of IS sensors is on the rise and they are, therefore, becoming less costly, the orbital IS sensors are still not operating as Seeking

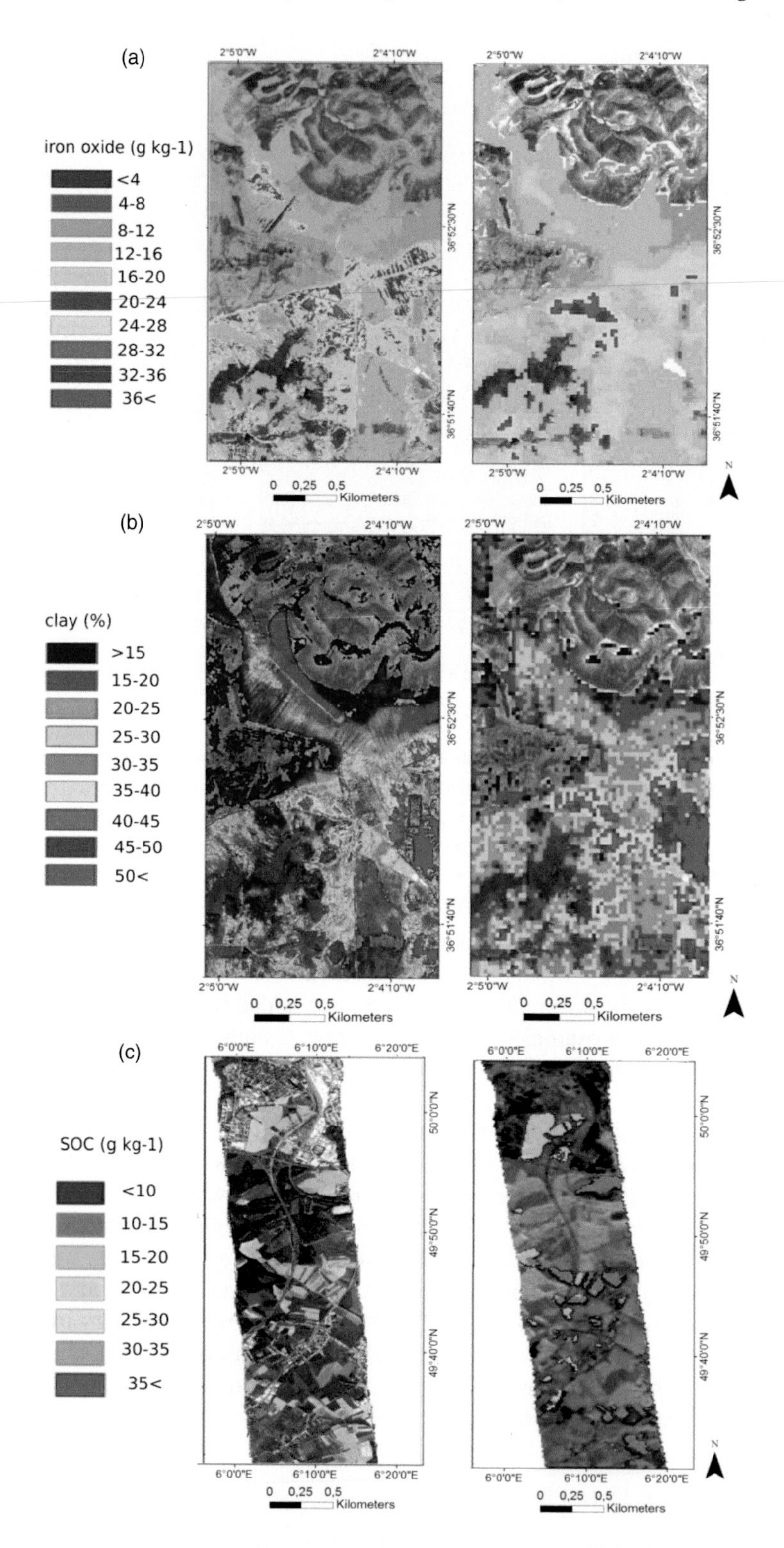

FIGURE 8.28 Maps of predicted soil properties: (a) Iron oxide, Cabo de Gata test site; (b) Clay content, Cabo de Gata test site; and (c) SOC content, Luxembourg test site. Left: Airborne imagery (a,b) HyMAP (4.5 m); and (c) AHS (2.6 m). Right: Spaceborne simulated EnMAP soil products based on EnMAP end-to-end simulated images (30 m). (Steinberg A. et al. 2016. *Remote Sensing* 8(7):613).

for more detailed information, Demattê and Terra (2014) observed important relationship between soil spectral properties and pedogenetic alterations, showing its relevance on this matter expected. The orbital sensor, with frequent temporal coverage of the entire globe, is important for monitoring changes in the soil environment. Moreover, it enables tracking after vegetation-free areas (soil) in seasons when photosynthetic activity is not relevant (Guerschman et al., 2009). To that end, the idea to track after favorable temporal coverage provided by orbital sensors to find free soil pixels that are temporally covered by vegetation is interesting and reached important results by Demattê et al. (2016). The authors developed a technique to detect bare soils that are seasonally covered by vegetation using time series analysis of satellite multispectral images (Demattê et al., 2018). The idea was to find bare soil using the temporal information, which allow to "see" soils as a continuum and thus, reach a deterministic and quantified information of soil in each pixel. Despite this, a limitation is that IS capability for soils has not yet been fully recognized by many end users. The relatively low signal-to-noise ratio, atmospheric attenuation, varying FOV for every pixel, spectral instability, low integration time for a given pixel, spectral mixing problems, optical shifts from one pixel to another, and BRDF effects are only a few of the problems that still call for practical solutions. The relatively low spatial resolution, mostly from the space domain, with low geolocation accuracy may hinder accurate mapping with the IS approach, but with advancing technology, this is expected to improve (e.g. the SHALOM orbital sensor intend to provide a pixel size of about 10 m). IS data require fine pre-processing of the raw data prior to any advanced correction. Highly skilled personnel are needed to meet all of these requirements, with experience, knowledge, and a well-equipped infrastructure (software and field measurements). In addition, it should be remembered that the soil surface is not always flat, smooth, or homogeneous and, therefore, sample preparation (as performed in the laboratory) is almost impossible. This leads to textural problems such as variations in particle size, adjacency, BRDF effects, and the need to develop methodologies that will represent a pixel on the ground and in the IS sensor from both the chemical and spectral perspectives. In the face of all of these obstacles, one should remember that although optical remote sensing does not go beneath the surface, it can eventually produce precise soil (profile) maps, by combining more electromagnetic methods with IS, and by using smart approaches, such as the spectral penetrating-probe assembly presented by Ben-Dor et al. (2008a,b). Another important problem is the validation-stage assessment. Since the pixel size cannot really represent point measurements, at least 3×3 pixels have to be averaged for both true spectral ground measurements and chemical analysis. Since a field may generate a non-homogeneous presentation, this can cause problems that, if not estimated properly, will bias the final "spectral-based" map. Within the validation stage, it should be pointed out that field spectral measurements suffer from systematics and non-systematic instabilities that can cause problems when comparing IS and field data. Problems such as weak radiation, invisible cirrus clouds during the field measurements, and varying geometry by different operators may be significant. Relying on laboratory spectral measurements (e.g., SSL) of selected soils that were sampled in the field and brought to the laboratory may also be problematic. It is extremely important that the undisturbed soil surface be measured for validation. Sampling the soil may harm its surface properties (e.g., physical crust, hydrophobicity, dust accumulation, surface structure). This may prevent a proper way to exploit the SSLs available today. Although improvement to that end has been demonstrated by Ben-Dor et al. (2017) in developing the SoilPRO® apparatuses to measure soil in the field under laboratory conditions, scaling up laboratory-based spectral data to airborne data still requires further attention. A special attention has to be given to the field moisture condition that may render the extraction of the main chromophores in the soil.

8.11 SPECTRAL PROXYMATION OF SOIL USING POINT AND IMAGE DOMAINS: FUTURE NOTES

For soil applications, air- and spaceborne imaging spectrometers should consist of a reasonable number of spectral channels across the entire VIS–NIR–SWIR region which will cover the spectrally active regions of all chromophores with a reasonable bandwidth and sampling interval. Price (1990) believes that a relatively low number of spectral channels (15–20) with bandwidths of 0.04–0.10 μm and high signal-to-noise ratio promise better remote-sensing capabilities for soils. Goetz and Herring (1989) prefer more spectral channels (192) but a wider bandwidth (about 10 nm) to permit diagnostic evaluation of specific features across the entire VIS–NIR–SWIR region. We believe that for quantitative analysis of soil spectra, the optimal bandwidth and number of channels may be strongly dependent upon the soil population and the property being examined. There is no doubt, however, that high signal-to-noise ratio is a crucial factor in quantitative analysis of soil spectra derived from both air and space measurements. It is also important that the air and space borne data be well calibrated and the atmosphere attenuation be accurately removed. Brook and Ben-Dor (2010) demonstrated a practical way of retrieving accurate reflectance information from IS data, by minimizing major uncertainty factors via application of a "smart vicarious calibration" method. This method permits better reflectance retrieval from the soil surface. As the IS technology holds promising capability for soil mapping, especially if accurate reflectance is extracted, use of all know-how in the laboratory can be adopted with small modifications, driving soil spectroscopy forward (Ben-Dor et al., 2008a,b). The innovative progress in quantitative analyses of soils from a spectral perspective, and the recent advances in IS sensors, will create a valuable environment for innovative studies and practical applications in the soil sciences. The remarkable achievements in IS sensor manufacturing are evidenced by the many relatively low-cost IS sensors which are now commercially available. The introduction of unmanned airborne vehicles (UAVs) with simple operational capabilities and ground-sensor availability has become a driving force in IS technology. Separation between pre-processing of IS data (including atmospheric rectification) and quantitative analysis is recommended. This can be achieved by collaboration or by purchasing the former service in advance. Similarly, developing full-chain capability from raw reflectance data is strongly recommended if IS is to become a major tool for the potential user. The current drawback of high operational costs is diminishing as the technology develops. Software to analyze IS data is available today for simple or complex applications and for spectral and spatial analyses, atmospheric correction, BRDF, and geometrical rectification. The problem of sensing the soil profile can be solved by merging IS technology with other remote-sensing techniques, as was done by Ben-Dor et al. (2008a,b) for soil salinity using ground-penetrating radar (GPR) and frequent domain electromagnetic (FDEM) antennas, as well as by developing new approaches, such as combining the IS information with point spectroscopy in the field. Another point of note is that the spectral range available today (VIS–NIR–SWIR) can be expanded to both the UV (<300 nm) and TIR (2500–14,000 nm) spectral regions, providing more quantitative information on soils (e.g., Viscarra Rossel et al., 2008; Notesco et al., 2014; Padilla et al., 2014). Recent papers on this issue have been published by Kopačková et al. (2017) who showed the added value of the TIR region to the VIS–NIR–SWIR region in predicting several attributes in a group of soils from the Czech Republic. In addition, Notesco et al. (2016) pointed out that the TIR region can account for Si-bearing minerals, the dominant chemical constituent of soil minerals. As Si cannot be seen by the traditional VIS–NIR–SWIR regions of the IS technology, TIR IS seems to be very promising. Promising results regarding several soil attributes in the TIR region have also been demonstrated by Hewson et al. (2012). Merging data from other sensors such as LiDAR is also a promising option.

Moving from airborne to ground IS sensors may also create new applications for soil mapping. A new initiative to put an IS sensor into orbit (Shen-En, 2016) is more good news, as it will generate high signal-to-noise spectral information on soils with high temporal resolution and wide spatial coverage. Several more orbital initiatives, such as Copernicus Hyperspectral Imaging Mission for the Environment (CHIME, also known as Sentinel 10) of ESA, EMIT of NASA or SHALOM of ISA-ASI, are advocating soil as an important application among other well known ones such as water, geology, and forestry. We thus anticipate a new era for the spectral sensing of soils. Ground IS sensors that are available today to all (with respect to both cost and size) are also a new and promising direction. The minimization of sensor size (mostly in the VIS–NIR spectral region) is resulting in lightweight IS cameras that can be carried on UAVs (e.g., AisaKestral of SPECIM and of Mjolnir V-1240 of HySPEX). We anticipate that as technology advances, sensors in the SWIR region will also be minimized, enabling the full optical range from UAVs. In summary, it can be said that IS for soil applications has progressed significantly since the last edition of this chapter, and has proven to hold great promise as an innovative vehicle to study soils from afar in a quantitative domain for more and more applications.

8.12 CONCLUSIONS

Based on the accumulated knowledge presented in this chapter, it is only obvious that soil spectral information is a treasure chest for mapping and recognizing soils remotely. An illustration that sums up the capability of soil spectroscopy as was discussed throughout this chapter is provided in Figure 8.29.

In this figure, the role of the Soil Sensing Cycle can be seen, from which all information brought from soils by different sensors can feed each other and create new interpretation perspectives of the same object. Indeed, it also shows the role of SSL with airborne and spaceborne IS sensors for soil mapping as well as the capability to assess soil surface and profile characteristics in situ. Over the past decade, a strong foundation has been established for the practical use of soil spectroscopy, despite the complexity of the soil matrix. Many researchers have generated intersecting studies that utilize an (chromometric) analytical approach to retrieve many soil properties solely from the reflectance measurements. Significant success has been achieved under controlled laboratory conditions and the method has become well accepted by the soil science community. It is very clear that moving from the point to spatial remote-sensing domain will advance the impact of spectral capability for practical applications. In the IS domain, more (cognitive) information is being generated, opening new frontiers in the field of soil science. However, this technology, especially when used from the air

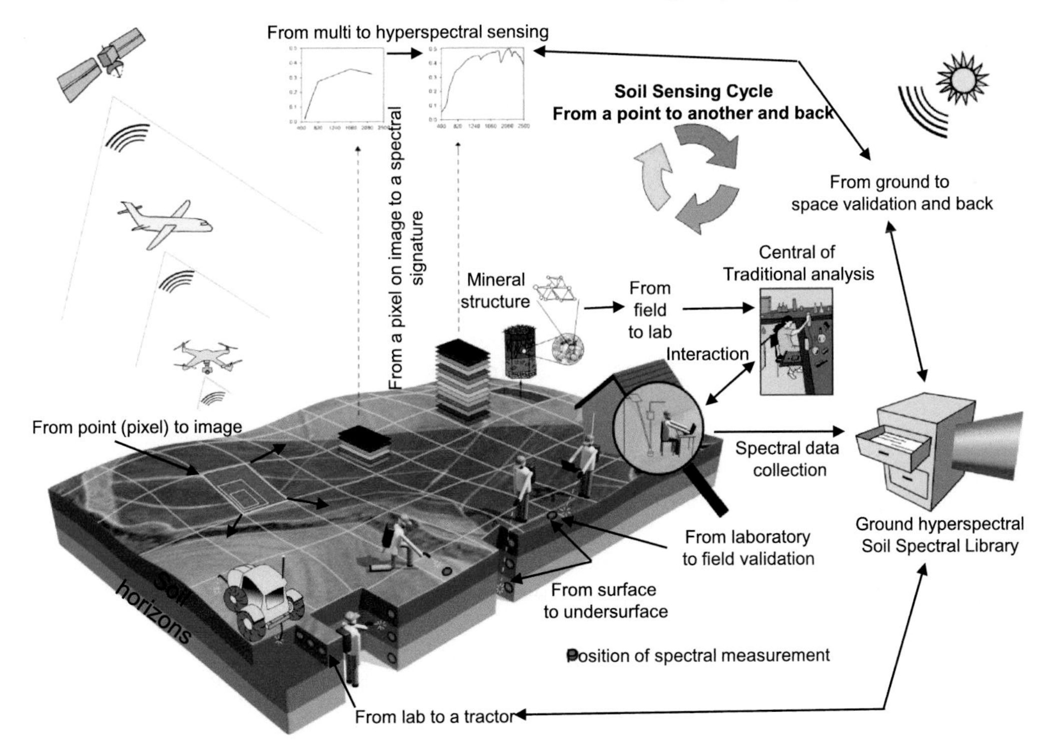

FIGURE 8.29 Illustration of the Soil Sensing Cycle: From soil spectral library, field measurements in 3D and remote sensing from drone to satellite.

and space domains, has encountered severe difficulties in providing accurate soil reflectance values. These problems include atmospheric attenuation, mixed pixels, low signal-to-noise ratios, geometric and optical distortions, BRDF effects, and sensing only the first 50 μm of the soil body.

Attempts to overcome these problems are being explored scientifically and it is believed that soon, spectral information from the air and space domain will be as accurate as that obtained in the laboratory. This will permit the implementation of spectral models originally developed for point spectrometry in the spatial (IS) domain. Several reports have shown innovative results using both point and image spectrometry to map soils, such as on soil degradation (salinity and erosion), soil genesis and formation, soil water content, soil contamination, soil formation, soil sealing, and soil expansion. Soil mapping and classification, which require soil profile information, have also become possible with the combined use of point and image spectrometry. The new (penetrating spectral) device, which can provide soil profile composition within a short time (using reflectance spectroscopy and NIRS), may replace the traditional trencher soil survey scheme. Although there is no doubt that soil spectroscopy in general and IS in particular harbor high potential for soil applications, so far, not many users have adopted this technology on a routine basis. This is mainly because of the relatively high price of operating airborne IS sensors and the need for skilled personnel to process the data; it is also because of failure to disseminate the technology to other end users. The new generation of IS sensors that can work on the ground may also open up a new frontier in soil spectroscopy applications. All of this leads to the promise of a bright future for soil spectroscopy technique. Electro-optic technology is now well developed and can offer infrastructure and never-before-seen capabilities. As progress continues, it is estimated that soil spectral technology will soon be fully commercialized. The success of soil spectroscopy relies on active chromophores in the soil matrix and on the development of sophisticated analytical approaches to extract the highly correlated wavelength of the attribute in question. The quantitative approach using NIRS technology (as developed in other disciplines) enables the extraction of spectral models that can be used simply and at low cost. NIRS is a well-established technique in soil science and is a successful means of determining several soil attributes, such as total C, organic C, total N, cation-exchange capacity, and moisture (in "as is" soil), clay content, free iron oxides, carbonates, and more. The field of soil NIRS is dynamic, with ongoing developments in field portability, software capability, sample presentation, and interfacing of instruments with GPS/GIS for mapping and monitoring soil. Soil reflectance spectroscopy has commercial applications and has been adopted by several companies. Nonetheless, soil spectroscopy still requires marketing and educating farmers, decision-makers, and other end users (e.g., soil scientists). It is believed that as soon as a good IS sensor becomes operational from orbit, a new era in soil mapping will begin. In summary, it can be concluded that reflectance spectroscopy is a powerful tool for quantitative and qualitative applications with soil material in the laboratory, field, and far-distance domains and is moving toward practical usage. Adding the TIR region to the IS technology, developing new tools to cope with quantitative analyses of the "big (soil) data," and the new initiative to place IS sensors in orbit (with particular attention to soil applications) are very promising. We strongly feel that if all spectral domains (point and image) are thoroughly researched and other active remote-sensing methods will be merged and combined, the applications are unlimited. Therefore, we believe that this review will shed light on future activities in the soil discipline. We hope that this will spark the imagination of both scientists and end users on how to utilize this treasure and simple technology for better and more effective management of the world's number one production resource—soil.

REFERENCES

Aase J.K. and Tanaka D.L. 1983. Effect of tillage practices on soil and wheat spectral reflectance. *Journal of Agronomy* 76:814–818.

Aber J., Wessman C.A., Peterson D.L., Mellilo J.M. and Fownes J.H. 1990. Remote sensing of litter and soil organic matter decomposition in forest ecosystems. In: *Remote Sensing of Biosphere Functioning*, Hobbs R.J. (ed.). pp. 87–101.

Agdu P.A., Fehrenbacher J.D. and Jansen I. 1990. Soil property relationships with SPOT satellite digital data in the East central Illinois. *Soil Science Society of American Journal* 54:807–812.

Ahmadian N., Demattê J.A., Xu D., Borg E. and Zölitz R. 2016. A new concept of soil line retrieval from Landsat 8 images for estimating plant biophysical parameters. *Remote Sensing* 8(9):738–730.

Al-Abbas H.H., Swain H.H. and Baumgardner M.F. 1972. Relating organic matter and clay content to multispectral radiance of soils. *Soils Science* 114:477–485.

Ammer U., Koch B., Schneider T. and Wittmeier H. 1991. High resolution spectral measurements of vegetation and soil in field and laboratory. *Proceedings of the 5th International Colloquium, Physical Measurements and Signatories in Remote Sensing*, Courchevel, France. I: pp. 213–218.

Banin A. and Amiel A. 1970. A correlation of the chemical physical properties of a group of natural soils of Israel. *Geoderma* 3:185–198.

Bartholomeus H., Epema G. and Schaepman M. 2007. Determining iron content in Mediterranean soils in partly vegetated areas, using spectral reflectance and imaging spectroscopy. *International Journal of Applied Earth Observation and Geoinformation* 9:194–203.

Bartholomeus H.M., Schaepman M.E., Kooistra L., Stevens A., Hoogmoed W.B. and Spaargaren O.S.P. 2008. Spectral reflectance based indices for soil organic carbon quantification. *Geoderma*, 145(1-2):28–36.

Bartholomeus H., Kooistra L., Stevens A., Leeuwen M.V., Wesemael B.V., Ben-Dor E. and Tychon B. 2010. Soil organic carbon mapping of partially vegetated agricultural fields. *International Journal of Applied Earth Observation and Geoinformation* 13(1):81–88.

Baumgardner M.F., Kristof S.J., Johannsen C.J. and Zachary A.L. 1970. Effects of organic matter on multispectral properties of soils. *Proceedings of the Indian Academy of Science* 79:413–422.

Baumgardner M.F., Silva L.F., Biehl L.L. and Stoner E.R. 1985. Reflectance properties of soils. *Advances in Agronomy* 38:1–44.

Beck R.H., Robinson B.F., McFee W.H. and Peterson J.B. 1976. Information Note 081176. *Laboratory Application of Remote Sensing*, Purdue University, West Lafayette, Indiana.

Bedidi A., Cervelle B. and Madeira J. 1991. Moisture effects on spectral signatures and CIE-color of lateritic soils. *Proceedings of the 5th International Colloquium, Physical Measurements and Signatures in Remote Sensing*, Courchevel, France. I: pp. 209–212.

Bedidi A., Cervelle B., Madeira J. and Pouget M. 1990. Moisture effects on spectral characteristics (visible) of lateritic soils. *Soil Science* 153:129–141.

Bellinaso H., Demattê J.A.M. and Araújo S. 2010. Soil spectral library and its use in soil classification. *R. Bras. Ci. Solo* 34:861–870.

Ben-Dor E. 2002. Quantitative remote sensing of soil properties. *Advances in Agronomy* 75:173–243.

Ben-Dor E. and Banin A. 1990. Near infrared reflectance analysis of carbonate concentration in soils. *Applied Spectroscopy* 44(6):1064–1069.

Ben-Dor E. and Banin A. 1994. Visible and near infrared (0.4–1.1 mm) analysis of arid and semiarid soils. *Remote Sensing of Environment* 48:261–274.

Ben-Dor E. and Banin A. 1995a. Near infrared analysis (NIRA) as a rapid method to simultaneously evaluate, several soil properties. *Soil Science Society of American Journal* 59:364–372.

Ben-Dor E. and Banin A. 1995b. Near infrared analysis (NIRA) as a simultaneously method to evaluate spectral featureless constituents in soils. *Soil Science* 159:259–269.

Ben-Dor E. and Banin A. 1995c. Quantitative analysis of convolved TM spectra of soils in the visible, near infrared and short-wave infrared spectral regions (0.4–2.5 mm). *International Journal of Remote Sensing* 18:3509–3528.

Ben-Dor E., Irons J.R. and Epema, G.F. 1999. Soil reflectante. *In Manual of Remote Sensing: Remote Sensing for Earth Science*, Rencz A.N., Ryerson R.A. (Eds), John Wiley & Sons, New York, pp. 111–187.

Ben-Dor, E., Chabrillat, S., Demattê, J.A.M., Taylor, G.R., Hill J., Whiting M.L. and Sommer S. 2009. Using imaging spectroscopy to study soil properties. *Remote Sensing of Environment* 113:S38–S55.

Ben-Dor E., Goldshalager N., Braun O., Kindel B., Goetz A.F.H., Bonfil D., Agassi M., Margalit N., Binayminy Y. and Karnieli A. 2004a. Monitoring of infiltration rate in semiarid soils using airborne hyperspectral technology. *International Journal of Remote Sensing* 25:1–18.

Ben-Dor E., Goldshleger N., Eshel M., Mirablis V. and Bason U. 2008a. Combined active and passive remote sensing methods for assessing soil salinity. In: *Remote Sensing of Soil Stalinization: Impact and Land Management*, Metternicht G. and Zinck A. (Eds.), CRC Press, Boca Raton, USA.

Ben-Dor E., Granot A. and Notesco G. 2017. A simple apparatus to measure soil spectral information in the field under stable conditions. *Geoderma* 306:73–80.

Ben-Dor E., Heller D. and Chudnovsky A. 2008b. A novel method of classifying soil profiles in the field using optical means. *Soil Science Society of American Journal* 72:1–13.

Ben-Dor E., Inbar Y. and Chen Y. 1997. The reflectance spectra of organic matter in the visible near infrared and short wave infrared region (400–2,500 nm) during a control decomposition process. *Remote Sensing of Environment* 61:1–15.

Ben-Dor E., Irons J.A. and Epema A. 1998. Soil spectroscopy. In: *Manual of Remote Sensing*, Third Edition, Rencz A. (ed.), J. Wiley & Sons Inc., New York, Chichester, Weinheim, Brisbane, Singapore, Toronto. pp. 111–189.

Ben-Dor E., Kindel B. and Goetz A.F.H. 2004b. Quality assessment of several methods to recover surface reflectance I using synthetic imaging spectroscopy (IS) data. *Remote Sensing of Environment* 90:389–404.

Ben-Dor E. and Kruse F.A. 1995. Surface mineral mapping of Makhtesh Ramon Negev, Israel using GER 63 channel scanner data. *International Journal of Remote Sensing* 18:3529–3553.

Ben-Dor E., Levin N., Singer A., Karnieli A., Braun O. and Kidron G.J. 2005. Quantitative mapping of the soil rubification process on sand dunes using an airborne hyperspectral sensor. *Geoderma* 131:1–21.

Ben-Dor E., Ong C. and Lau I.C. 2015. Reflectance measurements of soils in the laboratory: Standards and protocols. *Geoderma* 245:112–124.

Ben-Dor E., Patkin K., Banin A. and Karnieli A. 2002. Mapping of several soil properties using DAIS-7915 hyperspectral scanner data. A case study over clayey soils in Israel. *International Journal of Remote Sensing* 23:1043–1062.

Ben-Dor E. and Singer A. 1987. Optical density of vertisol clays suspensions in relation to sediment volume and dithionite-citrate-bicarbonate extractable iron. *Clays and Clay Minerals* 35:311–317.

Ben-Dor E., Taylor R.G., Hill J., Demattê J.A.M., Whiting M.L., Chabrillat S. and Sommer S. 2007. Imaging spectrometry for soil applications. *Agronomy Journal* 97:323–381.

Ben-Gera I. and Norris K.H. 1968. Determination of moisture content in soybeans by direct spectrophotometry. *Israeli Journal of Agriculture Research* 18:124–132.

Bishop J.L., Pieters C.M. and Edwards J.O. 1994. Infrared spectroscopic analyses on the nature of water in montmorillonite. *Clays and Clay Minerals* 42:702–716.

Bowers S. and Hanks R.J. 1965. Reflectance of radiant energy from soils. *Soil Science* 100:130–138.

Brodský L., Klement A., Penízek V., Kodesová R. and Borüvka L. 2011. Building soil spectral library of the Czech soils for quantitative digital soil mapping. *Soil and Water Research* 6(4):165–172.

Brook A. and Ben-Dor E. 2010. Supervised vicarious calibration (SCV) of hyperspectral remote sensing data. *Remote Sensing of Environment* 115(6):1543–1555.

Brown D.J., Shepherd K.D., Walsh M.G., Dewayne Mays M. and Reinsch T.G. 2006. Global soil characterization with VNIR diffuse reflectance spectroscopy. *Geoderma* 132:273–290.

Buol S.W., Hole F.D. and McCracken R.J. 1973. *Soil Genesis and Classification*. The Iowa State University Press, Ames p. 360.

Cambule A.H., Rossiter D.G., Stoorvogel J.J. and Smaling E.M.A. 2012. Building a near infrared spectral library for soil organic carbon estimation in the Limpopo National Park, Mozambique. *Geoderma* 183–184:41–348.

Cariati F., Erre L., Micera G., Piu P. and Gessa C. 1981. Water molecules and hydroxyl groups in montmorillonites as studied by near infrared spectroscopy. *Clay and Clay Minerals* 29:157–159.

Cariati F., Erre L., Micera G., Piu P. and Gessa C. 1983. Polarization of water molecules in phyllosilicates in relation to exchange cations as studied by near infrared spectroscopy. *Clays and Clay Minerals* 31:155–157.

Carmon N. and Ben-Dor E. 2017. An advanced analytical approach for spectral-based modelling of soil properties. *IJREAT* 7:9097.

Castaldi F., Palombo A., Santini F., Pascucci S., Pignatti S. and Casa R. 2016. Evaluation of the potential of the current and forthcoming multispectral and hyperspectral imagers to estimate soil texture and organic carbon. *Remote Sensing of Environment* 179:54–9065.

Chabrillat S., Foerster S., Steinberg A. and Segl K. 2014. Prediction of common surface soil properties using airborne and simulated EnMAP hyperspectral images: Impact of soil algorithm and sensor characteristic. *Geoscience and Remote Sensing Symposium (IGARSS)*, 2014 IEEE International, pp. 2914–2917.

Chabrillat S. and Goetz A.F.H. 1999. The search for swelling clays along the Colorado Front Range: The role of AVIRIS resolution in detection. *Summaries of the 8th JPL Airborne Earth Science Workshop*.

Chabrillat S., Goetz A.F.A., Krosley L. and Olsen H.W. 2002. Use of hyperspectral images in the identification and mapping of expansive clay soils and the role of spatial resolution. *Remote Sensing of Environment* 82:431–445.

Chappell A., Strong C., McTainsh G. and Leys J. 2006. Detecting induced in situ erodibility of dust-producing playa in Australia using a bi-directional soil spectral reflectance model. *Remote Sensing of Environment* 106:508–524.

Chappell A., Zobeck T.M. and Brunner G. 2005. Using on-nadir spectral reflectance to detect soil surface changes induced by simulated rainfall and wind tunnel abrasion. *Earth Surface Processes and Landforms* 30:489–511.

Chen Y. and Inbar Y. 1994. Chemical and spectroscopical analysis of organic matter transformation during composting in relation to compost maturity. In: *Science and Engineering of Composting: Design, Environmental, Microbiology and Utilization Aspects*, Hoitink H.A.J. and Keener H.M. (Eds.), Renaissance Publications, Worthington, OH, USA, pp. 551–600.

Cierniewski J. 1987. A model for soil surface roughness influence on the spectral response of bare soils in the visible and near infrared range. *Remote Sensing of Environment* 23:98–115.

Cierniewski J. 1989. The influence of the viewing geometry of bare soil surfaces on their spectral response in the visible and near infrared. *Remote Sensing of Environment* 27:135–142.

Cipra J.E., Franzmeir D.P., Bauer M.E. and Boyd R.K. 1980. Comparison of multispectral measurements from some nonvegetated soils using Landsat digital data and a spectroradiometer. *Soil Science Society of American Journal* 44:80–84.

Clark R.N. 1981. The reflectance of water-mineral mixtures at low temperatures. *Journal of Geophysical Research* 86:3074–3086.

Clark R.N. 1998. Spectroscopy of rocks and minerals, and principles of spectroscopy. In: *Remote Sensing for the Earth Sciences Manual of Remote Sensing*, Third Edition, Vol. 3. Rencz A.N. (ed.), John Wiley & Sons, Inc., New York, pp. 3–58.

Clark R.N. and Roush T.L. 1984. Reflectance spectroscopy: Quantitative analysis techniques for remote sensing applications. *Journal of Geophysical Research* 89:6329–6634.

Colin C., Collings K., Drummond P. and Lund E. 2004. A mobile sensor platform for measurement of soil pH and buffering. *American Society of Agricultural and Biological Engineers.*

Condit H.R. 1970. The spectral reflectance of American soils. *Photogrammetric Engineering* 36:955–966.

Coyne L.M., Bishop J.L., Sacttergood T., Banin A., Carle G. and Orenberg J. 1989. Near-infrared correlation spectroscopy. Quantifying iron and surface water in series of variably cation-exchanged montmorillonite clays. *Spectroscopic Characterization of Minerals and Their Surfaces.* ACS Symposium Series No. 415.

Csillag F., Pasztor L. and Biehl L.L. 1993. Spectral band selection for the characterization of salinity status of soils. *Remote Sensing of Environment* 43:231–242.

Da-Costa L.M. 1979. Surface soil color and reflectance as related to physicochemical and mineralogical soil properties. Ph.D. Dissertation, University of Missouri, Columbia.

Dalal R.C. and Henry R.J. 1986. Simultaneous determination of moisture, organic carbon and total nitrogen by near infrared reflectance spectroscopy. *Soil Science Society of America Journal* 50:120–123.

Davies A.M. and Grant A. 1987. Review: Near infrared analysis of food. *International Journal of Food Science and Technology* 22:191–207.

Davies A.M.C. 1987. Near infrared spectroscopy: Time for the giant to wake up. *European Spectroscopy News* 73.

Davies T. 1998. The history of near infrared spectroscopic analysis: Past, present and future "From sleeping technique to the morning star of spectroscopy". *Analusis* 26(4), 17–19.

Dekker A.G., Brando V.E., Anstee J.M., Pinnel N., Kutser T., Hoogenboom H.J., Pasterkamp R. et al. 2001. Imaging spectrometry of water. Chapter 11 In: *Imaging Spectrometry: Basic Principles and Prospective Applications: Remote Sensing and Digital Image Processing*, vol. IV, Kluwer Academic Publishers, Dordrecht, pp. 307–335.

Demattê J.A.M., Campos R.C. and Alves M.C. 1999. Evaluation of soil survey by spectral reflectance. In: *International Conference on Applied Geologic Remote Sensing.* Vol. 13, pp. 126–133.

Demattê J.A.M., Alves M.R., Terra F.S., Bosquila R.W.D., Fongaro C.T. and Barros P.P.S. 2016. Is it possible to classify topsoil texture using a sensor located 800 km away from the surface? *Brazilian Journal of Soil Science* 440:1–13.

Demattê J.A.M., Demattê J.L.I., Alves E.R., Negrão R. and Morelli J.L. 2014. Precision agriculture for sugarcane management: A strategy applied for Brazilian conditions. *Acta Scientiarum* 36(1):111–117.

Demattê J.A.M. and Garcia G.J. 1999. Alteration of soil properties through a weathering sequence as evaluated by spectral reflectance. *Soil Science Society of American Journal* 63:327–342.

Demattê J.A.M. and Terra F.S. 2014. Spectral pedology: A new perspective on evaluation of soils along pedogenetic alterations. *Geoderma* 217:190–200.

Demattê J.A.M., Horák-Terra I., Terra F.S., Marques K.P., Fongaro C.T., Silva A.A.C. and Torrado P.V. 2017. Genesis and properties of wetland soils by VIS-NIR-SWIR as a technique for environmental monitoring. *Journal of Environmental Management* 197:50–62.

Demattê J.A.M., Sousa A.A., Alves, M.C. and Nanni, M.R. 2006. Determining soil water status and other soil characteristics by spectral proximal sensing. *Geoderma* 135:179–195.

Demattê J.A.M., Fongaro C.T., Rizzo R. and Safanelli J.L. 2018. Geospatial soil sensing system (GEOS3): A powerful data mining procedure to retrieve soil spectral reflectance from satellite images. *Remote Sensing of Environment* 212: 161–175.

Dixon J.B. and Weed S.B. 1989. *Minerals in Soil Environments*. Soil Science Society of Soil Science Society of America Publishing, Madison WI.

Downey G. and Byrne P. 1986. Prediction of moisture and bulk density in milled peat by near infrared reflectance. *Journal of Food and Agriculture Science* 37:231–238.

Dunn B.W., Batten G.D., Beecher H.G. and Ciavarella S. 2002. The potential of near-infrared reflectance spectroscopy for soil analysis—A case study from the Riverine Plain of south-eastern Australia. *Australian Journal of Experimental Agriculture* 42(5):607–614.

Dwivedi R.S. and Sreenivas K. 1998. Image transforms as a tool for the study of soil salinity and alkalinity dynamics. *International Journal of Remote Sensing* 19:605–619.

Escadafal R. 1993. Remote sensing of soil color: Principles and applications. *Remote Sensing Reviews* 7:261–279.

Escadafal R., Girard M. and Courault D. 1989. Munsell soil color and soil reflectance in the visible spectral bands of Landsat MSS and TM data. *Remote Sensing of Environment* 27:37–46.

Escadafal R. and Hute A.R. 1991. Influence of the viewing geometry on the spectral properties (high resolution visible and NIR) of selected soils from Arizona. *Proceedings of the 5th International Colloquium, Physical Measurements and Signatures in Remote Sensing*, Courchevel, France. I: pp. 401–404.

FAO Year Book. 1994. *Production*, Vol. 48. *FAO Statistical Series #125*, Food and Agriculture organization of the Untied Nations, Rome, p. 3.

Farifteh J., Farshad A. and George R.J. 2006. Assessing salt-affected soils using remote sensing, solute modeling, and geophysics. *Geoderma* 130:191–206.

Feingersh T., Ben-Dor E. and Filin S. 2010. Correction of reflectance anisotropy: A multi-sensor approach. *International Journal of Remote Sensing* 31:49–74.

Feingersh T., Ben-Dor E. and Portugali J. 2007. Construction of synthetic spectral reflectance imagery for monitoring of urban sprawl. *Environmental Modeling and Software* 22:335–348.

Formaggio A.R., Epihanio J.C.E., Valeriano M.M. and Oliveira J.B. 1996. *Comportamento espectral (450–2450 nm) de solos tropicais de São Paulo* (Soil spectral behavior from tropical sol in São Paulo State). *Brazilian Journal of Soil Science* 20:467–474.

Franceschini M.H.D., Demattê J.A.M., Kooistra L., Bartholomeus H., Rizzo R., Fongaro C.T. and Molin J.P. 2017. Effects of external factors on soil reflectance measured on-the-go and assessment of potential spectral correction through orthogonalisation and standardisation procedures. *Soil and Tillage Research.*

Frazier B.E. and Cheng Y. 1989. Remote sensing of soils in the Eastern Palouse region with Landsat Thematic Mapper. *Remote Sensing of Environment* 28:317–325.

Gaffey S.J. 1985. Reflectance spectroscopy in the visible and near infrared (0.35–2.55 μm): Applications in carbonate petrology. *Geology* 13:270–273.

Gaffey S.J. 1986. Spectral reflectance of carbonate minerals in the visible and near infrared (0.35–2.55 μm): Calcite, aragonite and dolomite. *American Mineralogist* 71:151–162.

Gaffey S.J. and Reed K.L. 1987. Copper in calcite: Detection by visible and near infra-red reflectance. *Economic Geology* 82:195–200.

Galvao L.S., Vitorello I. and Paradella W.R. 1995. Spectroradiometric discrimination of laterites with principle components analysis and additive modeling. *Remote Sensing of Environment* 53:70–75.

Gausman H.W., Allen W.A., Cardenas R. and Bowen R.L. 1970. Color photos, cotton leaves and soil salinity. *Photogrammertic Engineering and Remote Sensing* 36:454–459.

Gerbermann A.H. and Neher D.D. 1979. Reflectance of varying mixtures of a clay soil and sand. *Photgrammertic Engineering and Remote Sensing* 45:1145–1151.

Gholizadeh A., Carmon N., Klement A., Ben-Dor E. and Borůvka L. 2018. Prediction of Czech Agricultural Soil Properties: Effects of Spectral Measurement Protocol and Data Mining Technique. *Remote Sensing* 9(10):1078.

Gholizadeh A., Carmon N., Klement A., Ben-Dor E. and Borůvka L. 2017. Agricultural soil spectral response and properties assessment: Effects of measurement protocol and data mining technique. *Remote Sensing* 9(10):107–100.

Goetz A.F.A. 1992. Principles of narrow band spectrometry in the visible and IR: Instruments and data analysis. In: *Imaging Spectroscopy: Fundamentals and Prospective Applications*, Toselli F. and Bodechtel J. (Eds.), ECSE, EEC, EAEC, Brussels and Luxembourg. pp. 21–32.

Goetz A.F.H., Hauff P., Shippert M. and Maecher A.G. 1991. Rapid detection and identification of OH-bearing minerals in the 0.9–1.0 mm region using new portable field spectrometer. *Proceeding of the 8th Thematic Conference on Geologic Remote Sensing*, Denver, Colorado. I: pp. 1–11.

Goetz A.F.H. and Herring M. 1989. The high resolution imaging spectrometer (HIRIS) for EOS. *IEEE Transactions on Geoscience and Remote Sensing* 27.2:136–144.

Gogé F., Joffre R., Jolivet C., Ross I. and Ranjard L. 2012. Optimization criteria in sample selection step of local regression for quantitative analysis of large soil NIRS database. *Chemometrics and Intelligent Laboratory Systems* 110(1):168–176.

Goldshlager N., Ben-Dor E., Chudnovsky A. and Agassi M. 2010. Soil reflectance as a generic tool for assessing infiltration rate induced by structural crust for heterogeneous soils. *European Journal of Soil Science* 60:1038–1951.

Gomez C., Oltra-Carrió R., Bacha S., Lagacherie P. and Briottet X. 2015. Evaluating the sensitivity of clay content prediction to atmospheric effects and degradation of image spatial resolution using Hyperspectral VNIR/SWIR imagery. *Remote Sensing of Environment* 164:1–15.

Gomez C., Viscarra Rossel R.A. and McBratney A.B. 2008. Soil organic carbon prediction by hyperspectral remote sensing and field vis-NIR spectroscopy: An Australian case study. *Geoderma* 146:403–411.

Guanter L., Kaufmann H., Segl K., Foerster S., Rogass C., Chabrillat S. and Straif C. 2015. EnMAP spaceborne imaging spectroscopy mission for earth observation. *Remote Sensing* 7(7):8830–8857.

Guerschman J.P., Hill M.J., Renzullo L.J., Barrett D.J., Marks A.S. and Botha E.J. 2009. Estimating fractional cover of photosynthetic vegetation, non-photosynthetic vegetation and bare soil in the Australian tropical savanna region upscaling the EO-1 Hyperion and MODIS sensors. *Remote Sensing of Environment* 113(5):928–945.

Hapke B.W. 1981a. Bidirectional reflectance spectroscopy I. Theory. *Journal of Geophysical Research* 86:3039–3054.

Hapke B.W. 1981b. Bidirectional reflectance spectroscopy: 2 Experiments and observation. *Journal of Geophysical Research* 86:3055–3060.

Hapke B.W. 1984. Bidirectional reflectance spectroscopy: Correction for macroscopic roughens. *Icarus* 59:41–59.

Hapke B.W. 1986. Bidirectional reflectance spectroscopy 4: The extinction coefficient and the opposition effect. *Icarus* 67:264–280.

Hapke B.W. 1993. *Theory of Reflectance and Emittance Spectroscopy*. Cambridge University Press, New York.

Hardisky M.A., Klemas V. and Smart R.M. 1983. The influence of soil salinity, growth form and leaf moisture on the spectral radiance of Spartina Alterniflora canopies. *Photogrammetric Engineering and Remote Sensing* 49:77–83.

Haubrock S.-N., Chabrillat S., Kuhnert M., Hostert P. and Kaufmann H. 2008a. Surface soil moisture quantification and validation based on hyperspectral data and field measurements. *Journal Application of Remote Sensing* 2: 023552.

Haubrock S.-N., Chabrillat S., Lemmnitz C. and Kaufmann H. 2008b. Surface soil moisture quantification models from reflectance data under field conditions. *International Journal of Remote Sensing* 29:3–29.

Herold M., Gardner M., Hadley B. and Roberts D. 2002. The spectral dimension in urban land cover mapping from high-resolution optical remote sensing data. *Proceedings of the 3rd Symposium on remote Sensing of Urban Areas*, Vol. 6, p. 2002.

Herold M., Roberts D.A., Gardner M.E. and Dennison P.E. 2004. Spectrometry for urban area remote sensing—Development and analysis of a spectral library from 350 to 2400 nm. *Remote Sensing of Environment* 91(3):304–2319.

Hewson R.D., Cudahy T.J., Jones M. and Thomas M. 2012. Investigations into soil composition and texture using infrared spectroscopy (2–14 m). *Applied and Environmental Soil Science.*

Hick R.T. and Russell W.G.R. 1990. Some spectral considerations for remote sensing of soil salinity. *Australian Journal of Soil Research* 28:417–431.

Hirschfeld T. 1985. Salinity determination using NIRA. *Applied Spectroscopy* 39:740–741.

Huete A.R. 1988. Soil adjusted vegetation index (SAVI). Remote sensing of environment 25:47–57. *Proceedings of the 5th International Colloquium, Physical Measurements and Signatures in Remote Sensing*, Courchevel, France. I: pp. 419–422.

Huete A.R., Chehbouni A., Leeuwen W. and Hua G. 1991. Normalization of multidirectional red and NIR reflectance with the SAVI. In: *Proceedings of the 5th International Colloquium, Physical Measurements and Signatures in Remote Sensing*. Courchevel, France, I:419–422.

Hummel J.W., Sudduth K.A. and Hollinger S.E. 2001. Soil moisture and organic matter prediction of surface and subsurface soils using an NIR soil sensor. *Computers and Electronics in Agriculture* 32:149–165.

Hunt G.R., 1980. Spectoscopic Properties of Rock and Minerals In: Stewart, C.R. (Ed.), *Handbook of Physical Properties Rocks*. CRC Press, p. 295.

Hunt G.R. and Salisbury J.W. 1970. Visible and near infrared spectra of minerals and rocks: I: Silicate minerals. *Modern Geology* 1:283–300.

Hunt G.R. and Salisbury J.W. 1971. Visible and near infrared spectra of minerals and rocks: Carbonates. *Modern Geology* 2:23–30.

Hunt G.R. and Salisbury J.W. 1976. Visible and near infrared spectra of minerals and rocks: XI Sedimentary rocks. *Modern Geology* 5:211–217.

Hunt G.R., Salisbury J.W. and Lenhoff A. 1971a. Visible and near-infrared spectra of minerals and rocks: III Oxides and hydroxides. *Modern Geology* 2:195–205.

Hunt G.R., Salisbury J.W. and Lenhoff C.J. 1971b. Visible and near-infrared spectra of minerals and rocks: Sulfides and sulfates. *Modern Geology* 3:1–14.

Hunt G.R., Salisbury J.W. and Lenhoff C.J. 1971c. Visible and near-infrared spectra of minerals and rocks: Halides, phosphates, arsenates, vandates and borates. *Modern Geology* 3:121–132.

Idso S.B., Jackson R.D., Reginato R.J., Kimball B.A. and Nakama F.S. 1975. The dependence of bare soil albedo on soil water content. *Journal of Applied Meteorology* 14:109–113.

Irons J.R., Weismiller R.A. and Petersen G.W. 1989. Soil reflectance In: *Theory and Application of Optical Remote Sensing*, Asrar G. (ed.), Willey Series in Remote Sensing, John Wiley & Sons, New York, pp. 66–106.

Ishida T., Ando H. and Fukuhara M. 1991. Estimation of complex refractive index of soil particles and its dependence on soil chemical properties. *Remote Sensing of Environment* 38:173–182.

Jackson R.D., Moran S., Slater P.N. and Biggar S.F. 1987. Field calibration of reflectance panels. *Remote Sensing of Environment* 22:145–158.

Jackson R.D., Teillet P.M., Slater P.N., Fedosjsvs G., Jasinski M.F., Aase J.K. and Moran M.S. 1990. Bidirectional measurements of surface reflectance for view angle corrections of oblique imagery. *Remote Sensing of Environment* 32:189–202.

Jacquemoud S., Baret F. and Hanocq J.F. 1992. Modeling spectral and bidirectional soil reflectance. *Remote Sensing of Environment* 41:123–132.

Karmanova L.A. 1981. Effect of various iron compounds on the spectral reflectance and color of soils. *Soviet Soil Science* 13:63–60.

Karnieli A., Kidron G.J., Glaesser C. and Ben-Dor E. 1999. Spectral characteristics of cyanobacteria soil crust in semiarid environments. *Remote Sensing of Environment* 69:67–75.

Karnieli A. and Tsoar H. 1994. Spectral reflectance of biogenic crust developed on desert dune sand along the Israel-Egypt border. *International Journal of Remote Sensing* 16:369–374.

Kemper T. and Sommer S. 2002. Estimate of heavy metal contamination in soils after a mining accident using reflectance spectroscopy. *Environmental Science Technology* 36:2742–2747.

Kierein-Young K.S. and Kruse F.A. 1989. Comparison of Landsat thematic mapper and geophysical and environmental research imaging spectrometer data for the cuprite mining district, Esmeralda and Nye Counties, Nevada. *Geoscience and Remote Sensing Symposium, 1989. IGARSS'89. 12th Canadian Symposium on Remote Sensing, 1989 International*, Vol. 2. IEEE.

Knadel M., Deng F., Thomsen A. and Greve M.H. 2012. Development of a Danish national Vis–NIR soil spectral library for soil organic carbon determination. In: Minasny B., Malone B.P., & McBratney A.B. *Digital Soil Assessments and Beyond: Proceedings of the 5th Global Workshop on Digital Soil Mapping.* 2012, Sydney, Australia. London: Taylor & Francis Group. Chapter 69, pp. 403–408.

Kneizys F.X., Abdersen G.P., Shettle E.P., Gallery W.O., Abreu L.W., Selby J.E.A., Chetwynd J.H. and Clough S.A. 1988. Users guide to LOWTRAN-7. Air Force Geophysics Laboratory, Hanscom AFB, Massachusetts AFGL-TR-88-0177.

Kokaly R., Rockwell B.W.M., Haire S. and King T.V.V. 2007. Characterization of post fire surface cover, soils and burn severity at the Cerro Grande Fire, New Mexico, using hyperspectral and multispectral remote sensing. *Remote Sensing of Environment* 106:305–325.

Kopačková V., and Ben-Dor, E. 2016. Normalizing reflectance from different spectrometers and protocols with an internal soil standard. *International Journal of Remote Sensing* 37(6):1276–1290.

Kopačková V., Ben-Dor E., Carmon N. and Notesco G. 2017. Modelling diverse soil attributes with visible to longwave infrared spectroscopy using PLSR employed by an automatic modelling engine. *Remote Sensing* 9(2):134.

Kosmas C.S., Curi N., Bryant R.B. and Franzmeier D.P. 1984. Characterization of iron oxide minerals by second derivative visible spectroscopy. *Soil Science Society of American Journal* 48:401–405.

Krishnan P., Alexander J.D., Bulter B.J. and Hummel J.W. 1980. Reflectance technique for predicting soil organic matter. *Soil Science Society of American Journal* 44:1282–1285.

Kristof S.F., Baumgardner M.F. and Johannsen. 1971. Spectral mapping of soil organic matter. *Journal Paper No.5390, Agricultural Experiment Station*, Purdue University, West Lafayette, Indiana.

Kruse F.A., Kierein-Young K.S. and Boardman J.W. 1990. Mineral mapping of Cuprite, Nevada with a 63-channel imaging spectrometer. *Photogrammetric Engineering and Remote Sensing* 56:83–92.

Kruse F.A., Thiry M. and Hauff P.L. 1991. Spectral identification (1.2–2.5 mm) and characterization of Paris basin kaolinite/smectite clays using a field spectrometer. *Proceedings of the 5th International Colloquium, Physical Measurements and Signatures in Remote Sensing*, Courchevel, France. I: pp. 181–184.

Latz K., Weismiller R.A. and Van Scoyoc G.E. 1981. *A study of the spectral reflectance of selected eroded soils of Indiana in relationship to their chemical and physical properties.* LARS Technical Report 082181.

Leger R.G., Millette G.J.F. and Chomchan S. 1979. The effects of organic matter, iron oxides and moisture on the color of two agricultural soils of Quebec. *Canadian Journal of Soil Science* 59:191–202.

Lee C.M., Cable M.L., Hook S.J., Green R.O., Ustin S.L., Mandl D.J. and Middleton E.M. 2015. An introduction to the NASA Hyperspectral InfraRed Imager (HyspIRI) mission and preparatory activities. *Remote Sensing of Environment* 167:6–19.

Lekner J. and Dorf M.C. 1988. Why some things are darker when wet. *Applied Optics* 27:1278–1280.

Lesaignoux A., Fabre S., Briottet X. and Olioso A. 2009. Soil moisture impact on lab measured reflectance of bare soils in the optical domain [0.4–15 μm]. *Geoscience and Remote Sensing Symposium, 2009 IEEE International, IGARSS 2009*, Vol. 3, pp. III–522.

Liang S. and Townshend R.G. 1996. A modified Hapke model for soil bidirectional reflectance. *Remote Sensing of Environment* 55:1–10.

Liu H.Q. and Huete A. 1995. A feedback based modification of the NDVI to minimize canopy background and atmospheric noise. *IEEE* 33.

Livne I., Goldshleger N., Ben-Dor E., Mirlas V. and Ben-Binyamin R. 2009. Monitoring soil salinity in agricultural lands using combined hyperspectral data and chemical measurements. *EARSeL SIG Imaging Spectroscopy Workshop*, 6th, Tel Aviv. Mar 16–19, 2009. Tel Aviv Univ., Tel Aviv, Israel.

Lobell D.B. and Asner G.P. 2002. Moisture effects on soil reflectance. *Soil Science Society of America Journal* 66:722–727.

Lugassi R., Ben-Dor E. and Eshel G. 2010. A spectral-based method for reconstructing spatial distributions of soil surface temperature during simulated fire events. *Remote Sensing of Environment* 114:322–331.

Madeira Netto J.S. 1996. Spectral reflectance properties of soils. *Photo Interp* 34:59–70.

Malec S., Rogge D., Heiden U., Sanchez-Azofeifa A., Bachmann M. and Wegmann M. 2015. Capability of spaceborne hyperspectral EnMAP mission for mapping fractional cover for soil erosion modeling. *Remote Sensing* 7(9):11776–11800.

Malley D.F., Martin P.D., and Ben-Dor E. 2004. Application in analysis of soils. Chapter 26 In: *Near Infrared Spectroscopy in Agriculture*, Craig R., Windham R. and Workman J. (Eds.), A three Societies Monograph (ASA, SSSA, CSSA) 44:729–784.

Matsunaga T., Iwasaki A., Tsuchida S., Iwao K., Tanii J., Kashimura O. and Nakamura R. et al., 2017. Current status of Hyperspectral Imager Suite (HISUI) onboard the International Space Station (ISS). In: *Proceedings of IGARSS 2017, IEEE International Geoscience and Remote Sensing Symposium*, Fort Worth, Texas, USA, July 23–28, 2017.

Mathews H.L., Cunningham R.L. and Peterson G.W. 1973. Spectral reflectance of selected Pennsylvania soils. *Proceedings of the Soil Science Society of America Journal* 37:421–424.

McBratney A.B., Mendonca Santos M.L. and Minasny B. 2003. On digital soil mapping. *Geoderma* 117:2–52.

Metternich G., Ben-Dor E., Goldshleger N. and Basson U. 2008. Sensors/platforms and popular classification algorithms. In: *Remote Sensing of Soil Stalinization: Impact and Land Management*, Metternicht G. and Zinck A. (Eds.), CRC Press, Boca Raton, USA.

Metternicht G.I. and Zink J.A. 1997. Spatial distribution of saline and sodium affected soil surfaces. *International Journal of Remote Sensing* 18:2571–2586.

Milton E.J. 1987. Review article principles of field spectroscopy 1987. *International Journal of Remote Sensing* 8:1807–1827.

Montgomery O.L. 1976. An investigation of the relationship between spectral reflectance and the chemical, physical and genetic characteristics of soils. *PhD thesis*, Purdue University, West Lafayette, Indiana (Libr. Congr. no 79-32236).

Montgomery O.L. and Baumgardner M.F. 1974. The effects of the physical and chemical properties of soil and the spectral reflectance of soils. *Information Note 1125 Laboratory for Applications of Remote Sensing, Purdue University*, West Lafayette, Indiana.

Moran M.S., Jackson R.D., Slater P.N. and Teillet P.M. 1992. Evaluation of simplified procedures for retrieval of land surface reflectance factors from satellite sensor output. *Remote Sensing of Environment* 41(2–3):169–184.

Morin Y., Benyamini Y. and Michaeli A. 1981. The dynamics of soil crusting by rainfall impact and the water movement in the soil profile. *Journal of Hydrology* 52:321–335.

Morra M.J., Hall M.H. and Freeborn L.L. 1991. Carbon and nitrogen analysis of soil fractions using near-infrared reflectance spectroscopy. *Soil Science Society of American Journal* 55:288–291.

Mouazen A.M., Maleki M.R., De Baerdemaeker J. and Ramon H. 2007. On-line measurement of some selected soil properties using a VIS–NIR sensor. *Soil and Tillage Research* 93(1):13–27.

Mougenot B., Epema G.F. and Pouget M. 1993. Remote sensing of salt-affected soils. *Remote Sensing Review* 7:241–259.

Muller E. and Dećamps H. 2000. Modeling soil moisture-reflectance. *Remote Sensing Environment* 76:173–180.

Murphy R.J. and Wadge G. 1994. The effects of vegetation on the ability to map soils using imaging spectrometer data. *International Journal of Remote Sensing* 15:63–86.

Nanni M.R. and Dematte J.A.M. 2006. Spectral reflectance methodology in comparison to traditional soil analysis. *Soil Science Society of America Journal* 70:393–407.

Nguyen T.T., Janik L.J. and Raupach M. 1991. Diffuse reflectance infrared Fourier transform (DRIFT) spectroscopy in soil studies. *Australian Journal of Soil Research* 29: 49–67.

Notesco G., Kopačková V., Rojík P., Schwartz G., Livne I. and Ben-Dor E. 2014. Mineral classification of land surface using multispectral LWIR and hyperspectral SWIR remote-sensing data. A case study over the Sokolov lignite open-pit mines, the Czech Republic. *Remote Sensing* 6(8):7005–7025.

Notesco G., Ogen Y. and Ben-Dor E., 2016. Integration of hyperspectral shortwave and longwave infrared remote-sensing data for mineral mapping of Makhtesh Ramon in Israel. *Remote Sensing* 8: 318. doi:10.3390/rs8040318.

Obukhov A.I. and Orlov D.S. 1964. Spectral reflectance of the major soil groups and the possibility of using diffuse reflection in soil investigations. *Soviet Soil Science* 2:174–184.

Odlare M., Svensson K. and Pell M. 2005. Near infrared reflectance spectroscopy for assessment of spatial soil variation in an agricultural field. *Geoderma* 126:193–202.

O'neill A.L. 1994. Reflectance spectra of microphytic soil crusts in semi-arid Australia. *International Journal of Remote Sensing* 15:675–681.

Orlov D.C. 1966. Quantitative patterns of light reflectance on soils I: Influence of particles (aggregate) size on reflectivity. *Soviet Soil Science* 13:1495–1498.

Otterman J., Brakke T. and Marshak A. 1995. Scattering by Lambertian-leaves canopy: Dependents of leaf-area projections. *International Journal of Remote Sensing* 16:1107–1125.

Padilla J.E., Calderón F.J., Acosta-Martinez V., Van Pelt S., Gardner T., Baddock M. and Noveron J.C. 2014. Diffuse-reflectance mid-infrared spectroscopy reveals chemical differences in soil organic matter carried in different size wind eroded sediments. *Aeolian Research* 15:193–201.

Palmer J.M. 1982. Field standards of reflectance. *Photogrametirc Engineerings and Remote Sensing* 48:1623–1625.

Peterson J.B., Robinson B.F. and Beck R.H. 1979. Predictability of change in soil reflectance on wetting. *Purdue e-Pubs LARAS Technical Reports*. Paper 73: 264-273. (http://docs.lib.purdue.edu/larstech/73)

Peterson J.B. 1980. Use of spectral data to estimate the relationship between soil moisture tension and their corresponding reflectance. *Annual Report*, OWRT Purdue University, pp. 1–18.

Pieters C.M. 1983. Strength of mineral absorption features in the transmitted component of near-infrared reflected light. First results from RELAB. *Journal of Geophysical Research* 88:9534–9544.

Pignatti S., Angelo P., Simone P., Filomena R., Federico S., Tiziana S. and Stefania M. 2013. The PRISMA hyperspectral mission: Science activities and opportunities for agriculture and land monitoring. *Geoscience and Remote Sensing Symposium (IGARSS), 2013 IEEE International*, IEEE. (pp. 4558–4561).

Pimstein A., Notesco G. and Ben-Dor E. 2010. Performance of three identical spectrometers in retrieving soil reflectance under laboratory conditions. *Soil Science Society of America Journal* 75(2):746–759.

Pinker R. and Karnieli A. 1995. Characteristics spectral reflectance of semi-arid environment. *International Journal of Soil Science* 16:1341–1363.

Pinty B., Verstraete M.M. and Dickson R.E. 1989. A physical model for prediction bidirectional reflectance over bare soil. *Remote Sensing of Environment* 27:273–288.

Pinty B., Verstraete M.M. and Gobron N. 1998. The effect of soil anisotropy on the radiance field emerging from vegetation canopies. *Geophysical Research Letters* 25:797–800.

Price J.C. 1990. On the information content of soil reflectance spectra. *Remote Sensing of Environment* 33:113–121.

Price J.C. 1995. Examples of high resolution visible to near-infrared reflectance spectra and a standardized collection for remote sensing studies. *International Journal of Remote Sensing* 16:993–1000.

Qi J., Chehbouni A., Huete A.R., Kerr Y.H. and Sorooshian S. 1994. A modified soil adjusted vegetation index. *Remote Sensing of Environment* 48(2):119–126.

Rao B.R.M., Ravi Sankar T., Dwivedi R.S., Thammappa S.S. and Venkataratnam L. 1995. Spectral behavior of salt-affected soils. *International Journal of Remote Sensing* 16:2125–2136.

Ray T.W. and Murray B.C. 1996. Nonlinear spectral mixing in desert vegetation. *Remote Sensing of Environment* 55:59–79.

Richardson A.J., Wiegand C.L., Gausman H.W., Cullar J.A. and Gerbermann A.H. 1975. Plant, soil and shadow reflectance components of raw crops. *Photogammetric Engineering and Remote Sensing* 41:1401–1407.

Richter R. and Schläpfer D. 2002. Geo-atmospheric processing of airborne imaging spectrometry data. Part 2: Atmospheric/topographic correction. *International Journal of Remote Sensing* 23:2631–2649.

Roberts D.A., Smith M.O. and Adams J.B. 1993. Green vegetation, nonphotosynthetic vegetation, and soils in AVIRIS data. *Remote Sensing of Environment* 44:255–269.

Rogers R.W. and Langer R.T. 1972. Soil surface lichens in arid and subarid south-eastern Australia. Introduction and floristics. *Australian Journal of Botany* 20:197–213.

Romero D.J., Ben-Dor E., Demattê J.A., e Souza A.B., Vicente L.E., Tavares T.R. and Gallo B.C., 2018. Internal soil standard method for the Brazilian soil spectral library: Performance and proximate analysis. *Geoderma* 312:95–103.

Rondeaux G., Steven M. and Baret F. 1996. Optimization of soil-adjusted vegetation indices. *Remote Sensing of Environment* 55:95–107.

Ronov A.A. and Yaroshevsky A.A. 1971. Chemical composition of the earth's crust. In: *The Earth's Crust and Upper Mantle*, Hart P.J. (Ed.). American Geophysical Union, Washington DC, pp. 37–57.

Schläpfer D., Richter R. and Feingersh T. 2010. Operational BRDF effects correction for wide-field-of-view optical scanners (BREFCOR). *IEEE Transactions on Geoscience and Remote Sensing* 53(4):1855–1864.

Schnitzer M. and Khan S.U. 1978. *Soil Organic Matter*. Elsvier Publication, Amsterdam.

Schreier H. 1977. *Proceeding of the 4th Canadian Symposium on Remote Sensing* I: pp. 106–112.

Schwartz G., Ben-Dor E. and Eshel G. 2013. Quantitative assessment of hydrocarbon contamination in soil using reflectance spectroscopy: A "Multipath" approach. *Applied Spectroscopy* 67(11):1323–1331.

Schwertmann U. 1988. Occurrence and formation of iron oxides in various pedoenvironments. In: *Iron In Soils and Clay Minerals*, Stucki J.W., Goodman B.A. and Schwertmann U. (Eds.), NATO ASI Series, Reidel Publishing Company, Dordrecht, Boston, Lancaster, Tokyo, pp. 267–308.

Shainberg I. 1992. Chemical and mineralogical components of crusting. In: *Soil Crusting*, Sumenr M.E. and Stewart B.A. (Eds.), Lewis Publications, Ann-Arbor, USA.

Sharma R.C. and Bhargava G.P. 1988. Landsat imagery for mapping saline soils and wetlands in north-west India. *International Journal of Remote Sensing* 9:39–44.

Shen-En Q. 2016. *Optical Payloads for Space Missions*. Wile Press, p. 1008.

Shepherd K.D. and Walsh M.G. 2002. Development of reflectance spectral libraries for characterization of soil properties. *Soil Science Society of America Journal* 66(3):988–1998.

Shi Z., Wang Q.L., Peng J., Ji W.J., Li H.J., Li X., and Viscarra Rossel R.A. 2014a. Development of a national VNIR soil-spectral library for soil classification and prediction of organic matter concentrations. *China Earth Sciences* 57(7):1671–1680.

Shi Z., Wang Q., Peng J., Ji W., Liu H., Li X. and Rossel R.A.V. 2014b. Development of a national VNIR soil-spectral library for soil classification and prediction of organic matter concentrations. *Science China Earth Sciences* 57(7):1671.

Shields J.A., Paui E.A., Arnaud R.J. and Head W.K. 1968. Spectrophotometric measurement of soil color and its relation to moisture and organic matter. *Canadian Journal of Soil Science* 48:271–1280.

Siegal B.S. and Goetz A.F.H. 1977. Effect of vegetation on rock and soil type discrimination. *Photogrammetric Engineering and Remote Sensing* 43:191–196.

Singh A.N. 1994. Monitoring change in the extent of salt-affected soils in northern India. 16:3173–3182.

Smith M.O., Ustin S.L., Adams J.B. and Gillespie A.R. 1990. Vegetation in desert: I A regional measure of abundances from multispectral images. *Remote Sensing of Environment* 31:1–26.

Soileau J.M. and McCraken R.J. 1967. Free iron and coloration in certain well-drained Coastal Plain soils in relation to their other properties and classification. *Soil Science Society of American Proceedings* 31:248–255.

Soriano-Disla J.M., Janik L.J., Viscarra Rossel R.A., MacDonald L.M. and McLaughlin M.J. 2014. The performance of visible, near-, and mid-infrared reflectance spectroscopy for prediction of soil physical, chemical, and biological properties. *Applied Spectroscopy Reviews* 49(2):139–186.

Staenz K., Mueller A. and Heiden U. 2013. Overview of terrestrial imaging spectroscopy missions. *Geoscience and Remote Sensing Symposium (IGARSS), 2013 IEEE International*, IEEE, pp. 3502–3505.

Steinberg A., Chabrillat S., Stevens A., Segl K. and Foerster S. 2016. Prediction of common surface soil properties based on Vis-NIR airborne and simulated EnMAP imaging spectroscopy data: Prediction accuracy and influence of spatial resolution. *Remote Sensing* 8(7):613.

Stevens A., Wesemael B.V., Bartholomeus H., Rosillon D., Tychon B. and Ben-Dor E. 2008. Laboratory, field and airborne spectroscopy for monitoring organic carbon content in agricultural soils. *Geoderma Geoderma* 144:395–640.

Stevenson F.J. 1982. *Humus Chemistry.* John Wiley & Sons Inc., New York.

Stoner E.R. 1979. Physicochemical, site and bidirectional reflectance factor characteristics of uniformly-moist soils. *PhD thesis*, Purdue University.

Stoner E.R. and Baumgardner M.F. 1981. Characteristic variations in reflectance of surface soils. *Soil Science Society of American Journal* 45:1161–1165.

Stoner E.R., Baumgardner M.F., Weismiller R.A., Biehl L.L. and Robinson F. 1980. Extension of laboratory soil spectra to field conditions. *Soil Science Society of American Journal* 44:572–574.

Strobl P., Mueller A.A., Schlaepfer D. and Schaepman M.E. 1997. Laboratory calibration and inflight validation of the Digital Airborne Imaging Spectrometer DAIS 7915. In: *Algorithms for Multispectral and Hyperspectral Imagery III*, Vol. 3071, pp. 225–237. International Society for Optics and Photonics.

Stuffler T., Förster K., Hofer S., Leipold M., Sang B., Kaufmann H., Penné B., Mueller A. and Chlebek C. 2009. Hyperspectral imaging—An advanced instrument concept for the EnMAP mission (Environmental Mapping and Analysis Programme). *Acta Astronautica* 65:7–8, 1107–1112.

Szilagyi A. and Baumgardner M.F. 1991. Salinity and spectral reflectance of soils. *Proceedings of ASPRS Annual Convention*, Baltimore. pp. 430–438.

Tanre D., Deroo C., Duhaut P., Herman M., Morcrette J.J., Perbos J. and Deschamps P.Y. 1986. Simulation of the Satellite Signal in the Solar Spectrum (5S), User Gide (UST de Lille, 59655 Villenueve D'asc, France: Laboratory d'Optique Atmospherique.

Taranik D.L. and Kruse F.A. 1990. Iron mineral reflectance in Geophysical and Environmental Research Imaging Spectrometer(GERIS) data. In: *Thematic Conference on Remote Sensing for Exploration Geology- Methods, Integration, Solutions, 7th.* Calgary, Canada, pp. 445–458.

Terra F.S., Demattê J.A. and Rossel R.A.V. 2015. Spectral libraries for quantitative analyses of tropical Brazilian soils: Comparing vis–NIR and mid-IR reflectance data. *Geoderma* 255:81–93.

Tkachuk R. and Law D.P. 1978. Near infrared diffuse reflectance standards. *Cereal Chemistry* 55:981–995.

Tóth G., Jones A. and Montanarella L. 2013. The LUCAS topsoil database and derived information on the regional variability of cropland topsoil properties in the European Union. *Environmental Monitoring and Assessment* 185(9):7409–7425.

Toth T., Csillag F., Biehl L.L. and Micheli E. 1991. Characterization of semivegetated salt-affected soil by means of field remote sensing. *Remote Sensing of Environment* 37:167–180.

Tucker C.J. and Miller L.D. 1977. Soil spectra contributions to grass canopy spectral reflectance. *Photogrammetric Engineering and Remote Sensing* 43:721–726.

Tueller P.T. 1987. Remote sensing science application in arid environment. *Remote Sensing of Environment* 23:143–154.

Twomey S.A., Bohren C.F. and Mergenthaler J.L. 1986. Reflectance and albedo differences between wet and dry surfaces. *Applied Optics* 25:431–437.

Ustin S.L. and Schaepman, M.E. 2009. Imaging spectroscopy: Special issue. *Remote Sensing of Environment* 113:1–3.

Van der Marel H.W. and Beutelspacher H. 1976. Clay and related minerals. In: Van der Marel, H.W.., Beutelspacher, H. (Eds.), *Atlas of Infrared Spectroscopy of Clay Minerals and Their Admixtures.* Elsevier Scientific, Amsterdam, The Netherlands, p. 396.

Van-der-Meer F. 1995. Spectral reflectance of carbonate mineral mixture and bidirectional reflectance theory: Quantitative analysis techniques for application in remote sensing. *Remote Sensing Reviews* 13:67–94.

Vane G., Goetz A.F.H. and Wellman J.B. 1984. Airborne imaging spectrometer: A new tool for remote sensing. *IEEE Trans Geosciences Remote Sensing* 22(6):546–549.

Vane G., Reimer J.H., Chrien T.G., Enmark H.T., Hansen E.G. and Porter W.M. 1993. Airborne visible/infrared imaging spectrometer (AVIRIS). *Remote Sensing of Environment* 44:127–143.

Veraverbeke S., Somers B., Gitas I., Katagis T., Polychronaki A. and Goossens R. 2012. Spectral mixture analysis to assess post-fire vegetation regeneration using Landsat Thematic Mapper imagery: Accounting for soil brightness variation. *International Journal of Applied Earth Observation and Geoinformation* 14(1):1–11.

Verma K.S., Saeena R.K., Barthwal A.K. and Deshmukh S.N. 1994. Remote sensing technique for mapping salt affected soils. *International Journal for Remote Sensing* 15:1901–1914.

Vinogradov B.V. 1981. Remote sensing of the humus content of soils. *Soviet Soil Science* 11:114–123.

Viscarra Rossel R.A. 2009. The soil spectroscopy group and the development of a global soil spectral library. *Geophysical Research Abstracts* 11:EGU2009-14021.

Viscarra-Rossel R.A. and Behrens T. 2010. Using data mining to model and interpret soil diffuse reflectance spectra. *Geoderma* 158: 46–54.

Viscarra Rossel R.A., Behrens T., Ben-Dor E., Brown D.J., Demattê J.A.M., Shepherd K.D., Shi Z. et al. 2016. A global spectral library to characterize the world's soil. *Earth-Science Reviews* 155:198–230.

Viscarra Rossel R.A., Jeon Y.S., Odeh I.O.A. and McBratney A.B. 2008. Using a legacy soil sample to develop a mid-IR spectral library. *Soil Research* 46(1):1–16.

Viscarra Rossel R.A., Lobsey C.R., Sharman C., Flick P. and McLachlan G. 2017. Novel proximal sensing for monitoring soil organic C stocks and condition. *Environment Science & Technology* 51(10):5630–5641.

Viscarra-Rossel R.A., Walvoort D.J.J., McBratney A.B., Janik L.J. and Skjemstad J.O. 2006. Visible, near infrared, mid infrared or combined diffuse reflectance spectroscopy for simultaneous assessment of various soil properties. *Geoderma* 131:59–75.

Viscarra Rossel R.A. and Webster R. 2012. Predicting soil properties from the Australian soil visible near infrared spectroscopic database. *European Journal of Soil Science* 63:848–860.

Weindner V.R. and Hsia J.J. 1981. Reflection properties of presses polytetrafluoroethylene powder. *Journal of Optical Society of America* 71:856–862.

West N.E. 1990. Structure and function of microphytic soil crust in wildland ecosystems of arid to semi-arid regions. *Advances in Ecological Research* 20:179–223.

Whiting M.L., Li L. and Ustin S.L. 2004a. Predicting water content using Gaussian model on soil spectra. *Remote Sensing of Environment* 89:535–552.

Whiting M.L., Li L. and Ustin S.L. 2004b. Correcting mineral abundance estimates for soil moisture. In: *13th Annual JPL Airborne Earth Science Workshop*, Green R.O. (ed.), Pasadena, California, March 30–April 3, 2004, JPL Publication 05-3-1.

Whiting M.L., Palacios-Orueta A., Li L. and Ustin S.L. 2005. Light absorption model for water content to improve soil mineral estimates in hyperspectral imagery. In *Pecora 16: Global Priorities in Land Remote Sensing*, Sioux Falls, South Dakota. October 23–27, 2005. American Society of Photogrammetry and Remote Sensing.

Wiegand C.L., Rhoades J.D., Escobar D.E. and Everitt J.H. 1994. Photographic and videographic observations for determining and mapping the response of cotton to soil salinity. *Remote Sensing of Environment* 49:212–223.

Young E.R., Clark K.C., Bennett R.B. and Houk T.L. 1980. Measurements and parameterization of the bidirectional reflectance feature of $BaSO_4$ paint. *Applied Optics* 19(20):3500–3505.

Zaady E., Karnieli A. and Shachak M. 2007. Applying a field spectroscopy technique for assessing successional trends of biological soil crusts in a semi-arid environment. *Journal of Arid Environments* 70:463–477.

Zabcic N., Rivard B., Ong C. and Mueller A. 2014. Using airborne hyperspectral data to characterize the surface pH and mineralogy of pyrite mine tailings. *International Journal of Applied Earth Observation and Geoinformation* 32:152–162.

Section IV

Hyperspectral Data Mining, Data Fusion, and Algorithms

9 Spaceborne Hyperspectral EO-1 Hyperion Data Pre-Processing

Methods, Approaches, and Algorithms

Itiya P. Aneece, Prasad S. Thenkabail, John G. Lyon, Alfredo Huete, and Terrance Slonecker

CONTENTS

9.1 INTRODUCTION TO PRE-PROCESSING OF HYPERSPECTRAL REMOTE SENSING DATA

Increases in global populations and urbanization and associated increases in food and nutrition demand are leading to rapid changes in global land use and land cover (LULC). Over the last 50 years, remote sensing has been widely used to monitor such changes over space and time. Most of these studies were conducted using data from multispectral broadband sensors. Multispectral broadband (MBB) data such as from the Landsat series, Système Probatoire d'Observation de la

Terre or SPOT series, and the Indian Remote Sensing (IRS) series have been successfully used in a wide range of applications. Limitations of MBBs include the broad waveband range of each band (e.g., 630–690 nm red band), very few wavebands (e.g., 4–7 bands) over long spectral range (400–13,000 nm), and the absence of targeted wavebands (e.g., band centered at 970 nm to measure plant moisture).

In contrast, hyperspectral sensors have tens, hundreds, or thousands of wavebands representing a continuous spectrum of electromagnetic data, each waveband has a narrow bandwidth (e.g., 1 nm, 5 nm, or 10 nm), and the wavebands are in targeted portions of the spectrum (e.g., 680 nm with 20 nm bandwidth). Hyperspectral data have been used for mapping and monitoring invasive plant species (He et al., 2011; Underwood et al., 2003, 2007), estimating biodiversity (Carlson et al., 2007), discriminating and mapping vegetation species (Andrew and Ustin, 2006; Zhang et al., 2006; Shafri et al., 2007; Banskota et al., 2011; Cho et al., 2012; Naidoo et al., 2012), mapping land cover classes (Zomer et al., 2009), scaling biochemical characteristics from leaf to canopy level (He and Mui, 2010), assessing the effects of pollution (Kefauver et al., 2012; Peterson et al., 2015; Beland et al., 2016), mapping crop residues (Daughtry et al., 2005), and monitoring various agricultural components (Smith and Blackshaw, 2003; Yang et al., 2008; Sahoo et al., 2015).

More specifically, hyperspectral Hyperion satellite data, with a spatial resolution of 30 m and 242 bands, have been used for mapping land cover classes (Asner and Heidebrecht, 2003; Thenkabail et al., 2004; Jafari and Lewis, 2012), detecting alteration minerals (Oskouei and Babakan, 2016), classifying rocks and land formations (Dadon et al., 2011), discriminating among plant and tree species (Thenkabail et al., 2004; Christian and Krishnayya, 2009; Papes et al., 2010; Somers and Asner, 2013a, 2014; George et al., 2014), monitoring invasive plant and tree species (Somers and Asner, 2013b; Walsh et al., 2008), mapping distribution of vegetation functional types (Huemmrich et al., 2013), monitoring mine waste mineralogy (Mielke et al., 2014), estimating foliar nutrition to inform fertilizer application (Sims et al., 2013), assessing fire danger (Roberts et al., 2003), assessing post-fire effects on vegetation (Mitri and Gitas, 2010), and assessing photosynthetic activity in forests (Zhang et al., 2013). There have also been several agricultural applications of Hyperion data including mapping crop residue (Bannari et al., 2015; Sonmez and Slater, 2016), monitoring tillage intensity (Sonmez and Slater, 2016), classifying a crop by age (Bhojaraja et al., 2015), classifying soybean varieties (Breunig et al., 2011), estimating crop planting area (Pan et al., 2013), differentiating coffee crop conditions after harvest (Lamparelli et al., 2012), mapping spatial variability of chlorophyll and nitrogen content in rice (Moharana and Dutta, 2016), estimating total canopy chlorophyll and leaf area index in crops (Houborg et al., 2016), estimating crop biophysical characteristics (Thenkabail et al., 2013), and estimating crop biomass and productivity (Mariotto et al., 2013; Marshall and Thenkabail, 2015a).

Nevertheless, hyperspectral narrowbands (HNBs) can have some limitations relative to MBBs especially in handling data volumes and massive data processing. HNBs also typically have lower signal-to-noise ratios than MBBs due to the reduced energy collected by sensors from the narrower spectral bands; this is especially the case in the shortwave infrared region, where solar irradiance is low (Castaldi et al., 2016). However, the advantages of using HNBs over MBBs far outweigh any limitation, especially at a time when cloud computing and machine learning algorithms offer powerful capabilities in data mining and data handling. For example, Mariotto et al. (2013) found that even the sub-meter to 5-m high spatial resolution (HSR) data such as QuickBird and IKONOS were not enough for predicting crop productivity without the availability of HNB data. Similarly, Marshall and Thenkabail (2015a) found that HSR WorldView-2 data were not able to successfully discriminate among crop types, except for rice. In contrast, HNB data have been found to explain 25% more variability in crop discrimination when compared with MBB data (Mariotto et al., 2013). Thenkabail et al. (2004) also found that Hyperion models explained 36%–83% more variability in rainforest biomass and had 45%–52% greater LULC classification accuracies than IKONOS, Advanced Land Imager (ALI), and Landsat. HNBs are also less likely to become saturated from rainforest vegetation than are MBBs (Thenkabail et al., 2004). Thus, hyperspectral remote sensing can lead to further advances in studying vegetation.

Given the huge advantages of HNB data over MBB and HSR examples, one would expect wider application of hyperspectral data. However, this has not happened due to limited availability, especially repetitive coverage of the world, and general lack of knowledge of the characteristics of hyperspectral data over the broader audience. Above all, one of the greatest difficulties in handling hyperspectral data is limited availability of standardized pre-processing steps, procedures, and algorithms. Whereas a number of authors have provided pre-processing steps (Barry, 2001; Beck, 2003; Datt et al., 2003; Goodenough et al., 2003; Khurshid et al., 2006; Dadon et al., 2010a; Richter et al., 2011; San and Suzen, 2011; Farifteh et al., 2013; Scheffler and Karrasch, 2014), there is no known comprehensive documentation of hyperspectral pre-processing steps along with complete description and implementation of various methods, procedures, and algorithms. Here, we define pre-processing as any processing done to an image before data analysis (such as the calculation of vegetation indices, band math, and so on).

Therefore, the overarching goal of this chapter is to outline various pre-processing steps involved in hyperspectral data, describe them, establish standardized methods and code to solve the problems and seamlessly generate products, and apply the algorithms on a collection of numerous images. The outcome is standardized and harmonized hyperspectral data products that are ready for applications. We will be describing these pre-processing steps using Earth Observing-1 (EO-1) Hyperion data, the world's first commercial hyperspectral sensor data that are available free of cost from USGS Earth Explorer for the 2001–2015 time-period.

9.2 DATA USED FOR PRE-PROCESSING

9.2.1 Description of Hyperion Images

The National Aeronautics Space Administration's (NASA's) New Millennium Program's Earth Observing 1 (EO-1) with the hyperspectral Hyperion sensor was launched on November 21, 2000 (Barry, 2001), in collaboration with the United States Geological Survey (USGS). The spacecraft carried the hyperspectral instrument Hyperion, the Advanced Land Imager (ALI), and an Atmospheric Corrector (AC; Barry, 2001). EO-1 was launched for a 1-year baseline mission to validate advanced remote sensing technology that would be used in future satellites, such as Landsat 7 (Flick et al., 2007; USGS, 2013). The extended mission, which continued until March 2017, was to promote infusion of EO-1 technology by increasing the use of on-orbit resources and reduce operational costs through the Continuous Improvement Program (Flick et al., 2007).

EO-1 had a sun-synchronous orbit at a height of 705 km. Hyperion images have a spatial resolution of 30 m, and a temporal resolution of 16 days. They have a spectral range of 356–2576 nm, with a spectral bandwidth of 10 nm and spectral resolution of 242 bands (220 unique bands; Barry, 2001). Hyperion images are 7.65 km by 185 km (Barry, 2001; Table 9.1). Signal-to-noise ratio is 190 to 40 (Beck, 2003). More than 70,000 of these images are freely available through the US Geological Survey Earth Explorer. Many of these images are already ingested into the Google Earth Engine (GEE) cloud computing platform, which facilitates global studies. Hyperion has been decommissioned as of March 2017 because of fuel depletion, but there are other hyperspectral satellite sensors in development, including NASA's HyspIRI, and Germany's EnMAP (Table 9.1).

9.2.2 Levels of Processing

Once data are collected from satellite or airborne remote sensing platforms, they are managed through the Earth Observing System Data and Information System (EOSDIS; Berrick, 2016). EOSDIS, through the Earth Science Mission Operations (ESMO) Project, assisted the Earth Observing System (EOS) satellite mission with command and control, scheduling, data collection, and processing to Level 0 (Berrick, 2016). EOSDIS also helped generate higher level data products for the EOS mission

TABLE 9.1
Comparison between Hyperspectral Satellite Sensors

	Hyperion	HyspIRI	EnMAP
Status	Decommissioned in 2017	Unknown expected launch date	Expected to launch in 2019
Spatial resolution	30 m	30 m for VNIR, 60 m for SWIR	30 m
Temporal resolution	16 days	19 days for VSWIR, 5 days for TIR	4 or 27 days depending on view angle
Spectral resolution	242 bands	221 bands	244 bands
Spectral range	356–2577 nm	380–2510 nm, 3–13 μm	420–2450 nm
Spectral bandwidth	10 nm	10 nm	6.5 nm in VNIR, 10 nm in SWIR
Swath width	7.75 km	153 km for VSWIR, 600 km for TIR	30 km

Abbreviations: VNIR: Visible and near-infrared; SWIR: Shortwave infrared; VSWIR: Visible, near-infrared, and shortwave infrared; TIR: Thermal infrared.

through the Earth Science Data and Information System (ESDIS) Project (Berrick, 2016). The Earth Resources Observation and Science (EROS) Center through the US Geological Survey (USGS) then functions as a Distributed Active Archive Center (DAAC), for processing, archiving, and distributing EOS data (USGS, 2013).

Historically, there were several levels of Hyperion image processing. Barry (2001) described them as follows:

- *Hyperion Level 0 data* are downlinked from EO-1, decoded, separated into files, and verified for data integrity and instrument performance. The visible and near-infrared (VNIR) and the shortwave-infrared (SWIR) datasets are combined and then converted into an HDF (Barry, 2001).
- *Hyperion Level 1 data* are generated when smear correction and then echo removal is performed on the Level 0 data (Barry, 2001). Echo is when a feature is anomalously repeated 11 lines offset from that feature due to "electronic cross-talk" (Pearlman, 2003), while smear is when a pixel is influenced by signal from an adjacent pixel due to spatial and scan-velocity mismatch (Lomheim et al., 1999; Pearlman, 2003). The data are then calibrated using pre- and post-image dark frames (Barry, 2001). The background is removed, and then the data are radiometrically calibrated (Barry, 2001). Known bad pixels are then repaired (Barry, 2001).
- *Hyperion Level 1A data* output is in signed integers, as opposed to the unsigned integers in the Level 1 product (Barry, 2001).
- *Hyperion Level 1B data* include the spatial image-to-image co-registration of the VNIR and SWIR datasets (Barry, 2001).

In both Level 1A and 1B products, saturated pixels were first identified using Level 0 data (Barry, 2001). The USGS Earth Explorer site also has Level 1Gst (radiometrically and geometrically corrected) and Level 1T (radiometrically corrected, and geometrically corrected with ground control points) Hyperion products.

9.2.3 Hyperion and Other Pushbroom Sensors: An Overview of Specific Pre-Processing Issues

Hyperion and other pushbroom sensors have specific pre-processing requirements. One is destriping. Striping in images generated by pushbroom sensors is caused by differences in calibration between the detector arrays and the fact that each column and band has a separate detector (Goodenough et al., 2003). Another requirement is a smile correction. The spectral curvature, or smile, effect is a change in band width and center wavelength going from the nadir point of the sensor or center of an image to the edge of the image (Aktaruzzaman, 2008; Dadon et al., 2010a). Similarly, the keystone effect is a change in pixel size going from the center to the edge (Dadon et al., 2010a). The maximum spatial shift due to the keystone effect is 0.5 pixels (Scheffler and Karrasch, 2014) and is not discussed further here. The smile effect is smaller in the SWIR than in the VNIR (San and Suzen, 2011). The first 12 VNIR bands and several SWIR bands were most affected (Scheffler and Karrasch, 2014). Lastly, there is one more correction that needs to be made specific to Hyperion images. Hyperion uses different algorithms for reading out data from two spectrometers, one measuring VNIR and one measuring SWIR (Scheffler and Karrasch, 2014). This leads to spatial differences in the VNIR and SWIR datasets (Scheffler and Karrasch, 2014). The right half of the SWIR image (between pixels 128 and 129) is one pixel lower than the left half (Khurshid et al., 2006). The SWIR is also counterclockwise rotated by 0.22° compared with the VNIR (Khurshid et al., 2006). The VNIR and SWIR regions measured by the two spectrometers have a few overlapping bands that can be used for spatial co-registration (Staenz et al., 2002a). It should be noted that although issues such as striping and smile/ keystone effects exist in whiskbroom sensors, they affect those sensors much less than they affect pushbroom sensors (Schlapfer et al., 2007).

9.3 PRE-PROCESSING OF HYPERION IMAGES: WORKFLOW

Several approaches to image pre-processing exist for Hyperion imagery (Barry, 2001; Asner and Heidebrecht, 2003; Beck, 2003; Datt et al., 2003; Goodenough et al., 2003; Khurshid et al., 2006; Aktaruzzaman, 2008; Miglani et al., 2008; Dadon et al., 2010a; Richter et al., 2011; San and Suzen, 2011; Perkins et al., 2012; Farifteh et al., 2013; Scheffler and Karrasch, 2014), though with no established standard protocol (Scheffler and Karrasch, 2014). Pre-processing of Hyperion data includes spatial shift of SWIR, destriping, angular shift correction, noise reduction, keystone detection and correction, smile detection and correction, atmospheric correction, and post-processing (Khurshid et al., 2006; Scheffler and Karrasch, 2014). Pre-processing of VNIR and SWIR separately is recommended, combining them after pre-processing (Datt et al., 2003).

However, different steps have been included in protocols by different researchers. For example, Staenz et al. (2002a) pre-processed Hyperion imagery by removing the SWIR line shift, removing striping, removing the smile effect, co-aligning VNIR and SWIR, and then performing atmospheric correction using MODTRAN4.2. Farifteh et al. (2013) pre-processed the imagery by performing an angular shift, destriping, smile correction, FLAASH atmospheric correction, and post-processing via filtering and spectral smoothing. Khurshid et al. (2006) used MODTRAN for atmospheric correction. A proposed workflow (Figure 9.1) may include:

1. VNIR and SWIR separation
2. Converting digital numbers to radiance
3. Removing problematic bands
4. Smile correction separately on VNIR and SWIR
5. Merging VNIR and SWIR
6. Destriping
7. Atmospheric correction
8. Cloud removal and spectral polishing/smoothing

9.4 HYPERSPECTRAL PRE-PROCESSING STEPS, DESCRIBED IN DETAIL

The pre-processing steps outlined in Section 9.3 (Figure 9.1) are described in detail in subsections below (Sections 9.4.1 through 9.4.8). These steps are computed in the GEE cloud platform, which is described in Section 9.5.

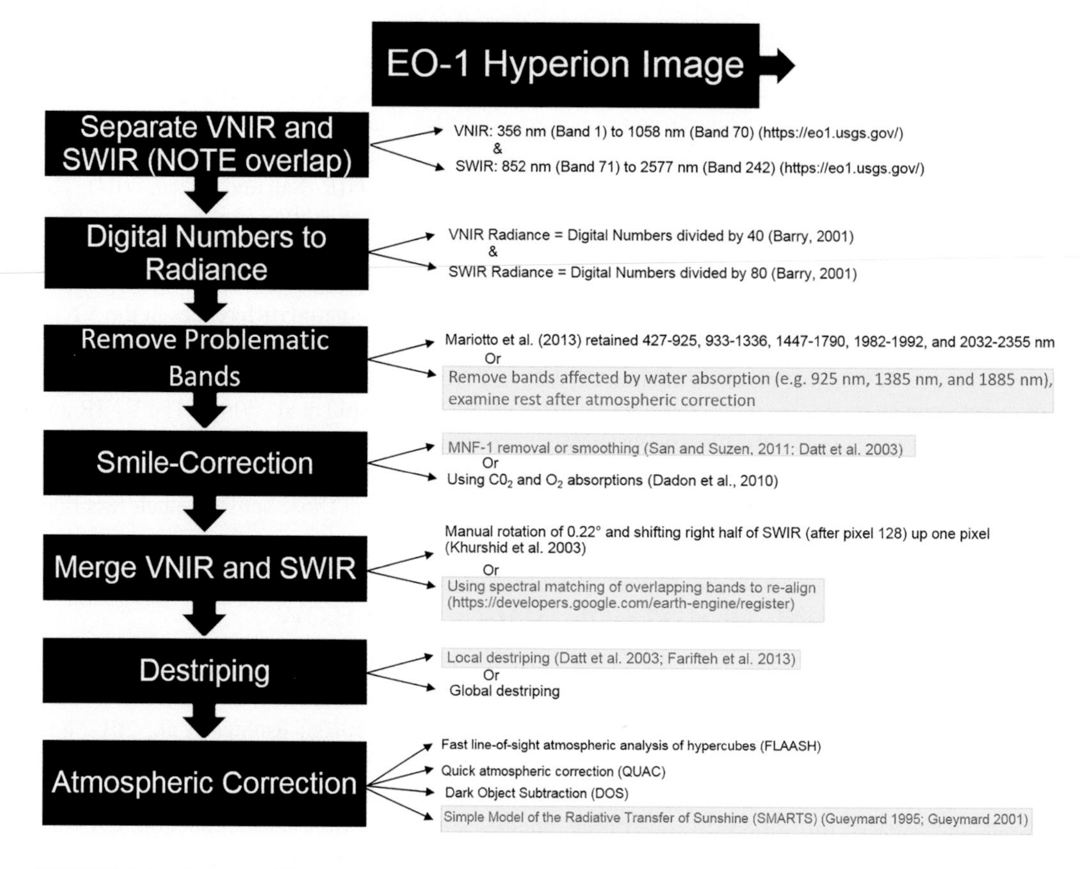

FIGURE 9.1 Suggested EO-1 Hyperion pre-processing workflow. These are the most common methods available for pre-processing steps; other methods exist. Highlighted methods are those the authors recommend.

9.4.1 VNIR and SWIR Separation

As mentioned above, the VNIR and SWIR data were obtained from separate spectrometers, which differ in calibration requirements. Therefore, the VNIR data, bands 1–70, and the SWIR data, bands 71–242 should be separated to make subsequent steps easier, such as converting to radiance (different scaling factors), smile correction (different severities of the smile effect), and VNIR-SWIR alignment.

9.4.2 Converting Digital Numbers to Radiance

In Hyperion imagery (Figure 9.2), digital numbers in the VNIR bands are converted to radiance by dividing the digital numbers by 40. Digital numbers in the SWIR bands are converted to radiance by dividing by 80 (Barry, 2001). The script example of how this can be done in the GEE API (application programming interface) are available in GitHub (https://github.com/ianeece/hyperion_preproc_gee_code), along with separation of the VNIR and SWIR regions.

9.4.3 Removing Problematic Bands

The 242 bands of Hyperion consist of approximately 40 times greater data volume compared to 6 non-thermal Landsat bands. Although there are 220 unique bands in Hyperion data, many of those bands are noisy. Slightly different sets of Hyperion bands have been retained by different researchers. Farifteh et al. (2013) retained bands 1–92 (426–1346 nm), 103–136 (1457–1790 nm),

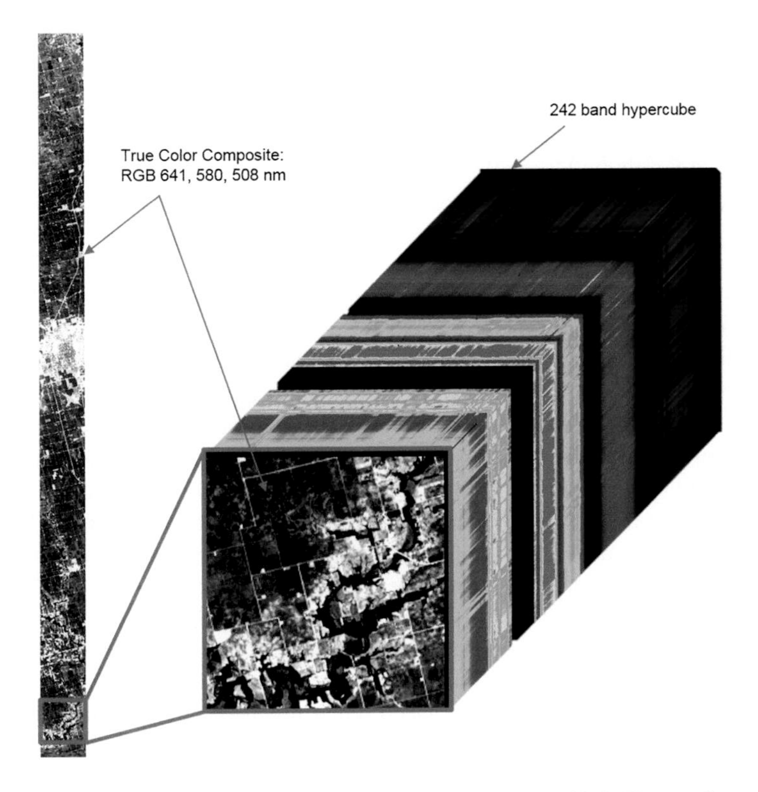

FIGURE 9.2 EO-1 Hyperion image over Champaign, IL, USA, July 12, 2012. Hypercube composed of raw data from 242 bands; surface of hypercube true color composite: RGB 641, 580, 508 nm.

and 154–193 (1971–2365 nm) for a total of 166 bands. San and Suzen (2011) used 141 bands: 437–922, 925–1295, 1477–1739, 2012–2375 nm. Thenkabail et al. (2004) used 157 bands: 427.55–851, 924–925.85, 932.72–1305, 1438–1789, and 1993–2364 nm. Mariotto et al. (2013), like Thenkabail et al. (2004), retained 158 bands: 8–57 (426.82–925.41 nm), 79–119 (932.64–1336.15 nm), 133–164 (1477.43–1790.19 nm), 183–184 (1981.86–1991.96 nm), and 188–220 (2032.35–2355.21 nm). Datt et al. (2003) retained 176 bands: 447–925, 953–1114, 1155–1336, 1488–1790, and 1972–2365 nm. These choices were made based on atmospheric absorption features and signal to noise ratios. Some bands also have more striping than other bands in Hyperion images. Of course, the study objectives also play a role in which bands can be removed, and which need to be retained.

For example, Mariotto et al. (2013) and Marshall and Thenkabail (2015b) determined which bands were the most useful for predicting crop biomass and yield, and which are best from crop discrimination. For crop productivity modeling, 53.6% of the best hyperspectral narrowbands (HNBs) were in the shorter wavelengths of the mid-infrared (EMIR, 1301–1990 nm) and far mid-infrared (FMIR, 1901–2350 nm), especially 1371, 1731, 2121, 2181, and 2191 nm (Mariotto et al., 2013). More specifically, Mariotto et al. (2013) found that the 933, 2052, and 2285 nm bands were best for estimating cotton wet biomass; in terms of 2-band indices, bands at 2194 and 1528 nm were best for linear modeling of wet biomass. The bands at 427, 437, 973, and 1165 nm were most important for estimating maize biomass, while the 1033 and 722 nm bands were best for linear modeling of maize wet biomass with a 2-band index (Mariotto et al., 2013). The best two-band normalized difference hyperspectral vegetation indices (HVIs) for estimating crop productivity were (1) 742 and 1175 nm, (2) 1296 and 1054 nm, (3) 1225 and 697 nm, and (4) 702 and 1104 nm (Mariotto et al., 2013). Additionally, the best three bands for crop type discrimination were 885, 943, and 2143 nm (Mariotto et al., 2013). Marshall and Thenkabail (2015b) found that the best two bands for estimating rice growth were 963 and 993 nm. For cotton, the best two bands were 1134 and 539 nm (Marshall and Thenkabail, 2015b). The best three bands for alfalfa were

631, 468, and 438 nm, while those for maize were 865, 845, and 794 nm (Marshall and Thenkabail, 2015b). Marshall and Thenkabail (2015a) found that the best bands for estimating biomass for rice using two-band vegetation indices were at 549 and 752 nm. For alfalfa, the best bands were at 925 and 1104 nm; for cotton, they were at 722 and 732 nm; for maize, they were at 529 and 895 nm (Marshall and Thenkabail, 2015a). Similarly, Marshall et al. (2016) assessed which hyperspectral vegetation indices were best for estimating evapotranspiration and its components, transpiration and soil evaporation.

Additionally, there are certain spectral regions that are most useful for specific applications. The 400–700 nm region is useful for estimating photosynthetic pigments, 680 nm for chlorophyll absorption, 700–750 nm for photosynthetic activity, 1080–1170 nm for the liquid water inflection point on the spectral curve, 1700–1780 nm for leaf waxes and oils, 2100 nm for cellulose, 2100–2300 nm for soil properties, and 2280–2290 nm for nitrogen and protein content (Datt et al., 2003). The near-infrared region is useful for predicting crop biophysical characteristics and crop productivity while the far-infrared and bands in the shorter wavelengths of the mid-infrared (1301–1900 nm) are useful for predicting biomass and leaf area index (Mariotto et al., 2013). The early mid-infrared and far mid-infrared (1901–2500 nm) are useful for studying lignin, cellulose, and starch content, geometric structure of canopies, and optical soil properties (Mariotto et al., 2013). Thenkabail et al. (2004) found that the 675 and 680 nm regions were useful for crop discrimination; the 1710, 1467, and 2052 nm regions were useful for estimating starches, 1467 nm for water, 2052 nm for protein, a broad absorption feature at 2050–2140 nm for lignin, and 783 and 895 nm for biomass. Somers and Asner (2013a,b, 2014) have also developed innovative techniques to remove correlated bands and select the bands most useful for classifying species by using the Separability Index. They also developed a modified multiple end-member spectral mixture analysis technique (MESMA) to select particular spectral subsets for classifying species in a given mixed pixel (Somers and Asner, 2013a,b, 2014). Thenkabail et al. (2000, 2004, 2013) have developed Lambda versus Lambda correlation analyses to reduce data dimensionality while retaining the most useful spectral information.

Thus, band-selection in terms of which bands to retain is highly dependent on application. Band removal also varies by study. However, three regions are very commonly removed: areas around 925, 1385, and 1885 nm. These regions are affected by water vapor absorption (San and Suzen, 2011) and have low signal-to-noise ratio (Thenkabail et al., 2004). The authors removed bands at 933–963, 1114–1155, 1336–1498, 1780–2042, and 2365–2396 nm, which were most affected by water vapor absorption (Figure 9.3).

9.4.4 Smile Correction

The keystone and smile effects in Hyperion imagery add spatial noise to the data. The keystone effect is the inter-band spatial mis-registration in Hyperion imagery, caused by geometric distortions and/or chromatic aberrations (Bannari et al., 2015). However, this effect has been found to be minor, and thus is often not incorporated into Hyperion pre-processing chains (Bannari et al., 2015). The smile effect is a wavelength shift due to spatial distortions or aberrations (Bannari et al., 2015). There are several methods used to correct for the smile effect including moving linear fitting and interpolation, column mean adjusted in radiance space, and column mean adjusted in Minimum Noise Fraction (MNF) space. The first works for AVIRIS but not completely for Hyperion data; however, it does retain spectral fidelity to the original data. The second assumes the image is homogenous, which is usually not the case, and thus creates false spectra. The third also creates false spectra when rotated from MNF back to radiance space (Goodenough et al., 2003). An MNF transform is essentially two principal components analyses (PCA) done back-to-back (Christian and Krishnayya, 2009; Underwood et al., 2007). The first MNF component displays variation due to the smile effect, but is also influenced by the keystone effect (Dadon et al., 2010a). This first component also represents variance due to albedo, brightness, topography, sun angle, and image mis-calibration (Underwood et al., 2007). This MNF-1 component can be corrected through polynomial smoothing or entirely removed and the original image then reconstructed to remove the smile effect (Dadon et al., 2010a) (Figure 9.4). Another method

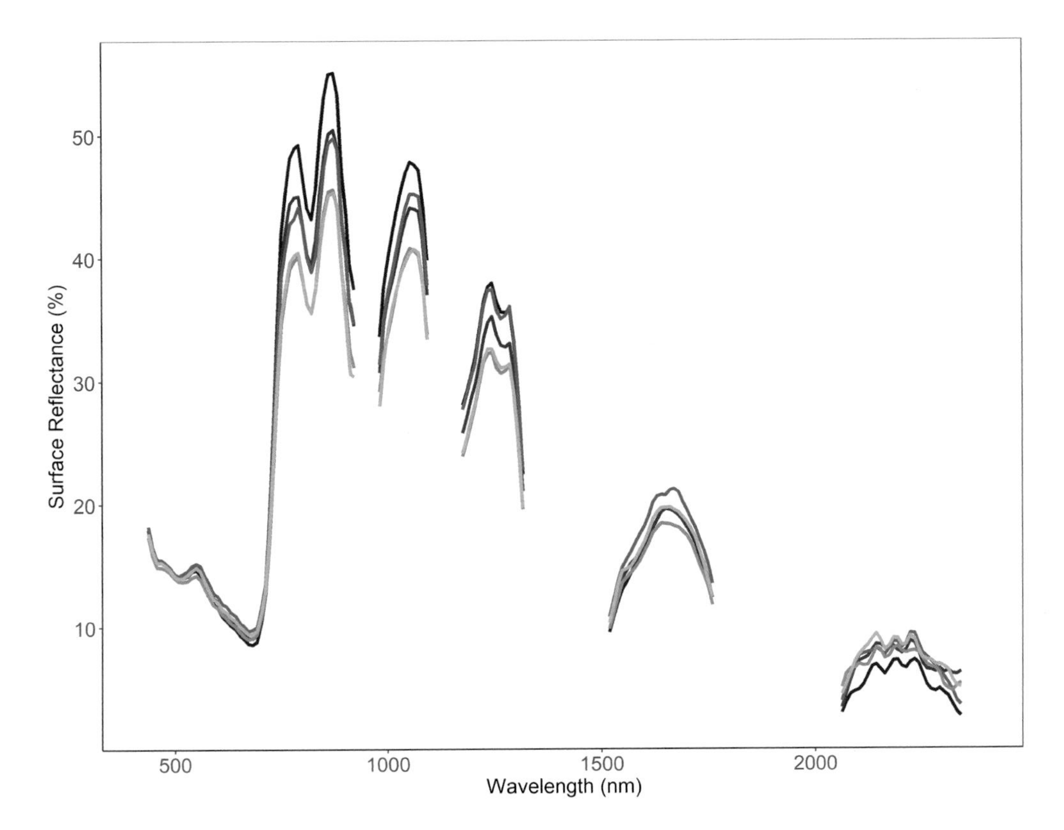

FIGURE 9.3 Example corn spectra from Hyperion image over Champaign, IL, USA, July 12, 2012. Problematic bands at 933–963, 1114–1155, 1336–1498, 1780–2042, and 2365–2396 nm were removed and spectra were smoothed using a moving average across 3 bands.

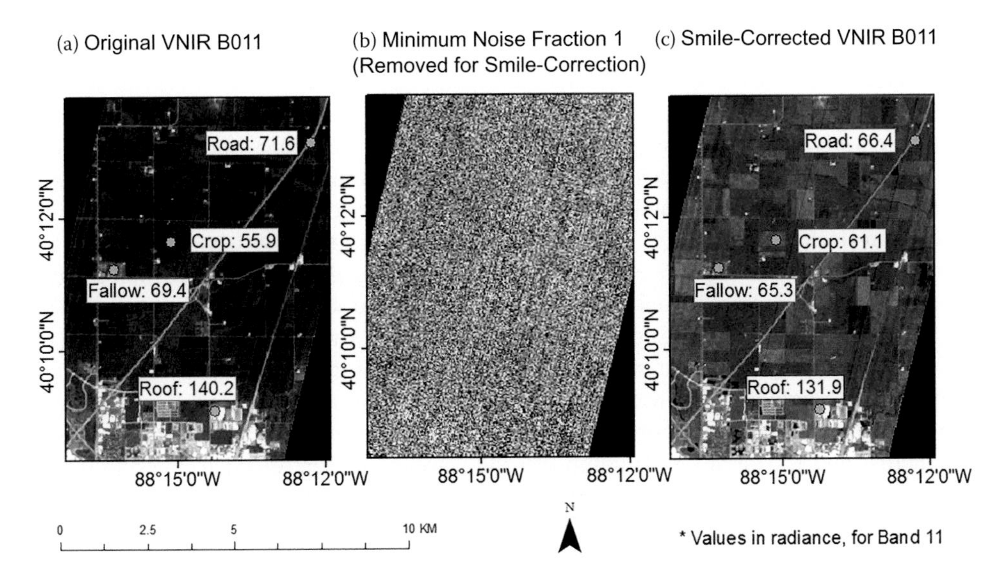

FIGURE 9.4 Smile correction of EO-1 Hyperion image band 11 (447 nm) over Champaign, IL, USA, July 12, 2012, by removing minimum noise fraction 1 (MNF-1) and reconstructing the image using two inverse principal component analyses. (a) Original, uncorrected radiance image; (b) MNF-1, which represents noise from smile effect; (c) smile-corrected radiance image.

involves conducting a trend-line smile correction, which uses the first MNF transform to remove spectral noise caused by the smile, but using a trendline to isolate spectral effects from spatial effects (Dadon et al., 2010a). Yet another method was used by San and Suzen (2011), in which a column mean adjustment procedure in MNF-transformed space outlined by Datt et al. (2003) was used for noise correction. This was done by first calculating the average value for each column and band. A line of best fit or a nonlinear polynomial function of best fit was then created using the CTIC algorithm (San and Suzen, 2011). New digital numbers (DN) were calculated by summing the difference between computed values and original DNs or by multiplying the ratio of the computed value to original DN (San and Suzen, 2011). Dadon et al. (2010a) used the O_2 absorption near 760 nm to detect the smile effect in the VNIR and CO_2 absorption at 2012 nm to detect that in the SWIR. The smile correction can be incorporated into the atmospheric correction (Richter et al., 2011), such as by using FLAASH-IDL (Perkins et al., 2012).

An example method of conducting a smile correction in GEE is available in GitHub (https://github.com/ianeece/hyperion_preproc_gee_code). In this example, two PCAs are conducted back-to-back to obtain MNF results. The first minimum noise fraction component, which represents noise from the smile effect, is removed, and two inverse principal components analyses are conducted to reconstruct the original image without the smile effect. However, the authors find that this method does not lead to a removal of only the smile effect, and thus is not the best method of smile-correction. This code has been provided for anyone interested in improving it to obtain more acceptable results.

9.4.5 VNIR-SWIR Alignment

VNIR and SWIR can be re-aligned by shifting the right half of the SWIR image one pixel up, and rotating the VNIR counterclockwise by 0.22° (Khurshid et al., 2006; Figure 9.5). Alternatively, Bands 56 and 57 from the VNIR image, which spectrally overlap with Bands 77 and 78 from the SWIR image, can be used to re-align the two images using a displacement tool in GEE.

9.4.6 Destriping

There are also several methods for destriping Hyperion images. Scheffler and Karrasch (2014) tested six destriping methods: ENVI Hyperion Tools, ENVI-SPEAR-Tools Vertical stripe removal, ENVI General Purpose Utilities-Destripe, methods outlined by Datt et al. (2003) (local destriping), ERDAS Periodic Noise Removal, and wavelet Fourier adaptive filtering. Local destriping is often used because it often outperforms global destriping (Datt et al., 2003). Local destriping can be done by matching the mean and standard deviation of each column to match the mean and standard deviation of the corresponding band; global destriping, not as good, uses the entire image rather than a column (Farifteh et al., 2013). One advantage of using global destriping, however, is that it can minimize the smile effect mentioned earlier (Datt et al., 2003). Local destriping has been found to substantially decrease striping (Datt et al., 2003); some stripes, especially several adjacent to each other, were not removed, but may be removed using a wider filter kernel (Scheffler and Karrasch, 2014). During local destriping, only a few columns are recalculated, and those that exceed an established threshold of difference from neighboring columns are addressed. Recalculation also preserves data more than other destriping techniques (Scheffler and Karrasch, 2014). On the other hand, Scheffler and Karrasch (2014) found that the wavelet Fourier Adaptive Filtering technique removed more stripes than did local destriping, but with greater alteration of the original data. Destriping and smile correction can result in negative pixel values. These negative values can then be replaced by the average of the surrounding eight pixels (Farifteh et al., 2013).

An example of destriping in GEE is shown in Figure 9.6. In the script, a kernel was constructed to calculate the mean values within a potential stripe. Then another kernel was constructed to calculate the mean values within the neighboring pixels of the potential stripe. The difference between the two regions was then compared, and a threshold difference of the standard deviation divided by 3.5

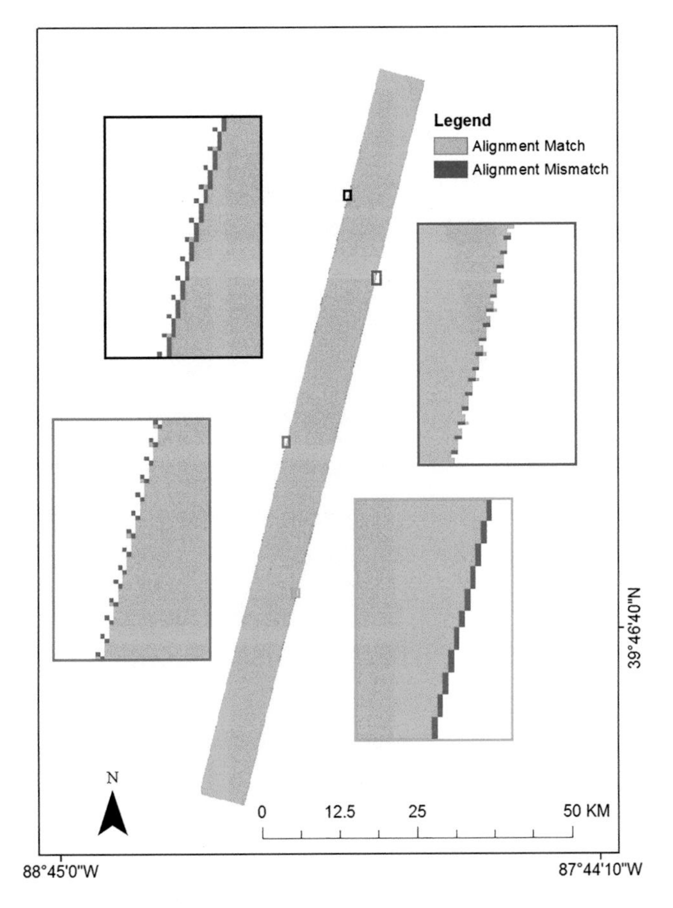

FIGURE 9.5 A comparison between the original shortwave infrared (SWIR) image footprint and the realigned SWIR footprint of an EO-1 Hyperion Image over Champaign, IL, USA, July 12, 2012. Orange represents where there is alignment match and blue highlights alignment mismatch, often less than 1 pixel in width.

was used to determine the stripes within the image. This local destriping is done on each individual band, since stripe locations and intensities vary by band. Thus, a pixel that might be masked out in one band may not be masked out in another adjacent band. The authors found that the value of a pixel that was masked out in one band was often not different from the value of that pixel in adjacent bands where it was not masked out. Thus, the authors find that after bad bands are removed, much of the striping issue is also resolved, without further need for destriping. The subtlety of these stripes is illustrated in Figure 9.6.

9.4.7 Atmospheric Correction

Atmospheric correction of satellite imagery is important to compare data from imagery across different dates, locations, and atmospheric conditions. Top of atmosphere reflectance (TOA) can be calculated using Equation 9.1 (Gueymard, 2001), where L is at-satellite radiance in W m^{-2} sr^{-1} nm^{-1}, d is Earth-Sun distance in astronomical units, E_{sun} is solar irradiance in W m^{-2} sr^{-1} nm^{-1}, and θ_z is zenith angle in radians. For surface reflectance to be calculated, we need to remove atmospheric effects. Assuming no haze, surface reflectance can be calculated using Equation 9.2 (Chavez, 1996), where θ_v is viewing angle in radians, T is transmittance (unitless) affected by various atmospheric elements, and E_{down} is diffuse downwelling irradiance. Transmittance can be calculated using Equation 9.3 (Gueymard, 2001), where $T_{R\lambda}$ is Rayleigh transmittance (unitless) dependent on

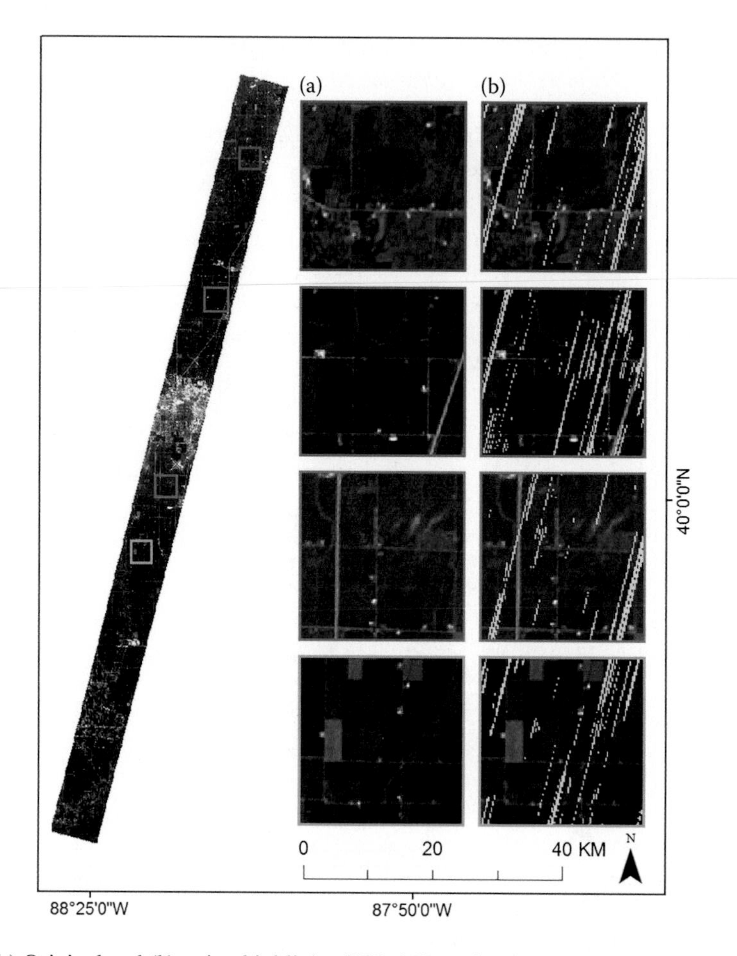

FIGURE 9.6 (a) Original and (b) stripe-highlighted EO-1 Hyperion image over Champaign, IL, USA, July 12, 2012, Band 11 (447 nm). Different colors of boxes serve as guides to match location of inserts on Hyperion image. This figure illustrates how subtle the striping is on non-noisy bands, so much so that they are difficult to see in the original image but able to be detected by an algorithm, as seen in the stripe-highlighted inserts.

wavelength, λ, $T_{o\lambda}$ is ozone transmittance (unitless), $T_{n\lambda}$ is nitrogen dioxide transmittance (unitless), $T_{g\lambda}$ is uniformly mixed gas transmittance (unitless), $T_{w\lambda}$ is water vapor transmittance (unitless), and $T_{a\lambda}$ is aerosol transmittance (unitless). Details for calculating these terms can be found in Gueymard (1995) and Gueymard (2001). E_{down} is calculated using Equation 9.4, where τ_D is the dimensionless transmission coefficient for direct solar radiation (Liu and Jordan, 1960). Details for calculating τ_D have been described by Gopinathan and Polokoana (1986).

$$\mathrm{TOA} = \frac{\pi * L * d^2}{E_{sun} * \cos(\theta_z)} \tag{9.1}$$

$$\mathrm{SR} = \frac{\pi * L}{\cos\theta_V * (E_{sun} * \cos\theta_Z * T + E_{down})} \tag{9.2}$$

$$T = T_{R\lambda} * T_{o\lambda} * T_{n\lambda} * T_{g\lambda} * T_{w\lambda} * T_{a\lambda} \tag{9.3}$$

$$E_{down} = E_{sun} * \tau_D \tag{9.4}$$

There are many papers comparing the use of various atmospheric models including MODTRAN (Moderate Resolution Atmospheric Transmittance), ATREM (Atmospheric Removal), ACORN (Atmospheric Correction Now), FLAASH (Fast-Line-of-sight-Atmospheric Analysis of Hypercubes), DISORT (Discrete Ordinate Radiative Transfer), A/CHEAT (Auscover/Curtin Hyperion Enhancement and Atmospheric Correction Technique), and SMARTS (Simple Model of the Radiative Transfer of Sunshine; Gueymard, 1995, 2001; Staenz et al., 2002b; Kruse, 2004; Gao et al., 2009; San and Suzen, 2011). Such models solve the radiative transfer equations to assess the influence of atmospheric components on incoming radiation. Details for the radiative transfer equations behind many of these methods are described by Bird and Hulstrom (1981), Fraser et al. (1989), Staenz et al. (2002b), Lenot et al. (2009), Berk et al. (1998), Perkins et al. (2012), and Vermote and Vermeulen (1999). Detailed descriptions of MODTRAN codes and algorithms have been provided by Berk et al. (1987, 1998, 2000, 2008) and Matthew et al. (2000). San and Suzen (2011) provide an example of the successful use of ACORN in mineral classification. Examples and descriptions of FLAASH, which uses MODTRAN (Farifteh et al., 2013), are provided by Farifteh et al. (2013), Felde et al. (2003), and Adler-Golden et al. (1998). Stamnes et al. (2000) provide codes for DISORT. The algorithms behind the SMARTS model are described by Gueymard (1995, 2001). Many of these models need a priori information about the location and environmental conditions in which these images were taken in order to do atmospheric correction (San and Suzen, 2011).

However, many times, this information is not available (Song et al., 2001). There are ways to estimate such parameters as atmospheric water content (Gao and Goetz, 1990) and visibility (Peng and Li, 2016) from existing imagery. However, in situations where values for all needed parameters are not available, there are scene-based techniques that can be used for more basic calibration, including dark object subtraction, flat field, internal average relative reflection, log residual modified model, and empirical line calibration (Gao et al., 2009; San and Suzen, 2011; Ganesh et al., 2013). These techniques are useful but not as reliable as models based on radiative transfer (Ganesh et al., 2013).

An example of atmospheric correction in GEE is illustrated in Figure 9.7. Within this script, the influence of atmospheric components on solar radiation going through the atmosphere towards the Earth's surface was determined using the SMARTS model (Gueymard, 1995, 2001). Seidel et al. (2010) have found that the SMARTS model is 25 times faster than the 6S model, with only a 5% difference in spaceborne data processing results. In GEE, we first defined image characteristics, such as zenith angle, which are in the metadata included in GEE. Digital numbers were then converted to radiance values. Then we defined several coefficients needed to calculate the influence of several atmospheric components. Additionally, we built a lookup table of coefficients that varied by wavelength. After defining the coefficients, we built a function to calculate the transmittance influenced by various atmospheric components including Rayleigh transmittance, ozone transmittance, NO_2 transmittance, mixed gas transmittance, water vapor transmittance, and aerosol transmittance. Site-specific relative humidity, pressure, and visibility information was needed for calculating transmittance. We used the National Oceanic and Atmospheric Administration (NOAA) Integrated Surface Hourly (ISH) data (NOAA, 2017) to obtain visibility, University of Idaho's Gridded Surface Meteorological (GRIDMET) dataset (Abatzoglou, 2011) for relative humidity, and NASA's North American Land Data Assimilation System 2 (NLDAS-2) data (Pearce et al., 2017) for site-level pressure. We then calculated surface reflectance using total transmittance and diffuse irradiance.

9.4.8 Cloud Removal and Spectral Smoothing

Clouds and cloud shadows can affect spectral values and spectral profile shapes, and thus pixels affected by these artifacts need to be removed. Clouds can be detected using threshold reflectance in the blue band and the sum of reflectance in the red, green, and blue bands since clouds usually have very high reflectance in the visible region. Cloud shadows can be detected using the sum of bands in the NIR and SWIR regions, because shadows often have much lower reflectance in those regions. Thresholds vary by image and the degree to which the user wants to mask out clouds and shadows. The cloud shadow mask also masks water bodies; a more complex mask would be needed if the project involves study of water.

Atmospheric correction often makes spectral profiles noisier so that they need spectral smoothing or polishing. This can be done using a Savitsky-Golay local polymer filter in the "signal" package in R (Marshall and Thenkabail, 2015a), by using the Empirical Flat Field Optimal Reflectance Transformation (EFFORT; Farifteh et al., 2013), by using MNF smoothing (Datt et al., 2003), and other methods. Datt et al. (2003) found that MNF smoothing after atmospheric correction almost always improved accuracy of biochemical and biophysical estimations. The authors used a 3-band moving window to average the spectral profiles in R.

9.5 CLOUD COMPUTING

With rapid advancements in remote sensing technology and the associated increase in data volume, we face new processing challenges that need to be addressed (Agapiou, 2017; Plaza et al., 2011). As an example, Sentinels 2A and 2B collect over 400 terabytes annually (Agapiou, 2017). Landsat 7 acquires 300 images daily, which is more than 1 petabyte of data (Padarian et al., 2015). Cloud computing, such as through the use of GEE, can help ameliorate those challenges. GEE stores petabytes of satellite images, digital elevation models, atmospheric data, meteorological data, and land cover data using Google's infrastructure (Padarian et al., 2015; Tang et al., 2016; Agapiou, 2017). With the use of GEE, users can work with terabytes of data without having to download them to a personal computer (Navulur et al., 2013; Tang et al., 2016). They can visualize data using the Explorer interface, or run analyses with the Code Editor using JavaScript, or a Python API (Huntington et al., 2016; Agapiou, 2017).

Cloud computing reduces the need for personal storage space for storing imagery and lessens demands for processing power on individual computers (Navulur et al., 2013). The parallel processing capability of GEE greatly speeds up processing time for large datasets (Dong et al., 2016). For example, Padarian et al. (2015) analyzed over 654 thousand Landsat images for their study, which amounted to 707 terabytes of data, in approximately 100 hours; without GEE, it would have taken 1,000,000 hours. These advantages of using GEE allow for the expansion of studies to larger spatial and temporal extents (Goldblatt et al., 2016; Huntington et al., 2016). GEE also makes it possible to include more images in a study without drastically increasing processing time. For example, the use of GEE allowed Soulard et al. (2016) to use continuous Landsat imagery rather than annual or decadal imagery, which then allowed them to detect dormant season fire effects and seasonal NDVI patterns that enabled the discrimination of deciduous meadow vegetation and evergreen conifers.

GEE has been used for various types of research, including archeology (Agapiou, 2017), wetland inundation (Tang et al., 2016), rice paddy mapping (Dong et al., 2016), groundwater change (Huntington et al., 2016; Carroll et al., 2017), agricultural land use suitability (Yalew et al., 2016), river avulsion (Edmonds et al., 2016), soil mapping (Padarian et al., 2015), crop yield mapping (Lobell et al., 2015), urban boundary mapping (Goldblatt et al., 2016), fire effects (Soulard et al., 2016), and oil palm cultivation changes (Okoro et al., 2016). GEE is still under "beta-testing" and thus still has limitations, but continues to develop and provides big opportunities for soil and environment science studies (Padarian et al., 2015).

9.6 DISCUSSION

9.6.1 Pre-Processing Summary

Above, we outlined the steps for pre-processing hyperspectral Hyperion satellite data and discussed various methods available to perform each step. Below, we will offer recommendations of methods

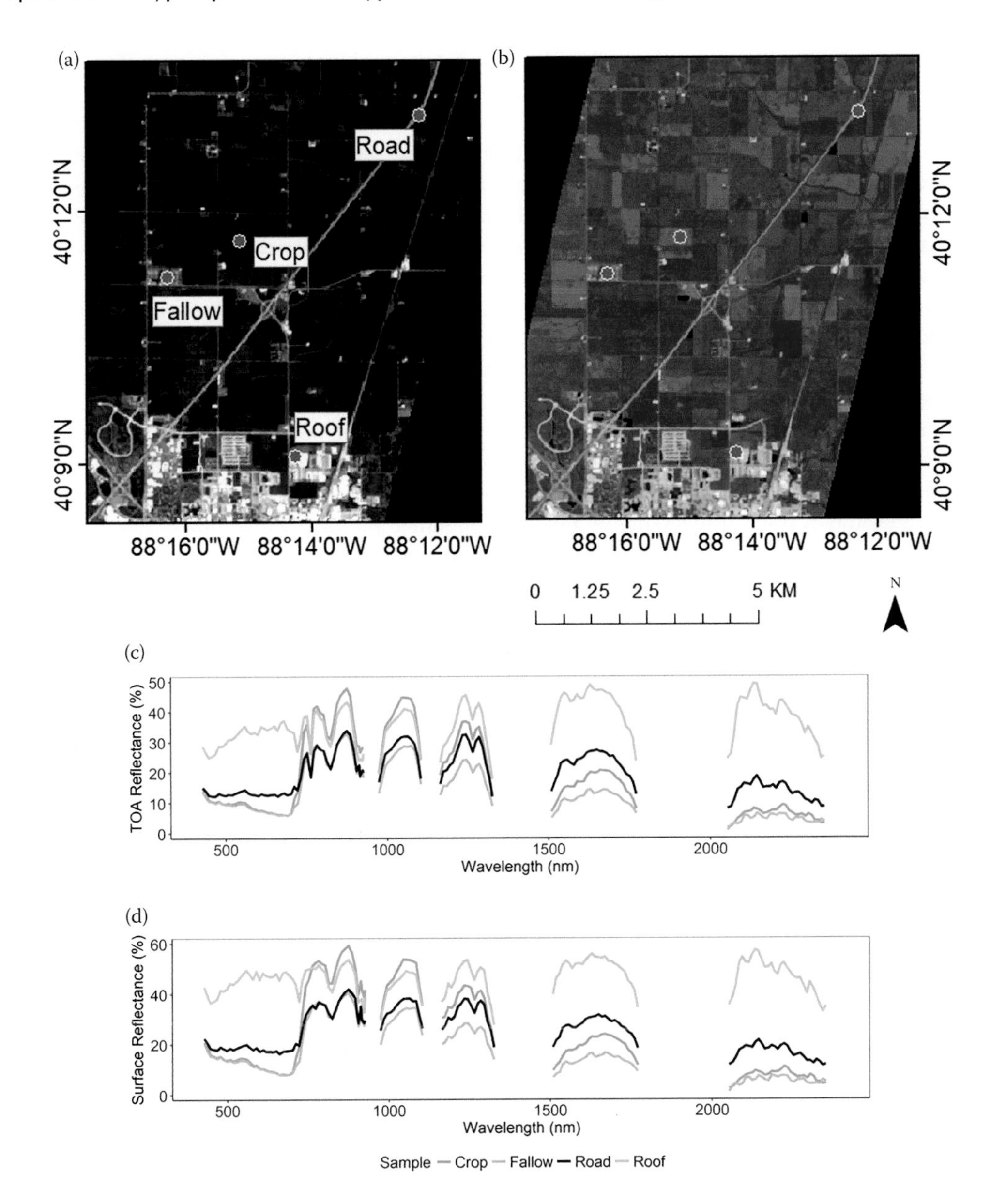

FIGURE 9.7 EO-1 Hyperion image over Champaign, IL, USA, July 12, 2012. Surface reflectance in (a) true color composite of RGB bands 29, 23, 16 (641, 579, 508 nm) and (b) false color composite of RGB bands 50, 23, 16 (854, 579, 508 nm). Spectral profiles showing (c) top of atmosphere reflectance and (d) atmospherically corrected surface reflectance.

for each step that had multiple methods mentioned above. For example, as mentioned earlier, bands retained vary across studies. The regions that are most often removed include bands around 925, 1385, and 1885 nm. However, there are many other regions that can be noisy. The benefit to removing all possible noisy bands is a clean dataset from the outset. However, this may lead to removal of bands that are not noisy in a particular dataset and thus to elimination of potentially useful data. Therefore, the authors recommend retention of as many bands as possible (perhaps removing the three regions just mentioned) throughout pre-processing, and evaluation of other problematic bands after pre-preprocessing. Of the smile correction methods discussed previously, MNF-1 removal or smoothing are recommended because they do not depend on in situ data as do some other methods. Merging of VNIR and SWIR by spectral matching of overlapping bands is preferable in GEE. There are also software programs that manually re-align the two images, like the Imaging Spectrometer Data Analysis System (ISDAS) developed by the Canada Centre for Remote Sensing. This software also conducts other pre-processing steps for hyperspectral data analysis. However, the use of this software for hundreds of Hyperion images may be time-consuming. The authors recommend local destriping over global because there are fewer assumptions made across the entire image. The SMARTS model is recommended for atmospheric correction because of the ease and speed of implementation as well as comparability with other more complex atmospheric correction as demonstrated by Seidel et al. (2010). Alternatively, the Py6S algorithm has recently become available in the GEE Python API. This more complex atmospheric correction is easy and fast to implement, although it makes a few assumptions that the authors are not as comfortable making for hyperspectral data.

Many researchers have proposed workflows for pre-processing of Hyperion imagery. However, many of those studies focus on one aspect of the workflow rather than the whole process. This chapter summarizes findings of best practices for the entire Hyperion pre-processing workflow, including examples of pre-processing in a cloud-computing platform, GEE. Since this platform is still in the development stage, there are still certain limitations to using it for hyperspectral data analysis (Padarian et al., 2015). For example, the statistical tools available for analyzing spectral data are still limited (Padarian et al., 2015). However, as users continue to contribute to the GEE community, the platform will become more robust and increase in its usefulness for hyperspectral data analysis.

9.6.2 Applications

There are many advantages of using hyperspectral data, and they have already proven useful for many applications. Hyperion itself has not only provided data useful for many applications, it has also paved the way for advanced technology in the Landsat mission, and in future hyperspectral sensors. Many researchers have compared Hyperion data with other sensors. Several researchers have compared the performance of multispectral sensors, especially Landsat, with Hyperion and have found that the increased spectral resolution was useful (Thenkabail et al., 2000, 2004; Numata et al., 2011; Thenkabail et al., 2013; Mariotto et al., 2013; Marshall and Thenkabail, 2015a; Bostan et al., 2016; Pervez et al., 2016; Sonmez and Slater, 2016).

Marshall and Thenkabail (2015a) showed that hyperspectral narrowband vegetation indices explained 3%–33% greater variability in modeling above ground wet biomass (AWB) when compared with multispectral broadbands. Their study also established the importance of the red-edge (700–740 nm), which is absent in most MBBs, and identified six specific HNBs within the 400–2500 nm range that explained the most AWB variability: 539, 722, 758, 914, 1130, and 1320 nm (bandwidth of 10 nm). Mariotto et al. (2013) showed that the best HNB models explained variability in biophysical parameters overwhelmingly above 93%, which is approximately 25% higher than the best multispectral broadband models. They recommended 29 hyperspectral narrowbands centered at waveband centers of Hyperion: 447, 508, 579, 651, 681, 722, 803, 824, 885, 933, 943, 953, 963, 983, 1064, 1084, 1094, 1124, 1134, 1144, 1195, 1205, 1488, 1528, 1982, 2123, 2143, 2264, and 2274 nm. Thenkabail et al. (2000, 2004, 2013) also recommended hyperspectral optimal wavebands in the study of agriculture and vegetation. Their studies demonstrated approximately 30% increase in classification accuracies when using selected optimal hyperspectral narrowbands when compared with Landsat broadbands.

Castaldi et al. (2016) found a slight improvement with hyperspectral data, but not enough for studying soil texture. Hamzeh et al. (2016) found that Landsat 7 was better at categorically classifying salinity stress in sugarcane, while Hyperion was better at quantifying it. Daughtry et al. (2006) found that both Landsat 8 and Hyperion data were successful in mapping crop residue and classifying tillage intensity. On the other hand, comparisons with AVIRIS data resulted in a poorer performance by Hyperion due to its low signal to noise ratio (Asner and Heidebrecht, 2003; Roberts et al., 2003; Kruse et al., 2006). Hyperion data have also been combined with data from other sensors. For example, Song et al. (2014) combined Advanced Synthetic Aperture Radar (ASAR) with Hyperion data to estimate soil moisture in vegetated areas.

Hyperion data have also been used to simulate data of other sensors. For example, these data have been used to simulate HyspIRI data to assess the ability to use future HyspIRI data for specific applications (Zhang et al., 2012; Abrams et al., 2013). Hill (2013) used Hyperion data to simulate Sentinel-2, MODIS, and VIIRS data to assess the ability to use indices from Sentinel-2 data to study grasslands and savannahs, and the ability to combine Sentinel-2, MODIS, and VIIRS data for the study. Thus, Hyperion data have many applications, and a standardized pre-processing protocol for these data would aid research in various fields.

9.7 CONCLUSION

In this chapter, we reviewed the pre-processing of hyperspectral Hyperion data and proposed some standardized methods and procedures for Hyperion pre-processing workflow. Hyperion pre-processing steps using the GEE cloud-computing platform are illustrated. The chapter clearly demonstrates the ability of handling massive hyperspectral data volumes through the GEE cloud. The hyperspectral pre-processing steps presented and demonstrated in this chapter can become standard in not only pre-processing Hyperion data but other upcoming sensors such as NASA's HyspIRI.

REFERENCES

Abatzoglou, J., 2011. University of Idaho Gridded Surface Meteorological Dataset (UofI METDATA), http://metdata.northwestknowledge.net/

Abrams, M., Pieri, D., Realmuto, V., Wright, R., 2013. Using EO-1 Hyperion data as HyspIRI preparatory data sets for volcanology applied to Mt Etna, Italy. *IEEE Journal of Selected Topics in Applied Earth Observations and Remote Sensing*, 6(2), 375–385.

Adler-Golden, S., Berk, A., Bernstein, L., Richtsmeier, S., Acharya, P., Matthew, M., Anderson, G., Allred, C., Jeong, L., Chetwynd, J., 1998. *FLAASH, a MODTRAN4 atmospheric correction package for hyperspectral data retrievals and simulations.* Technical report, Spectral Sciences, Inc; Phillips Laboratory, Geophysics Directorate, Burlington and Hanscom AFB, MA.

Agapiou, A., 2017. Remote sensing heritage in a petabyte-scale: Satellite data and heritage Earth Engine applications. *International Journal of Digital Earth*, 10(1), 85–102.

Aktaruzzaman, M., 2008. Simulation and correction of spectral smile effect and its influence on hyperspectral mapping. MSc thesis.

Andrew, M., Ustin, S., 2006. Spectral and physiological uniqueness of perennial pepperweed (Lepidium *latifolium*). *Weed Science*, 54, 1051–1062.

Asner, G., Heidebrecht, K., 2003. Imaging spectroscopy for desertification studies: Comparing AVIRIS and EO-1 Hyperion in Argentina drylands. *IEEE Transactions on Geoscience and Remote Sensing*, 41(6), 1283–1296.

Bannari, A., Staenz, K., Champagne, C., Khurshid, K., 2015. Spatial variability mapping of crop residue using Hyperion (EO-1) hyperspectral data. *Remote Sensing*, 7, 8107–8127.

Banskota, A., Wynne, R., Kayastha, N., 2011. Improving within-genus tree species discrimination using the discrete wavelet transform applied to airborne hyperspectral data. *International Journal of Remote Sensing*, 32(13), 3551–3563.

Barry, P., 2001. *EO-1/ Hyperion Science Data User's Guide*, Level 1_b. Pub. Release, L1_B HYP.TO.01.077, TRW Space, Defense and Information Systems, Redondo Beach, CA.

Beck, R., 2003. *EO-1 User Guide v. 2.3.* Technical report, University of Cincinnati, Department of Geography, Cincinnati, OH.

Beland, M., Roberts, D., Peterson, S., Biggs, T., Kokaly, R., Piazza, S., Roth, K., Khanna, S., Ustin, S., 2016. Mapping changing distributions of dominant species in oil-contaminated salt marshes of Louisiana using imaging spectroscopy. *Remote Sensing of Environment*, 182(2016), 192–207.

Berk, A., Acharya, P., Bernstein, L., Anderson, G., Chetwynd, J., Hoke, M., 2000. *Reformulation of the MODTRAN band model for higher spectral resolution*. SPIE AeroSense meeting SSI Report No. sr133, Orlando, FL.

Berk, A., Anderson, G., Acharya, P., Shettle, E., 2008. *MODTRAN 5.2.0.0 User's Manual*. Technical Report.

Berk, A., Bernstein, L., Anderson, G., Acharya, P., Robertson, D., Chetwynd, J., Adler-Golden, S., 1998. MODTRAN cloud and multiple scattering upgrades with application to AVIRIS. *Remote Sensing of Environment*, 65, 367–375.

Berk, A., Bernstein, L., Robertson, D., 1987. *MODTRAN: A moderate resolution model for LOWTRAN7*. Scientific Report No. 1 AFGL-TR-87-0220, Spectral Sciences, Inc., Burlington, MA.

Berrick, S., 2016. About EOSDIS: An overview of EOSDIS.

Bhojaraja, B., Shetty, A., Nagaraj, M.K., Manju, P., 2015. Age-based classification of Arecanut crops: A case study of Channagiri, Karnataka, India. *Geocarto International*, 1–11.

Bird, R., Hulstrom, R., 1981. A simplified clear sky model for direct and diffuse insolation on horizontal surfaces. Technical Report SERI/TR-642-761.

Bostan, S., Ortak, M., Tuna, C., Akoguz, A., Sertel, E., Ustundag, B., 2016. Comparison of classification accuracy of co-located hyperspectral and multispectral images for agricultural purposes. In *Fifth International Conference on Agro-Geoinformatics*, China, p. 4.

Breunig, F., Galvao, L., Formaggio, A., Epiphanio, J., 2011. Classification of soybean varieties using different techniques: Case study with Hyperion and sensor spectral resolution simulations. *Journal of Applied Remote Sensing*, 5, 053533-1–053533-15.

Buckingham, R., Staenz, K., 2008. Review of current and planned civilian space hyperspectral sensors for EO. *Canadian Journal of Remote Sensing*, 34, S187–S197.

Carlson, K., Asner, G., Hughes, R., Ostertag, R., Martin, R., 2007. Hyperspectral remote sensing of canopy biodiversity in Hawaiian lowland rainforests. *Ecosystems*, 10, 536–549.

Carroll, R., Huntington, J., Snyder, K., Niswonger, R., Morton, C., Stringham, T., 2017. Evaluating mountain meadow groundwater response to Pinyon-Juniper and temperature in a great basin watershed. *Ecohydrology*, p. 10.

Castaldi, F., Palombo, A., Santini, F., Pascucci, S., Pignatti, S., Casa, R., 2016. Evaluation of the potential of the current and forthcoming multispectral and hyperspectral imagers to estimate soil texture and organic carbon. *Remote Sensing of Environment*, 179, 54–65.

Chavez, Jr., P., 1996. Image-based atmospheric corrections-revisited and improved. *Photogrammetric Engineering and Remote Sensing*, 62(9), 1025–1036.

Cho, M., Mathieu, R., Asner, G., Naidoo, L., van Aardt, J., Ramoelo, A., Debba, P. et al., 2012. Mapping tree species composition in South African savannas using an integrated airborne spectral and LiDAR system. *Remote Sensing of Environment*, 125, 214–226.

Christian, B., Krishnayya, N., 2009. Classification of tropical trees growing in a sanctuary using Hyperion (EO-1) and SAM algorithm. *Current Science*, 96(12), 1601–1607.

Dadon, A., Ben-Dor, E., Beyth, M., Karnieli, A., 2011. Examination of spaceborne imaging spectroscopy data utility for stratigraphic and lithologic mapping. *Journal of Applied Remote Sensing*, 5, 053507-1–053507-14.

Dadon, A., Ben-Dor, E., Karnieli, A., 2010a. Use of derivative calculations and minimum noise fraction transform for detecting and correcting the spectral curvature effect (smile) in Hyperion images. *IEEE Transactions on Geoscience and Remote Sensing*, 48(6), 2603–2612.

Dadon, A., Karnieli, A., Ben-Dor, E., 2010b. Detecting and correcting the curvature effect (smile) in Hyperion images by use of MNF. In *Proc. Hyperspectral 2010 Workshop*, Frascati, Italy.

Datt, B., McVicar, T., Van Niel, T., Jupp, D., Pearlman, J., 2003. Preprocessing EO-1 Hyperion hyperspectral data to support the application of agricultural indexes. *IEEE Transactions on Geoscience and Remote Sensing*, 41(6), 1246–1259.

Daughtry, C., Doraiswamy, P., Hunt Jr, E., Stern, A., McMurtrey III, J., Prueger, J., 2006. Remote sensing of crop residue cover and soil tillage intensity. *Soil and Tillage Research*, 91, 101–108.

Daughtry, C., Hunt Jr, E., Doraiswamy, P., McMurtrey III, J., 2005. Remote sensing the spatial distribution of crop residues. *Agronomy Journal*, 97, 864–871.

Dong, J., Xiao, X., Menarguez, M., Zhang, G., Qin, Y., Thau, D., Biradar, C., Moore III, B., 2016. Mapping paddy rice planting area in northeast Asia with Landsat 8 images, phenology-based algorithm and Google Earth Engine. *Remote Sensing of Environment*, 185, 142–154.

Edmonds, D., Hajek, E., Downton, N., Bryk, A., 2016. Avulsion flow-path selection on rivers in foreland basins. *Geology*, 44(9), 695–698.

Farifteh, J., Nieuwenhuis, W., Garcia-Melendez, E., 2013. Mapping spatial variations of iron oxide by-product minerals from EO-1 Hyperion. *International Journal of Remote Sensing*, 34(2), 682–699.

Felde, G., Anderson, G., Cooley, T., Matthew, M., Adler-Golden, S., Berk, A., Lee, J., 2003. Analysis of Hyperion data with the FLAASH atmospheric correction algorithm. In *Proc. 2003 IEEE International*, vol. 1, Toulouse, France, pp. 90–92.

Flick, M., Mandl, D., Kane, L., Sabia, S., 2007. EO-1 validation report. Technical Report, National Aeronautics and Space Administration.

Fraser, R., Ferrare, R., Kaufman, Y., Mattoo, S., 1989. Algorithm for atmospheric corrections of aircraft and satellite imagery. Technical Report NASA TM-100751, National Aeronautics and Space Administration, Greenbelt, MD.

Ganesh, B., Aravindan, S., Raja, S., Thirunavukkarasu, A., 2013. Hyperspectral satellite data (Hyperion) preprocessing- a case study on banded magnetite quartzite in Godumalai Hill, Salem, Tamil Nadu, India. *Arabian Journal of Geosciences*, 6, 3249–3256.

Gao, B., Goetz, A., 1990. Column atmospheric water vapor and vegetation liquid water retrievals from airborne imaging spectrometer data. *Journal of Geophysical Research*, 95(D4), 3549–3564.

Gao, B., Montes, M., Davis, C., Goetz, A., 2009. Atmospheric correction algorithms for hyperspectral remote sensing data of land and ocean. *Remote Sensing of Environment*, 113, S17–S24.

George, R., Padalia, H., Kushwaha, S., 2014. Forest tree species discrimination in western Himalaya using EO-1 Hyperion. *International Journal of Applied Earth Observation and Geoinformation*, 28, 140–149.

Goldblatt, R., You, W., Hanson, G., Khandelwal, A., 2016. Detecting the boundaries of urban areas in India: A dataset for pixel-based image classification in Google Earth Engine. *Remote Sensing*, 8(634), p. 28.

Goodenough, D., Dyk, A., Niemann, K., Pearlman, J., Chen, H., Han, T., Murdoch, M., West, C., 2003. Processing Hyperion and ALI for forest classification. *IEEE Transactions on Geoscience and Remote Sensing*, 41(6), 1321–1331.

Gopinathan, K. Polokoana, P., 1986. Estimation of hourly beam and diffuse solar radiation. *Solar and Wind Technology*, 3(3), 223–229.

Gueymard, C., 1995. SMARTS2, a simple model of the atmospheric radiative transfer of sun- shine: Algorithms and performance assessment. Technical Report FSEC-PF-270-95, Florida Solar Energy Center; University of Central Florida, Cocoa, FL.

Gueymard, C., 2001. Parameterized transmittance model for direct beam and circumsolar spectral irradiance. *Solar Energy*, 71(5), 325–346.

Hamzeh, S., Naseri, A., AlaviPanah, S., Bartholomeus, H., Herold, M., 2016. Assessing the accuracy of hyperspectral and multispectral satellite imagery for categorical and quantitative mapping of salinity stress in sugarcane fields. *International Journal of Applied Earth Observation and Geoinformation*, 52, 412–421.

He, K., Rocchini, D., Neteler, M., Nagendra, H., 2011. Benefits of hyperspectral remote sensing for tracking plant invasions. *Diversity and Distributions*, 17, 381–392.

He, Y., Mui, A., 2010. Scaling up semi-arid grassland biochemical content from the leaf to the canopy level: Challenges and opportunities. *Sensors*, 10, 11072–11087.

Hill, M.J., 2013. Vegetation index suites as indicators of vegetation state in grassland and savanna: An analysis with simulated Sentinel 2 data for a North American transect. *Remote Sensing of Environment*, 137, 94–111.

Houborg, R., McCabe, M., Angel, Y., Middleton, E., 2016. Detection of chlorophyll and leaf area index dynamics from sub-weekly hyperspectral imagery. *Remote Sensing for Agriculture, Ecosystems, and Hydrology XVIII*, 9998, 999812-1–999812-11.

Huemmrich, K., Gamon, J., Tweedie, C., Campbell, P., Landis, D., Middleton, E., 2013. Arctic tundra vegetation functional types based on photosynthetic physiology and optical properties. *IEEE Journal of Selected Topics in Applied Earth Observations and Remote Sensing*, 6(2), 265–275.

Huntington, J., McGwire, K., Morton, C., Snyder, K., Peterson, S., Erickson, T., Niswonger, R., Carroll, R., Smith, G., Allen, R., 2016. Assessing the role of climate and resource management on groundwater dependent ecosystem changes in arid environments with the Landsat archive. *Remote Sensing of Environment*, 185, 186–197.

Jafari, R., Lewis, M., 2012. Arid land characterization with EO-1 Hyperion hyperspectral data. *International Journal of Applied Earth Observation and Geoinformation*, 19, 298–307.

Kefauver, S., Penuelas, J., Ustin, S., 2012. Applications of hyperspectral remote sensing and GIS for assessing forest health and air pollution. In *Geoscience and Remote Sensing Symposium (IGARSS), 2012 IEEE International*, Munich, Germany.

Khurshid, K., Staenz, K., Sun, L., Neville, R., White, H., Bannari, A., Champagne, C., Hitchcock, R., 2006. Preprocessing of EO-1 Hyperion data. *Canadian Journal of Remote Sensing*, 32(2), 84–97.

Kruse, F., 2004. Comparison of ATREM, ACORN, and FLAASH Atmospheric corrections using low-altitude AVIRIS data of Boulder, CO. In *Summaries of 13th Jet Propulsion Laboratory (JPL) Airborne Geoscience Workshop*, Pasadena, CA, p. 10.

Kruse, F., Perry, S., Caballero, A., 2006. District-level mineral survey using airborne hyper- spectral data, Los Menucos, Argentina. *Annals of Geophysics*, 49(1), 83–92.

Lamparelli, R., Johann, J., Dos Santos, E., Esquerdo, J., Rocha, J., 2012. Use of data mining and spectral profiles to differentiate condition after harvest of coffee plants. *Engenharia Agricultura Jaboticabal*, 32(1), 184–196.

Lenot, X., Achard, V., Poutier, L., 2009. SIERRA: A new approach to atmospheric and topographic corrections for hyperspectral imagery. *Remote Sensing of Environment*, 113, 1664–1677.

Liu, B., Jordan, R., 1960. The interrelationship and characteristic distribution of direct, diffuse and total solar radiation. *Solar Energy*, 4(3), 1–19.

Lobell, D., Thau, D., Seifert, C., Engle, E., Little, B., 2015. A scalable satellite-based crop yield mapper. *Remote Sensing of Environment*, 164, 324–333.

Lomheim, T., Kwok, J., Dutton, T., Shima, R., Johnson, J., Boucher, R., Wrigley, C., 1999. Imaging artifacts due to pixel spatial sampling smear and amplitude quantization in two-dimensional visible imaging arrays. *SPIE Proceedings*, 3701, p. 25.

Mariotto, I., Thenkabail, P., Huete, A., Slonecker, T., Platonov, A., 2013. Hyperspectral versus multispectral crop-productivity modeling and type discrimination for the HyspIRI mission. *Remote Sensing of Environment*, 139, 291–305.

Marshall, M., Thenkabail, P., 2015a. Advantage of hyperspectral EO-1 Hyperion over multispectral IKONOS, GeoEye-1, WorldView-2, Landsat ETM+, and MODIS vegetation indices in crop biomass estimation. *ISPRS Journal of Photogrammetry and Remote Sensing*, 108, 205–218.

Marshall, M., Thenkabail, P., 2015b. Developing in situ non-destructive estimates of crop biomass to address issues of scale in remote sensing. *Remote Sensing*, 7, 808–835.

Marshall, M., Thenkabail, P., Biggs, T., Post, K., 2016. Hyperspectral narrowband and multispectral broadband indices for remote sensing of crop evapotranspiration and its components (transpiration and soil evaporation). *Agricultural and Forest Meteorology*, 218–219, 122–134.

Matthew, M., Adler-Golden, S., Berk, A., Richtsmeier, S., Levine, R., Bernstein, L., Acharya, P. et al., 2000. Status of atmospheric correction using a MODTRAN4-based algorithm. In *Proc. SPIE 4049*, Orlando, FL.

Mielke, C., Boesche, N., Rogass, C., Kaufmann, H., Gauert, C., de Wit, M., 2014. Space-borne mine waste mineralogy monitoring in South Africa, applications for modern push-broom missions: Hyperion/OLI and EnMAP/Sentinel-2. *Remote Sensing*, 6, 6790–6816.

Miglani, A., Ray, S., Pandey, R., Parihar, J., 2008. Evaluation of EO-1 Hyperion data for agricultural applications. *Journal of the Indian Society of Remote Sensing*, 36, 255–266.

Mitri, G., Gitas, I., 2010. Mapping postfire vegetation recovery using EO-1 Hyperion imagery. *IEEE Transactions on Geoscience and Remote Sensing*, 48(3), 1613–1618.

Moharana, S., Dutta, S., 2016. Spatial variability of chlorophyll and nitrogen content of rice from hyperspectral imagery. *ISPRS Journal of Photogrammetry and Remote Sensing,* 122, 17–29.

Naidoo, L., Cho, M., Mathieu, R., Asner, G., 2012. Classification of savanna tree species, in the Greater Kruger National Park Region, by integrating hyperspectral and LiDAR data in a Random Forest data mining environment. *ISPRS Journal of Photogrammetry and Remote Sensing*, 69, 167–179.

Navulur, K., Lester, D., Marchetti, A., Hammann, G., 2013. Demystifying cloud computing for remote sensing applications. *Earth Imaging Journal*, 14–19.

NOAA, 2017. NOAA integrated surface database, https://www.ncdc.noaa.gov/isd

Numata, I., Cochrane, M., Galvao, L., 2011. Analyzing the impacts of frequency and severity of forest fire on the recovery of disturbed forest using Landsat time series and EO-1 Hyperion in the Southern Brazilian Amazon. *Earth Interactions*, 15(13), 1–17.

Okoro, S., Schickhoff, U., Bohner, J., Schneider, U., 2016. A novel approach in monitoring land-cover change in the tropics: Oil palm cultivation in the Niger Delta, Nigeria. *Die Erde*, 147(1), 40–52.

Oskouei, M., Babakan, S., 2016. Detection of alteration minerals using Hyperion data analysis in Lahroud. *Journal of the Indian Society of Remote Sensing*, 44(5), 713–721.

Padarian, J., Minasny, B., McBratney, A., 2015. Using Google's cloud-based platform for digital soil mapping. *Computers and Geosciences*, 83, 80–88.

Pan, Z., Huang, J., Wang, F., 2013. Multi range spectral feature fitting for hyperspectral imagery in extracting oilseed rape planting area. *International Journal of Applied Earth Observation and Geoinformation*, 25, 21–29.

Papes, M., Tupayachi, R., Martinez, P., Peterson, A., Powell, G., 2010. Using hyperspectral satellite imagery for regional inventories: A test with tropical emergent trees in the Amazon Basin. *Journal of Vegetation Science*, 21, 342–354.

Pearce, S., Rodell, M., Beaudoing, H., Mocko, D., McNaly, A., Jasinski, M., 2017. NASA LDAS land data assimilation systems, https://ldas.gsfc.nasa.gov/nldas/nldas2forcing.php

Pearlman, J., 2003. Hyperion validation report. Technical Report Boeing Report Number 03-ANCOS-001, Phantom Works, The Boeing Company, Kent, WA.

Peng, P., Li, C., 2016. Visibility measurements using two-angle forward scattering by liquid droplets. *Applied Optics*, 55(15), 3903–3908.

Perkins, T., Adler-Golden, S., Matthew, M., Berk, A., Bernstein, L., Lee, J., Fox, M., 2012. Speed and accuracy improvements in FLAASH atmospheric correction of hyperspectral imagery. *Optical Engineering*, 51(11), 111707-1–111707-7.

Pervez, W., Uddin, V., Khan, S., Khan, J., 2016. Satellite-based land use mapping: Comparative analysis of Landsat-8, Advanced Land Imager, and big data Hyperion imagery. *Journal of Applied Remote Sensing*, 10(2), 026004-1–026004-20.

Peterson, S., Roberts, D., Beland, M., Kokaly, R., Ustin, S., 2015. Oil detection in the coastal marshes of Louisiana using MESMA applied to band subsets of AVIRIS data. *Remote Sensing of Environment*, 159(2015), 222–231.

Plaza, A., Plaza, J., Paz, A., Sanchez, S., 2011. Parallel hyperspectral image and signal processing. *IEEE Signal Processing Magazine*, 119–126.

Richter, R., Schlapfer, D., Muller, A., 2011. Operational atmospheric correction for imaging spectrometers accounting for the smile effect. *IEEE Transactions on Geoscience and Remote Sensing*, 49(5), 1772–1780.

Roberts, D., Dennison, P., Gardner, M., Hetzel, Y., Ustin, S., Lee, C., 2003. Evaluation of the potential of Hyperion for fire danger assessment by comparison to the airborne visible/infrared imaging spectrometer. *IEEE Transactions on Geoscience and Remote Sensing*, 41(6), 1297–1310.

Sahoo, R., Ray, S., Manjunath, K., 2015. Hyperspectral remote sensing of agriculture. *Current Science*, 108(5), 848–859.

San, B., Suzen, M., 2011. Evaluation of cross-track illumination in EO-1 Hyperion imagery for lithological mapping. *International Journal of Remote Sensing*, 32(22), 7873–7889.

Scheffler, D., Karrasch, P., 2014. Destriping of hyperspectral image data: An evaluation of different algorithms using EO-1 Hyperion data. *Journal of Applied Remote Sensing*, 8, 083645-1–083645-18.

Schlapfer, D., Nieke, J., Ittens, K., 2007. Spatial PSF nonuniformity effects in airborne push- broom imaging spectrometry data. *IEEE Transactions on Geoscience and Remote Sensing*, 45(2), 458–468.

Seidel, F., Kokhanovsky, A., Schaepman, M., 2010. Fast and simple model for atmospheric radiative transfer. *Atmospheric Measurement Techniques*, 3, 1129–1141.

Shafri, H., Suhaili, A., Mansor, S., 2007. The performance of maximum likelihood, spectral angle mapper, neural network and decision tree classifiers in hyperspectral image analysis. *Journal of Computer Science*, 3(6), 419–423.

Sims, N., Culvenor, D., Newnham, G., Coops, N., Hopmans, P., 2013. Towards the operational use of satellite hyperspectral image data for mapping nutrient status and fertilizer requirements in Australian plantation forests. *IEEE Journal of Selected Topics in Applied Earth Observations and Remote Sensing*, 6(2), 320–328.

Smith, A., Blackshaw, R., 2003. Weed-crop discrimination using remote sensing: A detached leaf experiment. *Weed Technology*, 17, 811–820.

Somers, B., Asner, G., 2013a. Mapping tropical rainforest canopies using multi-temporal spaceborne imaging spectroscopy. *Remote Sensing for Agriculture, Ecosystems, and Hydrology XV*, 8887, 888704-1–888704-8.

Somers, B., Asner, G., 2013b. Multi-temporal hyperspectral mixture analysis and feature selection for invasive species mapping in rainforests. *Remote Sensing of Environment*, 136, 14–27.

Somers, B., Asner, G., 2014. Tree species mapping in tropical forests using multi-temporal imaging spectroscopy: Wavelength adaptive spectral mixture analysis. *International Journal of Applied Earth Observation and Geoinformation*, 31, 57–66.

Song, C., Woodcock, C., Seto, K., Lenney, M., Macomber, S.A., 2001. Classification and change detection using Landsat TM data: When and how to correct atmospheric effects? *Remote Sensing of Environment*, 75, 230–244.

Song, X., Ma, J., Li, X., Leng, P., Zhou, F., Li, S., 2014. First results of estimating surface soil moisture in the vegetated areas using ASAR and Hyperion data: The Chinese Heihe river basin case study. *Remote Sensing*, 6, 12055–12069.

Sonmez, N., Slater, B., 2016. Measuring intensity of tillage and plant residue cover using remote sensing. *European Journal of Remote Sensing*, 49(1), 121–135.

Soulard, C., Albano, C., Villarreal, M., Walker, J., 2016. Continuous 1985–2012 Landsat monitoring to assess fire effects on meadows in Yosemite National Park, CA. *Remote Sensing*, 8(371), p. 16.

Staenz, K., Neville, R., Clavette, S., Landry, R., White, H., 2002a. Retrieval of surface reflectance from Hyperion radiance data. p. 4.

Staenz, K., Secker, J., Gao, B., Davis, C., Nadeau, C., 2002b. Radiative transfer codes applied to hyperspectral data for the retrieval of surface reflectance. *ISPRS Journal of Photogrammetry and Remote Sensing*, 57, 194–203.

Stamnes, K., Tsay, S., Wiscombe, W., Laszlo, I., 2000. DISORT, a general-purpose Fortran program for discrete-ordinate-method radiative transfer in scattering and emitting layered media: Documentation of methodology (version 1.1, Mar 2000). Technical Report DISORT Report v1.1, Stevens Institute of Technology Department of Physics and Engineering Physics; NASA Goddard Space Flight Center Climate and Radiation Branch; University of Maryland Department of Meteorology.

Tang, Z., Li, Y., Gu, Y., Jiang, W., Xue, Y., Hu, Q., LaGrange, T., Bishop, A., Drahota, J., Li, R., 2016. Assessing Nebraska playa wetland inundation status during 1985–2015 using Landsat data and Google Earth Engine. *Environmental Monitoring and Assessment*, 188(654), p. 14.

Thenkabail, P., Enclona, E., Ashton, M., Legg, C., de Dieu, M., 2004. Hyperion, IKONOS, ALI, and ETM+ sensors in the study of African rainforests. *Remote Sensing of Environment*, 90, 23–43.

Thenkabail, P., Mariotto, I., Gumma, M., Middleton, E., Landis, D., Huemmrich, K., 2013. Selection of hyperspectral narrowbands (HNBs) and composition of hyperspectral two band vegetation indices (HVIs) for biophysical characterization and discrimination of crop types using field reflectance and Hyperion/ EO-1 data. *IEEE Journal of Selected Topics in Applied Earth Observations and Remote Sensing*, 6(2), 427–439.

Thenkabail, P., Smith, R., Pauw, E., 2000. Hyperspectral vegetation indices and their relationships with agricultural crop characteristics. *Remote Sensing of Environment*, 71, 158–182.

Underwood, E., Ustin, S., DiPietro, D., 2003. Mapping nonnative plants using hyperspectral imagery. *Remote Sensing of Environment*, 86, 150–161.

Underwood, E., Ustin, S., Ramirez, C., 2007. A comparison of spatial and spectral image resolution for mapping invasive plants in coastal California. *Journal of Environmental Management*, 39, 63–83.

USGS, 2013. Nasa partnerships, https://eros.usgs.gov/nasa-partnerships

Vermote, E., Vermeulen, A., 1999. Atmospheric correction algorithm: Spectral reflectances (MOD09) (Version 4.0, April 1999). Technical Report NAS5-96062. University of Maryland, Department of Geography, College Park, MD.

Walsh, S., McCleary, A., Mena, C., Shao, Y., Tuttle, J., Gonzalez, A., Atkinson, R., 2008. QuickBird and Hyperion data analysis of an invasive plant species in the Galapagos islands of Ecuador: Implications for control and land use management. *Remote Sensing of Environment*, 112, 1927–1941.

Yalew, S., van Griensven, A., van der Zaag, P., 2016. AgriSuit: A web-based GIS-MCDA framework for agricultural land suitability assessment. *Computers and Electronics in Agriculture*, 128, 1–8.

Yang, C., Everitt, J., Bradford, J., 2008. Yield estimation from hyperspectral imagery using spectral angle mapper (SAM). *Transactions of the ASABE*, 51(2), 729–737.

Zhang, J., Rivard, B., Sanchez-Azofeifa, A., Castro-Esau, K., 2006. Intra- and inter-class spectral variability of tropical tree species at La Selva, Costa Rica: Implications for species identification using HYDICE imagery. *Remote Sensing of Environment*, 105, 129–141.

Zhang, Q., Middleton, E., Cheng, Y., Landis, D., 2013. Variations of foliage chlorophyll fAPAR and foliage non-chlorophyll fAPAR (fAPARchl, fAPARnon-chl) at the Harvard forest. *IEEE Journal of Selected Topics in Applied Earth Observations and Remote Sensing*, 6(5), 2254–2264.

Zhang, Q., Middleton, E., Gao, B., Cheng, Y., 2012. Using EO-1 Hyperion to simulate HyspIRI products for a coniferous forest: The fraction of par absorbed by chlorophyll (fAPARchl) and leaf water content (LWC). *IEEE Transactions on Geoscience and Remote Sensing*, 50(5), 1844–1852.

Zomer, R., Trabucco, A., Ustin, S., 2009. Building spectral libraries for wetlands land cover classification and hyperspectral remote sensing. *Journal of Environmental Management*, 90, 2170–2177.

10 Hyperspectral Image Data Mining

Sreekala G. Bajwa, Yu Zhang, and Alimohammad Shirzadifar

CONTENTS

10.1 INTRODUCTION

Why data mining? In hyperspectral remote sensing, data are collected in numerous (hundreds to thousands) narrow wavebands in one or more regions of the electromagnetic spectrum. From satellite to Unmanned Aerial Systems (UAS), hyperspectral sensors can potentially generate data in terabytes to petabytes or exabytes of data. For example, the Rikola hyperspectral sensor (Senop, Lievestuore, Finland) from a UAS platform flying at 100 m above ground will cover a 65 m × 65 m area in one image. The camera can acquire data in up to 300 bands with a pixel dynamic range of 32 bits, generating 57 GB per image if all 300 bands were used. However, the sensor is often configured to collect data in fewer than 300 bands. This can translate to petabytes or exabytes of data for a state or region. While hyperspectral data offer tremendous potential for various new and existing applications, such huge datasets offer a variety of challenges on data transfer, storage, and processing that must be addressed to obtained actionable intelligence from the raw data.

Each hyperspectral dataset can take up hundreds of megabytes to gigabytes of storage space. For example, one still shot from a Rikola sensor is compared to a part of a scene from the AVIRIS sensor (Figure 10.1). The hyperspectral images are represented as hyperspectral cubes with two spatial dimensions of length and width of the scene, with the third dimension being the spectral axis containing different bands. The Rikola image has 47 image bands along the spectral axis spanning the wavelengths of 490–900 nm, and 7040 × 6780 pixels in the spatial dimension, each pixel representing 32-bits data per band. In comparison, the AVIRIS image has 220 bands in the wavelength range of 400–2500 nm, with 145 × 145 pixels in the spatial dimension, each representing 16-bits data per band. The AVIRIS image is a subset from a larger scene collected in 1992, and is calibrated to have pixel values proportional to radiance. The original AVIRIS images have 224 bands in the 360–2500 nm range, with 677 × 512 pixels. The twin Otter aircraft carrying the AVIRIS sensor collects data from 4 km above ground level. The images also show the high spatial resolution of hyperspectral data collected from a UAS platform, where each tree is clearly visible, compared to the AVIRIS data. In addition to the high dimensionality, hyperspectral data are also multi-collinear or redundant along the spectral axis, and autocorrelated spatially. It means that the information in each hyperspectral band is not unique, but very similar to that in the adjacent bands, and there is a spatial contextual relation. The high dimensionality, data size,

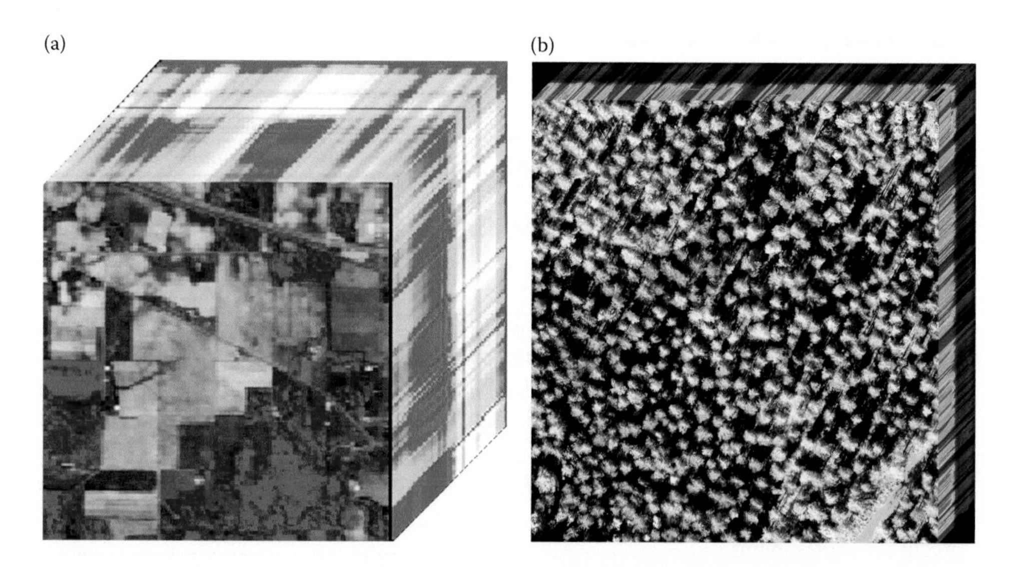

FIGURE 10.1 A side-by-side view of the hyperspectral image cube of (a) AVIRIS scene 92AV3C from June 12, 1992 (https://engineering.purdue.edu/~biehl/MultiSpec/hyperspectral.html), and (b) a high-resolution Rikola image collected from a UAS platform.

and information redundancy in hyperspectral data could pose major difficulties for the users of such data. For example, for a user interested in mapping a certain species of vegetation (or other similar applications), it may be difficult to know what bands to analyze, or how to extract relevant information from such a large dataset.

Some of the problems associated with the use of hyperspectral remote sensing are listed below:

a. *Data transfer/storage/processing issues*: The users of hyperspectral should have the capability to store and handle large size datasets in their own computer or through cloud services. They would require high performance computer with large storage and processing capacity, and network capable of transferring large datasets. Although data storage is becoming less of a problem with the decreasing cost of storage media or cloud services, data transfer and processing of such large datasets are still problematic.
b. *Data redundancy problems*: The information contained in each band of the hyperspectral image is not unique. In addition, there is considerable spatial correlation in the data. Hyperspectral data redundancy can be visualized through covariance or correlation between bands. The band correlation surface shown in Figure 10.2 reveals high positive correlations between adjacent bands. It also shows strong correlation between bands in different spectral regions. For example, the correlation between visible and near-infrared (NIR) bands are strongly negative ($r < -0.8$). Such high data redundancy indicates that data size could be significantly reduced by removing the redundant information.
c. *The curse of dimensionality*: As the number of bands in an image increases, the number of observations required to train a classifier increases exponentially to maintain the same classification accuracies. This is called the Hughes phenomenon, and it refers to the loss of classifiability of an image with the same fixed number of training samples when the dimensionality of data increases [1]. Often, it is not possible to increase the size of training data due to time and budgetary constraints. In such cases, the data dimensionality must be reduced using an appropriate feature selection method.

What is data mining? Data mining refers to extracting and preserving previously unknown, potentially useful, and reliable information or patterns, also called actionable intelligence, contained in a hyperspectral image that are relevant to a specific application.

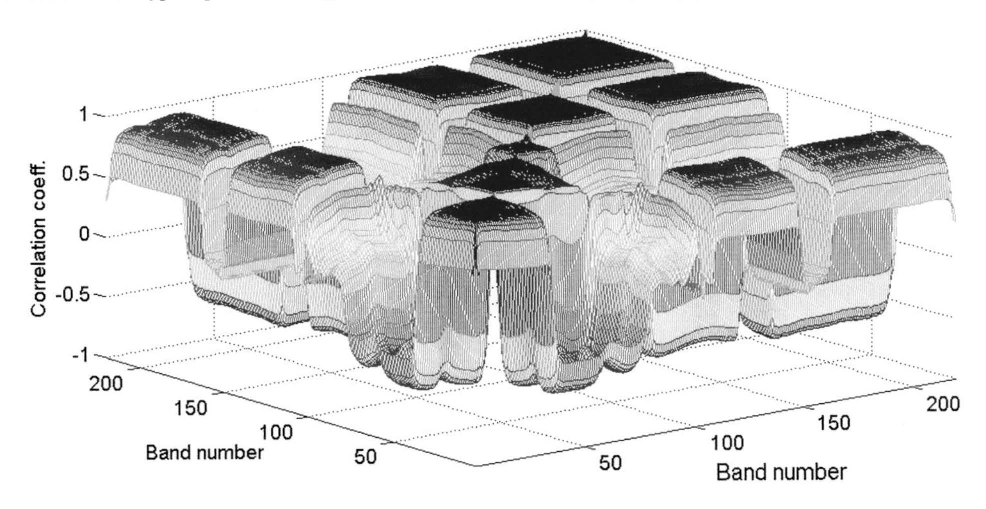

FIGURE 10.2 Correlation between the bands of hyperspectral image shown in Figure 10.1. Only the alternate bands are used to compute the correlation. The colors represent −1 for dark blue to +1 for the dark red.

10.2 DATA MINING METHODS

The potential applications of big data in many domains have led to the development of new data analytics and deep learning methods to discover the information or patterns buried in those datasets. Most hyperspectral image data mining procedures include a process of feature selection, followed by a process of information extraction. The word "feature" is used throughout this chapter since band selection implies only selection of a subset of bands from the original image bands, and do not explicitly include other features that are developed by transforming two or more bands, or spatial features.

Both feature selection and information extraction methods could be either supervised or unsupervised. Figure 10.3 gives a summary of the supervised and unsupervised methods for feature selection/extraction and information extraction. The unsupervised methods do not necessary require ground data or prior knowledge of the target characteristics or phenomenon of interest to the user. Most unsupervised methods for feature selection are based on one or more feature characteristics such as variance, entropy, correlation, covariance, similarity measures, divergence, and natural grouping within the data [2].

Supervised methods use prior knowledge or training data on target characteristics or a phenomenon of interest to identify a group of best features and to train a classifier. Some of the supervised feature selection methods include divergence measures, correlation between features and ground data, support vector machines, and so on. The output from both supervised and unsupervised methods of information extraction is usually a map indicating spatial patterns of interest or a table.

Data mining methods could also be either parametric or non-parametric. Generally speaking, parametric methods assume a data distribution, usually a normal distribution. A majority of the methods used for data mining have an underlying assumption that the data are normally distributed. On the other hand, non-parametric methods make no such assumptions, and may be more appropriate for some applications.

10.3 FEATURE SELECTION/EXTRACTION METHODS

A feature refers to an individual band or the result of a spectral or spatial transformation. For example, vegetation indices (VI) and principal components are features derived by spectral transformation.

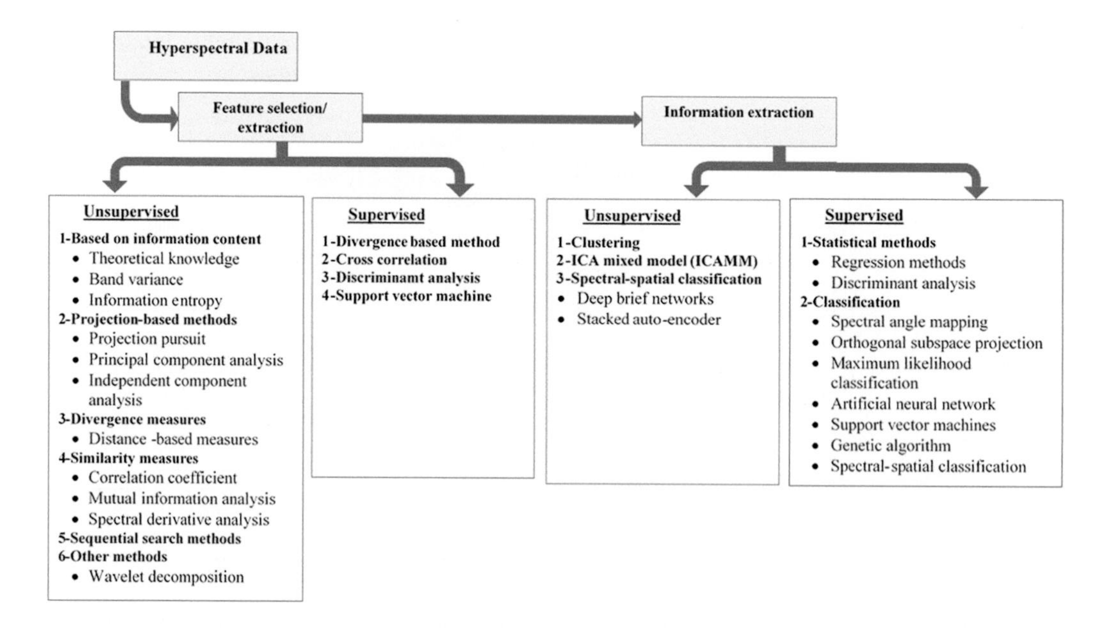

FIGURE 10.3 A summary of hyperspectral image data mining methods.

Object oriented features such as pixel morphology and texture are examples of features derived by spatial transformation. With hyperspectral data, spectral transformation is more commonly used. Hyperspectral data can potentially have hundreds to thousands of bands. Many thousands of features can be derived from these bands. Therefore, the feature space of a hyperspectral image could potentially have tens of thousands of features.

Interestingly, all the features available from a hyperspectral image may not be sensitive to the target variable or phenomenon of interest. Also, inclusion of all of these features in a data mining process such as classification can make the process extremely slow, less accurate (as explained by the Hughes phenomenon), and computationally expensive. Therefore, feature selection/extraction is an important step in hyperspectral image data mining.

There are many methods available for feature selection or extraction. A feature selection method results in a subset of the original features, whereas a feature extraction method would provide a combination of new and/or reduced set of features that may include original features as well as newly extracted ones. An ideal feature selection/extraction method should be able to identify the optimal set of features that would obtain the information of interest with the highest possible accuracy and reliability with the least amount of time, computational effort, and cost.

Feature extraction typically involves evaluation of a superset of features such as all bands and their specific transformations using an index of performance. Then, a subset of these features is selected based on one or more indices of performance. An index of performance is a measure of the capability of the band to supply the information of interest, called actionable intelligence.

There are many methods available for feature selection from hyperspectral data. The choice of a method for a specific application may depend on availability of ground data as well as the scale of the problem. If the feature space has up to 19 features, it is called a small-scale problem, whereas a medium scale has 20–49 features and large scale has 50 to an infinite number of features in the feature space [3].

10.3.1 Feature Selection Based on Information Content

This group of feature selection methods uses various measures of information content of individual features to select a subset most appropriate for extracting target characteristics of interest. All methods listed under this group are unsupervised methods, meaning they do not require any training data. Although there are numerous methods scientists have experimented with, only the most commonly used methods are listed below.

10.3.1.1 Feature Selection Based on Theoretical Knowledge

Feature extraction and feature selection can be made on theoretical knowledge of how the phenomenon or target characteristics of interest interact with radiation. For example, in optical remote sensing, a biochemical molecule that is indicative of the target characteristic of interest may have a fundamental frequency, one or more overtone frequencies, and combination frequencies. If the fundamental frequency for excitation of the molecule is known, it is easy to calculate the overtones and combination frequencies.

Information on the location of such frequencies could be used for extracting features from hyperspectral images in optical remote sensing. For example, the fundamental wavelength for three different vibrational modes of excitation for water vapor are 6270, 2738, and 2662 nm, with overtones at 3173 and 906 nm, and combination bands at 1876, 1135, 942, 906, 823, 796, and 652 nm [4]. Similarly, overtone or combination bands of other compounds include 1940 nm for liquid water, 2270 nm for lignin, 2336 nm for cellulose, and so on. Many primary and secondary minerals and biochemical constituents of vegetation exhibit electronic (in visible wavelengths) and vibrational spectral features in IR and NIR bands.

Several features such as VI have been derived based on the knowledge of the overtones or combination frequencies of chemical constituents of interest. In vegetation, chlorophyll a and b, carotenoids, phytochromes, cellulose, nitrogen, canopy water content, and so on, could be used as indicators of vegetation condition including growth stage, diseases, water stress, nutrient stresses, and other plant conditions. An example would be choosing Physiological Reflectance Index (PRI) as a feature of interest for detecting yellow rust fungus in wheat [5] as the rust disrupts the foliar photosynthetic activity, and PRI has been shown to indicate photochemical and radiation use efficiency of plants [6]. Almeida and De Souza Filho [7] have developed several VIs using the knowledge of overtone and combination bands of specific molecules of interest. These VI include R_{461}/R_{422} and R_{807}/R_{638} for chlorophyll a, R_{520}/R_{470} and R_{807}/R_{648} for chlorophyll b, R_{520}/R_{442} for α-carotene, R_{539}/R_{490} and R_{807}/R_{490} for carotenoids, R_{510}/R_{530} for anthocyanin, R_{845}/R_{730} for phytochrome P730, R_{778}/R_{658} for phytochrome P660, R_{1028}/R_{2101} for lignin, R_{2211}/R_{2400} for cellulose, R_{1731}/R_{1691} for nitrogen, and R_{1066}/R_{1452} for leaf water content.

10.3.1.2 Band Variance

In hyperspectral data, the variance of digital numbers (DN) or reflectance of an image band can be considered as a simple measure of its information content. Therefore, feature variance can be used as a basis for selecting hyperspectral features [8]. The variance of DN within an image band can be calculated with Equation 10.1.

$$\text{Variance of a sample population,} \quad s^2 = \frac{\sum_{i=1}^{N} (\text{DN}_i - \overline{\text{DN}})^2}{N-1} \tag{10.1}$$

where, DN_i is the digital number of ith pixel in the image band, $\overline{\text{DN}}$ is the average value of DN, and N is the number of pixels in the image band. The same equation can be applied to any feature, with the DN replaced by the pixel values of the feature.

The purpose of remote sensing is to understand a spatially dynamic phenomenon or variable through image data. A feature sensitive to the spatial variability in a target should represent that information through a corresponding variability in its values. Therefore, selecting a set of features or bands with the most variability is a simple method of feature selection. Using principal component analysis (PCA), or partial least square (PLS) factors, as features in many applications is based on the fact that these features capture the most variability in the image bands through specific transformation [9]. For example, the band variance of the AVIRIS hyperspectral image shown in Figure 10.1 indicates that the bands in the far red and NIR region have the highest variance (Figure 10.4).

A high information content in red and NIR is expected for an image scene consisting primarily of soil and vegetation. However, these adjacent bands can be very similar to each other since they are

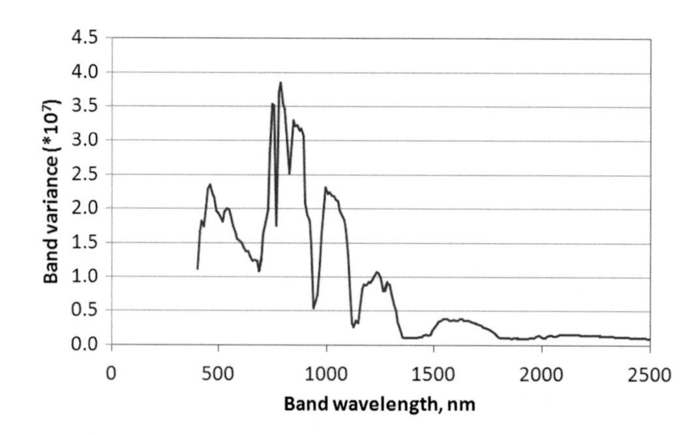

FIGURE 10.4 Band variance of the AVIRIS image shown in Figure 10.1.

highly correlated (Figure 10.2). This brings up a major drawback of this method—it does not exclude redundant bands. Therefore, the information carried by the features selected with this method could be highly redundant. Also, if the variable or phenomenon of interest is not the cause of the dominating variability in the image scene but causes rather subtle variability, this method may not capture the features carrying such information. PLS factors eliminate such cases of unrelated variability in image bands by maximizing the covariance between transformed image bands and the target characteristics of interest. Therefore, it is a supervised method of feature selection while PCA is an unsupervised method.

10.3.1.3 Information Entropy

Information entropy is a simple measure of information content of an image band [8]. It is calculated based on the probability of occurrence of each distinct DN. Entropy of an image band can be calculated using Equation 10.2.

$$\text{Information Entropy,} \quad H(X) = -\sum_{i=1}^{N} p_i \ln(p_i) \tag{10.2}$$

where, p_i is the probability of occurrence of the ith DN in the hyperspectral image band X, and N is the number of distinct DNs in this band.

If the entropy of an image band is high, then that band is considered to have high information content. For example, the entropy of a hyperspectral image of an agricultural field [8] indicates that the red bands have the highest entropy (Figure 10.5).

The entropy measure is very similar to variance measure in both benefits and drawbacks. Both methods are simple and easy to use, but do not consider redundancy of information in hyperspectral image bands.

The entropy measure can be expanded or combined with other measures to account for information redundancy between bands. Bruce and Reynolds [10] applied conflict data filtering based on mutual entropy as a measure for multiple feature extraction and fusion from hypertemporal hyperspectral data for ground cover classification. Zhao et al. [11] developed a variation index by combining pixel entropy, a measure of spectral information, with kernel spatial attraction, a measure of spatial similarity, to identify endmembers in a hyperspectral image classification problem to identify land cover classes.

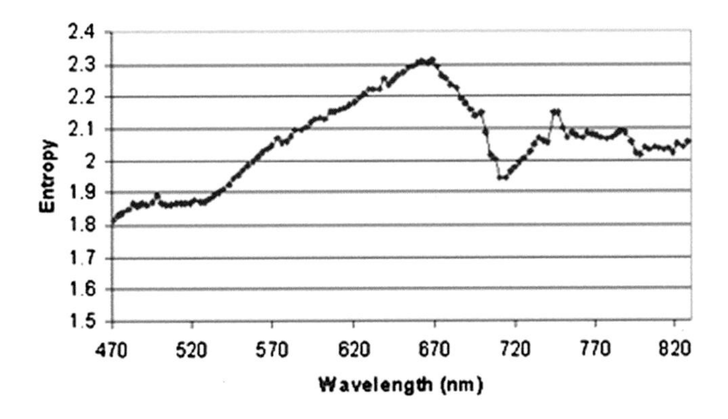

FIGURE 10.5 Entropy of a hyperspectral image of a corn field with 120 bands in the visible and near-infrared region. The image was acquired early in the season while the seeds were still germinating. (From Lavanya, A., and Sanjeevi, S., *Journal of Indian Society of Remote Sensing*, 41(2), 199–211, 2013, copyright [9].)

10.3.2 Projection-Based Methods

This group of methods is different from others based on information content of the feature in that it transforms or projects the high-dimensional data into a low dimensional space based on some constraint. A majority of the projection-based methods use projected bands that are linear combinations of the original bands.

10.3.2.1 Projection Pursuit (PP)

High-dimensional data space is mostly empty and, therefore, parametric methods are not very effective in image classification. High-dimensional data can be projected in numerous ways to lower dimensions. In the case of projections of structured data into one or two dimensions, normality is not a stringent requirement because the data in the lower dimension represent a shadow of the high-dimensional data with a certain distribution. However, a parametric PP is usually used for projecting hyperspectral data to a reasonable lower dimension with higher dimensionality than two features for subsequent classification.

PP is a common method of obtaining lower dimensional and interesting projections of high-dimensional data. It is an unsupervised method that can be used for dimensionality reduction, and also to detect anomalies or patterns in hyperspectral imagery [13]. Both PCA and discriminant analysis could be considered as special cases of PP.

The PP uses linear combinations of original features, image bands in this case, to maximize some index of performance. This is equivalent to rotating the feature space. Scientists have used different types of PP index for different purposes for which the data were projected.

For classifying an image, the PP index should measure the ability of the projected data to form meaningful clusters. On the other hand, the PP index should help in identifying the outliers if the purpose of PP is to detect anomalies in a hyperspectral image. The types of PP index reported include a relative entropy image, Shannon entropy, Legendre's index, ratios of moments of a standard normal distribution, and the Bhattacharyya distance [12–14].

10.3.2.2 Principal Component Analysis (PCA)

The PCA is a multivariate method commonly used for reducing data redundancy and dimensionality. Principal components (PCs) are linear combinations of the original bands or features such that the PCs are orthogonal to each other, and the information content is maximized in each principal component. The principal components are sorted based on their variance such that the first PC has the highest variance, and it diminishes in successive PCs. Thus, a majority of the information contained in hundreds of bands of a hyperspectral image is captured in a few PCs, thus, achieving data dimensionality reduction. The decorrelated PCs also ensure that there is no data redundancy, which is otherwise a problem in hyperspectral images. The PCA can be applied to all or a subset of original bands or transformed bands such as band ratios and other vegetation indices [7,8] in order to understand a scene characteristic [15]. Variations of PCA such as hyperspectral data slicing and segmented PCA have also been developed for dimensionality reduction [15,16].

The PCA is an unsupervised method that will capture the major variability in the image in the first few PCs. If the user is interested in a phenomenon or variable that causes subtle differences in target reflectance, then, PCA is not the best method for feature selection. For example, in an agricultural field with bare soil or sparse vegetation coverage, soil electrical conductivity can cause subtle differences in reflectance. When a hyperspectral image of such a field with 120 bands in the visible and NIR region was subjected to PCA, the first three PC represented 99.6% of variability in the scene. However, soil electrical conductivity showed the highest cross-correlation with PC5 and PC4 (Figure 10.6) [8]. The PC5 and PC4 represented only 0.2% and 0.1% of variability in the image, respectively. The PCA is not ideal for detecting small classes that are represented by relatively fewer pixels. In both these cases, the target characteristics of interest generated relatively small variability in band DN compared to the dominant variability in the image. In such cases, the PCA may not be able to correctly preserve the information of interest.

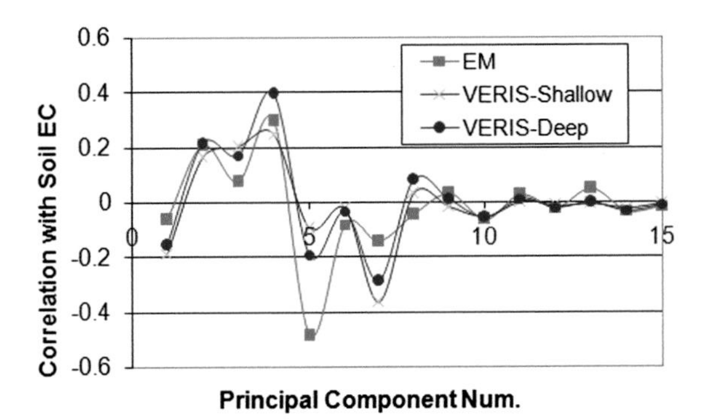

FIGURE 10.6 Correlation between principal components of a hyperspectral image of an agricultural field and apparent soil electrical conductivity at two different depths. (From Bajwa S.G. et al., *Transactions of the ASABE*, 47, 895–907, 2004, copyright used with permission [8].)

10.3.2.3 Independent Component Analysis (ICA)

Independent component analysis is a method superficially similar to PCA in that both methods reduce data redundancy and dimensionality. The ICA is different from PCA in that the PCA decorrelates the image bands whereas the ICA uses independence of components to estimate them.

The ICA reveals hidden factors that underlie a set of measurements or signals by assuming that each band is a linear mixture of independent hidden components. It then proceeds to recover the original factors or independent features through a linear unmixing operation [17]. The ICA can be performed on hyperspectral data transformed with PCA or on the original data under a minimal mutual information (MMI) framework to ensure that the independence assumption in component analysis, and the abundance assumption in linear mixed models, are met [18]. If the target classes of interest are not a major source of variability in the data, ICA on the original bands is recommended.

The limitation of ICA is that it can only separate linearly mixed models. It also assumes that the independent components are non-Gaussian. This would mean that the bands, which are linear combinations of non-Gaussian independent components could be Gaussian distributed according to the Central Limit Theorem. Another limitation of ICA is that neither the variances (energies) nor the order of the independent components can be determined. The advantage of ICA is that even when the components are not entirely independent, it finds a transformation that maximizes the degree of independence between the components.

10.3.3 Divergence Measures

A majority of the divergence measures are supervised methods that require training data. Even in multispectral remote sensing, the supervised classification methods start with a separability analysis on training data, which are based on divergence measures. Measures of divergence are mostly distance-based.

10.3.3.1 Distance-Based Measures

Although distance measures are commonly used for supervised feature selection, they can also be used for unsupervised cluster analysis. Distance-based measures usually use training data to select a subset of features that have the best discrimination based on the distance between the mean class vectors.

There are many measures of distance between pixel vectors, or between a pixel vector and class mean, or between class means. The most commonly used distance measures include Euclidean, "city block," Angular, Mahalanobis, divergence, Bhattacharyya, Kolmogorov variational distance, and Jeffries-Matusita distance (Table 10.1). Among these distance measures, the city block, Euclidean, and Angular measurements ignore the covariance C of the classes and do not make assumptions

TABLE 10.1
Distance Measures between Two Class Distributions *a* and *b*, and the Equations to Calculate These Distance Measures

Distance Measure	Equation
City block distance	$L_{CB} = \| \mu_a - \mu_b \| = \sum_{i=1}^{k} \| m_{ai} - m_{bi}$
Euclidean distance	$L_E = \| \mu_a - \mu_b \| = [(\mu_a - \mu_b)^T (\mu_a - \mu_b)]^{1/2}$
Angular distance	$\theta = a\cos\left[\frac{\mu_a^T \mu_b}{\| \mu_a \| \| \mu_b \|}\right]$
Normalized city block distance	$L_{NCB} = \sum_{i=1}^{k} \frac{\| m_{ai} - m_{bi} \|}{\left(\sqrt{c_{ai}} + \sqrt{c_{bi}}\right)/2}$
Mahalanobis distance	$L_M = \left[(\mu_a - \mu_b)^T \left(\frac{C_a + C_b}{2}\right)^{-1} (\mu_a - \mu_b)\right]^{1/2}$
Divergence	$D = \frac{1}{2}\text{tr}[(C_a - C_b)(C_b^{-1} - C_a^{-1})] + \text{tr}[(C_a^{-1} + C_b^{-1})(\mu_a - \mu_b)(\mu_a - \mu_b)^T]$
Transformed divergence	$D^t = 2(1 - e^{-D/8})$
Kolmogorov variational distance	$K_a = \int \| P_a(\omega) - P_b(\omega) \| d\omega$, where $P_a(\omega)$ and $P_b(\omega)$ are joint conditional cumulative distribution functions of class *a* and *b*.
Bhattacharyya distance	$L_B = \frac{L_M}{8} + \frac{1}{2}\ln\left[\frac{\|(C_a + C_b)/2\|}{(\| C_a \| \| C_b \|)^{1/2}}\right]$
Jeffries-Matusita	$L_{JM} = [2(1 - e^{-L_B})]^{1/2}$

Note: The two class means are $\mu_a = (m_{a1}, \ldots, m_{ak})$ and $\mu_b = (m_{b1}, \ldots, m_{bk})$, where k is the number of features, and C is the covariance matrix.

about the distribution of the classes. The remaining methods are considered more robust since they rely either on probability distribution or covariance.

A factor analysis method in combination with target signature separability measures such as Euclidean distance, divergence, transformed divergence, and Jeffries-Matusita distance have also been effective in dimensionality reduction [9].

In addition to the above measures, there are scatter matrix-based measures. Consider that S_w, S_b, and S_t are the within-class, between-class, and total scatter matrices that are computed as shown below:

$$S_w = \sum_{i=1}^{L} P(\omega_i)\, C_i \tag{10.3}$$

$$S_b = \sum_{i=1}^{L} P(\omega_i)(\mu_i - \mu_0)(\mu_i - \mu_0)^T \tag{10.4}$$

$$S_t = S_w + S_b \tag{10.5}$$

where, μ_i is the mean of *i*th class, μ_0 is the mean of all classes together, C_i is the class covariance, L is the number of classes, and $P(\omega_i)$ is the class probability. Two divergence measures based on scatter matrix include $\text{tr}(S_w^{-1}S_b)$and $\det(S_w^{-1}S_b)$. These measures are also used for feature selection.

10.3.4 Similarity Measures

Similarity measures use an index of performance that measures the degree of similarity between pairs of bands or features. Measures of similarity includes correlation coefficients, mutual information entropy, and spectral derivative analysis.

10.3.4.1 Correlation Coefficient

If the correlation between two bands is high, they are considered redundant and hence one band may be adequate to represent the information [8,9]. The correlation coefficient is a measure of linear dependency between two variables x and y, and is widely used as the statistical measure of similarity between the two spectral bands x and y. It is defined as:

$$\rho(x,y) = \frac{C(x,y)}{\sqrt{\sigma(x)\sigma(y)}} \tag{10.6}$$

where, ρ is the correlation coefficient between bands x and y, $C(x, y)$ is the covariance between x and y, and σ is the variance.

Correlation between hyperspectral features can be used as a measure of their common information content or redundancy. The correlation coefficient can vary from -1 to $+1$. A correlation coefficient of 0 indicates no linear dependency where as a $+1$ or -1 indicates a 100% dependency. For a pair of features that are highly correlated, one can be eliminated without losing significant information.

In supervised feature selection applications, the cross-correlation between the target characteristics and the superset of features can be used as an index for selection of the best set of features for target characterization [19]. If a feature shows high positive or negative correlation with the target characteristic, it should be included in the final set of features for estimating those characteristics. This method has been adopted for estimating vegetation biophysical variables from hyperspectral images, and for crop classification [19,20].

The drawbacks of correlation coefficients include their sensitivity to rotation of the scatter plot diagram in the (X, Y) plane. It is also invariant to scaling and translation of the variables. Because of these two properties, the correlation coefficient is somewhat unsuitable for feature selection in some applications. In spite of these drawbacks, cross-correlation is a widely adopted method for selecting features for developing simple or multivariate regression models to estimate target characteristics from hyperspectral data.

10.3.4.2 Mutual Information Analysis (MIA)

The MIA is a modification of the entropy-based method, and it considers the mutual information in pairs of bands in a hyperspectral image [9,18,21]. If band X has M levels of DN and band Y has N levels of DN, then the mutual information entropy can be calculated as:

$$H(X,Y) = -\sum_{i=1}^{M}\sum_{j=1}^{N} p_{i,j}\ln(p_{i,j}) \tag{10.7}$$

$$I(X,Y) = N\ln(N) - \sum_{i=1}^{M}\sum_{j=1}^{N} F_{i,j}\ln(F_{i,j}) - \sum_{i=1}^{M} F_{i,+}\ln(F_{i,+}) - \sum_{j=1}^{N} F_{+,j}\ln(F_{+,j}) \tag{10.8}$$

$$\mathrm{MI} = \frac{I(X,Y)}{I(Y)} \tag{10.9}$$

where, $p_{i,j}$ is the probability of occurrence of pixels with a DN of i in band X and j in band Y, $F_{i,+}$ is marginal summary of level i, and $F_{+,j}$ is marginal summary of level j, $I(X,Y)$ is the mutual information between two bands X and Y, and MI is the percentage of mutual information expressed by band X with respect to band Y.

If X and Y are probabilistically independent, $I(X,Y)$ would be zero. But, X and Y have perfect association, then $I(X,Y) = I(X) = I(Y)$. Using this method, bands with highest entropy and minimal mutual information can be selected. This method was shown to work well in identifying hyperspectral image bands [9] by eliminating up to 55% of bands without losing classification accuracy [21] and outperform entropy-based and correlation-based methods [22].

10.3.4.3 Spectral Derivative Analysis

The full width half maximum (FWHM), commonly referred to as bandwidth, can vary among bands in hyperspectral sensor design. The spectral derivative method explores the bandwidth variable as a function of added information. It is apparent that if two adjacent bands do not differ much, then the underlying geospatial phenomenon can be characterized with only one of the two bands.

Although higher order derivatives can be calculated, the first- and second-order derivatives are commonly used in identifying spectral features. The mathematical descriptions of the first and second spectral derivatives are illustrated in Equations 10.10 and 10.11.

$$\text{First derivative,} \quad D_{1\lambda} = \frac{\partial(\mathrm{DN}(x,\lambda))}{\partial\lambda} \tag{10.10}$$

$$\text{Second derivative,} \quad D_{2\lambda} = \frac{\partial^2(\mathrm{DN}(x,\lambda))}{\partial\lambda^2} \tag{10.11}$$

where DN represents the digital number of a pixel in the hyperspectral image, x is the spatial location of the pixel, and λ is the central wavelength.

If D_1 is equal to zero, then one of the two bands is redundant. In general, adjacent bands that differ significantly should be preserved for characterization, while adjacent bands similar to a specific band can be eliminated. The second derivative identifies bands that can be represented by a linear combination of adjacent bands. Thus, if two adjacent bands can linearly interpolate the third band, then the third band is redundant. The larger the deviation from a linear model, the higher the information value of the band. The drawback of the derivative analysis is that it usually compares only adjacent bands although higher order derivatives can be implemented to include a larger number of bands.

Spectral derivatives can be combined with other feature selection methods such as divergence measures to identify useful derivative features. Using spectral derivatives as features can result in high accuracies for spectral signature discrimination and classification [23]. Derivative analysis has been successfully utilized for mapping biodiversity in a tropical forest from hyperspectral image data [24], and to estimate biophysical characteristics such as leaf area index (LAI) in crops [25].

10.3.5 Sequential Search Methods

Selection of features with most discriminatory power from a superset of features may include a search method and a criterion for selection. An important index of performance that is commonly used in multispectral remote sensing is the error of estimation of the target characteristics, or classification accuracy or clustering ability, depending on the purpose of feature extraction and application of interest.

Estimation of the classification accuracy (or error of estimation) for all combinations of features in hyperspectral remote sensing is a prohibitive task because of the tens of thousands of potential features. If there are L features in a dataset, there will be $(2^L - 1)$ combination of features that will need to be evaluated. For example, a feature space with a mere 100 features, there could be $1.3*10^{30}$ feature combinations. An appropriate search methodology can reduce some of the computational needs. Sequential search algorithms explore the search methodology rather than the index of performance.

Most popular sequential search methodologies include sequential forward selection (SFS) and sequential backward selection (SBS) [26]. In the SFS method, the search starts with an empty feature space. Each feature is added one by one, based on an objective function, until a desired cardinality is obtained. The Firefly algorithm for band selection is an evolutionary algorithm that employs the SFS method for band selection on separability-based measures such as Jeffries-Matusita, and minimum estimated abundance covariance [27]. In the SBS method, the search starts with the full feature space, and features are removed one at a time based on an objective function until a desired cardinality is achieved. The supervised search methods use a similarity measure or a separability measure as the objective function. A Gaussian process regression method incorporates SBS based on regression coefficients as the objective function within a machine learning regression algorithm and selects 4–9 bands that achieved optimal accuracies for estimating crop biophysical variables [28].

Both SFS and SBS methods are suboptimal with the serious drawback that they cannot remove a feature already selected in a previous step (in case of SFS) or reconsider a feature that has been removed from the feature subset (in case of SBS). They could also be highly time consuming and computationally inefficient. The sequential floating selection applied to SFS or SBS by dynamically changing the number of features included or removed in a step overcomes these drawbacks. The sequential floating selection is computationally more efficient and allows to add a feature that has been removed in the backward selection process or remove a feature that has been previously added in the forward selection process.

Other sequential selection methods such as genetic algorithms (GA) were also proposed for feature selection from high-dimensional datasets [29,30]. Sequential search methods are appropriate for small and medium-scale problems, whereas GA is appropriate for large-scale problems [3].

10.3.6 Other Methods

There are many other less frequently used methods for feature selection/extraction and dimensionality reduction. One such method is increasing the bandwidth by combining adjacent bands of a hyperspectral image that were selected based on one of the band selection methods. Another such method is applying spectral filters on the image data to remove noise and less useful features. Wavelet decomposition is another important method for identifying important features from a hyperspectral image.

10.3.6.1 Wavelet Decomposition Method

The wavelet decomposition is a kernel-based method that works with the frequency components of hyperspectral signals. Therefore, it is an ideal approach for feature extraction where a multi-resolution approach is desirable [31]. The inherent ability of the wavelet function to vary the width of the operator allows it to separate fine-scale and large-scale information in a hyperspectral dataset. The fundamental operator used in wavelet transform (WT) is referred to as a *mother wavelet*. Any function $\psi(\lambda)$ can be used as a mother wavelet provided it satisfies the admissibility condition below.

$$\int_{-\infty}^{+\infty} \frac{|\Im(\psi(\lambda))|^2}{|\omega|} d\omega < \infty \tag{10.12}$$

where, $\Im$ indicates the Fourier transform of function $\psi(\lambda)$ and ω is the Fourier domain variable.

In other words, the mother wavelet function must oscillate with an average value of zero, while exhibiting an exponential decay and compact support. There are many mother wavelets available for one to choose. Some of the more common ones include biorthogonal spline, Haar, Daubechies, Gaussian, Symlet, Meyer, and Coiflet, to name a few.

After choosing a mother wavelet, the wavelet transformation can be applied to a one-dimensional or two-dimensional signal using a discrete or continuous WT where the width and scale can be systematically changed. Therefore, WT is a useful tool to separate and identify features that are associated with a phenomenon or variable that may present in different scales.

For example, in a hyperspectral image scene, the land covers (crops, forests, grass, soil, water, impervious, and so on) may cause the large-scale variation. However, if the user is interested in tree species or evapotranspiration, the variation may be occurring at a finer scale than that caused by land cover types.

Wavelet transform offers an efficient method for hyperspectral band selection [32], textural feature selection [33], and derivative analysis on hyperspectral data [31]. The three-dimensional (3D) wavelet transform can identify combined spectral and spatial features to improve classification accuracy [34]. The 3D Gabor wavelet feature extraction method combined with the pruning search based on a mutual information-based objective function successfully eliminated redundant information and increased classification accuracy [34].

10.4 INFORMATION EXTRACTION METHODS

Important information such as biophysical variables related to vegetation and vegetation types can be obtained using two major groups of information extraction methods. The first group includes deterministic or stochastic radiative transfer models (RTMs). These are process-based models that utilize our knowledge about the optical interaction of the target materials to inference on characteristics we are interested in mapping. Typically, a model is trained with known data on the optical properties of the target, and then, it is inversed to develop an inverse model to estimate the target property of interest [35,36]. The RTM are usually complex and difficult to implement.

The second group of methods for information extraction is based on statistical or heuristic methods. These methods typically utilize a subset of features to develop a quantitative model or a classification protocol. Two major groups of statistical/heuristic information extraction methods include:

a. *Modelling*: If the target characteristic to quantify is numerical and the user prefers quantitative output, a common method used is development of a decision support system with a mathematical relationship. Examples include estimation of variables such as LAI, biomass, canopy nitrogen content or evapotranspiration from hyperspectral data.
b. *Classification*: If the target characteristic of interest is categorical, then a classification approach is most appropriate for information extraction. If the variable of interest is distribution of different vegetation species, or identification of a specific invasive species, or land cover types, then classification is an appropriate strategy.

10.4.1 Statistical Methods

10.4.1.1 Regression Methods

Simple, multivariate, and partial least squares regression (PLSR) methods can be used for developing empirical models of a variable of interest based on hyperspectral image data. Simple regression is usually used when a vegetation index derived from hyperspectral data is used estimate a vegetation characteristic. Multivariate regression can be only used when the number of observations in the ground data or training data is considerably larger than the number of features used as independent variables. Multivariate regression uses the method of least square to estimate the parameters (intercept and regression coefficient) of regression in a model of the form:

$$y = b_0 + b_1 x_1 + b_2 x_2 + b_3 x_3 + \cdots + b_k x_k \tag{10.13}$$

where, y is the variable you are interested in estimating (e.g., biomass, evapotranspiration, canopy nitrogen content, and so on), k is the number of features, x_i is the ith feature, and b_i is the ith coefficient of regression with b_0 being the intercept.

One problem with multivariate regression on redundant image features is that it tends to exaggerate the goodness of fit. Using decorrelated or orthogonal features such as principal components or PLS factors eliminate this problem. Simple and multivariate regressions are extensively used for estimating biophysical variables such as LAI, biomass, foliar biochemical concentration, and so on, from decorrelated features such as PCA or vegetation index.

PLSR is a popular method of modeling optically active chemical constituents of a target material with spectroscopy. It is also ideal for developing quantitative models with high-dimensional data such as hyperspectral image data.

In PLSR, the extracted features are called PLS factors. The original bands are transformed to independent PLS factors such that the covariance between the dependent variable and the PLS factors in the training dataset is maximized. This transformation is similar to PCA with the only difference that the covariance is maximized here instead of the variance. The PLS factors are uncorrelated to each other. After transforming the data, a least square regression model is developed between the dependent variable and the PLS factors. An advantage of this method is that more than one y variable could be modeled in one step. The PLSR has been applied on both spectroscopic and hyperspectral image data to estimate biophysical variables relating to vegetation [25,37,38]. A modification of PLSR for capturing nonlinear relationships is the kernel-based PLSR (KPSLR). In this method, the hyperspectral data are transformed using a nonlinear mapping function called the kernel function, where the relationship between the transformed features and the target variable is linear [38].

Both the regression methods described here assume normal distribution of the data, and that the error term in regression is homoscedastic or has constant variance for all X since they come from the same population. However, spatially dynamic phenomena (that are often represented in an image) are heteroscedastic, with error variance dependent on the spatial lag (distance) between observations. Therefore, another approach for modeling spatially dynamic phenomena is by incorporating the spatial dependence into the model by using spatial regression models or autoregressive integrated moving average models.

In cases of non-parametric data, the machine learning regression algorithms (MLRA) such as bagging/boosting regression trees, random forest (RF), and kernel ridge regressions are computationally efficient methods that can lead to good model accuracies [24,37,38]. The RF method grows multiple decision trees on "bootstrap" samples from a training dataset known as the "in-bag" data. Each tree is grown on a bootstrap sample with replacement from the original data. A random subset of variables is used to split the tree, and the variable that results in the lowest impurity (error) is chosen for splitting the samples at each node. The tree is grown until each node holds samples of same group or characteristics. The average response from the various nodes or majority vote is used to make the final estimate of the class or value of the dependent variable. This method has shown the robustness to estimate vegetation characteristics [37–39].

10.4.1.2 Discriminant Analysis

Discriminant analysis works fast with high-dimensional hyperspectral data, and finds features that maximize the separation of the classes of interest. It is widely used in pattern recognition applications. The most commonly used discriminant analysis method is Fisher's linear discriminant analysis (FLDA), which employs Fisher's ratio, the ratio of between-class matrix to within-class scatter matrix (Equations 10.3 and 10.5).

To group hyperspectral data into L classes, the discriminant analysis identifies $L - 1$ eigenvectors as discriminant features. If there are k bands in the hyperspectral data, each pixel can be represented as a vector X with a $k \times 1$ dimension. If A is defined as a $k \times L - 1$ matrix with its columns representing the discriminant features, the original dataset X can be projected to onto the discriminant feature as $Y = A^T X$. The columns of A or the discriminant features are the eigenvectors of $(S_b^{-1} S_w)$ with nonzero eigenvalues, where S_b is the between-class scatter matrix and S_w within-class scatter matrix, which are calculated as shown in Equations 10.3 through 10.5.

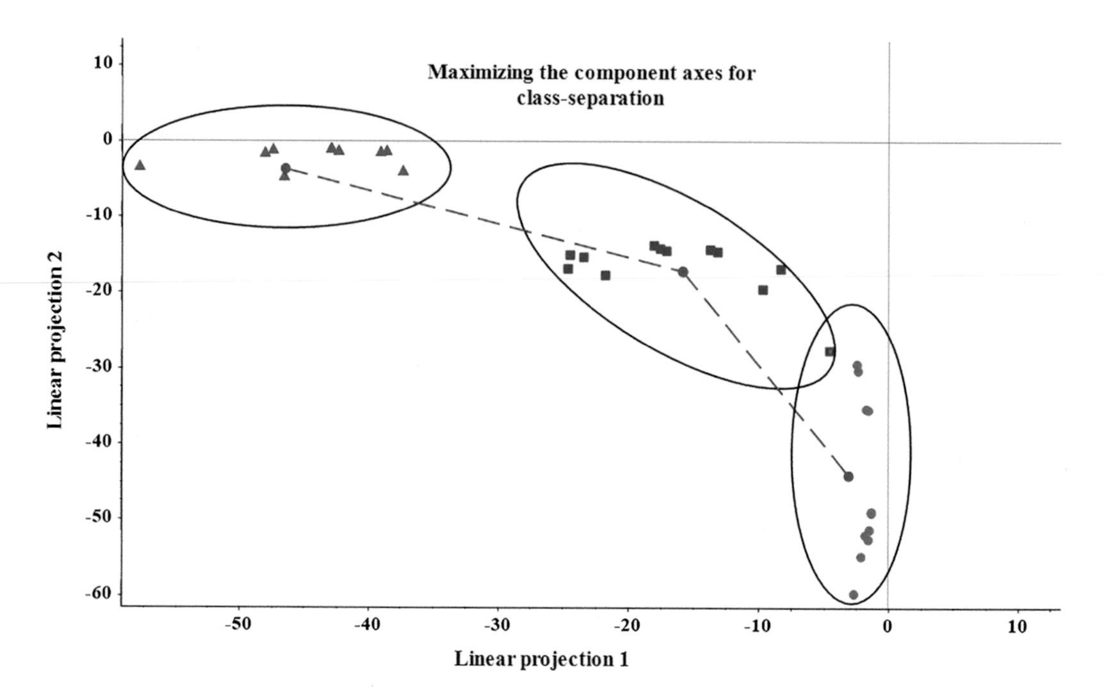

FIGURE 10.7 A figure depicting discriminant analysis of pixels on a scatter plot between two features. The linear projections maximize the separability between the three classes. The three classes represent canopy reflectance of sugar beet plants under disease infestation levels.

For example, Figure 10.7 shows the scatter plot of canopy reflectance of sugar beet plant under three different disease infestation levels. The linear projections of reflectance bands provide very high separability between the three disease levels. The classes did not show good separability if plotted against individual bands, even if those bands showed the highest correlation with the disease levels. The discriminant function identifies a linear projection that would provide the best discrimination between the target classes of interest.

The major problem with discriminant analysis is that it will identify only $L - 1$ features. In other words, the number of features cannot equal or exceed the number of classes. In hyperspectral images, the number of bands often exceeds the number of classes of interest. Therefore, a feature selection or dimensionality reduction method is usually employed before discriminant analysis. An example would be to use PLS for feature selection and to apply discriminant analysis for vegetation species identification [40]. A stepwise discriminant analysis can be used for identifying a subset of bands or features that can then be used for estimating target characteristics or class [13,41]. A stepwise band selection method and a subset of transformed features such as vegetation indices selected based on their correlation were used as input in an LDA and logistic discriminant analysis to identify plant diseases [41]. Discriminant analysis does not work well for distinguishing classes that have subtle differences, or very similar mean vectors. In many applications such as classifying similar species of vegetation, or different plant diseases or discriminating different types of stresses in the vegetation, the differences in reflectance is often very subtle. Such differences may not be easily distinguished with discriminant analysis since the class means may be very similar and discriminant analysis can use only relatively fewer ($L - 1$) features, resulting in low classification accuracy [41].

Fisher LDA can be combined with constrained energy minimization methods to develop linear constrained distance-based discriminant analysis [42]. If the dependent data are canonical in nature (interval variables), a canonical discriminant analysis (CDA) can be used. The CDA is similar to PCA in that it develops a linear combination of the independent variable (hyperspectral bands in this case) that has the highest multiple correlation with the dependent variable to form the first canonical variable. The process is repeated to develop multiple canonical variables. The CDA can be successfully used for vegetation classification from high-dimensional hyperspectral data [43]. Although discriminant analysis is mostly implemented as a supervised information extraction method, it can also be implemented for unsupervised information extraction using an automatic target detection method called target generation process [42].

10.4.2 Unsupervised Classification Methods

Unsupervised classification methods identify patterns of interest in an image data. This group of methods does not require training data or a priori knowledge of the true class labels. The output of classification is a class label for each pixel. For example, a regression-based information extraction system can provide a quantitative map of biomass or other target properties of interest. In contrast, unsupervised classification results in a class number that can be used as such, or labeled post-classification. The post-classification labeling is supervised process and requires ground data on true labels for sample areas. As in the previous category, unsupervised classification methods could use a set of features selected by the feature selection methods to classify the image into patterns of interest.

10.4.2.1 Clustering

Clustering uses an iterative optimization method to classify pixel vectors to a number of groups based on a similarity measure. The similarity between pixels can be measured using one of the distance measures such as Euclidean distance, Mahalanobis, distance, city block distance, and so on. The clustering algorithm can divide pixels into a user-specified number of groups using a k-means clustering, or to a flexible number of groups within a specified range using the ISODATA clustering method. In both cases, it is possible to delineate clusters in multiple ways. For example, assume that a user is interested in classifying the pixel vectors shown in Figure 10.8 into two classes. All three decision boundaries or lines shown in Figure 10.8 could potentially classify the pixels into two groups. The clustering methods usually adopt a clustering quality indicator such as "sum of squared errors" to select the best decision boundaries.

The clustering algorithm will check the distance between cluster means and a pixel vector that needs to be classified, and will assign a pixel to the closest cluster. In one iteration, all the pixels will be assigned to its nearest cluster and the cluster means will be recalculated. This process will be repeated until it meets one of the user-specified criteria for stopping the process. The user-specified criteria could include the number of classes, number of iterations, percentage of pixels migrating

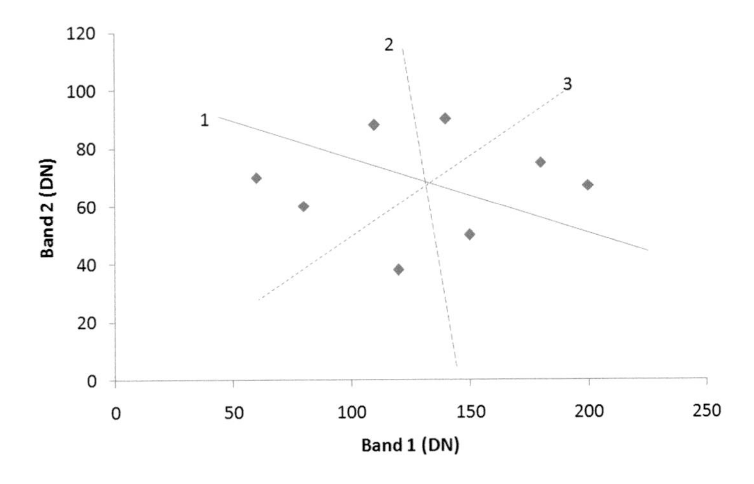

FIGURE 10.8 Three clustering options for a set of pixel vectors in a two-dimensional space. A line represents a decision boundary, the two sides of which represent the two clusters.

or changing its class label meets a user-defined criterion, and so on. A subset of hyperspectral image bands or a selection of features derived from the original image can be used as the input for unsupervised classification. Decorrelated bands and PCAs are commonly used as the input for unsupervised classification of hyperspectral data.

The most recent developments in clustering incorporates spatial-spectral features in the classification process, or unsupervised endmember identification and pixel unmixing, particularly when dealing with images of lower spatial resolution. Spatial features such as neighboring union histogram (indicating regional statistics) [44] and spatial textural features have been incorporated into clustering successfully. Unsupervised endmember identification using methods such as Vertex Component Analysis (VCA), where endmembers are chosen as orthogonal pixel vectors with the maximum length, followed by pixel unmixing process to generate classification maps [45] is another example of unsupervised classification.

10.4.2.2 ICA Mixed Model (ICAMM) Classification

The ICA is an unsupervised method that can be used for deriving the independent components that contribute to the mixed signals in a hyperspectral image when the source signals and the mixing information are missing. In a classification problem, the independent components can be viewed as a set of mutually exclusive classes. The ICAMM considers the hyperspectral image as a mixture of these mutually exclusive classes with non-Gaussian probability densities. The ICAMM first estimates these independent components as a linear transformation of the original hyperspectral bands using the statistical independency criterion and calculates the class component density [46]. It then calculates the class membership probability of each pixel using the Bayes theorem. Each pixel is assigned to a class based on the Bayes decision rule. Additional explanations of the theory behind ICA-based mixed model classification were given by [46,47].

The ICAMM is promulgated as a more appropriate method for unsupervised classification of hyperspectral images transformed with ICA. Because a majority of the classification methods assume normality of data and use second-order statistics for classification, they are not considered complimentary to ICA that employs higher order statistics. Classification using ICAMM has been shown to be superior to K-means classification when used in combination with various feature selection methods such as PCA [46]. Many modifications of ICAMM such as unmixing through sparse regression [48] and nonlinear mixing methods have improved classification accuracy [47,49]. An enhanced ICAMM method that incorporated the second derivative of the objective function also improved classification performance [49].

10.4.3 Supervised Classification

The first step in supervised classification is obtaining ground data or class signatures. Ground data refers to areas or pixels within an image with known class labels. The actual reflectance or DN of a given class for all wavebands in the image is referred to as the class signature. Surveying the scene is a common method for collecting ground data for a current image scene. Other sources of ground data should be used if a field survey is unavailable due to budgetary or time constraints. In such cases, maps, high resolution aerial photographs or images, farm records, archived spectral signatures of the classes of interest, and other historical records can be used.

During ground data collection, care should be taken to select areas representative of the class. Once ground data are developed, it is often divided into a training set and test set. A training set is used to train the classification algorithm to identify similar pixels in the entire image while test data are used for independent validation of the performance of the classifier.

The next step in supervised classification is to select or extract features from the hyperspectral image that needs to be classified. Either a supervised or unsupervised method can be used to select the features. If a supervised feature selection/extraction method is used, the ground data under the training sets are used to select/extract the features.

After feature selection/extraction, the next step is to train the classifier using the preselected features and the training data. The training process iteratively estimates the classification parameters for the specific classification algorithm until one of the user-specified classification criteria is achieved. The next step is to classify the entire image using the trained classifier. Depending on the classification method used, the output could be a hard classification, where each pixel is assigned one class, or a soft classification, where the membership of each pixel to each of the class is estimated as the output.

The last step in classification is a post-classification accuracy assessment using the ground data under the independent validation set to ensure that classification result is of acceptable quality. If the validation accuracies, represented by error matrix and kappa coefficient are acceptable, the process stops here. Otherwise, the classification process is repeated.

There are many methods available for supervised classification of hyperspectral data. Most methods available for classifying multispectral images can technically be used on hyperspectral data if the number of features has been reduced to similar levels as the multispectral data.

Usually, even after feature selection, the input space for classification of hyperspectral data could be considerably larger than that of multispectral data. Therefore, some of the multispectral classification methods are not optimal for hyperspectral data because of the large number of training samples required for obtaining acceptable classification accuracy. Therefore, methods developed specifically for the high-dimensional hyperspectral data are recommended information extraction for acceptable classification accuracies. The most commonly used supervised classification methods for hyperspectral data follow.

10.4.3.1 Spectral-Angle Mapper (SAM)

In an *N*-band (or *N*-feature) hyperspectral image, each pixel can be considered as an *N*-dimensional vector. Therefore, each vector defines a set of angles with the coordinates representing the band or features. In the SAM method of classification, the angular distance between pixel vectors is considered as the measure of distance (Figure 10.9) [50]. Each pixel is assigned to the class that is closest to it based on the angular distance. For example, pixel x is closer to reference spectrum A in Figure 10.9, and will be assigned to class A, while pixel y will be assigned to class B, based on their angular distance from those classes. In this respect, the SAM method is similar to the nearest neighbor classification method, except that the SAM uses angular distance. This method is developed for classifying hyperspectral data. It can easily handle high-dimensional data and large number of pixels in training set data as it reduces the dimensionality to the axes.

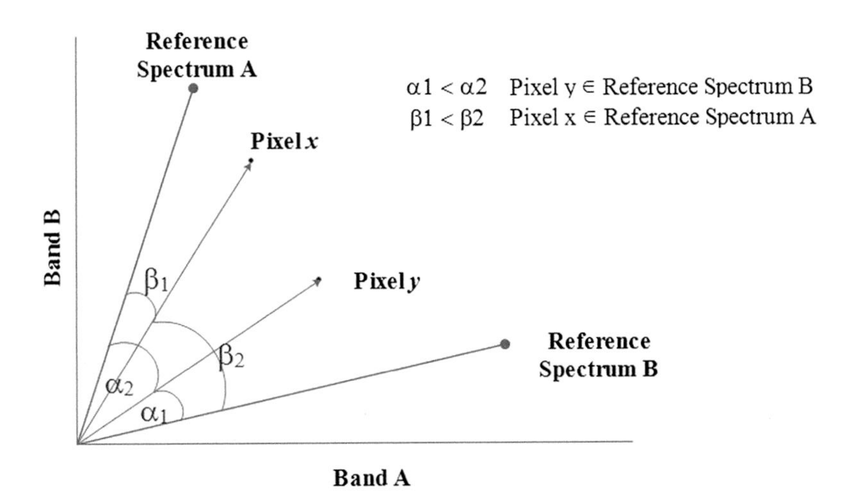

FIGURE 10.9 Depiction of an image pixel classification with Spectral Angle Mapper (SAM). The angular distance of a pixel from reference spectra or known class center is used to assign the pixel to the nearest class.

The advantage of the SAM is that it is insensitive to the magnitude of the pixel vectors since only the angular distance between vectors are used in establishing class membership. Therefore, it holds special significance for classifying vegetation.

Topographic shading usually interferes with vegetation signals in remote sensing. Since this interference tends to distribute the pixels along the same angular direction with different magnitude, it does not affect the angular orientation of the pixel vectors. Therefore, the SAM is a good classification method to use when topographic shading is suspected. However, SAM does not handle variability in the angular direction well. Combining spectral derivatives with the original spectra, and the use of multiple reference spectra, are shown to address this problem [51]. An extended the SAM method that incorporated a VCA to iteratively project data to an orthogonal subspace spanned by endmembers to identify all endmembers performed better than K-means and Mahalanobis distance classification methods [52].

10.4.3.2 Orthogonal Subspace Projection (OSP)

OSP is very effective in detecting and classifying constituent materials in a mixed pixel while suppressing undesired signatures [53]. Another benefit of OSP is its ability to reduce data dimensionality. Because of these properties, OSP is useful for class signature detection and discrimination as well as sub-pixel classification.

In this method, the classifier projects an unlabeled pixel vector onto a particular class vector (or subspace) of interest which is orthogonal to undesired signatures or other class vectors. Since the pixel vector is projected orthogonal to all other class vectors, their effect on the pixel vector under consideration is nullified. The basic idea used in OSP is that if a pixel vector with an unknown label is projected to each of the class vectors while nullifying influences from all other class signatures, it will provide the highest membership with the class where it belongs the best.

If an image is being classified into L classes of interest using supervised classification, there will be L corresponding mean class signatures. Matrix E represents the $K \times L$ class signatures or endmembers. The matrix containing the first $L - 1$ columns or endmembers is called U, and the last column containing the endmember of interest is called d. The OSP classification operator q^T is defined as:

$$q^T = d^T(I - UU^\#) \tag{10.14}$$

$$U^\# = (U^T U)^{-1} U^T \tag{10.15}$$

$$\alpha_p = \beta q^T \mathrm{DN} \tag{10.16}$$

$$\beta = (d^T P d)^{-1} \tag{10.17}$$

where, $U^\#$ is the pseudoinverse of U, $(1 - UU^\#)$ is the projection matrix P, DN is the pixel vector that needs to be classified, α_p is the projection of the pixel vector (with unknown label) to the specific endmember d, and β is a scalar normalizing factor. If the pixel vector with the unknown label belongs to the endmember d, the value of α_p will be the highest. Large values of α_p indicate better membership in the class with the signature of d.

This method is useful for pure pixel classification as well as mixed pixel classification. Although OSP is typically used for supervised classification, it can also be used for unsupervised classification. A recursive OSP, proposed by Song and Chang [54], performs OSP recursively without inverting the undesired signature matrix. The recursive OSP reduces the computational complexity without sacrificing the effectiveness.

10.4.3.3 Maximum Likelihood Classification (MLC)

The MLC is by far the most commonly used method of supervised classification of multispectral data when data are Gaussian distributed. The maximum likelihood classification can be performed based on the Bayes classification rule, which uses the conditional probabilities of pixel vectors to determine their class designation.

The likelihood function estimates the conditional probabilities $p(i|w)$ of a pixel with feature vector w belonging to class i by using the training data. Then, the maximum likelihood decision rule assigns this pixel to the class to which it has the highest probability of belonging.

For example, the probability of pixel a in Figure 10.10 belonging to class A is p_1, and class B is p_2. Since $p_1 > p_2$, pixel a will be assigned to class A. If the data distribution is multivariate normal and there are adequate numbers of training samples, maximum likelihood classification can result in high accuracy of classification. However, the number of training pixels required to maintain reasonable accuracy could be quite large in a hyperspectral dataset.

The accuracy of MLC depends of the accuracy of the mean vector and covariance matrix estimated for the classes. If N features used for classification, the training set for each class must contain a minimum of $N + 1$ pixels in order to calculate the sample covariance matrix, although Hoffbeck and Landgrebe [55] have reported a leave-one-out covariance estimation method in case of limited training data. As a rule of thumb, the number of pixels per class per feature is set around 10 to 100 for obtaining acceptable accurate class statistics. For example, if one wants to classify a hyperspectral image with 100 bands into 10 different classes, a minimum of 1000 to 10,000 training pixels per class will be needed for reasonable estimation of mean vector and covariance matrix. In other words, 10,000 to 100,000 total training pixels will be required. This requirement is without considering the accuracy degradation in hyperspectral images due to the Hughes phenomenon.

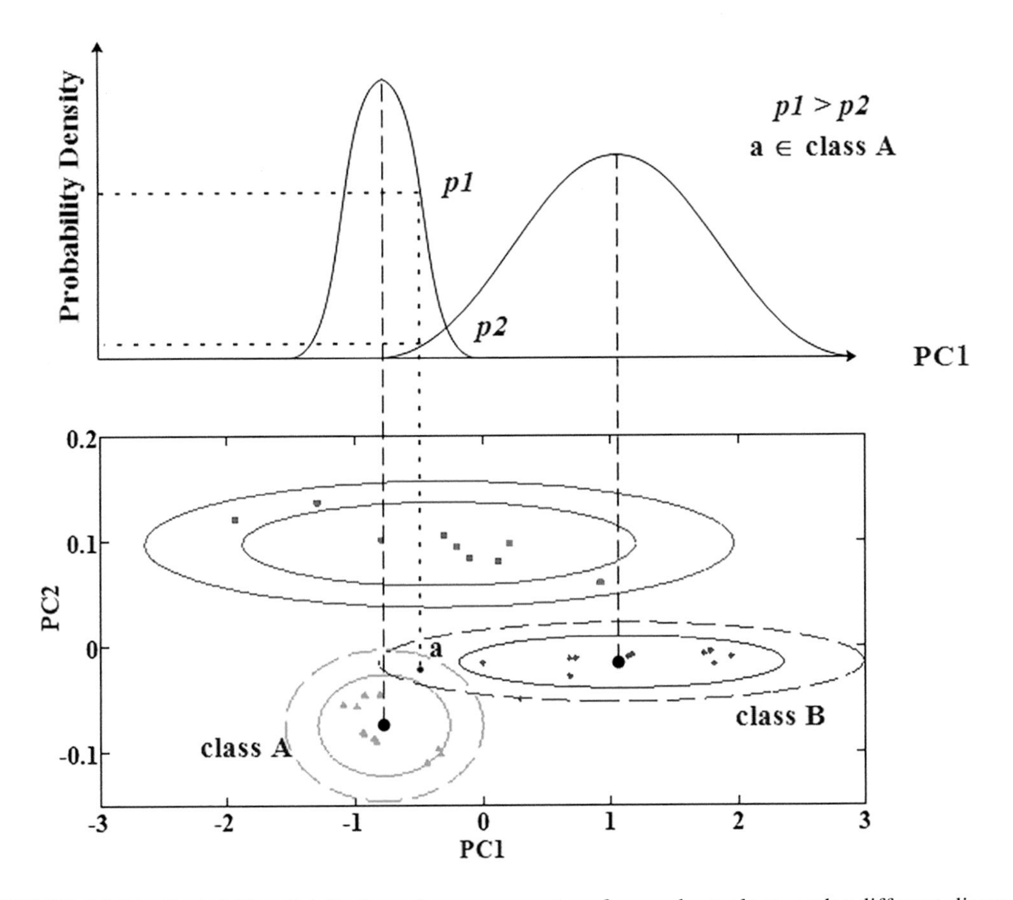

FIGURE 10.10 Probability distribution of canopy spectra of sugar beet plants under different disease infestation levels. Maximum likelihood classification of a new pixel based on the probability distribution of two of the classes.

Additionally, the number of training pixels required for achieving a specific accuracy also increases as the pixel variability within a class increases.

In summary, a large number of training samples are required in each class to obtain reasonable accuracy [1]. In a hyperspectral image with many features, such large numbers may not be always achievable, especially for classes with limited spatial extent. However, dimensionality reduction methods can be applied effectively to reduce the number of features before implementing MLC. For example, a comparison of MLC, SAM, artificial neural network, and decision tree classifiers found that MLC had the highest accuracy [56].

10.4.3.4 Artificial Neural Network (ANN)

The ANN is a non-parametric method of classification in that the decision boundaries of the classes are not determined by a deterministic rule. Rather, the decision boundaries are assigned in an iterative fashion, to minimize the error of labeling the training data.

A neural network is a simplified representation of how the human brain works in identifying objects and patterns. It contains data processing elements called neurons communicating through synaptic connections. Although there are many types of neural networks, the most commonly used the ANN model for hyperspectral image classification include multilayer perceptron (MLP), radial basis function, self-organizing networks, and AdaBoost models [56–59]. Both RBF and AdaBoost can be used with and without regularization.

A typical MLP neural network for image classification will have one input layer, one output layer, and one or more hidden layers (Figure 10.11). The output layer and hidden layers contains neurons where data are processed. The output layer can have one or more neurons depending on the application. The most common method of training an MLP network is using error back-propagation. In this method, the data are fed from input layer to output layer, where it is processed by each successive neuron, while the error is propagated back from output layer to input layer.

The input of a network for image classification is typically the features selected using one of the feature selection methods. Initially the model is trained using the training data with known output labels. The model training involved adjusting the weights associated with each synaptic connector and biases associated with each neuron to minimize the error between predicted output and target output. In other words, the network is trained until it learns the input patterns associated with a specific output with a

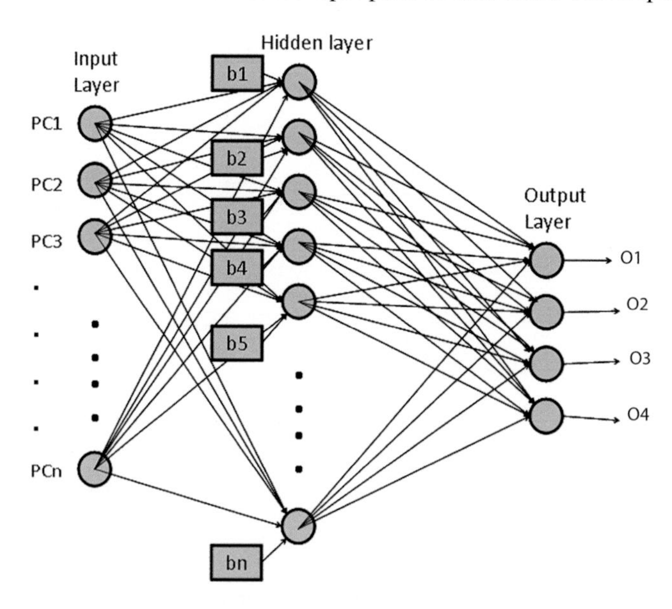

FIGURE 10.11 An example of an MLP network showing input, hidden, and output layers, all with multiple neurons. Here, principal components (PC) are listed as the input.

user-specified accuracy. There are also other criteria used for stopping training, which include inability of the network to learn any further, or reaching a maximum number of training epochs specified by the user.

Each processing node contains a summation operator and a transformation operator, which together processes the inputs into a weighted sum and then transforms it into the output (Equations 10.18 through 10.20) using a transformation function called activation function.

$$S_{j=\sum_i w_{ji} p_i} \tag{10.18}$$

$$O_j = f(S_j) \tag{10.19}$$

$$f(S) = \frac{1}{1+e^{-S}} \tag{10.20}$$

where, p_i represents the ith inputs to the jth neuron in a specific layer (either hidden or output), w_{ij} represents the weight of the synaptic connection from ith input from the previous layer to jth neuron in the current layer, O_j is the output from the jth neuron in the current layer, and f represents the transformation function.

Although there are many transformation functions that are available for use in a neuron, the most common are sigmoid functions, with the general form indicated in Equation 10.20. The selection of transformation function can affect the rate of convergence in a neural network. Since the ANN classifier is capable of developing highly nonlinear and non-parametric relationships even with noisy data, a properly trained network can provide highly accurate classification results. Some ANN-based classification regimes have shown similar accuracy to support vector machines, at reduced computational cost [58].

10.4.3.5 Support Vector Machines (SVMs)

Support vector machines (SVMs) represent a machine learning method that works well with high-dimensional data with limited training samples. The SVM can be used for feature selection, predictive modeling, and classification [39,60]. For linearly separable classes, the SVM tries to find the optimal separation surface between classes based on the training data. If the classes are not linearly separable, then the SVM uses a kernel-based method to find a nonlinear projection of the data where the classes are linearly separable. It is effective in separating classes with means very close to each other.

The SVM method is most commonly used for two-class separation problems, although it can be extended to multi-class separation as well. In a two-class problem, assume that (y_i, x_i) for $i = 1, 2, \ldots, N$ represents N training samples, where y_i is the label of the ith observation with values $+1$ or -1, and x_i is the corresponding feature vector with n features. The hyperplane separating the two classes has to be located such that the class labels $+1$ and -1 lies on either side of the hyperplane, and the minimum distance of the sample vectors to either side is maximized (Figure 10.12). The hyperplane is defined as:

$$wx + b = 0 \tag{10.21}$$

where, w and b are parameters of the hyperplane. Here, the intercept b is a scalar whereas the feature vector, x is L-dimensional. The vectors on either side of the hyperplane satisfy the condition that $wx + b \gtrless 0$. Therefore, the classifier can be expressed as:

$$f(x,\alpha) = \text{sign}\ (wx + b) \tag{10.22}$$

Therefore, the support vectors lie on two hyperplanes with equations:

$$wx + b = \pm 1 \tag{10.23}$$

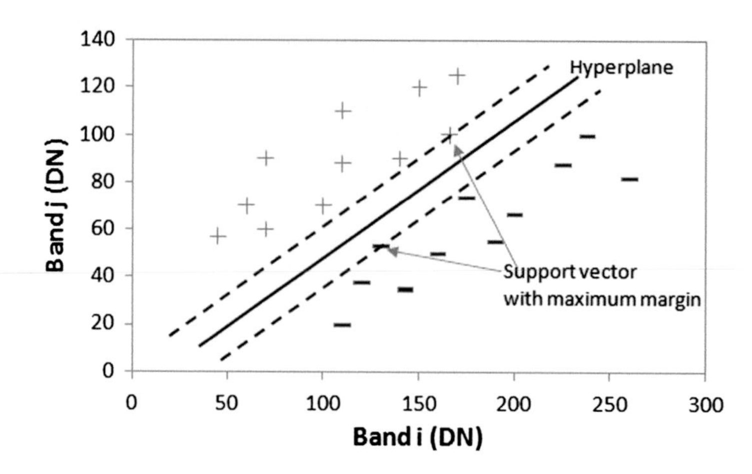

FIGURE 10.12 An example distribution of pixel vectors in two-dimensional, indicating the support vectors with maximum margin and the hyperplane defined as the decision boundary.

The optimal hyperplane is defined such that the margins (distance to support vector) is maximized. This constraint for optimizing the parameters of hyperplane can be expressed as:

$$\min\left\{\frac{1}{2}\|w\|^2\right\} \quad \text{with } y_i(wx+b)\geq 1, \quad i=1,2,\ldots,N \tag{10.24}$$

If the classes are not linearly separable, a regularization parameter C and error variable ε_i are introduced into the constraint in Equation 10.24.

The kernels used in SVMs are functions based on the quadratic distance between support vectors. The common kernel functions used include local kernels such as radial basis, kernel with moderate decreasing (KMOD) and inverse multi-quadratic, and global kernels such as linear, polynomial, and sigmoid. Spectral kernels were also defined, specifically for classifying hyperspectral data that uses the local kernel function with spectral angle as the measure of distance [61]. The spectral kernel function tended to decrease false classification caused by shadows when classical kernel was used. Also, SVM-based classification performed better than classification and regression trees (CART) and neural network classifiers for land use classification [59]. SVMs can be combined with optimization models such as genetic algorithm to obtain high accuracy of classification [60].

10.4.3.6 Genetic Algorithm (GA)

The genetic algorithm (GA) is an evolutionary approach that mimics the natural selection process in biological evolution to search and optimize a solution. It works well for both constrained and unconstrained optimization problems. With a data space containing known values for a set of spectral features and their corresponding response variable, the GA continually searches the solution space. When hyperspectral data are used in combination with the genetic algorithm, it treats each spectral feature as a gene. At each step, the algorithm randomly selects samples from the training set. It then uses crossover and mutation to identify the samples for the next iteration. Over many iterations, the population evolves towards the optimal solution.

Genetic algorithm in combination with feature reduction methods such as PLSR and derivative analysis has been successfully employed to estimate crop biophysical features such as LAI [25]. Genetic algorithm can also be used for feature selection method [62], and it works well in combination with discriminant analysis and a classification model [63].

10.4.4 Spectral-Spatial Classification and Deep Learning Methods

Spectral-spatial classification methods combine spatial contextual information with spectral features for classifying target features of interest. Incorporating both spatial and spectral information has shown to improve classification accuracies considerably. There are several types of methods for spectral-spatial classification. One group of spatial-spectral classification methods utilizes the spatial and spectral contextual information derived independently. For example, the spatial contextual information can be derived by mathematical morphological transforms on hyperspectral features to obtain spatial structural information. Spatial characteristics such as size, shape, or texture of a pixel's neighborhood, or spatial attribute profiles can provide spatial contextual information. Other morphological profile descriptors include geodesic operators [64], spatial segmentation methods, random multiscale representation [65], and wavelet analysis [66]. These spatial classification results are then combined with spectral classification to perform a pixel-based final classification. The concept of combining spatial and spectral classification information to perform a spectral-spatial classification is depicted in Figure 10.13. In this case, two classification files are created: the first, based on spatial features, and the second, based on spectral features. These classification results are then combined at the pixel level to generate a spectral-spatial classification.

The second group of method for spectral-spatial classification combined spatial features with spectral features to produce joint features. Fauvel et al. [64] describes the use of morphological profile that combine geodesic opening and closing operators and adaptive neighborhoods to identify spatial features. They proposed a method called extended morphological profile (EMP) that combined the morphological profile for each pixel with the spectral features selected with a feature selection method such as PCA or ICA, and then performed classification using a kernel classifier such as an SVM. This method resulted in high classification accuracies for land cover classification with hyperspectral data.

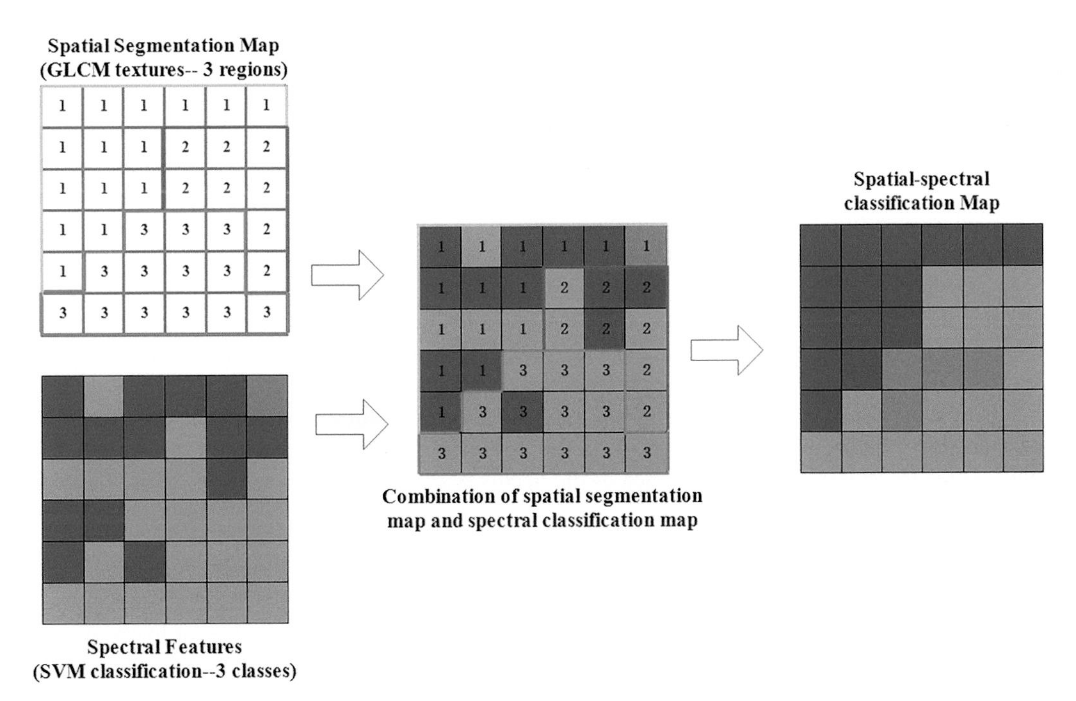

FIGURE 10.13 Depiction of spectral-spatial classification where a spectral classification method is combined with a spatial classification method to generate new classification results that use both information.

10.4.4.1 Deep Learning Methods

In the recent past, many deep learning methods have been developed to incorporate spectral and spatial features. These methods include both unsupervised and supervised methods, and are characterized by their deep learning features. For example, the image classification methods such as deep brief networks (DBF) and stacked auto-encoder (SAE) are unsupervised methods that are able to use spectral-spatial features [67,68]. The deep convoluted neural networks (CNN) is a supervised method that can incorporate deep learning using spectral-spatial features as inputs.

A deep CNN is a type of neural network composed of alternatively stacked convolution layers and spatial pooling layers. It is a kernel-based method where the convolution kernels are either pre-identified with a method such as k-means (called k-mean net), or the network automatically learns it from training data. The input data and the convolution kernel are major factors in CNN. The input data could be the hyperspectral bands or transformed features such as principal components. The convolution kernel could be 2D or 3D, depending on the type of analysis. The size of the kernel is also an important parameter. The 3D kernels can be derived from the hyperspectral data cube, or from the transformed feature data cube such as the principal components. The 3D CNN employing 3D kernels for spatial-spectral analysis have been shown to improve classification accuracy considerably [69,70] compared to 2D CNN, DBF, and SAE. The CNN networks can generate very high classification accuracy, better than most other methods reported for hyperspectral image classification.

10.5 ACCURACY ASSESSMENT

Information extraction procedures should be followed by an accuracy assessment of the product before they can be utilized for any real applications. In case of decision support models for estimating a numerical variable, the accuracy of the model should be assessed using an independent set of data (data not used for model development) called validation data. In such cases, the model accuracy can be expressed with several performance measures including the coefficient of determination (or R^2 value), root mean square error (RMSE) of prediction and cross-validation, and standard error of prediction.

In case of classification problems, error can be expressed as the error matrix based on an independent validation dataset that was not used for training. The error matrix can provide information on total number or percentage of correctly classified pixels as well as the errors of omission and commission for each class. Error of omission for a specific class is the number of pixels in that class that are classified as something else or given a wrong label, while error of commission for a certain class is the number of pixels from other classes labeled as this class under consideration.

In addition to the error matrix, classification accuracy can also be expressed using kappa coefficient. Kappa coefficient compares the classification results with respect to that of random assignment. A value of kappa coefficient of 1 indicates a perfectly accurate classification. More details on accuracy assessment can be obtained from [71].

For unsupervised classification, the common method of accuracy assessment requires post-classification labeling of the classified image based on information available on the actual class types. Once the classified image is labeled, the accuracy of labeling can be assessed in the same fashion as the supervised classification.

10.6 DISCUSSIONS

Many of the traditional multispectral data processing methods may be unsuitable for hyperspectral data because of the high dimensionality, data redundancy, and computation cost. In the past couple of decades, many data mining techniques for extracting useful information from hyperspectral data have been developed and tested. A characteristic of these data mining methods is the feature selection or dimensionality reduction step that usually results in independent features. Feature selection is usually followed by an information extraction step such as classification. The information extraction methods discussed in this chapter are specifically suitable for hyperspectral data. This chapter covered both supervised and unsupervised feature selection and information extraction methodologies.

A review of the published literature indicates that for characterizing a target property that can be expressed as a numerical variable (e.g., various biophysical properties), empirical models are the most commonly used. The most widely used empirical models are regression models (simple linear or multivariate or PLSR) on a subset of features selected or extracted with PCA or PLS factors, cross-correlation, or features selected based on knowledge of the absorbance characteristics of the target of interest. For characterizing a categorical variable (such as forest species, land cover, and so on), the most commonly adopted methods were classification with the SAM, ANN, and SVM on a set of reduced features, and the CNN for end-to-end classification. The features were selected with PCA, discriminant analysis, derivative analysis, ICA, and so on. The onset of new technologies such as unmanned aerial systems, miniaturized hyperspectral sensors, and gathering of big data with high spatial and spectral resolutions led to the development of new methods suited for high-dimensional data. Many of these methods utilize deep learning techniques that can exploit the high-level patterns in the data. Methods such as SVMs and deep learning methods such as CNNs incorporating spectral-spatial features are shown to be highly process efficient and accurate for hyperspectral data classification.

10.7 CONCLUSIONS

In summary, one may choose a specific combination of feature selection and information extraction methods depending on the application and objectives, the scale of the problem, skill level available, availability of training data, time, and budget constraints. With the proper selection of a suite of data mining techniques, it is possible to extract unique information from hyperspectral images that may not be possible with multispectral image data. The future direction in hyperspectral data is larger datasets with very high spatial and spectral resolutions. Therefore, the new methods in hyperspectral data mining should focus on process-efficient spectral-spatial methods that can handle large datasets and address spatial and spectral collinearity issues.

REFERENCES

1. Shahshahani, B.M., and Landgrebe, D.A., The effect of unlabeled samples in reducing the small sample size problem and mitigating the Hughes phenomenon, *IEEE Transactions on Geosciences and Remote Sensing*, 32, 1087–95, 1994.
2. Mitra, P., Murthy, C.A., and Pal, S.K., Unsupervised feature selection using feature similarity, *IEEE Transactions on Pattern Analysis and Machine Intelligence*, 24, 301–12, 2002.
3. Kudo, M., and Sklansky, J., Comparison of algorithms that select features for pattern classifiers, *Pattern Recognition*, 33, 25–41, 2000.
4. Rencs, A.N., *Remote Sensing for the Earth Sciences—Manual of Remote Sensing*, 3rd ed., vol. 3, John Wiley, New York, 1999.
5. Zhang, J., Huang, W., Li, J., Yang, G., Luo, J., Gu, X., and Wang, J., Development, evaluation and application of spectral knowledge base to detect yellow rust in winter wheat, *Precision Agriculture*, 12, 716–31, 2011.
6. Gamon, J.A., Serrano, L., and Surfus, J.S., The photochemical reflectance index: An optical indicator of photosynthetic radiation use efficiency across species, function types, and nutrient levels, *Oecologia*, 112(4), 492–501, 1997.
7. Almeida, T.I.R., and De Souza Filho, C.R., Principal component analysis applied to feature oriented band ratios of hyperspectral data: A tool for vegetation studies, *International Journal of Remote Sensing*, 25, 5005–23, 2004.
8. Bajwa, S.G., Bajcsy, P., Groves, P., and Tian, L.F., Hyperspectral image data mining for band selection in agricultural applications, *Transactions of the ASABE*, 47, 895–907, 2004.
9. Lavanya, A., and Sanjeevi, S., An improved band selection technique for hyperspectral data using factor analysis, *Journal of Indian Society of Remote Sensing*, 41(2), 199–211, 2013.

10. Bruce, L.M., and Reynolds, D., Game theory based data fusion for precision agriculture applications. In *Proceedings of Geoscience and Remote Sensing Symposium (IGARSS), 2016 IEEE International, IEEE*, 3563–66, 2016.
11. Zhao, C., Tian, M., Qi, B., and Wang, Y.A., Variation pixels identification method based on kernel spatial attraction model and local entropy for robust endmember extraction, *Journal of Central South University of Technology*, 23, 2016.
12. Malpica, J.A., Rejas, J.G., and Alonso, M.C., A projection pursuit algorithm for anomaly detection in hyperspectral imagery, *Pattern Recognition*, 41, 3313–27, 2008.
13. Ifarraguerri, A., and Chang, C., Unsupervised hyperspectral image analysis with projection pursuit, *IEEE Transactions on Geoscience and Remote Sensing*, 38, 2529–38, 2000.
14. Jones, M.C., and Sibson, R., What is projection pursuit? *Journal of the Royal Statistical Society. A (Statistics in Society)*, 150, 1–36, 1987.
15. Ren, J., Zabalza, J., Marshall, S., and Zheng, J., Effective feature extraction and data reduction in remote sensing using hyperspectral imaging, *IEEE Signal Processing Magazine*, 31(4), 149–54, 2014.
16. Tsai, F., Lin, E.K., and Yoshino, K., Spectrally segmented principal component analysis of hyperspectral imagery for mapping invasive plant species, *International Journal of Remote Sensing*, 28, 1023–39, 2007.
17. Falco, N., Bruzzone, L., and Benediktsson, J.A., An ICA based approach to hyperspectral image feature reduction. In *Geoscience and Remote Sensing Symposium (IGARSS), 2014 IEEE International, IEEE*, 3470–73, 2014.
18. Wang, N., Du, B., Zhang, L., and Zhang, L., An abundance characteristic-based independent component analysis for hyperspectral unmixing, *IEEE Transactions on Geoscience and Remote Sensing*, 53(1), 416–28, 2015.
19. Bajwa, S.G., Mishra, A.R., and Norman, R.J., Canopy reflectance response to plant nitrogen accumulation in rice, *Precision Agriculture*, 11(5), 488–506, 2010.
20. Thenkabail, P.S., Mariotto, I., Gumma, M.K., Middleton, E.M., Landis, D.R., and Huemmerich, K.F., Selection of hyperspectral narrowbands (HNBs) and composition of hyperspectral twoband vegetation indices (HVIs) for biophysical characterization and discrimination of crop types using field reflectance and Hyperion/EO-1 data, *IEEE Journal of Selected Topics in Applied Earth Observations and Remote Sensing*, 6(2), 427–39, 2013.
21. Guo, B., Damper, R.I., Gunn, S.R., and Nelson, J.D.B., Improving hyperspectral band selection by constructing an estimated reference map, *Journal of Applied Remote Sensing*, 8(1), 083692, 2014.
22. Guo, B., Gunn, S.R., Damper, R.I., and Nelson, J.D.B., Band selection for hyperspectral image classification using mutual information, *IEEE Geoscience and Remote Sensing Letters*, 3, 522–26, 2006.
23. Chang, C.-I., Chakravarty, S., Chen, H.M., and Ouyan, Y.-C., Spectral derivative feature coding for hyperspectral signature analysis, *Pattern Recognition*, 42, 395–408, 2009.
24. Laurin, G.V., Chan, J.C.W., Chen, Q., Lindsell, J.A., Coomes, D.A., Guerriero, L., and Valentini, R., Biodiversity mapping in a tropical West African forest with airborne hyperspectral data, *PloS One*, 9(6), e97910, 2014.
25. Thorp, K.R., Wang, G., Bronson, K.F., Badaruddin, M., and Mon, J., Hyperspectral data mining to identify relevant canopy spectral features for estimating durum wheat growth, nitrogen status, and grain yield, *Computers and Electronics in Agriculture*, 136, 1–12, 2017.
26. Jain, A., and Zongkar, D., Feature selection: Evaluation, application and small sample performance, *IEEE Transactions of Pattern Analysis and Machine Intelligence*, 19, 153–58, 1997.
27. Su, H., Yong, B., and Du, Q., Hyperspectral band selection using improved firefly algorithm, *IEEE Geoscience and Remote Sensing Letters*, 13(1), 68–72, 2016.
28. Verrelst, J., Rivera, J.P., Gitelson, A., Delegido, J., Moreno, J., and Camps-Valls, G., Spectral band selection for vegetation properties retrieval using Gaussian processes regression, *International Journal of Applied Earth Observation and Geoinformation*, 52, 554–67, 2016.
29. Paul, A., Bhattacharya, S., Dutta, D., Sharma, J.R., and Dadhwal, V.K., Band selection in hyperspectral imagery using spatial cluster mean and genetic algorithms, *GIScience & Remote Sensing*, 52(6), 643–59, 2015.
30. Hong, J.-H., and Cho, S.-B., Efficient huge-scale feature selection with speciated genetic algorithm, *Pattern Recognition Letters*, 27, 143–50, 2006.
31. Bruce, L.M., and Li, J., Wavelets for computationally efficient hyperspectral derivative analysis, *IEEE Transactions on Geoscience and Remote Sensing*, 39, 1540–46, 2001.
32. Feng, S., Itoh, Y., Parente, M., and Duarte, M.F., Hyperspectral band selection from statistical wavelet models, *IEEE Transactions on Geoscience and Remote Sensing*, 55(4), 2111–23, 2017.

33. Wang, L., Weijing, S., and Peng, L., Link the remote sensing big data to the image features via wavelet transformation, *Cluster Computing*, 19(2), 793–810, 2016.
34. Zhu, Z., Jia, S., He, S., Sun, Y., Ji, Z., and Shen, L., Three-dimensional Gabor feature extraction for hyperspectral imagery classification using a memetic framework, *Information Sciences*, 298, 274–87, 2015.
35. Verhoef, W., and Back, H., Simulation of hyperspectral and directional radiance images using coupled biophysical and atmospheric radiative transfer models, *Remote Sensing of Environment*, 87, 23–41, 2003.
36. Meroni, M., Colombo, R., and Panigada, C., Inversion of a radiative transfer model with hyperspectral observations for LAI mapping in poplar plantations, *Remote Sensing of Environment*, 92, 195–206, 2004.
37. Rivera-Caicedo, J.P., Verrelst, J., Muñoz-Marí, J., Camps-Valls, G., and Moreno, J., Hyperspectral dimensionality reduction for biophysical variable statistical retrieval, *ISPRS Journal of Photogrammetry and Remote Sensing*, 132, 88–101, 2017.
38. Pullanagari, R.R., Kereszturi, G., and Yule, I.J., Mapping of macro and micro nutrients of mixed pastures using airborne AisaFENIX hyperspectral imagery, *ISPRS Journal of Photogrammetry and Remote Sensing*, 117, 1–10, 2016.
39. Abdel-Rahman, E.M., Mutanga, O., Adam, E., and Ismail, R., Detecting *Sirex noctilio* grey-attacked and lightning-struck pine trees using airborne hyperspectral data, random forest and support vector machines classifiers, *ISPRS Journal of Photogrammetry and Remote Sensing*, 88, 48–59, 2014.
40. Peerbhay, K.Y., Mutanga, O., and Ismail, R., Commercial tree species discrimination using airborne AISA Eagle hyperspectral imagery and partial least squares discriminant analysis (PLS-DA) in KwaZulu–Natal, South Africa, *ISPRS Journal of Photogrammetry and Remote Sensing*, 79, 19–28, 2013.
41. Bajwa, S.G., Rupe, J., and Mason, J., Soybean disease monitoring with leaf reflectance, *Remote Sensing*, 9(2), 127, 2017.
42. Du, Q., and Chang, C.-I., A linear constrained distance-based discriminant analysis for hyperspectral image classification, *Pattern Recognition*, 34, 361–73, 2000.
43. Alonzo, M., Roth, K., and Roberts, D., Identifying Santa Barbara's urban tree species from AVIRIS imagery using canonical discriminant analysis, *Remote Sensing Letters*, 4(5), 513–21, 2013.
44. Yang, W., Hou, K., Liu, B., Yu, F., and Lin, L., Two-stage clustering technique based on the neighboring union histogram for hyperspectral remote sensing images, *IEEE Access*, 5, 5640–47, 2017.
45. Villa, A., Chanussot, J., Benediktsson, J.A., Jutten, C., and Dambreville, R., Unsupervised methods for the classification of hyperspectral images with low spatial resolution, *Pattern Recognition*, 46(6), 1556–68, 2013.
46. Shah, C.A., Varshney, P.K., and Arora, M.K., ICA mixture model algorithm for unsupervised classification of remote sensing imagery, *International Journal of Remote Sensing*, 28, 1711–31, 2007.
47. Bioucas, J.M., Plaza, A., Camps-Valls, G., Scheunders, P., Nasrabadi, N.M., and Chanussot, J., Hyperspectral remote sensing data analysis and future challenges, *IEEE Geoscience and Remote Sensing Magazine*, 1(2), 6–36, 2013.
48. Ertürk, A., Iordache, M.D., and Plaza, A., Sparse unmixing with dictionary pruning for hyperspectral change detection, *IEEE Journal of Selected Topics in Applied Earth Observations and Remote Sensing*, 10(1), 321–30, 2017.
49. Oliveira, P.R., and Romero, R.A., Improvements on ICA mixture models for image pre-processing and segmentation, *Neurocomputing*, 71(10–12), 2180–93, 2008.
50. van der Meer, F., The effectiveness of spectral similarity measures for the analysis of hyperspectral imagery, *International Journal of Applied Earth Observation and Geoinformation*, 8, 3017, 2006.
51. Zhang, X., and Li, P., Lithological mapping from hyperspectral data by improved use of spectral angle mapper, *International Journal of Applied Earth Observation and Geoinformation*, 31, 95–109, 2014.
52. Li, H., Lee, W.S., Wang, K., Ehsani, R., and Yang, C., Extended spectral angle mapping (ESAM) for citrus greening disease detection using airborne hyperspectral imaging, *Precision Agriculture*, 15(2), 162–83, 2014.
53. Harsanyi, J.C., and Chang, C.-I., Hyperspectral image classification and dimensionality reduction: An orthogonal subspace projection, *IEEE Transactions on Geoscience and Remote Sensing*, 32, 779–85, 1994.
54. Song, M., and Chang, C.I., A theory of recursive orthogonal subspace projection for hyperspectral imaging, *IEEE Transactions on Geoscience and Remote Sensing*, 53(6), 3055–72, 2015.
55. Hoffbeck, J.P., and Landgrebe, D.A., Covariance matrix estimation and classification with limited training data, *IEEE Transactions on Pattern Analysis and Machine Intelligence*, 18, 763–7, 1996.
56. Shafri, H.Z.M., Suhaili, A., and Mansor, S., The performance of maximum likelihood, spectral angle mapper, neural network and decision tree classifiers in hyperspectral image analysis, *Journal of Computer Science*, 3, 419–23, 2007.

57. Camps-Valls, G., and Bruzzone, L., Kernel-based methods for hyperspectral image classification, *IEEE Transactions on Geoscience and Remote Sensing*, 43, 1351–62, 2005.
58. Li, J., Du, Q., and Li, Y., An efficient radial basis function neural network for hyperspectral remote sensing image classification, *Soft Computing*, 20(12), 4753–9, 2016.
59. Shao, Y., and Lunetta, R.S., Comparison of support vector machine, neural network, and CART algorithms for the land-cover classification using limited training data points, *ISPRS Journal of Photogrammetry and Remote Sensing*, 70, 78–87, 2012.
60. Pal, M., Support vector machine-based feature selection for land cover classification: A case study with DAIS hyperspectral data, *International Journal of Remote Sensing*, 27, 2877–94, 2006.
61. Mercier, G., and Lennon, M., Support vector machines for hyperspectral image classification with spectral based kernels, *Proc. IEEE International Geoscience and Remote Sensing Symposium*, 1, 288–90, 2003.
62. Stavrakoudis, D.G., Galidaki, G.N., Gitas, I.Z., and Theocharis, J.B., A genetic fuzzy-rule-based classifier for land cover classification from hyperspectral imagery, *IEEE Transactions on Geoscience and Remote Sensing*, 50(1), 130–48, 2012.
63. Cui, M., Prasad, S., Li, W., Bruce, L.M., Locality preserving genetic algorithms for spatial-spectral hyperspectral image classification, *IEEE Journal of Selected Topics in Applied Earth Observation and Remote Sensing*, 6(3), 1688–97, 2013.
64. Fauvel, M., Tarabalka, Y., Benediktsson, J.A., Channusol, J., and Tilton, J.C., Advances in spectral-spatial classification of hyperspectral images, *Proceedings of IEEE*, 101(3), 652–75, 2013.
65. Liu, J., Wu, Z., Li, J., Xiao, L., Plaza, A., and Benediktsson, J.A., Spatial-spectral hyperspectral image classification using random multiscale representation, *IEEE Journal of Selected Topics in Applied Earth Observation and Remote Sensing*, 9(9), 4129–41, 2016.
66. Zhou, X., Prasad, X., and Crawford, M.M., Wavelet domain multiview active learning for spatial-spectral image classification, *IEEE Journal of Selected Topics in Applied Earth Observation and Remote Sensing*, 9(9), 4047–59, 2016.
67. Chen, Y., Lin, Z., Zhao, X., Wang, G., and Gu, Y., Deep learning-based classification of hyperspectral data, *IEEE Journal of Selected Topics in Applied Earth Observation and Remote Sensing*, 7, 2094–107, 2014.
68. Chen, Y., Zhao, X., and Jia, X., Spectral-spatial classification of hyperspectral data based on deep belief network, *IEEE Journal of Selected Topics in Applied Earth Observation and Remote Sensing*, 8, 2381–92, 2015.
69. Li, Y., Zhang, H., and Shen, Q., Spectral-spatial classification of hyperspectral imagery with 3D convoluted neural networks, *Remote Sensing*, 9(1), 67, 2017.
70. Chen, Y., Lin, Z., Zhao, X., Wang, G., and Gu, Y., Deep learning-based classification of hyperspectral data, *IEEE Journal of Selected Topics in Applied Earth Observations and Remote Sensing*, 7(6), 2094–107, 2014.
71. Congalton, R.G., and Green, K., *Assessing the Accuracy of Remotely Sensed Data—Principles and Practices*, CRC Press, Boca Raton, Florida, 1999.

11 Hyperspectral Data Processing Algorithms

Antonio Plaza, Javier Plaza, Gabriel Martín, and Sergio Sánchez

CONTENTS

11.1 INTRODUCTION

Hyperspectral imaging is concerned with the measurement, analysis, and interpretation of spectra acquired from a given scene (or specific object) at a short, medium, or long distance by an airborne or satellite sensor [1]. The concept of hyperspectral imaging originated at NASA's Jet Propulsion Laboratory in California with the development of the Airborne Visible InfraRed Imaging Spectrometer (AVIRIS), able to cover the wavelength region from 400 to 2500 nanometers using more than 200 spectral channels, at a nominal spectral resolution of 10 nanometers [2]. As a result, each pixel vector collected by a hyperspectral instrument can be seen as a spectral signature or fingerprint of the underlying materials within the pixel.

The special characteristics of hyperspectral datasets pose different processing problems [3], which must be necessarily tackled under specific mathematical formalisms, such as classification, segmentation, image coding, or spectral mixture analysis [4]. These problems also require specific dedicated processing software and hardware platforms. In most studies, techniques are divided into full-pixel and mixed-pixel techniques, where each pixel vector defines a spectral signature or fingerprint that uniquely characterizes the underlying materials at each site in a scene [5]. Mostly based on previous efforts in multispectral imaging, full-pixel techniques assume that each pixel vector measures the response of one single underlying material. Often, however, this is not a realistic assumption. If the spatial resolution of the sensor is not fine enough to separate different pure signature classes at a macroscopic level, these can jointly occupy a single pixel, and the resulting spectral signature will be a composite of the individual pure spectra, called endmembers in hyperspectral terminology [6]. Mixed pixels can also result when distinct materials are combined into a homogeneous or intimate mixture, which occurs independently of the spatial resolution of the sensor. To address these issues, spectral unmixing approaches have been developed under the assumption that each pixel vector measures the response of multiple underlying materials [7].

Our main goal here is to provide a seminal view on recent advances in techniques for full-pixel and mixed-pixel processing of hyperspectral images, taking into account both the spectral and spatial properties of the data. Due to the small number of training samples and the high number of features available in remote sensing applications, reliable estimation of statistical class parameters is a challenging goal [4]. As a result, with a limited training set, classification accuracy (in a full-pixel sense) tends to decrease as the number of features increases. This is known as the Hughes phenomenon. Furthermore, high-dimensional spaces are mostly empty, thus making density estimation more difficult. One possible approach to handle the problem of dimensionality is to consider the geometrical properties rather than the statistical properties of the classes. In this regard, it is important to develop techniques able to select the most highly informative training samples from the available training set [8]. The good classification performance already demonstrated by techniques such as kernel methods and support vector machines (SVMs) in remote sensing applications [9], using spectral signatures as input features, has been further increased using intelligent training sample selection algorithms [10].

It should be noted that most available hyperspectral data processing techniques (including both full-pixel and mixed-pixel techniques) focused on analyzing the data without incorporating information on the spatially adjacent data, that is, hyperspectral data are usually not treated as images, but as unordered listings of spectral measurements with no particular spatial arrangement. In certain applications, however, the incorporation of spatial and spectral information is mandatory to achieve sufficiently accurate mapping and/or classification results [11–13]. To address the need for developments able to exploit *a priori* information about the spatial arrangement of the objects in the scene in order to complement spectral information, this chapter also presents several techniques for spatial-spectral data processing in the context of a mixed-pixel classification scenario.

11.2 SUPPORT VECTOR MACHINES

Supervised classification is one of the most common analyses of remotely sensed hyperspectral data. The output of a supervised classification is effectively a thematic map that provides a snapshot representation of the spatial distribution of a particular theme of interest such as land cover. Research has indicated the considerable potential of SVM-based approaches for the supervised classification of remotely sensed hyperspectral data [14]. Comparative studies have shown that classification by an SVM can be more accurate than techniques such as neural networks, decision trees, and probabilistic classifiers such as maximum-likelihood classification [9]. This is due to the superior performance of SVMs when analyzing high-dimensional data (particularly in the presence of limited training samples) which generally results in higher relative accuracies than those reported for other classification methods. SVMs were designed for binary classification, but various methods exist to extend the binary approach to multiclass classification, such as the one-versus-rest and the one-versus-one strategies [15].

In essence, the SVM classification is based on fitting an optimal separating hyperplane between classes by focusing on the training samples that lie at the edge of the class distributions; which are the support vectors (Figure 11.1, reproduced from [9]). All of the other training samples are effectively discarded as they do not contribute to the estimation of hyperplane location. In this way, not only is an optimal hyperplane fitted, in the sense that it is expected to be generalizable to a large degree, but also a high accuracy may be obtained with the use of a small training set. It should be noted that the SVM used with a kernel function is a nonlinear classifier, where the nonlinear ability is included in the kernel. Different kernels lead to different SVMs. The most used kernels are the polynomial kernel, the Gaussian kernel, or the spectral angle mapper kernel, among many others [9].

Recently, innovative kernel-based algorithms with enhanced properties have been developed. These include semi-supervised or transductive SVMs (TSVMs) learning procedures [16], which are used to exploit both labeled and unlabeled pixels in the training stage, or contextual SVMs [17], in which spatial and spectral information is incorporated by means of the use of proper kernel

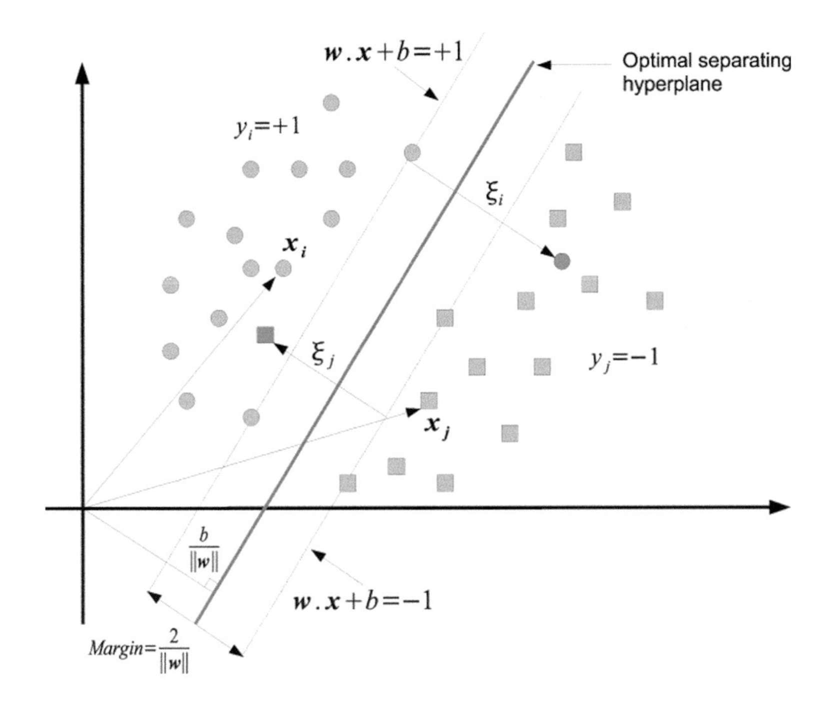

FIGURE 11.1 Classification of a nonlinearly separable case by a support vector machine (SVM).

functions. The capability of semi-supervised SVMs to capture the intrinsic information present in the unlabeled data can further mitigate the Hughes phenomenon, and contextual SVMs can address the issues related to the nonstationary behavior of the spectral signatures of classes in the spatial domain.

11.3 SPECTRAL UNMIXING OF HYPERSPECTRAL DATA

Spectral mixture analysis (also called spectral unmixing) has been an alluring exploitation goal from the earliest days of hyperspectral imaging [1] to the present [18]. No matter the spatial resolution, the spectral signatures collected in natural environments are invariably a mixture of the signatures of the various materials found within the spatial extent of the ground instantaneous field view of the imaging instrument [7]. The availability of hyperspectral imagers with a number of spectral bands that exceeds the number of spectral mixture components [2] has cast the unmixing problem in terms of an overdetermined system of equations, in which a given set of pure spectral signatures called endmembers, lies the actual unmixing work to determine apparent pixel abundance fractions that can be defined in terms of a numerical inversion process.

A standard technique for spectral mixture analysis is linear spectral unmixing [19], which assumes that the collected spectra at the spectrometer can be expressed in the form of a linear combination of endmembers weighted by their corresponding abundances. It should be noted that the linear mixture model assumes minimal secondary reflections and/or multiple scattering effects in the data collection procedure, and hence the measured spectra can be expressed as a linear combination of the spectral signatures of materials present in the mixed pixel (Figure 11.2a).

Although the linear model has practical advantages such as ease of implementation and flexibility in different applications [3], nonlinear spectral unmixing may best characterize the resultant mixed spectra for certain endmember distributions, such as those in which the endmember components are randomly distributed throughout the field of view of the instrument [10,20]. In those cases, the mixed spectra collected at the imaging instrument is better described by assuming that part of the source radiation is multiply scattered before being collected at the sensor (Figure 11.2b).

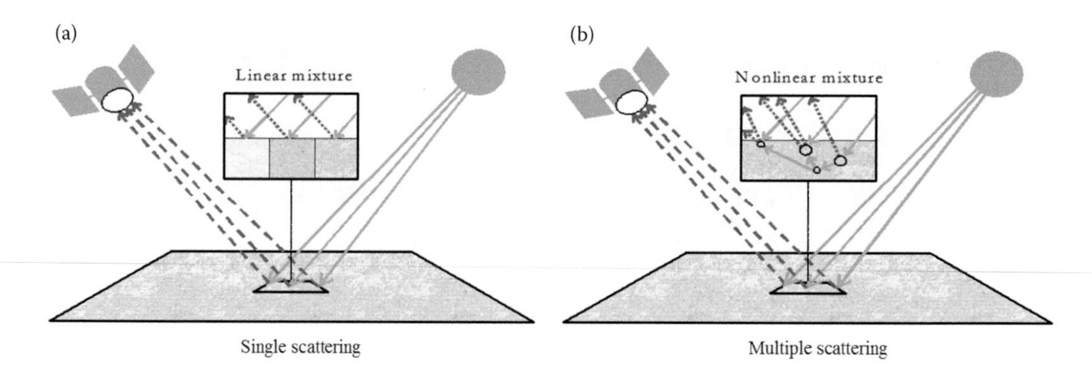

FIGURE 11.2 Graphical interpretation of the linear (a) versus the nonlinear (b) mixture model.

11.3.1 Linear Spectral Unmixing

In order to be able to correctly unmix a hyperspectral dataset using the linear model, two requirements are needed:

1. A successful estimation of the number of endmembers (spectrally distinct pure signatures) present in the input hyperspectral scene.
2. The correct determination of a set of endmembers and their correspondent abundance fractions at each pixel.

In order to address the first requirement, two successful techniques have been demonstrated including the virtual dimensionality (VD) [21] and HySime approaches [22]. The VD concept formulates the issue of whether a distinct signature is present or not in each of the spectral bands as a binary hypothesis testing problem, where a so-called Neyman–Pearson detector is generated to serve as a decision-maker based on a prescribed false alarm probability. In light of this interpretation, the issue of determining an appropriate value for the number of endmembers is further simplified and reduced to setting a specific value of the false alarm probability. In turn, the HySime uses a minimum mean squared error-based approach to determine the signal subspace in hyperspectral imagery.

Regarding the second requirement for successful implementation of the linear mixture model, several algorithms have been developed in recent years for automatic or semi-automatic extraction of spectral endmembers [6]. Classic techniques include the pixel purity index (PPI) [23], N-FINDR [24–26], iterative error analysis (IEA) [27], optical real-time adaptive spectral identification system (ORASIS) [28], convex cone analysis (CCA) [29], vertex component analysis (VCA) [30], and an orthogonal subspace projection (OSP) technique [31]. Other advanced techniques for endmember extraction have been recently proposed, but few of them consider spatial adjacency. However, one of the distinguishing properties of hyperspectral data is the multivariate information coupled with a two-dimensional (pictorial) representation amenable to image interpretation.

Subsequently, most endmember extraction algorithms listed above could benefit from an integrated framework in which both the spectral information and the spatial arrangement of pixel vectors are taken into account. An example is given in Figure 11.3, in which a hyperspectral data cube collected over an urban area (high spatial correlation) is modified by randomly permuting the spatial coordinates of the pixel vectors (i.e., removing the spatial correlation). In both scenes, the application of a spectral-based processing method would yield the same analysis results, while it is clear that a spatial-spectral technique could incorporate the spatial information present in the original scene into the process.

To the best of our knowledge, only a few attempts exist in the literature aimed at including the spatial information in the process of extracting spectral endmembers. Extended morphological

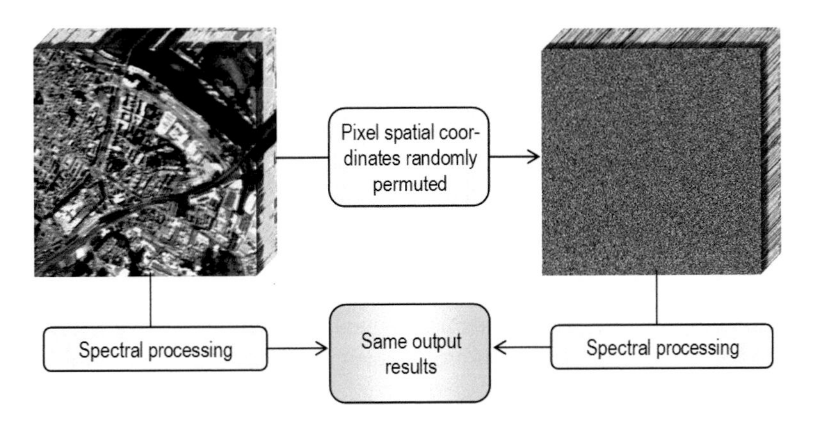

FIGURE 11.3 The importance of including spatial information in hyperspectral data processing.

operations [13] have been used as a baseline to develop an automatic morphological endmember extraction (AMEE) algorithm [32] for spatial-spectral endmember extraction. Also, spatial averaging of spectrally similar endmember candidates found via singular value decomposition (SVD) was used in the development of the spatial-spectral endmember extraction (SSEE) algorithm [33].

A spatial preprocessing (SPP) algorithm [34] has been proposed. A spatially-derived factor is used by this technique to weight the importance of the spectral information associated to each pixel in terms of its spatial context. The SPP is intended as a preprocessing module that can be used in combination with an existing spectral-based endmember extraction algorithm.

Once a set of endmembers have been extracted, their corresponding abundance fractions in a specific pixel vector of the scene can be estimated (in a least-squares sense) by using the unconstrained and constrained techniques [35]. It should be noted that the fractional abundance estimations obtained in an unconstrained fashion do not satisfy the abundance sum-to-one (ASC) and the abundance non-negativity (ANC) constraints that should hold in order for the linear mixture model to be physically meaningful (i.e., the derived endmember set should be complete and that the negative abundance estimations lack physical interpretation). Imposing the ASC and ANC constraints leads to a more complex optimization problem, which has been solved (in a least-squares sense) in the literature [36].

11.3.2 Nonlinear Spectral Unmixing

In a nonlinear model, the interaction between the endmembers and their associated fractional abundances is given by a nonlinear function which is not known *a priori*. Various machine learning techniques have been proposed in the literature to estimate this function. In particular, artificial neural networks have demonstrated great potential to decompose mixed pixels due to their inherent capacity to approximate complex functions [37]. Although many neural network architectures exist, for decomposition of mixed pixels in terms of nonlinear relationships, mostly feed-forward networks of various layers, such as the multi-layer perceptron (MLP), have been used [10,38,39]. It has been shown in the literature that MLP-based neural models, when trained accordingly, generally outperform other nonlinear models such as regression trees or fuzzy classifiers [40].

A variety of issues have been investigated in order to evaluate the impact of training in mixed-pixel classification accuracy, including the size and location of training sites, and the composition of training sets, but most of the attention has been paid to the issue of training set size, that is, the number of training samples required for the learning stage [41]. Sometimes, the smallness of a training set represents a major problem. This is especially apparent for analyses using hyperspectral sensor data, where the requirement of large volumes of training sites is a serious limitation [42].

Even if the identities of endmembers participating in mixtures in a certain area are known, proportions of these endmembers on a per-pixel basis are difficult to be estimated *a priori*. Therefore, one of the most challenging aspects in the design of neural network-based techniques for spectral mixture analysis is to reduce the need for very large training sets. Studies have investigated a range of issues [43], including the use of feature selection and feature extraction methods to reduce the dimensionality of the input data [38], the use of unlabeled and semi-labeled samples [42], the accommodation of spatial dependence in the data to define an efficient sampling design [33], or the use of statistics derived on other locations [44].

Our speculation (and that of many thoughtful investigators over the past 40 years [42,43]) is that the problem of mixed-pixel interpretation demands intelligent training sample selection algorithms, able to seek out the most informative training samples, thus optimizing the compromise between estimation accuracy (to be maximized) and ground-truth knowledge (to be minimized). In this sense, several efforts in the literature have been oriented toward the selection of mixed (border) training samples using previous work developed by Foody [43], as well as core (pure) training samples developed by simple endmember extraction algorithms.

In our experience, machine learning techniques such as MLP neural networks or SVMs can produce stable results when trained accordingly, a fact that leads us to believe that training can indeed be more important than the choice of a specific network architecture in mixture analysis applications.

11.4 EXPERIMENTAL RESULTS

11.4.1 Analysis of Supervised Hyperspectral Data Classification Using SVMs

The hyperspectral scene used for experiments in this subsection was gathered by AVIRIS over the Indian Pines test site in Northwestern Indiana, a mixed agricultural/forested area, early in the growing season, and consisted of 1939 × 677 pixels and 204 spectral bands in the wavelength range of 400–2500 nanometers (523 MB in size). Twenty AVIRIS bands (151–170) were removed from the original scene prior to analysis due to low signal-to-noise ratio (SNR) in those bands. The AVIRIS Indian Pines dataset represents a very challenging classification problem dominated by similar spectral classes and mixed pixels. Specifically, the primary crops of the area, mainly corn and soybeans, were very early in their growth cycle with only about 5% canopy cover. This fact makes most of the scene pixels highly mixed in nature. Discriminating among the major crops under these circumstances can be very difficult, a fact that has made this scene an extensively used benchmark to validate classification accuracy of hyperspectral imaging algorithms. For illustrative purposes, Figure 11.4a shows a randomly selected spectral band (587 nanometers) of the original scene and Figure 11.4b shows the corresponding ground-truth map, displayed in the form of a class assignment for each labeled pixel, with 30 mutually exclusive ground-truth classes. Part of these data, including the ground truth, are available online from Purdue University (http://dynamo.ecn.purdue.edu/~biehl/MultiSpec).

In the following, three types of kernels were used in experiments: polynomial, Gaussian, and spectral angle mapper. Small training sets, composed of 1%, 2%, 4%, 6%, 8%, 10%, and 20% of the ground-truth pixels available per class, were extracted using pure (core) and mixed (border) training sample selection algorithms [10], and also using a random selection procedure. The SVM was trained with each of these training subsets and then evaluated with the remaining test set. Each experiment was repeated five times in order to guarantee statistical significance, and the mean accuracy values were reported. Table 11.1 summarizes the overall classification results obtained using the three selected kernels and training sample selection algorithms.

From Table 11.1, it can be seen that SVMs generalize quite well: with only 1% of training pixels per class, almost 90% overall classification accuracy was reached by all kernels when trained using border training samples. In all cases, classification accuracies decreased when random and

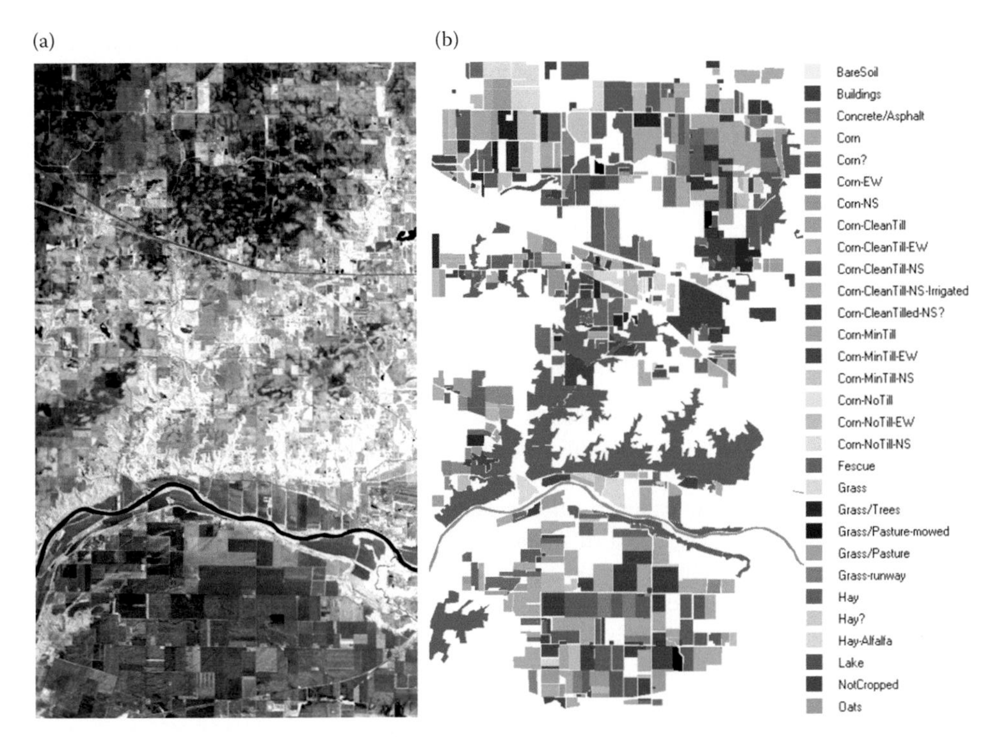

FIGURE 11.4 (a) Spectral band at the 587 nanometer wavelength of an AVIRIS scene comprising agricultural and forest features at Indian Pines region. (b) Ground-truth map with 30 mutually exclusive land cover classes.

pure samples were used for the training site. This confirms the fact that kernel-based methods in general and SVMs in particular are less affected by the Hughes phenomenon. It is also clear from Table 11.1 that the classification accuracy is generally correlated with the training set size. However, when border training samples were used, higher classification accuracies were achieved with less training samples. The above results indicate the importance of including mixed pixels at the border of class boundaries in the training set, as these border patterns are most efficient to determining the hyperplane between two classes.

TABLE 11.1
Overall Classification Accuracies (in Percentage) Achieved by the SVM Classifier after Applying Polynomial, Gaussian, and Spectral Angle Mapper Kernels to the AVIRIS Indian Pines Dataset, Using Different Strategies for Training Sample Selection (Random, Pure, Border Patterns)

Polynomial kernel	Random	82.33	82.94	83.21	83.82	85.34	86.12	86.52
	Pure	81.23	82.06	82.80	83.00	84.03	84.45	85.57
	Border	83.44	84.23	84.45	84.96	86.27	87.44	89.96
Gaussian kernel	Random	87.94	88.23	88.78	88.96	89.45	89.48	90.77
	Pure	86.53	87.02	87.64	87.93	88.12	88.26	88.55
	Border	89.45	90.25	91.24	92.08	92.93	93.04	93.67
Spectral angle kernel	Random	85.90	86.22	86.49	87.03	87.56	88.09	88.72
	Pure	85.12	85.67	86.08	86.45	86.97	87.13	87.81
	Border	86.05	86.93	87.57	88.12	89.30	90.12	90.57

Finally, it can be seen in Table 11.1 that the best classification scores were generally achieved for the Gaussian kernel, in which the overall accuracy obtained with 1% of the training pixels per class was only 4.22% lower than the overall accuracy obtained with 20% of the training pixels per class (extracted using border training sample selection). On the other hand, the spectral angle mapper kernel gives slightly degraded classification results. However, with accuracies above 85% in a challenging classification problem, this kernel also provides promising results. Finally, the polynomial kernel needs more training samples than the two other kernels to perform appropriately, as can be seen from the relatively poor results obtained by this kernel for a very limited number of training samples.

11.4.2 Analysis of Unsupervised Linear Unmixing of Hyperspectral Data

The hyperspectral scene used for experiments is the well-known AVIRIS Cuprite dataset, available online in reflectance units (http://aviris.jpl.nasa.gov/html/aviris.freedata.html) after atmospheric correction. This scene has been widely used to validate the performance of endmember extraction algorithms. The portion used in experiments corresponds to a 350 × 350-pixel subset of the sector labeled as "f970619t01p02_r02_sc03.a.rfl" in the online data. The scene (displayed in Figure 11.5a) comprised 224 spectral bands between 400 and 2500 nanometers, with full width at half maximum of 10 nanometers and spatial resolution of 20 meters per pixel.

Prior to the analysis, several bands (1–3, 150–170, and 217–224) were removed due to water absorption and low SNR in those bands, leaving a total of 192 reflectance channels to be used in the experiments. The Cuprite site is well understood mineralogically [45,46], and has several exposed minerals of interest included in a spectral library compiled by the U.S. Geological Survey (USGS) available online (http://speclab.cr.usgs.gov/spectral-lib.html). A few selected spectra from the USGS library, corresponding to several highly representative minerals in the Cuprite mining district (Figure 11.5b), are used in this work to substantiate endmember signature purity.

Two different metrics were used to compare the performance of endmember extraction and spectral unmixing algorithms in the AVIRIS Cuprite scene. The first metric was the spectral angle [3,19] between each extracted endmember and the set of available USGS ground-truth spectral signatures. Low spectral angle scores mean high spectral similarity between the compared vectors. This spectral similarity measure is invariant in the multiplication of pixel vectors by constants and, consequently, is invariant before unknown multiplicative scalings that may arise due to differences in illumination and angular orientation. In our experiments, the spectral angle allows us to identify

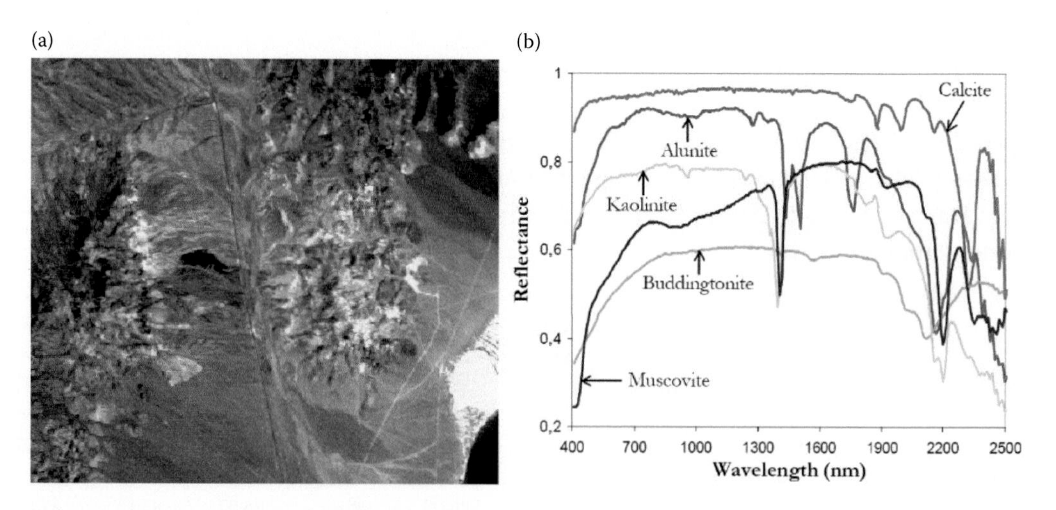

FIGURE 11.5 (a) False color composition of the remote sensing image used in experiments. (b) Reference spectral signatures provided by USGS and used for validation purposes.

the USGS signature which is most similar to each endmember automatically extracted from the scene by observing the minimum spectral angle reported for such endmember across the entire set of USGS signatures.

A second metric employed to evaluate the goodness of the reconstruction was the root mean square error (RMSE) obtained in the reconstruction of the hyperspectral image (using the derived endmembers and their corresponding abundance fractions). This metric is based on the assumption that a set of high-quality endmembers (and their corresponding estimated abundance fractions) may allow reconstruction of the original hyperspectral scene with higher precision than a set of low-quality endmembers. In this case, the original hyperspectral image is used to measure the fidelity of the reconstructed version of the same scene on a per-pixel basis.

Table 11.2 tabulates the spectral angles (in degrees) obtained after comparing the USGS library spectra of five highly representative minerals in the Cuprite mining district (alunite, buddingtonite, calcite, kaolinite, and muscovite) with the corresponding endmembers extracted by several different algorithms (listed in Section 11.3.1) from the AVIRIS Cuprite scene. In all cases, the input parameters of the different endmember extraction methods tested have been carefully optimized so that the best performance for each method was reported. Again, the smaller the spectral angles across the five minerals in Table 11.2, the better the results. It should be noted that Table 11.2 only displays the smallest spectral angle scores of all endmembers with respect to each USGS signature for each algorithm.

For reference, the mean spectral angle values across all five USGS signatures were also reported. In all cases, the number of endmembers to be extracted was set to 14 after using a consensus between the VD concept and the HySime method. Table 11.2 reveals that the AMEE provides very good results (all spectral angle values scores below 10 degrees), with the SSEE and the SPP+OSP (where SPP indicates spatial preprocessing prior to the classic OSP procedure for endmember extraction) are the algorithms that can provide comparable—but slightly worse—results. Table 11.2 also reveals that, in this real example, spatial preprocessing generally improves the signature purity of the endmembers extracted by spectral-based algorithms.

On the other hand, Figure 11.6 graphically represents the per-pixel RMSE obtained after reconstructing the AVIRIS Cuprite scene using 14 endmembers extracted by different methods. It can be seen that the methods using spatial preprocessing (SPP+OSP, SPP+N-FINDR, SPP+VCA) improve their respective spectral-based versions in terms of the quality of image reconstruction, while both AMEE and SSEE also provide lower reconstruction errors than OSP, N-FINDR, and VCA. These results suggest the advantages of incorporating spatial information into the automatic extraction of image endmembers from the viewpoint of obtaining more spatially representative spectral signatures, which can be used to describe other mixed signatures in the scene.

TABLE 11.2
Spectral Angle Scores (in Degrees) between the USGS Mineral Spectra and Their Corresponding Endmember Pixels Produced by Several Endmember Extraction Algorithms

Algorithm	Alunite	Buddingtonite	Calcite	Kaolinite	Muscovite	Mean
OSP	4.81	4.16	9.62	11.14	5.41	7.03
N-FINDR	9.96	7.71	12.08	13.27	5.24	9.65
VCA	10.73	9.04	6.36	14.05	5.41	9.12
SPP+OSP	4.95	4.16	9.96	10.90	4.62	6.92
SPP+N-FINDR	12.81	8.33	9.83	10.43	5.28	9.34
SPP+VCA	12.42	4.04	9.37	7.37	6.18	7.98
AMEE	4.81	4.21	9.54	8.74	4.61	6.38
SSEE	4.81	4.16	8.48	11.14	4.62	6.64

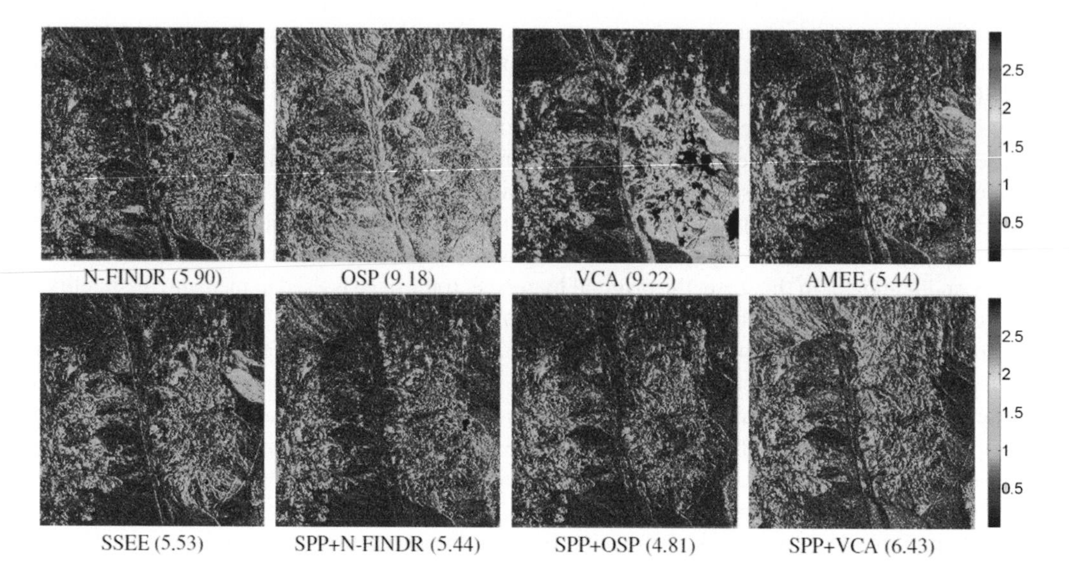

FIGURE 11.6 RMSE reconstruction errors (in percentage) for various endmember extraction algorithms after reconstructing the AVIRIS Cuprite scene.

11.4.3 Analysis of Supervised Nonlinear Unmixing of Hyperspectral Data Using MLPs

In the Iberian Peninsula, dehesa systems are used for a combination of livestock, forest, and agriculture activity [47]. The outputs of these systems include meat, milk, wool, charcoal, cork bark, and grain. Around 12%–18% of the area is harvested on a yearly basis. The crops are used for animal feed or for cash cropping, depending on the rainfall of the area. Determination of fractional land cover using remote sensing techniques may allow for a better monitoring of natural resources in dehesa agro-ecosystems.

Our choice of this type of landscape for evaluating nonlinear unmixing techniques was made on several accounts. The first one is the availability of hyperspectral image datasets with accurate geo-registration for a real dehesa test site in Caceres, southwest Spain. Data was collected simultaneously in July 2001 by two instruments operating at multiple spatial resolutions; the Digital Airborne Imaging Spectrometer (DAIS) 7915 and Reflective Optics Spectrographic Imaging System (ROSIS), and operated by the German Aerospace Agency (DLR). A second major reason is the simplicity of the dehesa landscape, which greatly facilitates the collection of reliable field data for model validation purposes. It is also important to emphasize that the scenes were collected in summertime, so atmospheric interferers were greatly minimized. Before describing our experiments, we first provide a comprehensive description of the datasets used and ground-truth activities in the study area.

The data used in this study consisted of two main components: image data and field measurements of land cover fractions, collected at the time of image data acquisition. The image data were formed by a ROSIS scene collected at high spatial resolution, with 1.2-meter pixels, and its corresponding DAIS 7915 scene, collected at low spatial resolution with 6-meter pixels. The spectral range from 504 to 864 nanometers (consisting of a total of 112 spectral bands) was selected for experiments, not only because it is adequate for analyzing the spectral properties of the landscape under study, but also because this spectral range is well covered by the two sensors through narrow spectral bands. Figure 11.7 shows the full flightline of the ROSIS scene, which comprises a dehesa area located between the facilities of the University of Extremadura in Caceres (leftmost part of the flightline) and Guadiloba water reservoir at the center of the flightline. Figure 11.8a shows the dehesa test site selected for experiments, which corresponds to a highly representative dehesa area that contains

FIGURE 11.7 Flightline of a ROSIS hyperspectral scene collected over a dehesa area in Caceres, Spain.

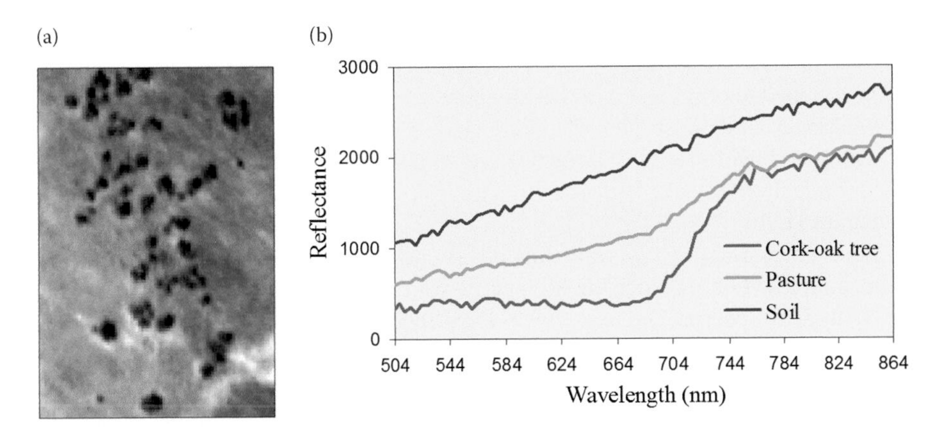

FIGURE 11.8 (a) Spectral band (584 nm) of a ROSIS dehesa subset selected for experiments. (b) Endmember signatures of soil, pasture, and cork-oak tree extracted by the AMEE algorithm, where scaled reflectance values are multiplied by a constant factor.

several cork-oak trees (appearing as dark spots) and several pasture (gray) areas on a bare soil (white) background. Several field techniques were applied to obtain reliable estimates of the fractional land cover for each DAIS 7915 pixel considered in the dehesa test site:

1. First, the ROSIS image was roughly classified into the three land cover components above using a maximum-likelihood supervised classification approach based on image-derived spectral endmembers, where Figure 11.8b shows the three endmembers used for mapping that were derived using the AMEE algorithm. Our assumption was that the pixels in the ROSIS image were sufficiently small to become spectrally simple to analyze.
2. Then, the classified ROSIS image was registered with the DAIS 7915 image using a ground control point-based method with sub-pixel accuracy [48].
3. The classification map was then associated with the DAIS 7915 image to provide an initial estimation of land cover classes for each pixel at the DAIS 7915 image scale. For that purpose, a 6 $\times$ 6-meter grid was overlaid on the 1.2 $\times$ 1.2-meter classification map derived

FIGURE 11.9 Ground measurements in the dehesa area of study located in Caceres, Spain. (a) Spectral sample collection using an ASD FieldSpec Pro spectroradiometer. (b) High-precision GPS geographic delimitation. (c) Field spectral measurements at different altitudes.

from the ROSIS scene, where the geographic coordinates of each pixel center point were used to validate the registration with sub-pixel precision.

4. Next, fractional abundances were calculated within each 6 × 6-meter grid as the proportion of ROSIS pixels labeled as cork-oak tree, pasture, and soil located within that grid, respectively.
5. Most importantly, the abundance maps at the ROSIS level were thoroughly refined using field measurements (Figure 11.9a) before obtaining the final proportions. Several approaches were developed to refine the initial estimations:
   a. Fractional land cover data were collected on the ground at more than thirty evenly distributed field sites within the test area. These sites were delineated during the field visit as polygons, using high-precision global positioning system (GPS) coordinates (see Figure 11.9b).
   b. Land cover fractions were estimated at each site using a combination of various techniques. For instance, field spectra were collected for several areas using an Analytical Spectral Devices (ASD) FieldSpec Pro spectroradiometer. Of particular interest were field measurements collected on top of tree crowns 9(c), which allowed us to model different levels of tree crown transparency.
   c. On the other hand, the early growth stage of pasture during the summer season allowed us to perform ground estimations of pasture abundance in selected sites of known dimensions, using pasture harvest procedures supported by visual inspection and laboratory analyses.

After following the abovementioned sequence of steps, we obtained a set of approximate fractional abundance labels for each pixel vector in the DAIS 7915 image. Despite our effort to conduct a reliable ground estimation of fractional land cover in the considered semi-arid environment, absolute accuracy was not claimed. We must emphasize, however, that the combined use of imagery data at different resolutions, sub-pixel ground control-based image registration, and extensive field work including high-precision GPS field work, spectral sample data collection, and expert knowledge, represents a novel contribution in the area of spectral mixture analysis validation, in particular, for dehesa-type ecosystems.

We evaluated the accuracy of linear spectral unmixing in the considered application using scatterplots, and Figure 11.10 shows measured versus linearly estimated fractional abundances (using linear spectral unmixing with the ASC and ANC constraints imposed). Displayed are the three land cover materials in the DAIS 7915 (low spatial resolution) image dataset, where the diagonal represents a perfect match and the two flanking lines represent the plus/minus 20% error bound. Here, the three spectral endmembers were derived using the AMEE algorithm, which incorporated spatial information into the endmember extraction process.

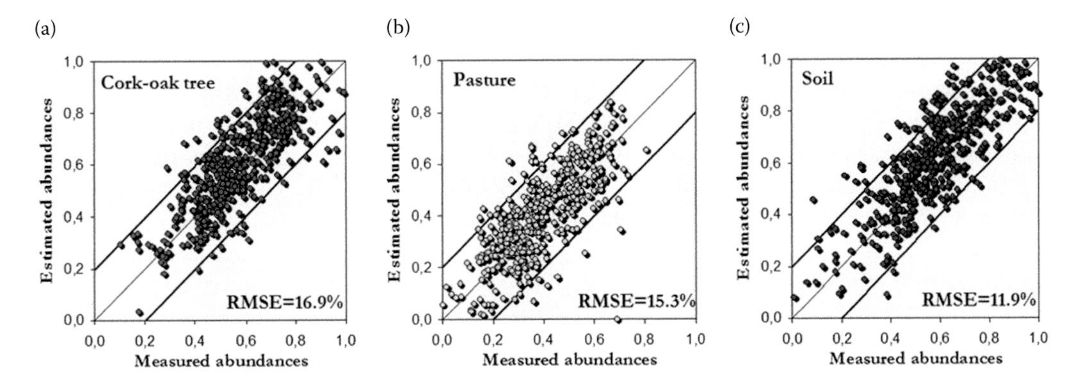

FIGURE 11.10 Abundance estimations of cork-oak tree (a), pasture (b), and soil (c) by the fully constrained linear mixture model from the DAIS 7915 image.

As expected, the flatness of the test site largely removed topographic influences in the remotely sensed response of soil areas. As a result, most linear predictions for the soil endmember fall within the 20% error bound (see Figure 11.10a). On the other hand, the multiple scattering within the pasture and cork-oak tree canopies (and from the underlying surface in the latter case) complicated the spectral mixing in a nonlinear fashion, which resulted in a generally higher number of estimations lying outside the error bound, as illustrated in Figure 11.10b and c. Also, the RMSE scores in abundance estimation for the soil (11.9%), pasture (15.3%), and cork-oak tree (16.9%) were all above 10% estimation error in percentage, which suggested that linear mixture modeling was not flexible enough to accommodate the full range of spectral variability throughout the landscape.

In order to characterize the dehesa ecosystem structure better than linear models did, we used nonlinear spectral unmixing to better characterize nonlinear mixing effects. For this purpose, we applied a mixed (border) training sample selection algorithm to automatically locate highly descriptive training sites in the DAIS 7915 scene and then used the obtained samples (and the ground-truth information associated to those samples) to train the MLP-based neural network model described in Section 11.3.2. Figure 11.11 shows the scatterplots of measured versus predicted fractional abundances for soil, pasture, and cork-oak tree by the proposed MLP-based model, first trained with the three pure training samples by the AMEE algorithm (Figure 11.11b) plus 40 additional training samples selected by an algorithm designed to seek out the most highly mixed training samples [10]. This represented less than 1% of the total number of pixels in the DAIS 7915 scene. These samples were excluded from the testing set made up of all remaining pixels in the scene. From Figure 11.11, it is clear that the utilization of intelligently selected training samples resulted

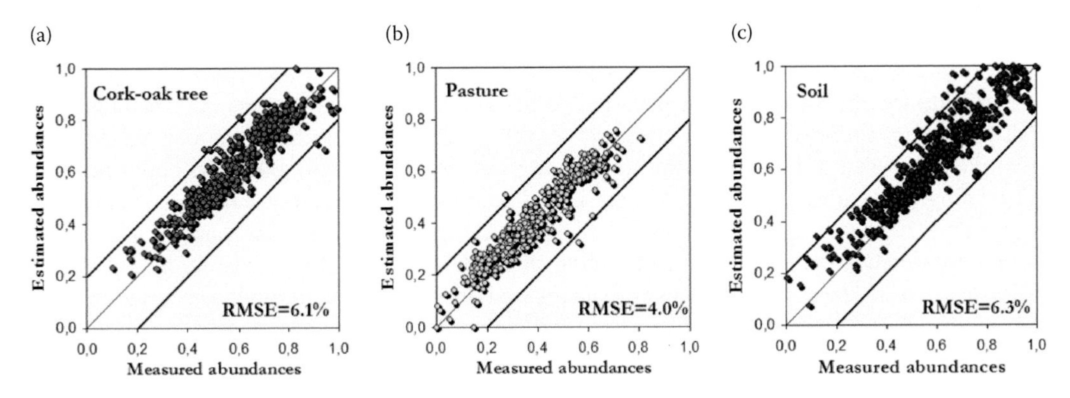

FIGURE 11.11 Abundance estimations of cork-oak tree (a), pasture (b), and soil (c) by the MLP-based mixture model, trained using mixed (border) samples, from the DAIS 7915 image.

in fewer points outside the two 20% difference lines, most notably, for both pasture and cork-oak abundance estimates.

The pattern of the scatter plots obtained for the soil predictions (Figure 11.11a) was similar (in particular, when the soil abundance was high). Most importantly, the RMSE scores in the abundance estimation were significantly reduced (with regards to the experiment using fully constrained linear unmixing) for the soil (6.1%), pasture (4%), and cork-oak tree (6.3%). These results confirmed our intuition that nonlinear effects in dehesa landscapes mainly result from multiple scattering effects in vegetation canopies. It is worth noting that, although the ASC and ANC constraints were not imposed in our proposed MLP-based learning stage, negative and/or unrealistic abundance estimations (which usually indicate a bad fit of the model and reveal inappropriate endmember/ training data selection) were very rarely found in our experiments.

The experimental validation carried out in this subsection indicated that the intelligent incorporation of mixed training samples can enable a more accurate representation of nonlinearly mixed signatures. It was apparent from experimental results that the proposed neural network-based model was able to generate abundance estimates that were close to abundance values measured in the field, using only a few intelligently generated training samples. The need for mixed training data does, however, require detailed knowledge on abundance fractions for the considered training sites. In practice, these data are likely to be derived from imagery acquired at a finer spatial resolution than the imagery to be classified, for example, using datasets acquired by sensors operating simultaneously at multiple spatial resolutions as it was the case of the DAIS 7915 and ROSIS instruments considered in this experiment. Such multi-resolution studies may also incorporate prior knowledge or ancillary information, which can be used to help target the location of training sites, and to focus training site selection activities on regions likely to contain the most informative training samples.

11.5 CONCLUSIONS AND FUTURE PERSPECTIVES

This chapter focused on hyperspectral data processing algorithms that included: (a) support vector machine (SVM) techniques for supervised classification using limited training samples; and (b) development of linear and nonlinear spectral unmixing techniques, with some of them integrating the spatial and the spectral information. The special characteristics of hyperspectral images pose new processing problems, not to be found in other types of remotely sensed data:

1. The high-dimensional nature of hyperspectral data introduces important limitations in supervised, full-pixel classifiers, such as the limited availability of training samples or the inherently complex structure of the data (leading to the Hughes phenomenon).
2. There is a need to integrate the spatial and spectral information to take advantage of the complementarities that both sources of information can provide, in particular, for unsupervised mixed-pixel classifiers.

In this regard, the SVM experiments reported in our quantitative assessment demonstrated that, with only 1% of training pixels per class, almost 90% overall classification accuracy was reached by all kernels when trained using border training samples. This highlighted the opportunity of overcoming the Hughes phenomenon using kernel approaches. On the other hand, our unmixing experiments indicated that new trends in algorithm design (such as the joint use of spatial and spectral information in linear spectral unmixing, or the development of nonlinear unmixing models based on machine learning techniques with an appropriate exploitation of limited training samples) can significantly improve the accuracy in the estimation of fractional abundances in real analysis scenarios.

As demonstrated by our experimental results and the determination of the accuracy of these approaches, techniques are rapidly changing from hard classifiers to soft classifiers. In this regard, we anticipate that the full adaptation of soft classifiers to mixed-pixel classification problems (e.g., via multi-regression and robust training sample selection algorithms) may push the frontiers of hyperspectral data classification into new application domains. Further developments on the joint exploitation of the spatial and the spectral information in the input data are also needed to complement initial approximations to the problem of interpreting the data in an unsupervised fashion, thus being able to cope with the dramatically enhanced spatial and spectral capabilities expected in the design of future imaging spectrometers. Advances in high-performance computing [49], including clusters of computers and distributed grids, as well as specialized hardware modules such as field programmable gate arrays (FPGAs) or graphics processing units (GPUs), will also be crucial to help increase algorithm efficiency and meet timeliness needs in many remote sensing applications.

ACKNOWLEDGMENT

This work has been supported by the European Community's Marie Curie Research Training Networks Programme under reference MRTN-CT-2006-035927, Hyperspectral Imaging Network (HYPER-I-NET), and also been supported by the Spanish Ministry of Science and Innovation (HYPERCOMP/EODIX project, reference AYA2008-05965-C04-02). Gabriel Martín and Sergio Sánchez were sponsored by research fellowships with references BES-2009-017737 and PTA2009-2611-P, respectively, both associated with the aforementioned project. Funding from Junta de Extremadura (local government) under project PRI09A110 is also gratefully acknowledged. The authors thank Andreas Mueller for his lead of the DLR project that allowed us to obtain the DAIS 7915 and ROSIS hyperspectral datasets over dehesa areas in Extremadura, Spain. We also thank David Landgrebe at Purdue University for making the AVIRIS Indian Pines scene available to the scientific community, and Robert O. Green at NASA/JPL for also making the AVIRIS Cuprite scene available to the scientific community. Last but not least, the authors would like to take this opportunity to gratefully acknowledge the Editors of this volume for their very kind invitation to contribute a chapter and for all their support and encouragement during the different stages of the production process for this monograph.

REFERENCES

1. A. F. H. Goetz, G. Vane, J. E. Solomon, and B. N. Rock, "Imaging spectrometry for earth remote sensing," *Science*, vol. 228, pp. 1147–1153, 1985.
2. R. O. Green, "Imaging spectroscopy and the airborne visible-infrared imaging spectrometer (AVIRIS)," *Remote Sensing of Environment*, vol. 65, pp. 227–248, 1998.
3. C.-I. Chang, "Hyperspectral imaging: Techniques for spectral detection and classification," Kluwer Academic and Plenum Publishers, New York, 2003.
4. D. A. Landgrebe, "Signal theory methods in multispectral remote sensing," John Wiley and Sons, Hoboken, NJ, 2003.
5. J. A. Richards, "Analysis of remotely sensed data: the formative decades and the future," *IEEE Transactions on Geoscience and Remote Sensing*, vol. 43, pp. 422–432, 2005.
6. A. Plaza, P. Martinez, R. Perez, and J. Plaza, "A quantitative and comparative analysis of endmember extraction algorithms from hyperspectral data," *IEEE Transactions on Geoscience and Remote Sensing*, vol. 42, pp. 650–663, 2004.
7. J. B. Adams, M. O. Smith, and P. E. Johnson, "Spectral mixture modeling: a new analysis of rock and soil types at the Viking Lander 1 site," *Journal of Geophysical Research*, vol. 91, pp. 8098–8112, 1986.
8. G. M. Foody and A. Mathur, "Toward intelligent training of supervised image classifications: directing training data acquisition for svm classification," *Remote Sensing of Environment*, vol. 93, pp. 107–117, 2004.
9. A. Plaza, J. A. Benediktsson, J. Boardman, J. Brazile, L. Bruzzone, G. Camps-Valls, J. Chanussot et al., "Recent advances in techniques for hyperspectral image processing," *Remote Sensing of Environment*, vol. 113, pp. 110–122, 2009.
10. J. Plaza, A. Plaza, R. Perez, and P. Martinez, "On the use of small training sets for neural network-based characterization of mixed pixels in remotely sensed hyperspectral images," *Pattern Recognition*, vol. 42, pp. 3032–3045, 2009.

11. P. Gamba, F. Dell'Acqua, A. Ferrari, J. A. Palmason, and J. A. Benediktsson, "Exploiting spectral and spatial information in hyperspectral urban data with high resolution," *IEEE Geoscience and Remote Sensing Letters*, vol. 1, pp. 322–326, 2004.
12. J. A. Benediktsson, J. A. Palmason, and J. R. Sveinsson, "Classification of hyperspectral data from urban areas based on extended morphological profiles," *IEEE Transactions on Geoscience and Remote Sensing*, vol. 42, pp. 480–491, 2005.
13. A. Plaza, P. Martinez, J. Plaza, and R. Perez, "Dimensionality reduction and classification of hyperspectral image data using sequences of extended morphological transformations," *IEEE Transactions on Geoscience and Remote Sensing*, vol. 43, no. 3, pp. 466–479, 2005.
14. G. Camps-Valls and L. Bruzzone, "Kernel-based methods for hyperspectral image classification," *IEEE Transactions on Geoscience and Remote Sensing*, vol. 43, pp. 1351–1362, 2005.
15. K. R. Muller, S. Mika, G. Ratsch, K. Tsuda, and B. Scholkopf, "An introduction to kernel-based learning algorithms," *IEEE Transactions on Neural Networks*, vol. 12, pp. 181–202, 2001.
16. L. Bruzzone, M. Chi, and M. Marconcini, "A novel transductive SVM for the semisupervised classification of remote sensing images," *IEEE Transactions on Geoscience and Remote Sensing*, vol. 44, pp. 3363–3373, 2006.
17. G. Camps-Valls, L. Gomez-Chova, J. Munoz-Mari, J. Vila-Frances, and J. Calpe-Maravilla, "Composite kernels for hyperspectral image classification," *IEEE Geoscience and Remote Sensing Letters*, vol. 3, pp. 93–97, 2006.
18. M. E. Schaepman, S. L. Ustin, A. Plaza, T. H. Painter, J. Verrelst, and S. Liang, "Earth system science related imaging spectroscopy—an assessment," *Remote Sensing of Environment*, vol. 113, pp. 123–137, 2009.
19. N. Keshava and J. F. Mustard, "Spectral unmixing," *IEEE Signal Processing Magazine*, vol. 19, pp. 44–57, 2002.
20. K. J. Guilfoyle, M. L. Althouse, and C.-I. Chang, "A quantitative and comparative analysis of linear and nonlinear spectral mixture models using radial basis function neural networks," *IEEE Transactions on Geoscience and Remote Sensing*, vol. 39, pp. 2314–2318, 2001.
21. C.-I. Chang and Q. Du, "Estimation of number of spectrally distinct signal sources in hyperspectral imagery," *IEEE Transactions on Geoscience and Remote Sensing*, vol. 42, pp. 608–619, 2004.
22. J. M. Bioucas-Dias and J. M. P. Nascimento, "Hyperspectral subspace identification," *IEEE Transactions on Geoscience and Remote Sensing*, vol. 46, pp. 2435–2445, 2008.
23. J. W. Boardman, F. A. Kruse, and R. O. Green, "Mapping target signatures via partial unmixing of AVIRIS data," *Proceedings JPL Airborne Earth Sci. Workshop*, pp. 23–26, 1995.
24. M. E. Winter, "N-FINDR: An algorithm for fast autonomous spectral endmember determination in hyperspectral data," *Proceedings of SPIE*, vol. 3753, pp. 266–277, 1999.
25. M. E. Winter, "A proof of the N-FINDR algorithm for the automated detection of endmembers in a hyperspectral image," *Proceedings of SPIE Algorithms and Technologies for Multispectral, Hyperspectral, and Ultraspectral Imagery X*, vol. 5425, pp. 31–41, 2004.
26. M. Zortea and A. Plaza, "A quantitative and comparative analysis of different implementations of N-FINDR: A fast endmember extraction algorithm," *IEEE Geoscience and Remote Sensing Letters*, vol. 6, pp. 787–791, 2009.
27. R. A. Neville, K. Staenz, T. Szeredi, J. Lefebvre, and P. Hauff, "Automatic endmember extraction from hyperspectral data for mineral exploration," *Proceedings of 21st Canadian Symposium on Remote Sensing*, pp. 21–24, 1999.
28. J. H. Bowles, P. J. Palmadesso, J. A. Antoniades, M.M. Baumback, and L. J. Rickard, "Use of filter vectors in hyperspectral data analysis," *Proceedings of SPIE Infrared Spaceborne Remote Sensing III*, vol. 2553, pp. 148–157, 1995.
29. A. Ifarraguerri and C.-I. Chang, "Multispectral and hyperspectral image analysis with convex cones," *IEEE Transactions on Geoscience and Remote Sensing*, vol. 37, no. 2, pp. 756–770, 1999.
30. J. M. P. Nascimento and J. M. Bioucas-Dias, "Vertex component analysis: A fast algorithm to unmix hyperspectral data," *IEEE Transactions on Geoscience and Remote Sensing*, vol. 43, no. 4, pp. 898–910, 2005.
31. J. C. Harsanyi and C.-I. Chang, "Hyperspectral image classification and dimensionality reduction: An orthogonal subspace projection," *IEEE Transactions on Geoscience and Remote Sensing*, vol. 32, no. 4, pp. 779–785.
32. A. Plaza, P. Martinez, R. Perez, and J. Plaza, "Spatial/spectral endmember extraction by multidimensional morphological operations," *IEEE Transactions on Geoscience and Remote Sensing*, vol. 40, pp. 2025–2041, 2002.

33. D. M. Rogge, B. Rivard, J. Zhang, A. Sanchez, J. Harris, and J. Feng, "Integration of spatial–spectral information for the improved extraction of endmembers," *Remote Sensing of Environment*, vol. 110, pp. 287–303, 2007.
34. M. Zortea and A. Plaza, "Spatial preprocessing for endmember extraction," *IEEE Transactions on Geoscience and Remote Sensing*, vol. 47, pp. 2679–2693, 2009.
35. D. Heinz and C.-I. Chang, "Fully constrained least squares linear mixture analysis for material quantification in hyperspectral imagery," *IEEE Transactions on Geoscience and Remote Sensing*, vol. 39, pp. 529–545, 2001.
36. C.-I. Chang and D. Heinz, "Constrained subpixel target detection for remotely sensed imagery," *IEEE Transactions on Geoscience and Remote Sensing*, vol. 38, pp. 1144–1159, 2000.
37. C. M. Bishop, "Neural networks for pattern recognition." Oxford: Oxford University Press, 1995.
38. J. Plaza and A. Plaza, "Spectral mixture analysis of hyperspectral scenes using intelligently selected training samples," *IEEE Geoscience and Remote Sensing Letters*, vol. 7, pp. 371–375, 2010.
39. A. Baraldi, E. Binaghi, P. Blonda, P. A. Brivio, and P. Rampini, "Comparison of the multilayer perceptron with neuro-fuzzy techniques in the estimation of cover class mixture in remotely sensed data," *IEEE Transactions on Geoscience and Remote Sensing*, vol. 39, pp. 994–1005, 2001.
40. W. Liu and E. Y. Wu, "Comparison of non-linear mixture models," *Remote Sensing of Environment*, vol. 18, pp. 1976–2003, 2004.
41. X. Zhuang, B. A. Engel, D. F. Lozano, R. B. Fernndez, and C. J. Johannsen, "Optimization of training data required for neuro-classification," *International Journal of Remote Sensing*, vol. 15, pp. 3271–3277, 1999.
42. M. Chi and L. Bruzzone, "A semilabeled-sample-driven bagging technique for ill-posed classification problems," *IEEE Geoscience and Remote Sensing Letters*, vol. 2, pp. 69–73, 2005.
43. G. M. Foody, "The significance of border training patterns in classification by a feedforward neural network using backpropagation learning," *International Journal of Remote Sensing*, vol. 20, pp. 3549–3562, 1999.
44. C. C. Borel and S. A. W. Gerslt, "Nonlinear spectral mixing models for vegetative and soil surfaces," *Remote Sensing of Environment*, vol. 47, pp. 403–416, 1994.
45. R. N. Clark, G. A. Swayze, K. E. Livo, R. F. Kokaly, S. J. Sutley, J. B. Dalton, R. R. McDougal, and C. A. Gent, "Imaging spectroscopy: Earth and planetary remote sensing with the USGS tetracorder and expert systems," *Journal of Geophysical Research*, vol. 108, pp. 1–44, 2003.
46. G. Swayze, R. N. Clark, F. Kruse, S. Sutley, and A. Gallagher, "Ground-truthing AVIRIS mineral mapping at Cuprite, Nevada," *Proceedings of the JPL Airborne Earth Science Workshop*, pp. 47–49, 1992.
47. F. J. Pulido, M. Diaz, and S. J. Hidalgo, "Size structure and regeneration of Spanish holm oak quercus ilex forests and dehesas: Effects of agroforestry use on their long-term sustainability," *Forest Ecology and Management*, vol. 146, pp. 1–13, 2001.
48. A. Plaza, J. L. Moigne, and N. S. Netanyahu, "Morphological feature extraction for automatic registration of multispectral scenes," *Proceedings of the IEEE International Geoscience and Remote Sensing Symposium*, vol. 1, pp. 421–424, 2007.
49. A. Plaza and C.-I. Chang, "High performance computing in remote sensing." Boca Raton: CRC Press, 2007.

12 Methods for Linking Drone and Field Hyperspectral Data to Satellite Data

Muhammad Al-Amin Hoque and Stuart Phinn

CONTENTS

12.1 INTRODUCTION

The objective of this chapter is to explain how drone and field spectrometer datasets are currently collected and linked to airborne and satellite image datasets. Given the proliferation of drones that carry various sensors (Cracknell, 2017), there is a great need to standardize data acquisition from these platforms to deliver properly calibrated data that can be linked to airborne and spaceborne image data and image derived data products and measurements. We also identify future requirements for improving our ability to use these datasets, linking traditional field- and plot-scale observations and measurements to airborne and satellite image datasets. The spatial and temporal scale of data collected by drones provides a unique and globally critical resource enabling traditional plot and transect scale manual measurements, to be made from an aerial perspective, and in a form that can be linked to satellite datasets. We will follow the convention of most aerospace authorities around the world, and use the terms drone or remotely piloted airborne system (RPAS) instead of Unmanned Aircraft System (UAS) or Unmanned Aerial Vehicle (UAV; IJRS, 2017).

The chapter begins by first identifying the range of imaging and non-imaging spectrometers that are currently being flown on drone platforms, including the range of platform-sensor combinations that are most commonly available. The nature and quality of the datasets produced from these platform-sensor combinations are then reviewed, as this is still an "experimental" area, given the relatively small sensor payloads able to be carried on all but larger military drones at present, and restrictions on civilian operations of large drones in most countries. Field-based spectrometer setups, including non-imaging and imaging sensors are next presented. Next, the types of processing required for linking field spectrometer and drone-based datasets to airborne and satellite image data are outlined. This covers necessary spectral and spatial resampling, and geo-referencing where exact field to drone image, and drone to satellite image match-ups are required. The final section reviews examples of field-spectrometer and drone-based sensor datasets linked to satellite image data, using: (1) modelling and simulation exercises; (2) calibration of empirical models; and (3) validation of the outputs from satellite-based data processing. Limitations of these approaches and clear directions for future developments are presented to conclude.

12.2 DRONE PLATFORMS

The platform types carrying hyperspectral sensors for civilian use include: fixed wing and multi-rotor small systems (<2 kg) and fixed wing and multi-rotor large systems (>2 kg). Significantly larger platforms are used for military work (e.g., GlobalHawk, ScanEagle), and are slowly moving to commercial applications in some countries where civil aviation authorities approve beyond line-of-sight operations.

Small systems (<2 kg) are typically user-friendly platforms which are often cheaper, safer, and comparatively easier to fly in terms of physical and regulatory issues (Colomina and Molina, 2014; Suomalainen et al., 2014). There are two types of small systems, fixed wing or multi-rotor. Example fixed wing small systems include SenseFly eBee, RTD X5, Parrot Disco, Lehmann Aviation LA500, and Baaz Flying Wing. These small systems can fly 3–5 kilometres and their maximum flying time is up to 1 hour. The payload of these systems is limited (<2 kg). The DJI Phantom® 4, 3DR Solo Quadcopter®, Topcon Falcon 8®, SenseFly eXom®, and Yuneec Typhoon 4K® are examples of multi-rotor systems. With maximum flying times of 30 minutes, these systems can fly up to 2 km. The general payload size is less than 1 kg. Fixed wing small systems are able to fly a longer distance and map larger areas than multi-rotors. Fixed wing systems often require a large (hundreds of metres) corridor to launch and land. In comparison, multi-rotor small systems are easy to operate, requiring only a few square meters of free space for their vertical take-off and landing. Despite being highly flexible platforms, the limitations of multi-rotor systems are their endurance (<20 minutes) and speed, restricting them to a relatively small area (<5 km^2).

Large drone (>2 kg) platforms are being used more commonly for civilian purposes, due to their long endurance (>30 minutes), stability in higher wind speed, sustained speeds, and relatively heavy payload capacity (>1 kg) (Salamí et al., 2014; Pádua et al., 2017). Deployments of these systems over longer durations, increased flying heights, beyond line-of-sight, and to cover larger areas, are commonly restricted by drone flight regulations in each country (Cracknell, 2017). Examples of commonly used fixed wing, large systems are QuestUAV Q-Pod, Trimble UX5, PrecisionHawk Lancaster, and Penguin B. The flying endurance of these systems ranges from 1 hour to 25 hours, covering ranges up to 100 km. These systems can carry larger payloads, from 1 to 10 kg. There are very few published examples of applications from large multi-rotor systems due to their commercial concerns, for example, MicroKopter 3500-Geo®, Yuneec H920 Tornado®, DJI Matrice 600®, and Freefly Systems Alta 8®. These systems can fly up to 32 minutes with 2–5 km coverage. These types of platforms are capable of carrying camera payload approximately between 0.5 and 9 kg. These systems only require a small launch ramp, while relatively large space is required for fixed wing systems (Pádua et al., 2017).

12.3 SENSORS USED ON DRONES

Two types of hyperspectral sensors are carried on drones: non-imaging and imaging systems, and the technology in this area is changing rapidly (Cracknell, 2017).

Non-imaging spectrometer operations are confined to single, low altitude measurements over set targets, or transects over features. A number of these systems are provided by Ocean Optics Inc: STS-VIS, USB4000-VIS-NIR, and Flame-NIR. The STS-VIS sensor has 1024 bands, 3 nm spectral resolution with spectral range between 350–800 nm. The USB4000-VIS-NIR sensor provides data within visible to near-infrared range (350–1000 nm) with 1.5–2.3 nm spectral resolution including 3648 spectral bands. The Flame-NIR sensor features a spectral range from 950–1650 nm with 10 nm spectral resolution. Datasets from these point or line samples are used for a variety of applications, including vegetation condition monitoring (Von Bueren et al., 2015), water quality monitoring, agricultural crops monitoring and diseases identification (Burkart et al., 2014), snow albedo monitoring, and land cover feature characterization (Garzonio et al., 2017).

Imaging spectrometer applications from drone platforms still appear to be in an experimental phase, with commercial providers demonstrating that their systems can collect accurately and precisely geo-corrected data with radiometrically sound and verifiable measurements of at-surface radiance and reflectance and at-sensor irradiance. The majority of drone imaging spectrometers operate in the visible to infrared portion of the spectrum (400–1100 nm), with band placements and bandwidths comparable to airborne and satellite systems. RPAS-based hyperspectral systems do not currently extend into the shortwave infrared (1100–2600 nm) portion of the spectrum due to the need for more complex non-silicon, cooled detectors, which are physically large and require power to cool them. This also applies to narrow-band thermal sensors.

Examples of currently available commercial imaging hyperspectral sensors able to be flown on drones include: Micro-Hyperspec X Series NIR®, Nano-Hyperspec VNIR®, Rikola Hyperspectral camera®, SOC710-GX®, Specim ImSpector V10 2/3®, OCI-UAV-1000®, and MicroHSI 410-SHARK®. The details of these sensors are provided in Table 12.1.

12.4 FIELD SPECTROMETER DATASETS

Unlike drone-based hyperspectral systems, field-based systems do not have stringent weight and power restrictions, unless they are being operated in remote locations on instrumented towers and platforms. Field hyperspectral systems are operated on a mobile, as needs basis, or permanently mounted on towers, platforms, and ships. Field systems can be non-imaging or imaging, they typically cover the full visible—shortwave infrared range (350–2600 nm), with narrow spectral bandwidths (<5 nm), a large number of spectral bands (>1000), high-signal-noise ratios, and traceable measurement units and calibration procedures. This last attribute is conspicuously absent from most commercially available drone-based hyperspectral sensors (imaging and non-imaging) at present.

Field-based non-imaging spectrometer datasets are widely used for calibrating and validating satellite image data and derived products (Matese et al., 2015). Commonly used non-imaging

TABLE 12.1
Currently Available Commercial Imaging Hyperspectral Sensors for Drones

Sensor	Manufacturer	Number of Bands	Bandwidths (nm)	Spectral Range (nm)	References
Micro-Hyperspec X Series NIR	Headwall Photonics Inc.	62	12.9	900–1700	http://www.headwallphotonics.com/spectral-imaging/hyperspectral/micro-hyperspec
Micro-Hyperspec VNIR	Headwall Photonics Inc.	270	5	380–1000	Lucieer et al. (2014)
Rikola Hyperspectral camera	Rikola Ltd.	40	10	500–900	Mozgeris et al. (2016)
SOC710-GX	Surface Optics Corp.	120	4.2	400–1000	https://surfaceoptics.com/products/hyperspectral-imaging/710-gx/
Specim ImSpector V10 2/3	Spectral Imaging Ltd.	101	9	400–1000	Gevaert et al. (2015)
OCI-UAV-1000	BaySpec Inc.	100	5	600–1000	http://www.bayspec.com/spectroscopy/oci-uav-hyperspectral-camera/
MicroHSI 410-SHARK	Corning Inc.	154	2	400–1000	https://www.corning.com/au/en/products/advanced-optics/product-materials/spectral-sensing.html

spectrometers include: ASD Field Spec (ASD Inc.,), Ocean Optic USB4000 (Ocean Optics Inc), and UniSpec DC Spectrometer Analysis System (PP systems). These spectrometers provide data at VNIR (325–1075 nm spectral range) with <3 nm spectral resolution, SWIR (1000–2500 nm spectral range) with 10 nm spectral resolution, and VIS/NIR (310–1100 nm spectral range) with <10 nm spectral resolution. These datasets are linked with satellite and airborne data for either calibration or validation purposes. This can include vicarious calibration of at-surface radiance and reflectance data, and subsequent validation of these standard products. In addition, field spectrometer data and coincident field survey are used to calibrate algorithms for estimating a wide range of biophysical properties of soils, water, vegetation, snow/ice, and the atmosphere. Examples of biophysical properties mapped include: leaf area index (LAI) (Fang et al., 2003), vegetation species (Laliberte et al., 2011), biomass (Wilson et al., 2011; Sibanda et al., 2015), chlorophyll a concentration (Zhou et al., 2014), vegetation community mapping (Davidson et al., 2016) and non-structural carbohydrates (NSC) concentration in forest canopies (Asner and Martin, 2015).

The uses of field-based imaging spectrometer datasets for linkage with satellite data are limited compared to non-imaging spectrometers (Table 12.2). Examples of field-based imaging spectrometers are Hyperspec RECON (Headwall Inc.®), GaiaField Hyperspectral Imaging System (Scientific Photonics®), OCI-2000 Snapshot Handheld Hyperspectral Imager (BaySpec Inc.®), and Specim ImSpector V10 (Specim Spectral Imaging Inc.®). The datasets from these imaging spectrometers cover spectral ranges 400–1000 nm at VNIR with spectral resolution <4 nm and 600–1000 nm at VNIR with spectral resolution <12 nm. These datasets have been integrated with satellite data for mapping vegetation communities (Zhang et al., 2015), mineral composition, land cover types, and agricultural crop types (Abd-Elrahman et al., 2011).

12.5 REQUIREMENTS FOR LINKING FIELD SPECTROMETER AND RPAS DATA TO AIRBORNE AND SATELLITE DATA

Linking field and drone-based spectrometer datasets to satellite datasets can be accomplished using a number of approaches, depending on the end goal of the satellite image data processing. We identify and then describe four commonly used approaches:

1. *Simulation modelling with radiative transfer equations to assess performance of planned satellites sensors*: Simulation modelling with radiative transfer equations, also known as physical-based models, is used to assess the performance of planned satellite sensors (Verrelst et al., 2015). Hedley et al. (2012) used a forward radiative transfer model (RTM) inversion to assess the capability of the planned Sentinel-2 sensor for mapping several attributes of coral reef. The detailed processing flow of their analysis is presented in Figure 12.1. Similarly, the Sentinel-2 sensors ability for estimating the LAI of sugar beet and maize were simulated by Richter et al. (2009) using an inversion of a physically based RTM (SAILH+PROSPECT). In another study, Justice et al. (1998) assessed the Moderate Resolution Imaging Spectroradiometer (MODIS) sensor performance for the retrieval of LAI where they used a 3-D RTM to generate spectral and angular biome specific signatures of vegetation canopies and other variables.
2. *Empirical modelling to link field-based or drone hyperspectral data to satellite data for radiance/reflectance measurement*: Empirical modelling is a common approach to link field-based or drone hyperspectral data to satellite data for radiance/reflectance measurement. A linear non-parametric empirical model is commonly applied due to its simplicity and ability to be optimized (Verrelst et al., 2015). Yue et al. (2017) used the partial least squares regression (PLSR) to relate field-based hyperspectral data to satellite data for reflectance measurement. The linear regression co-efficient approach was used by Laliberte et al. (2011) to link drone hyperspectral data to satellite data for reflectance measurement of vegetation and soils (Figure 12.2).

3. *Empirical modelling to link biophysical properties derived from field-based or drone hyperspectral data to satellite data for mapping biophysical properties*: Empirical modelling is widely used to link biophysical properties derived from field-based or drone hyperspectral data to satellite data for mapping biophysical properties. PLSR is one of the most commonly used linear non-parametric models for mapping biophysical properties and including field datasets (Verrelst et al., 2015). The popularity of PLSR has increased due to its mapping speed and availability in most of the imaging processing packages. The PLSR were used for mapping non-structural carbohydrates (NSC) in forest canopies (Asner and Martin, 2015), leaf mass area (LMA) and foliar traits (Asner et al., 2015; Chadwick and Asner, 2016), above ground biomass (Sibanda et al., 2015), as well as LAI and leaf nitrogen concentrations of crops (Cho et al., 2007; Im et al., 2009). In comparison, simple regression models were also used for the estimation of pasture (Numata et al., 2008), biomass (Messinger et al., 2016) and chlorophyll a concentrations (Zhou et al., 2014) integrating field and satellite data. The applications of some non-linear non-parametric models are also evident in the literature. For example, Skidmore et al. (2010) applied Artificial Neural Network (ANN) for mapping foliar protein and polyphenols for trees and grass; Malenovský et al. (2017) used support vector regression (SVR) for estimation of chlorophyll a + b content (Cab) and effective leaf density (ELD) of Antarctic mosses from reflectance measurements. Satellite and drone image derived Cab and ELD estimates were linked together for validation purposes (Figure 12.3). Further, Adam et al. (2012) applied random forest (RF) to discriminate vegetation species, while Wilson et al. (2011) used the Bayesian networks for biomass mapping. These models, also known as machine learning regression algorithms, are becoming popular due to their flexibility to link different data sources into the analysis.
4. *Validation of satellite-based estimates of biophysical properties using maps of biophysical properties derived from field-based or drone hyperspectral data*: Satellite-based estimates of biophysical properties can also be validated using maps of biophysical properties derived from field-based or drone hyperspectral data. However, limited applications have been reported in the literature. Davidson et al. (2016) validated the Arctic tundra vegetation community map produced from WorldView-2 satellite data using the map which was developed from field spectroscopy measurements. In another study, Malenovský et al. (2017) validated the health map of fragile polar vegetation developed from WorldView-2 satellite data using the map that was produced from the drone-based hyperspectral data.

12.6 CURRENT CAPABILITIES FOR LINKING FIELD SPECTROMETER AND RPAS DATA TO AIRBORNE AND SATELLITE DATA

The approaches presented in the previous section have a number of limitations related to the sensors used, platform and acquisition methods, and linking method.

1. *Simulation modelling with radiative transfer equations to assess performance of planned satellites sensors*: Simulation modelling with RTM to assess performance of planned satellite sensor has a number of limitations. This approach is computationally demanding because of per-pixel-based analysis. Accurate parameterization and optimization procedure are required in this approach since in most of the cases reference data cannot be seen in reality. Moreover, the accuracy of output variables of this approach depends on the quality of RTMs, prior knowledge, and regularizations.
2. *Empirical modelling to link field-based or drone data hyperspectral data to satellite data for radiance/reflectance measurement*: Empirical modelling, which is used to link field hyperspectral data to satellite data for radiance/reflectance measurement, is site-,

TABLE 12.2
Summary of Methods for Linking UAV/UAS/RPAS/Drone and Field Hyperspectral Data to Satellite Data

Platform	Sensor Type	Remote Sensing Data Produced	Information Product, for example, Output Spatial Data (Mapped Variable e.g., Biomass, Spatial, and Temporal Dimension)
		UAV/UAS datasets:	Vegetation community mapping
UAV (fixed wing)	UAV (fixed wing):	Spatial: 5 cm to 1m	Above ground carbon density mapping
UAV (Multi-rotor)	Cannon S110 RGB (Multispectral)	Spectral: 101 bands with 5 nm spectral resolution (spanning from 450–900 nm)	Chla spatial distribution map
UAS (fixed wing)	Specim ImSpector V10 2/3 (Hyperspectral)	3648 spectral bands with 1.5 nm spectral resolution (spanning from 350–1000 nm)	Crop condition mapping
UAS (Multi-rotor)			Forest canopy traits
Satellite Airborne Handheld spectrometer	Sony DSC-WX200 (multispectral)	10 nm spectral resolution (450–915 nm)	Mapping Non-Structural Carbohydrates (NSC) in forest canopies
	UAV (Multi-rotor):	Radiometric: 10 bit, 12 bit	Mapping grape yields
	Tetracam ADC Lite (Multispectral)	Temporal: Same day	Biomass mapping
	UAS (fixed wing):	Field spectrometer datasets: Spectral range: VNIR: 325–1000 nm	Estimating canopy-level foliar N concentration
	Multispectral 4C camera	SWIR: 1000–2500 nm	Monitoring crop traits
	UAS (Multi-rotor):	Spectral resolution: 10 nm in the SWIR region	Land cover mapping
	PhotonFocus SM2-D1312 camera – P (Hyperspectral)	1.5 nm in the VNIR region <3 nm in the VNIR region	Health mapping of fragile polar vegetation
	USB 400 Spectrometer (Hyperspectral)	1.4 nm for the 350–1000 nm spectral region 2 nm for the 1000–2500 spectral regions.	

Transformations Applied to Link UAS or Field Spectrometer to Scale of Satellite Data or Satellite Data Product	Known Limitations of Method Identified	Directions for Future Work	References
(A) Data mining/machine learning: (1) Artificial Neural Networks (ANN) (2) Random forest (RF) (3) Support Vector Regressions (SVR) (B) Empirical approaches: (1) Partial least squares regression (PLSR) (2) Hierarchical Bayesian statistical model (3) linear regression coefficient (4) The Pearson correlation (5) Sparse partial least squares regression (SPLSR)	(A) Data mining/ machine learning: ANN cannot efficiently run on large datasets and it is also sensitive to noise or over-fitting The RF algorithm does not automatically select the optimal number of variables that could yield the lowest error rate. SVR requires a specific three-dimensional upscaling mechanism. (B) Empirical approaches: PLSR treats each wavelength as independent, which incorporate noise created by non-informative wavelengths. It's not also computationally efficient.	(A) Data mining/ machine learning: Need an approach which can run on large datasets, not sensitive to noise and can handle thousands of input variables and required few parameters. (B) Empirical approaches: New method is required that can eliminate useless wavebands in PLS analysis. Need a computationally efficient method. Need a model which is transferable. Targets should be spectrally stable over time, the measurements need not be concurrent with data acquisition Need more automated approach	Fang et al. (2003) Martin et al. (2008) Dorigo et al. (2009) Skidmore et al. (2010) Wilson et al. (2011) Laliberte et al. (2011) Adam et al. (2012) Zhou et al. (2014) Asner et al. (2015) Sibanda et al. (2015) Gevaert et al. (2015) Asner and Martin (2015) Matese et al. (2015) Chadwick and Asner (2016)

(*Continued*)

TABLE 12.2 (*Continued*)
Summary of Methods for Linking UAV/UAS/RPAS/Drone and Field Hyperspectral Data to Satellite Data

Platform	Sensor Type	Remote Sensing Data Produced	Information Product, for example, Output Spatial Data (Mapped Variable e.g., Biomass, Spatial, and Temporal Dimension)
	Micro-hyperspectral (Hyperspectral)	Radiometric: 16 bit, 11 bit.	Mapping Canopy Foliar Traits
	Satellite: WorldView-2, (Multispectral) HJ-1A (Hyperspectral)	Airborne: VSWIR spectrometer (480 bands spanning from 252–2648 nm) 5 nm spectral resolution.	Mapping wheat breading nurseries Species level vegetation classification
	Formosat-2 (Multispectral) RapidEye (Multispectral)	HyMap (500–2400 nm spectral range, 96 bands 15 nm bandwidth, 4.2 m spatial resolution).	Mapping vegetation biophysical and biochemical variables
	Multispectral (Landsat7ETM) Landsat 8 OLI (Multispectral) Sentinel 2 MSI (Multispectral)	Satellite datasets: WorldView-2 (8 spectral bands, 1.84 m spatial resolution, 1.1 days revisit time and 11 bit).	Leaf area index mapping (LAI) Discriminating vegetation species Above ground biomass mapping
	Airborne:	HJ-1A (100 m spatial resolution, 115 narrow-bands in visible and NIR wavelength and four days revisit time).	
	LiDAR: CAO AtoMS CAO AtoMS (VSWIR And waveform LiDAR)	Formosat-2 (8 m spatial resolution, 4 bands, daily revisit capacity)	
	CAO VSWIR (Multispectral) HyMap (Hyperspectral)	Landsat7 ETM (8 bands, 30 m resolution, 8 bit)	
	AVIRIS (Hyperspectral) HiFIS	Landsat 8 (11 spectral band with 30 m spatial resolution, 12 bit, 16 days temporal resolution)	
		Sentinel 2 (12 spectral bands with 10 to 60 m spatial resolution, 5 days temporal resolution, 12 bit) RapidEye (5 m spatial resolution with 6 spectral band, 16 bit. 5 to 6 days temporal resolution	

Transformations Applied to Link UAS or Field Spectrometer to Scale of Satellite Data or Satellite Data Product	Known Limitations of Method Identified	Directions for Future Work	References
(B)Physical/deterministic: (1) Integrating Canopy radiative transfer model (RT) and Genetic algorithm optimization technique (2) CRASh (Canopy variable Retrieval Approach based on PROSPECT and SAILh)	The accuracy of Hierarchical Bayesian statistical depends on sample size/available data and site characteristics and condition. Careful selection and characterization of calibration targets is required for linear regression coefficient. Transferability is limited by site-specific regression coefficient estimation SPLSR does not work well in multi category classification (C) Physical/ deterministic: Accurate parameterization and optimization procedure are required. The approach is computationally demanding. The accuracy of output variables depends on the quality of radiative transfer model.	(C)Physical/ deterministic: Development is required in the physical-based models to make them computationally efficient and simple.	Haghighattalab et al. (2016) Davidson et al. (2016) Messinger et al. (2016) Ahmed et al. (2017) Domingues Franceschini et al. (2017) Garzonio et al. (2017) Malenovský et al. (2017)

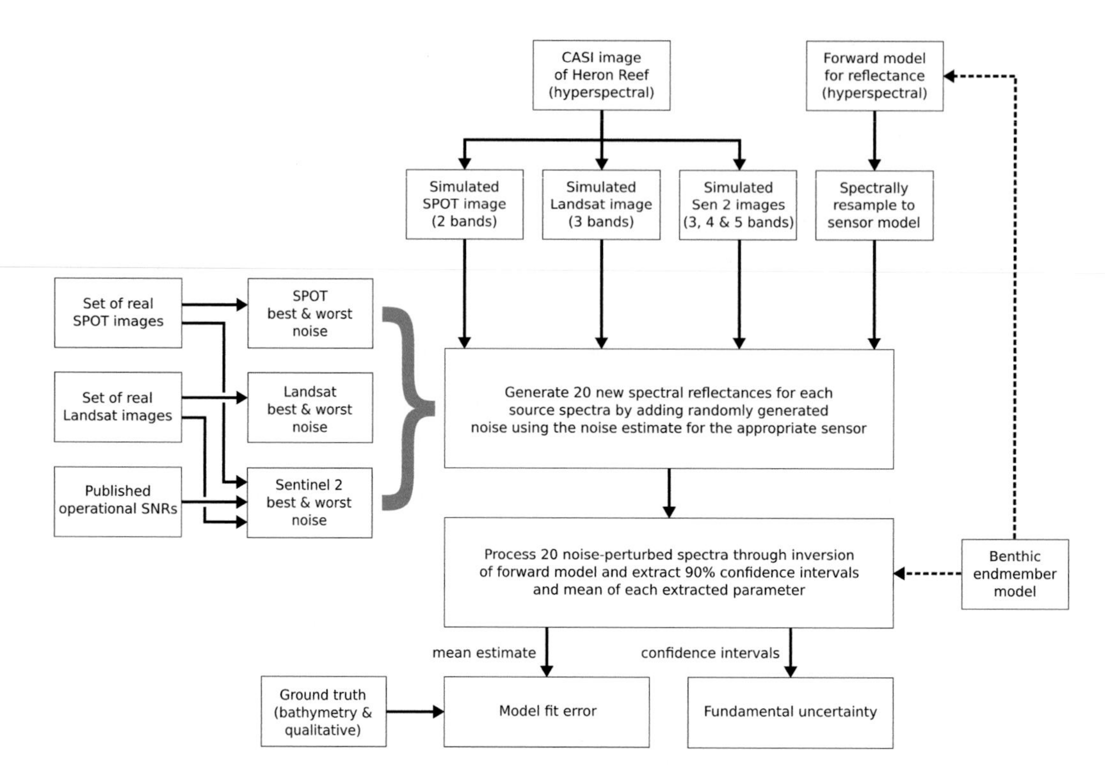

FIGURE 12.1 Processilng flow used to assess performance of planned satellite sensor Sentinel 2 using a radiative transfer model inversion. (From Hedley J et al. 2012. *Remote Sensing of Environment* 120:145–155.)

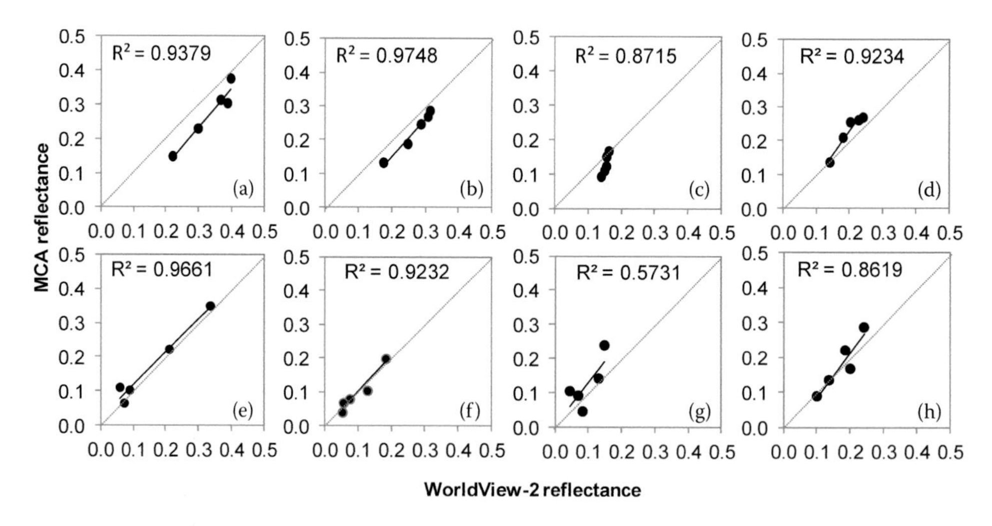

FIGURE 12.2 Correlation of reflectance measurements for eight vegetation/soil targets extracted from WorldView-2 and Drone (MCA) image in the five bands. (Blue, Green, Red, Red-Edge, Near-Infrared) (From Laliberte AS et al. 2011. *Remote Sensing* 3:2529–2551.)

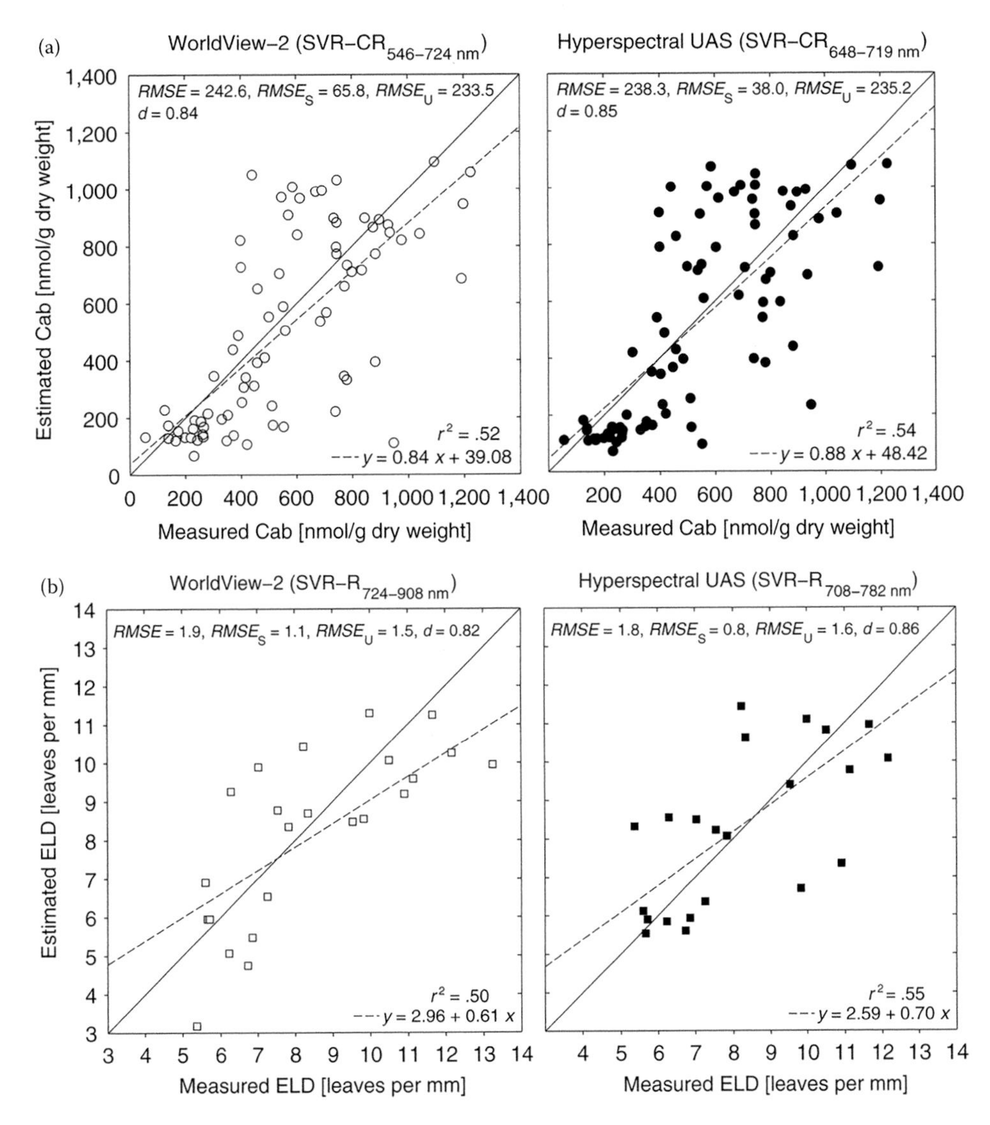

FIGURE 12.3 Support Vector Regression (SVR) estimating (a) Chlorophyll a + b content (Cab) and (b) effective leaf density (ELD) of Antarctic mosses from reflectance measurements convolved to spectral bands of satellite WorldView-2 and Drone to link each other. (From Malenovský Z et al. 2017. *Methods in Ecology and Evolution*.)

data-, and environment-specific. As a result, this approach is not transferable. Selection and characterization of calibration targets are required to perform carefully prior to field-based reflectance/radiance data collection. Expert knowledge is also essential for tuning to improve the better agreement between field and satellite-based reflectance measurement.

3. *Empirical modelling to link biophysical properties derived from field-based or drone hyperspectral data to satellite data for mapping biophysical properties*: several limitations are evident with the empirical approaches as well as data quality and collection procedures to relate biophysical properties derived from field-based or drone hyperspectral to satellite data for mapping biophysical properties. In terms of field and remote sensing data, the mismatch between field data point or small-scale plot measurement and large-scale remote sensing signal are a great challenge to link each other using empirical approaches. Moreover, upscaling and downscaling of spectrometer data and measurement density in

time and space also has great effects in accuracy to integrate field-based data to satellite data. Empirical linking approaches lack generalization and reproducibility as these are site and sensor specific. Moreover, some of the empirical linking algorithms cannot run efficiently on large datasets and require more parameters, for example, the ANN model. The model PLSR treats each wavelength as independent, which incorporates noise generated by non-informative wavelength. This model is also not computationally efficient. In addition, the Bayesian network model is not appropriate for wider application since the accuracy of this model depends on the sample size as well as site characteristics and conditions.

4. *Validation of satellite-based estimates of biophysical properties using maps of biophysical properties derived from field-based or drone hyperspectral data*: The accurate and appropriate scaling up field-based hyperspectral data to satellite data scales for producing biophysical properties maps for the purpose of satellite data derived map validation is a significant challenge (Lucieer et al., 2014).

12.7 CONCLUSIONS AND FUTURE REQUIREMENTS FOR LINKING FIELD SPECTROMETER AND DRONE DATA TO AIRBORNE AND SATELLITE DATA

This chapter shows that our current capabilities for linking hyperspectral data from field and drone datasets to airborne and satellite images is limited to field spectrometry with well-established methods and protocols. These include the application of simulation modelling with radiative transfer equations to assess performance of planned satellites sensors; empirical modelling to link field-based hyperspectral data to satellite data for radiance/reflectance measurement; empirical modelling to link biophysical properties derived from field-based or drone hyperspectral data to satellite data for mapping biophysical properties; and validation of satellite-based estimates of biophysical properties using maps of biophysical properties derived from field-based hyperspectral data. Drone-based hyperspectral data have significant global potential to be used extensively in this domain; however, the lack of significant numbers of peer reviewed publications indicate that these systems are not yet delivering research and operations ready systems.

To move forward, there are some immediate challenges we need to tackle with closer collaboration between the application/user community and private industry making hyperspectral systems, and these include:

- Develop standard protocols for collecting (field and flight standards), correcting, and processing field or drone-based hyperspectral data to allow direct integration with satellite datasets, such as the "best practice standards" outlined by Held et al. (2015) and various CEOS Working Groups.
- Develop high-quality, lightweight, operational, hyperspectral sensors with appropriate geometric and radiometric corrections—in a turn-key ready-to-use format, like the fully operational multispectral systems (e.g., Parrot Sequoia®).
- Development of best practice workflows that can ensure collected field-based or drone-based hyperspectral data that are corrected to appropriate levels (geometric, radiometric, and atmospheric) to be linked with fully corrected datasets.
- More effective linking models to handle large datasets and variables for mapping biophysical properties. Development is required in the physical-based models to make them computationally efficient, as well as simple. Empirical models need to be improved in terms of their handling large datasets, accuracy, wider applications, and computational efficiency. The appropriate and effective upscaling and downscaling techniques are also required to integrate field-based or drone-based hyperspectral data to satellite data scale in an accurate way to reduce the loss of spectral information (Table 12.3).

TABLE 12.3
Detailed Summary of Reviewed Literature for Linking UAV/UAS/RPAS/Drone and Field Hyperspectral Data to Satellite Data

Platform	Sensor Type	Remote Sensing Data Produced	Information Product, for example, Output Spatial Data (Mapped Variable e.g., Biomass, Spatial, and Temporal Dimension)	Transformations Applied to Link UAS or Field Spectrometer to Scale of Satellite Data or Satellite Data Product	Known Limitations of Method Identified	Directions for Future Work	References
Field data: Handheld spectrometer Satellite	Satellite data: WorldView-2 (Multispectral)	Spectrometer: 256 discrete bands spanning from 450 to 1040 nm, on a single date. Satellite: Worldview-2 multispectral data with 2 m spatial resolution, 11 bit.	Vegetation community mapping	Rescaling narrow-band data to multispectral satellite data by averaging band reflectance	Loss of spectral information prior to aggregating spectral data across a wider bandwidth, scaling from plot to landscape.	The use of airborne hyperspectral imagery or unmanned aerial vehicles (UAV) with very high spatial resolution (sub-centimetre for UAV), acquired at the same time of field data could improve the accuracy of vegetation mapping	Davidson et al. (2016)

(Continued)

TABLE 12.3 (*Continued*)
Detailed Summary of Reviewed Literature for Linking UAV/UAS/RPAS/Drone and Field Hyperspectral Data to Satellite Data

Platform	Sensor Type	Remote Sensing Data Produced	Information Product, for example, Output Spatial Data (Mapped Variable e.g., Biomass, Spatial, and Temporal Dimension)	Transformations Applied to Link UAS or Field Spectrometer to Scale of Satellite Data or Satellite Data Product	Known Limitations of Method Identified	Directions for Future Work	References
UAV (fixed wing) Aircraft for LiDAR data	UAV: Cannon S110 RGB (Multispectral) LiDAR: CAO AtoMS	UAV: Georeferenced orthoimages and digital canopy Model (DCM) using UAV data, the ground sample distance (GSD) of the imagery was 6.2–7.7 cm DTM and canopy height model (CHM) was created using LiDAR data. The density of LiDAR data of one point per 1.25 m^2	Above ground carbon density mapping	Structure from motion (SFM) method was used to register DCM with LiDAR product CHM	Poor accuracy of CHM geo-referencing resulting from the use of consumer-grade GPS is the primary limitation to the method.	The use of higher-precision differential GPS should eliminate the need for manual registration by reducing error in X, Y, and Z to significantly less than 1 m.	Messinger et al. (2016)

(*Continued*)

TABLE 12.3 *(Continued)*
Detailed Summary of Reviewed Literature for Linking UAV/UAS/RPAS/Drone and Field Hyperspectral Data to Satellite Data

Platform	Sensor Type	Remote Sensing Data Produced	Information Product, for example, Output Spatial Data (Mapped Variable e.g., Biomass, Spatial, and Temporal Dimension)	Transformations Applied to Link UAS or Field Spectrometer to Scale of Satellite Data or Satellite Data Product	Known Limitations of Method Identified	Directions for Future Work	References
Boat for field spectral data collection	ASD FieldSpec Spectrophotometer	Spectrometer: Sensitivity range from 350 to 1075 nm with spectral sampling of 1.5 nm, collected in 512 bands.	Chla spatial distribution map	Plotting data and correlation coefficient of three bands of model	In comparing the selected bands location for the spectra-based model and HJ-1A HSI-based model, the first and third bands of the two models are located nearly in the same range, yet the second band shifted about two bands width.	The three bands of model may work better using MODIS or MERIS data	Zhou et al. (2014)
Satellite	Satellite: HJ-1A (Hyperspectral)	Satellite: 100 m spatial resolution, a field of view of 50 km, 115 narrow spectral bands in visible and NIR wavelengths and the four days revisit time.					

(Continued)

TABLE 12.3 (*Continued*)
Detailed Summary of Reviewed Literature for Linking UAV/UAS/RPAS/Drone and Field Hyperspectral Data to Satellite Data

Platform	Sensor Type	Remote Sensing Data Produced	Information Product, for example, Output Spatial Data (Mapped Variable e.g., Biomass, Spatial, and Temporal Dimension)	Transformations Applied to Link UAS or Field Spectrometer to Scale of Satellite Data or Satellite Data Product	Known Limitations of Method Identified	Directions for Future Work	References
UAV (fixed wing) Satellite	UAV: Specim ImSpector V10 2/3 (Hyperspectral) Satellite: Multispectral (Formosat -2)	UAV: 1 m spatial resolution at four dates, 101 bands every 5 nm between 450 and 900 nm Satellite: 8 images with 8 m spatial resolution, 4 bands, collected during 2013 growing season Acquired in different date	Crop condition mapping	Spectral temporal response surface (STRSs) method using Bayesian theory The UAV imagery taken were convolved to the Formosat-2 spectral bands using the Formosat-2 spectral response function and radiometrically normalized.	The used method can accurately interpolate a limited number of hyperspectral measurements to daily observations during an entire growing season	Future studies could combine additional sensor, providing surface reference data at the spectral and temporal intervals.	Gevaert et al. (2015)

(*Continued*)

TABLE 12.3 (*Continued*)
Detailed Summary of Reviewed Literature for Linking UAV/UAS/RPAS/Drone and Field Hyperspectral Data to Satellite Data

Platform	Sensor Type	Remote Sensing Data Produced	Information Product, for example, Output Spatial Data (Mapped Variable e.g., Biomass, Spatial, and Temporal Dimension)	Transformations Applied to Link UAS or Field Spectrometer to Scale of Satellite Data or Satellite Data Product	Known Limitations of Method Identified	Directions for Future Work	References
Airborne Handheld spectrometer	Airborne: CAO AtoMS (VSWIR) And waveform LiDAR)	The VSWIR spectrometer measures spectral radiance in 480 channels spanning the 252–2648 nm wavelength range in 5 nm increments (full-width at half-maximum). The LiDAR has a beam divergence set to 0.5 mrad, and was operated at 200 kHz with 17° scan half-angle from nadir, providing swath coverage similar to the VSWIR spectrometer, DSM and DCM (Digital Canopy Model)	Forest canopy traits	Data fusion approach and matching by 1-hectare resolution grid cell with the spectral signature of each 1-hectare cell was derived by averaging the spectra of all 2-meter resolution VSWIR measurements that passed the filtering criteria like NDVI, vegetation height and minimum intra canopy shade	Inter-annual variation in canopy chemistry and phenology would thus certainly have affected the validation process, especially in the context of model performances	Further testing of the methods presented here will be required before the approach can be made operational with other airborne and future spaceborne imaging spectrometers.	Asner et al. (2015)

(*Continued*)

TABLE 12.3 (*Continued*)

Detailed Summary of Reviewed Literature for Linking UAV/UAS/RPAS/Drone and Field Hyperspectral Data to Satellite Data

Platform	Sensor Type	Remote Sensing Data Produced	Information Product, for example, Output Spatial Data (Mapped Variable e.g., Biomass, Spatial, and Temporal Dimension)	Transformations Applied to Link UAS or Field Spectrometer to Scale of Satellite Data or Satellite Data Product	Known Limitations of Method Identified	Directions for Future Work	References
Handheld spectrometer Airborne	Airborne: CAO VSWIR (Multispectral)	Handheld spectrometer: The spectra were collected with a field spectrometer in the 400–2500 nm wavelength range. Airborne: The VSWIR spectrometer measures spectral radiance in 480 channels spanning the 252–2648 nm wavelength range in 5-nm increments (full-width at half-maximum). All the data were collected in the same season to keep consistency	Mapping Non-Structural Carbohydrates (NSC) in forest canopies	Chemometric method, called partial least squares regression (PLSR) analysis was used to estimate NSC at leaf, modeled-canopy and actual forest stand scales. Radiative transfer model was used to transfer the spectra to the canopy level.	This method occurs canopy scale absorption. Precision and accuracy did suffer under such extreme canopy structural variability compared to the leaf-level analyses.	Insufficient evidence to recommend that structural variability in forest canopies will critically impair NSC estimation from imaging spectroscopy.	Asner and Martin (2015)

(*Continued*)

TABLE 12.3 (*Continued*)
Detailed Summary of Reviewed Literature for Linking UAV/UAS/RPAS/Drone and Field Hyperspectral Data to Satellite Data

Platform	Sensor Type	Remote Sensing Data Produced	Information Product, for example, Output Spatial Data (Mapped Variable e.g., Biomass, Spatial, and Temporal Dimension)	Transformations Applied to Link UAS or Field Spectrometer to Scale of Satellite Data or Satellite Data Product	Known Limitations of Method Identified	Directions for Future Work	References
UAV (Multi-rotor) Satellite	UAV : Tetracam ADC Lite (Multispectral) Satellite: RapidEye (Multispectral)	UAV: 0.05 m spatial resolution with 3spectral wavebands (between 520–900 nm), 10 bit RAW. All images were taken between 12:00 and 13:00 in clear sky condition in the same day. Satellite: 5 m spatial resolution with 6 spectral band (between 440–850 nm), 16 bit. 5 to 6 days temporal resolution.	Mapping grape yields	Resampling UAV image to satellite image using block-averaging function (Quadrant decomposition) A FieldSpec Pro spectroradiometer (ASD) was used to conduct a radiometric calibration in field for UAV images	This method doesn't work in the case of heterogeneity.	Need a method that can account heterogeneity.	Matese et al. (2015)

(*Continued*)

TABLE 12.3 (*Continued*)
Detailed Summary of Reviewed Literature for Linking UAV/UAS/RPAS/Drone and Field Hyperspectral Data to Satellite Data

Platform	Sensor Type	Remote Sensing Data Produced	Information Product, for example, Output Spatial Data (Mapped Variable e.g., Biomass, Spatial, and Temporal Dimension)	Transformations Applied to Link UAS or Field Spectrometer to Scale of Satellite Data or Satellite Data Product	Known Limitations of Method Identified	Directions for Future Work	References
Handheld spectrometer Airborne	Airborne: HyMap (Hyperspectral)	Spectrometer: Field sampling were conducted in the middle of the day (i.e., 2 h before and after true midday) to collect high quality spectra using a field spectrometer. Airborne: HyMap hyperspectral imagery (500–2400 nm spectral range, 96 bands, 15 nm bandwidth, 4.2 m spatial resolution, mounted on a Zeiss stabilized platform). The imagery was acquired between 11:00 and 12:00 to coincide with the time of field data collection.	Biomass mapping	Artificial Neural Networks (ANN) machine learning algorithm	It cannot efficiently run on large datasets, It is sensitive to noise or over-fitting, More parameters are required	Need an approach which can run on large datasets, not sensitive to noise and can handle thousands of input variables and required few parameters.	Skidmore et al. (2010)

(*Continued*)

TABLE 12.3 (*Continued*)

Detailed Summary of Reviewed Literature for Linking UAV/UAS/RPAS/Drone and Field Hyperspectral Data to Satellite Data

Platform	Sensor Type	Remote Sensing Data Produced	Information Product, for example, Output Spatial Data (Mapped Variable e.g., Biomass, Spatial, and Temporal Dimension)	Transformations Applied to Link UAS or Field Spectrometer to Scale of Satellite Data or Satellite Data Product	Known Limitations of Method Identified	Directions for Future Work	References
Handheld spectrometer Airborne	Airborne: AVIRIS (Hyperspectral)	Spectrometer: analysis for foliar nitrogen concentration was performed using a FOSS NIR 6500 spectrometer. Satellite: The AVIRIS sensor captures upwelling spectral radiance in 224 contiguous spectral bands for wavelengths from 400 to 2500 nm, with a 10 nm nominal bandwidth. Pixel size is approximately 17 m.	Estimating canopy-level foliar nitrogen concentration	Partial least squares regression (PLSR) was used to relate satellite image spectra to field-measured nitrogen concentration	PLSR treats each wavelength as independent, which incorporate noise created by non-informative wavelengths It's not computationally efficient.	New method is required that can eliminate useless wavebands in PLSR analysis. Need a computationally efficient method.	Martin et al. (2008)

(*Continued*)

TABLE 12.3 (*Continued*)
Detailed Summary of Reviewed Literature for Linking UAV/UAS/RPAS/Drone and Field Hyperspectral Data to Satellite Data

Platform	Sensor Type	Remote Sensing Data Produced	Information Product, for example, Output Spatial Data (Mapped Variable e.g., Biomass, Spatial, and Temporal Dimension)	Transformations Applied to Link UAS or Field Spectrometer to Scale of Satellite Data or Satellite Data Product	Known Limitations of Method Identified	Directions for Future Work	References
Handheld spectrometer UAS (Multi-rotor)	UAS: PhotonFocus SM2-D1312 camera – P (Hyperspectral)	Spectrometer: canopy reflectance was measured using a handheld Cropscan Multispectral Radiometer that performs readings in 16 spectral bands in the visible, near-infrared and shortwave infrared (from 490 to 870 nm). UAV: Images were collected on different days during the growing season, corresponding to distinct crop growth stages. During data acquisition the sensor system was positioned at 80 m altitude resulting in an approximate spatial resolution of 0.2 m and 10 nm spectral resolution between 450 and 915 nm, with 10 nm spectral resolution.	Monitoring crop traits	Geo-referencing the image pixel to field plot using edge detection approach. The spectral from field and image were averaged at plot level for spectral resampling.	The decision about the final boundaries was made visually which add bias to the process. Illumination conditions to reflectance and spatial resolution differences affects in the results	Simple linear transfer functions could be used to combine outputs of different sensors.	Domingues Franceschini et al. (2017)

(*Continued*)

TABLE 12.3 (*Continued*)
Detailed Summary of Reviewed Literature for Linking UAV/UAS/RPAS/Drone and Field Hyperspectral Data to Satellite Data

Platform	Sensor Type	Remote Sensing Data Produced	Information Product, for example, Output Spatial Data (Mapped Variable e.g., Biomass, Spatial, and Temporal Dimension)	Transformations Applied to Link UAS or Field Spectrometer to Scale of Satellite Data or Satellite Data Product	Known Limitations of Method Identified	Directions for Future Work	References
Handheld spectrometer UAS (Multi-rotor)	UAS: USB 400 Spectrometer (Hyperspectral)	UAS: The USB4000 is a non-imaging fiber-optic spectrometer, which measures 3648 spectral bands from 350 to 1000 nm with a spectral resolution about 1.5 nm. Handheld spectrometer: FieldSpec HandHeld (HH) spectrometer (ASD) was use to acquire field reflectance data. Field spectra were collected in the same time of UAS data collection.	Land cover mapping	Radiometric transformation: Spectra are resampled to a common web length–grid trough linear interpolation using dual spectrometer method.	Dual instrument transfer function is a valid for short time frame. this method requires a fine spectral and radiometric inter-calibration between the two spectrometers together with the use of the empirical transfer function	Low-flying UAS could offer a more practical approach to collect spatially-distributed fluorescence measurements within an acceptable level of accuracy.	Garzonio et al. (2017)

(*Continued*)

TABLE 12.3 (*Continued*)
Detailed Summary of Reviewed Literature for Linking UAV/UAS/RPAS/Drone and Field Hyperspectral Data to Satellite Data

Platform	Sensor Type	Remote Sensing Data Produced	Information Product, for example, Output Spatial Data (Mapped Variable e.g., Biomass, Spatial, and Temporal Dimension)	Transformations Applied to Link UAS or Field Spectrometer to Scale of Satellite Data or Satellite Data Product	Known Limitations of Method Identified	Directions for Future Work	References
Handheld spectrometer Satellite	Satellite: Multispectral (Landsat7ETM)	Spectrometer: Hyperspectral reflectance from 380 to 1000 nm using a portable spectrometer with a 2-m fiber-optic cord mounted to a pole and held 1 m from the canopy. Satellite: Landsat7 ETM with 30 m resolution, 8 bit. Both data were collected within much closed dates.	Biomass mapping	Hierarchical Bayesian statistical model to relate the plot level field data to satellite data	The number and size of plots needed for this type of satellite validation will depend on the homogeneity of the vegetation at the scale of the satellite data and the parameters of interest. This model has limitation for wider scale application as the accuracy of this model depends on sample size/ available data and site characteristics.	Need a model which is transferable	Wilson et al. (2011)

(*Continued*)

TABLE 12.3 (*Continued*)

Detailed Summary of Reviewed Literature for Linking UAV/UAS/RPAS/Drone and Field Hyperspectral Data to Satellite Data

Platform	Sensor Type	Remote Sensing Data Produced	Information Product, for example, Output Spatial Data (Mapped Variable e.g., Biomass, Spatial, and Temporal Dimension)	Transformations Applied to Link UAS or Field Spectrometer to Scale of Satellite Data or Satellite Data Product	Known Limitations of Method Identified	Directions for Future Work	References
UAS (Multi-rotor) Satellite	UAS: Micro-hyperspectral (Hyperspectral) Satellite: Multispectral (WorldView-2)	UAS: 5.0 cm spatial resolution with 162 spectral wavebands (between 361–961 nm), 12 bit RAW Satellite: 1.84 m spatial resolution with 8 spectral band (between 350–1050 nm) 11 bit.	Health mapping of fragile polar vegetation	Machine learning Support Vector Regressions (SVR)	This method requires a specific three-dimensional upscaling mechanism,	Modern approaches like canopy radiative transfer models could solve the three-dimensional mechanism requirement.	Malenovský et al. (2017)
Handheld spectrometer Airborne	Airborne HiFIS	Airborne: The VSWIR spectrometer measures spectral radiance in 427 channels spanning the 350–2610 nm wavelength range in 6-nm spectral resolution.	Mapping Canopy Foliar Traits	Partial least squares (PLSR) regression was used to relate satellite image spectra to field-measured nitrogen concentration	PLSR is considered limited because it treats each wavelength as independent, which incorporate noise created by non-informative wavelengths	New method is required that can eliminate useless wavebands in PLS analysis	Chadwick and Asner et al. (2016)

(*Continued*)

TABLE 12.3 (*Continued*)
Detailed Summary of Reviewed Literature for Linking UAV/UAS/RPAS/Drone and Field Hyperspectral Data to Satellite Data

Platform	Sensor Type	Remote Sensing Data Produced	Information Product, for example, Output Spatial Data (Mapped Variable e.g., Biomass, Spatial, and Temporal Dimension)	Transformations Applied to Link UAS or Field Spectrometer to Scale of Satellite Data or Satellite Data Product	Known Limitations of Method Identified	Directions for Future Work	References
Handheld spectrometer UAS (fixed wing)	UAS: Multispectral 4C camera	UAS: UAS image contains four distinct bands with no spectral overlap (530–810 nm), and is controlled by the eBee Ag autopilot during the flight. Handheld spectrometer: Field spectral measurements were taken using ASD VNIR handheld point-based spectroradiometer (ASD Inc.) with a wavelength range of 325–1075 nm, and a spectral resolution of <3 nm at 700 nm. The instrument digitizes spectral values to 16 bits.	Mapping wheat breading nurseries	The Pearson correlation between the average VI values of the sample plots and field spectra for each platform was Calculated.	Transferability is limited by site-specific regression coefficient estimation	Need more generic and transferable approach	Haghighattalab et al. (2016)

(*Continued*)

TABLE 12.3 (*Continued*)

Detailed Summary of Reviewed Literature for Linking UAV/UAS/RPAS/Drone and Field Hyperspectral Data to Satellite Data

Platform	Sensor Type	Remote Sensing Data Produced	Information Product, for example, Output Spatial Data (Mapped Variable e.g., Biomass, Spatial, and Temporal Dimension)	Transformations Applied to Link UAS or Field Spectrometer to Scale of Satellite Data or Satellite Data Product	Known Limitations of Method Identified	Directions for Future Work	References
Handheld spectrometer UAS (fixed wing) Satellite	UAS: Multispectral (Mini MCA-6) Satellite: Multispectral (WorldView-2)	UAS: 14 cm spatial resolution with 100 spectral wavebands (between 450–850 nm), 10 bit RAW. Satellite: 1.84 m spatial resolution with 8 spectral band (between 350–1050 nm) 11 bit. Single date and two weeks difference after the acquisition of satellite image Handheld spectrometer: The FieldSpec Pro collects data in multiple narrow bands (3–10 nm spectral resolution) with a spectral range of 350–2500 nm.	Species level vegetation classification	Regression coefficient to relate spectral reflectance between satellite image and UAS image	Careful selection and characterization of calibration targets is required especially where the calibration targets may contain small proportions of highly contrasting materials	Targets should be spectrally stable over time, the measurements need not be concurrent with data acquisition	Laliberte et al. (2011)

(*Continued*)

TABLE 12.3 (*Continued*)
Detailed Summary of Reviewed Literature for Linking UAV/UAS/RPAS/Drone and Field Hyperspectral Data to Satellite Data

Platform	Sensor Type	Remote Sensing Data Produced	Information Product, for example, Output Spatial Data (Mapped Variable e.g., Biomass, Spatial, and Temporal Dimension)	Transformations Applied to Link UAS or Field Spectrometer to Scale of Satellite Data or Satellite Data Product	Known Limitations of Method Identified	Directions for Future Work	References
Handheld spectrometer UAV (fixed wing)	UAV: Sony DSC-WX200 (multispectral)	UAV: a 1.2 megapixel Parrot Sequoia that captures ~50 nm wide bands in the green (~550 nm), red (~660 nm), and near-infrared (~790 nm) regions, as well as a ~10 nm wide band in the red-edge (~735 nm) region. Single date. Handheld spectrometer: ASD Field Spectrometer at multiple light and dark calibration target locations was collected throughout the same day.	Land cover and vegetation mapping	The raw UAV reflectance data were calibrated using an empirical line calibration method	Need sufficient knowledge of the calibration procedure, the site under investigation and the spectral properties of proposed calibration targets, erroneous results can be produced.	Need more automated approach	Ahmed at al. (2017)

(*Continued*)

TABLE 12.3 (*Continued*)

Detailed Summary of Reviewed Literature for Linking UAV/UAS/RPAS/Drone and Field Hyperspectral Data to Satellite Data

Platform	Sensor Type	Remote Sensing Data Produced	Information Product, for example, Output Spatial Data (Mapped Variable e.g., Biomass, Spatial, and Temporal Dimension)	Transformations Applied to Link UAS or Field Spectrometer to Scale of Satellite Data or Satellite Data Product	Known Limitations of Method Identified	Directions for Future Work	References
Handheld spectrometer Airborne	Hyperspectral (HyMap)	Airborne: Electromagnetic spectrum between 0.438 and 2.483 μm with a total of 126 contiguous bands having a spectral resolution between 11 and 22 nm Handheld spectrometer: In each sample plot, 10 spectroradiometric measurements were taken from nadir using a portable Fieldspec PRO FR spectrometer.	Mapping vegetation biophysical and biochemical variables	CRASh (Canopy variable Retrieval Approach based on PROSPECT and SAILh) is based on the combined leaf reflectance model PROSPECT and the canopy radiative transfer model SAILh.	The structure of the approach need a validated land-use map or the definition of plausible phenological properties.		Dorigo et al. (2009)

(*Continued*)

TABLE 12.3 (*Continued*)
Detailed Summary of Reviewed Literature for Linking UAV/UAS/RPAS/Drone and Field Hyperspectral Data to Satellite Data

Platform	Sensor Type	Remote Sensing Data Produced	Information Product, for example, Output Spatial Data (Mapped Variable e.g., Biomass, Spatial, and Temporal Dimension)	Transformations Applied to Link UAS or Field Spectrometer to Scale of Satellite Data or Satellite Data Product	Known Limitations of Method Identified	Directions for Future Work	References
Handheld spectrometer Satellite	Satellite: Multispectral (Landsat ETM+)	Satellite: 8 multispectral band with 30 m spatial resolution, 8 bit. Handheld spectrometer: The surface reflectance was taken with the ASD. In each field, about 50–100 points along several random transactions were measured.	Leaf area index mapping	Integrating canopy radiative transfer model (RT) and genetic algorithm optimization technique	A large number of iterations are needed to converge toward appropriate solutions.	To solve this problem, more efficient GA optimization algorithms and GA-RT coupling methods are needed.	Fang et al. (2003)

(*Continued*)

TABLE 12.3 (*Continued*)

Detailed Summary of Reviewed Literature for Linking UAV/UAS/RPAS/Drone and Field Hyperspectral Data to Satellite Data

Platform	Sensor Type	Remote Sensing Data Produced	Information Product, for example, Output Spatial Data (Mapped Variable e.g., Biomass, Spatial, and Temporal Dimension)	Transformations Applied to Link UAS or Field Spectrometer to Scale of Satellite Data or Satellite Data Product	Known Limitations of Method Identified	Directions for Future Work	References
Handheld spectrometer Airborne	Hyperspectral (HyMap)	Airborne: HyMAP comprising 126 wavelengths, operating over the spectral range 436.5–2485 nm, with average spectral resolutions of 15 nm (437–1313 nm), 13 nm (1409–1800 nm) and 17 nm (1953–2485 nm) Handheld spectrometer: The wavelengths ranging from 350 to 2500 nm with a sampling interval of 1.4 nm for the 350 to1000 nm spectral region and a 2.0 nm sampling interval for the 1000 to 2500 nm spectral region using ASD spectrometer.	Discriminating vegetation species	Random forest (RF) machine learning algorithm	The RF algorithm does not automatically choose the optimal number of variables that could yield the lowest error rate.	Need an approach that can select the optimal number of variables to minimize the error rate.	Adam et al. (2012)

(*Continued*)

TABLE 12.3 (*Continued*)
Detailed Summary of Reviewed Literature for Linking UAV/UAS/RPAS/Drone and Field Hyperspectral Data to Satellite Data

Platform	Sensor Type	Remote Sensing Data Produced	Information Product, for example, Output Spatial Data (Mapped Variable e.g., Biomass, Spatial, and Temporal Dimension)	Transformations Applied to Link UAS or Field Spectrometer to Scale of Satellite Data or Satellite Data Product	Known Limitations of Method Identified	Directions for Future Work	References
Handheld spectrometer Satellite	Satellite: Multispectral (Landsat 8 OLI) and Sentinel 2 MSI	Satellite: Landsat 8 having 11 spectral bands with 30 m spatial resolution, 12 bit, 16 days temporal resolution. Sentinel 2 having 12 spectral bands with 10 to 60 m spatial resolution, 5 days temporal resolution, 12 bit. Handheld spectrometer: The ASD spectrometer acquire radiation at 1.4 nm intervals for the spectral region 350–1000 and 2 nm intervals for the spectral region 1000–2500 nm. Spectral measurements were taken under clear sky conditions between 10 AM and 2 PM.	Aboveground biomass mapping	Sparse partial least squares regression (SPLSR) Sentinel 2 and Landsat 8 data were calibrated using field-plot-based reflectance data	It does not work well in multi-category classification Higher sensitivity in variable selection in multi- category classification.	Need an approach that can deal with multi-category classification.	Sibanda et al. (2015)

REFERENCES

Abd-Elrahman A, Pande-Chhetri R, Vallad G. 2011. Design and development of a multi-purpose low-cost hyperspectral imaging system. *Remote Sensing* 3:570–586

Adam E, Mutanga O, Rugege D, Ismail R. 2012. Discriminating the papyrus vegetation (*Cyperus papyrus L.)* and its co-existent species using random forest and hyperspectral data resampled to HYMAP. *International Journal of Remote Sensing* 33:552–569

Ahmed OS, Shemrock A, Chabot D, Dillon C, Williams G, Wasson R, Franklin SE. 2017. Hierarchical land cover and vegetation classification using multispectral data acquired from an unmanned aerial vehicle. *International Journal of Remote Sensing* 38:2037–2052

Asner GP, Martin RE. 2015. Spectroscopic Remote Sensing of Non-Structural Carbohydrates in Forest Canopies. *Remote Sensing* 7:3526–3547. doi:10.3390/rs70403526

Asner GP, Martin RE, Anderson CB, Knapp DE. 2015. Quantifying forest canopy traits: Imaging spectroscopy versus field survey. *Remote Sens Environ* 158:15–27. doi:10.1016/j.rse.2014.11.011

Burkart A, Cogliati S, Schickling A, Rascher U. 2014. A novel UAV-based ultra-light weight spectrometer for field spectroscopy. *IEEE Sensors Journal* 14:62–67

Chadwick KD, Asner GP. 2016. Organismic-scale remote sensing of canopy foliar traits in lowland tropical forests. *Remote Sensing* 8. doi:10.3390/rs8020087

Cho MA, Skidmore A, Corsi F, Van Wieren SE, Sobhan I. 2007. Estimation of green grass/herb biomass from airborne hyperspectral imagery using spectral indices and partial least squares regression. *International Journal of Applied Earth Observation and Geoinformation* 9:414–424

Colomina I, Molina P. 2014. Unmanned aerial systems for photogrammetry and remote sensing: A review. *ISPRS Journal of Photogrammetry and Remote Sensing* 92:79–97

Cracknell AP. 2017. UAVs: Regulations and law enforcement. *International Journal of Remote Sensing* 38:3054–3067. doi:10.1080/01431161.2017.1302115

Davidson SJ, Santos MJ, Sloan VL, Watts JD, Phoenix GK, Oechel WC, Zona D. 2016. Mapping Arctic Tundra vegetation communities using field spectroscopy and multispectral satellite data in North Alaska, USA. *Remote Sensing* 8:978

Domingues Franceschini MH, Bartholomeus H, van Apeldoorn D, Suomalainen J, Kooistra L. 2017. Intercomparison of unmanned aerial vehicle and ground-based narrow band spectrometers applied to crop trait monitoring in organic potato production. *Sensors* 17:1428

Dorigo W, Richter R, Baret F, Bamler R, Wagner W. 2009. Enhanced automated canopy characterization from hyperspectral data by a novel two step radiative transfer model inversion approach. *Remote Sensing* 1:1139–1170

Fang H, Liang S, Kuusk A. 2003. Retrieving leaf area index using a genetic algorithm with a canopy radiative transfer model. *Remote Sens Environ* 85:257–270

Garzonio R, Di Mauro B, Colombo R, Cogliati S. 2017. Surface reflectance and sun-induced fluorescence spectroscopy measurements using a small hyperspectral UAS. *Remote Sensing* 9:472

Gevaert CM, Suomalainen J, Tang J, Kooistra L. 2015. Generation of spectral–temporal response surfaces by combining multispectral satellite and hyperspectral UAV imagery for precision agriculture applications. *IEEE Journal of Selected Topics in Applied Earth Observations and Remote Sensing* 8:3140–3146

Haghighattalab A et al. 2016. Application of unmanned aerial systems for high throughput phenotyping of large wheat breeding nurseries. *Plant Methods* 12:35

Hedley J, Roelfsema C, Koetz B, Phinn S. 2012. Capability of the Sentinel 2 mission for tropical coral reef mapping and coral bleaching detection. *Remote Sensing of Environment* 120:145–155

Held A., Phinn S., Soto-Berelov M., Jones S. (Eds.). 2015. *AusCover Good Practice Guidelines: A Technical Handbook Supporting Calibration and Validation Activities of Remotely Sensed Data Products*. Version 1.2. TERN AusCover, ISBN 978-0-646-94137-0

IJRS (ed.). 2017. Unmanned Aerial Vehicles for Environmental Applications. *International Journal of Remote Sensing* 38:2029–2036. doi:10.1080/01431161.2017.1301705

Im J, Jensen JR, Coleman M, Nelson E. 2009. Hyperspectral remote sensing analysis of short rotation woody crops grown with controlled nutrient and irrigation treatments. *Geocarto International* 24:293–312

Justice CO et al. 1998. The Moderate Resolution Imaging Spectroradiometer (MODIS): Land remote sensing for global change research. *IEEE Transactions on Geoscience and Remote Sensing* 36:1228–1249

Laliberte AS, Goforth MA, Steele CM, Rango A. 2011. Multispectral remote sensing from unmanned aircraft: Image processing workflows and applications for rangeland environments. *Remote Sensing* 3:2529–2551

Lucieer A, Malenovský Z, Veness T, Wallace L. 2014. HyperUAS—Imaging spectroscopy from a multirotor unmanned aircraft system. *Journal of Field Robotics* 31:571–590

Malenovský Z, Lucieer A, King DH, Turnbull JD, Robinson SA. 2017. Unmanned aircraft system advances health mapping of fragile polar vegetation. *Methods in Ecology and Evolution*

Martin ME, Plourde LC, Ollinger SV, Smith ML, McNeil BE. 2008. A generalizable method for remote sensing of canopy nitrogen across a wide range of forest ecosystems. *Remote Sensing of Environment* 112:3511–3519. doi:10.1016/j.rse.2008.04.008

Matese A et al. 2015. Intercomparison of UAV, aircraft and satellite remote sensing platforms for precision viticulture. *Remote Sensing* 7:2971–2990

Messinger M, Asner GP, Silman M. 2016. Rapid assessments of Amazon forest structure and biomass using small unmanned aerial systems. *Remote Sensing* 8. doi:10.3390/rs8080615

Mozgeris G, Gadal S, Jonikavičius D, Straigytė L, Ouerghemmi W, Juodkienė V. 2016. Hyperspectral and color-infrared imaging from ultralight aircraft: Potential to recognize tree species in urban environments. In: *2016 8th Workshop on Hyperspectral Image and Signal Processing: Evolution in Remote Sensing (WHISPERS)*. IEEE, pp. 1–5.

Numata I, Roberts DA, Chadwick OA, Schimel JP, Galvao LS, Soares JV. 2008. Evaluation of hyperspectral data for pasture estimate in the Brazilian Amazon using field and imaging spectrometers. *Remote Sensing of Environment* 112:1569–1583. doi:10.1016/j.rse.2007.08.014

Pádua L, Vanko J, Hruška J, Adão T, Sousa JJ, Peres E, Morais R. 2017. UAS, sensors, and data processing in agroforestry: A review towards practical applications. *International Journal of Remote Sensing* 38:2349–2391

Richter K, Atzberger C, Vuolo F, Weihs P, d'Urso G. 2009. Experimental assessment of the Sentinel-2 band setting for RTM-based LAI retrieval of sugar beet and maize. *Canadian Journal of Remote Sensing* 35:230–247

Salamí E, Barrado C, Pastor E. 2014. UAV flight experiments applied to the remote sensing of vegetated areas. *Remote Sensing* 6:11051–11081

Sibanda M, Mutanga O, Rouget M. 2015. Examining the potential of Sentinel-2 MSI spectral resolution in quantifying above ground biomass across different fertilizer treatments. *ISPRS Journal of Photogrammetry and Remote Sensing* 110:55–65

Skidmore AK et al. 2010. Forage quality of savannas—Simultaneously mapping foliar protein and polyphenols for trees and grass using hyperspectral imagery. *Remote Sensing of Environment* 114:64–72. doi:10.1016/j.rse.2009.08.010

Suomalainen J et al. 2014. A lightweight hyperspectral mapping system and photogrammetric processing chain for unmanned aerial vehicles. *Remote Sensing* 6:11013–11030

Verrelst J, Camps-Valls G, Muñoz-Marí J, Rivera JP, Veroustraete F, Clevers JG, Moreno J. 2015. Optical remote sensing and the retrieval of terrestrial vegetation bio-geophysical properties—A review. *ISPRS Journal of Photogrammetry and Remote Sensing* 108:273–290

Von Bueren S, Burkart A, Hueni A, Rascher U, Tuohy M, Yule I. 2015. Deploying four optical UAV-based sensors over grassland: Challenges and limitations. *Biogeosciences* 12:163

Wilson AM, Silander JA, Gelfand A, Glenn JH. 2011. Scaling up: Linking field data and remote sensing with a hierarchical model. *International Journal of Geographical Information Science* 25:509–521

Yue J, Yang G, Li C, Li Z, Wang Y, Feng H, Xu B. 2017. Estimation of winter wheat above-ground biomass using unmanned aerial vehicle-based snapshot hyperspectral sensor and crop height improved models. *Remote Sensing* 9:708

Zhang L, Sun X, Wu T, Zhang H. 2015. An analysis of shadow effects on spectral vegetation indexes using a ground-based imaging spectrometer. *IEEE Geoscience and Remote Sensing Letters* 12:2188–2192

Zhou LG, Roberts DA, Ma WC, Zhang H, Tang L. 2014. Estimation of higher chlorophyll a concentrations using field spectral measurement and HJ-1A hyperspectral satellite data in Dianshan Lake, China. *Isprs Journal of Photogrammetry and Remote Sensing* 88:41–47. doi:10.1016/j.isprsjprs.2013.11.016

13 Integrating Hyperspectral and LiDAR Data in the Study of Vegetation

Jessica J. Mitchell, Nancy F. Glenn, Kyla M. Dahlin, Nayani Ilangakoon, Hamid Dashti, and Megan C. Maloney

CONTENTS

13.1 INTRODUCTION

This chapter provides fundamental concepts of integrating hyperspectral and LiDAR data (Light Detection and Ranging) for studying terrestrial ecological processes, and vegetation in general, from satellite, airborne, and ground-based platforms. We describe data integration approaches and commonly used sensors. Following this introduction are a number of application areas, including plant diversity mapping and modeling, the characterization of sparse vegetation in dryland systems, and ground-based remote sensing techniques that support advances in validation, scaling, and hyperspectral/LiDAR data processing and integration.

Hyperspectral data provide narrowbands containing rich spectral information that can describe vegetation species, function, structure, and processes. Hyperspectral data are extensively used for vegetation type/species classification as well as for biophysical (leaf area index [LAI]), biochemical (carbon [C], nitrogen [N], chlorophyll content, and so on), and stress (plant water content) parameter estimation. Remote sensing parameters that describe vegetation are typically derived using statistical methods (multivariate statistics or machine learning algorithms) or physical models (radiative transfer models [RTMs]).

Complementary to hyperspectral data is the active remote sensing technique LiDAR which provides detailed 3-dimensional (3D) structure and some radiometric properties of vegetation using information from the backscatter of laser pulses (waveforms) that are sampled either discretely or near-continuously (full waveform LiDAR [FWF]). The sampled laser returns can be visually displayed in a 3D point cloud format.

Hyperspectral and LiDAR data fusion is the integration of the two datasets to form a rich, comprehensive dataset. Data fusion retains the significant information of the component datasets and is enriched with the additional dimensions of information. Data fusion can be performed at multiple stages of data processing and using a variety of approaches. However, it is typically essential to have an accurate geometric alignment of the hyperspectral and LiDAR datasets before fusion. Data fusion approaches that have been applied to hyperspectral and LiDAR data are widespread and include land cover land-use classification as well as biomass and carbon balance monitoring and modeling (e.g., Alonzo et al., 2014).

The potential to couple hyperspectral and LiDAR data allows scientists to use multiple dimensions of radiometric and structural information for retrieving vegetation parameters. In addition, vegetation structural parameters (e.g., height, biomass, LAI) and reflectance have been found to be statistically correlated depending on the biophysical environment, thus including 3D information about plant structure further enables hyperspectral sensing and vice versa.

13.1.1 Hyperspectral and LiDAR Data Integration for Vegetation Studies

The integration of different remote sensing sources to improve the utility of individual sensors or develop new products using multiple sensor measurements is not a new idea and is widely discussed in the remote sensing community. This multisensor data fusion is commonly referred to as a "technology to enable combining information from several sources to form a unified picture" (Schmitt and Zhu, 2016). The first attempts at data fusion in remote sensing were the fusion of different images (i.e., image fusion or pan-sharpening) to improve spatial resolution. Image fusion has been implemented with optical, LiDAR, radar, and other remote sensing sensors. More recent methods for combining hyperspectral and LiDAR data for estimating vegetation parameters can be divided into three broad categories (Figure 13.1): (1) multivariate statistical analysis, usually through high dimensional regression analysis, machine learning, segmentation, and other common statistical methods; (2) physical modeling using radiative transfer concepts; and (3) ecological or vegetation dynamic modeling using data assimilation. In all of these methods, feature extraction may be combined into a single feature space for use in different integration methods or LiDAR and optical features can be used separately in different steps of an integration algorithm. Before integration, one

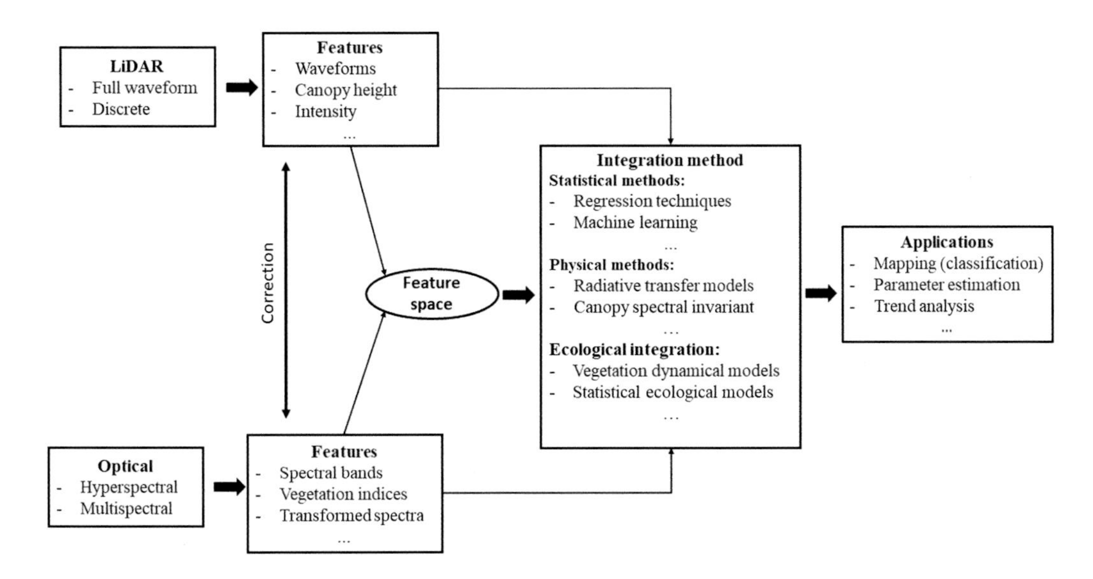

FIGURE 13.1 General scheme of LiDAR and optical sensor integration.

might also use LiDAR for the correction of optical data and vice versa. In the following sections, we discuss integration methods in more detail.

13.1.1.1 Multivariate Statistical Analysis

Statistical methods represent the most common approach to integrating LiDAR and hyperspectral data. Optical reflectance data (i.e., individual bands) or band ratios calculated from the optical data (e.g., narrowband indices) are combined with LiDAR-derived variables such as height, intensity, and cross section. A multivariate statistical analysis is then applied to the combined dataset and used to develop a relationship between these remotely sensed variables and vegetation properties sampled on the ground. Common statistical methods include ordinary least squares regression (Hudak et al., 2002), support vector machines (SVMs; Dalponte et al., 2008; Chen and Hay, 2011), random forests (Zald et al., 2016), and partial least square regression (Balzotti et al., 2016). While these models are relatively simple to develop and fast to implement, there are some challenges associated with them. The main drawback is difficulty in developing accurate generalized models across multiple collection sites and/or dates. For example, a statistical model developed for the estimation of LAI in a forest ecosystem may not produce strong results in another ecosystem or at another data acquisition time period. Another issue is that statistical methods are subject to misinterpretation as the relationships are an implicit representation of the optical-LiDAR-vegetation system. Despite these challenges, statistical methods are useful techniques that can enable analysts to provide rapid response to natural hazards such as wildfire (e.g., Mutlu et al., 2008) or quickly quantify resources such as surface water volume (e.g., Irwin et al., 2017).

Statistical data fusion of co-registered hyperspectral and LiDAR data can be implemented at the pixel, feature, or decision level. LiDAR-derived variables (e.g., height, intensity) are combined with hyperspectral data as individual reflectance bands, narrowband indices, or bands of reduced dimensionality using transformations such as principal component analysis (PCA) or minimum noise fraction (MNF). Using pixel level data fusion, both hyperspectral and LiDAR data are rasterized to have equal pixel sizes. Typically, the pixel size is defined based on the point density of the LiDAR, the native pixel size of the hyperspectral data, and the ability to achieve an accurate co-registration. To enhance the information by means of fusion, ideally the point density of the LiDAR data is dense enough such that rasterization can occur at the same or higher resolution than the hyperspectral data. LiDAR variables (e.g., height statistics) are often calculated by processing individual LiDAR point returns classified as either ground or vegetation at a resolution that is compatible with the nominal pixel resolution of the hyperspectral imagery. The rasterized LiDAR variables (bands) are then combined with hyperspectral data with varying degrees of image enhancement. The most common pixel level fusion methods are layer stacking LiDAR and hyperspectral bands (with or without the application of data reduction techniques such as PCA and MNF transformations). Mundt et al. (2006) used the layer stacking technique to improve shrub classifications where soil and shrubs are spectrally similar; LiDAR vegetation roughness provided an added level of information to the sub-pixel abundances determined from hyperspectral data. Luo et al. (2015) obtained an improved land cover classification accuracy using layer stacking of hyperspectral reflectance bands with LiDAR-derived images of digital terrain model (DTM), digital surface model (DSM), normalized digital surface model (nDSM), and intensity. In another study, classification accuracy of tea and palms increased from 83% to 91% using MNF and PCA fused hyperspectral and waveform LiDAR features compared to hyperspectral imagery without LiDAR (Chu et al., 2016).

LiDAR data can also be represented in volumetric pixels (rather than 2D pixels) referred to as voxels (Stoker, 2009). Using voxels that have a base area equal to the hyperspectral pixel size can facilitate data fusion (Buddenbaum et al., 2013; Miltiadou et al., 2015; Wang and Glennie, 2015). The use of 3D voxels preserves the additional dimension in the LiDAR data and can provide a voxel volume measurement for measures such as biomass, as well as be used with ray tracing to describe the 3D light environment of individual plants (e.g., Van der Zande et al., 2009).

At the feature level, LiDAR are treated as point clouds and aggregated by geometric boundaries in 2D or 3D space. One approach, object-based feature extraction, has the advantage of preserving the radiometric and structural features of the hyperspectral and LiDAR data, whereas pixel-based classifications are often coarser. With object-based feature extraction, the fusion step can be performed either before or after feature extraction. In the case of *fusing data before feature extraction*, LiDAR point clouds are colored using information from hyperspectral and/or LiDAR variables, then feature extraction operations are performed. This approach was used by Gerke and Xiao (2014), where voxels were assigned textural and other variable information from color-infrared images. A segmentation approach was then applied to the voxels, which were subsequently classified into feature classes of interest. More recently, "super-voxels," which are essentially a low-level grouping or clustering of voxels, have been used in object-oriented segmentation. Super-voxels are used to decrease redundancy in information in high density point clouds and thus increase computational efficiency (Ramiya et al., 2016). In the case of *fusing data after feature extraction*, rasterized LiDAR variables (e.g., intensity, DSM) are used to extract or delineate in an automated fashion the boundaries of real-world objects in a scene of interest (e.g., trees, buildings, roads). The point cloud data contained within the boundaries of extracted features are then colored by hyperspectral and/or LiDAR images to extract features information (Zhang and Qiu, 2012).

At the decision level, hyperspectral and LiDAR datasets are processed independently and the final results are combined for decision-making using their respective spatial references. For example, Sarrazin et al. (2010) derived waveform LiDAR and hyperspectral data matrices independently and applied a stepwise discriminant analysis (SDA) approach for species-level structural assessment in a savanna ecosystem. Swatantran et al. (2011) used FWF LiDAR metrics and sub-pixel estimates of vegetation from hyperspectral imagery in a linear regression model to estimate biomass in a forested ecosystem. In a similar approach, Dalponte et al. (2008) used LiDAR and hyperspectral metrics with SVMs to classify a forested ecosystem.

13.1.1.2 Physical Modeling Using Radiative Transfer Concepts

Physical modeling methods estimate vegetation parameters using the concept of a waveform or pulse that propagates through the canopy. In this approach, the absorption and scattering of electromagnetic waves and different components of vegetation are simulated using radiative transfer models (RTMs). Optical RTMs includes models that have been developed at both leaf (e.g., PROSPECT; Jacquemoud and Baret, 1990) and canopy (e.g., SAIL; Verhoef and Bach, 2007) scales. LiDAR recorded in full waveform resolution can improve the robustness of RTM inversion estimates and reduce model uncertainty by providing canopy structural information (Koetz et al., 2007). Geometric Optical-Radiative Transfer (GORT) and Discrete Anisotropic Radiative Transfer (DART) are two commonly used LiDAR RTMs (Ni-Meister et al., 2001; Gastellu-Etchegorry et al., 2015) The DART model has been widely used for integration of LiDAR and multispectral data (Zhou et al., 2010), but, to our knowledge, DART has not yet been implemented with hyperspectral data. In the forward mode, RTMs use input parameters such as sensor-target geometry and vegetation characteristics to simulate canopy reflectance spectra or full waveform LiDAR returns. Thus, in order to obtain vegetation characteristics, an inverse problem should be solved. It has been shown that the integration of LiDAR and optical RTMs can significantly improve the estimation of vegetation canopy characteristics (Koetz et al., 2007; Bye et al., 2017). The main challenges associated with physically based models are the process of inversion and the many input parameters that need to be provided in order to run the model. The inverse problem is usually a time-consuming process to solve and solutions are not always unique (i.e., ill-posed problem). Moreover, it is not always possible to provide all of the parameters required by the model, and parameters left as free exacerbate the ill-posed problem. In this situation, yet another source of uncertainty is introduced when impacts from soil, understory, and atmosphere on simulated waveforms are removed.

13.1.1.3 Ecological Modeling Using Data Assimilation

In general, the main application of remote sensing of vegetation is in ecological studies. Thus, an ecological model provides a meaningful framework for the integration of hyperspectral and LiDAR data. Integrating vegetation dynamic models (VDMs) with remotely sensed data adds value to the data and is commonly known as data assimilation (Liang, 2005). In this approach, remotely sensed data are used to initialize or tune VDM parameters by imposing additional constraints (Figure 13.2). For example, vegetation height information is extracted from terrestrial laser scanning (TLS) and LiDAR data to initialize a VDM,

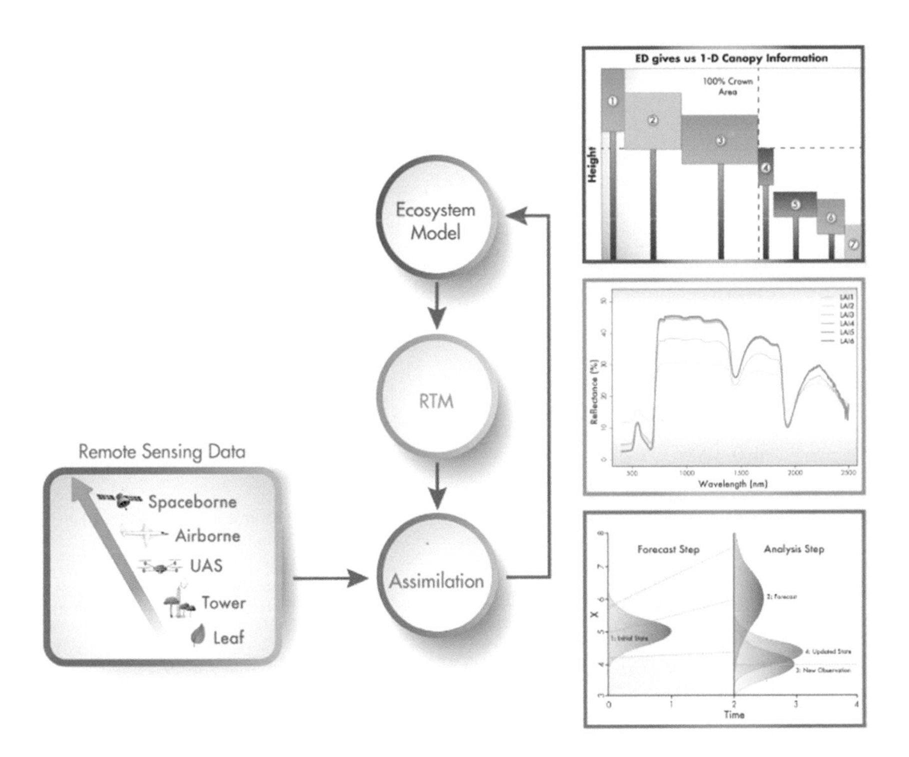

FIGURE 13.2 Remote sensing data from multiple platforms can be integrated to tune vegetation dynamic model (VDM) parameters by imposing additional constraints. (From Brookhaven National Lab, https://www.bnl.gov/envsci/testgroup/nasa-te.php.)

and then due to the cheaper accessibility of optical data, the data are used to tune the model during the simulation period. In other words, the integration happens inside the model through a mechanistic approach. There are different VDMs and integration frameworks used for ecological studies (Smolander and Stenberg, 2005; Detto et al., 2015; Fisher et al., 2015; Stark et al., 2015; Widlowski et al., 2015).

13.1.2 Sensors

Scientists have a suite of sensors available for hyperspectral and LiDAR data integration (Table 13.1). Primarily plant spectral and structural information obtained from these sensors are paired, but a temporal pairing can also be equally advantageous. Available and upcoming sensors can be deployed

TABLE 13.1
Example Operational and Future Hyperspectral and LiDAR Sensors Suitable for Integration

Sensor	Optical Range (nm)	Number of Spectral Channels	Spatial Resolution (m)	References
Hyperspectral				
AVIRIS	400–2500	224	Variable	Alonzo et al. (2014)
AVIRIS-NG	280–2510	432	Variable	https://aviris-ng.jpl.nasa.gov/
HyMAP	450–2500	126	3.5–10	Molan et al. (2014)
Hyperion	400–2500	220	30	Petropoulos et al. (2012)
DESIS[a]	400–1000	240	Variable	http://www.dlr.de/
EnMAP[a]	420–2450.	230	30	Guanter et al. (2015)
HyspIRI[a]	400–2500	213	30	Lee et al. (2015), Mitchell et al. (2015)
HISUI[a]	400–2500	185	20 × 30	http://www.jspacesystems.or.jp/en_/
LiDAR				
LVIS	1064 (waveform)	1	10–25	Blair et al. (1999)
ICESat-GLAS	532 and 1064 (waveform)	2	70	Wang et al. (2011)
ICESat-2 ATLAS[a]	532 (photon-counting laser)	1	~17	Abdalati et al. (2010), Herzfeld et al. (2014)
GEDI[a]	1064 (waveform)	1	~25	Qi and Dubayah (2016), Stavros et al. (2017)
Multisensor[b]				
G-LiHT	Hyperspectral: 400–1000	409	Variable	Cook et al. (2013)
	LiDAR: 1550 and 905	2		
CAO-2 AToMS	Hyperspectral: 380–2510	427	Variable	Asner et al. (2015)
	LiDAR: 1064	1		
NEON-AOP	Hyperspectral:380–2510	428	Variable	Kampe et al. (2010)
	LiDAR: 1064	1		
NASA-JPL-ASO	Hyperspectral:370–1050	144 or 288	Variable	Painter et al. (2016)
	LiDAR:1064	1		

[a] Future launch.

[b] In addition to hyperspectral and LiDAR, multisensor systems may also consist of irradiance spectrometers, thermal and RGB cameras, and so on.

via ground-based (e.g., terrestrial laser scanning, field spectroscopy), unmanned aerial, airborne, and satellite platforms.

Much of the previous work on data integration has occurred using airborne hyperspectral imagery (e.g., AVIRIS, HyMAP) with small-footprint (e.g., commercial narrow beam, discrete return) and large-footprint (e.g., NASA's Land, Vegetation, and Ice Sensor, [LVIS]) LiDAR (e.g., Mundt et al., 2006; Swatantran et al., 2011; Latifi et al., 2012). The 30-m spatial resolution satellite-based Hyperion hyperspectral imagery has also been combined with airborne LiDAR (Ghosh et al., 2014); however, the instrument is onboard the Earth Observing-1 (EO-1) satellite mission, which was launched in November 2000 and decommissioned in March 2017.

Notably, there are several airborne platforms available to the scientific community in which hyperspectral and LiDAR are collected co-incidentally. These include, but are not limited to, NASA Goddard's LiDAR, Hyperspectral and Thermal (G-LiHT) Airborne Imager (Cook et al., 2013), the Carnegie Airborne Observatory (CAO; Asner et al., 2007; and CAO-2; Asner et al., 2012), the National Ecological Observatory Network's (NEON) Airborne Observation Platform (Kampe et al., 2010), and NASA Jet Propulsion Laboratory's (JPL) Airborne Snow Observatory (ASO; Painter et al., 2016). The type of imaging spectrometer and LiDAR instruments that are flown on these systems vary as do their applicability to ecological mapping capabilities.

In the coming decade, an increase in both hyperspectral and LiDAR spaceborne missions will enable new data fusion opportunities in ecology. For example, the upcoming Ice, Cloud, and Land Elevation Satellite-2 Advanced Topographic Laser Altimeter System (ICESat-2 ATLAS) extends the previous ICESat Geoscience Laser Altimeter System (GLAS) mission with a photon-counting LiDAR. The Global Ecosystem Dynamics Investigation LiDAR (GEDI), a large-footprint full waveform system on the International Space Station (ISS), is designed to provide new 3D structural measures of vegetation. Additional ISS technologies that are planned or operational and can be coupled with structural information to assess plant health include a commercial hyperspectral sensor in the visible-near-infrared (DESIS, DLR Earth Sensing Imaging Spectrometer), a hyperspectral sensor in the visible, near-infrared, and shortwave infrared (HISUI, Hyperspectral Imager Suite) and the thermal radiometer, ECOsystem Spaceborne Thermal Radiometer Experiment on Space Station (ECOSTRESS).

A German hyperspectral satellite, the Environmental Mapping and Analysis Programme (EnMAP) and an Italian hyperspectral satellite, the PRecursore IperSpettrale della Missione Applicativa (PRISMA; Labate et al., 2009) are also planned. EnMAP will provide 30 m resolution data with a 4-day revisit opportunity with tasking (Guanter et al., 2015). Similarly, NASA's Hyperspectral Infrared Imager (HyspIRI) mission study is ongoing with intent for a future launch to provide global imaging spectrometer data (Lee et al., 2015).

Emerging technologies with unmanned aerial systems (UAS) platforms, including lighter and higher-calibrated sensors, are now enabling hyperspectral and point cloud data fusion. Commercial sensors such as the Resonon PIKA II (Hruska et al., 2012) and Headwall's Micro-Hyperspec (Zarco-Tejada et al., 2012) are used for species discrimination and plant health analysis. Vegetation structural measures have been quantified using LiDAR sensors for UAS platforms including the IBEO Lux LiDAR (Wallace et al., 2014) and Riegl's VUX-1UAV (Brede et al., 2017). Studies such as Aasen et al. (2015) are using lightweight hyperspectral cameras along with structure from motion (SfM) to derive 3D hyperspectral information for ecology. Similarly, ground-based systems such as hyperspectral LiDAR have been developed (Hakala et al., 2012).

13.2 STRUCTURAL AND FUNCTIONAL DIVERSITY MAPPING AND MODELING

Documenting and understanding patterns of biodiversity is a central issue in macroecology and, to date, continental to global scale plant diversity mapping has been based on occurrence data, biased by uneven sampling and focused on species richness (Barthlott et al., 2007). Vegetation diversity or phytodiversity occurs along a range of temporal, biological, and spatial scales, from genes to species, ecosystems, and global biodiversity. Patches or corridors of biodiversity across a local landscape can reflect areas where fine-scale patterns of diversity are mechanistically related to broad-scale dynamics as represented by environmental variables. With the integration of hyperspectral and LiDAR data, there is new potential for mapping finer-scale diversity variables such as plant structural and functional diversity more frequently and across larger geographic extents.

13.2.1 What Is Plant Structural and Functional Diversity?

Plant functional traits and plant structure are critical to our understanding of ecosystem processes from local to global scales. Plant functional traits are measurable aspects of plant physiology that influence their growth and overall success (Figure 13.3). Structure here refers to the physical arrangement of the vegetation—this can be a representation of plant functional types (e.g., grassland versus shrubs versus forest) but it can also refer to the structural variation within a given functional type, like the branch architecture or gap distribution of a forest. Many aspects of plant structural and functional diversity can be mapped with remote sensing (Tables 13.2 and 13.3).

Much of the work on plant functional traits has focused on the idea of a leaf economic spectrum—that tradeoffs in life history strategies lead to a limited set of combinations of traits. Díaz et al. (2016) used principle components to show that most of the plants in the TRY database (Kattge et al., 2011) fell along just a few axes of variation, mostly relating to growth strategy; some plants grow fast and inexpensively, while others grow more slowly and with more expensive tissues. Some, though certainly not all, plant functional traits can be mapped using imaging spectroscopy (Asner and Martin, 2009; Singh et al., 2015). Leaf chemical constituents (N, C, phosphorus [P], chlorophyll, and so on), and some leaf structural properties (leaf mass per area [LMA], leaf water content [LWC]) are now relatively trivial to map with airborne imaging spectroscopy and statistical techniques like

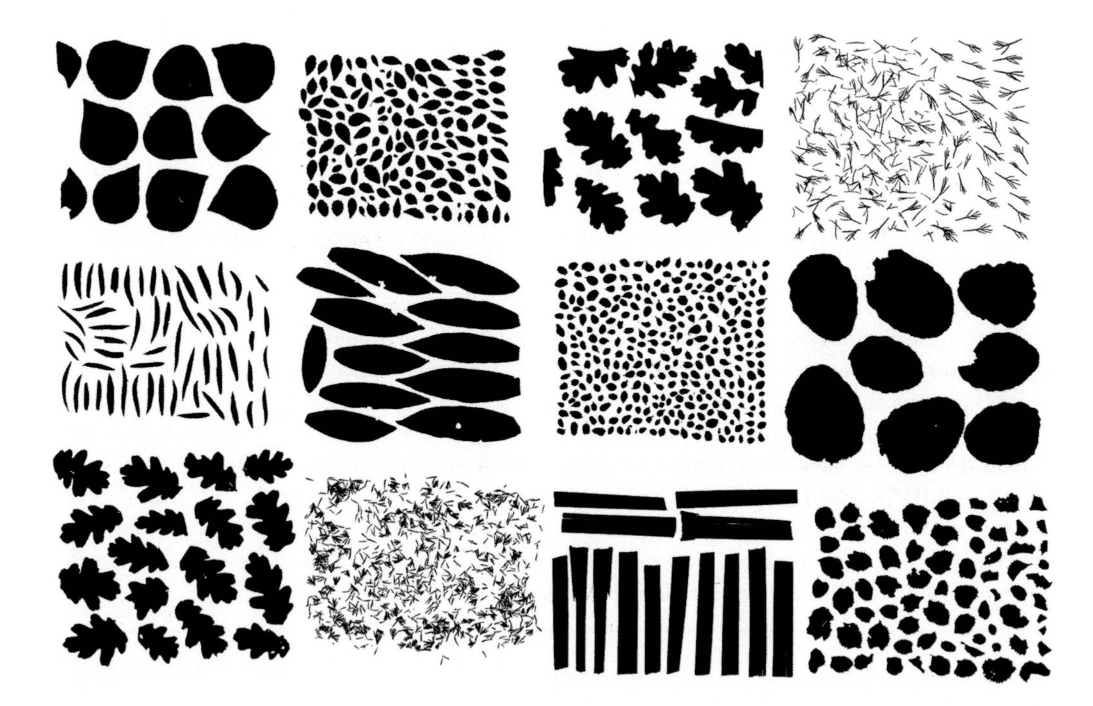

FIGURE 13.3 Scans of leaves collected from Jasper Ridge Biological Preserve (California, USA) illustrating the variety of different shapes and sizes of leaves. LMA is one of many leaf traits that correlates with plant function—thinner, cheaper leaves tend to grow on faster growing, early successional species, while thicker, more expensive leaves tend to grow on slower growing, longer lived species. (From Kyla Dahlin.)

TABLE 13.2
Plant Traits That Can Be Mapped Using Remote Sensing, the Types of Observations That Are Often Used, and an Example of Each One

Plant trait(s)	Observation Type	Examples
Leaf chemistry & morphology	Hyperspectral	Singh et al. (2015)
	Hyperspectral + LiDAR	Asner et al. (2017)
Species identification	Hyperspectral	Roth et al. (2012)
	Hyperspectral + LiDAR	Roth et al. (2012)
	Multi-wavelength ground-based LiDAR	Budei et al. (2018)
Canopy structure	Airborne LiDAR	Stark et al. (2015)
	Ground-based LiDAR	Hardiman et al. (2014)
	Hyperspectral + LiDAR	Mitchell et al. (2015)
Aboveground biomass	LiDAR	Sherrill et al. (2008)
	Hyperspectral + LiDAR	Dahlin et al. (2012)

TABLE 13.3
Example Studies in Which a Comparison of the Accuracies Was Made between Using Hyperspectral, LiDAR, and Both Hyperspectral and LiDAR for Leaf and Plant Trait Remote Sensing

Leaf/ Plant Trait(s)	Example	Hyperspectral only (R^2 or Percent Accuracy[a])	Lidar only (R^2 or Percent Accuracy[a])	Hyperspectral and Lidar (R^2 or Percent Accuracy[a])	Literature
Leaf and canopy chemistry and morphology	Canopy chlorophyll	0.78–0.79	Not evaluated	0.84	Thomas et al. (2008)
	LMA, N, P, C, water, Lignin, Phenols	Not evaluated separately		✓	Asner et al. (2017)
Species identification	15 urban tree species	51.1[a]	61[a]	70[a]	Liu et al. (2017)
	Forest stands	55.9[a]	61.7[a]	72.2[a]	Buddenbaum et al. (2013)
Canopy structure	Canopy cover	Not evaluated	0.49	0.58	Mitchell et al. (2015)
	Clumping index	0.86	0.81	0.90	Thomas et al. (2008)
Biomass	Aboveground biomass—African tropical forest	0.36	0.64	0.70	Laurin et al. (2014)
	Aboveground biomass—Sierra National Forest, USA	0.60	0.77	0.80	Swatantran et al. (2016)

[a] Percent classification accuracy

partial least squares regression (PLSR). There are many important plant functional traits we cannot map (e.g., wood density, bark thickness, fruit characteristics), but many of these likely correlate with those we can map, at least within biomes. Recent work has also begun to consider whether hyperspectral data by itself can be used to look at functional trait diversity even in the absence of ground data (Rocchini et al., 2008; Dahlin, 2016).

Vegetation structure is readily mapped with LiDAR allowing for a 3D view of the landscape. In a heterogeneous environment, LiDAR permits the separation of different plant functional types (Dahlin et al., 2013), which is by itself a measure of functional diversity. Within a given functional type, like forests, LiDAR information allows us to understand something about the structure of the forest, and therefore the light environment (Parker 1995), which can be a significant contributor to productivity (Hardiman et al., 2014). Since vertically structured ecosystems also have vertically structured nutrient distributions (Niinemets et al., 2015) knowing something about this vertical structure can help inform studies of the distribution of nutrients through plant canopies.

Studies combining hyperspectral imaging and LiDAR for diversity mapping and modeling are only beginning to emerge, largely due to a lack of available data. However, as more combined hyperspectral data and LiDAR systems come online, we expect more work in this area to emerge. In the next section, we describe the types of models these data can and do inform, some current uses of hyperspectral data and LiDAR in isolation, and the potential for future analyses capitalizing on the complementarity of these two data types for plant structural and functional diversity mapping.

13.2.2 Why Do We Want to Map and Model Structural and Functional Diversity?

13.2.2.1 Biodiversity Monitoring and Prediction

An international focus has recently been placed on biodiversity monitoring and predicting future biodiversity change (Jetz et al., 2016). To that end, organizations like the Group on Earth Observation Biodiversity Observation Network (GEO BON) have begun to develop "Essential Biodiversity Variables" (EBVs) which are suggested measurements that can capture, directly or indirectly, the current status and change over time of global biodiversity (Figure 13.4; Pereira et al., 2013). Since mapping individual species is not possible with spatially coarse data or over large spatial extents, and, even if plant species could be mapped, much of the focus of biodiversity research is on other taxonomic groups, many of the suggested EBVs capture plant structural and functional diversity, with the idea that these EBVs should correlate with other forms of biodiversity.

As of 2018, candidate EBVs that can be remotely sensed include phenology, physiological traits, primary productivity, disturbance regime, habitat structure, ecosystem extent and fragmentation,

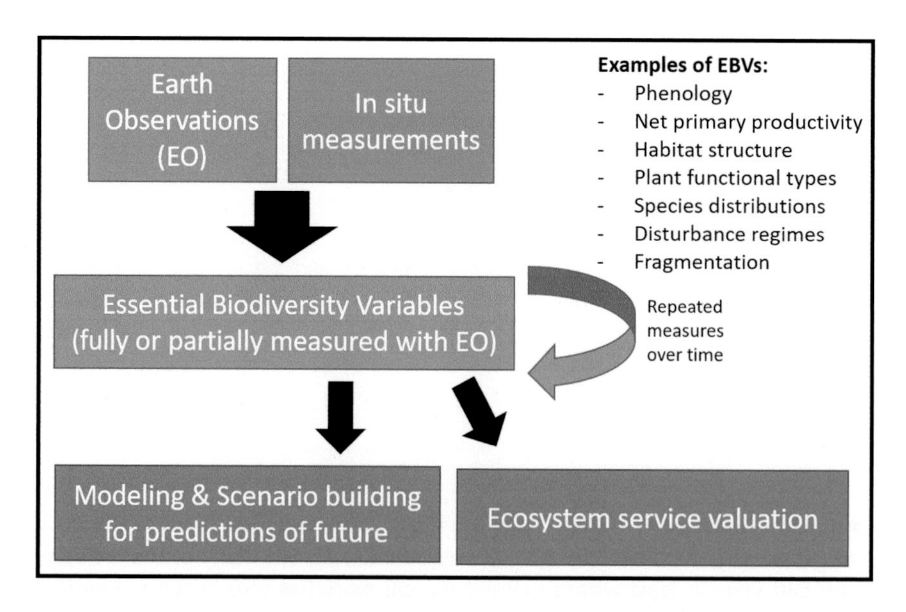

FIGURE 13.4 An illustration of the link between remote sensing data, Essential Biodiversity Variables (EBVs), indicators, and future scenarios and calculations of ecosystem services. (From Kyla Dahlin.)

and ecosystem composition by functional type. The aim of GEO BON is to look at continental to global scales, meaning there are not currently hyperspectral or LiDAR datasets that can contribute to this effort. However, future hyperspectral and LiDAR missions like the European Space Agency's Environmental Mapping and Analysis Program (EnMAP, scheduled to launch in 2018-2020), Japan Space System's Hyperspectral Imager Suite (HISUI, scheduled to launch in 2020), NASA's Global Ecosystem Dynamics Investigation (GEDI LiDAR, scheduled to be installed on the International Space Station end of 2018), and ICESat-2 (scheduled to launch September 2018), and Hyperspectral Infrared Imager (HyspIRI, in study stage as of 2018) have the potential to significantly contribute to EBV mapping efforts. In combination and over long periods, these structural and functional EBVs will reveal different aspects of biodiversity and the distribution and pace of biodiversity change over time.

13.2.2.2 Ecosystem and Land Surface Modeling

At the local scale, there are two commonly used types of models that incorporate plant structural and functional diversity and can be informed by hypserspectral/LiDAR fusion. Forest gap models are models that essentially track every tree or small patch in a landscape from germination through canopy emergence and senescence (Bugmann, 2001). These models are often parameterized using field inventory data (stem counts and measurements) and tested against other field data. Though these models often assume no intraspecific heterogeneity in traits, they can be trained with hyperspectral data and LiDAR data to include trait variation, size distributions, and potentially demographics (Shugart et al., 2015). A recent advance in this genre of models is the Ecosystem Demography (ED2) model (Moorcroft et al., 2001; Medvigy et al., 2009) which uses partial differential equations to represent populations of trees across a landscape. Incorporating traits (Fisher et al., 2015), hyperspectral and LiDAR (Antonarakis et al., 2014) data, into ED2 is an active area of research. Canopy radiative transfer models like the Forest Light Environment Simulator (FLiES; Kobayashi and Iwabuchi, 2008) also have the potential to incorporate LiDAR and hyperspectral trait data.

Biogeochemical or ecosystem models abstract the components of ecosystems into pools and fluxes of matter and energy that move through an integrated system. Models like CENTURY (Parton et al., 1987) include pools of leaves and their chemical constituents (e.g., C, N, P), which move from tree leaves to branches to the forest floor and into the soil, and so on, hyperspectral data, and LiDAR could be used to inform these models by refining estimates of leaf nutrients and forest biomass that would typically be scaled up from field plots.

In contrast to local scale gap and ecosystem models, land surface models (LSMs) attempt to represent the land surface at the global scale, typically with very coarse spatial resolution (Lawrence et al., 2011). These models are built to couple to atmosphere, ocean, and ice models to form Earth System Models (ESMs) which simulate the entire global environment (Hurrell et al., 2013). While these models are very coarse and simple in XY space (typically about $1° \times 1°$) with fractional cover of simple plant functional types like "temperate broadleaf deciduous trees," the importance and inclusion of canopy vertical complexity has been recognized for decades (Sellers et al., 1996). The Community Land Model (CLM; Lawrence et al., 2011), for example, includes a two-layer canopy with different leaf traits in the upper and lower canopies. Hyperspectral and LiDAR data have not yet been incorporated into LSMs but recent efforts to use hyperspectral data to model plant physiological traits like Vcmax (Serbin et al., 2015) suggest that this will be a fruitful avenue of research as more hyperspectral data become available.

13.2.3 How Are Hyperspectral Data and LiDAR Currently Being Used in Diversity Mapping?

13.2.3.1 Functional Diversity Mapping

Mapping plant functional diversity is an increasingly important part of ecological monitoring programs. Jetz et al. (2016) showed that despite the large amount of field data in databases like

TRY (Kattge et al., 2011), these still only represent ~2% of total plant diversity, with the largest discrepancies between plant diversity and trait information in the tropics. The difficulty of collecting these data in the field has led to an emphasis on the need for more hyperspectral data (from satellites and other platforms) to fill in these gaps. Jetz et al. (2016) also emphasized that the combination of LiDAR data with hyperspectral could yield even more information, as the structural diversity data can help differentiate between different plant functional types. One avenue for biodiversity monitoring that is currently being explored is the use of "spectral diversity" (Rocchini et al. 2008) as a proxy for plant species or functional diversity, based on the assumption that spectral variation should correlate to plant trait variation across a landscape.

The Spectral Variation Hypothesis (SVH) states that there should be some relationship between spectral heterogeneity and biodiversity. This concept was introduced and tested by Palmer et al. (2002), but using a panchromatic image and small prairie plots in Oklahoma, USA, they did not find strong correlations. Nevertheless, Palmer et al. (2002) state that they expect this concept to prove more fruitful with multispectral data and larger spatial scales, and since then it has been tested or used several times.

Rocchini et al. (2008) showed correlations between spectral diversity using Landsat ETM+ imagery and field quadrats. Working across all of Switzerland using plots from the Swiss Biodiversity Monitoring Programme, they calculated species rarefaction curves for the plots. To calculate spectral diversity, they then created spectral "species" by using PCA. The first Principal Component (PC) of the Landsat data divided into 256 equal bins as "species." They found that the estimated spectral and species richness were significantly correlated once a good number of quadrats and areas was considered in the analysis, confirming the importance of spatial scale in considering the relationship between biodiversity and spectral diversity. A review by Rocchini et al. (2010) described the most common ways of estimating spectral diversity from multispectral data as looking at measures of dispersion in NDVI, each band, or one principal component, or using a method that retains the multispectral information like calculating the distance from a spectral centroid in n-dimensional space.

Dahlin (2016) used AVIRIS data collected over the Kellogg Biological Station (Michigan, USA) and compared spectral diversity between plots set in closed canopy forest patches and green agricultural fields. To estimate spectral diversity, Dahlin first isolated vegetation-only pixels using thresholds for NDVI and NDWI. She then used PCA to reduce the dimensionality of the hyperspectral data. She calculated spectral diversity as both the sum of the variance in the first three PCs of the AVIRIS data as the convex hull volume (CHV) formed by those three PCs. She then developed true and predicted spectral diversity area relationships for the different sets of plots and found that the forest plots were closer to random, with high alpha spectral diversity but low beta diversity, while the agricultural plots had lower alpha diversity but higher beta diversity (Figure 13.5). One key point of this research is that it was essential to have closed canopy vegetation to make diversity predictions, otherwise soil or water reflectance could have altered the results. While not used in these studies, LiDAR could have been incorporated to eliminate areas with little to no vegetation, or to provide another axis of diversity (e.g., vegetation height) in the diversity calculations.

Another approach to understanding biodiversity and its relationship to ecosystem function involves the use of experimental plots. The BioDIV experiment at the Cedar Creek Ecosystem Science Reserve in Minnesota, USA, is one such experiment. There are 168 plots of planted prairie grasses and forbs that have been maintained since 1994, each containing a mix of species from 1 to 16 per plot (Tilman et al., 1996). Recent work has used the BioDIV experiment plots to test spectral diversity relationships using a combination of handheld spectroradiometers, a spectroradiometer mounted on a tram, and airborne hyperspectral (400–970 nm) imagery (Wang et al., 2017). The researchers found that there was a relationship between spectral diversity and alpha biodiversity, but that the resolution necessary to detect this diversity properly was much finer than current or near-future satellites could collect.

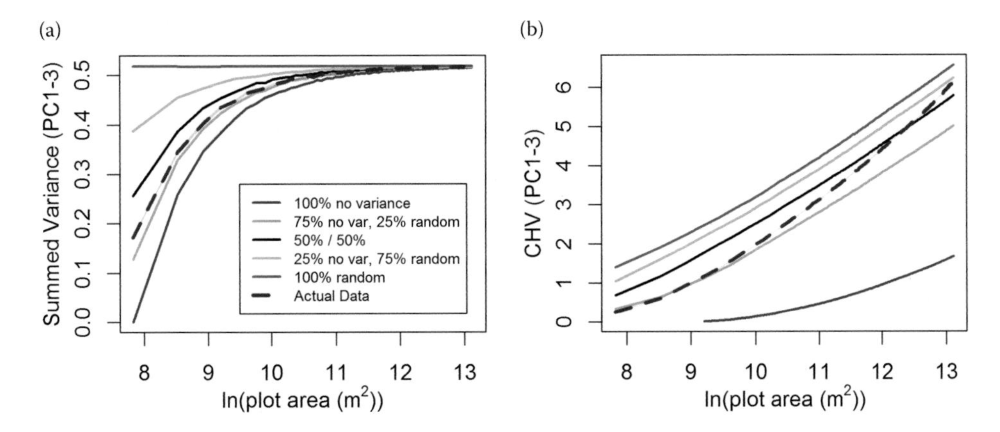

FIGURE 13.5 Spectral diversity area relationships (SDAR) using AVIRIS hyperspectral imagery from southwestern Michigan. Two metrics of spectral diversity are calculated—the summed variance of the first three principal components of hyperspectral data and the convex hull volume (CHV). Real data is compared to synthetic combinations of spectra. (From Dahlin K.M., 2016. *Ecological Applications*, 26, 2756–2766.)

The science of linking spectral diversity to biodiversity is still in its infancy; however, as more hyperspectral and LiDAR data become available there will undoubtedly be more explorations into this realm.

13.2.3.2 Physiology Mapping

In addition to mapping plant chemical constituents with good success (Asner and Martin, 2009; Serbin et al., 2014), recent work has shown that hyperspectral imaging has the potential to map physiological parameters like leaf age (Chavana-Bryant et al., 2017), photosynthetic capacity (Serbin et al., 2015), and growth rates (Caughlin et al., 2016).

While physiological traits are often treated as fixed for an individual or even for a species, Chavana-Bryant et al. (2017) demonstrated both how leaf traits like water content, N, and P vary over the lifespan of a leaf. They also showed that these variations are detectable with spectroscopy using a handheld spectrometer on individual leaves. At remotely sensed scales of meters or more, the issue of mixed pixels will make identifying leaf ages more challenging; however, these results suggest that leaf age could be an important confounding factor in linking ground observations to satellites.

While leaf traits such as N and LMA are often identified as important for understanding plant physiology, other scientists have worked to use hyperspectral data to map important physiological variables directly to inform photosynthesis models. Serbin et al. (2015) focused on two important physiological variables—Vcmax and Ev. Vc is the rate that rubisco works in chloroplasts, which is the rate limiting step in photosynthesis. Vcmax is the maximum rate for a given leaf, which is temperature dependent. Vc is usually estimated at a reference temperature, then activation energy is also quantified (Ev) allowing Vc measurements to be extrapolated to a maximum value (Vcmax). Working in irrigated monospecific agricultural fields in California, USA, Serbin et al. (2015) showed that Vc30 and Ev could be estimated from AVIRIS data even at relatively coarse spatial resolutions (18 m) and without canopy closure. This avenue of research opens the door to direct estimates of plant physiology which could be linked to plant structure using LiDAR to estimate total plant photosynthesis.

Other researchers have stretched these methods even further, associating hyperspectral data with growth rates directly. Caughlin et al. (2016) showed that in tropical tree monoculture plantings, growth rates (changes in diameter at breast height) could be estimated from a single hyperspectral image, along with LiDAR-derived elevation, with reasonable accuracy. This result suggests that hyperspectral data could be used to monitor plantation growth in the future, and, if extended to natural ecosystems, could provide more direct information about plant growth over time.

The 2012–2016 drought in California provided an excellent testbed for mapping drought stress using hyperspectral imagery and LiDAR. Several recent studies have focused on the impacts of drought on giant sequoias (*Sequoiadendron giganteum*) in the southern Sierra Nevada of California, using the Carnegie Airborne Observatory. One study used hyperspectral imagery to identify sequoia stands and canopy dieback in Sequoia National Park (Paz-Kagan et al., 2017a). They then used a digital elevation model derived from LiDAR to map different terrain measures and features, showing that a higher probability of mortality was found in areas where reduced access to water would be expected. Another study (Paz-Kagan et al., 2017b) used hyperspectrally-derived canopy water content (CWC) to assess plant stress over 2 years of drought, showing that trees declined in CWC over the 2 years in areas related, again, to topographic variation.

13.2.3.3 Landscape Controls on Functional Diversity

One of the key ecological questions that hyperspectral and LiDAR remote sensing have the potential to help address is how much control environmental gradients (soil, solar radiation, hydrology, and so on) exert over plant communities. In 2013, Dahlin et al. used hyperspectral-derived plant functional traits (N, C, and LWC) to assess the importance of environmental gradients on the distribution of plant traits within and across plant communities (Figure 13.6). LiDAR was incorporated in this study first to mask out shaded pixels from the analysis (Figure 13.7), then the LiDAR digital elevation model was used to calculate gradients like slope, aspect, and incident solar radiation. They found that while environmental gradients could explain some of the variation in these traits, much of that variation

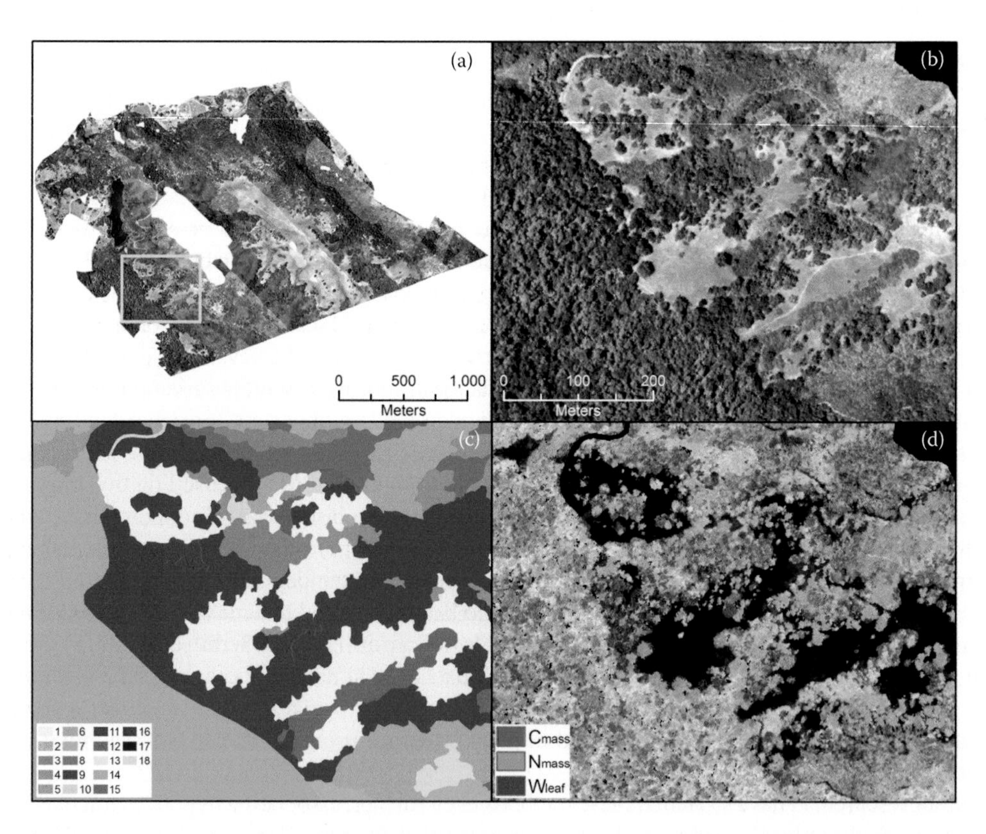

FIGURE 13.6 Trait mapping at Jasper Ridge Biological Preserve (JRBP) using the Carnegie Airborne Observatory. (a) True color image of JRBP; (b) true color image of focal area (orange box in a); (c) vegetation map based on field classification; (d) RGB image where red is foliar carbon, green is foliar nitrogen, and blue is LWC. (From K.M. Dahlin et al. 2013. *Proceedings of the National Academy of Sciences of the United States of America*, 110, 6895–900.)

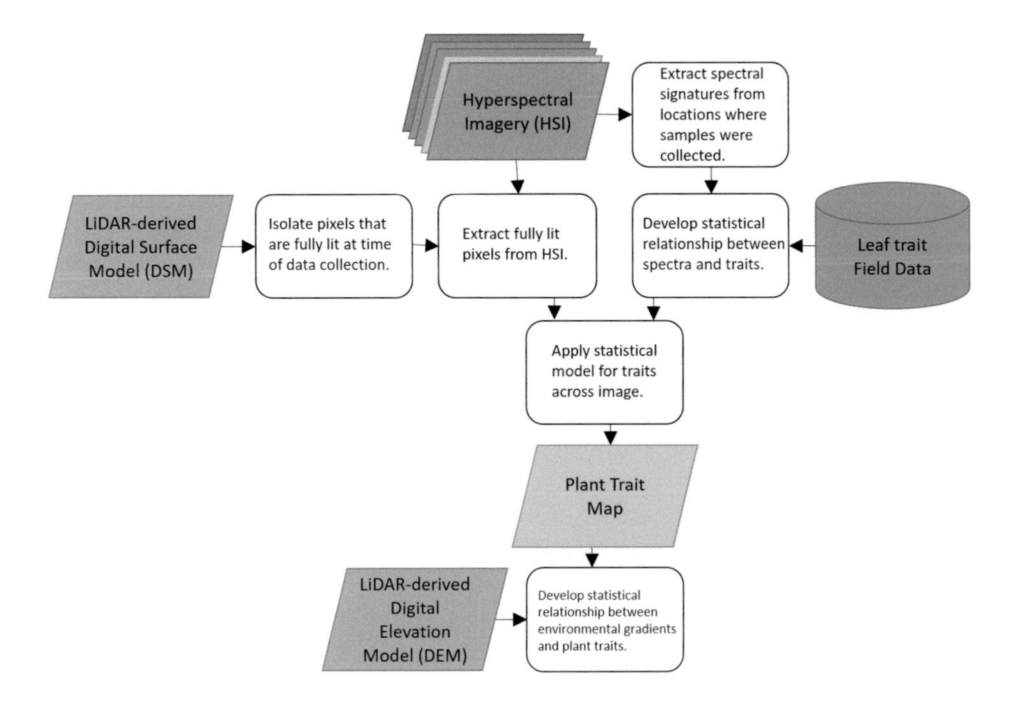

FIGURE 13.7 Example workflow for combining hyperspectral and LiDAR imagery to map plant traits and understanding plant trait x environment relationships.

could also be explained using a plant community classification map. This result suggested that changes in plant community (driven by environment as well as other factors like land-use history) were more important to determining the trait patterns across the landscape than the environmental gradients alone.

Across the Peruvian Amazon, Asner et al. (2017) showed that the forest functional groups they mapped with hyperspectral imagery could be related to known geomorphological and topo-edaphic features across the landscape, showing how at the regional scale, variations in environmental gradients play an important role in determining the trait composition of the Amazon. Across the Upper Midwest and Mid-Atlantic states, Singh et al. (2015) showed that foliar N and LMA correlated with latitude and N deposition was a significant predictor of leaf N.

13.2.3.4 Structural Heterogeneity and the Carbon Cycle

The high uncertainty in estimates of the terrestrial carbon sink in land surface models (Friedlingstein et al., 2014) is frequently attributed to the lack of functional diversity represented in these models. However, a lack of structural diversity may also be to blame. While land surface models typically use a "big leaf" or possibly a two-layer model of forest canopies, in reality, a forest canopy is a cluster of microenvironments with different amounts of light, temperature, and humidity regimes (Parker, 1995). LiDAR remote sensing has allowed researchers to begin to quantify this variation and show its impact on the forest carbon budget. Using a portable canopy LiDAR (ground-based) system, Hardiman et al. (2014) showed that in even-aged northern hardwood forests, LiDAR-measured canopy structural complexity was a strong predictor of wood net primary productivity (Figure 13.8). Canopy structural complexity may be one reason why older forests remain more productive than would otherwise be expected (Gough et al., 2016).

13.2.3.5 Forest Structure in Demographic Models

Currently, a major focus in global ecosystem modeling is incorporating size and age structure, along with functional diversity, into ecosystem models that can be coupled to atmospheric and ocean

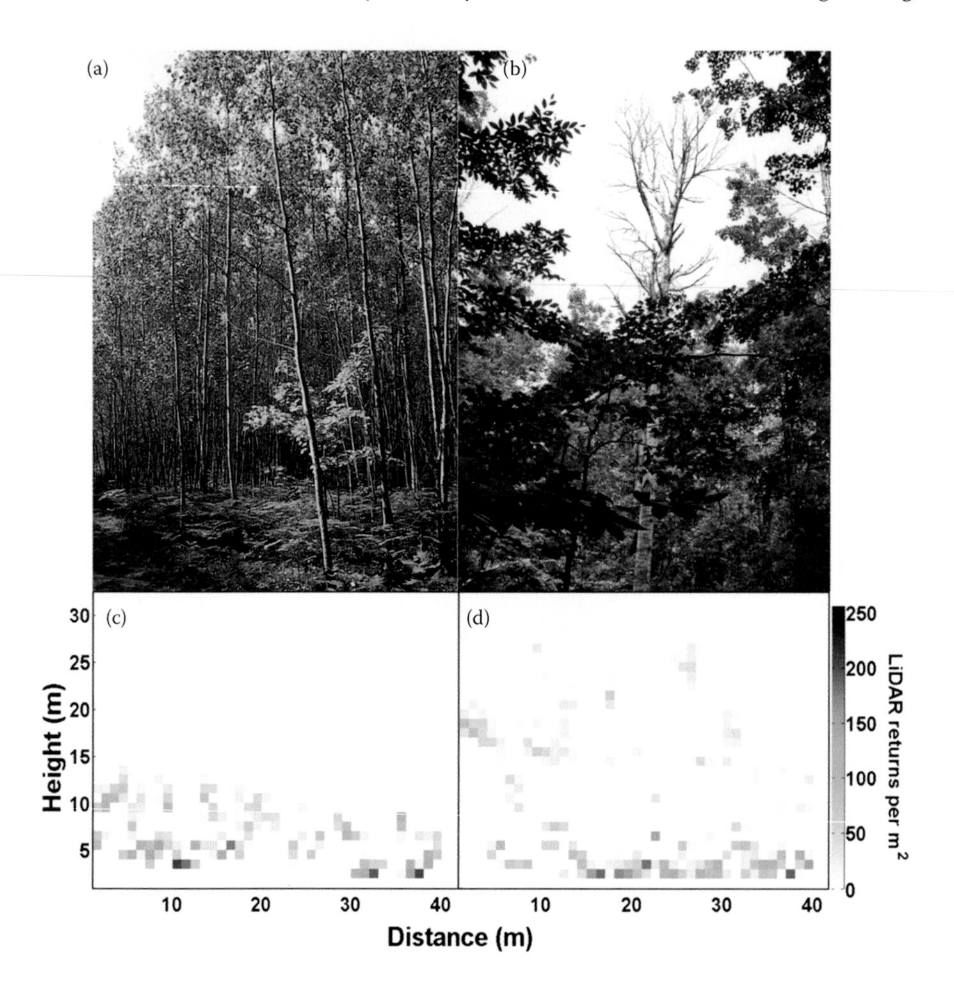

FIGURE 13.8 Terrestrial LiDAR-derived measures of canopy structural complexity in a young (30-year-old) forest (a and c) and a more structurally complex, older forest (b and d). (From Gough C.M. et al. 2016. *Ecosphere*, 7, 1–15.)

models to build Earth system models (ESMs). One such model, the Ecosystem Demography model (ED2; Medvigy et al., 2009) uses partial differential equations to represent cohorts of trees with different physiological properties that compete for light and other resources. The 3D structure of these models (Figure 13.9) and their reliance on finely resolved functional types (e.g., early versus late successional) aligns them well with hyperspectral and LiDAR data. Antonarakis et al. (2014) incorporated AVIRIS and LiDAR data from Harvard Forest (Massachusetts, USA) into runs of ED2. They found that initializing the model with hyperspectral and LiDAR derived measurements of forest demographics performed as well or better than model runs performed with ground-based measurements. This suggests that in the future, models like ED2 could be parameterized directly from remotely sensed measurements, reducing the need for time-consuming forest inventories.

13.2.4 Future Directions for Mapping Diversity

While the concepts around mapping structural and functional diversity with hyperspectral and LiDAR remote sensing have been around for decades, it is only recently that data have become available to make these endeavors widespread. As more hyperspectral and LiDAR datasets become available, we expect these lines of enquiry to expand. One avenue that has not been explored much at all is variation in these measurements over time. The majority of airborne collections have been single

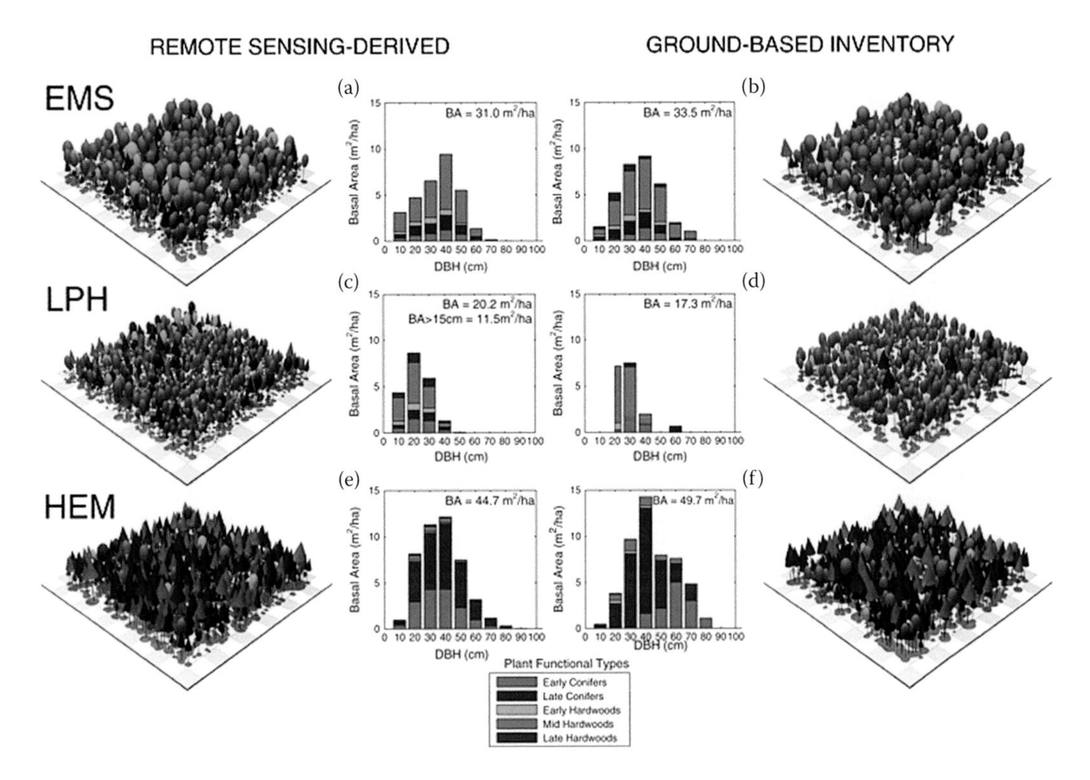

FIGURE 13.9 Remotely sensing-derived and ground-based inventory-derived estimates of forest structure and composition to inform uses of the Ecosystem Demography model. Actual sites are at Harvard Forest. (From Antonarakis A.S. et al. 2014. *Geophysical Research Letters*, 41, 2535–2542.)

campaigns, leaving many open questions about how ecosystems change in quality and structure over the course of a season or over years to decades. We expect that the increasing availability of manned aircraft collections as well as the miniaturization of hyperspectral and LiDAR sensors for use on drones will change this research area dramatically in the coming decade. Multisensor platforms like the Carnegie Airborne Observatory (Asner et al., 2017), the National Ecological Observatory Network's Airborne Observation Platform (Kampe et al., 2010), and NASA's G-LiHT (Cook et al., 2013) are making these fused datasets more common and available to address a wide variety of ecological questions.

13.3 CHARACTERIZING SPARSE VEGETATION IN DRYLAND ECOSYSTEMS

13.3.1 What Are Dryland Ecosystems and Why Are They Important to Map?

Dryland ecosystems are considered areas where water scarcity exists due to a low ratio of total annual precipitation to potential evapotranspiration and limits ecosystem services such as forage, wood, and agricultural production. Drylands are particularly sensitive to disturbance (e.g., changes in fire regime, overgrazing, invasive species, land-use change), and trending toward desertification, yet cover 41% of Earth's terrestrial surface (Figure 13.10). Overall, the sensitivities of drylands and their ability to drive climate change are poorly understood (e.g., Adeel et al., 2005; Ahlstrom, 2015), yet recent studies have indicated that the largest warming in the past 100 years occurred in drylands (Huang et al., 2017). Globally, these lands house 35.5% of our growing human population, 90% of whom live in developing countries and may be particularly vulnerable to environmental changes and dependent on ecosystem services (Adeel et al., 2005; Gilbert, 2011; United Nations Environment Management Group, 2011). Drylands provide habitat for important species, many of which are endangered, listed,

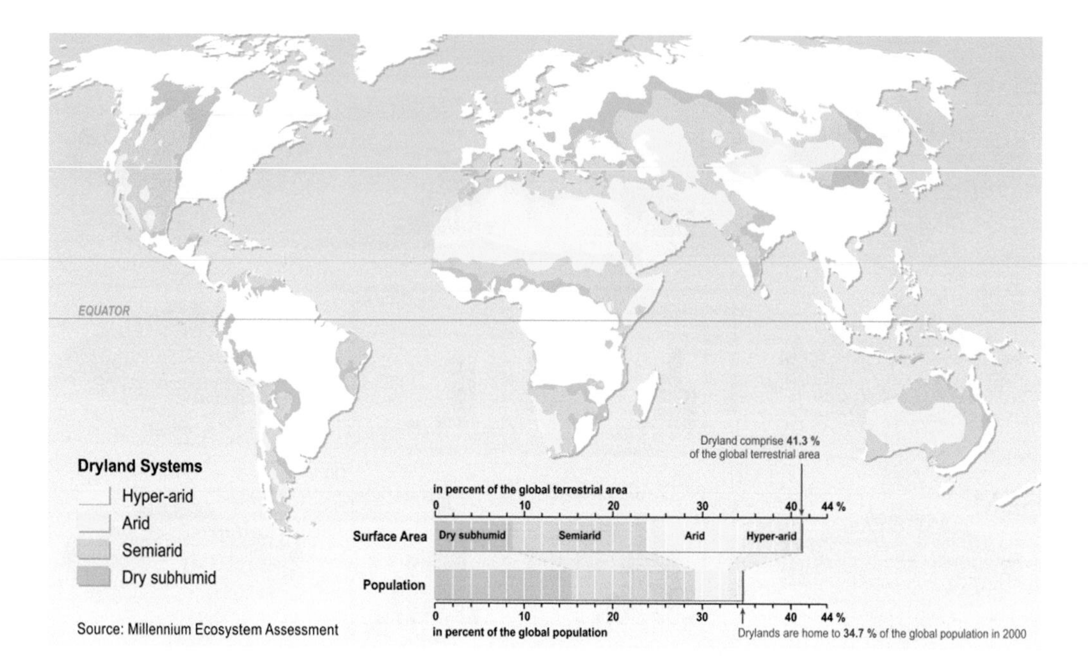

FIGURE 13.10 Global distribution of dryland systems, which include hyper-arid, arid, semi-arid, and dry sub-humid biomes, in relation to human population percentages. (From Adeel, Z. et al. 2005. *Ecosystems and Human Well-Being: Desertification Synthesis*. World Resources Institute, Washington, DC.)

or endemic. These species provide services related to pollination; ecotourism and recreation; food and water security; pharmaceuticals and medicinal research; and genetic resources that are important to adaptation and survival in a changing climate (United Nations Environment Management Group, 2011). Interconnected pressures with poorly understood feedback interactions are exacerbating land degradation and desertification trends. Synoptic and scalable remote sensing products are needed to quantify and monitor how drylands respond to and drive environmental variables.

Drylands store roughly 15% of Earth's total soil organic carbon (SOC) due to their spatial extent and high SOC pools (Eswaran et al., 2001; Lal, 2004; Noojipady et al., 2015). This is twice the organic carbon of forest ecosystems despite the sparse vegetation and low carbon sequestration rates of drylands (Noojipady et al., 2015). Drylands additionally have low, though increasing, net primary productivity (NPP); a global model derived from satellite observations from 1981–2000 estimated NPP levels to be 703 $\pm$ 44 g/m^2 compared to 869 $\pm$ 34 g/m^2 for forests and woodlands (Cao et al., 2004; Safriel and Adeel, 2008). New research is finding that drylands exert major influence on the interannual variability of the global carbon sink (Ahlström, 2015; Poulter et al., 2014). This influence is caused by not just expansive global coverage, but also the sensitivity of these systems to rainfall, temperature, and disturbances—all of which alter NPP and carbon fixation and emission (Poulter et al., 2014; Ahlstrom, 2015). For example, changes in the timing and amount of precipitation cause greater plant growth and carbon fixation and suppress fire; drier conditions reduce plant growth and carbon fixation and make fire-related emissions more likely (Poulter et al., 2014). Dryland ecosystems are historically subject to natural disturbance regimes including wildfire and drought (Field et al., 2014) and vulnerable to desertification and wind erosion. Globally, approximately 10–20% of drylands are considered degraded by desertification processes and a larger amount is at risk (Adeel et al., 2005). Additional factors that influence carbon emissions and fixation rates include increasing atmospheric carbon dioxide (CO_2) concentrations, and changes in temperature averages and extremes (Field et al., 2014; Ahlstrom, 2015). Direct land-use pressures and development also alter carbon emissions and fixation rates. Examples include land cover conversion to grazing and agriculture; suppression of regional rainfall through changes in albedo and evapotranspiration from alterations of the Earth's surface (e.g., urbanization and dust storms), and increases in sources of fire ignition (Pausas and Vallejo, 1999; Adeel et al., 2005; Belnap, 2005; Wendt et al., 2007; Sankey et al., 2011; Silvestrini et al., 2011; Rutherford et al., 2017; Saha et al., 2017). All of these factors are expected to intensify pressure on ecosystem services, increase the rate of habitat shift and loss; and alter the disturbance regimes of wildfire, droughts, flood, and erosion (Adeel et al., 2005; Field et al., 2014).

13.3.1.1 Vegetation Responses to Environmental Changes in Dryland Ecosystems

Remote sensing observations of increased greenness in drylands have been attributable to rising atmospheric carbon dioxide (CO_2), which could improve the efficiency of carbon fixation (Poulter et al., 2014). Under this scenario, plants keep their stomata closed more often and retain more water, which reserves soil water levels in a limited system where increased water use efficiency due to carbon fertilization is hypothesized to increase vegetative growth (Donohue et al., 2013). By contrast, in situ studies have shown that small temperature increases in dryland sites, such as those expected to occur as CO_2 rises, may reduce soil water and limit growth and productivity, thereby possibly favoring certain species and exerting controls on non-native plant invasion patterns (Wertin et al., 2017).

Vegetative presence and structure affects hydrology as well as resistance to wind erosion in drylands by, for example, creating microclimates with lower temperatures and evaporative rates under the canopy (Breshears et al., 1998). The distribution of woody vegetation across a landscape creates heterogeneous patterns of aeolian sediment transport, which can shape nutrient distribution patterns and drive land degradation and desertification processes (Mueller et al., 2007; Okin, 2008; Sankey et al., 2012). Post-fire wind erosion increases sediment transport but allows redistribution of soil nutrients (Ravi et al., 2009). Together, changing climate and vegetation affects nitrogen mineralization (Bobbink et al., 2010), soil structure, and soil microbial communities (Field et al., 2014).

In short, relationships that describe how dryland ecosystems drive and react to climate change are complex and interactive. Despite variation in greening and water retention, globally, the changes in climate and land use have resulted in overall increased aridity, drought, and wind erosion (Field et al., 2014). While vegetation is relatively sparse compared to other ecosystems, the global extent of these ecosystems and our heavy reliance on the services they provide make it important to improve our understanding of complex multi-scale processes and develop numerical decision support systems for land managers. Remote sensing provides the opportunity to monitor ecosystem trends and better understand interactions between disturbance, climate, nutrient cycling, and vegetation composition and structure. Land cover classification studies have generally been limited due to weak vegetation signal and lack of vertical structure; however, multispectral time series, hyperspectral imaging, and airborne LiDAR have been able to address some mapping challenges in dryland ecosystems.

13.3.2 Why Do We Need Hyperspectral and LiDAR Data to Map Dryland Vegetation?

In forested ecosystems, the integration of hyperspectral imaging with LiDAR data has been demonstrated to boost species cover, composition and biomass mapping performance, and improve analytical power by combining structural and biochemical information for understanding changes in forest functional and structural diversity patterns over time (see Section 13.2). Similar vegetation studies designed to discriminate species or estimate structural and biochemical information in dryland ecosystems are limited by multiple scattering, bright soil reflectance, penetrable canopies, and spectrally indiscriminate targets (e.g., Smith et al., 1990; Jakubauskas et al., 2001; Okin et al., 2001; Mirik et al., 2007). Despite the ability to unmix reflectance signatures of materials within hyperspectral pixels for estimating vegetation quantities such as cover, biomass, and leaf area index (LAI) in arid and semiarid environments (typically less than 50% vegetation cover), spectral mixture analysis has limited reliability when cover is below 30% or where there is little spectral contrast between vegetation and surrounding background materials (Okin et al., 2001). One ambiguity is the assumption that materials within a given pixel combine linearly; yet, there is a nonlinear mixing component, due in part to multiple scattering from non-photosynthetic vegetation (e.g., litter, woody

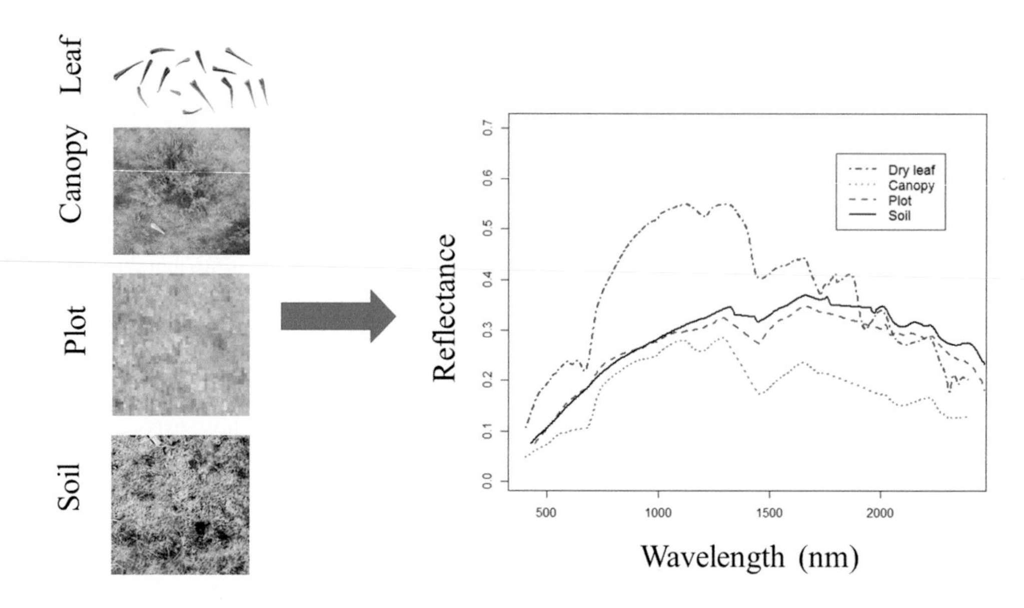

FIGURE 13.11 The spectral signatures of a sagebrush (*Artemisia tridentata*) leaf (dry), sagebrush canopy, and plot scales (10 × 10 m) that also include significant soil mixing. Unique challenges are related to high degrees of mixing and nonlinear scattering caused by bare ground, non-photosynthetic vegetation, and fine-scale heterogeneity. (From Boise Center Aerospace Lab, USA.)

shrubs), bare ground, and patterns of fine-scale heterogeneity (Figure 13.11; Roberts et al. 1993, Borel and Gerstl 1994, Ray and Murray 1996).

Small-footprint, discrete return LiDAR is not limited by many of these spectral challenges and has the potential for estimating shrub canopy characteristics at scales appropriate for landscape assessments (e.g., Ritchie et al., 1992; Mundt et al., 2006; Streutker and Glenn, 2006; Su and Bork, 2006, 2007; Riaño et al., 2007; Li et al., 2015, 2017a). However, separating LiDAR returns in low-height vegetation is difficult because the canopy returns are often close to ground returns in both space and time (Figure 13.12a). Furthermore, there are fewer vegetation returns in sparsely vegetated

FIGURE 13.12 While LiDAR can overcome some of the spectral challenges associated with hyperspectral remote sensing of dryland vegetation, vegetation returns are challenging to separate from ground returns and high canopy penetration (a), results in shrub height underestimation by roughly one-third of average shrub height in sagebrush steppe (b). (From Boise Center Aerospace Lab, USA.)

semiarid ecosystems than in more foliated ecosystems. A limited number of studies have evaluated the use of LiDAR in shrub environments (Hopkinson et al., 2005; Streutker and Glenn, 2006; Riaño et al., 2007; Su and Bork, 2007; Glenn et al., 2011; Mitchell et al., 2011; Spaete et al., 2011) and these studies have consistently found that small-footprint LiDAR systems underestimate shrub canopy height (Figure 13.12b). Shrub height underestimation is attributed to the low probability of the laser hitting the top of the canopy, or laser pulses penetrating the canopy, which generates return signals from material within the canopy (Weltz et al., 1994; Næsset and Økland, 2002; Gaveau and Hill, 2003; Clark et al., 2004).

Consequently, remote sensing researchers are faced with the challenge of improving the discrimination of dryland vegetation, including grasses and non-natives, for applications such as inventory mapping (e.g., structure, nutritional status, habitat quality) and ecological modeling (e.g., nutrient cycling, ecosystem productivity).

13.3.3 How Are Hyperspectral Data and LiDAR Currently Being Used in Dryland Ecosystem Mapping?

Widely available airborne LiDAR datasets delivered with average point densities around 9 pts/m^2, can contribute to shrub height underestimation and lack the resolution for automatic delineation of individual shrub crowns (Figure 13.13) for feature-level fusion techniques (see Section 13.1.1) and some ecological applications, particularly those dependent on shrub allometry for biomass estimation. In so far as metrics such as 2D canopy shape and height have been relatable to aboveground biomass for dominant shrubs in the western United States, such as sagebrush (*Artemisia tridentata*) and rabbitbrush (*Purshia tridentata*), using allometry at the individual shrub scale (Cleary et al., 2008; Olsoy et al., 2014a,b), several studies have tried to circumvent LiDAR challenges unique to dryland ecosystems by using statistical approaches such as random forests to test the integration of vegetation indices from hyperspectral imagery to improve cover estimation results (Mitchell et al., 2015) and develop bulk shrub biomass estimation approaches (Li et al., 2015, 2017a, 2017b).

Mitchell et al. (2015) found that combining narrowband vegetation indices derived from airborne hyperspectral imagery, such as the ratio of red to green bands (Gamon and Surfus, 1999) and anthocyanin reflectance index (Gitelson et al., 2001) with LiDAR-derived variables (mean absolution deviation from median height of all vegetation returns; interquartile range of height of all vegetation returns; and vegetation cover, defined as the ratio of vegetation returns greater than 15 cm height and total returns) boosted cover estimation results from $r^2 = 0.49$; root mean square error (RMSE) 8.19 using LiDAR alone to $r^2 = 0.58$; RMSE 7.35 using LiDAR and hyperspectral combined. Results from this study also indicated that spectral variables will play a dominant role in estimating grass cover and that upscaling this approach to regional cover mapping will be limited by sensitivity to ground thresholds, which tend to vary by collection site. Li et al. (2015) were able to apply a hierarchical scaling approach to biomass estimation by delineating individual shrub crowns using terrestrial laser scanning datasets with high point density instead of airborne LiDAR (5 points/m^2), which was instead used for extrapolating shrub biomass to the watershed scale using a ratio of vegetation and ground returns as a measure of canopy cover. Height did not explain biomass variation at the plot scale and biomass was estimated with an adjusted $r^2 = 0.55$, RMSE $= 4.01$ kg/5 m^2, and normalized RMSE $= 35\%$. By comparison, the lowest forest biomass estimation error is achieved with approaches that synergistically use LiDAR rather than various airborne or satellite platforms alone; normalized error ranges from approximately 22% to 27% normalized RMSE (Zolkos et al., 2013).

While hyperspectral and LiDAR combined can boost shrub cover estimation results, only a relatively small portion of the full spectrum of data contained within the hundred plus hyperspectral image bands are used to calculate relevant narrowband vegetation indices. A larger portion of the hyperspectral data can be leveraged to map shrub foliar canopy chemistry in dryland ecosystems. Foliar N, for example, tends to yield surprisingly successful results, despite challenges associated with spectrally discriminating vegetation targets from arid backgrounds using an empirical approach

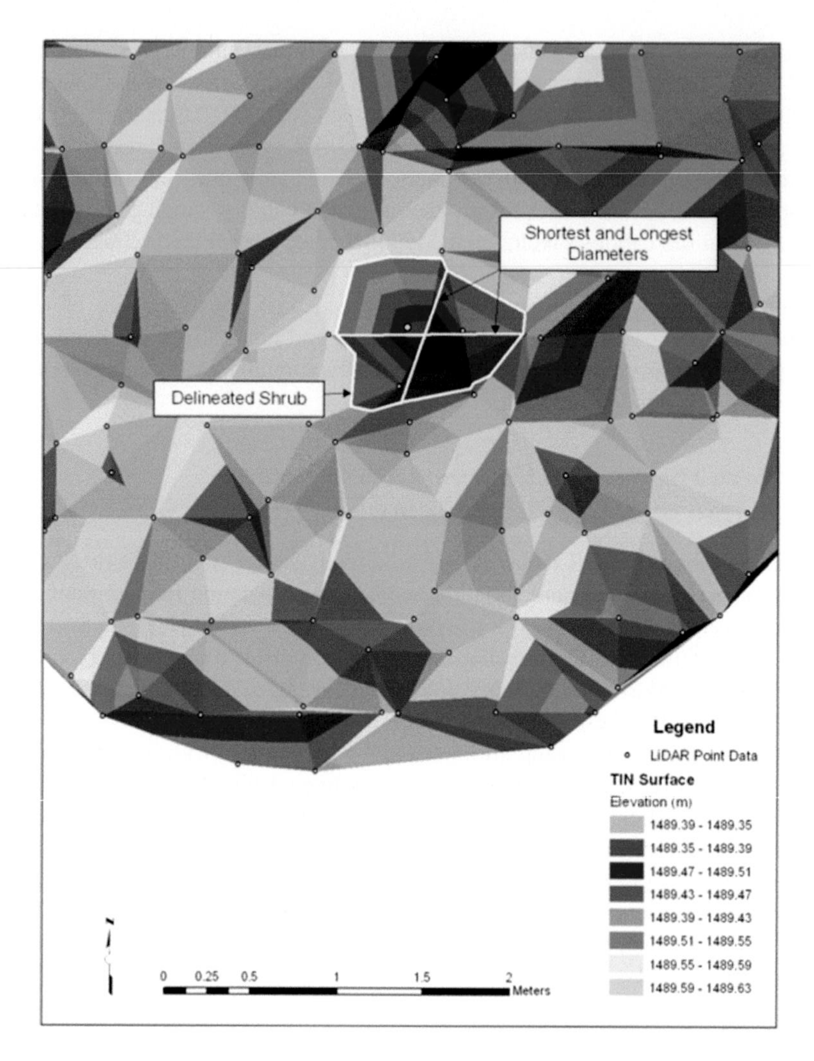

FIGURE 13.13 Discrete return LiDAR datasets with point densities of approximately 8–10 points/m^2 do not support the automated delineation of individual shrub canopies, which could be useful for feature-level fusion approaches and hierarchical scaling for applications such as shrub biomass estimation. (From Mitchell J. et al. 2011. *Photogrammetric Engineering and Remote Sensing*, 77(5), 521–530.)

and multivariate statistical analysis (e.g., Serrano et al., 2002; Huang et al., 2004; Mutanga et al., 2004; Skidmore et al., 2010; Mitchell et al., 2012). Techniques that are particularly useful for mapping foliar N in dryland systems include continuum removal transformations (Kokaly and Clark, 1999) to suppress background noise, focusing on areas of shrub cover greater than approximately 40%, and using structural metrics such as LMA, LAI, and biomass to scale from leaf to plot level field estimates of foliar N for building empirical relationships. Since foliar N equates to protein, ecological applications include mapping habitat quality and forage quantity, expanded rangeland health monitoring, and emerging areas in foraging ecology that consider not only nutrition, but also plant defensive chemistry and cover from predation (Olsoy et al., 2016b, 2017; Forbey et al., 2017).

13.3.4 Future Directions for Mapping Vegetation in Drylands

Global dryland ecosystem mapping products have been produced using aridity (precipitation, potential evapotranspiration (PET), and temperature) to delineate climatic zones (UNCCD, 2011), and land cover approaches such as percent tree cover per 500-meter MODIS pixel (Hansen et al., 2003). Degradation in drylands has been monitored at the continental to global scale using NDVI time series by relating greenness trend to rainfall or using NDVI to estimate NPP (e.g., Herrmann et al., 2005; Olsson et al., 2005; Bai et al., 2008; Vlek et al., 2008). Higher resolution studies are needed to understand drivers of land degradation and finer-scale ecosystem processes.

Future directions in the areas of shrub biomass and canopy chemistry mapping in dryland ecosystems should focus on constraining uncertainty error with attention to upscaling and the dominant factors that influence estimation accuracy across scales. Although the importance of shrub biomass in annual carbon storage fluctuations is becoming recognized in global DVMs, quantitative estimates and error bounds have not been established, as there is a need for both baseline data and identification of acceptable error levels for monitoring biomass responses to environmental change over time. There is opportunity to reduce shrub biomass estimation error at the individual and plot scales using terrestrial laser scanning to non-destructively improve allometric relationships, particularly for large, undersampled shrubs. In general, combining optical and LiDAR datasets reduces uncertainty in dryland vegetation cover estimates; although uncertainty increases as canopy cover decreases (e.g., Mitchell et al., 2015; Glenn et al., 2016). A recent study by Ilangakoon et al. (2018) demonstrate FWF LiDAR variables to differentiate low-height vegetation from ground, as well as different PFTs in drylands. Future considerations for constraining uncertainty and upscaling canopy chemistry mapping in dryland ecosystems should evaluate the value of adding LiDAR metrics, including FWF, to hyperspectral datasets. In addition, scaling from leaf to plot estimates of constituents such as foliar N could be improved using precise ground-based measurements of biomass rather than bulk approximations. Finally, it should be noted that tradeoffs between multivariate statistics for canopy estimated compared to physically based radiosity models need to be evaluated in the context of generalizing relationships across landscapes.

Satellite missions such as GEDI are expected to support future biomass and canopy chemistry mapping efforts in dryland ecosystems. Efforts that test tradeoffs between multispectral and hyperspectral sensors for dryland vegetation mapping applications should continue to be explored while improvements in LiDAR point density and range resolution from Geiger counting and full waveform LiDAR systems are expected to improve structural measurement errors, along with ground-based laser scanning.

13.4 TERRESTRIAL LASER SCANNING IN SUPPORT OF HYPERSPECTRAL/LiDAR VEGETATION MAPPING APPLICATIONS

Ground-based or terrestrial laser scanning is an active remote sensing technique that generates 3D point clouds of targets from low divergence, mono-spectral laser beams. The laser pulses operate in one to several specific wavelengths that sample in the visible and/or near-infrared (NIR) region(s) of the electromagnetic spectrum. Cost reductions; increased range, precision and accuracy; and availability of software with improved computational capabilities have expanded the operational use of TLS techniques to larger plot sizes (several hectares) and a variety of biomes, including grasslands and vertically complex tropical forests (e.g., Calders et al., 2015a). Since TLS instruments are able to accurately and efficiently collect fine-scale resolution data, the technology is also being adopted more frequently for several niche application areas in remote sensing of vegetation. One major role is to refine allometric relationships by supplementing destructively sampled ground truth data. A second major role is to characterize and model fine-scale vegetation biophysical and biochemical properties, either directly from TLS collections, or via statistical or physical models. In many cases, TLS datasets are collected and combined with other remote sensing datasets acquired from multispectral, hyperspectral, LiDAR, and synthetic aperture radar (SAR) sensors. Therefore, a third major TLS vegetation application area is to calibrate and validate other relatively large-footprint active and passive remote sensing datasets and help develop products such as canopy cover and height estimates derived from LiDAR full waveform signals.

13.4.1 Augmenting Ground Truth Data and Refining Allometry with TLS

Hyperspectral remote sensing has been extensively used in vegetation applications at a range of spatial scales to detect species type, composition, biophysical properties (structural properties such as LAI, biomass, yield, density), biochemical constituents (e.g., chlorophyll, N), environmental stresses (insect infestation, plant invasion, water stress, drought), moisture content (e.g., leaf moisture), light-use efficiency, and NPP. In order to enhance vegetation mapping estimates from hyperspectral data, accurate and effective ground truth data are needed to calibrate and validate hyperspectral products, and to integrate these products with other data sources across spatial scales. Ground truth data at leaf and plot scales play a crucial role in constraining uncertainties in remote sensing of vegetation and ecological modeling applications. Common ground truth measurements include 3D vegetation structure (height, area, volume, number of stems, LAI), biomass, leaf, and canopy biochemistry (water, chlorophyll, N), bare ground cover, and relative distribution of plant functional types (PFTs). Some of these ground truth data, such as biomass and LAI, require destructive sampling, which is not only time and labor intensive, but can also preclude future baseline comparisons. In the case of destructively sampled measurements such as biomass, a limited number of specimens are collected to develop allometric models based on empirical relationships between tree, shrub, and grass metrics and weight. Consequently, allometric biomass models remain highly uncertain for many ecosystems. Moreover, most allometric models depends on in situ measured vegetation structure parameters (e.g., height, diameter breast height (DBH), volume, and so on), which suffer from non-optimal sampling for reasons including incorrect measurements, undersampling (e.g., small or unrepresentative sample size), and bias (e.g., heteroscedasticity at the individual scale and occlusion of understory vegetation with larger trees at the plot scale). However, fine-scale vegetation structure is strongly linked to biomass, biodiversity, and dynamics with carbon and water fluxes (Baldocchi et al., 2004; Goetz and Dubayah, 2011; Calders et al., 2015a; Guo et al., 2017). Therefore, fine-scale, accurate, and efficient ground truth data are necessary for refining allometric relationships, reducing uncertainty in fine-scale vegetation products, and providing reference data for testing new metrics derived from airborne FWF LiDAR collections.

For TLS 3D point cloud datasets, each point is georeferenced to an accuracy on the order of millimeters to centimeters. Each laser return in the high density point cloud contains x, y, z GPS coordinate information, laser intensity and/or reflectance, red-green-blue (RGB) color (if camera is used), and a laser return index number. Vegetation height can be calculated from TLS data by (1) rasterizing point cloud metrics at sub-meter pixel resolution, (2) segmenting individual canopies then deriving height metrics, or (3) using height-based point cloud slicing methods. Similar techniques can also be applied to extract DBH, basal area, and stem metrics that strongly correlate to field measured data (Moskal and Zheng, 2011). Such vegetation structural parameters frequently appear in allometric models and thus can be easily used to estimate biomass at individual and plot scales.

Given the heterogeneity of plant structure, it is evident that canopy cover and volume are key components of biomass allometry. Manually measuring the area or volume of randomly distributed, irregular bodies like vegetation is challenging; however, the 3D volume of individual shrubs to tall trees can be reconstructed and calculated by applying techniques such as voxelization, convex hull, and quantitative structure modeling (QSM) to TLS point clouds. Using TLS-based vegetation volume approximations, new high predictive relationships have been generated to estimate biomass in semi-arid ecosystems (Olsoy et al., 2014a; Li et al., 2015), Arctic tundra (Greaves et al., 2017), and grasslands (Cooper et al., 2017). In addition to allometric models, reconstructed TLS-based vegetation tree volume and species-specific wood density (biomass = wood density * volume) have been successfully used to replace the allometric biomass models to calculate tree biomass without bias from tree morphology (Olschofsky et al., 2016; Gonzalez de Tanago et al., 2018).

In addition to replacing destructively sampled biomass measurements, TLS point clouds have also been used to estimate LAI, plant area index (PAI), gap fraction, and clumping index with strong correlations to ground reference measurements (Zheng et al., 2013; García et al., 2015; Olsoy et al., 2016a; Li et al., 2017a, 2017b). The high correlation of destructive field estimates and TLS estimates in above studies highlight the potential of TLS to replace ground reference measurements as validation data in vegetation remote sensing studies, without need for additional destructive harvest. Further, relationships between TLS and LiDAR-derived vegetation parameters are widely used to calibrate the retrieval of LiDAR vegetation metrics at regional scales (Vincent et al., 2015; Paris et al., 2017).

13.4.2 Using TLS to Quantify Fine-Scale Vegetation Structure and Biochemistry

The intensity, reflectance, incident angle, and other characteristics of TLS point clouds and waveforms are also used to estimate fine-scale properties of vegetation structure and biochemical properties. The dual-wavelength (1063 and 1545 nm) Salford Advanced Laser Canopy Analyser (SALCA; Danson et al., 2014) terrestrial laser scanner has been used to measure the influence of incident angle on leaf reflectance calculated from both wavelengths. In a recent study using full waveform terrestrial LiDAR signals, Zhu et al. (2017) were able to estimate the vertical distribution of LWC within a canopy ($r^2 = 0.66$) using incident angle-corrected backscatter coefficients. While leaf reflectance is a function of incident angle, it has been shown that observed differences between fresh and dry leaves and between leaf and bark are greater than reflectance differences related to incident angle (Hancock et al., 2017). Eitel et al. (2010) used a TLS instrument that operates in the green (532 nm) to simultaneously approximate the spatial distribution of chlorophyll a and b content, leaf area, and leaf angle. Changes in the intensity of TLS values in the green at different light exposures have been used to understand the patterns and mechanisms driving foliar photoprotection (Magney et al., 2014). A recent advance in terrestrial LiDAR systems is ultraviolet fluorescence LiDAR (UFL series), which is currently used to measure chlorophyll a in lake waters (Pelevin et al., 2017). Further, TLS time series data shows great potential to estimate the seasonal dynamics as a function of time and canopy height at fine scales in vertically complex forests (Calders et al., 2015b; Figure 13.14).

Due to the importance of integrating both spectral and structural components, terrestrial hyperspectral LiDAR has been introduced and is used to estimate the chlorophyll content of plants very precisely and accurately. The chlorophyll estimates of Scots pine derived from hyperspectral LiDAR linearly correlates with the laboratory analyzed chlorophyll concentrations, and are able to represent a range of shoot chlorophyll concentrations ($r^2 = 0.88$, RMSE = 0.10 mg/g; Figure 13.15). The results indicate that the hyperspectral instrument has the potential to estimate vegetation biochemical parameters such as the chlorophyll concentration. Having both spectral and structural components incorporated, this hyperspectral TLS instrument holds much potential in various environmental applications over single wavelength LiDAR or passive optical systems for environmental remote sensing (Nevalainen et al., 2014).

In addition to statistical approaches to estimate vegetation properties, TLS data can be integrated with other active and passive sensors via physically based radiative transfer models designed to calibrate sensor signals and retrieve target properties such as LAI, chlorophyll, and structure at leaf or canopy levels.

Fine-scale TLS point clouds of vegetation that span individual plants to plots over a hectare in size can be used to derive vegetation structure and biophysical attributes (e.g., LAI, height, DBH, crown width) that are generally used as radiative transfer model inputs. These inputs serve as common reference data that are used to calculate and control aspects of vegetation structure and simulated signals. The use of TLS data for physically based modelling applications allows rapid testing of various RTM assumptions in a real-world context (Vaccari et al., 2013; Gonzalez de Tanago et al., 2018).

13.4.3 Integration of TLS with Other Remote Sensing Data Sources

Terrestrial laser scanning applications have further expanded through fusion with terrestrial hyperspectral imagery and integration of ground-based scanning with other remote sensing data

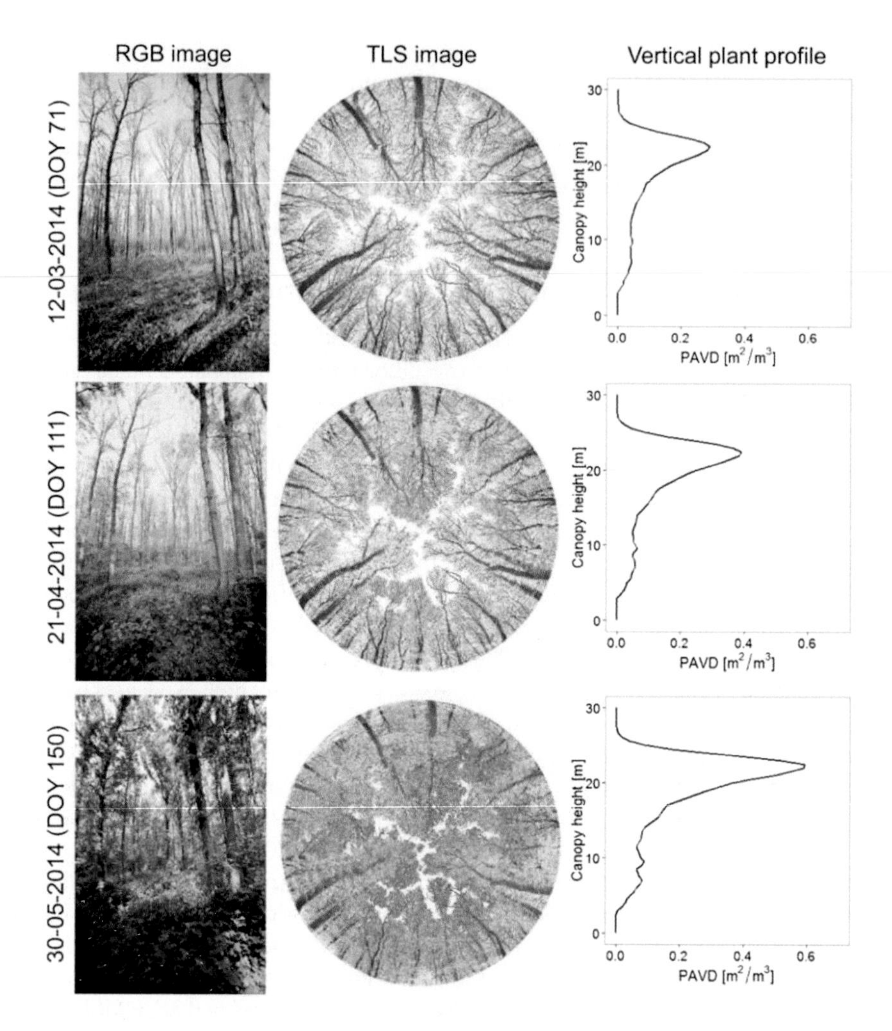

FIGURE 13.14 In this terrestrial laser scanning (TLS) time series, seasonal vegetation dynamics are captured for three dates using true color or red-green-blue (RGB) images (left), polar projected laser scans colored by apparent reflectance, where light green is low intensity and dark red is high intensity (center); and vertical plant profiles of plant area vertical density (PAVD) derived from the TLS data (right). (From Calders K. et al. 2015b. *Agricultural and Forest Meteorology*, 203, 158–168. https://doi.org/10.1016/j.agrformet.2015.01.009.)

sources. For example, locational and intensity information acquired from TLS points have been fused with hyperspectral camera images to correct for the confounding influence of shadow in images acquired from ground-based visible to near-infrared (VNIR) and shortwave infrared (SWIR) cameras. Hartzell et al. (2017) fused hyperspectral images with 3D laser point cloud returns by deriving a scaling factor from the ratio of hyperspectral reflectance to the TLS reflectance at the operational wavelength of the TLS system. The scale factor was then applied to the entire hyperspectral spectrum. The shadow-restored bands improved correlation with a validation image by 45% compared to existing shadow restoration methods. Although this method was demonstrated in a rock outcrop study, it could be extended to minimize the influence of shadow in remote sensing of vegetation studies. Another approach to fusing terrestrial LiDAR and hyperspectral datasets has been to combine collections from both single wavelength TLS and newly designed active hyperspectral laser scanning instruments (rather than imagery acquired from passive hyperspectral cameras) for the purpose of designing single instruments capable of acquiring both x, y, z locational information and reflectance data. Puttonen et al. (2010) fused data from these two types of instruments using

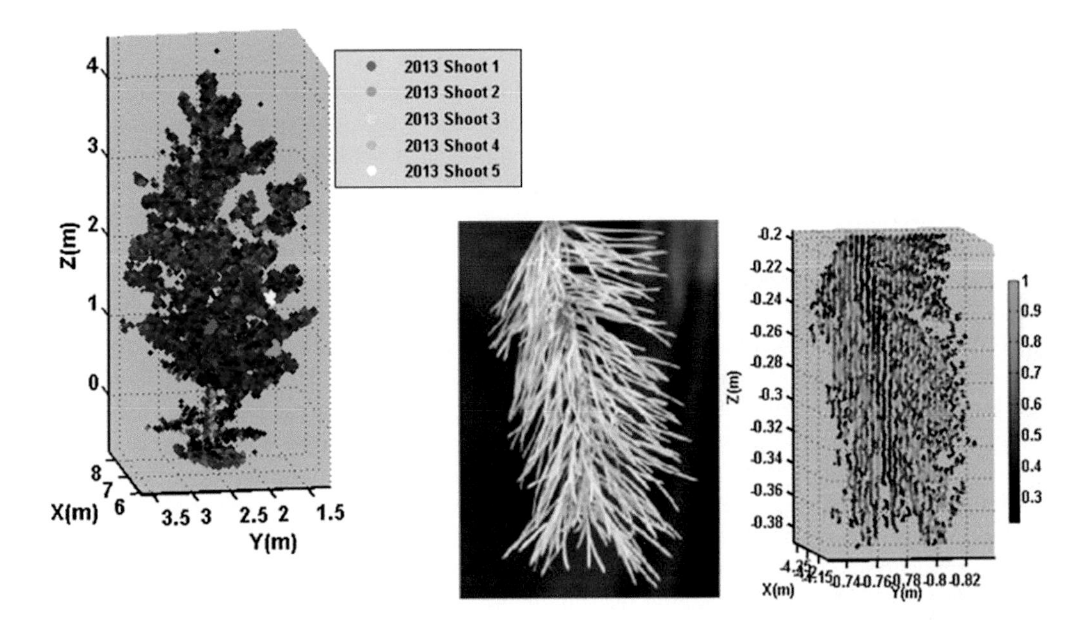

FIGURE 13.15 A 3D LiDAR/hyperspectral point cloud depicts a scanned Scots pine tree, including the locations of shoots that were sampled for chlorophyll content (left). Total chlorophyll concentration was estimated for every point in the scanned shoot (right) using a linear regression model that was developed as a function of the modified chlorophyll absorption ratio index (Wu et al., 2008), as estimated from return intensities for instrument channels closest to 750 and 705 nm wavelengths. (From Nevalainen O. et al. 2014. *Agricultural and Forest Meteorology*, 198, 250–258. https://doi.org/10.1016/j.agrformet.2014.08.018.)

simultaneously collected coincident points and dataset matching to generate a combined dataset that improved classification accuracy for three boreal tree species by using both reflectance and shape-derived classification variables.

Geographic position and intensity data acquired from TLS systems are also becoming increasingly used to calibrate, validate, and simulate LiDAR data acquisitions. Vierling et al. (2013) simulated discrete return airborne LiDAR point clouds using TLS data and estimated the location, height, and crown area of individual shrubs by using a 2D spatial wavelet analysis technique. The resultant accuracies were computed in comparison to field observations ($r^2 = 0.47–0.\ 94$) and demonstrated the use of airborne LiDAR point cloud data for automatic quantification of challenging shrub biophysical parameters across broad spatial scales. Liu et al. (2017) were able to relate TLS-derived vegetation understory volume estimates from voxels to an airborne LiDAR-derived understory point cloud distribution variable. This study confirmed the potential of TLS to advance airborne LiDAR understory mapping through its use as a tool for validating forest material volumes estimated at the landscape scale.

Despite high point cloud density, TLS is generally inadequate at detecting the top of canopy on account of its bottom up viewing within forest plots. With recent advancements in photogrammetric analysis techniques such as structure from motion (SfM), complimentary UAS platforms are emerging where the top of canopy can be retrieved from high density point clouds using low cost three and four band camera images. The accurate co-registration of point cloud data derived from UAS and TLS sources can provide complete, in-depth 3D coverage of a study area. A number of recent studies have combined coincident TLS- and UAS-derived point clouds to generate datasets that are sufficient in resolution to, for example, reconstruct a cultural heritage site (Xu et al., 2014) or to refine vegetation structure (height, volume) estimation errors obtained with TLS (Prandi et al., 2016). In the vegetation study, the UAS/TLS point cloud provided forest inventory and timber volume estimates with high levels of accuracy for large plots (over 12 ha), with detailed measurements that included individual stem detection and DBH.

13.4.4 Terrestrial Laser Scanning Limitations and Challenges

Many technological challenges associated with ground-based laser scanning technology are captured in instrument design parameters and measurement precision. Measurement precision is influenced not only by instrument specifications, but also by factors such as outdoor field setup, co-registration procedures, and computationally intensive workflows.

Many newer TLS instrument models can collect point clouds in discrete return and full waveform modes because the systems are equipped with waveform digitization capabilities. Most TLS instruments still operate in a single wavelength located in either the green (e.g., Leica—532 nm) or NIR (e.g., Riegl VZ series/ Echidna—1064 nm). Recently, multibeam (dual wavelength, e.g., DWEL and SALCA—1063 and 1545 nm), multispectral (e.g., Optech Titan—532, 1064, and 1550 nm), and hyperspectral (e.g., HSL of NTech Industries, Inc., Ukiah, CA—538–910 nm) TLS have been used in vegetation studies (Sun et al., 2017). Further, each of these instruments is configured with specific pulse repetition rate (PRF), pulse width (FWHM), beam divergence (in mrad), max and min range (m), scanning azimuth and zenith resolution (degrees) settings, which control point density, field of view, footprint size, detection range, and application (e.g., vegetation structure vs. biochemistry; snow and ice). Many TLS instruments are designed for specific applications, such as the Echidna Validation Instrument (EVI), which is used for used for close range vegetation 3D structural measurements. Specifications for instrument wavelength, PRF, pulse width, beam divergence, maximum detection range, zenith, and azimuth were: 1064 nm (waveform based), 2 kHz, 14.9 ns, 2–15 mrad, 130 m, ± 130°, and 180°. In addition to these types of design parameters, instrument weight and cost are also important considerations that can present challenges and influence measurement precision.

Field setup is another consideration that can influence TLS measurement precision. Scanners always require a stable and rigid platform (usually a tripod with known height) and level surface within a given tolerance before starting measurements. Occlusion (blockage of distant target by foreground targets) and edge effect (partial hit of the laser beams at the edge of the target) are considered major issues. To avoid occlusion, especially in forest plots, a collection strategy with multiple systematic scans per plot is implemented (Wilkes et al., 2017). However, it is essential to have accurate co-registration of multiple scans. The co-registration process requires accurate determination of scanner relative position within a common reference coordinate system. (Wilkes et al., 2017). To facilitate measurement accuracy in this co-registration task, retro-reflective targets (spheres, cylinders, or flat) of known size are commonly set up at fixed heights aboveground in locations that are visible from multiple scan stations. Aside from the standard reflective targets, ground objects such as building edges can also be used to derive additional control points for co-registration.

Some studies have experimented with the minimum number of scans needed to capture specific targets as well as field setup strategies for maximizing co-registration accuracy and precision. Anderson et al. (2017) have established that three to five scans can capture approximately 70% of one hectare plot dominated by either grass or short stature shrub. Wilkes et al. (2017) found that six or more scans are required to capture a large tree with dense canopy. However, it should be noted that these recommendations can be instrument-specific and that the distance between scans is a function of density of understory vegetation. For example, a grid layout with a 10 × 10 m TLS scan area is recommended for high density understory vegetation (Wilkes et al., 2017).

As each TLS produces enormous data volume, the multiple scans required to adequately cover a target of interest must consider optimal angular resolution so that the necessary number of scans can be collected timely and efficiently. Further, the type of data also determines the data volume (e.g., FWF collections may have ~6–7 times data volume as a discrete return point cloud). Another consideration during field setup is the weather condition. Rain and moist conditions affect laser transmission and also alter target scattering properties. Wind can alter target position and introduce uncertainty in co-registration.

In addition to instrument and field setup considerations, TLS post-acquisition data processing techniques need to be optimized based on the type of output parameter that is desired. The data derived from TLS falls into two broad categories: gap probability (based on gap fraction) and geometric modeling. Gap fraction-based techniques commonly use a voxel approach to derive vegetation canopy gaps, where voxel size plays a critical role in accurately deriving parameters such as LAI, vegetation volume, gap fraction, and plant area density volume. Geometric modeling includes the determination of vegetation structure (e.g., height, DBH, stems) and volume. Geometric modeling approaches, such as Quantitative Structure Models (QSM), fit hard geometric shapes to approximate the volume of features in point cloud datasets such as stems and branches. Both data processing approaches are computationally intensive and large dataset handling challenges may need to be identified and addressed.

13.5 CONCLUSION

The integration of hyperspectral and LiDAR data allows scientists to combine dimensions of complimentary plant structural (e.g., height, biomass, LAI) and biochemical information. This coupling is particularly useful for mapping and understanding controls on finer-scale diversity variables such as plant structural and functional diversity and for addressing mapping challenges associated with sparse, short, and open canopy vegetation in dryland ecosystems. Major approaches to combining hyperspectral and LiDAR datasets include multivariate statistical analysis, physical-based modeling using radiative transfer concepts, and ecological modeling at ecosystem and global scales. The integrated use of high resolution hyperspectral and LiDAR datasets, including TLS, boosts accuracies typically by 5–10%, and reduces uncertainty in vegetation parameter estimates. The use of TLS is particularly useful in reducing biomass uncertainty estimates by improving allometry and in development of new vegetation metrics from emerging remote sensing data sources (UAS, airborne, and satellite). Spatially and temporally coordinated observations across multiple remote sensing platforms have the potential to capture the structural heterogeneity and temporal frequency necessary to resolve cause and effect relationships between plant function and ecosystem processes globally (Stavros et al., 2017).

REFERENCES

Aasen, H., Burkart, A., Bolten, A. and Bareth, G. 2015. Generating 3D hyperspectral information with lightweight UAV snapshot cameras for vegetation monitoring: From camera calibration to quality assurance. *ISPRS Journal of Photogrammetry and Remote Sensing*, 108, 245–259.

Abdalati, W., Zwally, H.J., Bindschadler, R., Csatho, B., Farrell, S.L., Fricker, H.A. et al. 2010. The ICESat-2 laser altimetry mission. *Proceedings of the IEEE*, 98(5), 735–751.

Adeel, Z., Safriel, U., Niemeijer, D. and White, R., 2005. *Ecosystems and Human Well-Being: Desertification Synthesis*. World Resources Institute, Washington, DC.

Ahlstrom, A., Raupach, M.R., Schurgers, G., Smith, B., Arneth, A., Jung, M. et al. 2015. The dominant role of semi-arid ecosystems in the trend and variability of the land CO2 sink. *Science*, 348, 895–899. https://doi.org/10.1126/science.aaa1668.

Alonzo, M., Bookhagen, B. and Roberts, D.A. 2014. Urban tree species mapping using hyperspectral and LiDAR data fusion. *Remote Sensing of Environment*, 148, 70–83.

Anderson, K.E., Glenn, N.F., Spaete, L.P., Shinneman, D.J., Pilliod, D.S., Arkle, R.S. et al. 2017. Methodological considerations of terrestrial laser scanning for vegetation monitoring in the sagebrush steppe. *Environmental Monitoring and Assessment*, 189(11). https://doi.org/10.1007/s10661-017-6300-0

Antonarakis, A.S., Munger, J.W. and Moorcroft, P.R. 2014. Imaging spectroscopy- and LiDAR-derived estimates of canopy composition and structure to improve predictions of forest carbon fluxes and ecosystem dynamics. *Geophysical Research Letters*, 41, 2535–2542.

Asner, G.P. and Martin, R.E. 2009. Airborne spectranomics: Mapping canopy chemical and taxonomic diversity in tropical forests. *Frontiers in Ecology and the Environment*, 7, 269–276.

Asner, G.P., Knapp, D.E., Boardman, J., Green, R.O., Kennedy-Bowdoin, T., Eastwood, M., Martin, R.E., Anderson, C. and Field, C.B. 2012. Carnegie airborne observatory-2: Increasing science data dimensionality via high-fidelity multi-sensor fusion. *Remote Sensing of Environment*, 124, 454–465.

Asner, G.P., Knapp, D.E., Kennedy-Bowdoin, T., Jones, M.O., Martin, R.E., Boardman, J.W. and Field, C.B. 2007. Carnegie airborne observatory: In-flight fusion of hyperspectral imaging and waveform light detection and ranging for three-dimensional studies of ecosystems. *Journal of Applied Remote Sensing*, 1(1), 013536.

Asner, G.P., Martin, R.E., Anderson, C.B. and Knapp, D.E. 2015. Quantifying forest canopy traits: Imaging spectroscopy versus field survey. *Remote Sensing of Environment*, 158, 15–27.

Asner, G.P., Martin, R.E., Knapp, D.E., Tupayachi, R., Anderson, C.B., Sinca, F., Vaughn, N.R. and Llactayo, W. 2017. Airborne laser-guided imaging spectroscopy to map forest trait diversity and guide conservation. *Science*, 355, 385–389.

Bai, Z.G., Dent, D.L., Olsson, L. and Schaepman, M.E. 2008. Proxy global assessment of land degradation. *Soil Use and Management*, 24(3), 223–234.

Baldocchi, D.D., Xu, L. and Kiang, N. 2004. How plant functional-type, weather, seasonal drought, and soil physical properties alter water and energy fluxes of an oak–grass savanna and an annual grassland. *Agricultural and Forest Meteorology*, 123(1–2), 13–39. https://doi.org/10.1016/j.agrformet.2003.11.006.

Balzotti, C.S., Asner, G.P., Taylor, P.G., Cleveland, C.C., Cole, R., Martin, R.E., Nasto, M., Osborne, B.B., Porder, S., Townsend, A.R. 2016. Environmental controls on canopy foliar nitrogen distributions in a Neotropical lowland forest. *Ecol. Appl.*, 26, 2451–2464. doi:10.1002/eap.1408.

Barthlott, W., Hostert, A., Kier, G., Küper, W., Kreft, H., Mutke, J., Rafiqpoor, M.D. and Sommer, J.H. 2007. Geographic patterns of vascular plant diversity at continental to global scales (Geographische Muster der Gefäßpflanzenvielfalt im kontinentalen und globalen Maßstab). *Erdkunde*, 305–315.

Belnap, J., Welter, J.R., Grimm, N.B., Barger, N. and Ludwig, J.A. 2005. Linkages between microbial and hydrologic processes in arid and semiarid watersheds. *Ecology*, 86(2), 298–307.

Blair, J.B., Rabine, D.L. and Hofton, M.A. 1999. the laser vegetation imaging sensor: A medium-altitude, digitisation-only, airborne laser altimeter for mapping vegetation and topography. *ISPRS Journal of Photogrammetry and Remote Sensing*, 54(2–3), 115–122.

Bobbink, R., Hicks, K., Galloway, J., Spranger, T., Alkemade, R., Ashmore, M. et al. 2010. Global assessment of nitrogen deposition effects on terrestrial plant diversity: A synthesis. *Ecological Applications*, 20, 30–59.

Borel, C.C. and Gerstl, S.A. 1994. Nonlinear spectral mixing models for vegetative and soil surfaces. *Remote Sensing of Environment*, 47(3), 403–416.

Brede, B., Lau, A., Bartholomeus, H.M. and Kooistra, L. 2017. Comparing RIEGL RiCOPTER UAV LiDAR derived canopy height and DBH with terrestrial LiDAR. *Sensors*, 17(10), 2371.

Breshears, D.D., Nyhan, J.W., Heil, C.E. and Wilcox, B.P. 1998. Effects of woody plants on microclimate in a semiarid woodland: soil temperature and evaporation in canopy and intercanopy patches. *International Journal of Plant Sciences*, 159(6), 1010–1017.

Buddenbaum, H., Seeling, S. and Hill, J. 2013. Fusion of full-waveform LiDAR and imaging spectroscopy remote sensing data for the characterization of forest stands. *International Journal of Remote Sensing*, 34(13), 4511–4524. https://doi.org/10.1080/01431161.2013.776721.

Budei, B.C., St-Onge, B., Hopkinson, C., Audet, F.A. 2018. Identifying the genus or species of individual trees using a three-wavelength airborne lidar system. *Remote Sensing of Environment*, 204, 632–647. doi:10.1016/j.rse.2017.09.037.

Bugmann, H. 2001. A review of forest gap models. *Climatic Change*, 51, 259–305.

Bye, I.J., North, P.R.J., Los, S.O., Kljun, N., Rosette, J.A.B., Hopkinson, C., Chasmer, L. and Mahoney, C. 2017. Estimating forest canopy parameters from satellite waveform LiDAR by inversion of the FLIGHT three-dimensional radiative transfer model. *Remote Sensing of Environment*, 188, 177–189. https://doi.org/10.1016/j.rse.2016.10.048.

Calders, K., Newnham, G., Burt, A., Murphy, S., Raumonen, P., Herold, M. et al. 2015a. Nondestructive estimates of above-ground biomass using terrestrial laser scanning. *Methods in Ecology and Evolution*, 6(2), 198–208. https://doi.org/10.1111/2041-210X.12301.

Calders, K., Schenkels, T., Bartholomeus, H., Armston, J., Verbesselt, J. and Herold, M. 2015b. Monitoring spring phenology with high temporal resolution terrestrial LiDAR measurements. *Agricultural and Forest Meteorology*, 203, 158–168. https://doi.org/10.1016/j.agrformet.2015.01.009.

Cao, M., Prince, S.D., Small, J. and Goetz, S.J. 2004. Remotely sensed interannual variations and trends in terrestrial net primary productivity 1981–2000. *Ecosystems*, 7(3), 233–242.

Caughlin, T.T., Graves, S.J., Asner, G.P., Van Breugel, M., Hall, J.S., Martin, R.E., Ashton, M.S. and Bohlman, S.A. 2016. A hyperspectral image can predict tropical tree growth rates in single-species stands. *Ecological Applications*, 26, 2367–2373.

Chavana-Bryant, C., Malhi, Y., Wu, J., Asner, G.P., Anastasiou, A., Enquist, B.J. et al. 2017. Leaf aging of Amazonian canopy trees as revealed by spectral and physiochemical measurements. *New Phytologist*, 214(3), 1049–1063.

Chen, G. and Hay, G.J. 2011. A support vector regression approach to estimate forest biophysical parameters at the object level using airborne LiDAR transects and quickbird data. *Photogrammetric Engineering & Remote Sensing*, 77(7), 733–741.

Chu, H.J., Wang, C.K., Kong, S.J. and Chen, K.C. 2016. Integration of full-waveform LiDAR and hyperspectral data to enhance tea and areca classification. *GIScience & Remote Sensing*, 53(4), 542–559.

Clark, M.L., Clark, D.B. and Roberts, D.A. 2004. Small-footprint LiDAR estimation of sub-canopy elevation and tree height in a tropical rain forest landscape. *Remote Sensing of Environment*, 91(1), 68–89. 53(4), 542–559. https://doi.org/10.1080/15481603.2016.1177249.

Cleary, M.B., Pendall, E. and Ewers, B.E. 2008. Testing sagebrush allometric relationships across three fire chronosequences in Wyoming, USA. *Journal of Arid Environments*, 72(4), 285–301. https://doi.org/10.1016/j.jaridenv.2007.07.013.

Cook, B.D., Corp L.A., Nelson R.F., Middleton E.M., Morton D.C., McCorkel J.T., Masek J.G., Ranson K.J., Ly V. and Montesano P.M. 2013. NASA Goddard's LiDAR, hyperspectral and thermal (G-LiHT) airborne imager. *Remote Sensing*, 5, 4045–4066.

Cooper, S., Roy, D., Schaaf, C. and Paynter, I. 2017. Examination of the potential of terrestrial laser scanning and structure-from-motion photogrammetry for rapid nondestructive field measurement of grass biomass. *Remote Sensing*, 9(6), 531. https://doi.org/10.3390/rs9060531.

Dahlin, K.M. 2016. Spectral diversity area relationships for assessing biodiversity in a wildland-agriculture matrix. *Ecological Applications*, 26, 2756–2766.

Dahlin, K.M., Asner, G.P. and Field, C.B. 2012. Environmental filtering and land-use history drive patterns in biomass accumulation in a Mediterranean-type landscape. *Ecological Applications*, 22(1), 104–118.

Dahlin, K.M., Asner G.P. and Field C.B. 2013. Environmental and community controls on plant canopy chemistry in a Mediterranean-type ecosystem. *Proceedings of the National Academy of Sciences of the United States of America*, 110, 6895–900.

Dalponte, M., Bruzzone, L. and Gianelle, D. 2008. Fusion of hyperspectral and LIDAR remote sensing data for classification of complex forest areas. *IEEE Transactions on Geoscience and Remote Sensing*, 46(5), 1416–1427.

Danson, F.M., Gaulton, R., Armitage, R.P., Disney, M., Gunawan, O., Lewis, P., Pearson, G. and Ramirez, A.F. 2014. Developing a dual-wavelength full-waveform terrestrial laser scanner to characterize forest canopy structure. *Agricultural and Forest Meteorology*, 198, 7–14. https://doi.org/10.1016/j.agrformet.2014.07.007.

Detto, M., Asner, G.P., Muller-Landau, H.C. and Sonnentag, O. 2015. Spatial variability in tropical forest leaf area density from multireturn LiDAR and modeling. *Journal of Geophysical Research: Biogeosciences*, 120(2), 294–309. doi: 10.1002/2014JG002774.

Donohue, R.J., Roderick, M.L., McVicar, T.R. and Farquhar, G.D. 2013. Impact of CO2 fertilization on maximum foliage cover across the globe's warm, arid environments. *Geophysical Research Letters*, 40(12), 3031–3035.

Díaz, S., Kattge J., Cornelissen J.H.C., Wright I.J., Lavorel S., Dray S. et al. 2016. The global spectrum of plant form and function. *Nature*, 529, 167–171.

Eitel, J.U., Vierling, L.A. and Long, D.S. 2010. Simultaneous measurements of plant structure and chlorophyll content in broadleaf saplings with a terrestrial laser scanner. *Remote Sensing of Environment*, 114(10), 2229–2237.

Eswaran, H., Lal, R. and Reich, P.F. 2001. Land degradation: An overview. *Responses to Land Degradation*, 20–35.

Field, C.B., Barros, V.R., Intergovernmental Panel on Climate Change (Eds.), 2014. Climate Change 2014: Impacts, Adaptation, and Vulnerability: Working Group II Contribution to the Fifth Assessment Report of the Intergovernmental Panel on Climate Change. Cambridge University Press, New York, NY.

Fisher, R.A., Muszala, S., Verteinstein, M., Lawrence, P., Xu, C., McDowell, N.G. et al. 2015. Taking off the training wheels: The properties of a dynamic vegetation model without climate envelopes. *Geoscientific Model Development Discussions*, 8, 3293–3357.

Forbey, J.S., Patricelli, G.L., Delparte, D.M., Krakauer, A.H., Olsoy, P.J., Fremgen, M.R. et al. 2017. Emerging technology to measure habitat quality and behavior of grouse: Examples from studies of greater sage-grouse. *Wildlife Biology*, wlb-00238. https://doi.org/10.2981/wlb.00238.

Friedlingstein, P., Meinshausen, M., Arora, V.K., Jones, C.D., Anav, A., Liddicoat, S.K. and Knutti, R. 2014. Uncertainties in CMIP5 climate projections due to carbon cycle feedbacks. *Journal of Climate*, 27, 511–526.

Gamon, J.A. and Surfus, J.S. 1999. Assessing leaf pigment content and activity with a reflectometer. *The New Phytologist*, 143(1), 105–117.

García, M., Gajardo, J., Riaño, D., Zhao, K., Martín, P. and Ustin, S. 2015. Canopy clumping appraisal using terrestrial and airborne laser scanning. *Remote Sensing of Environment*, 161, 78–88. https://doi.org/10.1016/j.rse.2015.01.030.

Gastellu-Etchegorry, J.P., Yin, T., Lauret, N., Cajgfinger, T., Gregoire, T., Grau, E. et al. 2015. Discrete Anisotropic Radiative Transfer (DART 5) for modeling airborne and satellite spectroradiometer and LIDAR acquisitions of natural and urban landscapes. *Remote Sensing*, 7(2), 1667–1701. doi:10.3390/rs70201667.

Gaveau, D.L.A. and Hill, R.A. 2003. Quantifying canopy height underestimation by laser pulse penetration in small-footprint airborne laser scanning data. *Canadian Journal of Remote Sensing*, 29(5), 650–657.

Gerke, M. and Xiao, J. 2014. Fusion of airborne lasers canning point clouds and images for supervised and unsupervised scene classification. *ISPRS Journal of Photogrammetry and Remote Sensing*, 87, 78–92. https://doi.org/10.1016/j.isprsjprs.2013.10.011.

Ghosh, A., Fassnacht, F.E., Joshi, P.K. and Koch, B. 2014. A framework for mapping tree species combining hyperspectral and LiDAR data: Role of selected classifiers and sensor across three spatial scales. *International Journal of Applied Earth Observation and Geoinformation*, 26, 49–63. https://doi.org/10.1016/j.jag.2013.05.017.

Gilbert, N. 2011. Science enters desert debate. *Nature News*, 477, 262–262. https://doi.org/10.1038/477262a.

Gitelson, A.A., Merzlyak, M.N. and Chivkunova, O.B. 2001. Optical properties and nondestructive estimation of anthocyanin content in plant leaves. *Photochemistry and Photobiology*, 74(1), 38–45. https://doi.org/10.1562/0031-8655(2001)074<0038:OPANEO>2.0.CO;2.

Glenn, N.F., Neuenschwander, A., Vierling, L.A., Spaete, L., Li, A., Shinneman, D.J., Pilliod, D.S., Arkle, R.S. and McIlroy, S.K. 2016. Landsat 8 and ICESat-2: Performance and potential synergies for quantifying dryland ecosystem vegetation cover and biomass. *Remote Sensing of Environment*, 185, 233–242.

Glenn, N.F., Spaete, L.P., Sankey, T.T., Derryberry, D.R., Hardegree, S.P. and Mitchell, J.J. 2011. Errors in LiDAR-derived shrub height and crown area on sloped terrain. *Journal of Arid Environments*, 75(4), 377–382.

Goetz, S. and Dubayah, R. 2011. Advances in remote sensing technology and implications for measuring and monitoring forest carbon stocks and change. *Carbon Management*, 2(3), 231–244. https://doi.org/10.4155/cmt.11.18.

Gonzalez de Tanago, J., Lau, A., Bartholomeus, H., Herold, M., Avitabile, V., Raumonen, P. et al. 2018. Estimation of above-ground biomass of large tropical trees with terrestrial LiDAR. *Methods in Ecology and Evolution*, 9, 223–234. https://doi.org/10.1111/2041-210X.12904.

Gough, C.M., Curtis, P.S., Hardiman, B.S., Scheuermann, C. and Bond-Lamberty, B. 2016. Innovative Viewpoint: Disturbance, complexity, and succession of net ecosystem production in North America's temperate deciduous forests. *Ecosphere*, 7, 1–15.

Greaves, H.E., Vierling, L.A., Eitel, J.U.H., Boelman, N.T., Magney, T.S., Prager, C.M. and Griffin, K.L. 2017. Applying terrestrial LiDAR for evaluation and calibration of airborne LiDAR-derived shrub biomass estimates in Arctic tundra. *Remote Sensing Letters*, 8(2), 175–184. https://doi.org/10.1080/2150704X.2016.1246770.

Guanter, L., Kaufmann, H., Segl, K., Foerster, S., Rogass, C., Chabrillat, S. et al. 2015. The EnMAP spaceborne imaging spectroscopy mission for earth observation. *Remote Sensing*, 7(7), 8830–8857.

Guo, X., Coops, N.C., Tompalski, P., Nielsen, S.E., Bater, C.W. and John Stadt, J. 2017. Regional mapping of vegetation structure for biodiversity monitoring using airborne LiDAR data. *Ecological Informatics*, 38, 50–61. https://doi.org/10.1016/j.ecoinf.2017.01.005.

Hakala, T., Suomalainen, J., Kaasalainen, S. and Chen, Y. 2012. Full waveform hyperspectral LiDAR for terrestrial laser scanning. *Optics Express*, 20(7), 7119–7127.

Hancock, S., Gaulton, R. and Danson, F.M. 2017. Angular reflectance of leaves with a dual-wavelength terrestrial LiDAR and its implications for leaf-bark separation and leaf moisture estimation. *IEEE Transactions on Geoscience and Remote Sensing*, 55(6), 3084–3090. https://doi.org/10.1109/TGRS.2017.2652140.

Hansen, M.C., DeFries, R.S., Townshend, J.R.G., Carroll, M., Dimiceli, C. and Sohlberg, R.A. 2003. Global percent tree cover at a spatial resolution of 500 meters: First results of the MODIS vegetation continuous fields algorithm. *Earth Interactions*, 7(10), 1–15. https://doi.org/10.1175/1087-3562(2003)007<0001:GPTCAA>2.0.CO;2.

Hardiman, B.S., Bohrer, G., Gough, C.M., Vogel, C.S., Curtis, P.S., Vogel, S., Curtis, S. and Hardiman, S. 2014. The role of canopy structural complexity in wood net primary production of a maturing northern deciduous forest. *Ecology*, 92, 1818–1827.

Hartzell, P., Glennie, C. and Khan, S. 2017. Terrestrial hyperspectral image shadow restoration through LiDAR fusion. *Remote Sensing*, 9(5), 421. https://doi.org/10.3390/rs9050421.

Herrmann, S.M., Anyamba, A. and Tucker, C.J. 2005. Recent trends in vegetation dynamics in the African Sahel and their relationship to climate. *Global Environmental Change*, 15(4), 394–404. https://doi.org/10.1016/j.gloenvcha.2005.08.004.

Herzfeld, U.C., McDonald, B.W., Wallin, B.F., Neumann, T.A., Markus, T., Brenner, A. and Field, C. 2014. Algorithm for detection of ground and canopy cover in micropulse photon-counting lidar altimeter data in preparation for the ICESat-2 mission. *IEEE Transactions on Geoscience and Remote Sensing*, 52(4), 2109–2125.

Hopkinson, C., Chasmer, L.E., Sass, G., Creed, I.F., Sitar, M., Kalbfleisch, W. and Treitz, P. 2005. Vegetation class dependent errors in lidar ground elevation and canopy height estimates in a boreal wetland environment. *Canadian Journal of Remote Sensing*, 31(2), 191–206.

Hruska, R., Mitchell, J., Anderson, M. and Glenn, N.F. 2012. Radiometric and geometric analysis of hyperspectral imagery acquired from an unmanned aerial vehicle. *Remote Sensing*, 4(9), 2736–2752.

Huang, J., Li, Y., Fu, C., Chen, F., Fu, Q., Dai, A. et al. 2017. Dryland climate change: Recent progress and challenges. *Reviews of Geophysics*, 55.

Huang, Z., Turner, B.J., Dury, S.J., Wallis, I.R. and Foley, W.J. 2004. Estimating foliage nitrogen concentration from HYMAP data using continuum removal analysis. *Remote Sensing of Environment*, 93(1), 18–29. https://doi.org/10.1016/j.rse.2004.06.008.

Hudak, A.T., Lefsky, M.A., Cohen, W.B., Berterretche, M. 2002. Integration of LiDAR and Landsat ETM+ data for estimating and mapping forest canopy height. *Remote Sens. Environ.* 82, 397–416. https://doi.org/10.1080/014311698215748.

Hurrell, J.W., Holland, M.M., Gent, P.R., Ghan, S., Kay, J.E., Kushner, P.J. et al. 2013. The community earth system model: A framework for collaborative research. *Bulletin of the American Meteorological Society*, 94, 1339–1360.

Ilangakoon, N., Glenn, N.F., Dashti, H., Painter, T., Mikesell, D., Spaete, L., Mitchell, J., Shannon, K. 2018. Constraining plant functional types in a semi-arid ecosystem with waveform lidar. *Remote Sensing of Environment*, 209, 497–509.

Irwin, K., Beaulne, D., Braun, A. and Fotopoulos, G. 2017. Fusion of SAR, Optical imagery and airborne LiDAR for surface water detection. *Remote Sensing*, 9(9), 890. DOI:10.3390/rs9090890.

Jacquemoud, S., Baret, F. 1990. PROSPECT: A model of leaf optical properties spectra. *Remote Sensing of Environment*, 34, 75–91. doi:10.1016/0034-4257(90)90100-Z.

Jakubauskas, M., Kindscher, K. and Debinski, D. 2001. Spectral and biophysical relationships of montane sagebrush communities in multi-temporal SPOT XS data. *International Journal of Remote Sensing*, 22(9), 1767–1778.

Jetz, W., Cavender-Bares J., Pavlick R., Schimel D., Davis F.W., Asner G.P. et al. 2016. Monitoring plant functional diversity from space. *Nature Plants*, 2, 16024.

Kampe, T.U., Johnson, B.R., Kuester, M.A. and Keller, M. 2010. NEON: The first continental-scale ecological observatory with airborne remote sensing of vegetation canopy biochemistry and structure. *Journal of Applied Remote Sensing*, 4(1), 043510.

Kattge, J., Diaz, S., Lavorel, S., Prentice, I.C., Leadley, P., Bonisch, G. et al. 2011. TRY—a global database of plant traits. *Global Change Biology*, 17, 2905–2935.

Kobayashi, H. and Iwabuchi, H. 2008. A coupled 1-D atmosphere and 3-D canopy radiative transfer model for canopy reflectance, light environment, and photosynthesis simulation in a heterogeneous landscape. *Remote Sensing of Environment*, 112, 173–185.

Koetz, B., Sun, G., Morsdorf, F., Ranson, K.J., Kneubühler, M., Itten, K., Allgöwer, B. 2007. Fusion of imaging spectrometer and LIDAR data over combined radiative transfer models for forest canopy characterization. *Remote Sensing of Environment*, 106, 449–459.

Kokaly, R.F. and Clark, R.N. 1999. Spectroscopic determination of leaf biochemistry using band-depth analysis of absorption features and stepwise multiple linear regression. *Remote Sensing of Environment*, 67(3), 267–287. https://doi.org/10.1016/S0034-4257(98)00084-4.

Labate, D., Ceccherini, M., Cisbani, A., De Cosmo, V., Galeazzi, C., Giunti, L., Melozzi, M., Pieraccini, S. and Stagi, M. 2009. The PRISMA payload optomechanical design, a high performance instrument for a new hyperspectral mission. *Acta Astronautica*, 65(9–10), pp.1429–1436.

Lal, R., 2004. Soil carbon sequestration impacts on global climate change and food security. *Science*, 304(5677), 1623–1627.

Latifi, H., Fassnacht, F. and Koch, B. 2012. Forest structure modeling with combined airborne hyperspectral and LiDAR data. *Remote Sensing of Environment*, 121, 10–25.

Laurin, G.V., Chen, Q., Lindsell, J.A., Coomes, D.A., Del Frate, F., Guerriero, L., Pirotti, F. and Valentini, R. 2014. Above ground biomass estimation in an African tropical forest with lidar and hyperspectral data. *ISPRS Journal of Photogrammetry and Remote Sensing*, 89, 49–58.

Lawrence, D.M., Oleson, K.W., Flanner, M.G., Thornton, P.E., Swenson, S.C., Lawrence, P.J. et al. 2011. Parameterization improvements and functional and structural advances in Version 4 of the Community Land Model. *Journal of Advances in Modeling Earth Systems*, 3, M03001.

Lee, C.M., Cable, M.L., Hook, S.J., Green, R.O., Ustin, S.L., Mandl, D.J. and Middleton, E.M. 2015. An introduction to the NASA Hyperspectral InfraRed Imager (HyspIRI) mission and preparatory activities. *Remote Sensing of Environment*, 167, 6–19. https://doi.org/10.1016/j.rse.2015.06.012.

Li, A., Dhakal, S., Glenn, N.F., Spaete, L.P., Shinneman, D.J., Pilliod, D.S., Arkle, R.S. and McIlroy, S.K. 2017a. LiDAR aboveground vegetation biomass estimates in shrublands: Prediction, uncertainties and application to coarser scales. *Remote Sensing*, 9(9), 903.

Li, A., Glenn, N.F., Olsoy, P.J., Mitchell, J.J. and Shrestha, R. 2015. Aboveground biomass estimates of sagebrush using terrestrial and airborne LiDAR data in a dryland ecosystem. *Agricultural and Forest Meteorology*, 213, 138–147. https://doi.org/10.1016/j.agrformet.2015.06.005.

Li, Y., Guo, Q., Su, Y., Tao, S., Zhao, K. and Xu, G. 2017b. Retrieving the gap fraction, element clumping index, and leaf area index of individual trees using single-scan data from a terrestrial laser scanner. *ISPRS Journal of Photogrammetry and Remote Sensing*, 130, 308–316. https://doi.org/10.1016/j.isprsjprs.2017.06.006.

Liang, S. 2005.*Quantitative Remote Sensing of Land Surfaces*, vol. 30. John Wiley & Sons.

Liu, L., Pang, Y., Li, Z., Si, L. and Liao, S. 2017. Combining airborne and terrestrial laser scanning technologies to measure forest understorey volume. *Forests*, 8(4), 111. https://doi.org/10.3390/f8040111.

Luo, S., Wang, C., Xi, X., Zeng, H., Li, D., Xia, S. and Wang, P. 2015. Fusion of airborne discrete-return LiDAR and hyperspectral data for land cover classification. *Remote Sensing*, 8(1), 3. https://doi.org/10.3390/rs8010003.

Magney, T.S., Eusden, S.A., Eitel, J.U.H., Logan, B.A., Jiang, J. and Vierling, L.A. 2014. Assessing leaf photoprotective mechanisms using terrestrial LiDAR: Towards mapping canopy photosynthetic performance in three dimensions. *New Phytologist*, 201(1), 344–356. https://doi.org/10.1111/nph.12453.

Medvigy, D., Wofsy, S.C., Munger, J.W., Hollinger, D.Y. and Moorcroft, P.R. 2009. Mechanistic scaling of ecosystem function and dynamics in space and time: Ecosystem demography model version 2. *Journal of Geophysical Research: Biogeosciences*, 114, 1–21.

Miltiadou, M., Warren, M.A., Grant, M. and Brown, M. 2015. Alignment of hyperspectral imagery and full-waveform LiDAR data for visualization and classification purposes. *ISPRS—International Archives of the Photogrammetry, Remote Sensing and Spatial Information Sciences*, XL-7/W3, 1257–1264. https://doi.org/10.5194/isprsarchives-XL-7-W3-1257-2015.

Mirik, M., Norland, J.E., Biondini, M.E., Crabtree, R.L. and Michels, G.J. 2007. Relationships between remotely sensed data and biomass components in a big sagebrush (Artemisia tridentata) dominated area in Yellowstone National Park. *Turkish Journal of Agriculture and Forestry*, 31(2), 135–145.

Mitchell, J., Glenn, N., Sankey, T. and Derryberry, D., Anderson, M. and Hruska, R. 2011. Sagebrush canopy height and shape estimations using small footprint LiDAR. *Photogrammetric Engineering and Remote Sensing*, 77(5), 521–530.

Mitchell, J.J., Glenn, N.F., Sankey, T.T., Derryberry, D.R. and Germino, M.J. 2012. Remote sensing of sagebrush canopy nitrogen. *Remote Sensing of Environment*, 124, 217–223. https://doi.org/10.1016/j.rse.2012.05.002.

Mitchell, J.J., Shrestha, R., Spaete, L.P. and Glenn, N.F. 2015. Combining airborne hyperspectral and LiDAR data across local sites for upscaling shrubland structural information: Lessons for HyspIRI. *Remote Sensing of Environment*, 167, 98–110.

Molan, Y.E., Refahi, D. and Tarashti, A.H. 2014. Mineral mapping in the Maherabad area, eastern Iran, using the HyMap remote sensing data. *International Journal of Applied Earth Observation and Geoinformation*, 27, 117–127.

Moorcroft, P.R., Hurtt, G.C. and Pacala, S.W. 2001. A method for scaling vegetation dynamics: The ecosystem demography model (ED). *Ecological Monographs*, 71, 557–585.

Moskal, L.M. and Zheng, G. 2011. Retrieving forest inventory variables with terrestrial laser scanning (TLS) in urban heterogeneous forest. *Remote Sensing*, 4(12), 1–20. https://doi.org/10.3390/rs4010001.

Mueller, E.N., Wainwright, J. and Parsons, A.J. 2007. The stability of vegetation boundaries and the propagation of desertification in the American Southwest: A modelling approach. *Ecological Modelling*, 208(2), 91–101.

Mundt, J.T., Streutker, D.R. and Glenn, N.F. 2006. Mapping sagebrush distribution using fusion of hyperspectral and LiDAR classifications. *Photogrammetric Engineering & Remote Sensing*, 72(1), 47–54.

Mutanga, O., Skidmore, A.K. and Prins, H.H.T. 2004. Predicting in situ pasture quality in the Kruger National Park, South Africa, using continuum-removed absorption features. *Remote Sensing of Environment*, 89(3), 393–408. https://doi.org/10.1016/j.rse.2003.11.001.

Mutlu, M., Popescu, S.C. and Zhao, K. 2008. Sensitivity analysis of fire behavior modeling with LIDAR-derived surface fuel maps. *Forest Ecology and Management*, 256(3), 289–294.

Nevalainen, O., Hakala, T., Suomalainen, J., Mäkipää, R., Peltoniemi, M., Krooks, A. and Kaasalainen, S. 2014. Fast and nondestructive method for leaf level chlorophyll estimation using hyperspectral LiDAR. *Agricultural and Forest Meteorology*, 198, 250–258. https://doi.org/10.1016/j.agrformet.2014.08.018.

Ni-Meister, W., Jupp, D.L.B., Dubayah, R. 2001. Modeling LiDAR waveforms in heterogeneous and discrete canopies. *IEEE Transactions of Geoscience and Remote Sensing*, 39, 1943–1958. doi:10.1109/36.951085.

Niinemets, Ü., Keenan, T.F. and Hallik, L. 2015. A worldwide analysis of within-canopy variations in leaf structural, chemical and physiological traits across plant functional types. *New Phytologist*, 205(3), 973–993.

Noojipady, P., Prince, S.D., Rishmawi, K. 2015. Reductions in productivity due to land degradation in the drylands of the southwestern United States. *Ecosystem Health and Sustainability*, 1, 1–15. https://doi.org/10.1890/EHS15-0020.1.

Næsset, E. and Økland, T. 2002. Estimating tree height and tree crown properties using airborne scanning laser in a boreal nature reserve. *Remote Sensing of Environment*, 79(1), 105–115.

Okin, G.S. 2008. A new model of wind erosion in the presence of vegetation. *Journal of Geophysical Ressearch: Earth Surface*, 113(F2). https://doi.org/10.1029/2007JF000758.

Okin, G.S., Roberts, D.A., Murray, B., Okin, W.J. 2001. Practical limits on hyperspectral vegetation discrimination in arid and semiarid environments. *Remote Sensing of Environment*, 77, 212–225. https://doi.org/10.1016/S0034-4257(01)00207-3.

Olschofsky, K., Mues, V. and Köhl, M. 2016. Operational assessment of aboveground tree volume and biomass by terrestrial laser scanning. *Computers and Electronics in Agriculture*, 127, 699–707. https://doi.org/10.1016/j.compag.2016.07.030.

Olsoy, P.J., Glenn, N.F. and Clark, P.E. 2014a. Estimating sagebrush biomass using terrestrial laser scanning. *Rangeland Ecology & Management*, 67(2), 224–228. https://doi.org/10.2111/REM-D-12-00186.1.

Olsoy, P.J., Glenn, N.F., Clark, P.E. and Derryberry, D.R. 2014b. Aboveground total and green biomass of dryland shrub derived from terrestrial laser scanning. *ISPRS Journal of Photogrammetry and Remote Sensing*, 88, 166–173. https://doi.org/10.1016/j.isprsjprs.2013.12.006.

Olsoy, P.J., Griggs, T.C., Ulappa, A.C., Gehlken, K., Shipley, L.A., Shewmaker, G.E. and Forbey, J.S. 2016b. Nutritional analysis of sagebrush by near-infrared reflectance spectroscopy. *Journal of Arid Environments*, 134, 125–131. https://doi.org/10.1016/j.jaridenv.2016.07.003.

Olsoy, P.J., Mitchell, J.J., Levia, D.F., Clark, P.E. and Glenn, N.F. 2016a. Estimation of big sagebrush leaf area index with terrestrial laser scanning. *Ecological Indicators*, 61, 815–821. https://doi.org/10.1016/j.ecolind.2015.10.034.

Olsoy, P.J., Shipley, L.A., Rachlow, J.L., Forbey, J.S., Glenn, N.F., Burgess, M.A. and Thornton, D.H. 2017. Unmanned aerial systems measure structural habitat features for wildlife across multiple scales. *Methods in Ecology and Evolution*. https://doi-org.proxy006.nclive.org/10.1111/2041-210X.12919.

Olsson, L., Eklundh, L. and Ardö, J. 2005. A recent greening of the Sahel—trends, patterns and potential causes. *Journal of Arid Environments*, 63(3), 556–566. https://doi.org/10.1016/j.jaridenv.2005.03.008.

Painter, T.H., Berisford, D.F., Boardman, J.W., Bormann, K.J., Deems, J.S., Gehrke, F. et al. 2016. The airborne snow observatory: Fusion of scanning LiDAR, imaging spectrometer, and physically-based modeling for mapping snow water equivalent and snow albedo. *Remote Sensing of Environment*, 184, 139–152.

Palmer, M.W., Earls, P.G., Hoagland, B.W., White, P.S. and Wohlgemuth, T. 2002. Quantitative tools for perfecting species lists. *Environmetrics*, 13, 121–137.

Paris, C., Kelbe, D., van Aardt, J. and Bruzzone, L. 2017. A novel automatic method for the fusion of ALS and TLS LiDAR data for robust assessment of tree crown structure. *IEEE Transactions on Geoscience and Remote Sensing*, 55(7), 3679–3693. https://doi.org/10.1109/TGRS.2017.2675963.

Parker, G.G. 1995. Structure and microclimate of forest canopies. In: Lowman, M.D. and Nadkarni, N.M. (Eds.), *Forest Canopies*. First edition. Academic Press, San Diego, pp. 73–106.

Parton, W., Schimel, D., Cole, C. and Ojima, D. 1987. Analysis of factors controlling soil organic matter levels in Great Plains grasslands. *Soil Science Society of America Journal*, 51, 1173–1179.

Pausas, J.G., Vallejo, R.V. 1999. The role of fire in European Mediterranean ecosystems. In: Chuvieco, E. (Ed.), *Remote Sensing of Large Wildfires in the European Mediterranean Basin*. Springer-Verlag, pp. 3–16.

Paz-Kagan, T., Brodrick, P.G., Vaughn, N.R., Das, A.J., Stephenson, N.L., Nydick, K.R. and Asner, G.P. 2017a. What mediates tree mortality during drought in the southern Sierra Nevada? *Ecological Applications*, 27, 2443–2457.

Paz-Kagan, T., Vaughn, N.R., Martin, R.E., Brodrick, P.G., Stephenson, N.L., Das, A.J., Nydick, K.R. and Asner, G.P. 2017b. Landscape-scale variation in canopy water content of giant sequoias during drought. *Forest Ecology and Management*, 0–1.

Pelevin, V., Zlinszky, A., Khimchenko, E. and Toth, V. 2017. Ground truth data on chlorophyll- *a*, chromophoric dissolved organic matter and suspended sediment concentrations in the upper water layer as obtained by LIF LiDAR at high spatial resolution. *International Journal of Remote Sensing*, 38(7), 1967–1982. https://doi.org/10.1080/01431161.2016.1274446.

Pereira, H.M., Ferrier, S., Walters, M., Geller, G.N., Jongman, R.H.G., Scholes, R.J. et al. 2013. Essential Biodiversity Variables. *Science*, 339, 277–278.

Petropoulos, G.P., Arvanitis, K. and Sigrimis, N. 2012. Hyperion hyperspectral imagery analysis combined with machine learning classifiers for land use/cover mapping. *Expert Systems with Applications*, 39(3), 3800–3809.

Poulter, B., Frank, D., Ciais, P., Myneni, R.B., Andela, N., Bi, J. et al. 2014. Contribution of semi-arid ecosystems to interannual variability of the global carbon cycle. *Nature*, 509, 600–604. https://www.nature.com/articles/nature13376.

Prandi, F., Magliocchetti, D., Poveda, A., De Amicis, R., Andreolli, M. and Devigili, F. 2016. New Approach for forest inventory estimation and timber harvesting planning in mountain areas: The SLOPE project. *ISPRS—International Archives of the Photogrammetry, Remote Sensing and Spatial Information Sciences*, XLI-B3, 775–782. https://doi.org/10.5194/isprsarchives-XLI-B3-775-2016.

Puttonen, E., Suomalainen, J., Hakala, T., Räikkönen, E., Kaartinen, H., Kaasalainen, S. and Litkey, P. 2010. Tree species classification from fused active hyperspectral reflectance and LIDAR measurements. *Forest Ecology and Management*, 260(10), 1843–1852. https://doi.org/10.1016/j.foreco.2010.08.031.

Qi, W. and Dubayah, R.O. 2016. Combining Tandem-X InSAR and simulated GEDI LiDAR observations for forest structure mapping. *Remote Sensing of Environment*, 187, 253–266.

Ramiya, A.M., Nidamanuri, R.R. and Ramakrishnan, K. 2016. A supervoxel-based spectro-spatial approach for 3D urban point cloud labelling. *International Journal of Remote Sensing*, 37(17), 4172–4200. https://doi.org/10.1080/01431161.2016.1211348.

Ravi, S., D'Odorico, P., Wang, L., White, C.S., Okin, G.S., Macko, S.A., Collins, S.L. 2009. Post-fire resource redistribution in desert grasslands: A possible negative feedback on land degradation. *Ecosystems* 12, 434–444. https://doi.org/10.1007/s10021-009-9233-9.

Ray, T.W. and Murray, B.C. 1996. Nonlinear spectral mixing in desert vegetation. *Remote Sensing of Environment*, 55(1), 59–6.

Riaño, D., Chuvieco, E., Ustin, S.L., Salas, J., Rodríguez-Pérez, J.R., Ribeiro, L.M., Viegas, D.X., Moreno, J.M. and Fernández, H. 2007. Estimation of shrub height for fuel-type mapping combining airborne LiDAR and simultaneous color infrared ortho imaging. *International Journal of Wildland Fire*, 16(3), 341–348.

Ritchie, J.C., Everitt, J.H., Escobar, D.E., Jackson, T.J., and Davis, M.R. 1992. Airborne laser measurements of rangeland canopy cover. *Journal of Range Management*, 2, 189–193. 4.

Roberts, D.A., Smith, M.O. and Adams, J.B. 1993. Green vegetation, nonphotosynthetic vegetation, and soils in AVIRIS data. *Remote Sensing of Environment*, 44(2–3), 255–269.

Rocchini, D., Balkenhol, N., Carter, G.A., Foody, G.M., Gillespie, T.W., He, K.S. et al. 2010. Remotely sensed spectral heterogeneity as a proxy of species diversity: Recent advances and open challenges. *Ecological Informatics*, 5, 318–329.

Rocchini, D., Wohlgemuth, T., Ghisleni, S. and Chiarucci, A. 2008. Spectral rarefaction: Linking ecological variability and plant species diversity. *Community Ecolog*, 9, 169–176.

Roth, K.L., Dennison, P.E. and Roberts, D.A. 2012. Comparing endmember selection techniques for accurate mapping of plant species and land cover using imaging spectrometer data. *Remote Sensing of Environment*, 127, 139–152.

Rutherford, W.A., Painter, T.H., Ferrenberg, S., Belnap, J., Okin, G.S., Flagg, C. and Reed, S.C. 2017. Albedo feedbacks to future climate via climate change impacts on dryland biocrusts. *Scientific Reports*, 7.

Safriel, U. and Adeel, Z. 2008. Development paths of drylands: Thresholds and sustainability. *Sustainability Science*, 3(1), 117–123.

Saha, M.V., D'Odorico, P. and Scanlon, T.M. 2017. Albedo changes after fire as an explanation of fire-induced rainfall suppression. *Geophysical Research Letters*, 44(8), 3916–3923.

Sankey, J.B., Eitel, J.U., Glenn, N.F., Germino, M.J. and Vierling, L.A. 2011. Quantifying relationships of burning, roughness, and potential dust emission with laser altimetry of soil surfaces at submeter scales. *Geomorphology*, 135(1), 181–190.

Sankey, J.B., Ravi, S., Wallace, C.S., Webb, R.H. and Huxman, T.E. 2012. Quantifying soil surface change in degraded drylands: Shrub encroachment and effects of fire and vegetation removal in a desert grassland. *Journal of Geophysical Research: Biogeosciences*, 117(G2), 1–2.

Sarrazin, D., van Aardt, J., Asner, G.P., McGlinchy, J., Messinger, D.W. and Wu, J. 2010. Fusing waveform lidar and hyperspectral data for species-level structural assessment in savanna ecosystems. *In Laser Radar Technology and Applications XV* (Vol. 7684, p. 76841H). International Society for Optics and Photonics. https://doi.org/10.1117/12.849882

Schmitt, M. and Zhu, X.X. 2016. Data fusion and remote sensing: An ever-growing relationship. *IEEE Geoscience and Remote Sensing Magazine*, 4(4), 6–23. https://doi.org/10.1109/MGRS.2016.2561021.

Sellers, P.J., Randall, D.A., Collatz, G.J., Berry, J.A., Field, C.B., Dazlich, D.A., Zhang, C., Collelo, G.D. and Bounoua, L. 1996. A revised land surface parameterization (SiB2) for atmospheric GCMs. *Part I: Model Formulation.*

Serbin, S.P., Singh, A., Desai, A.R., Dubois, S.G., Jablonski, A.D., Kingdon, C.C., Kruger, E.L. and Townsend, P.A. 2015. Remotely estimating photosynthetic capacity, and its response to temperature, in vegetation canopies using imaging spectroscopy. *Remote Sensing of Environment*, 167, 78–87.

Serbin, S.P., Singh, A., McNeil, B.E., Kingdon, C.C. and Townsend, P.A. 2014. Spectroscopic determination of leaf morphological and biochemical traits for northern temperate and boreal tree species. *Ecological Applications*, 24, 1651–1669.

Serrano, L., Penuelas, J. and Ustin, S.L. 2002. Remote sensing of nitrogen and lignin in Mediterranean vegetation from AVIRIS data: Decomposing biochemical from structural signals. *Remote Sensing of Environment*, 81(2), 355–364. https://doi.org/10.1016/S0034-4257(02)00011-1.

Sherrill, K.R., Lefsky, M.A., Bradford, J.B. and Ryan, M.G. 2008. Forest structure estimation and pattern exploration from discrete-return lidar in subalpine forests of the central Rockies. *Canadian Journal of Forest Research*, 38(8), 2081–2096.

Shugart, H.H., Asner, G.P., Fischer, R., Huth, A., Knapp, N., Le Toan, T. and Shuman, J.K. 2015. Computer and remote-sensing infrastructure to enhance large-scale testing of individual-based forest models. *Frontiers in Ecology and the Environment*, 13, 503–511.

Silvestrini, R.A., Soares-Filho, B.S., Nepstad, D., Coe, M., Rodrigues, H. and Assunção, R. 2011. Simulating fire regimes in the Amazon in response to climate change and deforestation. *Ecological Applications*, 21(5), 1573–1590.

Singh, A., Serbin, S.P., McNeil, B.E., Kingdon, C.C. and Townsend, P.A. 2015. Imaging spectroscopy algorithms for mapping canopy foliar chemical and morphological traits and their uncertainties. *Ecological Applications*, 25(8), 2180–2197.

Skidmore, A.K., Ferwerda, J.G., Mutanga, O., Van Wieren, S.E., Peel, M., Grant, R.C., Prins, H.H., Balcik, F.B. and Venus, V. 2010. Forage quality of savannas—Simultaneously mapping foliar protein and polyphenols for trees and grass using hyperspectral imagery. *Remote Sensing of Environment*, 114(1), 64–72. https://doi.org/10.1016/j.rse.2009.08.010.

Smith, M.O., Ustin, S.L., Adams, J.B. and Gillespie, A.R. 1990. Vegetation in deserts: I. A regional measure of abundance from multispectral images. *Remote Sensing of Environment*, 31(1), 1–26.

Smolander, S. and Stenberg, P. 2005. Simple parameterizations of the radiation budget of uniform broadleaved and coniferous canopies. *Remote Sensing of Environment*, 94(3), 355–363.

Spaete, L.P., Glenn, N.F., Derryberry, D.R., Sankey, T.T., Mitchell, J.J. and Hardegree, S.P. 2011. Vegetation and slope effects on accuracy of a LiDAR-derived DEM in the sagebrush steppe. *Remote Sensing Letters*, 2(4), 317–326.

Stark, S.C., Enquist, B.J., Saleska, S.R., Leitold, V., Schietti, J., Longo, M., Alves, L.F., Camargo, P.B. and Oliveira, R.C. 2015. Linking canopy leaf area and light environments with tree size distributions to explain Amazon forest demography. *Ecology Letters*, 18(7), 636–645. doi:10.1111/ele.12440.

Stavros, E.N., Schimel, D., Pavlick, R., Serbin, S., Swann, A., Duncanson, L. et al. 2017. ISS observations offer insights into plant function. *Nature Ecology and Evolution*, 1. doi:10.1038/s41559-017-0194.

Stoker, J. 2009. Visualization of multiple-return LiDAR data: Using voxels. *Photogramm. Eng. Remote Sens*, 75(2), 109–112.

Streutker, D. and Glenn, N. 2006. LiDAR measurement of sagebrush steppe vegetation heights. *Remote Sensing of Environment*, 102, 135–145.

Su, J.G. and Bork, E.W. 2007. Characterization of diverse plant communities in Aspen Parkland rangeland using LiDAR data. *Applied Vegetation Science*, 10, 407–416.

Su, J. and Bork, E. 2006. Influence of vegetation, slope, and lidar sampling angle on DEM accuracy. *Photogrammetric Engineering & Remote Sensing*, 72(11), 1265–1274.

Sun, J., Shi, S., Gong, W., Yang, J., Du, L., Song, S. et al. 2017. Evaluation of hyperspectral LiDAR for monitoring rice leaf nitrogen by comparison with multispectral LiDAR and passive spectrometer. *Scientific Reports*, 7, 40362. https://doi.org/10.1038/srep40362.

Swatantran, A., Dubayah, R., Roberts, D., Hofton, M. and Blair, J.B. 2011. Mapping biomass and stress in the Sierra Nevada using LiDAR and hyperspectral data fusion. *Remote Sensing of Environment*, 115(11), 2917–2930.

Swatantran, A., Tang, H., Barrett, T., DeCola, P. and Dubayah, R., 2016. Rapid, high-resolution forest structure and terrain mapping over large areas using single photon lidar. *Scientific reports*, 6, 28277.

Thomas, R.Q., Hurtt, G.C., Dubayah, R. and Schilz, M.H. 2008. Using lidar data and a height-structured ecosystem model to estimate forest carbon stocks and fluxes over mountainous terrain. *Canadian Journal of Remote Sensing*, 34(sup2), S351–S363.

Tilman, D., Wedin, D. and Knops, J. 1996. Productivity and sustainability influenced by biodiversity in grassland ecosystems. *Nature*, 379(6567), 718.

UNCCD. 2011. *Desertification: A Visual Synthesis. United Nations Convention to Combat Desertification.* Zoi Environment Network, Bresson, France, p. 52. ISBN: 978-92-95043-49-7.

United Nations Environment Management Group. 2011. Global Drylands: A UN system-wide response [WWW Document]. Global Drylands: A UN system-wide response. URL http://www.unccd.int/Lists/SiteDocumentLibrary/Publications/Global_Drylands_Full_Report.pdf (accessed 30 October, 2016).

Vaccari, S., Calders, K., Herold, M., Bartholomeus, H. and van Leeuwen, M. 2013. Terrestrial laser scanning for 3D forest modeling and reflectance simulation through radiative transfer. In *Proceedings SilviLaser 2013, 13th International Conference on LiDAR Applications for Assessing Forest Ecosystems, Beijing, China, 9–11 October 2013*, pp. 18–25.

Van der Zande, D., Mereu, S., Nadezhdina, N., Cermak, J., Muys, B., Coppin, P. and Manes, F. 2009. 3D upscaling of transpiration from leaf to tree using ground-based LiDAR: Application on a Mediterranean Holm oak (Quercus ilex L.) tree. *Agricultural and Forest Meteorology*, 149(10), 1573–1583.

Verhoef, W., Bach, H. 2007. Coupled soil–leaf-canopy and atmosphere radiative transfer modeling to simulate hyperspectral multi-angular surface reflectance and TOA radiance data. *Remote Sensing of Environment*, 109, 166–182. doi:10.1016/j.rse.2006.12.013.

Vierling, L.A., Xu, Y., Eitel, J.U.H. and Oldow, J.S. 2013. Shrub characterization using terrestrial laser scanning and implications for airborne LiDAR assessment. *Canadian Journal of Remote Sensing*, 38(6), 709–722. https://doi.org/10.5589/m12-057.

Vincent, G., Antin, C., Dauzat, J., Grau, E. and Durrieu, S. 2015. September. Mapping plant area index of tropical forest by LiDAR: Calibrating ALS with TLS. In *SilviLaser 2015*, pp. 146–148.

Vlek, P.L., Le, Q.B. and Tamene, L. 2008. November. African land degradation in a world of global atmospheric change. In *Sustainable Development in Drylands–Meeting the Challenge of Global Climate Change*, vol. 7, p. 104).

Wallace, L., Lucieer, A. and Watson, C.S. 2014. Evaluating tree detection and segmentation routines on very high resolution UAV LiDAR data. *IEEE Transactions on Geoscience and Remote Sensing*, 52(12), 7619–7628.

Wang, H. and Glennie, C. 2015. Fusion of waveform LiDAR data and hyperspectral imagery for land cover classification. *ISPRS Journal of Photogrammetry and Remote Sensing*, 108, 1–11. https://doi.org/10.1016/j.isprsjprs.2015.05.012.

Wang, R., Gamon, J.A., Cavender-Bares, J., Townsend, P.A. and Zygielbaum, A.I. 2017. The spatial sensitivity of the spectral diversity-biodiversity relationship: An experimental test in a prairie grassland. *Ecological Applications.*

Wang, X., Cheng, X., Gong, P., Huang, H., Li, Z. and Li, X. 2011. Earth science applications of ICESat/GLAS: A review. *International Journal of Remote Sensing*, 32(23), 8837–8864.

Weltz, M.A., Ritchie, J.C. and Fox, H.D. 1994. Comparison of laser and field measurements of vegetation heights and canopy cover. *Water Resources Research*, 30, 1311–1320.

Wendt, C.K., Beringer, J., Tapper, N.J. and Hutley, L.B. 2007. Local boundary-layer development over burnt and unburnt tropical savanna: An observational study. *Boundary-Layer Meteorology*, 124(2), 291–304.

Wertin, T.M., Belnap, J. and Reed, S.C. 2017. Experimental warming in a dryland community reduced plant photosynthesis and soil CO2 efflux although the relationship between the fluxes remained unchanged. *Functional Ecology*, 31(2), 297–305.

Widlowski, J.L., Mio, C., Disney, M., Adams, J., Andredakis, I., Atzberger, C. et al. 2015. The fourth phase of the radiative transfer model intercomparison (RAMI) exercise: Actual canopy scenarios and conformity testing. *Remote Sensing of Environment*, 169, 418–437.

Wilkes, P., Lau, A., Disney, M., Calders, K., Burt, A., de Tanago, J.G., Bartholomeus, H., Brede, B. and Herold, M. 2017. Data acquisition considerations for terrestrial laser scanning of forest plots. *Remote Sensing of Environment*, 196, 140–153. https://doi.org/10.1016/j.rse.2017.04.030.

Wu, C., Niu, Z., Tang, Q. and Huang, W. 2008. Estimating chlorophyll content from hyperspectral vegetation indices: Modeling and validation. *Agricultural and Forest Meteorology*, 148(8), 1230–1241. https://doi.org/10.1016/j.agrformet.2008.03.005.

Xu, Z., Wu, L., Shen, Y., Li, F., Wang, Q. and Wang, R. 2014. Tridimensional reconstruction applied to cultural heritage with the use of camera-equipped UAV and terrestrial laser scanner. *Remote Sensing*, 6(11), 10413–10434. https://doi.org/10.3390/rs61110413.

Zald, H.S.J., Wulder, M.A., White, J.C., Hilker, T., Hermosilla, T., Hobart, G.W., Coops, N.C. 2016. Integrating Landsat pixel composites and change metrics with LiDAR plots to predictively map forest structure and aboveground biomass in Saskatchewan, Canada. *Remote Sensing of Environment*, 176, 188–201. doi:http://dx.doi.org/10.1016/j.rse.2016.01.015.

Zarco-Tejada, P.J., González-Dugo, V. and Berni, J.A. 2012. Fluorescence, temperature and narrow-band indices acquired from a UAV platform for water stress detection using a micro-hyperspectral imager and a thermal camera. *Remote Sensing of Environment*, 117, 322–337. https://doi.org/10.1016/j.rse.2011.10.007.

Zhang, C. and Qiu, F. 2012. Mapping individual tree species in an urban forest using airborne LiDAR data and hyperspectral imagery. *Photogrammetric Engineering & Remote Sensing*, 78(10), 1079–1087. https://doi.org/10.14358/PERS.78.10.1079.

Zheng, G., Moskal, L.M. and Kim, S.-H. 2013. Retrieval of effective leaf area index in heterogeneous forests with terrestrial laser scanning. *IEEE Transactions on Geoscience and Remote Sensing*, 51(2), 777–786. https://doi.org/10.1109/TGRS.2012.2205003.

Zhou, J., Proisy, C., Descombes, X., Hedhli, I., Barbier, N., Zerubia, J., Gastellu-Etchegorry, J.P., Couteron, P. 2010. Tree crown detection in high resolution optical and LiDAR images of tropical forest. In: Neale, C.M.U. and Maltese, A. (Eds.), *SPIE*, p. 78240Q. doi:10.1117/12.86506.8.

Zhu, X., Wang, T., Skidmore, A.K., Darvishzadeh, R., Niemann, K.O. and Liu, J. 2017. Canopy leaf water content estimated using terrestrial LiDAR. *Agricultural and Forest Meteorology*, 232, 152–162. https://doi.org/10.1016/j.agrformet.2016.08.016.

Zolkos, S.G., Goetz, S.J. and Dubayah, R. 2013. A meta-analysis of terrestrial aboveground biomass estimation using LiDAR remote sensing. *Remote Sensing of Environment*, 128, 289–298. https://doi.org/10.1016/j.rse.2012.10.017.

14 Fifty-Years of Advances in Hyperspectral Remote Sensing of Agriculture and Vegetation—Summary, Insights, and Highlights of Volume I

Fundamentals, Sensor Systems, Spectral Libraries, and Data Mining for Vegetation

Prasad S. Thenkabail, John G. Lyon, and Alfredo Huete

CONTENTS

Hyperspectral point or imaging spectroscopy data refers to a simultaneous acquisition of geometrically registered narrow spectral wavebands across the electromagnetic spectrum. The key factor here is that the data are acquired in numerous unique spectral bands across the entire length of the spectrum (e.g., 400–2500 nm; 2500–14,500 nm). Ideally, hyperspectral data are acquired continuously or near-continuously (e.g., every 1-nm) throughout the electromagnetic spectrum, resulting in a spectral signature (Figure 14.1; Hamzeh et al., 2016) from each pixel. In contrast, multispectral broadband sensors such as Landsat Enhanced Thematic Mapper Plus (ETM) only capture data from specific broad spectral bands (e.g., Figure 14.1). In Figure 14.1, hyperspectral

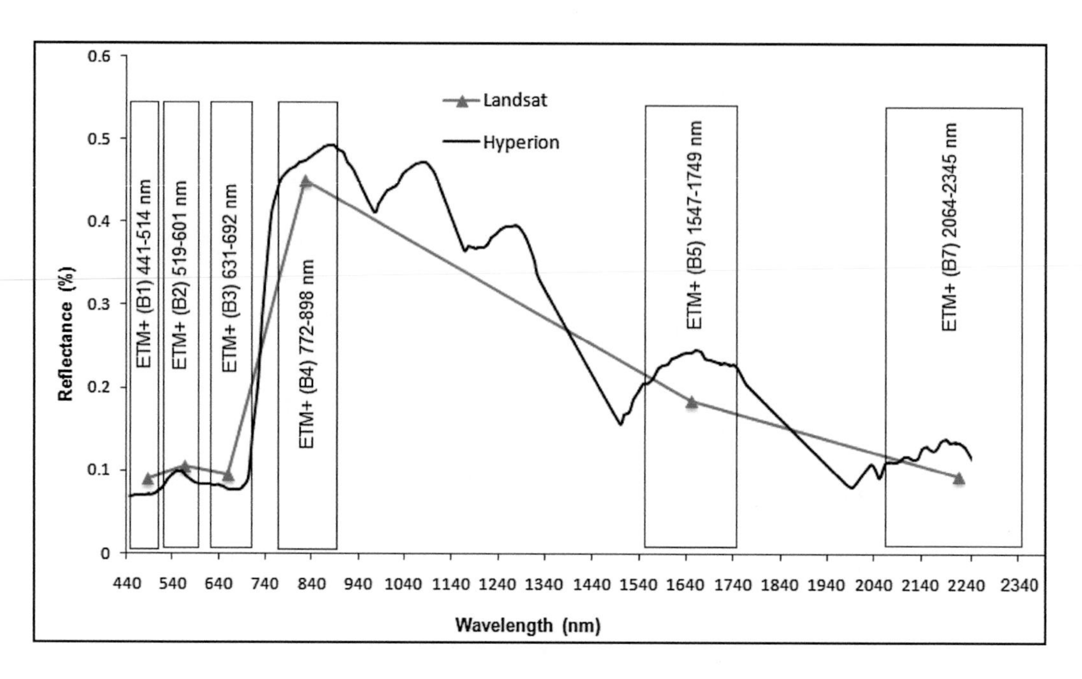

FIGURE 14.1 Sugarcane spectral profile from the hyperspectral Hyperion narrowband sensor versus the multispectral Landsat ETM broadband sensor. Sugarcane spectral curve, extracted from the hyperspectral narrowband Hyperion and multispectral broadband Landsat 7 ETM+ images with the overlap between the spectral wavelength coverage of these images. (From S. Hamzeh et al. 2016. *International Journal of Applied Earth Observation and Geoinformation*, Volume 52, 2016, Pages 412–421, https://doi.org/10.1016/j.jag.2010.01.007.)

data of sugarcane crop acquired from the Earth Observing-1 (EO-1) spaceborne Hyperion sensor consisting of 242 narrowbands (each of 10-nm bandwidths, across the 400–2500 nm spectral range) is compared with 6 broadbands of Landsat ETM across the optical spectral region. It is abundantly clear from Figure 14.1 that hyperspectral data capture spectral variability throughout the spectrum whereas Landsat ETM averages the spectrum over its broad wavelength ranges. Hyperspectral data can of course be captured in much narrower wavelengths (e.g., 1-nm or even less) if point spectroradiometers are used (Thenkabail et al., 2000, 2004a, 2004b) or even by using hyperspectral imaging data such as from the spaceborne EO-1 Hyperion which captures data in 10-nm bandwidths in 400–2500 nm (Pearlman et al., 2003). In addition, hyperspectral data (point or imaging) do not necessarily need to cover the entire spectral range or to have hundreds of bands. However, it is essential to have adequate numbers of narrowbands (<20-nm) spread across the electromagnetic spectrum such as visible (VIS), near-infrared (NIR), short-wave infrared (SWIR), mid-wave infrared (MWIR), and long-wave infrared (LWIR) ranges. As needed, hyperspectral data can be acquired across these entire wavelength ranges or in selective regions, but always as continuous or near-continuous data in discrete narrowbands along the spectrum under consideration.

Hyperspectral data are gathered from various platforms: ground-based (lab and field), drone-level flown, manned airborne (low and high altitude), and spaceborne platforms (Thenkabail, 2015). In each of these cases, data are acquired over a spatial dimension or pixel (e.g., Figures 14.2 and 14.3). To measure spectral properties of objects on Earth (e.g., Figure 14.2 showing crops versus bare soil), we not only need high spectral resolution, but we also need high spatial resolution. Often, in reality, every pixel in hyperspectral imaging data has spectral signatures of mixed objects. For example, EO-1 Hyperion data with 30-m x 30-m pixels (1 pixel = 0.09 hectares) can capture pure signatures of crops when fields are larger than 0.09 hectares, which is often the case for an overwhelming

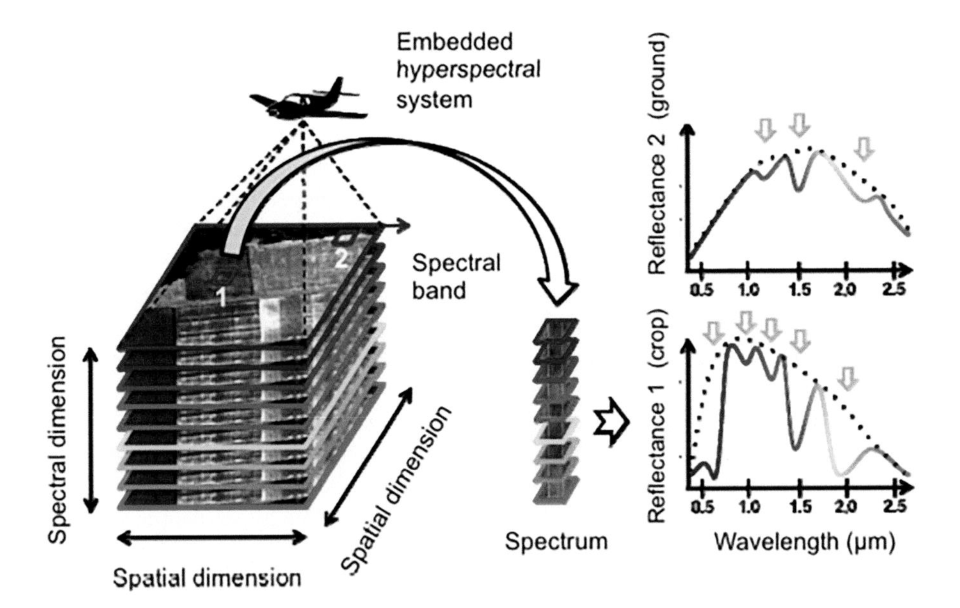

FIGURE 14.2 Diagram of a remote sensing hyperspectral image and acquiring hyperspectral data bank of features or themes. The reflectance spectrum of each of the two materials (crops and bare ground) is shown with its continuum (dotted line) and its absorption bands (arrows). (From X. Ceamanos and S. Valero, 2016. *Optical Remote Sensing of Land Surface,* edited by N. Baghdadi and M. Zribi, Elsevier, Pages 163–200, https://doi.org/10.1016/B978-1-78548-102-4.50004-1.)

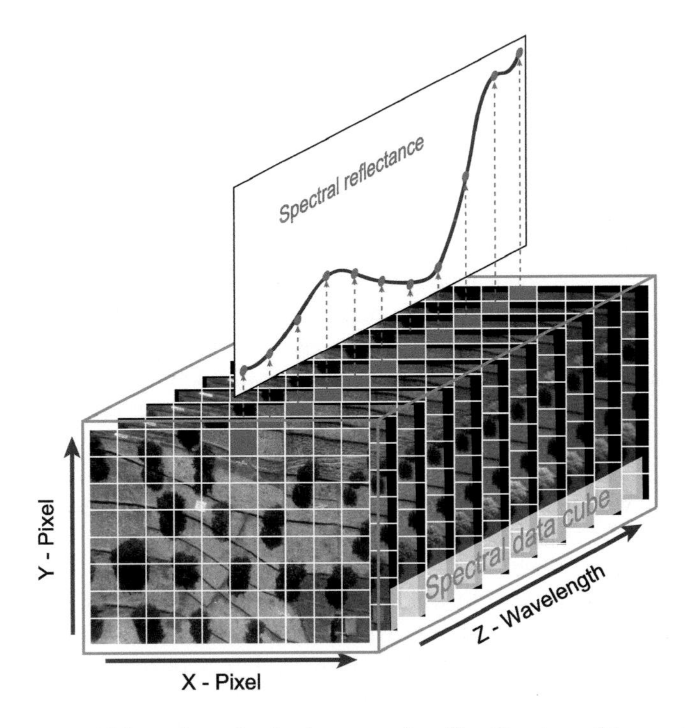

FIGURE 14.3 Hyperspectral data cube and gathering spectral profiles. Structure of the spectral data cube. In the field campaign, the liquid crystal tunable filter (LCTF) imager recurrently captured spectral images ranging from 460 to 780 nm at 10 nm intervals, which equates to 33 images per cycle. (From T. Ishida et al. 2018. *Computers and Electronics in Agriculture*, Volume 144, 2018, Pages 80–85, https://doi.org/10.1016/j.compag.2017.11.027.)

proportion of farms in the world. So, crop types will have pure crop signatures (by ensuring edge-pixels are ignored). However, in a 30-m x 30-m area of a tropical forest, for example, one can have several plant species within a single pixel. So, the hyperspectral data of a 30-m pixel in the forest case will contain a mixed tree species signature. So, ideally, we should be in a position to capture pure spectra of individual objects (e.g., a single plant). However, that is feasible only when high spatial and spectral resolution capabilities are available (e.g., a hyperspectral imaging sensor with spatial resolution of a few centimeters). For most studies pertaining to agriculture and vegetation, hyperspectral data captured at about a 1-m spatial resolution will be quite adequate. By meeting these conditions, hyperspectral imaging spectroscopy data will allow us to capture unique signatures of various objects (not necessarily vegetation) on Earth (e.g., Figure 14.2; Ceamanos and Valero, 2016), which are then stacked as a data cube (e.g., Figure 14.3; Ishida et al., 2018) for thematic data analysis and classification.

Today, many hyperspectral sensors are available worldwide. The widely used, operational airborne hyperspectral sensors include: (1) US National Aeronautics and Space Administration (NASA's) Airborne Visible/Infrared Imaging Spectrometer (AVIRIS), and AVIRIS-New Generation (AVIRIS NG), (2) FENIX 1 K sensor from SPACIM, (3) Australia's airborne hyperspectral imaging sensor (HyMap), and (4) many others (Panda et al., 2015). Characteristics of some of the most widely used ground-based spectroradiometers and spaceborne sensors (of past, present, near-future) are provided in Table 14.1. Readers are also encouraged to study Panda et al. (2015) for a more detailed listing and characterization of a multitude of hyperspectral sensors across platforms over the last 50 years.

14.1 ADVANCES IN HYPERSPECTRAL REMOTE SENSING OF AGRICULTURE AND VEGETATION

There is currently about 50-years' worth of hyperspectral research on agriculture and vegetation (Vane and Goetz, 1993; Thenkabail et al., 2000, 2002, 2004a, 2004b, 2012, 2013; Galvão et al., 2009; Goetz, 2009; Ustin et al., 2009; Mariotto et al., 2013; Thorp et al., 2013; Marshall and Thenkabail, 2014, 2015; Aasen et al., 2015; Roberts et al., 2015; Roth et al., 2015a, 2015b; Thenkabail, 2015; Middleton et al., 2016; Clark, 2017; Féret et al., 2017; Mutanga et al., 2017)). This body of knowledge has provided us with numerous strengths and limitations of hyperspectral data in understanding, modeling, mapping, and monitoring various biophysical, biochemical, and structural quantities of agricultural crops and a wide range of natural vegetation spread across forests, savannas, wetlands, and other ecosystems. Key highlights in these advances include: **(1) Data mining**: needed to reduce data redundancy and for efficient and effective computation of massively large datasets from imaging spectroscopy; **(2) Overcoming Hughes phenomenon:** the requirement to overcome "Hughes Phenomenon" or the curse of high dimensionality of hyperspectral data without which classifiers either require an exponential increase in training samples to maintain classification integrity or classifier accuracies suffer with increased dimensionality; **(3) Hyperspectral narrowbands (HNBs) and hyperspectral vegetation indices (HVIs)** to estimate specific biophysical, biochemical, and structural variables otherwise not possible using multispectral broadband (MBB) data; **(4) Advances by HNBs and HVIs in comparison with multispectral broadbands (MBBs):** significantly improve the modeling power of HNBs and HVIs as opposed to MBB data and the indices derived from them; **(5) Species and cultivar separability:** capability of HNBs to separate plant species, crop types, or vegetation types well beyond the capabilities of MBBs; **(6) Increased classification accuracies:** of crop types, species types, and vegetation types relative to MBBs; and **(7) Others**: a number of important advances discussed throughout this Second Edition of Hyperspectral Remote Sensing of Vegetation.

A common strength of hyperspectral remote sensing in agriculture and vegetation studies is the specific features or quantities that one can detect and study from specific wavelength regions throughout the electromagnetic spectrum (e.g., Figure 14.4; Thenkabail et al., 2013, 2015). Specific crop variables such as the pigments (anthocyanin, carotenoids, chlorophyll), biomass, leaf area index (LAI), water stress, moisture conditions, starch, lignin, cellulose, and, recently, even fluorescence

TABLE 14.1
Characteristics of Spaceborne Hyperspectral Sensors (Either in Orbit or Planned for Launch) for Ocean, Atmosphere, Land, and Water Applications Compared with ASD Spectroradiometer

Sensor, Satellite[a]	Spatial (meters)	Spectral (#)	Swath (km)	Band Range (µm)	Bandwidths (µm)	Irradiance (W $m^{-2}sr^{-1}µm^{-1}$)	Data Points (# per hectares)	Launch (Date)
I. Coastal Hyperspectral Spaceborne Imagers								
3. HICO, ISS USA	90	128	42	353–1080	5.7	See data in Neckel and Labs (1984).	0.81	2009-present
II. Atmosphere\Ozone Hyperspectral Spaceborne Imagers								
3. OMI, Aura USA	13,000 × 12,000	740	145	270–500	0.45–1	See data in Neckel and Labs (1984). Plot it	1/16,900	2004-present
3. SCIAMACHY, ENVISAT ESA	30,000 × 60,000	~2000	960	212–2384	0.2–1.5	See data in Neckel and Labs (1984).	1/180,000	2002-present
III. Land and Water Hyperspectral Spaceborne Imagers								
1. Hyperion, EO-1 USA	30	220 (196[b])	7.5	196 effective Calibrated bands VNIR (band 8 to 57 427.55 to 925.85 nm SWIR (band 79 to 224) 932.72 to 2395.53 nm	10 nm wide (approx.) for all 196 bands	See data in Neckel and Labs (1984). Plot it and obtain values for Hyperion bands	11.1	2000-present
2. CHRIS, PROBA ESA	25	19	17.5	200–1050	1.25–11	same as above	16	2001-present
3. HyspIRI VSWIR USA	60	210	145	210 bands in 380–2500 nm	10 nm wide (approx.) for all 210 bands	See data in Neckel and Labs (1984).	2.77	2020+
4. HyspIRI TIR USA	60	8	145	7 bands in 7500–12,000 nm and 1 band in 3000–5000 nm (3980 nm center)	7 bands in 7500–12000 nm	See data in Neckal and Labs (1984).	2.77	2020+
5. EnMAP Germany	30	92 108	30	420–1030 950–2450	5–10 10–20	same as above	11.1	2015+
6. PRISMA	30	250	30	400–2500	<10	same as above	11.1	2014+
7. Tiangong-1 China	20	64	30	1000–2500	~25	same as above	11.1	2011+

(*Continued*)

TABLE 14.1 (*Continued*)
Characteristics of Spaceborne Hyperspectral Sensors (Either in Orbit or Planned for Launch) for Ocean, Atmosphere, Land, and Water Applications Compared with ASD Spectroradiometer

Sensor, Satellite[a]	Spatial (meters)	Spectral (#)	Swath (km)	Band Range (μm)	Bandwidths (μm)	Irradiance (W $m^{-2}sr^{-1}\mu m^{-1}$)	Data Points (# per hectares)	Launch (Date)
I. Land and Water Hand-held Spectroradiometer								
7. ASD spectroradiometer	1134 cm^2 @ 1.2 m Nadir view 18-degree Field of view	~2100 bands 1 nm width between 400–2500 nm	N\A	~2100 effective bands	1 nm wide (approx.) in 400–2500 nm	See data in Neckel and Labs (1984). Plot it and obtain values for Hyperion bands	88,183	last 30+ years
8. Spectral Evolution SR-6500	1134 cm^2 @ 1.2 m Nadir view 18-degree Field of view	1.5 nm @ 700 nm 3.0 nm @ 1500 nm 3.8 nm @ 2100 nm 350–2500 nm	N\A	~1000 effective bands	1 nm wide (approx.) in 400–2500 nm	See data in Neckel and Labs (1984). Plot it and obtain values for Hyperion bands	88,183	last 5+ years

Source: Modified and adopted from S.S. Panda et al. 2015; P.S. Thenkabail, (Editor-in-Chief), 2015, "*Remote Sensing Handbook*" (Volume I): Remotely sensed data characterization, classification, and accuracies, Taylor and Francis Inc./CRC Press, Boca Raton, London, New York, Pages 3–60; P.S. Thenkabail, 2015. "*Remote Sensing Handbook*" (Volume II): Land Resources Monitoring, Modeling, and Mapping with Remote Sensing, Editor-in-Chief, P.S. Thenkabail, Taylor and Francis Inc./CRC Press, Boca Raton, London, New York, Pages 201–236; P.S. Thenkabail et al. 2012b. *Hyperspectral Remote Sensing of Vegetation*. CRC Press, Taylor &Francis Group, Boca Raton, London, New York, Page 781; P.S. Thenkabail, et al. 2014. *Photogrammetric Engineering and Remote Sensing*, Volume 80, Issue 4, Pages 697–709; Qi et al. 2012. Hyperspectral remote sensing in global change studies, Chapter 3. In *Hyperspectral Remote Sensing of Vegetation*. CRC Press, Taylor & Francis group, Boca Raton, London, New York. pp. 781.

Note: There are several other tables in this volume that show characteristics of other hyperspectral sensors.

[a] HICO = **Hyperspectral** Imager for the Coastal Ocean onboard International Space Station. OMI = Ozone Monitoring Instrument onboard AURA of NASA; SCIAMACHY (Scanning Imaging Absorption Spectrometer for Atmospheric CHartographY) of ESA; Hyperion EO-1 = hyperspectral sensor onboard EO-1 = Earth Observing-1; CHRIS PROBA = Compact High Resolution Imaging Spectrometer Project for On Board Autonomy satellite of ESA; HyspIRI VSWIR = Hyperspectral Infrared Imager Visible to Short Wavelength InfraRed of NASA; HyspIRI TIR = Hyperspectral Infrared Imager thermal infrared of NASA; Environmental Mapping and Analysis Program of Germany; PRISMA = PRecursore IperSpetlrale della Missione Applicativa of Italy.

[b] Of the 242 bands, 196 are unique and calibrated. These are: (A) Band 8 (427.55 nm) to band 57 (925.85 nm) that are acquired by visible and near-infrared (VNIR) sensor; and (B) Band 79 (932.72 nm) to band 224 (2395.53 nm) that are acquired by short-wave infrared (SWIR) sensor.

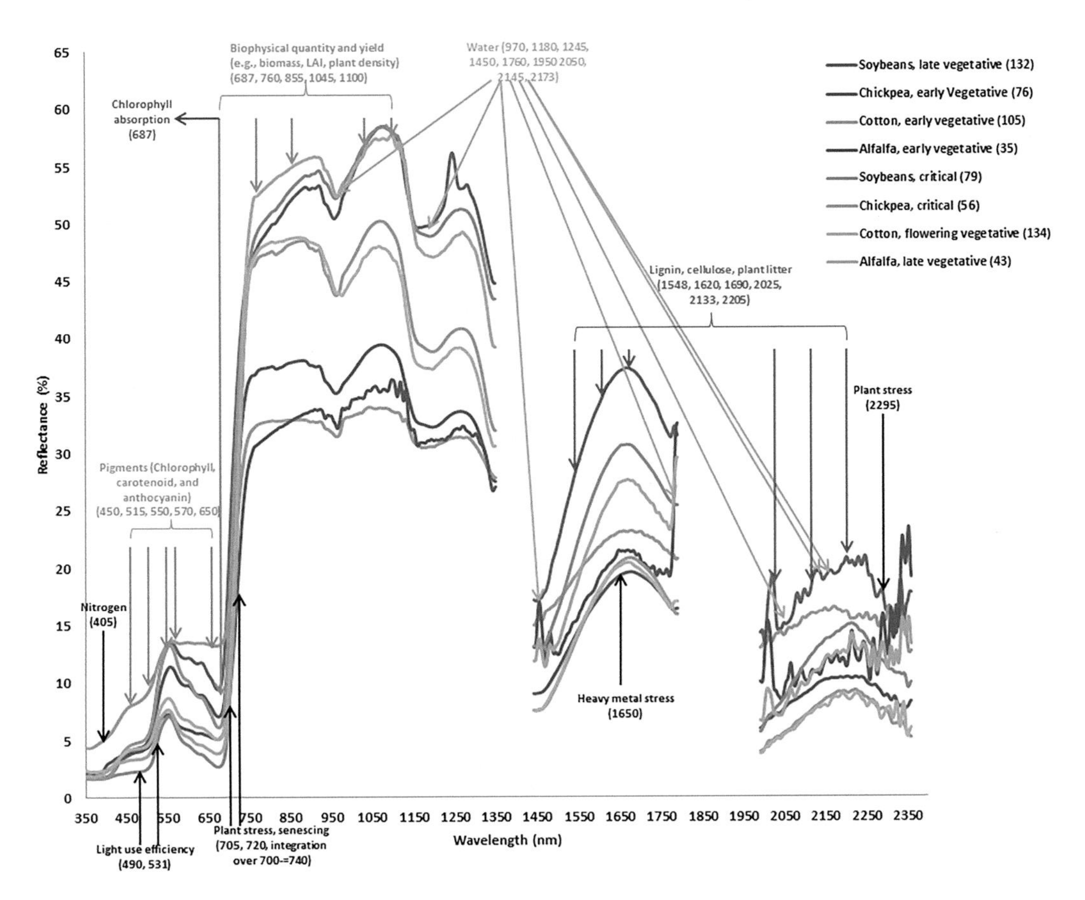

FIGURE 14.4 Hyperspectral data bank of some leading world crops. Hyperspectral data of some of the leading world crops and the critical crop features studied using data from specific portions of the spectrum.

are all best studied in specific wavelengths or specific combinations (e.g., involving indices) of wavelengths. This is fundamentally different from broadband sensors where, often, a normalized difference vegetation index (NDVI) becomes a proxy to study a wide range of Earth's features.

In Chapter 1, Dr. Prasad S. Thenkabail et al. provide a detailed specific focus on using hyperspectral data from ground-based and satellite-based sensors in the study of leading world crops as well as some of the other vegetation in forests and savannas spread across distinct agro-ecological zones of the world. The chapter discusses various methods of mining data specific to vegetation studies, approaches to reducing data redundancy, and overcoming the Hughes phenomenon. The highlight of the chapter is in determining optimal HNBs to study agricultural crop characteristics. Distinct HVIs were derived and discussed. Novel methods, such as the Lambda (λ) versus Lambda (λ) plot of computing thousands of HVIs to best model crop biophysical quantities, are presented in detail and discussed. These methods include ratio-based HVIs, normalized HVIs, derivative HVIs, and optimum multi-band HVIs (or OMBVs). The chapter illustrates various methods and approaches to model crop biophysical and biochemical quantities. The ability of hyperspectral data to classify various species and crop types and to do so with significantly higher classification accuracies relative to multispectral broadband data are presented and discussed.

14.2 HYPERSPECTRAL SENSORS: SPACEBORNE, AIRBORNE, HAND-HELD

Hyperspectral remote sensing sensors are used to acquire data from various platforms: hand-held, platform-mounted, drone-flown, airborne, and spaceborne. The most commonly used hyperspectral instruments in vegetation studies have been hand-held pointing spectroradiometers of various kinds (e.g., Castro-Esau et al., 2006; Thenkabail, 2015). These have been widely used to spectrally characterize vegetation and agricultural crops over the last 50 years (Roberts et al., 2003; Clark et al., 2005; Mariotto et al., 2013; Thenkabail et al., 2013; Marshall et al., 2014; Thenkabail, 2015). Airborne hyperspectral sensors are typically flown on many specialized missions and have played a great role in hyperspectral research of agriculture and vegetation (Toth and Jóźków, 2016; Bareth and Waldhoff, 2018). Hyperspectral data acquired from Unmanned Aerial System (UAS) platforms or drones are becoming more and more common (Toth and Jóźków, 2016; Bareth and Waldhoff, 2018). However, consistent, repeated, global coverage of hyperspectral data acquisition is only possible through various spaceborne systems (see some examples in Table 14.1). There are specialized hyperspectral spaceborne sensors to capture data of land, water, and atmosphere characteristics that are either currently in operation or have already completed their operation, or will be launched soon (Table 14.1).

In Chapter 2, Dr. Fred Ortenberg described hyperspectral remote sensing systems from all platforms except those from UAVs. The strength of the chapter is in providing technical details of the sensors including their physics and design. The differences between pushbroom scanners and whiskbroom scanners, so fundamental in hyperspectral sensor design, are discussed in detail. This is followed by detailed presentation of ground-based, airborne, and spaceborne sensors. For UAVs not covered, readers can refer to studies by Bareth and Waldhoff, (2018), Toth and Jóźków (2016), and Panda et al. (2015). Drone or UAV flown hyperspectral sensors are becoming ubiquitous. However, there is no standardization with different drone systems built by different countries and/or companies. This makes data acquired from these systems difficult to compare. Nevertheless, UAVs carrying hyperspectral sensors have huge potential in local and regional studies and can be deployed with short notice, especially over small study areas. However, their utility over global scales is very limited and will not happen at least in the foreseeable future, mainly due to security and legal issues and not because of any technical issues, which are quite mature. UAV-borne hyperspectral sensors are probably the best option given their ready local availability, local technical support, low costs, and ability to fly anytime, anywhere, as long as UAV capacity exists. They are also capable of flying under most types of cloud cover. New spaceborne hyperspectral sensors such as NASA's HyspIRI, and many others planned by many private players, are currently in various stages of deployment. Finally, Chapter 2 also discusses integration of LiDAR data with hyperspectral data. This is important because hyperspectral data are not a panacea. As we will learn from various chapters, hyperspectral data have their challenges (e.g., data redundancy, data volumes, computing speed) and their capabilities are best enhanced when integrated with other data such as LiDAR, which provides canopy height data in addition to biophysical, biochemical, and structural information obtained from hyperspectral sensors.

It is very likely in coming years that hyperspectral sensors will be put into space using micro or mini satellites (e.g., Cubesats and Smallsats) in constellations similar to existing hyperspatial systems such as PlanetScope, and will have the capability to cover the entire planet repeatedly via coverage either daily or every few days (e.g., Table 14.2). When that happens, the challenges will be in the domain of data processing and management, handling petabytes of data, and requiring capability to perform fast computing in the cloud using machine learning algorithms and artificial intelligence.

14.3 HYPERSPECTRAL SENSOR SYSTEMS AND DATA CHARACTERISTICS IN GLOBAL CHANGE STUDIES

The use of hyperspectral data in global change studies is an evolving subject. All global change studies using remote sensing, hitherto, have been conducted using multispectral broadband data acquired from systems such as Landsat-series satellites, Terra-Aqua-Suomi National Polar-orbiting Partnership (NPP) (formerly The National Polar-orbiting Operational Environmental Satellite System or NPOESS Preparatory Project or NPP) satellites, Sentinel satellites, and a multitude of other satellite

TABLE 14.2
Characteristics of the Twenty-First Century Sensors

	HyTI	PlanetScope	RapidEye	SkySat	Landsat 8	Sentinel-2
Number of satellites	1	175+	5	13	1	2
Spectral resolution	40 bands (8–10.7 μm)	4 bands (R, G. B. NIR)	5 bands (R, G, B, RE, NIR)	5 bands (R, G, B, NIR, pan)	11 bands (VNIR, SWIR, TIR, pan)	13 bands (VNIR. SWIR)
Spatial resolution	64 m	3–4 m	6.5 m	0.8–1 m	30 m multispectral, 15 m pan	10,20, or 60 m depending on band
Temporal resolution	5 days	Daily at nadir	Daily off-nadir, 5.5 days at nadir	4–5 days	16 days	5 days
Image dimension		24 by 7 km	77 km by up to 300 km	3.2 by 1.3 km	185 by 180 km	290 by 300 km
Radiometric resolution	16 bit	8 to 16 bit	8 to 16 bit	8 to 16 bit	12 to 16 bit	12 bit

Note: HyTI, PlanetScope, and other sensor characteristics for use in crop water productivity and irrigated area studies of the world crops.
Author would like to thank Dr. Itiya Aneece for helping prepare this table.

sensors providing data at coarse-resolution (≥250-m), moderate-resolution (>30-m to <250-m), and high-spatial resolution (>10-m to <30-m), the characteristics of which are discussed in detail by Panda et al. (2015). However, it is widely recognized that data acquired by hyperspectral narrowband sensors provide great advantages over multispectral broadband data (Thenkabail et al., 2014; Marshall et al., 2016; Clark, 2017). Nevertheless, except for a few hyperspectral technology demonstration satellite sensor systems such as Hyperion onboard NASA's EO-1 satellite, and the Compact High Resolution Imaging Spectrometer onboard Project for On Board European Space Agency's (ESA's) Autonomy satellite (CHRIS PROBA; see Table 14.1), there are no civilian hyperspectral systems onboard satellites in the past or at present time. Even existing satellite systems like Hyperion, CHRIS PROBA, and PRISMA only acquired data intermittently for selected places in the world without any operational coverage of the planet collected routinely and repeatedly to enable global change studies. In the coming years, only a few advanced hyperspectral systems onboard satellites are to be launched, such as HyspIRI by NASA (Table 14.1) and only a few others are planned that have operational capability with potential for coverage of the world every 19 days. Further, certain missions by private companies are beginning to take shape and probably in a decade or so, operational hyperspectral remote sensing of the Earth may become a reality much like the Landsat satellites of today.

Operational routine hyperspectral imaging of the world would certainly lead to great advancements in our understanding of Earth's vegetation when compared to existing studies using multispectral imaging (Belluco et al., 2006; Mariotto et al., 2013; Marshall et al., 2014; Thenkabail et al., 2014; Hamzeh et al., 2016; Stagakis et al., 2016). For example, Thenkabail et al. (2004a, 2004b, 2004c, 2013, 2014) clearly demonstrated that agricultural crop types were classified with as much as 30% higher accuracies using 15–20 narrowbands relative to the 6 multispectral broadband data provided by Landsat satellites.

In Chapter 3, Dr. Jiaguo Qi et al. discussed prominent global change related efforts involving satellites as used in the study of land, water, and atmospheric systems. For example, they discuss instruments such as: the Ozone Monitoring Instrument (OMI) on NASA's Earth Observing Systems (EOS's) Aura spacecraft designed to measure atmospheric composition; Hyperspectral Imager for the Coastal Ocean (HICO) used for coastal ocean studies; as well as land-focused instruments like the Moderate Resolution Imaging Spectroradiometer (MODIS). These integrated sensors (Table 14.1) enable planet-scale studies to be conducted repeatedly and routinely in a consistent manner (but with coarse spatial resolution). In addition, the chapter highlights methods and approaches required in global change studies using hyperspectral data including: support vector machines (SVMs), spectral libraries (for spectral matching techniques), hyperspectral narrowband indices, derivative indices, and neural networks. They examined the capability of hyperspectral data in detecting and mapping subtle changes that are difficult to detect or far less accurate when derived from broadband remote sensors. Some of the issues that hyperspectral data can identify and help with a study of global change include: (1) water quantity and quality, (2) carbon sequestration and fluxes, (3) greenhouse gas emissions, (4) atmospheric chemistry, (5) vegetation ecology, (6) vegetation biochemical properties, (7) invasive plant species detection, and (8) vegetation health. When hyperspectral imaging becomes operational, the data can be used to advance global change studies in virtually every aspect of issues pertaining to Earth's land, water, and atmosphere.

An important part of Chapter 3 is the recognition of the huge challenges of: handling massively large data volumes from hyperspectral remote sensing, overcoming data redundancy, ensuring efficient use of data for desired applications, and developing and advancing methods and approaches for various applications that in turn help global change studies. It is obvious that in the future, satellites will need to carry hyperspectral remote sensing systems for global mapping, modeling, and monitoring. For example, Hestir et al. (2015) compared the existing and near-future multispectral broadband and hyperspectral narrowband satellite sensors for global studies of freshwater ecosystems. They demonstrated that NASA's future HyspIRI supplies a unique measurement capability in both spatial and temporal resolutions (Figure 14.5), and provides significantly more spectral information than any other global mapper (Hestir et al., 2015).

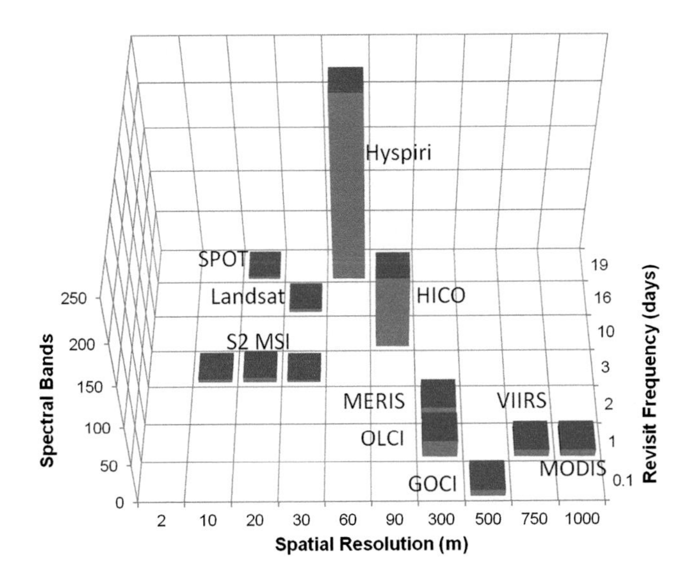

FIGURE 14.5 Characteristics of some of the global mapping sensors. The spectral, spatial, and temporal characteristics of current and near future global mappers commonly used for freshwater ecosystem measurements. (Credits: E.L Hestir et al. 2015. *Remote Sensing of Environment*, Volume 167, 2015, Pages 181–195, https://doi.org/10.1016/j.rse.2015.05.0; Creative Commons Attribution-Noncommercial-No Derivatives License.)

14.4 STUDYING VEGETATION DIVERSITY AND HEALTH THROUGH SPECTRAL TRAITS AND TRAIT VARIATIONS BASED ON HYPERSPECTRAL REMOTE SENSING

A wide range of vegetation types and their characteristics are best studied using a variety of remote sensing sensors that include multispectral, hyperspectral, RADAR, and light detection and ranging (LiDAR). Hyperspectral data are of specific focus since they provide unique opportunities for fast and reliable estimation of numerous characteristics associated with various structural, biochemical, and physiological traits (Sytar et al., 2017). Diverse and complex vegetation types and species are best studied using hyperspectral narrowband data relative to multispectral broadband data (e.g., Figure 14.6; Thenkabail et al., 2000, 2004a, 2004b). Vegetation diversity studies using broadband sensors such as Landsat-derived NDVI are insufficient as it only exploits spectral information from red and near-infrared bands and it does not consider canopy background conditions and hence it is affected by soil brightness which lowers its sensitivity to vegetation (Madonsela et al., 2017). As such NDVI is insufficient in explaining vegetation species diversity. However, use of SWIR region data from Landsat is related to plant properties (Madonsela et al., 2017) and improves vegetation diversity studies when used along with NDVI and/or related bands. Hyperspectral data further advance this understanding by explaining more of the variability encountered and resulting in as much as 30% greater accuracies when classifying complex vegetation types (Figure 14.7) relative to multispectral broadband data (Figure 14.7; Thenkabail et al., 2004a, 2004b, 2004c). Even further advances are possible when hyperspectral data are combined with other data such as LiDAR and are used through multisensor data integration called data fusion. For example, Laurin et al. (2014) showed hyperspectral data when combined with LiDAR data explained 70% of the variability in modeling aboveground biomass of tropical rainforests compared to 64% explained by hyperspectral data alone.

Vegetation condition or health is a function of numerous factors such as excess or deficient water in plants, pest or disease infestation, drought or floods, and lack of fertilizer or poor soils. Remote sensing detects these conditions of health, but requires field knowledge from certain sample locations for proper attribution of the vegetation condition or health status. Multispectral broadband data are widely used to

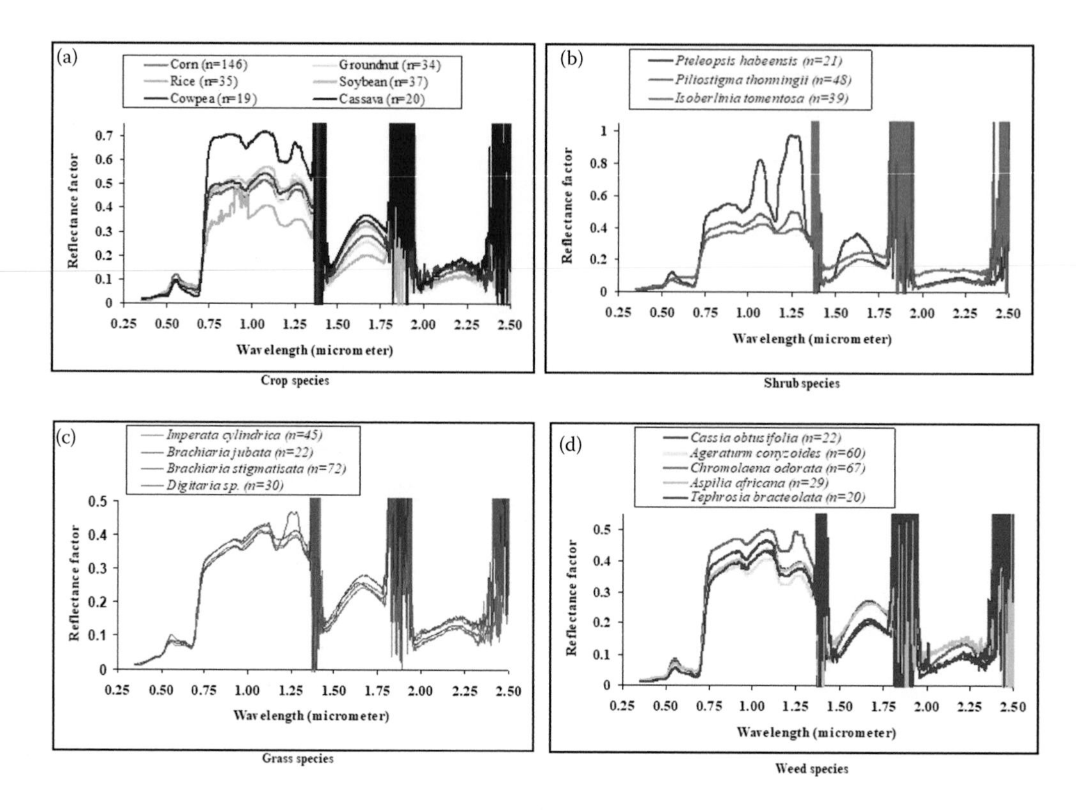

FIGURE 14.6 Hyperspectral Signature Bank of some of the world crops and vegetation collected using portable spectroradiometer. Hyperspectral datasets for vegetation and crop species where the mean spectral profile of major vegetation and crop species in the African savannas (a) crop species; (b) shrub species; (c) grass species; (d) weed species. (From P.S. Thenkabail et al. 2004a, 2004b, 2004c.)

study crop conditions or health such as in drought monitoring (Thenkabail et al., 2004c; Tadesse et al., 2005; Zhang et al., 2017a, 2017b). However, when specific conditions are targeted, specific narrowbands are needed. For example, soil arsenic content is best established through a narrowband index (Shi et al., 2016: $[R_{716}-R_{568}]/[R_{552}-R_{568}]$ and red-edge position [REP]), using reflectance at 552, 568, and 716 nm. Similarly, studies pertaining to plant species (Sonobe and Wang, 2018; also see Figure 14.6), crop water (Pasqualotto et al., 2018), and a host of other vegetation quantitative and qualitative characteristics (Thenkabail et al., 2000; Burud et al., 2016) are best performed using various hyperspectral narrowbands (HNBs) or hyperspectral vegetation indices (HVIs). Plant quantitative and qualitative characteristics studied using broadband remote sensing data are known to have relatively great uncertainties (Marshall and Thenkabail, 2014, 2015) resulting in significantly lower accuracies (Figure 14.7). Thereby, even though great advances have been made in the study of vegetation function and traits with regard to retrieving useful plant biochemical, physiological, and structural quantities across a range of spatial and temporal scales, translation of remote sensing data into meaningful descriptors of vegetation function and traits is still associated with large uncertainties. These uncertainties are due to complex interactions between leaf, canopy, soils, and atmospheric media, and significant challenges in the treatment of confounding factors in spectrum-trait relations (Houborg et al., 2015), which can only be addressed using a wide array of specific and targeted HNBs and HVIs.

In Chapter 4, Dr. Angela Lausch and Dr. Pedro Leitao provide approaches for monitoring vegetation diversity and its characteristics. They focus on the study of phylogenetic, taxonomic, structural, and functional vegetation characteristics using hyperspectral remote sensing. Spectral traits of vegetation are best studied by hyperspectral data and methods to capture their variations are presented and discussed.

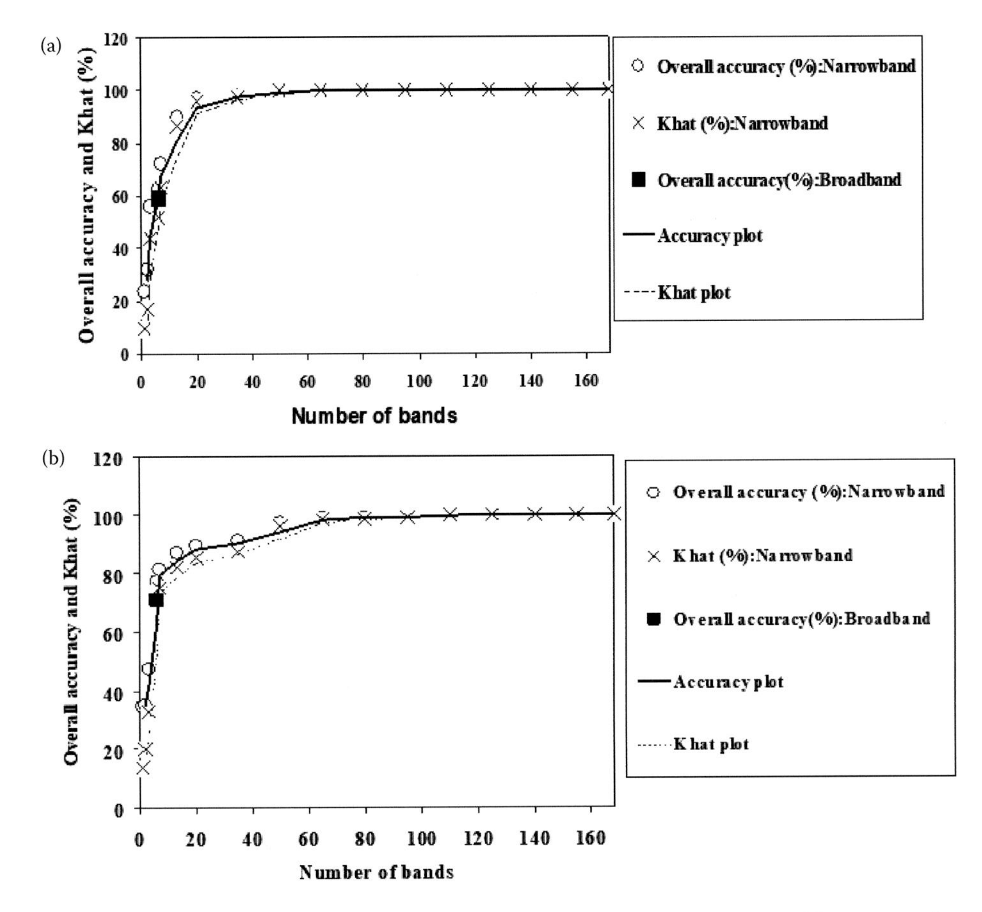

FIGURE 14.7 (a) Accuracy assessment using hyperspectral data obtained from portable spectroradiometer. Overall accuracies and Khat when classifying five weed species (*A. conyzoides*, *A. africana*, *C. obtusifolia*, *C. odorata*, and *T. bracteolata*) from the four ecoregions of the African savannas. (b) Accuracy assessment. Overall accuracies and Khat when classifying six crop species (cassava—*Manihot esculenta*, corn—*Zea mays*, cowpea—*Vigna unguiculata*, groundnut—*Arachis hypogaea*, rice—*Oryza sativ*, and soybean—*Glycine max*) from the four ecoregions of the African savannas. (From P.S. Thenkabail, 2003. *International Journal of Remote Sensing*, Volume 24(14), Pages 2879–2904.)

14.5 HYPERSPECTRAL PROXIMAL SENSING FOR PHENOTYPING PLANT BREEDING TRIALS

The future of plant breeding requires characterizing specific and all-encompassing phenological (observable) and genomic (genetic) traits. Hyperspectral proximal sensing is best suited to achieve this. A wide array of plant traits can be characterized using a surrogate or indirect variable to measure or "proximal sensing" such as the nitrogen (N) status of crops (Padilla et al., 2016), crop stress (Zhou et al., 2016), study of crop phenomics, including leaf water content, specific leaf mass, leaf chlorophyll, and LAI (Thorp et al., 2015). Further capabilities are elucidated such as detection of species types as demonstrated for grassland species (Lopatin et al., 2017), study of agricultural crops for their growth, water use, and leaf water content (Gao et al., 2009; Ge et al., 2016), and forest species detection (Asner et al., 2017). The four pillars of the crop breeding pipeline (environmental adaptation, phenotypic characterization, genetic diversity, and genetic information) are comprehensively studied through data from proximal sensing, imaging at frequent intervals, laboratory analyses of samples taken at specific intervals, and near-infrared spectroscopy (NIRS) as stated by Araus and Cairns (2014; Figure 14.8) and also discussed by White (2012).

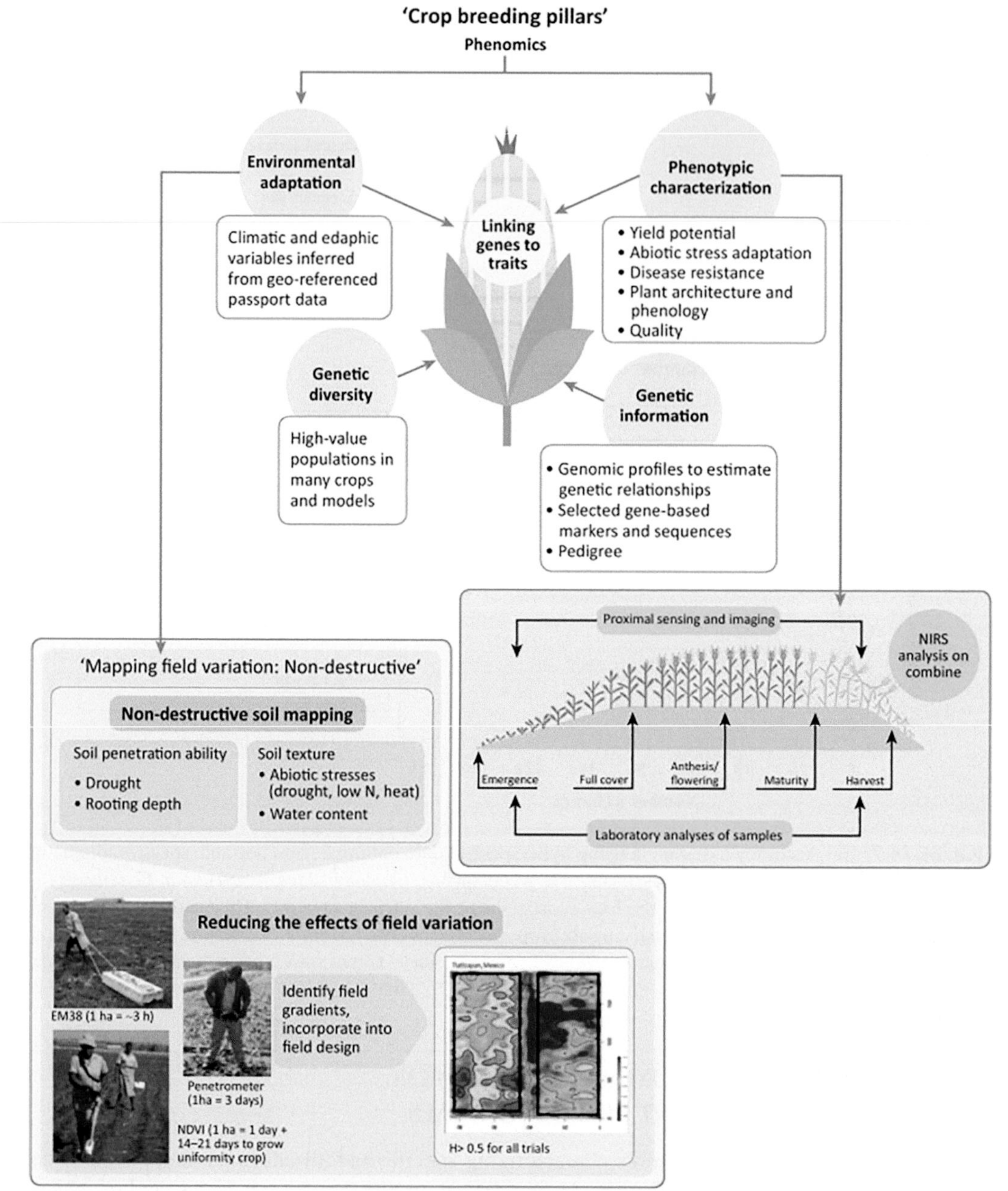

FIGURE 14.8 Phenotyping four pillars of crop breeding using hyperspectral data. (Upper) The four pillars of the crop breeding pipeline (environmental adaptation, phenotypic characterization, genetic diversity, and genetic information) and the implications of phenotyping. The importance of phenotyping is highlighted by its involvement in two of these pillars. (Lower left) Mapping field variability in a non-destructive manner implies the use of different methodological alternatives and its further integration. (Lower right) Diagram of the main categories of phenotyping techniques deployed over the life cycle of an annual seed crop. Types of data acquisition include: proximal sensing and imaging at frequent intervals, laboratory analyses of samples taken at specific intervals, and near-infrared spectroscopy (NIRS) of seed for oil or protein content during combine harvesting. (From J.L. Araus and J.E. Cairns, 2014. *Trends in Plant Science*, Volume 19, Issue 1, Pages 52–61, https://doi.org/10.1016/j.tplants.2013.09.008; J.W. White, 2012. *Field Crops Research*, Volume 133, Pages 101–112.)

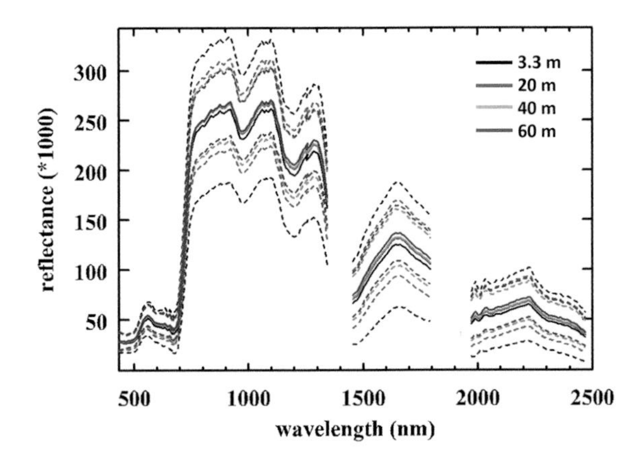

FIGURE 14.9 Studying spectral traits of a plant species using hyperspectral data acquired in different spatial resolutions. Mean (solid) and ± 1 standard deviation (dashed) spectra for *Abies concolor* across all four spatial resolutions. Variance decreases at each successive resolution. (From K.L. Roth et al. 2015a. *Remote Sensing of Environment*, Volume 167, Pages 135–151, https://doi.org/10.1016/j.rse.2015.05.007; K.L. Roth et al. 2015b. *Remote Sensing of Environment*, Volume 171, Pages 45–57, https://doi.org/10.1016/j.rse.2015.10.004.)

Phenotyping and genotyping most often require imaging spectroscopy (hyperspectral) data acquired at hyperspatial resolution (≤20 m) for land cover such as forest, wetland, and shrub species (Roth et al., 2015a, 2015b) and as fine as 1-m or less for certain vegetation species pertaining to agriculture, wetlands, grasslands, and forests. For example, a single 30-m x 30-m plot in a tropical rainforest can have tens or hundreds of species. Class discriminations are best achieved when we can minimize within-class spectral variance while at the same time maximizing between-class variance (Roth et al., 2015a, 2015b). Roth et al. (2015a, 2015b) found that the variance decreases at each successive resolution (Figure 14.9).

In Chapter 5, Dr. Andries Potgieter et al. used two proximal hyperspectral sensors with very high spectral resolution (0.1–2.6 nm across the entire spectral region) to study wheat and sorghum crops. In order to ensure very high spatial resolution, the hyperspectral sensors were carried on a high-throughput plant phenotyping (HTPP) system (e.g., Figure 14.10) which was mounted and driven very close to the canopy to capture data proximally from the hyperspectral sensor as well as from

FIGURE 14.10 High-throughput plant phenotyping (HTPP) hyperspectral system. The platform to mount the multisensor system for HTPP. (From G. Bai et al. 2016. *Computers and Electronics in Agriculture*, Volume 128, Pages 181–192, https://doi.org/10.1016/j.compag.2016.08.021.)

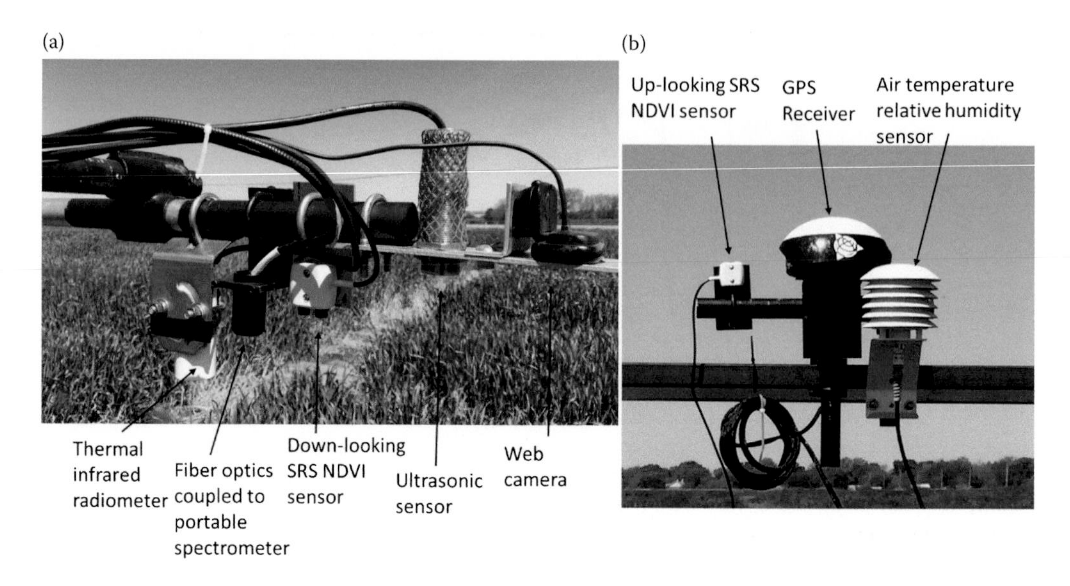

FIGURE 14.11 High-throughput plant phenotyping (HTPP) hyperspectral system. The five sensing modules mounted on a sensor bar to measure one crop row or plot (a); and the GPS receiver, air temperature/relative humidity sensor, and up-looking SRS NDVI sensor mounted at the center of the platform (b). (From G. Bai et al. 2016. *Computers and Electronics in Agriculture*, Volume 128, Pages 181–192, https://doi.org/10.1016/j.compag.2016.08.021.)

several other sensors (e.g., Figure 14.11). By combining the information obtained from these multiple sensors (e.g., Figures 14.10 and 14.11), a better understanding of phenotypes and genotypes was feasible. Chapter 5 focused on detecting genotype differences in phenology, chemical composition, photosynthetic capacity, water relations, and responses to environmental cues. They reported that a thorough and improved understanding of crop traits will help provide a step change which in turn will help foster increases in crop yield. In this chapter, they pre-selected three hyperspectral vegetation indices (HVIs) to study plant traits. However, readers should be aware that hyperspectral remote sensing allows us to compute thousands of HVIs. Even though an overwhelming number of them are redundant, there are at least 15–20 most valuable HVIs that can be exploited (see Thenkabail et al., 2013, 2014; Thenkabail, 2015) that can be used to study plant traits. The three HVIs used in Chapter 5 are valuable, but one should explore other HVIs to study specific traits (e.g., lignin, pigments, N, water, stress, chlorophyll, anthocyanins) of crops (see Galvão et al., 2009; Thenkabail et al., 2012a, 2012b, 2013; Mariotto et al., 2013; Thorp et al., 2013; Marshall and Thenkabail, 2014; Roberts et al., 2015; Thenkabail, 2015).

14.6 LINKING ONLINE SPECTRAL LIBRARIES WITH HYPERSPECTRAL TEST DATA

Hyperspectral libraries of agricultural crops and vegetation are currently evolving, but are of great importance for advancing science from imaging spectroscopy. For example, a large collection of hyperspectral databanks of different crops collected through field (point) spectroradiometers will help in providing "training data" for classifying or modeling data acquired from imaging spectroscopy collected by drone-based, aircraft-based, and space-based hyperspectral sensor platforms. This, for example, will help classify crop types and/or model their quantitative biophysical and biochemical characteristics (e.g., biomass, LAI, yield, chlorophyll, lignin, N, pigments) based on imaging spectroscopy data. The systematic development of hyperspectral libraries of crops (e.g., Figures 14.12 through 14.14) is quite complex in that the practice requires collection of spectra under a wide variety of

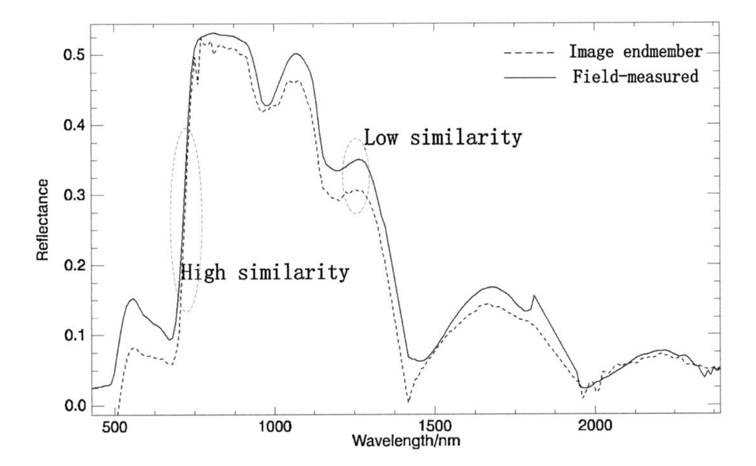

FIGURE 14.12 Similarities and differences in hyperspectral data acquired from an image versus field spectroradiometer. There are areas like the red-edge where they match perfectly and other places of the spectrum where they differ, often in magnitude of reflectivity and at times even in shape. Causes of these differences are discussed in the text. (From Z. Pan et al. 2013. *International Journal of Applied Earth Observation and Geoinformation*, Volume 25, Pages 21–29, https://doi.org/10.1016/j.jag.2013.03.002.)

conditions such as a different crop types and their varying growing stages (e.g., Figures 14.13 and 14.14), crop conditions, watering methods (irrigated versus rainfed), soil types in which crops are grown, management conditions (e.g., tillage, no-tillage, drainage, no-drainage), and inputs (e.g., N application, fertilizers, herbicides, pesticides). Since crops and vegetation grow in a wide array of conditions as mentioned above, developing a standardized spectral library of crops becomes quite complex. This requires a very systematic approach to developing vegetation or agricultural crop spectral libraries that consider data gathered of each crop in varying growth phases and conditions over space and time.

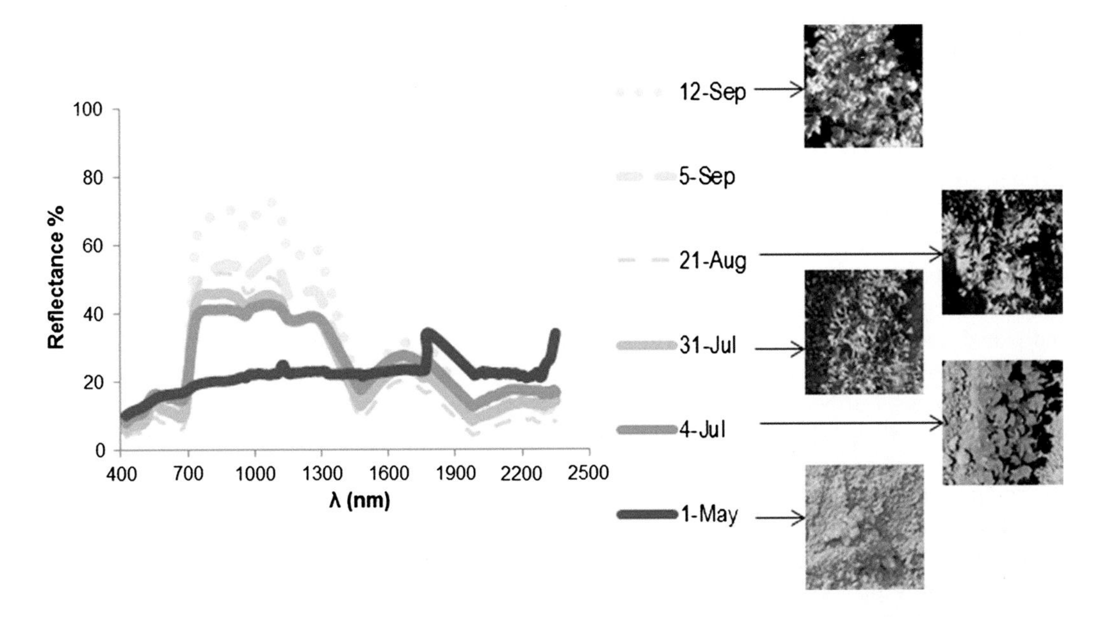

FIGURE 14.13 Field spectroradiometer measurements of cotton crops during different growth phases. Seasonal (May to September) vegetation spectra measured using ASD field spectroradiometer for a cotton field, and some photographs showing their respective biomass growth. (From I. Mariotto et al. 2013. *Remote Sensing of Environment*, Volume 139, Pages 291–305.)

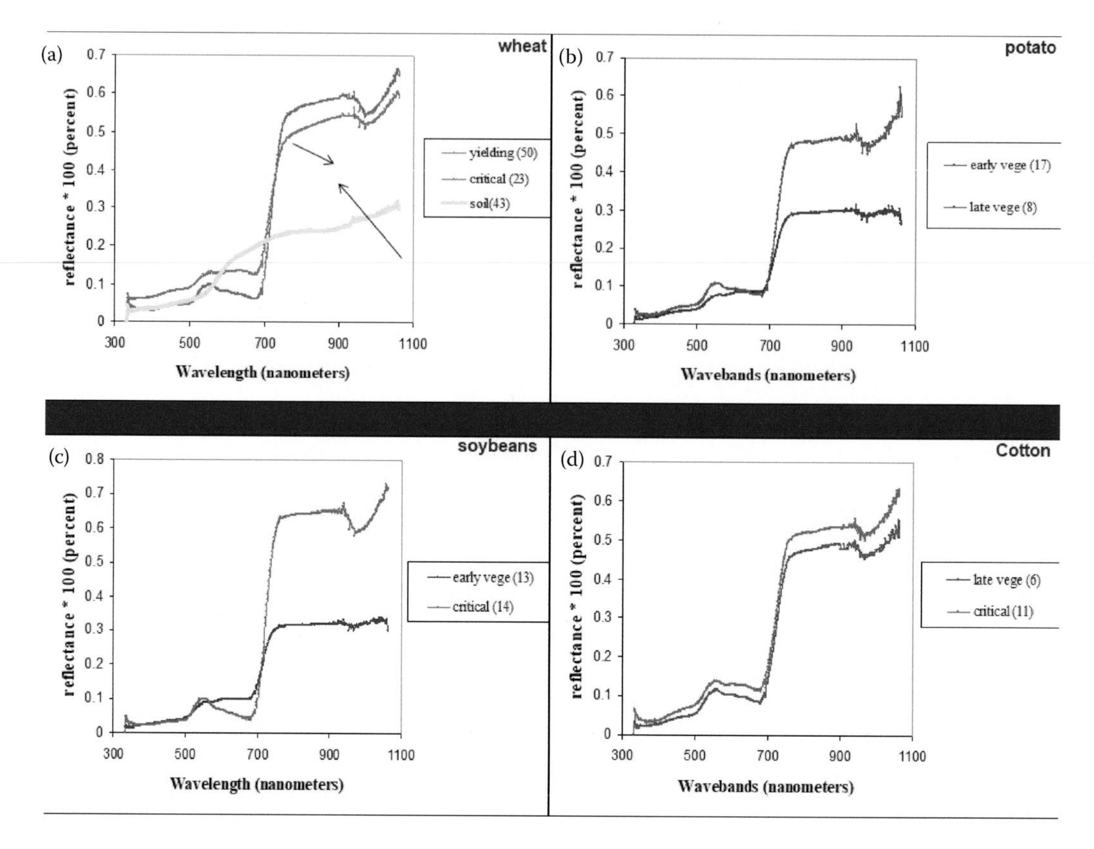

FIGURE 14.14 Field spectroradiometer measurements of crop types and their growth stages. Spectral reflectance characteristics of different crops at distinct growth stages. Sample sizes are shown inside the legend brackets of each crop. (a) cotton; (b) potato; (c) soybeans; (d) corn. (From P.S. Thenkabail et al. 2000. *Remote Sensing of Environment*, Volume 71, Issue 2, Pages 158–182, https://doi.org/10.1016/S0034-4257(99)00067-X.)

The goal of such standardization is multi-fold and includes developing spectral banks of crop traits, and building spectral libraries to "train" imaging spectroscopy datasets for use in modeling and mapping. Further, spectral libraries developed with data collected from various sensor systems and platforms will have their own variability (Figure 14.12) that needs to be accounted for and understood. Once robust spectral libraries are established, say for agriculture, one can begin to analyze and understand crop types, their growth stages, and their growing conditions (e.g., whether a crop is in stress or not).

However, good hyperspectral libraries of agricultural crops and vegetation do not exist currently. Individual researchers (e.g., Thenkabail et al., 2000; Mariotto et al., 2013) have built spectral libraries and used them in their research, but rarely have made them available online for public access in standard formats. Further, unless very careful efforts are made in collecting hyperspectral libraries of crops, there will be substantial difficulties in matching ground-based hyperspectral libraries with airborne and spaceborne imaging spectroradiometer data as implied in the research of Nidamanuri and Zbell (2011). They were able to show excellent relationships between the hyperspectral field data with HyMAP (which is an airborne hyperspectral imaging sensor developed in Australia and has 126 spectral bands in the 450–2500 nm spectral range) data in discriminating crops like winter rape and alfalfa. However, they found significant spectral confusion among crops like winter barley, winter rye, and winter wheat, thus raising questions about the existence of meaningful spectral matches between some field reflectance spectra with airborne hyperspectral imagery data.

In Chapter 6, Dr. Muhammad Al-Amin Hoque and Dr. Stuart Phinn bring to the reader's attention three forms of spectral libraries that are in existence including: (1) the Advanced Spaceborne Thermal

Emission and Reflection Radiometer (ASTER) spectral library that is associated with a particular project, (2) the Vegetation Spectral Library (VSL) which is a collection of measurements made by various instruments, and (3) a number of libraries like SPECCHIO (http://specchio.ch/) and ECOSIS (https://ecosis.org/) that are collections of measurements made for a single application or multiple applications that are stored in a common format. Whereas there appears to be several spectral libraries of vegetation and/or crops (Tables 6.1 and 6.2 of Chapter 6), in reality, all these spectral libraries are primarily based on a project in a specific area, and/or for specific crops in specific areas, and/or certain vegetation types over a time-period in small study areas, and/or measurements based on a wide range of instruments that are hard to inter-relate. What is definitely lacking is a systematic study of selected crops or vegetation over time and space and in different growing conditions that are collected using standard protocols and shared using standard formats. Hoque and Phinn highlighted the need for collecting hyperspectral libraries based on standard protocols, based on internationally accepted meta-data standards, and to publish these spectral libraries online like in SPECCHIO (http://specchio.ch/). This is of great importance at a time when there are several efforts to launch hyperspectral sensors onboard multiple satellites (e.g., EnMAP, HyspIRI, as well as many private players; also see Table 14.1).

14.7 USE OF SPECTRAL DATABASES FOR REMOTE SENSING OF CROPS

The use of spectral databases in the study of crops is now quite mature, widespread, and encompasses crop type mapping, modeling crop biophysical and biochemical quantities, estimating productivities, modeling yields, and monitoring of drought and disaster impacts on crops over space and time. Biophysical quantities have been modeled using remote sensing involving various approaches such as statistical models, physical models, and hybrid model inversions. Remote sensing is widely used in providing many crop variables critical to a wide range of crop models which in turn are crucial for crop yield modeling. These variables include (Gitelson et al., 2003; Ustin et al., 2009; Jin et al., 2018): absorbed photosynthetically active radiation (FAPAR), chlorophyll, other pigments (anthocyanin, carotenoids, xanthophylls, betalains), LAI, canopy cover, wet and dry biomass, leaf nitrogen (N) accumulation, evapotranspiration (ET), soil properties (e.g., soil moisture), crop phenology information, lignin, crop water content, leaf water content, plant height, crop stress, and a host of others (e.g., plant structural characteristics such as erectophile vs. planophile plants). Remote sensing plays a key role in providing parameters for many of the crop models (Figure 14.15). Multispectral broadband remote sensing data such as from the Landsat-series of satellites can characterize all of these crop variables, however, there will be limitations in the accuracies with which they can characterize them. Here is where hyperspectral narrowband data can provide substantial improvements (e.g., Figure 14.16). Further, whereas one can detect crop stress using remote sensing with a high degree of accuracy, attributing these stresses to an exact cause remains a challenge. Crop stress can occur due to various factors such as water deficit, water excess, pests, diseases, and a host of other reasons like poor management. So, to attribute a cause for the crop stress detected in remote sensing data, one needs substantial field knowledge in addition to image analysis. In this regard, Panda et al., 2015 demonstrated the inherent trade-offs between image resolution, spatial coverage, acquisition costs, optimal predictions, and high classification accuracies that hinder effective remote sensing applications in monitoring crop diseases and pests, especially in countries with poor economies. Again, hyperspectral narrowbands will help study these characteristics with substantially improved accuracies (Mariotto et al., 2013; Thenkabail et al., 2013; Thenkabail et al., 2014; Marshall and Thenkabail, 2015). Nevertheless, hyperspectral data collected during various times of the day, in various view angles, and various structural and other variables have great variability (Figures 14.17 and 14.18). Thereby, those collecting these data must keep in mind the need to understand these variations and to account for them to ensure a high degree of accuracy and low uncertainty in crop models (Figure 14.15) that predict yield or productivity.

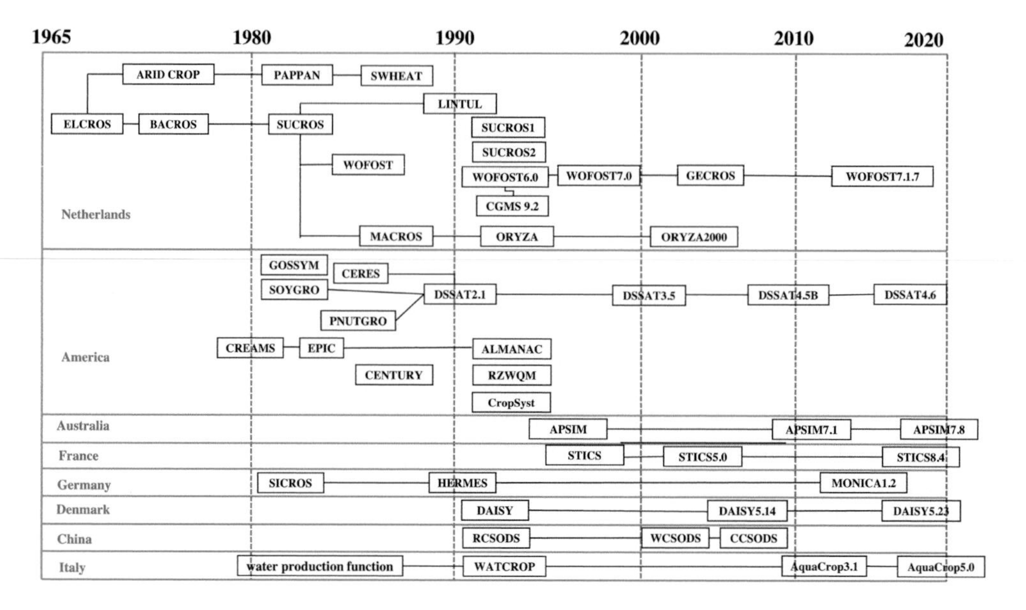

FIGURE 14.15 Crop models. Development of main crop models over time. Note: horizontal continuous lines indicate the development of new crop models. (From X. Jin et al. 2018. *European Journal of Agronomy*, Volume 92, Pages 141–152, https://doi.org/10.1016/j.eja.2017.11.002.)

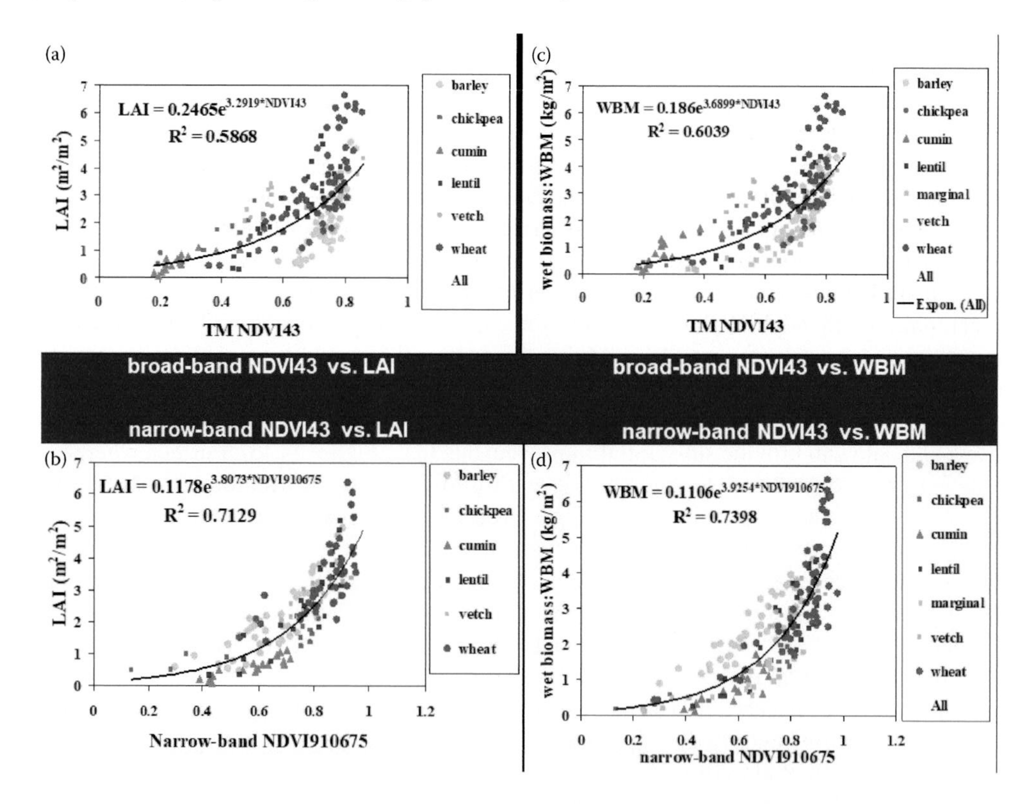

FIGURE 14.16 Leaf area index (LAI) and wet biomass (WBM) modeled using Landsat Thematic Mapper broadbands versus hyperspectral narrowbands. Comparison between the broadband and the narrowband soil-adjusted indices for potato WBM (a and b), and soybeans LAI (c and d). NIR and red-based indices of broadbands (a and c) versus visible band-based indices of narrowbands (b and d). (From P.S. Thenkabail et al. 2004a. *Remote Sensing of Environment*, Volume 90, Pages 23–43.)

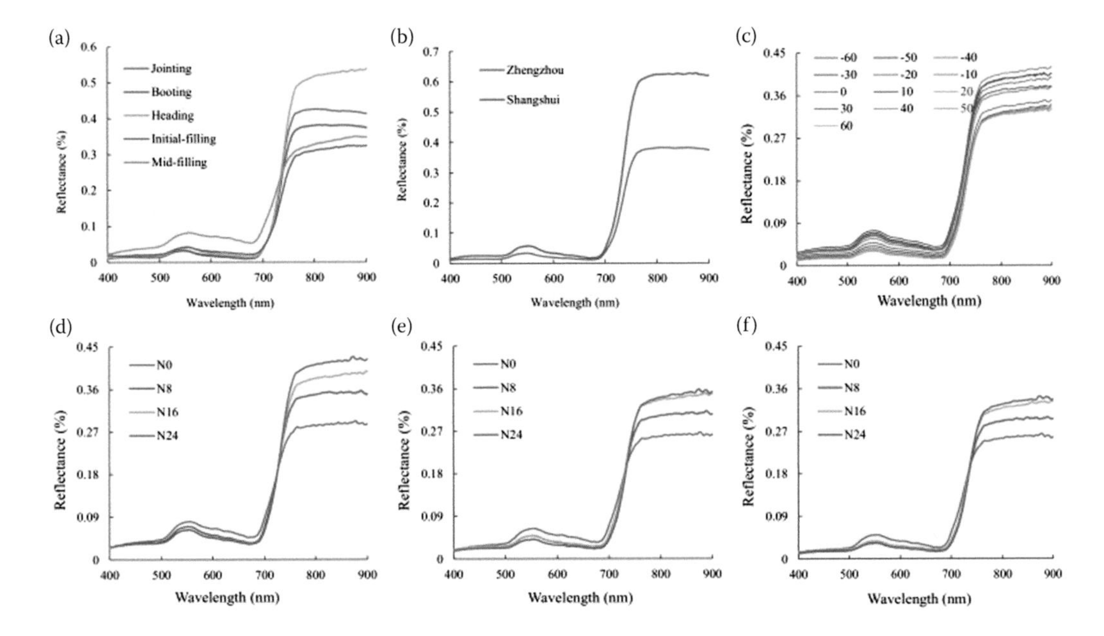

FIGURE 14.17 Some of the variables that cause variations in hyperspectral measurements. Example of changes in the reflectance with varied growth periods, sites, and viewing zenith angles: (a) growth periods; (b) sites; (c) viewing zenith angles; (d) – 20°; (e) nadir; (f) 20°. (From L. He et al. 2016. *European Journal of Agronomy*, Volume 73, Pages 170–185, https://doi.org/10.1016/j.eja.2015.11.017.)

In order to be able to widely use hyperspectral data in crop models (Figure 14.15) and to adapt their use in standardized formats across many sites, instruments, and a suite of conditions (e.g., Figure 14.17), especially to suit machine learning and artificial intelligence in analyzing big data, we need mature spectral libraries of crops. In Chapter 7, Dr. Andreas Hueni et al. highlight the need for these spectral libraries, their standardization, and potential for large-scale use with the advent of "big data" analyses through machine learning. Good spectral libraries are rare and the few good ones that exist, such as the United States Geological Survey's Spectral library (Kokaly et al., 2017) and the ASTER spectral library

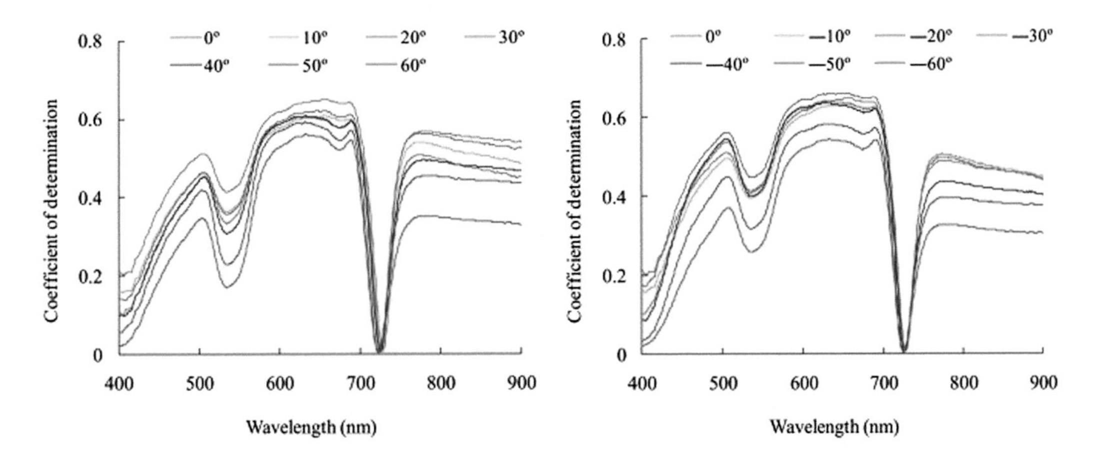

FIGURE 14.18 R-squared variability as a result of view angles of plant leaves. Coefficients of determination (r^2) between reflectance (R_x) and leaf nitrogen content (LNC) at 13 viewing zenith angles. (Left: forward scattering, right: backward scattering). (From L. He et al. 2016. *European Journal of Agronomy*, Volume 73, Pages 170–185, https://doi.org/10.1016/j.eja.2015.11.017.)

(Baldridge et al., 2009), are focused on minerals and rocks and not on vegetation or agricultural crops. Some excellent efforts in developing a spectral library that includes vegetation such as SPECCHIO (http://www.specchio.ch/), and EcoSIS (https://ecosis.org/) have made significant advances. However, their scope in documenting agricultural crops and vegetation is still very limited and, at best, site specific, plant specific, and/or instrument specific. As a result, the need for an exclusive agricultural crop spectral library of the world is much called for given its potential importance in food and water security (Thenkabail et al., 2010; Teluguntla et. al., 2015). Development of a hyperspectral library of agricultural crops of the world will be a major effort requiring large resources. However, it is a must in using data from hyperspectral sensors such as HyspIRI and other upcoming hyperspectral sensors (see Table 14.1) to quantify crop biophysical, biochemical, and structural properties that support rapid analysis of massive big datasets (e.g., www.croplands.org) using machine learning and artificial intelligence via cloud computing. Similarly, vegetation studies of various types and ecosystems require spectral libraries of various ecosystems such as tropical forests, savannahs, and wetlands as each one of these have very distinct vegetation types, species, and phenology.

14.8 CHARACTERIZING SOIL PROPERTIES USING REFLECTANCE SPECTROSCOPY

Specific soil properties such as salinity, organic carbon, soil moisture, and heavy metals have been widely studied over the years (Bilgili et al., 2010; Araújo et al., 2015; Ben-Dor et al., 2015; Knox et al., 2015; Gandariasbeitia et al., 2017; Mohamed et al., 2017; Sun and Zhang, 2017). Soil carbon fractions (e.g., Figure 14.19) were accurately modeled by Knox et al. (2015). Summers et al. (2011) demonstrated that it is possible to predict clay content, soil organic carbon, iron oxide content, and carbonate content using visible and near-infrared (VNIR) data. Brown et al. (2006) performed broad sampling of soils across the Americas, Africa, and Asia and found strong relationships between VNIR reflectance and soil properties (e.g., Figure 14.20) such as relative kaolinite content, relative montmorillonite content, clay content (pipette), cation exchange capacity (CEC), and soil organic carbon (SOC). VNIR technology has proven to be a relevant alternative for assessing and monitoring soil quality properties such as pH, soil organic matter (SOM), total-N (TN), and basal respiration or BR (Gandariasbeitia et al., 2017). Others (e.g., Bilgili et al., 2010) have shown VNIR spectroscopy accurately predicts important environmental soil variables such as clay and SOM content. Araújo et al. (2015) concluded that spectral data were useful in assessing the spatial distribution of some of the most important Amazonian dark soils. They showed that the mid-infrared (MIR) spectroscopy,

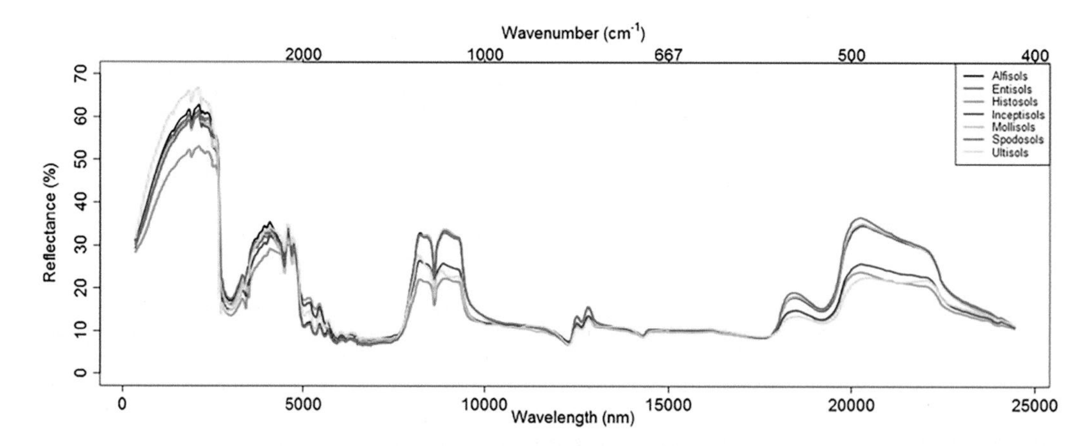

FIGURE 14.19 Spectral reflectivity variability along the spectrum based on soil orders. Mean combined VNIR–MIR spectrum for each of the soil orders collected in this study. (From N.M. Knox et al. 2015. *Geoderma*, Volumes 239–240, Pages 229–239, https://doi.org/10.1016/j.geoderma.2014.10.019.)

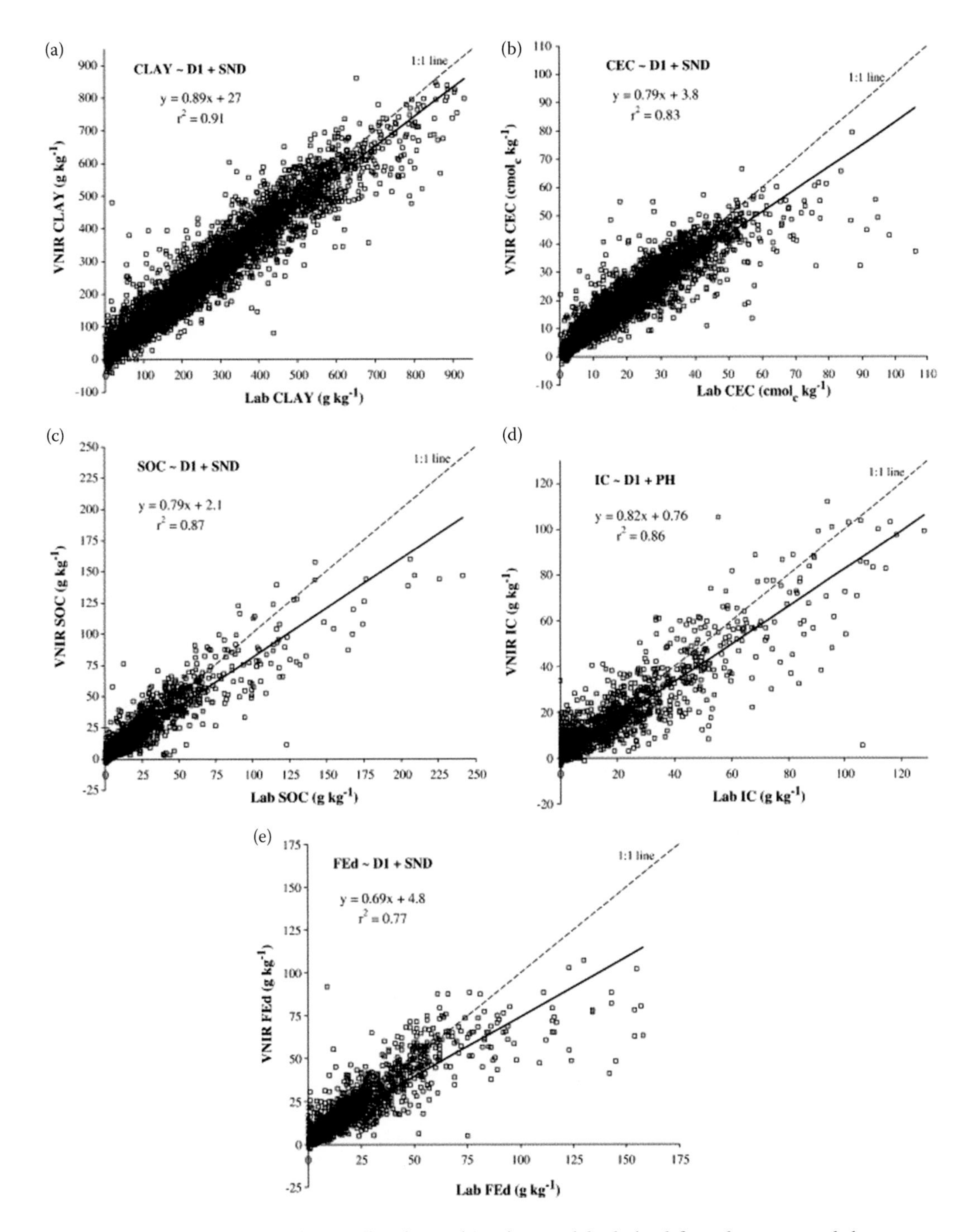

FIGURE 14.20 Soil properties predicted (y-axis) using models derived from hyperspectral data versus measured soil properties in the lab (x-axis). Predicted vs. measured: (a) clay content (pipette method); (b) CEC (NH_4 at pH = 7); (c) SOC (Walkley–Black); (d) IC (electronic manometer); (e) FEd (dithionite–citrate extraction) with 1/6th cross-validation using air-dry, 1st derivative VNIR reflectance spectra, Treenet®, and either sand content or pH ($CaCl_2$) as an auxiliary predictor. (From D.J. Brown et al. 2006. *Geoderma*, Volume 132, Issues 3–4, Pages 273–290, https://doi.org/10.1016/j.geoderma.2005.04.025.)

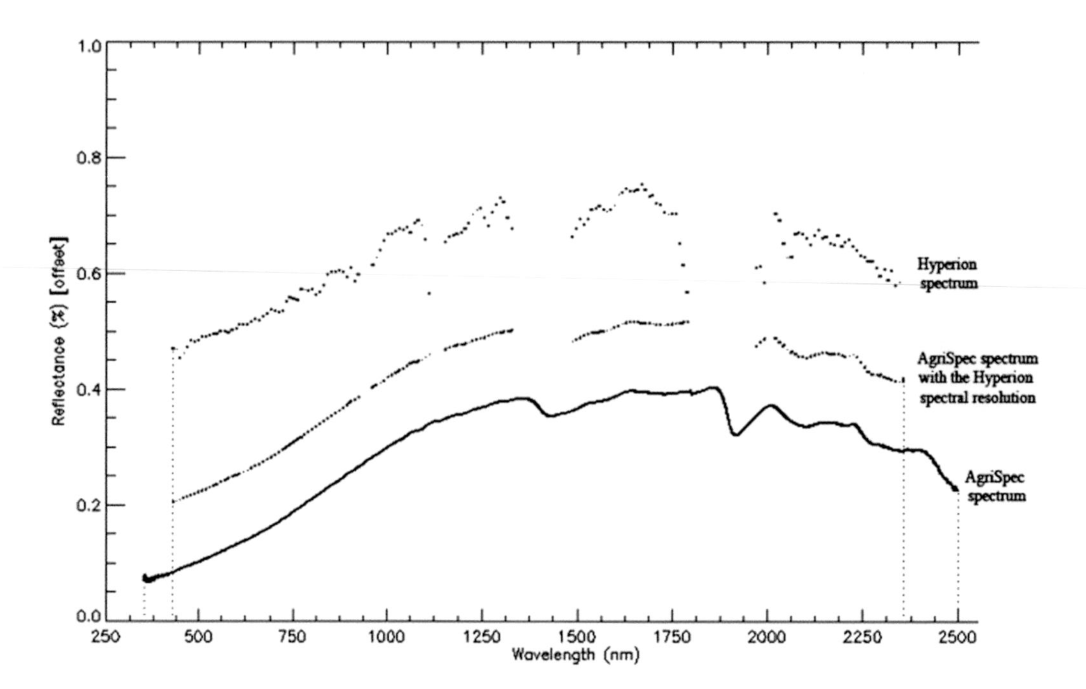

FIGURE 14.21 Hyperspectral data of soils measured with different spectral resolution and platforms. Plot of an AgriSpec high resolution spectrum (2151 spectral bands) of a soil sample on travelling stock route, an AgriSpec spectrum resampled to the low spectral resolution similar to that of the Hyperion data (152 spectral bands) of the same soil sample, and the Hyperion spectrum of the stock route. (From C. Gomez et al. 2008. *Geoderma*, Volume 146, Issues 3–4, Pages 403–411, https://doi.org/10.1016/j.geoderma.2008.06.011.)

in particular, has the potential to be an alternative to traditional soil analysis. Sun and Zhang (2017) demonstrated the huge potential of reflectance spectroscopy in estimating Zinc concentrations in soil and in monitoring soil contamination by heavy metals. The most common models used in the study of quantitative soil properties include stepwise multiple linear regression (SMLR), partial least squares regression (PLSR), multivariate adaptive regression splines (MARS), principal component regression (PCR), and artificial neural networks (ANN) (Mohamed et al., 2017). However, soil properties when studied from various platforms will have slight differences (e.g., Figure 14.21), which need to be well understood. Often, soils are covered by vegetation, at which time it is important to understand soil characteristics through vegetation signatures (e.g., Figure 14.22). Combined mapping of laboratory infrared soil spectral analysis, MODIS remote sensing data and statistical methods, soil properties such as SOC, pH, sand content, and sum of bases (SB), as well as root-depth restrictions provided maps with unprecedented accuracy and spatial resolution for the African continent (Vågen et al., 2016; Figure 14.23). Nevertheless, field spectroscopy data of soils needs to be better understood and correlated with spectral data of soils from airborne or spaceborne sensors. As large parts of the world are covered by various degrees of vegetation, we need to develop a full understanding of soils beneath vegetation cover.

In Chapter 8, Dr. Eyal Ben-Dor et al. highlighted the increasing importance of reflectance spectroscopy in soil characterization. They point to the fact that many soil properties are gathered solely by reflectance measurements. The reliability of these measurements from the ground or laboratory is very high. However, difficulties from airborne and spaceborne sensors are many due to cloud cover, haze, and other atmosphere effects, signal-to-noise issues, vegetation, and a host of other factors. These limitations have been significantly overcome with recent airborne and spaceborne sensors that are closing the gap in spectral, spatial, radiometric, and temporal frequency of data acquisition when compared with earlier generations of sensors. In a comprehensive coverage of both soil and spectral characteristics, they discuss soil properties such as clay minerals, carbonates, organic matter, water,

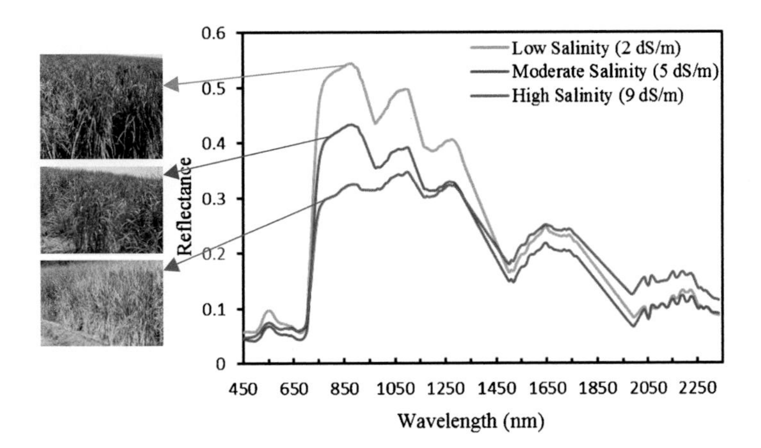

FIGURE 14.22 Hyperspectral data of vegetation covering soils with different salinity values. Spectral reflectance of a sugarcane canopy as affected by different salinity stress levels. (From S. Hamzeh et al. 2016. *International Journal of Applied Earth Observation and Geoinformation*, Volume 52, 2016, Pages 412–421, https://doi.org/10.1016/j.jag.2016.06.024.)

iron, and soil salinity. They also discuss the factors that are affecting soil reflectance such as: the biosphere, crust, soil moisture, and atmosphere. The heart of the chapter is soil spectroscopy from various sensors and with liberal illustrations. The lack of standards in building soil spectral libraries (Ben-Dor et al., 2015) is akin to the lack of standards in developing spectral libraries of crops. Ben-Dor et al. (2015) recommends the use of an internal soil standard (ISS; Pimstein et al., 2011) in developing a spectral library of crops in the field and in the laboratory and recommend using a specific measurement protocol, although the last recommendation is not mandatory unless the ISS is to be used with any protocol. They provide the "standard" soil spectral library vision, where all laboratories will be able to merge their data smoothly using the ISS procedure, and provide some examples for that. They also mention the added value of the MIR spectral region for soil proxy analysis.

14.9 HYPERSPECTRAL HYPERION DATA PREPROCESSING STEPS

One of the first steps of hyperspectral data analysis is to conduct a rigorous and comprehensive set of preprocessing protocol steps. This is required to overcome a wide variety of issues that include atmospheric correction (Gao et al., 2009), geometric correction (Jafari and Lewis, 2012), radiometric calibration (Jafari and Lewis, 2012), topographic resampling (Gao et al., 2009), removal of noisy or uncalibrated bands (Thenkabail et al., 2004a), band selection (Thenkabail et al., 2004a, 2004b), de-striping (Datt et al., 2003), "smile effect" correction (Goodenough et al., 2003), noisy pixel removal with techniques such as minimum noise fraction (Boardman and Kruse, 1994), and converting radiance to reflectance (Thenkabail et al., 2004a). Atmospheric corrections of hyperspectral data have evolved over the years and use a number of algorithms that include the Fast Line-of-sight Atmospheric Analysis of Spectral Hypercubes (FLAASH) algorithm (Matthew et al., 2000), Atmosphere Removal code (ATREM; Gao et al., 2009), the Atmosphere Correction Now (ACORN; Kruse, 2004), High-accuracy Atmospheric Correction for Hyperspectral Data (HATCH; Qu et al., 2003), and a few others as discussed in detail by Gao et al. (2009). Also, methods that comprised vicarious calibration targets coupled with the Atmospheric Correction (ATCOR) radiative transfer model were successfully used (e.g., Brook and Ben-Dor, 2001).

Hyperion is the first spaceborne civilian hyperspectral imaging spectrometer carried onboard the EO-1 satellite (Pearlman et al., 2003; Ungar et al., 2003). Thereby, much of the learning and development of hyperspectral data preprocessing has evolved through 16 years of handling and processing Hyperion data along with the significant development of airborne hyperspectral sensors

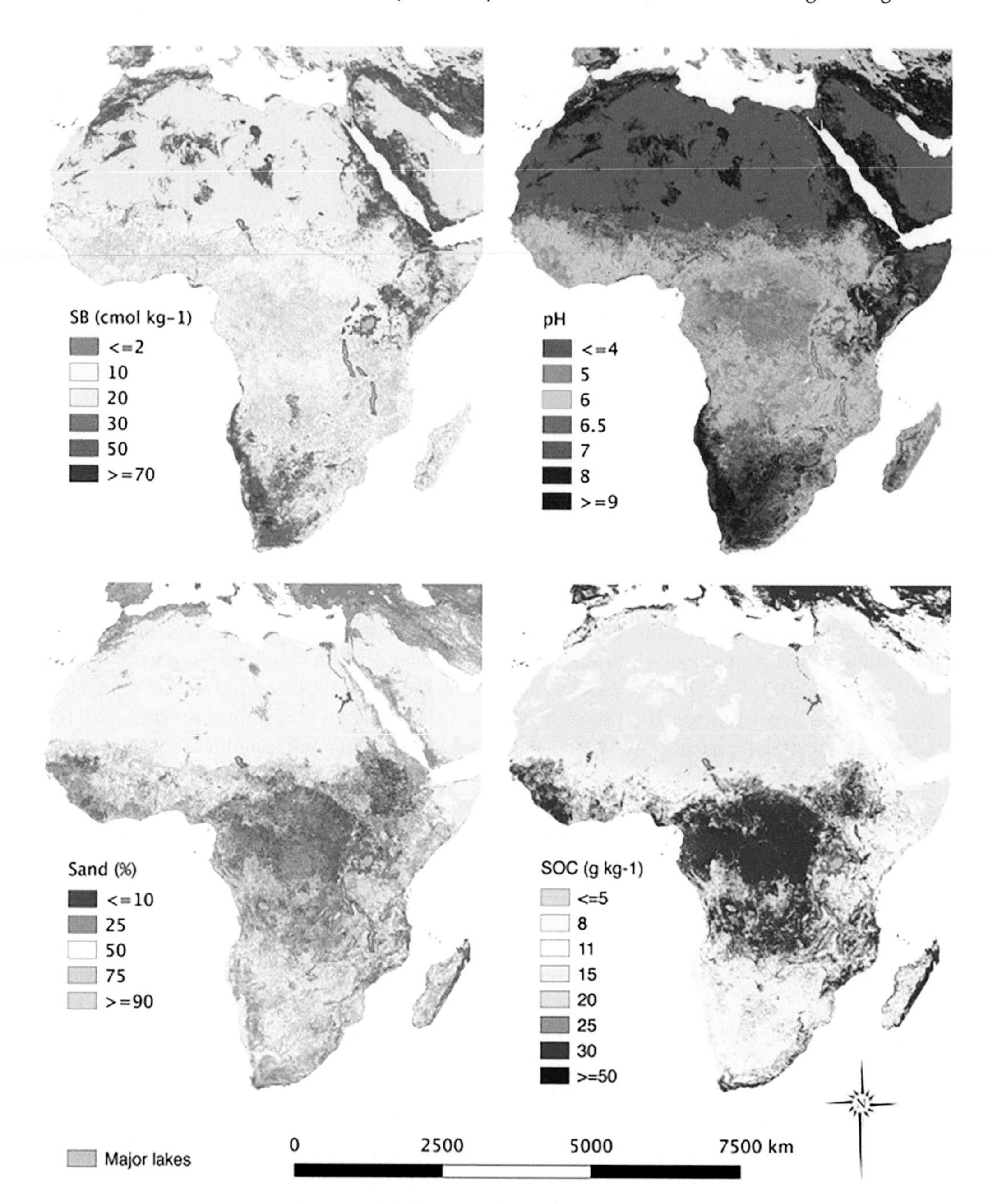

FIGURE 14.23 Soil maps of Africa showing different soil variables. Maps of sum of bases (SB), pH, SOC, and sand content using MODIS reflectance data for 2012. (From T.G. Vågen et al. 2016. *Geoderma*, Volume 263, 2016, Pages 216–225, https://doi.org/10.1016/j.geoderma.2015.06.0.)

by many vendors. Hyperion was launched in November 2000 and was decommissioned in March 2017. The entire archive of over 70,000 image cubes collected by Hyperion for the entire world is available free of charge through USGS' Earth Explorer (https://earthexplorer.usgs.gov/). Hyperion had two spectrometers, one for the VNIR and another for the SWIR. Hyperion gathered data from these two spectrometers in 242 narrowbands, each with 10-nm bandwidth, across a 357–2576-nm spectral region with a 16-bit dynamic range. However, due to noise effects in the lower and higher

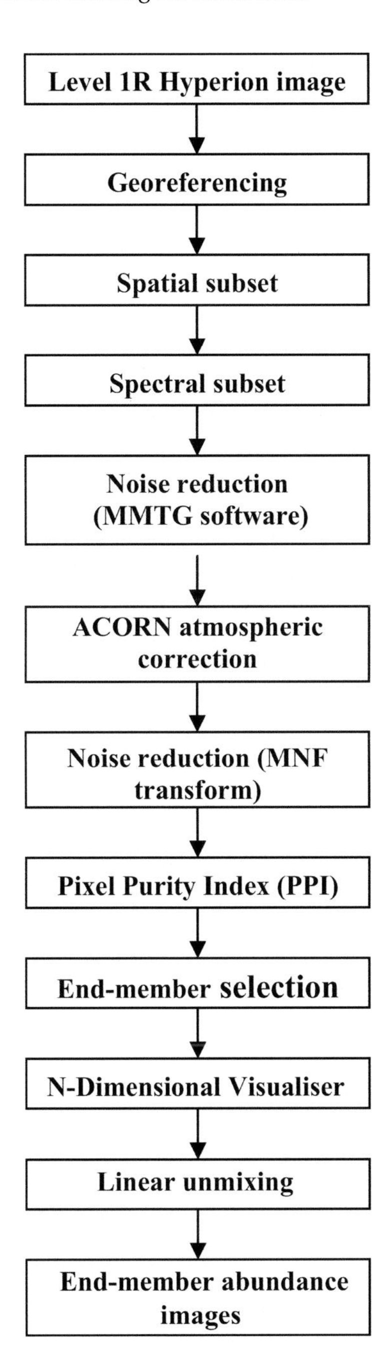

FIGURE 14.24 Hyperspectral Hyperion preprocessing steps. Flowchart of Hyperion image analysis. (From R. Jafari and M.M. Lewis, 2012. *International Journal of Applied Earth Observation and Geoinformation*, Volume 19, Pages 298–307, https://doi.org/10.1016/j.jag.2012.06.001.)

wavelength ranges, data were usually gathered in the 427–2365-nm range. A number of researchers (Thenkabail, 2004; Thenkabail et al., 2004a; Pignatti et al., 2009; Mariotto et al., 2013; Marshall and Thenkabail, 2015; Rautiainen and Lukeš, 2015; Zhang, 2016) provide details of these preprocessing steps (e.g., Figure 14.24) when applying the Hyperion data for a wide range of applications. Thenkabail et al. (2004a) found that of the original 242 Hyperion bands, only 196 were unique and calibrated: bands 8 (427.55 nm) to 57 (925.85 nm) from the VNIR sensors, and bands 79 (932.72 nm) to 224 (2395.53 nm) from the SWIR sensors. The redundant and uncalibrated bands were in the 357–417, 936–1068, and 852–923 nm ranges (Thenkabail et al., 2004a). They also found high levels of atmospheric and other noise and/or poor signals in the Hyperion data over the 1306–1437, 1790–1992, and 2365–2396 nm ranges. Hence, they recommended dropping bands in this range of the spectrum. This led to Thenkabail et al. (2004a) selecting 157 out of 242 bands as good, calibrated, and noise-free bands. Such careful steps for preprocessing the data are required prior to using them for various applications.

Nevertheless, what is lacking in the literature is a systematic and comprehensive workflow of methods and approaches of preprocessing hyperspectral Hyperion data. Such a workflow is provided in Chapter 9 by Dr. Itiya Aneece et al. The preprocessing steps are outlined and each and every step is then clearly analyzed. The strength of the chapter is four-fold: (1) detailed and systematic theoretical presentation of the steps of preprocessing Hyperion data, (2) practical illustration of the steps by processing Hyperion images, (3) coding and running these preprocessing steps on the Google Earth Engine cloud computing platform, and (4) making available these codes in the public domain. A unique feature of the chapter is also preprocessing of Hyperion keeping in view of agricultural crop and vegetation applications to ideally fit to the vision of this book. Keeping this in mind, Chapter 9 produces reflectance data, derived from Hyperion, for certain leading world crops. Overall, the chapter illustrated step-by-step the methods and procedures of deriving reflectance data of agricultural crops from Hyperion hyperspectral data by taking raw Hyperion images and demonstrating preprocessing steps. These methods and procedures will be invaluable for upcoming spaceborne hyperspectral sensors such as HyspIRI.

14.10 HYPERSPECTRAL DATA MINING

The high dimensionality of hyperspectral imaging spectroscopy data contains significant redundant information for any given application. High dimensionality leads to the "Hughes Phenomenon" or "curse of high dimensionality" (Liu et al., 2017a, 2017b). This happens because with a fixed number of training samples, the predictive power of a classifier is reduced as the dimensionality increases (Hughes, 1968; Liu et al., 2015, 2017a, 2017b). Hence, processing hyperspectral data without removing the redundant bands is a great strain on human and computing resources. So, it is essential to "mine" the hyperspectral data to eliminate redundant bands and to optimize and obtain the most useful bands for a given application (e.g., Figure 14.25).

Dimensionality reduction (DR) methods have been adopted to reduce hyperspectral data volumes leading to retention of only the most valuable non-redundant bands. Selecting a subset of the original bands and finding band transformations with high discriminative power are the common approaches. However, every DR method should preserve information of original features, not distort them or reduce their information content more than a very small, insignificant, percentage. DR methods include principal component analysis (PCA), independent component analysis (ICA), minimum noise fraction (MNF), discrete wavelet transform based dimensionality reduction (DWT-DR), optimal band selection (OBS), transformations such as the first-order derivative and hyperspectral vegetation indices (HVIs), partial least squares (PLS), neural networks, and extreme learning machines (Ksieniewicz et al., 2018). PCA, for example, performs an orthogonal transformation from the initial spectral space to another space of equal dimension showing no linear correlation between latent features (Tochon et al., 2015). The above process, through one or more methods, leads to optimal hyperspectral band selection before image classification (Yang et al., 2017). Rivera-Caicedo et al. (2017), for example, used 11 dimensionality reduction (DR) methods in combination with advanced machine learning regression algorithms (MLRAs) to show improved accuracies and computational efficiencies in retrieving the LAI biophysical variable. Thenkabail et al. (2000, 2004a, 2004b, 2013, 2014) presented a number of unique hyperspectral waveband reduction methods that worked best in the study of agriculture and vegetation (e.g., Figure 14.25). In Chapter 8, Dr. Eyal

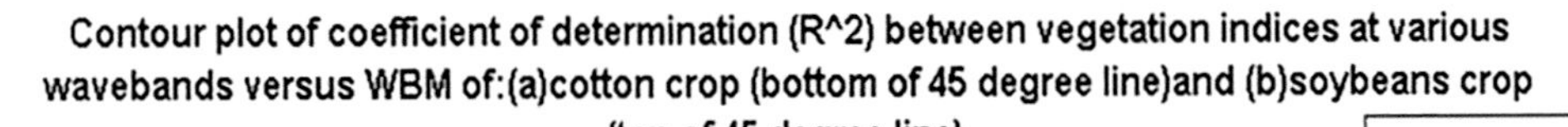

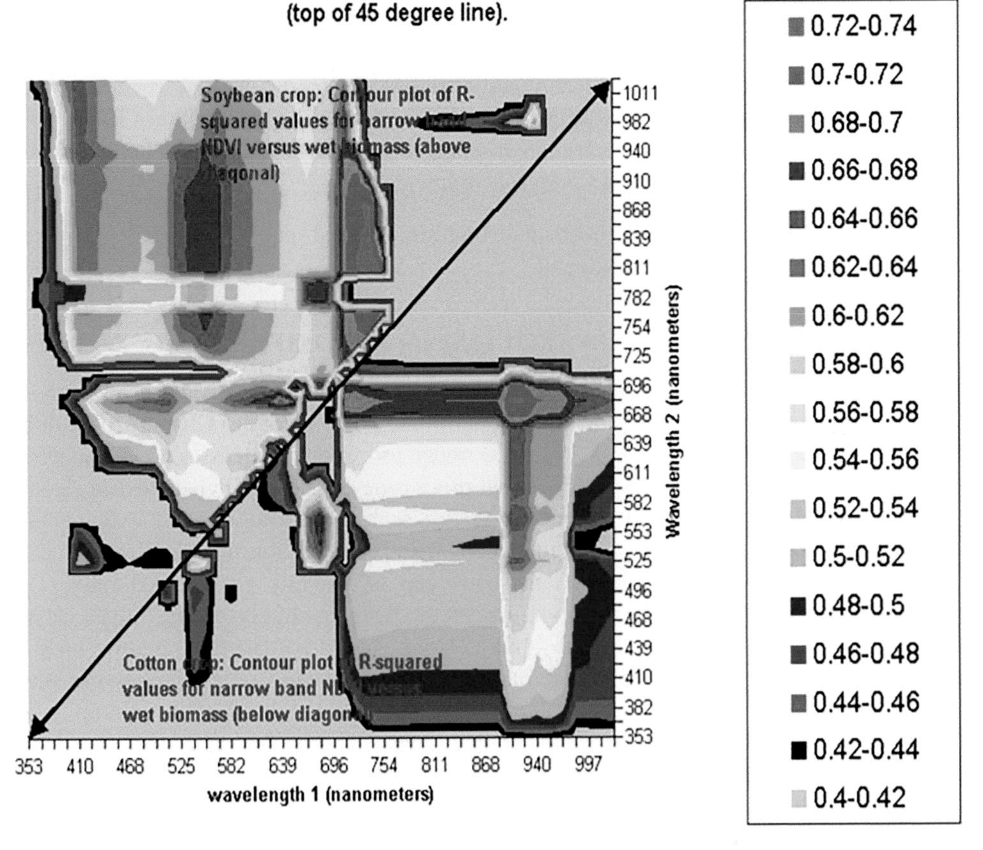

FIGURE 14.25 Lambda λ_1 (350–1050 nm) versus Lambda λ_2 (350–1050 nm) contour plots of R-square values to establish best hyperspectral narrowbands. Contour plot showing the correlation (R^2) between WBM and narrowband NDVI values calculated for 490 narrowbands spread across λ_1 (350–1050 nm) and λ_2 (350–1050 nm). The different areas of "bulls-eye" are the regions with high R^2 values, which were ranked and from which band centers (λ_1 versus λ_2) and bandwidths ($\Delta\lambda_1$ and $\Delta\lambda_2$) were calculated for the seven best indices of each crop variable. (From P.S. Thenkabail et al. 2000. *Remote Sensing of Environment*, Volume 71, Issue 2, Pages 158–182, https://doi.org/10.1016/S0034-4257(99)00067-X.)

Ben-Dor et al. presents an automatic data mining machine to extract the best prediction models out of the many available. This machine, termed PARACUDA II, does preprocessing of the data band by band while applying many data analyzers such as PLSR, Neural Networks, and others using a high parallel computing system and without the need of a human in the loop.

In Chapter 10, Dr. Sreekala Bajwa et al. presented hyperspectral image data mining methods in two groups: (1) feature selection methods (feature selection, projection, divergent measures, similarity measures, sequential search), and (2) information extraction methods (statistical, supervised and unsupervised classification, and spectral and spatial classification and deep learning).

In feature selection methods, they present:

A. Unsupervised methods of feature selection based on information content. Unsupervised methods do not require training data for classification.
B. Projection based methods such as PCA and ICA, which are transformations of original bands through some linear combinations.

C. Similarity measures such as the correlation coefficients and derivative spectral analysis.
D. Other methods of feature selection including wavelets.

In information extraction methods, they present:

E. Regression methods, linear discriminant analysis.
F. Unsupervised methods like clustering and ICA.
G. Supervised methods like spectral angle mapper, maximum likelihood classifier, artificial neural networks, and support vector machines.

Finally, deep learning methods that include both spectral and spatial information in hyperspectral data analysis are presented and discussed.

14.11 HYPERSPECTRAL DATA PROCESSING ALGORITHMS

When dealing with hyperspectral data processing, a differentiation needs to be made between hyperspectral image processing and hyperspectral signal processing. Chang (2013) defined the difference between the two as: "Hyperspectral image processing, processes a hyperspectral image as an image cube whereas hyperspectral signal processing considers a hyperspectral signature as a one-dimensional signal so that no sample correlation such as spectral correlation among pixels in a hyperspectral image cube can be taken into account and used for algorithm design."

Hyperspectral image classification methods can be broadly categorized into: (1) feature selection methods, and (2) pattern recognition methods. Every hyperspectral image classification goes through three major steps (Figure 14.26): (1) input of raw images, (2) feature preprocessing to eliminate redundant and noisy bands, and (3) remote sensing image processing such as for classification to determine classes in various themes (e.g., for land use/land cover, crop types, urban areas) or computation of hyperspectral vegetation indices, and for other purposes (e.g., monitoring droughts and disasters). Feature selection is based on dimensionality reduction and as such is a key requirement in hyperspectral image classification and helps avoid the Hughes phenomenon, which happens when there are only a fixed number of training samples in classifying high dimensional hyperspectral remote sensing images with hundreds or thousands of bands, resulting in the reduced predictive power (or reduced classification accuracies) of classes (Hughes, 1968; Liu et al., 2015; Liu et al., 2017a, 2017b). Acquiring a large number of training and validation data to overcome the Hughes phenomenon is resource intensive. In contrast, utilizing all available hyperspectral narrowbands is not optimal, includes many redundant, highly correlated bands, and requires very heavy computing resources. Nevertheless, working with only a limited number of samples in class classifications involving high dimensional hyperspectral data will lead to significant uncertainties in class precision as well as accuracies. As a result, researchers (Thenkabail et al., 2000, 2004a, 2004b, 2013; Mariotto et al., 2013; Marshall and Thenkabail, 2014, 2015; Thenkabail, 2015) have suggested selecting optimal hyperspectral narrowbands (HNBs) that will retain the most important bands and eliminate redundant HNBs. Establishing optimal HNBs is done by removing highly-correlated redundant HNBs and retaining selective optimal HNBs. In addition to removing redundant bands, one should also remove "noisy pixels or bands" which occur due to causes like instrument noise, cloud cover, or "foggy" atmospheric windows. Once optimal bands are selected, understanding and characterizing endmembers (e.g., Figure 14.27) are central to hyperspectral image analysis and processing. Approaches, methods, and algorithms for endmember analysis include four popular methods (Zhang et al., 2017a, 2017b): the Sequential Maximum Angle Convex Cone (SMACC), N-FINDR, Vertex Component Analysis (VCA), and Minimum Volume Constrained Non-Negative Matrix Factorization (MVC-NMF).

In Chapter 11, Dr. Antonio Plaza presents various hyperspectral data processing algorithms that include pixel-based supervised support vector machines (SVMs), linear unmixing that combines spectral and spatial information, and nonlinear unmixing techniques based on machine learning. Readers should

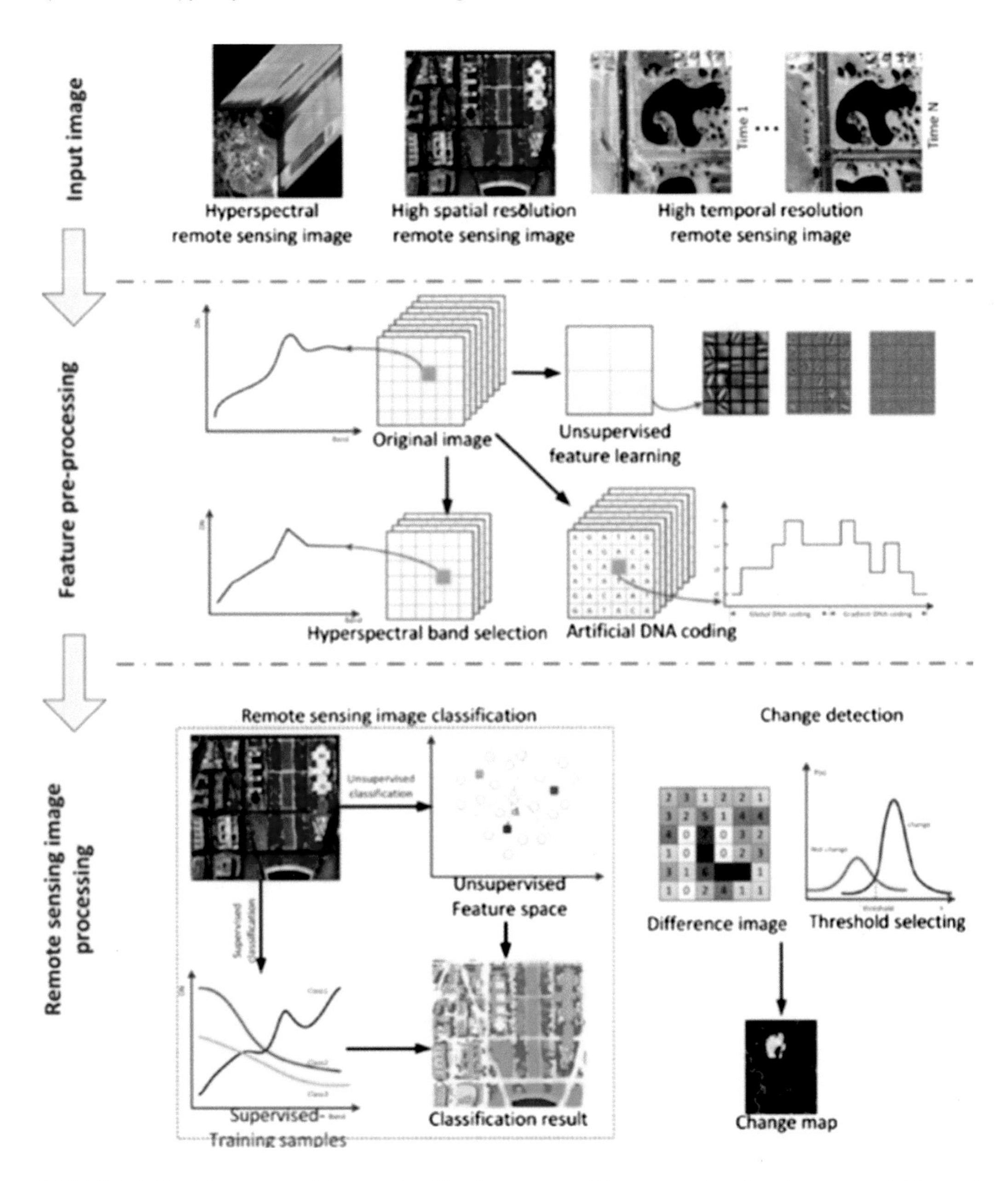

FIGURE 14.26 Hyperspectral image classification process. The process involves image acquisition and composition, preprocessing to ensure the geometric, radiometric, and quality synthesis and harmonization, and finally processing the image to obtain classes, or extract signals to model, map, and study various characteristics. (From Y. Zhong et al. 2018. *Applied Soft Computing*, Volume 64, 2018, Pages 75–93, https://doi.org/10.1016/j.asoc.2017.11.045.)

also explore machine learning algorithms such as random forest, and neural networks for hyperspectral image classification. In contrast to hyperspectral data processing, hyperspectral signal processing involves gathering near-continuous spectra (e.g., Figure 14.28) across a wavelength range and analyzing them. For example, Figure 14.28 illustrates assessment of crop stress, health, and senescence. To perform such analysis, one can gather hyperspectral signatures (signals) by sampling an image and then analyzing the same for such studies as vegetation health, stress, or senescence (Figure 14.28).

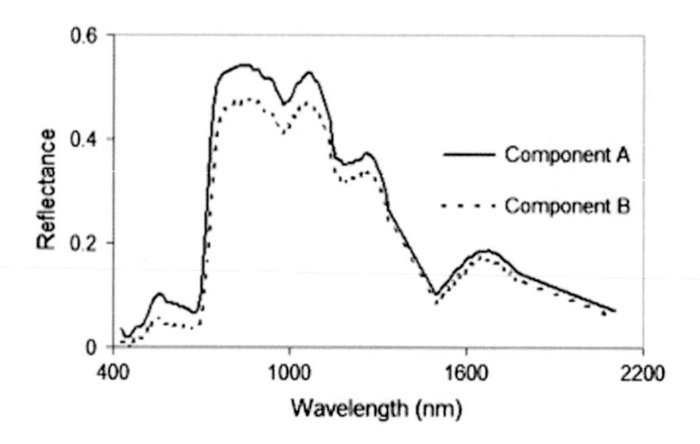

FIGURE 14.27 Endmember analysis of hyperspectral signatures of two crops. Spectral profile of averaged reflectance of endmembers in the two vegetation clusters, that is, Components A (poppy) and B (wheat), respectively. (From J.J. Wang et al. 2014. *Afghanistan, Science of The Total Environment*, Volume 476477, 2014, Pages 1–6, https://doi.org/10.1016/j.scitotenv.2014.01.006.)

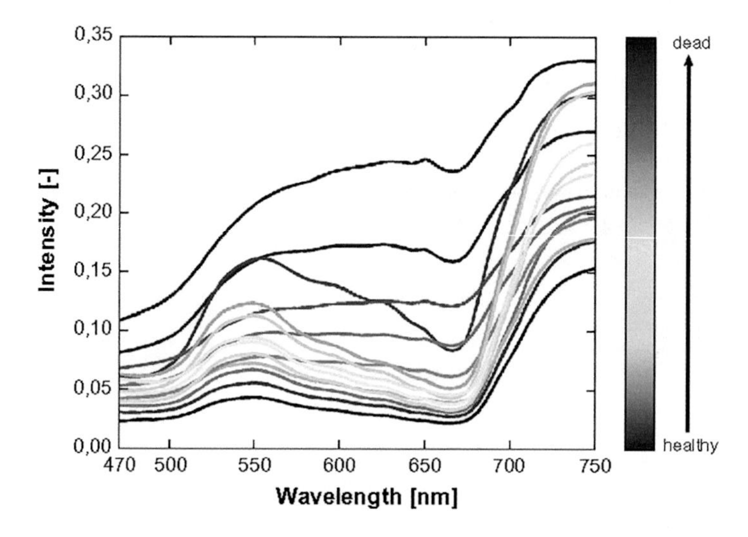

FIGURE 14.28 Vegetation health and stress as studies using hyperspectral signatures. The more stressed or senesced the vegetation, the greater the reflectivity in the 470–750 nm portion of the spectrum. (From J. Behmann et al. 2014. *ISPRS Journal of Photogrammetry and Remote Sensing*, Volume 93, Pages 98–111, https://doi.org/10.1016/j.isprsjprs.2014.03.016.)

14.12 METHODS OF LINKING UAV AND FIELD HYPERSPECTRAL DATA TO SATELLITE DATA

Imaging spectroscopy data are, primarily, gathered from four platforms: (1) ground-based, hand-held, (2) truck or platform mounted, (3) airborne (either aircrafts or drones), and (4) spaceborne. Depending on the requirements of the project, one should gather hyperspectral data from one or more platforms. Each platform has its own advantages and disadvantages. UAV platforms and miniaturized spectrometers (Zarco-Tejada et al., 2012; Burkart et al., 2014) provide unprecedented opportunities for high-spatial, spectral, and multi-angular field measurements (Cogliati et al., 2015). However, they have several processing challenges including atmospheric disturbances of the platform geometry, and challenges of coordinating with the team members within and across growing seasons. Ground-based and UAV-based spectroscopy or imaging spectroscopy help build

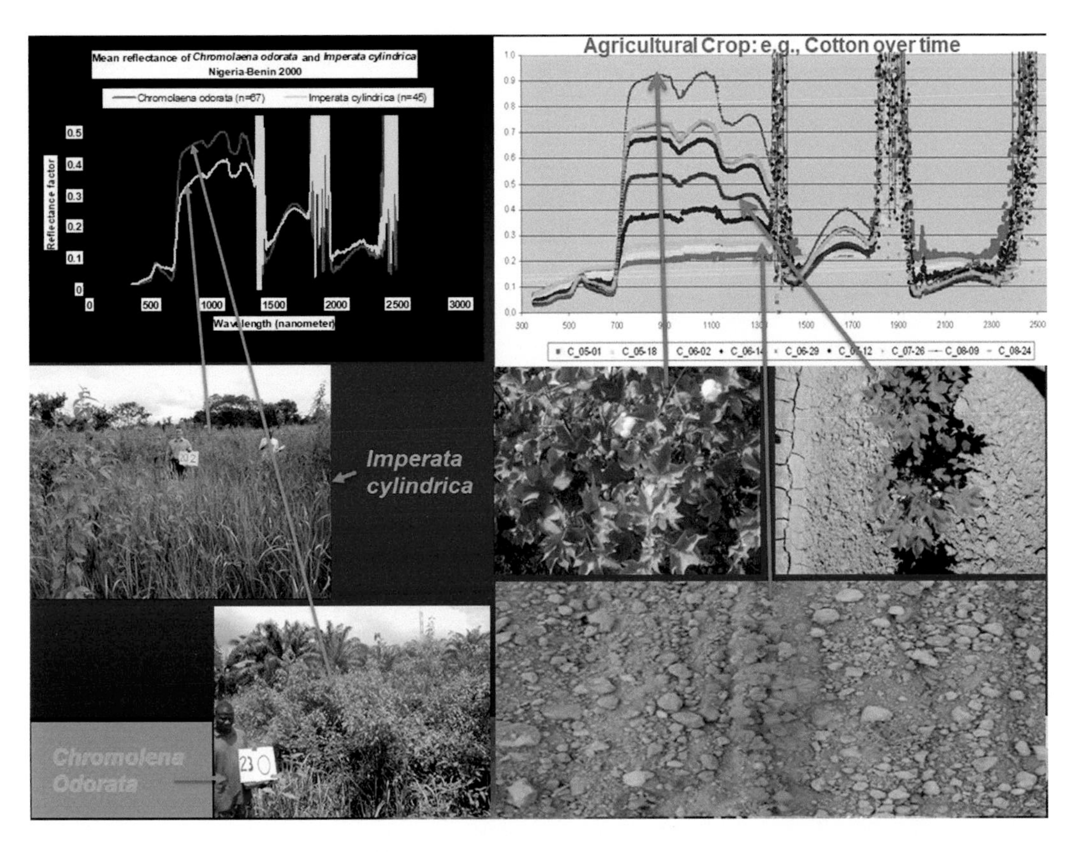

FIGURE 14.29 Ground-measured hyperspectral datasets for vegetation and crop species. Mean spectral profile of major vegetation and crop species in the African savannas (a) crop species; (b) shurb species; (c) grass species; (d) weed species.. (From P.S. Thenkabail et al. 2004a. *Remote Sensing of Environment*, Volume 90, Pages 23–43.)

spectral libraries of crops and vegetation (e.g., Figure 14.29) with better control of the situation (e.g., one can avoid gathering data during poor weather conditions) and help to collect spatial data that can be used to calibrate and/or verify airborne and spaceborne data collections. When large areas need to be covered routinely, spaceborne platforms are preferable. Once in orbit, data collected by spaceborne imaging spectroscopy are automatic, stable, and have the capability to gather data over every area of the Earth in a routine and repetitive manner. While drone platforms still suffer from carrying heavy payloads, piloted aircraft can do this job. Accordingly, they can carry sophisticated heavy sensors that cover the entire passive range from 0.4 to 12 μm. Although all are prone to cloud and other atmospheric effects, the drones that fly in very low altitude (e.g., 30 m) may obtain data in cloudy (uniform) days as well. Atmospheric attenuation is well addressed today by applying radiative transfer-based models or by empirical calibration methods.

Here, we show agriculture and vegetation spectra collected by a hand-held or portable spectroradiometer (Figure 14.29) and from the spaceborne Hyperion sensor (Figure 14.30). A direct comparison between the data collected from various platforms is only feasible if the spatial resolutions of the two platforms are the same. The hyperspectral signatures of crops shown in Figures 14.29 and 14.30 have many characteristic differences, except that both observe data roughly over the same wavelength range of 400–2500 nm. The ground-based, hand-held hyperspectral measurements (Figure 14.29) were collected from a spectroradiometer held 1.2 meters aboveground, with a FOV of 18 degrees. So, essentially, the ground spatial resolution was 1134 cm^2. In contrast, the agricultural crops data extracted from the spaceborne EO-1 Hyperion sensor (Figure 14.20) has a nominal

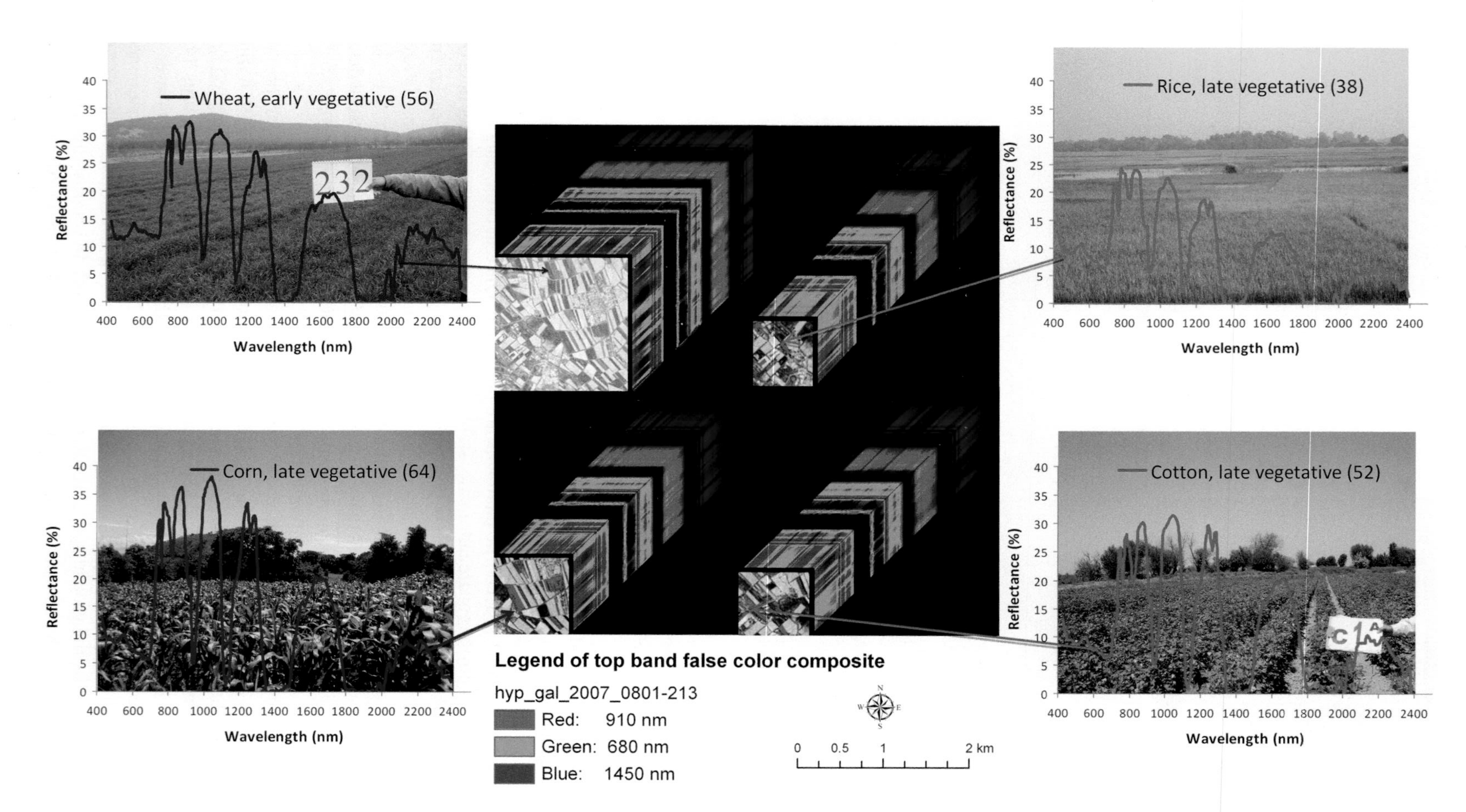

FIGURE 14.30 Spaceborne hyperspectral Hyperion data of agricultural crops. Mean spectral profile of some major world crops studied using EO-1 Hyperion hyperspectral data. (From Thenkabail et al. 2013. *IEEE Journal of Selected Topics in Applied Earth Observations and Remote Sensing*, Volume 6, Issue 2, Pages 427–439. doi: 10.1109/JSTARS.2013.2252601.

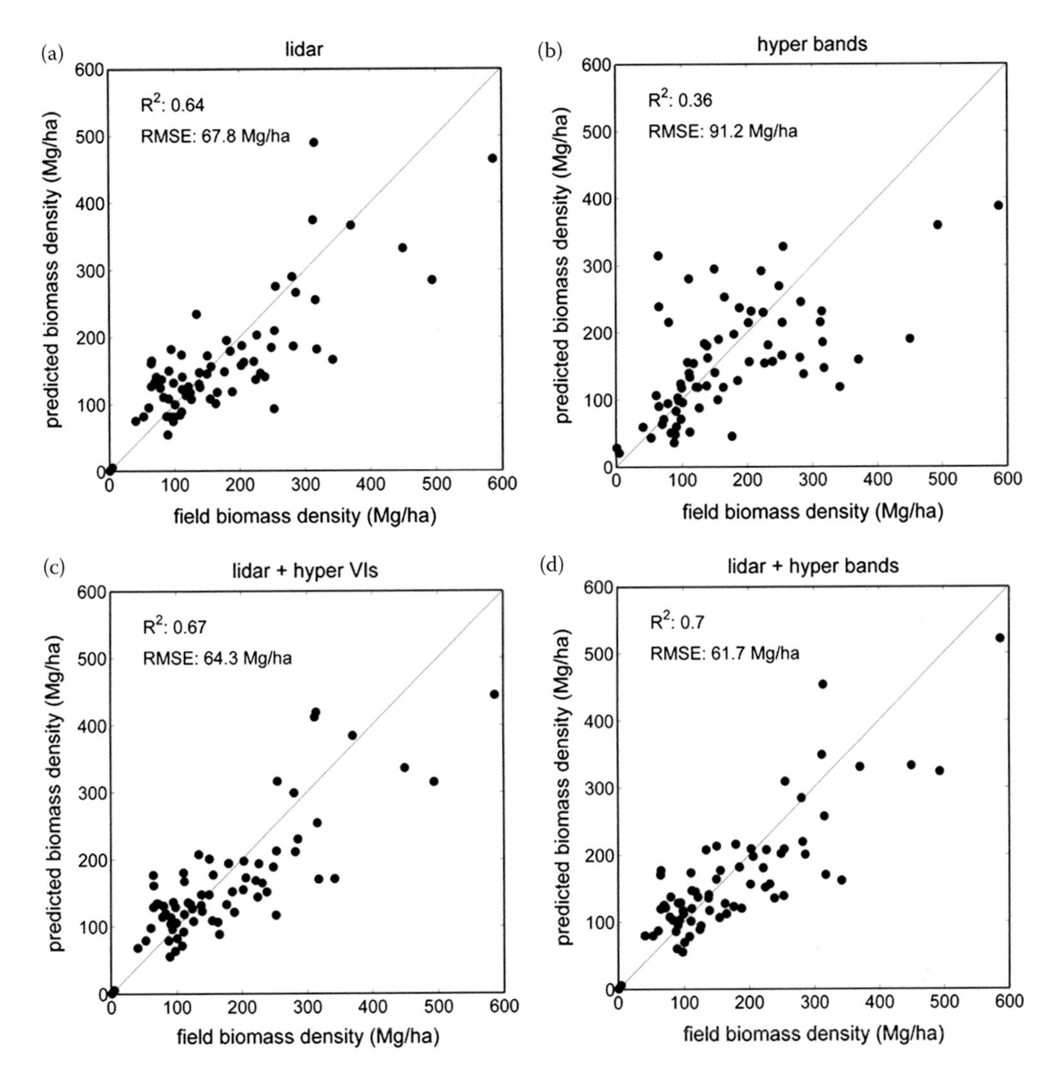

FIGURE 14.31 Aboveground biomass (AGB) models using hyperspectral data, LiDAR data, and a combination of the two datasets. Scatterplots of predicted vs. field observed AGB for the following inputs: (a) LiDAR metrics; (b) hyperspectral bands; (c) LiDAR metrics and VIs; (d) LiDAR metrics and hyperspectral bands. (From G.V. Laurin et al. 2014. *ISPRS Journal of Photogrammetry and Remote Sensing*, Volume 89, Pages 49–58, https://doi.org/10.1016/j.isprsjprs.2014.01.001.)

spatial resolution of 30-m. So, the Hyperion hyperspectral signature is an average of many plants in 30 m × 30 m pixel whereas the ground-based sensor often captures a signature focused on a single plant. Thus, a direct comparison of the hyperspectral signatures gathered from different platforms is not necessarily feasible unless attempts are made to synchronize spectral, spatial, and radiometric resolutions of the data captured from different platforms. Thereby, the spectral libraries established from different platforms need to be treated differently.

In Chapter 12, Dr. Muhammad Al-Amin Hoque, Dr. Stuart Phinn, and others provided various approaches, methods, and systems of collecting imaging and non-imaging spectroscopy data from various platforms. They clearly illustrated the possibilities and limitations of linking field spectroscopy and drone-based data with satellite sensor data using modeling and simulation exercises and by calibration through empirical models. Their specific illustrations in the study of biophysical properties are very relevant to vegetation studies found here. In the chapter, they also

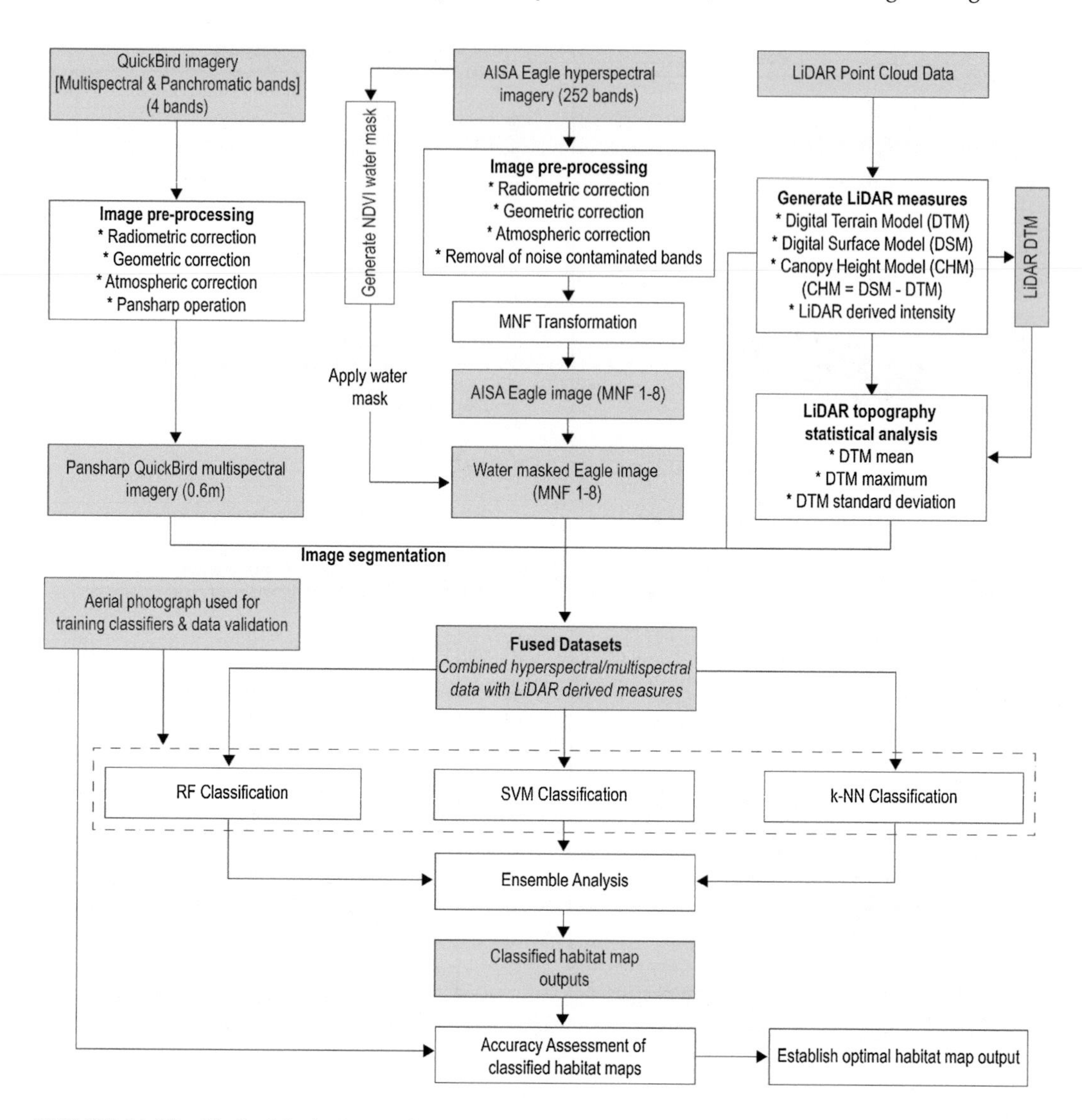

FIGURE 14.32 Methodological steps for integrating hyperspectral and LiDAR data in vegetation studies. Framework combining high resolution multispectral/hyperspectral imagery and LiDAR data for habitat vegetation mapping in the study area. (From A.O. Onojeghuo and A.R. Onojeghuo, 2017. *International Journal of Applied Earth Observation and Geoinformation*, Volume 59, Pages 79–91, https://doi.org/10.1016/j.jag.2017.03.007.)

provide a comprehensive table that provides state-of-the-art knowledge in linking UAV drone and field spectroscopy data with satellite sensor data.

14.13 INTEGRATING HYPERSPECTRAL AND LiDAR DATA IN THE STUDY OF VEGETATION

In a number of chapters, capabilities and limitations of hyperspectral data in the study of numerous agricultural crops and natural vegetation are well documented. Those studies clearly demonstrate the tremendous strengths of HNB data as opposed to MBB data. Nevertheless, combining hyperspectral data with other data sources such as LiDAR data further adds to its strength in classifying, identifying, and modeling vegetation characteristics (Puttonen et al., 2010; Laurin et al., 2014; Luo et al., 2017;

Goldbergs et al., 2018). Probably the best dataset to integrate with hyperspectral data is LiDAR, as it has many unique qualities for establishing vegetation characteristics such as canopy height, vegetation biomass, species detection, and a number of other features like topographic variations, and carbon assessments.

Many researchers have shown an increase in R^2 values and their significance in vegetation studies such as biomass estimation, tree height estimations, and species identification via applications conducted in forests, wetlands, savannas, and other ecosystems when using LiDAR data in combination with hyperspectral data (Puttonen et al., 2010; Laurin et al., 2014; Luo et al., 2017; Goldbergs et al., 2018), as opposed to using one of these data types alone. For example, most models explained about 6% greater variability in modeling vegetation biomass using both hyperspectral and LiDAR data as opposed to using only LiDAR or hyperspectral data as illustrated in an aboveground biomass (AGB) tropical forest study by Laurin et al. (2014; Figure 14.31).

A methodological framework for combining hyperspectral and LiDAR data is illustrated in Figure 14.32 (Onojeghuo and Onojeghuo, 2017). This process needs to be carefully implemented. The fusion of hyperspectral imagery (HI) and full waveform LiDAR (FWL) data needs to rely on: (1) processing the FWL data to a set of discrete returns; and (2) proper merging because the data structure and sampling interval of HI and FWL are distinctly different (Wang and Glennie, 2015).

The concepts, methods, and approaches of integrating hyperspectral data with LiDAR are presented in Chapter 13 by Dr. Jessica Mitchell et al. who provides a clear overview of many existing and upcoming land-based, airborne, and spaceborne hyperspectral and LiDAR sensors and their importance in vegetation studies. The various studies of structural (e.g., plant height, biomass, LAI) and biochemical properties using integrated hyperspectral and LiDAR data are highlighted. The authors point out that data acquired from a terrestrial laser scanner when combined with hyperspectral data reduces the uncertainty in biomass estimation significantly. The chapter puts particular emphasis on vegetation studies in dryland systems of the world.

REFERENCES

Aasen, H., Burkart, A., Bolten, A., Bareth, G. 2015. Generating 3D hyperspectral information with lightweight UAV snapshot cameras for vegetation monitoring: From camera calibration to quality assurance, *ISPRS Journal of Photogrammetry and Remote Sensing*, Volume 108, Pages 245–259, ISSN 0924-2716, https://doi.org/10.1016/j.isprsjprs.2015.08.002.

Araújo, S.R., Söderström, M., Eriksson, J., Isendahl, C., Stenborg, P., Demattê, J.A.M. 2015. Determining soil properties in Amazonian Dark Earths by reflectance spectroscopy, *Geoderma*, Volume 237238, Pages 308–317, ISSN 0016-7061, https://doi.org/10.1016/j.geoderma.2014.09.014.

Araus, J.L., and Cairns, J.E. 2014. Field high-throughput phenotyping: the new crop breeding frontier, *Trends in Plant Science*, Volume 19, Issue 1, Pages 52–61, ISSN 1360-1385, https://doi.org/10.1016/j.tplants.2013.09.008.

Asner, G.P., Martin, R.E., Tupayachi, R., Llactayo, W. 2017. Conservation assessment of the Peruvian Andes and Amazon based on mapped forest functional diversity, *Biological Conservation*, Volume 210, Part A, Pages 80–88, ISSN 0006-3207, https://doi.org/10.1016/j.biocon.2017.04.008.

Bai, G., Ge, Y., Hussain, W., Baenziger, P.S., Graef, G. 2016. A multi-sensor system for high throughput field phenotyping in soybean and wheat breeding, *Computers and Electronics in Agriculture*, Volume 128, Pages 181–192, ISSN 0168-1699, https://doi.org/10.1016/j.compag.2016.08.021.

Baldridge, A.M., Hook, S.J., Grove, C.I., Rivera, G. 2009. The ASTER Spectral Library Version 2.0, *Remote Sensing of Environment*, Volume 113, Pages 711–715.

Bareth, G., and Waldhoff, G. 2018. 2.01—GIS for mapping vegetation, In *Comprehensive Geographic Information Systems*, edited by B. Huang, Elsevier, Oxford, Pages 1–27, ISBN 9780128047934, https://doi.org/10.1016/B978-0-12-409548-9.09636-6.

Behmann, J., Steinrücken, J., Plümer, L. 2014. Detection of early plant stress responses in hyperspectral images, *ISPRS Journal of Photogrammetry and Remote Sensing*, Volume 93, Pages 98–111, ISSN 0924-2716, https://doi.org/10.1016/j.isprsjprs.2014.03.016.

Belluco, E., Camuffo, M., Ferrari, S., Modenese, L., Silvestri, S., Marani, A., Marani, M. 2006. Mapping salt-marsh vegetation by multispectral and hyperspectral remote sensing, *Remote Sensing of Environment*, Volume 105, Issue 1, Pages 54–67, ISSN 0034-4257, https://doi.org/10.1016/j.rse.2006.06.006.

Ben-Dor, E., Ong, C., Lau, I.C. 2015. Reflectance measurements of soils in the laboratory: Standards and protocols, *Geoderma*, Volumes 245–246, Pages 112–124, ISSN 0016-7061, https://doi.org/10.1016/j.geoderma.2015.01.002.

Bilgili, A.V., van Es, H.M., Akbas, F., Durak, A., Hively, W.D. 2010. Visible-near infrared reflectance spectroscopy for assessment of soil properties in a semi-arid area of Turkey, *Journal of Arid Environments*, Volume 74, Issue 2, Pages 229–238, ISSN 0140-1963, https://doi.org/10.1016/j.jaridenv.2009.08.011.

Boardman, J.W., and Kruse, F.A. 1994. Automated spectral analysis: a geological example using AVIRIS data, north Grapevine Mountains, Nevada. *Proceedings, ERIM Tenth Thematic Conference on Geologic Remote Sensing, Environmental Research Institute of Michigan*, Ann Arbor, MI, Pages I-407–I-418.

Brook, A., and Dor, E. B. 2011. Supervised vicarious calibration (SVC) of hyperspectral remote-sensing data, *Remote Sensing of Environment*, Volume 115, Issue 6, Pages 1543–1555.

Brown, D.J., Shepherd, K.D., Walsh, M.G., Mays, M.D., Reinsch, T.G. 2006. Global soil characterization with VNIR diffuse reflectance spectroscopy, *Geoderma*, Volume 132, Issues 3–4, Pages 273–290, ISSN 0016-7061, https://doi.org/10.1016/j.geoderma.2005.04.025.

Burkart, A., Cogliati, S., Schickling, A., Rascher, U. 2014. A novel UAV-Based ultra-light weight spectrometer for field spectroscopy, *IEEE Sensors Journal*, Volume 14, Issue 1, Pages 62–67, 10.1109/JSEN.2013.2279720.

Burud, I., Moni, C., Flo, A., Futsaether, C., Steffens, M., Rasse, D.P. 2016. Qualitative and quantitative mapping of biochar in a soil profile using hyperspectral imaging, *Soil and Tillage Research*, Volume 155, Pages 523–531, ISSN 0167-1987, https://doi.org/10.1016/j.still.2015.06.020.

Castro-Esau, K.L., Sánchez-Azofeifa, G.A., Rivard, B. 2006. Comparison of spectral indices obtained using multiple spectroradiometers, *Remote Sensing of Environment*, Volume 103, Issue 3, Pages 276–288, ISSN 0034-4257, https://doi.org/10.1016/j.rse.2005.01.019.

Ceamanos, X., and Valero, S. 2016. 4—Processing Hyperspectral Images, In *Optical Remote Sensing of Land Surface*, edited by N. Baghdadi and M. Zribi, Elsevier, Pages 163–200, ISBN 9781785481024, https://doi.org/10.1016/B978-1-78548-102-4.50004-1.

Chang, C.I. 2013. *Hyperspectral Data Processing: Algorithm Design and Analysis*. Wiley. USBN: 978-1-118-26977-0. Page 1071.

Cogliati, S., Rossini, M., Julitta, T., Meroni, M., Schickling, A., Burkart, A., Pinto, F., Rascher, U., Colombo, R. 2015. Continuous and long-term measurements of reflectance and sun-induced chlorophyll fluorescence by using novel automated field spectroscopy systems, *Remote Sensing of Environment*, Volume 164, Pages 270–281, ISSN 0034-4257, https://doi.org/10.1016/j.rse.2015.03.027.

Clark, M.L. 2017. Comparison of simulated hyperspectral HyspIRI and multispectral Landsat 8 and Sentinel-2 imagery for multi-seasonal, regional land-cover mapping, *Remote Sensing of Environment*, Volume 200, Pages 311–325, ISSN 0034-4257, https://doi.org/10.1016/j.rse.2017.08.028.

Clark, M.L., Roberts, D.A., Clark, D.B. 2005. Hyperspectral discrimination of tropical rain forest tree species at leaf to crown scales, *Remote Sensing of Environment*, Volume 96, Issues 3–4, Pages 375–398, ISSN 0034-4257, https://doi.org/10.1016/j.rse.2005.03.009.

Datt, B., McVicar, T.R., Van Niel, T.G., Jupp, D.L.B. 2003. Preprocessing EO-1 Hyperion hyperspectral data to support the application of agricultural indices, *IEE Transactions of Geoscience and Remote Sensing*, Volume 41, Issue 6, Pages 1246–1259.

Féret, J.-B., Gitelson, A.A., Noble, S.D., Jacquemoud, S. 2017. PROSPECT-D: Towards modeling leaf optical properties through a complete lifecycle, *Remote Sensing of Environment*, Volume 193, Pages 204–215, ISSN 0034-4257, https://doi.org/10.1016/j.rse.2017.03.004.

Galvão, L.S., Ponzoni, F.J., Liesenberg, V., Santos, J.R.D. 2009. Possibilities of discriminating tropical secondary succession in Amazônia using hyperspectral and multiangular CHRIS/PROBA data, *International Journal of Applied Earth Observation and Geoinformation*, Volume 11, Issue 1, Pages 8–14, ISSN 0303-2434, https://doi.org/10.1016/j.jag.2008.04.001.

Gandariasbeitia, M., Besga, G., Albizu, I., Larregla, S., Mendarte, S. 2017. Prediction of chemical and biological variables of soil in grazing areas with visible- and near-infrared spectroscopy, *Geoderma*, Volume 305, Pages 228–235, ISSN 0016-7061, https://doi.org/10.1016/j.geoderma.2017.05.045.

Gao, B.C., Montes, M.J., Davis, C.O., Goetz, A.F.H. 2009. Atmospheric correction algorithms for hyperspectral remote sensing data of land and ocean, *Remote Sensing of Environment*, Volume 113, Supplementary 1, Pages S17–S24, ISSN 0034-4257, https://doi.org/10.1016/j.rse.2007.12.015.

Ge, Y., Bai, G., Stoerger, V., Schnable, J.C. 2016. Temporal dynamics of maize plant growth, water use, and leaf water content using automated high throughput RGB and hyperspectral imaging, *Computers and Electronics in Agriculture*, Volume 127, Pages 625–632, ISSN 0168-1699, https://doi.org/10.1016/j.compag.2016.07.028.

Gitelson, A.A., Gritz, Y., Merzlyak, M.N. 2003. Relationships between leaf chlorophyll content and spectral reflectance and algorithms for non-destructive chlorophyll assessment in higher plant leaves, *Journal of Plant Physiology*, Volume 160, Issue 3, Pages 271–282, ISSN 0176-1617, https://doi.org/10.1078/0176-1617-00887.

Goodenough, G., Dyk, A., Niemann, K.O., Pearlman, J.S., Chen, H., Han, T. et al. 2003. Processing Hyperion and ALI for forest classification, *IEEE Transaction on Geoscience and Remote Sensing*, 41(6), Pages 1321–1331.

Goetz, A.F.H. 2009. Three decades of hyperspectral remote sensing of the Earth: A personal view, *Remote Sensing of Environment*, Volume 113, Supplement 1, Pages S5–S16, ISSN 0034-4257, https://doi.org/10.1016/j.rse.2007.12.014.

Gomez, C., Viscarra Rossel, R.A., McBratney, A.B. 2008. Soil organic carbon prediction by hyperspectral remote sensing and field vis-NIR spectroscopy: An Australian case study, *Geoderma*, Volume 146, Issues 3–4, Pages 403–411, ISSN 0016-7061, https://doi.org/10.1016/j.geoderma.2008.06.011.

Goldbergs, G., Levick, S.R., Lawes, M., Edwards, A. 2018. Hierarchical integration of individual tree and area-based approaches for savanna biomass uncertainty estimation from airborne LiDAR, *Remote Sensing of Environment*, Volume 205, Pages 141–150, ISSN 0034-4257, https://doi.org/10.1016/j.rse.2017.11.010.

Hamzeh, S., Naseri, A.A., Kazem, S., Panah, A., Bartholomeus, H., Herold, M. 2016. Assessing the accuracy of hyperspectral and multispectral satellite imagery for categorical and Quantitative mapping of salinity stress in sugarcane fields, *International Journal of Applied Earth Observation and Geoinformation*, Volume 52, Pages 412–421, ISSN 0303-2434, https://doi.org/10.1016/j.jag.2016.06.024.

Hestir, E.L., Brando, V.E., Bresciani, M., Giardino, C., Matta, E., Villa, P., Dekker, A.G. 2015. Measuring freshwater aquatic ecosystems: The need for a hyperspectral global mapping satellite mission, *Remote Sensing of Environment*, Volume 167, Pages 181–195, ISSN 0034-4257, https://doi.org/10.1016/j.rse.2015.05.0

Houborg, R., Fisher, J.B., Skidmore, A.K. 2015. Advances in remote sensing of vegetation function and traits, *International Journal of Applied Earth Observation and Geoinformation*, Volume 43, Pages 1–6, ISSN 0303-2434, https://doi.org/10.1016/j.jag.2015.06.001.

Hughes, G.F. 1968. On the mean accuracy of statistical pattern recognizers, *IEEE Transaction on Information Theory*, 14, 55–63.

Ishida, T., Kurihara, J., Viray, F.A., Namuco, S.B., Paringit, E.C., Perez, G.J., Takahashi, Y., Marciano, J.J. 2018. A novel approach for vegetation classification using UAV-based hyperspectral imaging, *Computers and Electronics in Agriculture*, Volume 144, Pages 80–85, ISSN 0168-1699, https://doi.org/10.1016/j.compag.2017.11.027.

Jafari, R., and Lewis, M.M. 2012. Arid land characterisation with EO-1 Hyperion hyperspectral data, *International Journal of Applied Earth Observation and Geoinformation*, Volume 19, Pages 298–307, ISSN 0303-2434, https://doi.org/10.1016/j.jag.2012.06.001.

Jin, X., Kumar, L., Li, Z., Feng, H., Xu, X., Yang, G., Wang, J. 2018. A review of data assimilation of remote sensing and crop models, *European Journal of Agronomy*, Volume 92, Pages 141–152, ISSN 1161-0301, https://doi.org/10.1016/j.eja.2017.11.002.

Knox, N.M., Grunwald, S., McDowell, M.L., Bruland, G.L., Myers, D.B., Harris, W.G. 2015. Modelling soil carbon fractions with visible near-infrared (VNIR) and mid-infrared (MIR) spectroscopy, *Geoderma*, Volumes 239–240, Pages 229–239, ISSN 0016-7061, https://doi.org/10.1016/j.geoderma.2014.10.019.

Kokaly, R.F., Clark, R.N., Swayze, G.A., Livo, K.E., Hoefen, T.M., Pearson, N.C. et al. 2017, USGS Spectral Library Version 7 Data: U.S. Geological Survey data release, https://dx.doi.org/10.5066/F7RR1WDJ.

Kruse, F.A. 2004. Comparison of ATREM, ACORN, and FLAASH atmospheric corrections using low-altitude AVIRIS data of Boulder, *CO. Summaries of 13th JPL Airborne Geoscience Workshop*, Jet Propulsion Lab, Pasadena, CA.

Ksieniewicz, P., Krawczyk, B., Woźniak, M. 2018. Ensemble of Extreme Learning Machines with trained classifier combination and statistical features for hyperspectral data, *Neurocomputing*, Volume 271, Pages 28–37, ISSN 0925-2312, https://doi.org/10.1016/j.neucom.2016.04.076.

Laurin, G.V., Chen, Q., Lindsell, J.A., Coomes, D.A., Frate, F.D., Guerriero, L., Pirotti, F., Valentini, R. 2014. Above ground biomass estimation in an African tropical forest with lidar and hyperspectral data, *ISPRS Journal of Photogrammetry and Remote Sensing*, Volume 89, Pages 49–58, ISSN 0924-2716, https://doi.org/10.1016/j.isprsjprs.2014.01.001.

Liu, S., Bruzzone, L., Bovolo, F., Du, P. 2015. Hierarchical change detection in multitemporal hyperspectral images, *IEEE Trans. Geosci. Remote Sens.*, Volume 53, Pages 244–260.

Liu, S., Du, Q., Tong, X., Samat, A., Pan, H., Ma, X. 2017a. Band Selection-Based Dimensionality Reduction for Change Detection in Multi-Temporal Hyperspectral Images, *Remote Sens.*, Volume 9, Page 1008.

Liu, N., Budkewitsch, P., Treitz, P. 2017b. Examining spectral reflectance features related to Arctic percent vegetation cover: Implications for hyperspectral remote sensing of Arctic tundra, *Remote Sensing of Environment*, Volume 192, Pages 58–72, ISSN 0034-4257, https://doi.org/10.1016/j.rse.2017.02.002.

Lopatin, J., Fassnacht, F.E., Kattenborn, T., Schmidtlein, S. 2017. Mapping plant species in mixed grassland communities using close range imaging spectroscopy, *Remote Sensing of Environment*, Volume 201, Pages 12–23, ISSN 0034-4257, https://doi.org/10.1016/j.rse.2017.08.031.

Luo, S., Wang, C., Xi, X., Pan, F., Qian, M., Peng, D., Nie, S., Qin, H., Lin, Y. 2017. Retrieving aboveground biomass of wetland *Phragmites australis* (common reed) using a combination of airborne discrete-return LiDAR and hyperspectral data, *International Journal of Applied Earth Observation and Geoinformation*, Volume 58, Pages 107–117, ISSN 0303-2434, https://doi.org/10.1016/j.jag.2017.01.016.

Madonsela, S., Azong Cho, M., Ramoelo, A., Mutanga, O. 2017. Remote sensing of species diversity using Landsat 8 spectral variables, *ISPRS Journal of Photogrammetry and Remote Sensing*, Volume 133, Pages 116–127, ISSN 0924-2716, https://doi.org/10.1016/j.isprsjprs.2017.10.008.

Mariotto, I., Thenkabail, P.S., Huete, H., Slonecker, T., Platonov, A., 2013. Hyperspectral versus Multispectral Crop- Biophysical Modeling and Type Discrimination for the HyspIRI Mission, *Remote Sensing of Environment*, Volume 139, Pages 291–305. IP-049224.

Marshall, M.T., and Thenkabail, P.S. 2014. Biomass modeling of four leading World crops using hyperspectral narrowbands in support of HyspIRI mission, *Photogrammetric Engineering and Remote Sensing*, Volume 80, Issue 4, Pages 757–772. IP-052043.

Marshall, M.T., and Thenkabail, P.S. 2015. Advantage of hyperspectral EO-1 Hyperion over multispectral IKONOS, GeoEye-1, WorldView-2, Landsat ETM, and MODIS vegetation indices in crop biomass estimation, *International Society of Photogrammetry and Remote Sensing (ISPRS) Journal of Photogrammetry and Remote Sensing (ISPRS P&RS)*, Volume 108, Pages 205–218. http://dx.doi.org/10.1016/j.isprsjprs.2015.08.001. IP-060745.

Marshall, M.T., Thenkabail, P.S., Biggs, T., Post, K. 2016. Hyperspectral narrowband and multispectral broadband indices for remote sensing of crop evapotranspiration and its components (transpiration and soil evaporation), *Agricultural and Forest Meteorology*, Volume 218–219, Pages 122–134. IP-065032.

Matthew, M.W., Adler-olden, S.M., Berk, A., Richtsmeier, S.C., Levine, R.Y., Bernstein, L.S., Miller, D.P. 2000. Status of atmospheric correction using a MODTRAN4-based algorithm, *Proc. SPIE*, Volume 4049, Page 199.

Middleton, E.M., Huemmrich, K.F., Landis, D.R., Black, T.A., Barr, A.G., McCaughey, J.H. 2016. Photosynthetic efficiency of northern forest ecosystems using a MODIS-derived Photochemical Reflectance Index (PRI), *Remote Sensing of Environment*, Volume 187, Pages 345–366, ISSN 0034-4257, https://doi.org/10.1016/j.rse.2016.10.021.

Mohamed, E.S., Saleh, A.M., Belal, A.B., Gad, A.A. 2017. Application of near-infrared reflectance for quantitative assessment of soil properties, *The Egyptian Journal of Remote Sensing and Space Science*, ISSN 1110-9823, https://doi.org/10.1016/j.ejrs.2017.02.001.

Neckel, H., and Labs, D. 1984, The solar radiation between 3300 and 12500 A, *Solar Physics*, Volume 90, Pages 205-258.

Nidamanuri, R.R., and Zbell, B. 2011. Use of field reflectance data for crop mapping using airborne hyperspectral image, *ISPRS Journal of Photogrammetry and Remote Sensing*, Volume 66, Issue 5, Pages 683–691, ISSN 0924-2716, https://doi.org/10.1016/j.isprsjprs.2011.05.001.

Onojeghuo, A.O., and Onojeghuo, A.R. 2017. Object-based habitat mapping using very high spatial resolution multispectral and hyperspectral imagery with LiDAR data, *International Journal of Applied Earth Observation and Geoinformation*, Volume 59, Pages 79–91, ISSN 0303-2434, https://doi.org/10.1016/j.jag.2017.03.007.

Padilla, F.M., Teresa Peña-Fleitas, M., Gallardo, M., Thompson, R.B. 2016. Proximal optical sensing of cucumber crop N status using chlorophyll fluorescence indices, *European Journal of Agronomy*, Volume 73, Pages 83–97, ISSN 1161-0301, https://doi.org/10.1016/j.eja.2015.11.001.

Pan, Z., Huang, J., Wang, F. 2013. Multi range spectral feature fitting for hyperspectral imagery in extracting oilseed rape planting area, *International Journal of Applied Earth Observation and Geoinformation*, Volume 25, Pages 21–29, ISSN 0303-2434, https://doi.org/10.1016/j.jag.2013.03.002.

Panda, S. S., Rao, M.N., Thenkabail, P.S., Fitzerald, J.E. 2015. Remote Sensing Systems – Platforms and Sensors: Aerial, Satellites, UAVs, Optical, Radar, and LiDAR, Chapter 1. In Thenkabail, P.S., (Editor-in-Chief),

2015. "*Remote Sensing Handbook" (Volume I): Remotely sensed data characterization, classification, and accuracies*. ISBN 9781482217865—CAT# K22125. Taylor and Francis Inc./CRC Press, Boca Raton, London, New York. Pages 3–60. IP-060641.

Pasqualotto, N., Delegido, J., Wittenberghe, S.V., Verrelst, J., Rivera, J.P., Moreno, J. 2018. Retrieval of canopy water content of different crop types with two new hyperspectral indices: Water Absorption Area Index and Depth Water Index, *International Journal of Applied Earth Observation and Geoinformation*, Volume 67, Pages 69–78, ISSN 0303-2434, https://doi.org/10.1016/j.jag.2018.01.002.

Pignatti, S., Cavalli, R.M., Cuomo, V., Fusilli, L., Pascucci, S., Poscolieri, M., Santini, F. 2009. Evaluating Hyperion capability for land cover mapping in a fragmented ecosystem: Pollino National Park, Italy, *Remote Sensing of Environment*, Volume 113, Issue 3, Pages 622–634, ISSN 0034-4257, https://doi.org/10.1016/j.rse.2008.11.006.

Pearlman, J.S., Barry, P.S., Segal, C.C., Shepanski, J., Beiso, D., Carman, S.L. 2003. Hyperion, a space-based imaging spectrometer, *IEEE Transactions on Geoscience and Remote Sensing*, Volume 41, Pages 1160–1173.

Pimstein, A., Ben-Dor, E., Notesko, G. 2011. Performance of three identical spectrometers in retrieving soil reflectance under laboratory conditions, *Soil Science Society of America. Journal*, Volume 75, Pages 110–174.

Puttonen, E., Suomalainen, J., Hakala, T., Räikkönen, E., Kaartinen, H., Kaasalainen, S., Litkey, P. 2010. Tree species classification from fused active hyperspectral reflectance and LIDAR measurements, *Forest Ecology and Management*, Volume 260, Issue 10, Pages 1843–1852, ISSN 0378-1127, https://doi.org/10.1016/j.foreco.2010.08.031.

Qi. J., Inouue, Y., and Wiangwang, N. 2012. Hyperspectral remote sensing in global change studies, Chapter 3. In *Hyperspectral Remote Sensing of Vegetation*. CRC Press, Taylor & Francis group, Boca Raton, London, New York. pp. 781.

Qu, Z., Kindel, B.C., Goetz, A.F.H. 2003. The high accuracy atmospheric correction for hyperspectral data (HATCH) model, *IEEE Transactions on Geoscience and Remote Sensing*, Volume 41, Pages 1223–1231.

Rautiainen, M., and Lukeš, P. 2015. Spectral contribution of understory to forest reflectance in a boreal site: an analysis of EO-1 Hyperion data, *Remote Sensing of Environment*, Volume 171, Pages 98–104, ISSN 0034-4257, https://doi.org/10.1016/j.rse.2015.10.009.

Rivera-Caicedo, J.P., Verrelst, J., Muñoz-Marí, J., Camps-Valls, G., Moreno, J. 2017. Hyperspectral dimensionality reduction for biophysical variable statistical retrieval, *ISPRS Journal of Photogrammetry and Remote Sensing*, Volume 132, Pages 88–101, ISSN 0924-2716, https://doi.org/10.1016/j.isprsjprs.2017.08.012.

Roberts, D.A., Dennison, P.E., Roth, K.L., Dudley, K., Hulley, G. 2015. Relationships between dominant plant species, fractional cover and Land Surface Temperature in a Mediterranean ecosystem, *Remote Sensing of Environment*, Volume 167, Pages 152–167, ISSN 0034-4257, https://doi.org/10.1016/j.rse.2015.01.026.

Roberts, D.A., Keller, M., Soares, J.V. 2003. Studies of land-cover, land-use, and biophysical properties of vegetation in the Large Scale Biosphere Atmosphere experiment in Amazônia, *Remote Sensing of Environment*, Volume 87, Issue 4, Pages 377–388, ISSN 0034-4257, https://doi.org/10.1016/j.rse.2003.08.012.

Roth, K.L., Roberts, D.A., Dennison, P.E., Alonzo, M., Peterson, S.H., Beland, M. 2015a. Differentiating plant species within and across diverse ecosystems with imaging spectroscopy, *Remote Sensing of Environment*, Volume 167, Pages 135–151, ISSN 0034-4257, https://doi.org/10.1016/j.rse.2015.05.007.

Roth, K.L., Roberts, D.A., Dennison, P.E., Peterson, S.H., Alonzo, M. 2015b. The impact of spatial resolution on the classification of plant species and functional types within imaging spectrometer data, *Remote Sensing of Environment*, Volume 171, Pages 45–57, ISSN 0034-4257, https://doi.org/10.1016/j.rse.2015.10.004.

Shi, T., Liu, H., Chen, Y., Wang, J., Wu, G. 2016. Estimation of arsenic in agricultural soils using hyperspectral vegetation indices of rice, *Journal of Hazardous Materials*, Volume 308, Pages 243–252, ISSN 0304-3894, https://doi.org/10.1016/j.jhazmat.2016.01.022.

Sonobe, R., and Wang, Q. 2018. Nondestructive assessments of carotenoids content of broadleaved plant species using hyperspectral indices, *Computers and Electronics in Agriculture*, Volume 145, Pages 18–26, ISSN 0168-1699, https://doi.org/10.1016/j.compag.2017.12.022.

Stagakis, S., Vanikiotis, T., Sykioti, O. 2016. Estimating forest species abundance through linear unmixing of CHRIS/PROBA imagery, *ISPRS Journal of Photogrammetry and Remote Sensing*, Volume 119, Pages 79–89, ISSN 0924-2716, https://doi.org/10.1016/j.isprsjprs.2016.05.013.

Sytar, O., Brestic, M., Zivcak, M., Olsovska, K., Kovar, M., Shao, H., He, X. 2017. Applying hyperspectral imaging to explore natural plant diversity towards improving salt stress tolerance, *Science of The Total Environment*, Volume 578, Pages 90–99, ISSN 0048-9697, https://doi.org/10.1016/j.scitotenv.2016.08.014.

Summers, D., Lewis, M., Ostendorf, B., Chittleborough, D. 2011. Visible near-infrared reflectance spectroscopy as a predictive indicator of soil properties, *Ecological Indicators*, Volume 11, Issue 1, Pages 123–131, ISSN 1470-160X, https://doi.org/10.1016/j.ecolind.2009.05.001.

Sun, W., and Zhang, X. 2017. Estimating soil zinc concentrations using reflectance spectroscopy, *International Journal of Applied Earth Observation and Geoinformation*, Volume 58, Pages 126–133, ISSN 0303-2434, https://doi.org/10.1016/j.jag.2017.01.013.

Tadesse, T., Brown, J.F., Hayes, M.J. 2005. A new approach for predicting drought-related vegetation stress: Integrating satellite, climate, and biophysical data over the U.S. central plains, *ISPRS Journal of Photogrammetry and Remote Sensing*, Volume 59, Issue 4, Pages 244–253, ISSN 0924-2716, https://doi.org/10.1016/j.isprsjprs.2005.02.003.

Teluguntla, P., Thenkabail, P.S., Xiong, J., Gumma, M.K., Giri, C., Milesi, C. et al. 2015. Global Cropland Area Database (GCAD) derived from Remote Sensing in Support of Food Security in the Twenty-first Century: Current Achievements and Future Possibilities, Chapter 6, In *"Remote Sensing Handbook" Volume II: Land Resources: Monitoring, Modeling, and Mapping: Advances over Last 50 Years and a Vision for the Future*, Editor-in-Chief, P.S. Thenkabail, Taylor and Francis Inc./CRC Press, Boca Raton, London, New York, Page 800.

Thenkabail, P.S. 2003. Biophysical and yield information for precision farming from near-real time and historical Landsat TM images, *International Journal of Remote Sensing*, Volume 24(14), Pages 2879–2904.

Thenkabail, P.S. 2015. Hyperspectral Remote Sensing for Terrestrial Applications, Chapter 9. In *"Remote Sensing Handbook" (Volume II): Land Resources Monitoring, Modeling, and Mapping with Remote Sensing*, Editor-in-Chief, P.S. Thenkabail, Taylor and Francis Inc./CRC Press, Boca Raton, London, New York, Pages 201–236. ISBN 9781482217957.

Thenkabail, P.S., Enclona, E.A., Ashton, M.S., Legg, C., Jean De Dieu, M. 2004a. Hyperion, IKONOS, ALI, and ETM sensors in the study of African rainforests, *Remote Sensing of Environment*, Volume 90, Pages 23–43.

Thenkabail, P.S., Enclona, E.A., Ashton, M.S., Van Der Meer, V. 2004b. Accuracy assessments of hyperspectral waveband performance for vegetation analysis applications, *Remote Sensing of Environment*, Volume 91, Issues 2–3, Pages 354–376.

Thenkabail, P.S., Gamage, N., Smakhin, V. 2004c. *The use of remote sensing data for drought assessment and monitoring in south west Asia*. IWMI Research Report # 85. Pages 25. IWMI, Colombo, Sri Lanka.

Thenkabail, P.S., Gumma, M.K., Teluguntla, P., Mohammed, I.A. 2014. Hyperspectral Remote Sensing of Vegetation and Agricultural Crops. Highlight Article, *Photogrammetric Engineering and Remote Sensing*, Volume 80, Issue 4, Pages 697–709. IP-052042.

Thenkabail, P.S., Hanjra, M.A., Dheeravath, V., Gumma, M. 2010. A holistic view of global croplands and their water use for ensuring global food security in the 21st century through advanced remote sensing and non-remote sensing approaches, *Remote Sens*, Volume 2, Pages 211–261.

Thenkabail, P.S., Knox, J.W., Ozdogan, M., Gumma, M.K., Congalton, R.G., Wu, Z., Milesi, C., Finkral, A., Marshall, M., Mariotto, I. 2012a. Assessing future risks to agricultural productivity, water resources and food security: how can remote sensing help? *Photogrammetric Engineering and Remote Sensing*, Volume 78, Issue 8, Pages 773–782.

Thenkabail, P.S., Lyon, G.J., Huete, A. (Editors) 2012b. *Hyperspectral Remote Sensing of Vegetation*. CRC Press- Taylor and Francis Group, Boca Raton, London, New York, Page 781.

Thenkabail, P.S., Mariotto, I., Gumma, M.K., Middleton, E.M., Landis, D.R., Huemmrich, F.K. 2013. Selection of hyperspectral narrowbands (HNBs) and composition of hyperspectral twoband vegetation indices (HVIs) for biophysical characterization and discrimination of crop types using field reflectance and Hyperion/EO-1 data, *IEEE Journal of Selected Topics in Applied Earth Observations and Remote Sensing*, Volume 6, Issue 2, April, Pages 427–439, doi: 10.1109/JSTARS.2013.2252601. (80%). IP-037139.

Thenkabail, P.S., Smith, R.B., De-Pauw, E. 2002. Evaluation of narrowband and broadband vegetation indices for determining optimal hyperspectral wavebands for agricultural crop characterization, *Photogrammetric Engineering and Remote Sensing*, Volume 68, Issue 6, Pages 607–621.

Thenkabail, P.S., Smith, R.B., Pauw, D.P. 2000. Hyperspectral vegetation indices and their relationships with agricultural crop characteristics, *Remote Sensing of Environment*, Volume 71, Issue 2, Pages 158–182, ISSN 0034-4257, https://doi.org/10.1016/S0034-4257(99)00067-X.

Thorp, K.R., Gore, M.A., Andrade-Sanchez, P., Carmo-Silva, A.E., Welch, S.M., White, J.W., French, A.N. 2015. Proximal hyperspectral sensing and data analysis approaches for field-based plant phenomics, *Computers and Electronics in Agriculture*, Volume 118, Pages 225–236, ISSN 0168-1699, https://doi.org/10.1016/j.compag.2015.09.005.

Thorp, K.R., French, A.N., Rango, A. 2013. Effect of image spatial and spectral characteristics on mapping semi-arid rangeland vegetation using multiple endmember spectral mixture analysis (MESMA), *Remote Sensing of Environment*, Volume 132, Pages 120–130, ISSN 0034-4257, https://doi.org/10.1016/j.rse.2013.01.008.

Tochon, G., Féret, J.B., Valero, S., Martin, R.E., Knapp, D.E., Salembier, P., Chanussot, J., Asner, G.P. 2015. On the use of binary partition trees for the tree crown segmentation of tropical rainforest hyperspectral images, *Remote Sensing of Environment*, Volume 159, Pages 318–331, ISSN 0034-4257, https://doi.org/10.1016/j.rse.2014.12.020.

Toth, C., and Jóźków, G. 2016. Remote sensing platforms and sensors: A survey, *ISPRS Journal of Photogrammetry and Remote Sensing*, Volume 115, Pages 22–36, ISSN 0924-2716, https://doi.org/10.1016/j.isprsjprs.2015.10.004.

Ungar, S.G., Pearlman, J.S., Mendenhall J., Reuter D. 2003. Overview of the Earth Observing 1 (EO-1) mission, *IEEE Transactions Geoscience and Remote Sensing*, Volume 41, Issue 6, Pages 1149–1159.

Ustin, S.L., Gitelson, A.A., Jacquemoud, S., Schaepman, M., Asner, G.P., Gamon, J.A., Zarco-Tejada, P. 2009. Retrieval of foliar information about plant pigment systems from high resolution spectroscopy, *Remote Sensing of Environment*, Volume 113, Supplement 1, Pages S67–S77, ISSN 0034-4257, https://doi.org/10.1016/j.rse.2008.10.019.

Vågen, T.G., Winowiecki, L.A., Tondoh, J.E., Desta, L.T., Gumbricht. T. 2016. Mapping of soil properties and land degradation risk in Africa using MODIS reflectance, *Geoderma*, Volume 263, Pages 216–225, ISSN 0016-7061, https://doi.org/10.1016/j.geoderma.2015.06.0

Vane, G., and Goetz, A.F.H. 1993. Terrestrial imaging spectrometry: Current status, future trends, *Remote Sensing of Environment*, Volume 44, Issues 2–3, Pages 117–126, ISSN 0034-4257, https://doi.org/10.1016/0034-4257(93)90011-L.

Wang, H., and Glennie, C. 2015. Fusion of waveform LiDAR data and hyperspectral imagery for land cover classification, *ISPRS Journal of Photogrammetry and Remote Sensing*, Volume 108, Pages 1–11, ISSN 0924-2716, https://doi.org/10.1016/j.isprsjprs.2015.05.012.

Wang, J.J., Zhang, Y., Bussink, C. 2014. Unsupervised multiple endmember spectral mixture analysis-based detection of opium poppy fields from an EO-1 Hyperion image in Helmand, *Afghanistan, Science of the Total Environment*, Volume 476477, Pages 1–6, ISSN 0048-9697, https://doi.org/10.1016/j.scitotenv.2014.01.006.

White, J.W. 2012. Field-based phenomics for plant genetics research, *Field Crops Research*, Volume 133, Pages 101–112.

Yang, R., Su, L., Zhao, X., Wan, H., Sun, J. 2017. Representative band selection for hyperspectral image classification, *Journal of Visual Communication and Image Representation*, Volume 48, Pages 396–403, ISSN 1047-3203, https://doi.org/10.1016/j.jvcir.2017.02.002.

Zarco-Tejada, P.J., González-Dugo, V., Berni, J.A.J. 2012. Fluorescence, temperature and narrow-band indices acquired from a UAV platform for water stress detection using a micro-hyperspectral imager and a thermal camera, *Remote Sensing of Environment*, Volume 117, Pages 322–337.

Zhang, C. 2016. Multiscale quantification of urban composition from EO-1/Hyperion data using object-based spectral unmixing, *International Journal of Applied Earth Observation and Geoinformation*, Volume 47, Pages 153–162, ISSN 0303-2434, https://doi.org/10.1016/j.jag.2016.01.002.

Zhang, L., Jiao, W., Zhang, H., Huang, C., Tong, Q. 2017a. Studying drought phenomena in the Continental United States in 2011 and 2012 using various drought indices, *Remote Sensing of Environment*, Volume 190, Pages 96–106, ISSN 0034-4257, https://doi.org/10.1016/j.rse.2016.12.010.

Zhang, C., Qin, Q., Zhang, T., Sun, Y., Chen, C. 2017b. Endmember extraction from hyperspectral image based on discrete firefly algorithm (EE-DFA), *ISPRS Journal of Photogrammetry and Remote Sensing*, Volume 126, Pages 108–119, ISSN 0924-2716, https://doi.org/10.1016/j.isprsjprs.2017.02.005.

Zhong, Y., Ma, A., Ong, Y.S., Zhu, Z., Zhang, L. 2018. Computational intelligence in optical remote sensing image processing, *Applied Soft Computing*, Volume 64, Pages 75–93, ISSN 1568-4946, https://doi.org/10.1016/j.asoc.2017.11.045.

Zhou, J., Khot, L.R., Bahlol, H.Y., Boydston, R., Miklas, P.N. 2016. Evaluation of ground, proximal and aerial remote sensing technologies for crop stress monitoring, *IFAC-PapersOnLine*, Volume 49, Issue 16, Pages 22–26, ISSN 2405-8963, https://doi.org/10.1016/j.ifacol.2016.10.005.

Index

A

B

C

D

N

O

T

U

V

W

Y